W9-BBY-236

SECOND EDITION

Molecular Biotechnology

*Principles and Applications
of Recombinant DNA*

SECOND EDITION

Molecular Biotechnology

*Principles and Applications
of Recombinant DNA*

**Bernard R. Glick and
Jack J. Pasternak**

Department of Biology, University of Waterloo
Waterloo, Ontario, Canada

ASM
PRESS
WASHINGTON, D.C.

Library of Congress Cataloging-in-Publication Data

Glick, Bernard R.
 Molecular biotechnology : principles and applications of
recombinant DNA / Bernard R. Glick and Jack J. Pasternak. — 2nd ed.
 p. cm.
 Includes bibliographical references and index.
 ISBN 1-55581-136-1
 1. Biotechnology. 2. Genetic engineering. 3. Molecular biology.
I. Pasternak, Jack J. II. Title.
TP248.2.G58 1998
660′.65—dc21

10 9 8 7 6 5 4 3 2 1

Cover and interior design: Susan Brown Schmidler
Cover illustration: Terese Winslow

To Marcia, for being Marcia
BRG

To Lili, for everything
JJP

Contents

PART THREE

Eukaryotic Systems 425

Preface

In the past 4 years, since the publication of the first edition of *Molecular Biotechnology: Principles and Applications of Recombinant DNA*, there has been an outpouring of new discoveries from biotechnology laboratories, and many gene-based innovations have been introduced into the marketplace. For example, human pharmaceuticals that are the products of cloned genes have been commercialized, while many other promising biotechnology drugs are being tested; immunological and PCR-based diagnostic assays for several disease-causing organisms and human disorders have become the standard feature of clinical testing procedures; novel recombinant vaccines that are based on gene products and cloned genes have been developed; many human disease genes have been identified and characterized; the number of human gene therapy clinical trials has increased 10-fold; detailed genetic and physical maps of human chromosomes have been constructed; the first mammal has been cloned from a differentiated somatic cell; and the first commercial transgenic plants have been used on a large scale.

In the first edition, we presented the principles and applications of molecular biotechnology, in a broad biological context, in a way that would be interesting and informative. Since then, we have received many helpful comments from colleagues, graduate students, and undergraduates from many different institutions around the world. With our original objectives in mind and taking into consideration the advice of a number of readers, we have updated, expanded, and reorganized *Molecular Biotechnology: Principles and Applications of Recombinant DNA*. We hope that the second edition conveys the excitement of molecular biotechnology, the scientific underpinnings of this area of study, the significance of recent discoveries, and how this information will be used to create "goods and services." Specifically, a new chapter that discusses the organisms that are commonly used by molecular biotechnologists has been added. As well, there is a complete chapter that describes the fundamental principles of molecular biology. The chapters dealing with human molecular genetics, human gene therapy, and

plant biotechnology have been extensively expanded to account for the recent advances in these areas. The chapters pertaining to diagnostic assays and vaccines have been thoroughly reorganized. The coverage of material has been updated throughout the book. In addition to these extensive changes, the numbers of figures and tables have been increased by approximately 50% to facilitate the understanding of both experimental strategies and important theoretical concepts. The glossary has also been updated and expanded. Notwithstanding all of the revising and reorganization, we have tried to preserve the same jargon-free and "reader-friendly" style that made the first edition popular with students.

We thank the crew at ASM Press, including Ken April, Susan Birch, Greg Payne, and Jeff Holtmeier for their support and encouragement. Susan Schmidler, as usual, did an excellent job with design of the book. Special thanks to Ken April, who handled the manuscript with diligence and thoroughness and us with patience.

Bernard R. Glick
Jack J. Pasternak

Preface to the First Edition

Molecular biotechnology emerged as a new research field that arose as a result of the fusion in the late 1970s of recombinant DNA technology and traditional industrial microbiology. Whether one goes to the movies to see *Jurassic Park* with its ingenious but scientifically untenable plot of cloning dinosaurs, reads in the newspaper about the commercialization of a new "biotech" tomato that has an extended shelf life, or hears one of the critics of molecular biotechnology talking about the possibility of dire consequences from genetic engineering, there is a significant public awareness about recombinant DNA technology. In this book, we introduce and explain what molecular biotechnology actually is as a scientific discipline, how the research in the area is conducted, and how this technology may realistically impact on our lives in the future.

We have written *Molecular Biotechnology: Principles and Applications of Recombinant DNA* to serve as a text for courses in biotechnology, recombinant DNA technology, and genetic engineering or for any course introducing both the principles and the applications of contemporary molecular biology methods. The book is based on the biotechnology course we have offered for the past 12 years to advanced undergraduate and graduate studies from the biological and engineering sciences at the University of Waterloo. We have written this text for students who have an understanding of basic ideas from biochemistry, molecular genetics, and microbiology. We are aware that it is unlikely that students will have had all of these courses before taking a course on biotechnology. Thus, we have tried to develop the topics in this text by explaining their broader biological context before delving into molecular details.

This text emphasizes how recombinant DNA technology can be used to create various useful products. We have, wherever possible, used experimental results and actual methodological strategies to illustrate basic concepts, and we have tried to capture the flavor and feel of how molecular biotechnology operates as a scientific venture. The examples that we have selected—from a vast and rapidly growing literature—were chosen as case studies that not only illustrate particular points but also provide the reader with a solid basis for understanding current research in specialized areas of

molecular biotechnology. Nevertheless, we expect that some of our examples will be out of date by the time the book is published, because molecular biotechnology is such a rapidly changing discipline.

For the ease of the day-to-day practitioners, scientific disciplines often develop specialized terms and nomenclature. We have tried to minimize the use of technical jargon and, in many instances, have deliberately used a simple phrase to describe a phenomenon or process that might otherwise have been expressed more succinctly with technical jargon. In any field of study, synonymous terms that describe the same phenomenon exist. In molecular biotechnology, for example, recombinant DNA technology, gene cloning, and genetic engineering, in a broad sense, have the same meaning. When an important term or concept appears for the first time in this text, it is followed in parentheses with a synonym or equivalent expression. An extensive glossary can be found at the end of the book to help the reader with the terminology of molecular biotechnology.

Each chapter opens with an outline of topics and concludes with a detailed summary and list of review questions to sharpen students' critical thinking skills. All of the key ideas in the book are carefully illustrated by the more than 200 full-color diagrams in the pedagogical belief that a picture is indeed worth a thousand words. After introducing molecular biotechnology as a scientific and economic venture in Chapter 1, the next five chapters (2 to 6) deal with the methodologies of molecular biotechnology. The chapters of Part I act as a stepping-stone for the remainder of the book. Chapters 7 to 12 in Part II present examples of microbial molecular biotechnology covering such topics as the production of metabolites, vaccines, therapeutics, diagnostics, bioremediation, biomass utilization, bacterial fertilizers, and microbial pesticides. Chapter 13 describes some of the key components of large-scale fermentation processes using genetically engineered (recombinant) microorganisms. In Part III, we deal with the molecular biotechnology of plants and animals (Chapters 14 and 15). The isolation of human disease-causing genes by using recombinant DNA technology and how, although it is in its early stages, genetic manipulation is being currently contemplated for the treatment of human diseases are presented in Chapters 16 and 17. The book concludes with coverage of the regulation of molecular biotechnology and patents in Part IV.

A brief mention should be made about the reference sections that follow each chapter. Within many of the chapters we have relied upon the published work of various researchers. In all cases, although not cited directly in the body of a chapter, the original published articles are noted in the reference section of the appropriate chapter. In some cases, we have taken "pedagogic license" and either extracted or reformulated data from the original publications. Clearly, we are responsible for any distortions or misrepresentations from these simplifications, although we hope that none has occurred. The reference sections also contain other sources that we used in a general way, which might, if consulted, bring the readers closer to a particular subject.

Acknowledgments

We express our appreciation to the following people who reviewed various parts of the manuscript as it was being developed. The comments of these expert scientists and teachers helped us immeasurably: Arthur I. Aronson, Purdue University; Ronald M. Atlas, University of Louisville; Fred

Ausubel, Massachusetts General Hospital; David R. Benson, University of Connecticut; Jean E. Brenchley, Pennsylvania State University; A. M. Chakrabarty, University of Illinois at Chicago; Stan Gelvin, Purdue University; Janet H. Glaser, University of Illinois at Urbana-Champaign; David Gwynne, Cambridge NeuroScience; George D. Hegeman, Indiana University; James B. Kaper, University of Maryland at Baltimore; Donald R. Lightfoot, Eastern Washington University at Cheney and Spokane; Cynthia Moore, Washington University; William E. Newton, Virginia Polytechnic University; Danton H. O'Day, University of Toronto in Mississauga; Richard D. Palmiter, University of Washington; David H. Persing, Mayo Clinic; William S. Reznikoff, University of Wisconsin; Campbell W. Robinson, University of Waterloo; Marc Siegel, University of Waterloo; Aaron J. Shatkin, Center for Advanced Biotechnology and Medicine at Rutgers University; Jim Schwartz, Genentech; Daniel C. Stein, University of Maryland at College Park; Dean A. Stetler, University of Kansas; and Robert T. Vinopal, University of Connecticut.

The following professionals at ASM Press worked on the book and deserve our thanks: Susan Birch, senior production editor; Ruth Siegal, developmental editor; Jodi Simpson, copy editor; Susan Schmidler, designer and art director; Peg Markow at Ruttle, Shaw & Wetherill, Inc., senior project manager; and Network Graphics, illustrators. Finally we are indebted to Patrick Fitzgerald, Director of ASM Press, who, in all possible ways, helped transform our original efforts into an acceptable final form. His encouragement as a persistent and friendly "torturer" was deeply appreciated.

Bernard R. Glick
Jack J. Pasternak

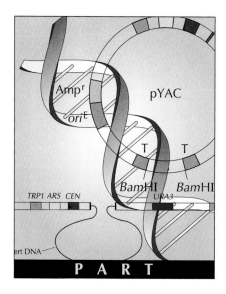

Fundamentals of Molecular Biotechnology

MOLECULAR BIOTECHNOLOGY is an exciting, revolutionary scientific discipline that is based on the ability of researchers to transfer specific units of genetic information from one organism to another. This conveyance of a gene(s) relies on the techniques of genetic engineering (recombinant DNA technology). The objective of recombinant DNA technology is often to produce a useful product or a commercial process. In Part I, the concept of molecular biotechnology, the organisms that are used by molecular biotechnologists, some fundamentals of molecular biology, and recombinant DNA procedures are presented. Essential molecular biotechnology laboratory techniques, including chemical synthesis of genes, the polymerase chain reaction (PCR), and DNA sequencing are discussed. In addition to isolating (cloning) genes, it is important that these genes function properly in a host organism. To this end, strategies for optimizing the expression of a cloned gene in either prokaryotic or eukaryotic cells are reviewed. Finally, procedures for modifying cloned genes by the introduction of specific nucleotide changes (in vitro mutagenesis) to enhance the properties of the target proteins are examined. Together, the chapters in Part I provide the conceptual and technical underpinnings for understanding the applications of molecular biotechnology that are described in the ensuing chapters.

The Molecular Biotechnology Revolution

Recombinant DNA Technology

On October 15, 1980, within 20 minutes of the start of trading on the New York Stock Exchange, the price of shares in the biotechnology company Genentech went from $35 to $89. At the time, this was the fastest increase in the value of any stock in the history of this market. When the market closed that day, Genentech stock was valued at $71.25 per share. The bidding for the 528,000 shares of Genentech stock was so frenetic that many investors who wanted "a piece of the action" never even got a chance to purchase a single share.

This may very well have been the first time that a major technological revolution was acclaimed by the clanging of stock exchange bells. At the time of their first public offering in 1980, Genentech was a 4-year-old California company specializing in gene splicing (also called recombinant DNA technology, genetic engineering, gene transplantation, or gene cloning). Two years previously, scientists at Genentech had successfully isolated portions of the gene (DNA sequence) that encodes human insulin and had transplanted them into genetic elements (cloning vectors) that could be maintained in the common bacterium *Escherichia coli*. These bacterial host cells acted as biological factories for the production of the two peptide chains of human insulin that, when combined, could then be purified and used as a medicine by those diabetics who were allergic to the commercially available porcine (pig) insulin. In the previous decade, this feat would have seemed absolutely impossible; by today's standards, however, this type of genetic engineering is considered commonplace.

The frenzy of buying Genentech shares was due to both an assessment of the potential of recombinant DNA technology and dreams of future possibilities that this unprecedented methodology engendered. Many people thought that recombinant DNA technology was the 20th-century version of the horn of plenty of Greek mythology that would fill itself with food and drink according to the wishes of the owner. Based on the enthusiasm of reports that appeared in newspapers, magazines, and television and of promotion by stockbrokers, the dreams, often tinged with science fiction fantasies, were limitless. Amazing biological menageries of manufactured

3

microorganisms, plants, and animals were conjured up. The dreamers predicted that some genetically engineered microorganisms would replace chemical fertilizers and others would eat up oil spills; plants with inherited resistance to a variety of pests and exceptional nutritional content would be created; and livestock would have faster growing times, more efficient feed utilization, and meat with low fat content. The dreamers were convinced that as long as a biological characteristic was genetically determined by one or a few genes (units of inheritance), organisms with novel genetic constitutions could be readily created. Today we see that, despite the commercial hype that dominated reality then, this infatuation with recombinant DNA technology was not totally unfounded. In a little more than 15 years, a number of the more sensible versions of the initial claims, although trimmed in scope, have become realities. How this has happened and what can be anticipated in the future from recombinant DNA technology form the basis of this book.

The conveyance of a functional unit of inheritance (a gene) from one organism to another was based on the strategy devised by Stanley Cohen and Herbert Boyer in 1973. It was clear to Cohen and Boyer and others that recombinant DNA technology had far-reaching possibilities. As Cohen noted at the time, "It may be possible to introduce in [the bacterium] *E. coli*, genes specifying metabolic or synthetic functions such as photosynthesis or antibiotic production indigenous to other biological classes."

One of the first scientific responses to this new technology was a voluntary moratorium on certain experiments that were thought to be potentially hazardous. This research ban was self-imposed by a group of molecular biologists, including Cohen and Boyer. They were concerned that combining genes from two different organisms might accidentally create a novel organism with undesirable and dangerous properties. Within a few years, however, these concerns were allayed as scientists gained laboratory experience with this technology and safety guidelines were formulated for recombinant DNA research. The temporary cessation of some recombinant DNA research projects did not dampen the enthusiasm for genetic engineering. In fact, the new technology continued to receive unprecedented attention from both the public and the scientific community.

The Cohen and Boyer strategy for gene cloning was an experiment "heard round the world." Once their concept was made public, many other researchers immediately appreciated the power of being able to clone genes; consequently, scientists created a large variety of experimental protocols that made identifying, isolating, characterizing, and utilizing genes more efficient and relatively easy. These technological developments have had a significant impact on generating new knowledge in practically all biological disciplines, including animal behavior, developmental biology, molecular evolution, cell biology, and human genetics. However, of all the scientific fields, recombinant DNA technology triggered the most profound changes in biotechnology.

Emergence of Molecular Biotechnology

In the early 1970s, traditional biotechnology was not a well-known scientific discipline; research in this area was centered in departments of chemical engineering and occasionally in specialized microbiology programs. In a broad sense, biotechnology is concerned with the production of commercial products generated by the metabolic action of microorganisms. More formally, biotechnology may be defined as "the application of scientific and engineering principles to the processing of material by biological agents to

provide goods and services." Thus, from a historical perspective, biotechnology dates back to the time when yeast was first deliberately used to ferment beer and bacteria were first used to make yogurt.

The term "biotechnology" was created in 1917 by a Hungarian engineer, Karl Ereky, to describe an integrated process for the large-scale production of pigs by using sugar beets as the source of food. According to Ereky, biotechnology was "all lines of work by which products are produced from raw materials with the aid of living things." This fairly precise definition was more or less ignored. For a number of years, biotechnology was used to describe two very different engineering disciplines. On one hand, it referred to industrial fermentation. On the other, it was used for the study of efficiency in the workplace—what is now called ergonomics. This ambiguity ended in 1961 when the Swedish microbiologist Carl Göran Hedén recommended that the title of a scientific journal dedicated to publishing research in the fields of applied microbiology and industrial fermentation be changed from the *Journal of Microbiological and Biochemical Engineering and Technology* to *Biotechnology and Bioengineering*. From that time on, biotechnology has clearly and irrevocably been associated with the study of "the industrial production of goods and services by processes using biological organisms, systems, and processes." And it was firmly grounded on expertise in microbiology, biochemistry, and chemical engineering.

An industrial biotechnology process that uses microorganisms for producing a commercial product typically has three key stages (Fig. 1.1):

1. Upstream processing: preparation of a raw material so that it can be used as a food source for the target microorganism.
2. Fermentation and transformation: growth (fermentation) of the target microorganism in a large bioreactor (usually >100 liters) with the consequent production (biotransformation) of a desired compound, which can be, for example, an antibiotic, an amino acid, or a protein.
3. Downstream processing: purification of the desired compound from either the cell medium or the cell mass.

Biotechnology research is dedicated to maximizing the overall efficiency of each of these steps and to finding microorganisms that make products that are useful as foods, food supplements, and drugs. During the 1960s and 1970s, this research focused on upstream processing, bioreactor design, and downstream processing. These studies led to enhanced bioinstrumentation for monitoring and controlling the fermentation process and to efficient large-scale growth facilities that increased the yields of various products.

The biotransformation component of the overall process was the most difficult phase to optimize. Commodity production by naturally occurring microbial strains on a large scale was often considerably less than optimum. Consequently, mutations induced by chemical mutagens or ultraviolet radiation were used to change the genetic constitution of existing strains in an effort to create variants (mutants) with enhanced product yields. However, the level of improvement that could be achieved in this way was often limited biologically. If a mutated strain, for example, synthesized too much of a compound, other metabolic functions often were impaired, thereby causing the strain's growth during large-scale fermentation to be less than desired. Despite this constraint, the traditional "induced mutagenesis and selection" strategies of strain improvement were extremely successful for a number of processes, such as the production of antibiotics.

Figure 1.1 Principal steps of a bioengineered biotechnology process. Parenthetically, Karl Ereky's scheme entailed using inexpensive sugar beets (raw material) to feed pigs (biotransformation) for the production of pork (downstream processing).

The traditional genetic improvement regimens were tedious, time-consuming, and costly because of the large numbers of colonies that had to be selected, screened, and tested. Moreover, the best result that could be expected with this approach was the improvement of an existing inherited property of a strain rather than the expansion of the genetic capabilities of a strain. Despite these limitations, by the late 1970s, effective processes for the mass production of a wide range of commercial products had been perfected.

However, the nature of biotechnology was changed forever by the development of recombinant DNA technology. With these techniques, the optimization of the biotransformation phase of a biotechnology process was achieved more directly. Genetic engineering provided the means to create, rather than merely isolate, highly productive strains. Microorganisms and

Table 1.1 Historical development of molecular biotechnology

Date	Event
1917	Karl Ereky coined the term *biotechnology*
1943	Penicillin produced on an industrial scale
1944	Avery, MacLeod, and McCarty demonstrated that DNA is the genetic material
1953	Watson and Crick determined the structure of DNA
1961	Journal *Biotechnology and Bioengineering* established
1961–1966	Entire genetic code deciphered
1970	First restriction endonuclease isolated
1972	Khorana and coworkers synthesized an entire tRNA gene
1973	Boyer and Cohen established recombinant DNA technology
1975	Kohler and Milstein described the production of monoclonal antibodies
1976	First guidelines for the conduct of recombinant DNA research issued
1976	Techniques developed to determine the sequence of DNA
1978	Genentech produced human insulin in *E. coli*
1980	U.S. Supreme Court ruled in the case of *Diamond v. Chakrabarty* that genetically manipulated microorganisms can be patented
1981	First commercial automated DNA synthesizers sold
1981	First monoclonal antibody-based diagnostic kit approved for use in the United States
1982	First animal vaccine produced by recombinant DNA technology approved for use in Europe
1983	Engineered Ti plasmids used to transform plants
1988	U.S. patent granted for a genetically engineered mouse susceptible to cancer
1988	Polymerase chain reaction (PCR) method published
1990	Approval granted in the United States for a trial of human somatic cell gene therapy
1990	Human Genome Project officially initiated
1994–1995	Detailed genetic and physical maps of human chromosomes published
1996	First recombinant protein, erythropoietin, exceeds 1 billion dollars (U.S.) in annual sales
1996	Complete DNA sequence of all the chromosomes of a eukaryotic organism, the yeast *Saccharomyces cerevisiae*, determined
1997	Nuclear cloning of a mammal, a sheep, with a differentiated cell nucleus

eukaryotic cells could be used as "biological factories" for the production of insulin, interferon, growth hormone, viral antigens, and a variety of other proteins. Recombinant DNA technology could also be used to facilitate the biological production of large amounts of useful low-molecular-weight compounds and macromolecules that occur naturally in minuscule quantities. Plants and animals became natural bioreactors producing new or altered gene products that could never have been created either by mutagenesis and selection or by crossbreeding. Finally, this new technology facilitated the development of radically new medical therapies and diagnostic systems.

The union of recombinant DNA technology with biotechnology created a vibrant, highly competitive field of study that has been called molecular biotechnology. This new research area, like molecular biology in its early stages in the 1960s, is a brash discipline in which claims and expectations sometimes exceed the researchers' abilities to produce the desired results. As well, the strategies and experimental foundations of molecular biotechnology undergo rapid changes. It is not unusual for various approaches to be superseded by different strategies within a short time. What is inevitable is that in the future, molecular biotechnology will be the standard method for developing living systems with novel functions and capabilities for the synthesis of important commercial products.

Most new scientific disciplines do not arise solely on their own. They are often formed by the amalgamation of knowledge from different areas of research. For molecular biotechnology, the biotechnology component was perfected by industrial microbiologists and chemical engineers, whereas the recombinant DNA technology portion owes much to discoveries in molecular biology, bacterial genetics, and nucleic acid enzymology (Table 1.1). In a broad sense, molecular biotechnology draws on knowledge from a diverse set of scientific disciplines to create a wide range of commercial products (Fig. 1.2).

Figure 1.2 Many scientific disciplines contribute to molecular biotechnology, which generates a wide range of commercial products.

Commercialization of Molecular Biotechnology

The ultimate objective of all biotechnology research is the development of commercial products. Consequently, molecular biotechnology is driven, to a great extent, by economics. Not only does financial investment currently sustain molecular biotechnology, but clearly the expectation of financial gain was responsible for the considerable interest and excitement during the initial stages of its development. By nightfall on October 15, 1980, the principal shareholders of Genentech stock were worth millions of dollars. The unprecedented enthusiastic public response to Genentech encouraged others to follow. Between 1980 and 1983, about 200 small biotechnology companies were founded in the United States with the help of tax incentives and funding from both stock market speculation and private investment. Like Herbert Boyer, who was first a research scientist at the University of California at San Francisco and then a vice president of Genentech, university professors started many of the early companies.

By 1985, there were over 400 biotechnology companies in the United States, including many with names that contained variants of the word "gene" to emphasize their expertise in gene cloning: Biogen, Amgen, Calgene, Engenics, Genex, and Cangene. Today, there are over 1,500 biotechnology companies in the United States and more than 3,000 worldwide. In addition, all large multinational chemical and pharmaceutical companies, including Monsanto, Du Pont, Upjohn, American Cyanamid, Eli Lilly, SmithKline Beecham, Merck, Novartis, and Hoffmann-LaRoche, to name but a few, have made significant research commitments to molecular biotechnology. During the rapid proliferation of the biotechnology business in the 1980s, small companies were absorbed by larger ones, strategic mergers took place, and joint ventures were undertaken. For example, in 1991, 60% of Genentech was sold to Hoffmann-LaRoche for $2.1 billion. And, inevitably, for various reasons, there were a number of bankruptcies. This state of flux is a characteristic feature of the biotechnology industry.

By the mid-1990s, more than a dozen new drugs that were produced by recombinant DNA technology had been certified for use with humans; over 100 are in the process of being tested in human trials, and more than 500 other recombinant pharmaceutical products are being developed. Similarly, many new molecular biotechnology products for enhancing crop and livestock yields have been created and are being marketed. The annual earnings of the molecular biotechnology industry have increased from about $6 million in 1986 to about $30 billion in 1996. It is estimated that by 2000, product sales for molecular biotechnology commodities will be more than $60 billion per year. However, generally speaking, profits for biotechnology companies have been elusive. Notwithstanding these losses, the continuing enthusiasm of investors for biotechnology indicates, at least in their estimation, that molecular biotechnology will have a pervasive impact on society.

New, independent molecular biotechnology companies have always been specialized and tend to stress the use of a particular aspect of recombinant DNA technology which is often reflected in their names. For example, after the formation of companies dedicated to the cloning of commercially important genes, several U.S. molecular biotechnology companies, including Immunex, ImmuLogic, ImmunoGen, Immunomedics, MedImmune, and Immune Response, were formed to produce genetically engineered antibodies for treating infectious diseases, cancer, and other disorders in humans.

MILESTONE

Construction of Biologically Functional Bacterial Plasmids In Vitro

S. N. COHEN, A. C. Y. CHANG, H. W. BOYER, and R. B. HELLING

Proc. Natl. Acad. Sci. USA **70**:3240–3244, 1973

The landmark study of Cohen et al. established the foundation for recombinant DNA technology by showing how genetic information from different sources could be joined to create a novel, replicatable genetic structure. In this instance, the new genetic entities were derived from bacterial autonomously replicating extrachromosomal DNA structures called plasmids. In a previous study, Cohen and Chang (*Proc. Natl. Acad. Sci. USA* **70**:1293–1297, 1973) produced a small plasmid from a large naturally occurring plasmid by shearing the larger plasmid into smaller random pieces and introducing the mixture of pieces into a host cell, the bacterium *E. coli*. By chance, one of the fragments that was about 1/10 the size of the original plasmid was perpetuated as a functional plasmid. To overcome the randomness of this approach and make the genetic manipulation of plasmids more manageable, Cohen and his coworkers decided to use an enzyme (restriction endonuclease) that cuts a DNA molecule at a specific site and produces a short extension at each end. The extensions of the cut ends of a restriction endonuclease-treated DNA molecule can combine with the extensions of another DNA molecule that was cleaved with the same restriction endonuclease. Consequently, when DNA molecules from different sources are treated with the same restriction endonuclease and mixed together, new DNA combinations that never existed before can be formed. In this way, Cohen et al. not only introduced a gene from one plasmid into another plasmid but demonstrated that the introduced gene was biologically active. To their credit, these authors fully appreciated that their strategy was "potentially useful for insertion of specific sequences from prokaryotic or eukaryotic chromosomes or extrachromosomal DNA into independently replicating bacterial plasmids." In other words, any gene from any organism could theoretically be cloned into a plasmid which, after introduction into a host cell, would be maintained indefinitely and, perhaps, produce the protein encoded by the cloned gene. By demonstrating the feasibility of gene cloning, Cohen et al. provided the experimental basis for recombinant DNA technology; established that plasmids could act as vehicles (vectors) for maintaining cloned genes; motivated others to pursue research in this area that rapidly led to the development of more sophisticated vectors and gene cloning strategies; engendered concerns about the safety and ethics of this kind of research that, in turn, was responsible for the establishment of official guidelines and governmental agencies for conducting and regulating recombinant DNA research, respectively; and contributed to the formation of the molecular biotechnology industry.

Much of the commercial development of molecular biotechnology has been centered in the United States. In other countries where investment capital is not readily available and where there is less emphasis on entrepreneurial activity, large corporations and governments have taken an active role in creating indigenous molecular biotechnology initiatives. In Japan, biotechnology was declared a "strategic industry" and made a national priority by the government. Instead of a plethora of small independent companies formed with venture capital, large Japanese corporations, which lacked basic researchers with training in recombinant DNA technology, initially formed collaborations with U.S. universities and companies. Now that they have acquired the necessary expertise, major Japanese corporations are more directly involved in molecular biotechnology research and product development.

The European biotechnology industry has been developing steadily. In 1995, there were over 600 European biotechnology companies. In economically less developed countries, governments have played a significant role as the initiators of molecular biotechnology. The motivation for this activity lies in the strong belief that molecular biotechnology is the "last great technological revolution of the twentieth century." No country wants to be left out of the gains expected from this technology.

Now that molecular biotechnology is nearing the end of its second decade, it has become the engine powering the development of many novel products. Although at first some scientists believed that molecular biotechnology was an esoteric technical development with marginal practical utility, now it is mainstream. Undoubtedly, the next 10 years of commercial molecular biotechnology will be exciting, although specific predictions are risky because new technical innovations rapidly replace existing technologies.

Concerns and Consequences

Molecular biotechnology ought to contribute unprecedented benefits to humanity. It should

- provide opportunities to accurately diagnose and prevent or cure a wide range of infectious and genetic diseases
- significantly increase crop yields by creating plants that are resistant to insect predation, fungal and viral diseases, and environmental stresses such as short-term drought and excessive heat
- develop microorganisms that will produce chemicals, antibiotics, polymers, amino acids, enzymes, and various food additives
- develop livestock and other animals that have enhanced genetically determined attributes
- facilitate the removal of pollutants and waste materials from the environment

Although it is exciting and important to emphasize the positive aspects of new advances, there are also social concerns and consequences that must be addressed. Because molecular biotechnology is so broadly based, its potential impact on society must be considered. For example,

- Will some genetically engineered organisms be harmful either to other organisms or to the environment?
- Will the development and use of genetically engineered organisms reduce natural genetic diversity?
- Should humans be genetically engineered?
- Will diagnostic procedures undermine individual privacy?
- Should genetically engineered animals be patentable?
- Will financial support for molecular biotechnology constrain the development of other important technologies?
- Will the emphasis on commercial success mean that the benefits of molecular biotechnology will be available only to the wealthy?
- Will agricultural molecular biotechnology undermine traditional farming practices?
- Will medical therapies based on molecular biotechnology supersede equally effective traditional treatments?
- Will the quest for patents inhibit the free exchange of ideas among research scientists?

These and many other issues have been considered by government commissions, discussed extensively at conferences, and thoughtfully debated and analyzed by individuals in popular and academic publications. On this basis, rules and regulations have been formulated, guidelines es-

tablished, and policies created. There has been an active and extensive participation by both scientists and the general public in deciding how molecular biotechnology should proceed, although some controversies still remain.

Molecular biotechnology, with much fuss and fanfare, has become a comprehensive scientific venture, both commercially and academically, in a remarkably short time. A number of new scientific and business publications are devoted to molecular biotechnology. Both graduate and undergraduate programs and courses have been created at many universities throughout the world to teach molecular biotechnology. The enthusiasm expressed in a 1987 U.S. Office of Technology Assessment document may not have been exaggerated in the statement that molecular biotechnology is "a new scientific revolution that could change the lives and futures of . . . citizens as dramatically as did the Industrial Revolution two centuries ago and the computer revolution today. The ability to manipulate genetic material to achieve specified outcomes in living organisms . . . promises major changes in many aspects of modern life."

SUMMARY

In 1973, Stanley Cohen, Herbert Boyer, and their coworkers devised a method for transferring genetic information (genes) from one organism to another. This procedure, which became known as recombinant DNA technology, enabled researchers to isolate specific genes and perpetuate them in host organisms. Recombinant DNA technology has been beneficial to many different areas of study. However, its impact on biotechnology has been extraordinary.

Biotechnology, for the most part, uses microorganisms on a large scale for the production of commercially important products. Before the advent of recombinant DNA technology, the most effective way of increasing the productivity of an organism was to induce mutations and then use selection procedures to identify organisms with superior traits. This process was not foolproof; it was time-consuming, labor-intensive, and costly, and only a small set of traits could be enhanced in this way. Recombinant DNA technology, however, provided a rapid, efficient, and powerful means for creating microorganisms with specific genetic attributes. Moreover, the tools of recombinant DNA technology enable not only microorganisms but also plants and animals to be genetically engineered. Combining recombinant DNA technology with biotechnology created a dynamic and exciting discipline called molecular biotechnology.

From its beginning, molecular biotechnology captured the imagination of the public. Many small companies dedicated to gene cloning (recombinant DNA technology) were established with funding from private investors. Although it took these biotechnology companies somewhat longer than originally expected to bring their products to the marketplace, a number of recombinant DNA-based products are currently available and many more are expected soon.

Because of its broad impact, molecular biotechnology has been scrutinized carefully for its potential effects on society. Some of the concerns that have been considered include its safety, possible negative effects on the environment, and the private or public ownership of genetically engineered organisms.

References

Anonymous. 1987. *New Developments in Biotechnology—Background Paper: Public Perceptions of Biotechnology.* Office of Technology Assessment, U.S. Congress, U.S. Government Printing Office, Washington, D.C.

Bud, R. 1991. Biotechnology in the twentieth century. *Soc. Stud. Sci.* **21:**415–457.

Bud, R. 1993. *The Uses of Life: a History of Biotechnology.* Cambridge University Press, Cambridge, United Kingdom.

Busch, L., W. B. Lacy, J. Burkhardt, and L. R. Lacy. 1992. *Plants, Power, and Profit: Social, Economic and Ethical Consequences of the New Biotechnologies.* Blackwell Publishers, Cambridge, Mass.

Cohen, S., A. C. Y. Chang, H. W. Boyer, and R. B. Helling. 1973. Construction of biologically functional bacterial plasmids in vitro. *Proc. Natl. Acad. Sci. USA* **70:**3240–3244.

Davis, B. D. (ed.). 1991. *The Genetic Revolution: Scientific Prospects and Public Perceptions.* The Johns Hopkins University Press, Baltimore, Md.

Grace, E. S. 1997. *Biotechnology Unzipped: Promises & Realities.* Trifolium Press, Inc., Toronto, Canada.

REVIEW QUESTIONS

1. What are the shortcomings of the "mutation and selection" method for developing enhanced organisms that have economic importance?

2. Why was the work reported by Cohen and Boyer and their coworkers in 1973 considered to be important?

3. What does Karl Ereky have to do with biotechnology?

4. Describe the basic steps of a bioengineered biotechnology process.

5. Distinguish between biotechnology and molecular biotechnology.

6. Discuss some of the concerns that have been raised about molecular biotechnology.

7. Discuss the statement "Molecular biotechnology is a diverse science."

8. What are some of the potential opportunities that are anticipated from molecular biotechnology?

9. Outline the history of the biotechnology industry in the last 30 years.

10. Read the News sections of the most recent issue of the journal *Nature Biotechnology*, which can be found in a university science library or by contacting http://biotech.nature.com on the World Wide Web, and discuss two of these topics in detail.

Molecular Biotechnology Biological Systems

Molecular biotechnologists use a very broad range of biological systems, including microorganisms; insect, plant, and mammalian cell lines; insect, plant, and mammalian viruses; and multicellular organisms such as plants, mice, and domestic animals, depending on the particular circumstances. In general, biological systems are extremely important for the biotransformation phase of a biotechnology process. And, in many instances, a genetically modified replicatable biological unit, whether it is a microorganism, virus, plant, or animal, is the actual commercial product. Of the many different replicatable biological units that form the repertoire for molecular biotechnology research, the bacterium *Escherichia coli*, the unicellular yeast *Saccharomyces cerevisiae*, and various animal cell lines are workhorses. In this context, *E. coli* is commonly used as the host cell for recombinant DNA technology, and all three of these systems are important for the production of proteins from cloned genes.

Various biological systems with highly specific functions will be described fully in ensuing chapters. For example, an insect virus–insect cell line system that is used for the production of authentic proteins from cloned genes and the genetic modification of domestic animals such as cows, sheep, and pigs are reviewed in Chapters 7 and 19, respectively. In this chapter, the more prominent molecular biotechnology biological systems, which are also discussed in later chapters, are considered briefly.

Prokaryotic and Eukaryotic Organisms

Living organisms are assigned to two major groups: prokaryotes and eukaryotes. A number of structural features are responsible for these designations. At various places in later chapters, the differences between prokaryotes and eukaryotes will be elaborated in more detail. Here, in broad terms, the main distinctions between these two classes of organisms are the presence or absence of a nucleus that contains the chromosomal DNA, the chemical components of the cell wall, and the presence or absence of subcellular cytoplasmic organelles. In a prokaryotic cell, such as a bacterium,

the chromosomal DNA is in direct contact with the cytoplasm, the rigid cell wall may contain peptidoglycan but never chitin or cellulose, and there are no subcellular cytoplasmic organelles. By contrast, in a eukaryotic cell, the chromosomal DNA is enclosed in a membrane to form a nucleus; the cell wall, if present, may contain chitin or cellulose but never peptidoglycan; and there are subcellular organelles in the cytoplasm such as mitochondria, Golgi vesicles, and, in plants, chloroplasts (Fig. 2.1).

Escherichia coli

The bacterium *E. coli* is one of the most studied organisms in the world. An immense amount of information about its genetics, molecular biology, biochemistry, physiology, and general biology has been gathered in the last 50 years. It is a gram-negative, nonpathogenic, short, motile, rod-shaped organism (<1 micrometer [μm] in length) that is naturally found in the intestines of humans but not normally in soil or water. Its ability to multiply by binary fission in the laboratory on a simple culture medium consisting of

Figure 2.1 Schematic representations of a prokaryotic bacterium **(A)** and a eukaryotic animal cell **(B)**.

A

B

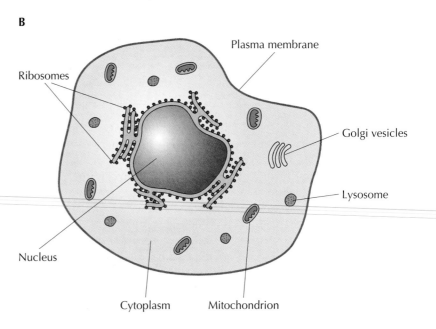

MILESTONE

Novel Genotypes in Mixed Cultures of Biochemical Mutants of Bacteria
J. LEDERBERG and E. L. TATUM
Cold Spring Harbor Symp. Quant. Biol. **11**:113–114, 1946

For nearly the first half of the 20th century, the prevalent view held by bacteriologists was that bacteria, unlike other organisms, did not rely on preexisting, chance genetic changes (mutations) for survival. Rather, it was believed that the environment, in most instances, triggered the appropriate physiological response that enabled bacteria to survive under extreme conditions. This doctrine was dispelled by S. E. Luria and M. Delbrück (*Genetics* **28**:491–511, 1943), who proved that resistance of *E. coli* to a bacterial virus (bacteriophage) was due to preexisting mutations, not to the effect of the bacteria being exposed to the bacteriophage. Other researchers substantiated this observation with different environmental agents. The Luria-Delbrück study marked the beginning of modern bacterial genetics.

The work of Lederberg and Tatum in 1946 catapulted *E. coli* into the forefront of genetic studies when they showed that genetic information could be exchanged between members of genetically mixed *E. coli* cultures and that new genetic combinations were formed, as in sexual organisms, by physical exchange between chromosomes (genetic recombination). Briefly, Lederberg and Tatum established lines of *E. coli* by inducing mutations and isolating cells that had multiple inherited metabolic defects. If, for example, one of the pure cultures could synthesize compounds A, B, and C but not D or E, it was designated ABCde. A pure culture with a complementary biological constitution that was abcDE was also created. When these two cultures were mixed, bacteria that were ABCDE were formed. The ABCDE type did not occur in either pure culture spontaneously. After other possibilities were experimentally eliminated, Lederberg and Tatum concluded that the ABCDE combination arose by the physical exchange of genetic information between two chromosomes. In other words, *E. coli* had a form of sexuality that facilitated the formation of new genetic combinations that was similar to the genetic systems of other organisms.

At the time, Luria claimed that this discovery might be "among the most fundamental advances in the whole history of bacteriological science." This prediction was, for the most part, on the mark. The findings of Lederberg and Tatum engendered a series of brilliant studies by Lederberg and others that established how the genetic system of *E. coli* functions and led to the extensive use of *E. coli* as a research organism for studying the molecular basis of the synthesis of nucleic acids and proteins as well as other important biological processes.

ions (Na^+, K^+, Mg^{2+}, Ca^{2+}, NH_4^+, Cl^-, HPO_4^{2-}, and SO_4^{2-}), trace elements, and a carbon source such as glucose has made it a favorite research organism. In complex (enriched) liquid culture medium that contains amino acids, vitamins, salts, trace elements, and a carbon source, the cell generation time at 37°C for *E. coli* in the logarithmic phase of growth is about 22 minutes.

Every organism grows within a particular temperature range. At the upper temperature limit, proteins begin to denature and other important cellular components are damaged, leading to the death of a cell. At the lower temperature limit, vital processes also cease because of changes in the structure of proteins. On the basis of the temperature range within which a microorganism proliferates, it can be classified as a thermophile (45 to >90°C), mesophile (10 to 47°C), psychrophile (−5 to 20°C), or psychrotroph (−5 to 35°C). A microorganism that has a particular temperature range may be helpful for a specific biotechnological purpose. For example, thermophiles are often used as sources of genes for enzymes that are stable at high temperatures and are used in either an industrial or a laboratory process, whereas psychrotrophs are genetically engineered to perform certain tasks such as the biodegradation of toxic pollutants in cold terrestrial and aquatic environments.

E. coli is a versatile microorganism that can be grown in either the presence (aerobically) or absence (anaerobically) of oxygen. However, for the

Table 2.1 Some microorganisms that have been genetically engineered to perform biotechnological processes

Acremonium chrysogenum
Bacillus brevis
Bacillus subtilis
Bacillus thuringiensis
Corynebacterium glutamicum
Erwinia herbicola
Escherichia coli
Pseudomonas spp.
Rhizobium spp.
Streptomyces spp.
Trichoderma reesei
Xanthomonas campestris
Zymomonas mobilis

optimal production of recombinant protein, *E. coli* and other host microorganisms are routinely grown aerobically. For the production of recombinant protein in a research laboratory, bacterial cultures are grown in a complex liquid culture medium in flasks. The growth temperature and extent of aeration are maintained by placing flasks with a fresh inoculum of bacteria into a shaking water bath. Under these conditions, the amount of aeration is sufficient for cell proliferation but not necessarily optimal for the production of a recombinant protein. The extent of growth and protein production is limited by the concentration of dissolved oxygen, which is approximately 9 parts per million of O_2 dissolved in 1 liter of water at 20°C, in the growth medium rather than by the level of the carbon or nitrogen sources. For the industrial production of recombinant protein by a microorganism, the amount of oxygen that is available to the cells is a critical factor. Special containers (fermenters) and aeration systems have been designed to maximize the output of large amounts of recombinant protein.

In addition to *E. coli*, a large number of different microorganisms are used for molecular biotechnology applications (Table 2.1). Generally, these microorganisms fall into two groupings: those that provide genes for specific functions and those that are genetically engineered to perform certain tasks more effectively. In the former category, for example, genes such as those encoding heat-resistant DNA polymerases that are essential for a widely used technique called the polymerase chain reaction have been isolated from thermophiles and cloned into *E. coli*. In the latter category, for example, various strains of *Corynebacterium glutamicum* have been genetically modified to improve the yield of industrially important amino acids.

Saccharomyces cerevisiae

S. cerevisiae is a nonpathogenic, single-celled (~5-μm diameter) yeast that in many respects is a eukaryotic version of *E. coli*. Its genetics, molecular biology, and metabolism have been studied extensively. As a laboratory organism, *S. cerevisiae*, which reproduces by the budding off of a sibling cell from a parent cell, readily proliferates in the same kind of simple culture medium that is used for growing *E. coli*. For centuries, its ability to convert sugars to ethanol and carbon dioxide has been used for the production of alcoholic beverages and bread. Today, more than a million tons of *S. cerevisiae* are used worldwide annually. In addition to its bioindustrial importance, *S. cerevisiae* is a significant research organism, in particular as a model system for other eukaryotic organisms, including humans. In the latter context, for example, many of the cell division cycle genes of *S. cerevisiae* are similar to those in humans; this knowledge has led to the isolation and characterization of human genes that are responsible for cancer. Also, a widely used yeast genetic system, the yeast artificial chromosome, is indispensable for studies of the physical organization of human DNA. In 1996, adding to its fame, *S. cerevisiae* became the first eukaryotic organism to have its complete set of chromosomes entirely sequenced.

Frequently, after a protein of a eukaryotic organism is synthesized, it is modified by the enzymatic addition of a low-molecular-weight compound(s). In many cases, these additions are required for the proper functioning of the protein. Unfortunately, *E. coli* and other prokaryotes do not possess the biosynthetic ability to perform these protein modifications. Consequently, *S. cerevisiae* has been used for the production of recombinant proteins from eukaryotic organisms. Other yeasts such as *Kluyveromyces lac-*

tis, Saccharomyces diastaticus, Schizosaccharomyces pombe, Yarrowia lipolytica, Pichia pastoris, and *Hansenula polymorpha* have also been used as host organisms for this purpose. Of all the yeasts, *P. pastoris* and *H. polymorpha* produce the largest amounts of correctly modified recombinant proteins.

Eukaryotic Cells in Culture

Although the specifics are quite different, very similar approaches and procedures are used for growing insect, plant, and mammalian cells in culture. Initially, a small sample of tissue is removed from an organism and the cells are released from the surrounding intercellular network (extracellular matrix) that holds them together by treatment with enzymes that digest proteins (proteolytic enzymes). With plant cells, additional enzymes are used to break down the cell wall. After the cells are freed of the extracellular material that mechanically prevents cell division in vitro, they are placed into a complex growth medium that contains amino acids, antibiotics, vitamins, salts, glucose, and growth factors. Under these conditions, the cells undergo cell divisions until the surface of the vessel is covered with a monolayer of cells. At this point, cell division ceases unless samples of the cells are collected, diluted, and transferred into fresh medium in new culture vessels. Usually, initial (primary) cell cultures can be transferred (passaged, or subcultured) and maintained for 50 to 100 cell generations before the cells lose the ability to divide and begin to die off. Cell growth for a definite period followed by cell death is characteristic of a primary cell culture. The cells of a primary cell culture retain some of the properties of the original cell type and therefore are used to study biochemical features of various tissues.

Frequently, during passages of primary cell cultures some of the cells undergo genetic changes that facilitate growth in cell culture. The cells that have this selective advantage grow indefinitely in vitro and are referred to as established cell lines. Some established cell lines retain a few biochemical characteristics of the original cells; others do not. Most of the cells that grow indefinitely have significant chromosome changes that include extra versions of some chromosomes and loss of others. For molecular biotechnology, established cell lines are used, on a small scale, for maintaining viruses and determining the protein that is encoded by a cloned DNA sequence and, on a large scale, for the commercial production of vaccines and proteins encoded by cloned genes (recombinant proteins).

SUMMARY

Molecular biotechnologists use a number of different biological systems for genetic manipulations and producing commercial products. Of these replicatable biological units, the most significant are the bacterium *Escherichia coli,* the yeast *Saccharomyces cerevisiae,* and insect, plant, and mammalian cell cultures.

References

Demain, A. L., and N. A. Solomon (ed.). 1985. *Biology of Industrial Organisms.* Benjamin/Cummings Publishing Co., Inc., Menlo Park, Calif.

Dujon, B. 1996. The yeast genome project: what did we learn? *Trends Genet.* **12:**263–270.

Lodish, H., D. Baltimore, A. Berk, S. L. Zipursky, P. Matsudaira, and J. Darnell. 1995. *Molecular Cell Biology,* 3rd ed. Scientific American Books, New York, N.Y.

O'Leary, W. M. (ed.). 1989. *Practical Handbook of Microbiology.* CRC Press, Inc., Boca Raton, Fla.

REVIEW QUESTIONS

1. Why do molecular biotechnologists use so many different biological systems?

2. What is a prokaryote?

3. What is a eukaryote?

4. What is *E. coli*?

5. What does "gram negative" signify?

6. What is *S. cerevisiae*?

7. What are the major components of a simple liquid culture medium?

8. What are the major components of a complex liquid culture medium?

9. What is a primary cell culture?

10. What is an established cell line?

CHAPTER 3

DNA, RNA, and Protein Synthesis

The information encoded in genetic material, deoxyribonucleic acid (DNA), is responsible for establishing and maintaining the cellular and biochemical functions of an organism. In most organisms, the DNA is an extended double-stranded polymer. The sequence of units (deoxyribonucleotides) of one DNA strand is complementary to the deoxyribonucleotides of the other strand. This complementarity enables new DNA molecules to be synthesized with the same linear array of deoxyribonucleotides in each strand as an original DNA molecule. The process of DNA synthesis is called replication. A specific order of deoxyribonucleotides determines the information content of an individual genetic element (gene). Some genes encode proteins, and others encode only RNA molecules. The protein-coding genes (structural genes) are decoded by two successive major cellular processes: RNA synthesis (transcription) and protein synthesis (translation). First, a messenger RNA (mRNA) molecule is synthesized from a structural gene. Second, an individual mRNA molecule interacts with other components, including ribosomes, transfer RNAs (tRNAs), and enzymes to produce a protein molecule. A protein consists of a precise sequence of amino acids which is essential for its activity.

Molecular biotechnologists use genes from a variety of organisms to create beneficial products. To understand how this is accomplished, it is helpful to know about the structure of DNA, replication, transcription, and translation.

Structure of DNA

The chemistry of DNA has been studied since 1868. By the 1940s, it was known that DNA was made up of individual units called nucleotides that are linked to each other to form long chains. A nucleotide consists of an organic base (base), a five-carbon sugar (pentose), and a phosphate group (Fig. 3.1A). The phosphate group and organic base are attached to the 5′ carbon and 1′ carbon atoms of the sugar moiety, respectively. In DNA, there are four kinds of organic bases: adenine (A), guanine (G), cytosine (C), and thymine (T) (Fig. 3.1B). The sugar of DNA is 2′-deoxyribose because it has

19

Figure 3.1 Chemical structures of the components of DNA. **A.** A representative nucleotide. The term "Base" denotes any of the four bases (adenine, guanine, cytosine, and thymine) that are found in DNA. The deoxyribose sugar is enclosed by dashed lines. The numbers with the primes mark the carbon atoms of the deoxyribose moiety. **B.** The bases of DNA. The circled nitrogen atom is the site of attachment of the base to the 1′ carbon atom of the deoxyribose moiety.

a hydroxyl (OH) group only on the 3′ carbon of the sugar moiety. By contrast, ribose, which is another five-carbon sugar, has hydroxyl groups at both the 2′ and 3′ carbons of the pentose ring. The nucleotides of DNA are joined by phosphodiester bonds, with the phosphate group of the 5′ carbon of one nucleotide linked to the 3′ OH group of the deoxyribose of the adjacent nucleotide (Fig. 3.2). A polynucleotide strand has a 3′ OH group at one end (3′ end) and a 5′ phosphate group at the other (5′ end).

In 1953, James Watson and Francis Crick, using X-ray diffraction analysis of crystallized DNA, discovered that native DNA consists of two long chains (strands) that form a double-stranded helix (Fig. 3.3). The coiled polynucleotide chains of DNA are held together by hydrogen bonds between the bases of the opposite strands. The bases occur as specific sets of

5′ end

Adenine

Phosphodiester
linkage

Thymine

Guanine

Cytosine

3′ end

Figure 3.2 Chemical structure of a single strand of DNA.

Figure 3.3 A rod-ribbon model of double-helical DNA. The rods represent the complementary base pairs, and the ribbons represent the deoxyribose-phosphate backbones.

complementary pairs (Fig. 3.4). A pairs only with T, and G pairs only with C. The A·T base pairs are held together by two hydrogen bonds, and the G·C base pairs are held together by three. The number of complementary base pairs (bp) is often used to characterize the length of a double-stranded DNA molecule. For DNA molecules with thousands or millions of base pairs, the designations are kilobase pairs (kb) or megabase pairs (Mb), respectively. For example, the DNA of human chromosome 1 is one double-stranded helix that has about 263 Mb.

The A·T and G·C base pairs lie within the interior of the molecule and the 5′-to-3′-linked phosphorus and deoxyribose components form the backbone of each strand (Fig. 3.4). The two strands of a duplex DNA molecule run in opposite directions to each other (antiparallel chains). One chain is oriented in a 3′-to-5′ direction, and the other is oriented in a 5′-to-3′ direction. Because of the base pairing requirements, when one strand of DNA has, for example, the base sequence 5′-TAGGCAT-3′, the complementary

Figure 3.4 Chemical structure of double-stranded DNA.

strand must be 3′-ATCCGTA-5′. In this case, the double-stranded form would be $\begin{smallmatrix}5'\text{-TAGGCAT-}3'\\3'\text{-ATCCGTA-}5'\end{smallmatrix}$. By convention, when DNA is drawn on a horizontal plane, the 5′ end of the upper strand is on the left.

In theory, genetic material has two major functions. It must be reproduced (replicated) with a high degree of accuracy and encode information for the production of proteins. The Watson-Crick model of DNA fully meets these important requirements. First, with base complementarity, each preexisting DNA strand acts as a template for the production of a new complementary strand. Consequently, after one round of replication, two daughter molecules are produced, with each having the same sequence of nucleotide pairs as the original DNA molecule. Second, the sequence of nucleotides of a gene provides the code for the production of a protein.

DNA Replication

As predicted by the Watson-Crick model of DNA, each strand of an existing DNA molecule acts as a template for the production of a new strand,

and the sequence of nucleotides of the synthesized (growing) strand is determined by base complementarity. During replication, the phosphate group of each incoming nucleotide is enzymatically joined by a phosphodiester linkage to the 3'-OH group of the last nucleotide that was incorporated in the growing strand (Fig. 3.5A). The nucleotides that are used for DNA replication are triphosphate deoxyribonucleotides which have three consecutive phosphate groups attached to the 5' carbon of the deoxyribose sugar moiety. The phosphate that is attached to the 5' carbon is designated the α phosphate, the next phosphate is the β phosphate, and the third one is the γ phosphate (Fig. 3.6). During the replication process, the β and γ phosphates are cleaved off as a unit, and the α phosphate is linked to the 3'-OH group of the previously incorporated nucleotide (Fig. 3.5B). The DNA synthesis machinery of prokaryotes and eukaryotes includes a large number of different proteins. Of these, DNA polymerases are responsible for binding deoxyribonucleotides, fitting the correct nucleotide into place according to the base pairing requirement of the template strand, and forming the phosphodiester linkage.

In bacteria, DNA replication is initiated at a specific region called the origin of replication (or origin). In eukaryotes, a chromosome has many different sites of initiation of replication. Because of these multiple origins of replication, part of the eukaryotic replication process includes enzymatically joining (ligating) segments of newly synthesized DNA together with phosphodiester bonds. As well, in eukaryotes, a special replication enzyme called telomerase is used for the completion of the ends (telomeres) of each chromosome.

Decoding Genetic Information: RNA and Protein

The vast majority of genes encode information for the production of protein chains. Proteins are essential polymers that are involved in almost all biological functions. They catalyze chemical reactions; transport molecules within cells; escort molecules between cells; control membrane permeability; give support to cells, organs, and body structures; cause movement; provide protection against infectious agents and toxins; and regulate the differential production of other gene products. A protein chain consists of a specific sequence of units called amino acids. All amino acids have the same basic chemical organization. There is a central carbon atom (the α carbon) that has a hydrogen (H), a carboxyl group (COO^-), an amino group (NH_3^+), and an R group attached to it (Fig. 3.7A). An R group can be any one of 20 different side chains (groups) that make up the 20 different amino acids found in proteins. When R, for example, is a methyl group (CH_3), then the amino acid is alanine. The amino acids of proteins are designated by either a three- or a one-letter notation (Table 3.1). In a protein, each amino acid is linked to an adjacent amino acid by a peptide bond that joins the carboxyl group of one amino acid to the amino group of the adjacent one (Fig. 3.7B). The first amino acid of a protein has a free amino group (N terminus), and the last amino acid in the polypeptide chain has a free carboxyl group (C terminus).

Proteins range in length from about 40 to more than 1,000 amino acid residues. A protein folds into a particular shape (configuration or conformation) depending on the location of specific amino acid residues and the overall amino acid composition. As well, many functional proteins consist of two or more polypeptide chains. In some cases, multiples of the same

Figure 3.5 DNA replication. **A.** The incoming nucleotide is a deoxyribonucleotide triphosphate that is directed by DNA polymerase to pair with the complementary base of the template strand. The letters A, G, C, and T denote the bases adenine, guanine, cytosine, and thymine, respectively.

B

Figure 3.5 **B.** The α phosphate of the incoming nucleotide forms a phosphodiester bond with the 3′ hydroxyl group of the growing strand. The next incoming nucleotide of the growing strand that is complementary to the nucleotide of the template strand is positioned by DNA polymerase.

Figure 3.6 Structure of a deoxyribonucleotide triphosphate. The phosphate units are designated α, β, and γ. The α phosphate is directly attached to the 5′ carbon of the deoxyribose sugar. During DNA synthesis, the α phosphate participates in the formation of the phosphodiester bond and the β and γ phosphate groups are released as a unit (pyrophosphate).

Figure 3.7 Generalized structures of an amino acid and a peptide bond. **A.** An amino acid. The R represents the location of the side chain. **B.** A peptide bond. The peptide bond is encircled, and R1 and R2 represent different side chains.

A

B

Peptide bond

polypeptide chain are required for an active protein molecule (homomeric protein). In other instances, a set of different protein chains (subunits) assembles to form a functional protein (heteromeric protein). Finally, large protein complexes that are made up of many different subunits often perform important cellular functions.

The decoding of genetic information is carried out through intermediary ribonucleic acid (RNA) molecules that are transcribed from regions of the DNA. RNA molecules are linear polynucleotide chains that differ from DNA in two important aspects. First, the sugar moiety of the nucleotides of RNA is ribose, which has hydroxyl groups on both the 2′ and 3′ carbons of the sugar. Second, instead of thymine, the base uracil (U) is found in RNA. Most RNA molecules are single stranded, although often there are segments of nucleotides within a single chain that are complementary to each

Table 3.1 The amino acids of proteins and their designations

Amino acid	Three-letter designation	Single-letter designation
Alanine	Ala	A
Arginine	Arg	R
Asparagine	Asn	N
Aspartic acid	Asp	D
Cysteine	Cys	C
Glutamic acid	Glu	E
Glutamine	Gln	Q
Glycine	Gly	G
Histidine	His	H
Isoleucine	Iso	I
Leucine	Leu	L
Lysine	Lys	K
Methionine	Met	M
Phenylalanine	Phe	F
Proline	Pro	P
Serine	Ser	S
Threonine	Thr	T
Tryptophan	Trp	W
Tyrosine	Tyr	Y
Valine	Val	V

other and form double-stranded regions (Fig. 3.8). The base pairing within a single RNA strand is the same as the base pairing between complementary sequences of DNA, except that uracil pairs with adenine. Base pairing can occur between two RNA molecules if they contain complementary sequences of base pairs.

The major kinds of RNA molecules that are essential for the decoding of genetic information are mRNA, ribosomal RNA (rRNA), and tRNA. The production of RNA from DNA is called transcription. In most prokaryotes, a single RNA polymerase is responsible for the transcription of all types of RNA. In eukaryotic organisms, mRNA, rRNA, and tRNA are each transcribed by a different RNA polymerase.

In many of its features, transcription resembles replication. Briefly, one strand of the DNA of a specific region acts as a template for transcription. RNA polymerase sequentially joins together with 3'-5' phosphodiester linkages a string of ribonucleotides that are complementary to the nucleotides of the template DNA strand (Fig. 3.9). As transcription proceeds, the newly synthesized RNA is released from the DNA and the DNA helix is reconstituted. Since only segments of DNA molecules are transcribed, sets of short stretches of base pairs (signal regions) within the DNA are required to ensure that transcription is initiated at the correct nucleotide and that it terminates at a specific nucleotide. The sequences that control the initiation of transcription usually precede the coding sequence, and the termi-

Figure 3.8 Secondary structure of an RNA molecule. The lines represent hydrogen bonding between complementary base pairs. The ribose-phosphate backbone is omitted.

Figure 3.9 Schematic representation of transcription. The arrow indicates the direction of transcription.

nation signal sequences follow it. The DNA segment that precedes a gene is called the 5′-flanking or upstream region, and the one following a gene is the 3′-flanking or downstream region.

From a molecular perspective, a gene is a specific nucleotide sequence that is transcribed into RNA. Structural genes, which make up the vast majority of transcribed DNA sequences, encode proteins. The functional transcription product of a structural gene is an mRNA. In prokaryotes, a contiguous DNA segment forms a structural gene. Prokaryotic transcription entails the binding of RNA polymerase to a promoter region, the initiation of transcription at the first nucleotide of the structural gene (the coding region), and the cessation of transcription at a termination sequence that lies downstream from the coding region (Fig. 3.10). In eukaryotic organisms, most structural genes consist of a set of coding regions (exons) that are separated by noncoding regions (introns, or intervening sequences). After an entire eukaryotic structural gene is transcribed, the introns are enzymatically cut out of this primary transcript and the exons, in the correct order, are linked (spliced) together to form a functional mRNA (Fig. 3.11; Fig. 3.12). In general, exons tend to be 150 to 200 bases long, and introns can vary from as few as 40 to over 10,000 bases long. A small number of eukaryotic structural genes lack introns. And, in a few instances, an mRNA may undergo legitimate alternative splicing events. For example, in one kind of tissue, all the exons of the primary transcript may be spliced together to form

Figure 3.10 Schematic representation of a prokaryotic structural gene. The promoter region (p), the site of initiation and direction of transcription (the right-angled arrow), and the termination sequence for RNA polymerase (t) are depicted. A prokaryotic structural gene is transcribed into mRNA and then directly into protein.

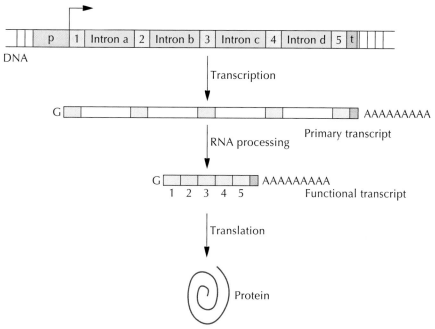

Figure 3.11 Schematic representation of a eukaryotic structural gene. The promoter region (p), the site of initiation and direction of transcription (the right-angled arrow), and the termination sequence for RNA polymerase (t) are depicted. The numbers 1 to 5 mark the exons of the structural gene, and the letters a to d mark the introns. The primary transcript is polyadenylated at the 5′ end and capped with a guanine (G) at the 3′ end. Processing of the primary transcript removes the introns. The functional RNA is translated into protein.

a functional mRNA, whereas in another tissue, the initial transcript may undergo a different pattern of exon splicing, with an exon being skipped during the process of intron removal and a novel functional mRNA being produced. The exon-skipping mechanism generates different gene products in different tissues from the same structural gene (Fig. 3.13).

Figure 3.12 Splicing of a eukaryotic primary RNA transcript. The bracketing arrows mark the sites that are spliced together after the removal of the intervening DNA regions. In this example, introns a and b are spliced out of the primary transcript and exons 1, 2, and 3 are spliced together to form a functional mRNA.

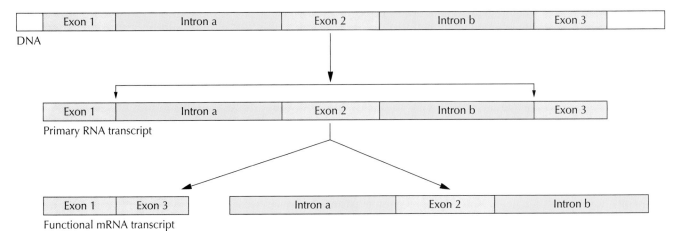

DNA

Primary RNA transcript

Functional mRNA transcript

Figure 3.13 Alternative splicing of a eukaryotic primary RNA transcript. The bracketing arrows mark the sites that are spliced together after the removal of the intervening DNA region. In this example, exon 2 flanked by introns 1 and 2 is spliced out of the primary transcript and exons 1 and 3 are spliced together to form a functional mRNA transcript.

In a metabolically active cell, about 3 to 5% of the cellular RNA is mRNA, about 90% is rRNA, and about 4% is tRNA. There can be dozens of different mRNAs in a cell. By contrast, there are two major types of rRNA. The larger of these rRNAs combines with a set of proteins to form a ribonucleoprotein complex called the large ribosomal subunit. The smaller rRNA combines with another set of proteins to form a small ribosomal subunit. During protein synthesis, one large ribosomal subunit and one small ribosomal subunit combine to form a ribosome. Both of the ribosomal subunits of eukaryotes are larger than those of prokaryotes.

In addition to thousands of ribosomes, a cell that is actively synthesizing proteins has about 60 different types of tRNA molecules. The tRNA molecules range in length from about 75 to 93 nucleotides. Because of intrastrand complementary segments of nucleotides, each tRNA molecule forms a folded, L-shaped structure (Fig. 3.14). A particular amino acid is enzymatically linked by its carboxyl end to the 3′ end of a specific tRNA. For example, the enzyme arginyl-tRNA synthetase adds the amino acid arginine to the tRNA^Arg molecule. There is at least one tRNA for each of the 20 amino acids found in proteins. After the binding of a particular amino acid to its tRNA, the tRNA is said to be "charged." In another part of the tRNA molecule, there are three unpaired nucleotides which together are called the anticodon sequence. This sequence plays a crucial role in the formation of the linear array of amino acids that constitute a protein.

Figure 3.14 Conformation of a charged tRNA. The amino acid is attached to the 3′ end of the tRNA, and the location of the anticodon region is enclosed by the dashed lines.

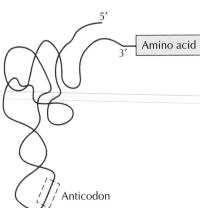

Translation

Translation requires the interaction of mRNA, charged tRNAs, ribosomes, and a large number of proteins (factors) that facilitate the initiation, elongation, and termination of the polypeptide chain. In prokaryotes, translation is initiated by the binding of the 3′-UAC-5′ anticodon of a specific initiator tRNA, fMet-tRNA^fMet, where f represents a formyl moiety that is bound to the methionine residue, to a 5′-AUG-3′ codon of an mRNA that is bound to a small ribosomal subunit. No other charged tRNA can bind to an mRNA–small ribosomal subunit complex. Proteins (initiation factors) facil-

itate the binding of the initiator tRNA to the mRNA–small ribosomal subunit complex. The binding of an mRNA to the small ribosomal subunit is mediated by base pairing between a sequence of about eight nucleotides (a Shine-Dalgarno [SD] sequence) that is located near the 5′ end of the mRNA and a complementary sequence near the 3′ end of the rRNA of the small ribosomal subunit. A large ribosomal subunit combines with the fMet-tRNA$_F^{Met}$–mRNA–small subunit complex to form the initiation complex (Fig. 3.15).

In eukaryotes, translation is initiated by the binding of a specific charged initiator tRNA, Met-tRNAMet, along with initiation factors, to a small ribosomal subunit. Next, the 5′ end of an mRNA associates with the initiator tRNA–small ribosomal subunit complex, and the complex migrates along the mRNA until an AUG sequence (initiator codon) is encountered. Then the UAC anticodon sequence of the initiator Met-tRNAMet base pairs with the AUG sequence of the mRNA, the migration stops, and the large ribosomal subunit joins the complex to form the initiation complex (Fig. 3.16).

The elongation and termination phases of translation are very similar in prokaryotes and eukaryotes. The elongation process entails the formation of a peptide bond between adjacent amino acids, with the order of the amino acids being determined by the order of codons of the mRNA (Fig. 3.17). More specifically, after the initiation complex is formed, the sec-

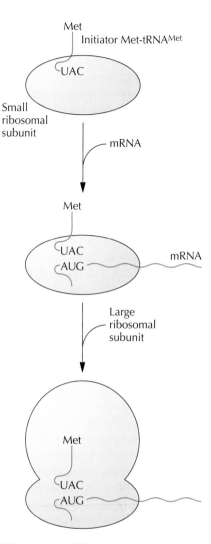

Figure 3.16 Schematic representation of the initiation of translation in eukaryotes. The initiator tRNA (Met-tRNAMet) binds to a small ribosomal subunit, and then the mRNA moves along the complex until the anticodon (UAC) of the initiator tRNA base pairs with the start codon (AUG) of the mRNA. The large ribosomal subunit combines with the mRNA–initiator tRNA–small ribosomal subunit complex to form the initiation complex.

Figure 3.15 Schematic representation of the initiation of translation in prokaryotes. The mRNA binds to the small ribosomal subunit. For some mRNAs, a sequence (Shine-Dalgarno sequence [SD]) near the 5′ end of the mRNA base pairs with a sequence near the 3′ end of the rRNA of the small ribosomal subunit. The anticodon (UAC) of the initiator fMet-tRNA$_F^{Met}$ base pairs with the start codon (AUG) of the mRNA. The large ribosomal subunit combines with the initiator tRNA–mRNA–small ribosomal subunit complex to form the initiation complex. The amino acid methionine of the initiator tRNA is formylated (CHO) at the amino group in prokaryotes (not depicted). After translation, the initial formyl-methionine is removed from the protein chain.

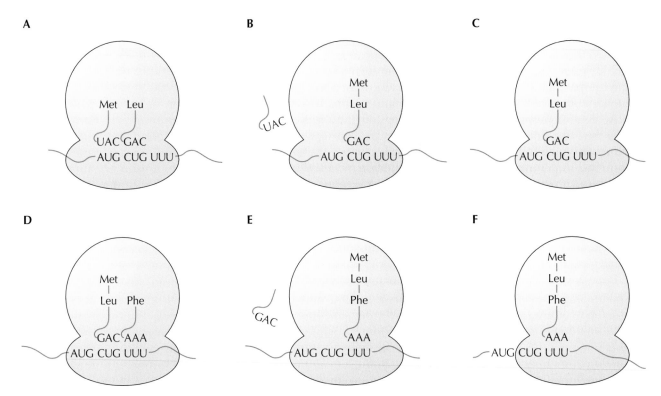

Figure 3.17 Schematic representation of the elongation phase of translation. **A.** The second codon (CUG) base pairs with the anticodon (GAC) of Leu-tRNA^Leu. **B.** The methionine of the initiator tRNA is joined to the leucine of Leu-tRNA^Leu by a peptide bond, and the uncharged initiator tRNA is ejected from the ribosome. **C.** Translocation of the peptidyl-tRNA and the mRNA to the peptidyl site from the aminoacyl site, which opens the aminoacyl site for the next codon (UUU). **D.** The third codon (UUU) base pairs with the anticodon (AAA) of Phe-tRNA^Phe. **E.** The leucine of the peptidyl-tRNA is joined to the phenylalanine of the Phe-tRNA^Phe by a peptide bond, and the uncharged Leu-tRNA^Leu is ejected from the ribosome. **F.** Translocation of the peptidyl-tRNA and mRNA to the peptidyl site from the aminoacyl site, which opens the aminoacyl site for the next codon and codon–anticodon interaction.

ond set of three nucleotides (triplet, or codon) in the mRNA that immediately follows the AUG codon dictates the anticodon sequence and, therefore, which charged tRNA will bind to the ribosome complex. Uncharged tRNAs do not bind efficiently to ribosomes. If the second nucleotide triplet in the mRNA is CUG, then the charged tRNA with the anticodon sequence 3'-GAC-5' will bind. This charged tRNA carries the amino acid leucine. Once this charged tRNA is in place, a peptide bond is formed between the carboxyl group of the methionine and the amino group of the leucine. The leucine remains bound to its tRNA. Peptide bond formation is catalyzed by enzymatic activity associated with the large ribosomal subunit. The formation of the peptide bond "uncharges" the initiator tRNA because the bond between the carboxyl group of methionine and its tRNA is cleaved to make the carboxyl group available for peptide bond formation. The uncharged tRNA is ejected from the ribosomal complex. The methionine–leucine–tRNA^Leu–mRNA combination shifts (translocates) along the ribosome to the site vacated by the initiator tRNA, and as a consequence, the next codon of the mRNA is available for binding by a charged tRNA with the appropriate anticodon sequence. If the third codon is UUU, then the charged

tRNA with an AAA anticodon will bind. In this case, the tRNA with an AAA anticodon carries the amino acid phenylalanine. Once this charged tRNA is in place, the linkage between the carboxyl group of leucine and its tRNA is broken and a peptide bond is formed between the carboxyl group of the leucine and the amino group of the phenylalanine. After ejection of the uncharged tRNALeu, the "peptidyl" tRNAPhe, in combination with the methonine-leucine-phenylanine amino acid sequence and the mRNA, is translocated, and the next codon is available for binding by the appropriately charged tRNA.

The succession of operations that includes binding of a charged tRNA by means of anticodon-codon pairing, peptide bond formation, ejection of an uncharged tRNA, and translocation continues until all the amino acids that are encoded by the mRNA are linked together. Translation occurs in a 5'-to-3' direction at a rate of about 15 amino acids per second. When the 5' end of the mRNA is free of a ribosome, it can combine with another initiation complex. A single mRNA can be translated simultaneously by a number of ribosomes, with each ribosome producing a polypeptide chain.

The elongation process continues until a UAA, UAG, or UGA codon (stop codon, or termination codon) is encountered (Fig. 3.18). There are no naturally occurring tRNAs with anticodons that are complementary to

Figure 3.18 Schematic representation of the termination of translation. The stop codon (UAG) interacts with a termination factor that leads to the termination of translation. The last tRNA is cleaved from the peptide chain and ejected. The mRNA and the finished peptide are released. The ribosomes are prepared for recycling by a ribosome releasing factor.

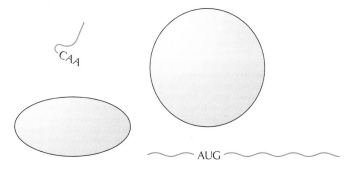

Table 3.2 The genetic code and codon usage in *E. coli* and humans

Codon	Amino acid	Frequency of use in:	
		E. coli	Humans
GGG	Glycine	0.13	0.23
GGA	Glycine	0.09	0.26
GGU	Glycine	0.38	0.18
GGC	Glycine	0.40	0.33
GAG	Glutamic acid	0.30	0.59
GAA	Glutamic acid	0.70	0.41
GAU	Aspartic acid	0.59	0.44
GAC	Aspartic acid	0.41	0.56
GUG	Valine	0.34	0.48
GUA	Valine	0.17	0.10
GUU	Valine	0.29	0.17
GUC	Valine	0.20	0.25
GCG	Alanine	0.34	0.10
GCA	Alanine	0.22	0.22
GCU	Alanine	0.19	0.28
GCC	Alanine	0.25	0.40
AAG	Lysine	0.24	0.60
AAA	Lysine	0.76	0.40
AAU	Asparagine	0.39	0.44
AAC	Asparagine	0.61	0.56
AUG	Methionine	1.00	1.00
AUA	Isoleucine	0.07	0.14
AUU	Isoleucine	0.47	0.35
AUC	Isoleucine	0.46	0.51
ACG	Threonine	0.23	0.12
ACA	Threonine	0.12	0.27
ACU	Threonine	0.21	0.23
ACC	Threonine	0.43	0.38
UGG	Tryptophan	1.00	1.00
UGU	Cysteine	0.43	0.42
UGC	Cysteine	0.57	0.58
UGA	Stop	0.30	0.61
UAG	Stop	0.09	0.17
UAA	Stop	0.62	0.22
UAU	Tyrosine	0.53	0.42
UAC	Tyrosine	0.47	0.58
UUU	Phenylalanine	0.51	0.43
UUC	Phenylalanine	0.49	0.57
UCG	Serine	0.13	0.06
UCA	Serine	0.12	0.15
UCU	Serine	0.19	0.17
UCC	Serine	0.17	0.23
AGU	Serine	0.13	0.14
AGC	Serine	0.27	0.25
CGG	Arginine	0.08	0.19
CGA	Arginine	0.05	0.10
CGU	Arginine	0.42	0.09
CGC	Arginine	0.37	0.19
AGG	Arginine	0.03	0.22
AGA	Arginine	0.04	0.21
CAG	Glutamine	0.69	0.73
CAA	Glutamine	0.31	0.27
CAU	Histidine	0.52	0.41
CAC	Histidine	0.48	0.59
CUG	Leucine	0.55	0.43
CUA	Leucine	0.03	0.07
CUU	Leucine	0.10	0.12
CUC	Leucine	0.10	0.20
UUG	Leucine	0.11	0.12
UUA	Leucine	0.11	0.06
CCG	Proline	0.55	0.11
CCA	Proline	0.20	0.27
CCU	Proline	0.16	0.29
CCC	Proline	0.10	0.33

these codons. However, a protein(s) (a termination factor, or release factor) recognizes a stop codon and binds to the ribosome. After binding of a termination factor, the bond between the last tRNA, which has the complete chain of amino acids linked to it, and its amino acid is broken, resulting in clearing of the uncharged tRNA, the complete protein, and the mRNA from the ribosome. In addition, a ribosome releasing factor separates the ribosomal subunits so that they can be reused for the translation of other mRNAs.

After translation, a protein may be modified in various ways. In most proteins, the methionine at the N terminus is cleaved off, leaving the second encoded amino acid as the N-terminal moiety. In eukaryotes, certain proteins are selectively cleaved at defined sites (processed) to make smaller protein chains that have discrete functions. In other instances, especially in eukaryotes, phosphate groups, lipids, carbohydrates, or other low-molecular-weight groups are enzymatically added to certain amino acids of a protein. These chemical additions create proteins that mediate specific cellular activities.

The complete genetic code consists of 64 codons. Three of these codons are reserved for stops, and one (AUG) is reserved for initiation (Table 3.2). When a methionine residue occurs internally in a protein, the codon AUG is recognized by another Met-tRNAMet which is neither formylated nor the initiator tRNA. There is one codon (UGG) for the amino acid tryptophan. For the rest of the amino acids that are found in proteins, there are at least two, usually four, and sometimes as many as six codons. For example, there are six codons (UUA, UUG, CUU, CUC, CUA, and CUG) for the amino acid leucine. Different codons are used to different extents in different organisms. Of the four codons for glycine, GGA is used about 26% of the time by human structural genes and about 9% of the time by the protein-coding genes of the bacterium *Escherichia coli*. The stop codons are also used to different extents in different organisms. In humans, the frequencies of usage of UAA, UAG, and UGA are 0.22, 0.17, and 0.61, respectively, whereas in *E. coli*, they are 0.62, 0.09, and 0.30, respectively. Notwithstanding the differences in codon usage, the genetic code, with a few rare exceptions, is the same in all organisms.

Regulation of mRNA Transcription in Bacteria

In bacteria, the production of amino acids, nucleotides, and other essential metabolites; replication; transcription; translation; cell growth; catabolic pathways; energy-generating systems; and responses to environmental changes all depend on proteins. However, the energy resources of a cell are not sufficient to support the transcription and translation (expression) of all of its structural genes at the same time. Consequently, only those genes that encode proteins that maintain basic cellular functions are expressed continuously. The transcription of the remaining structural genes is regulated. If a protein(s) is required by a cell, then a signaling system initiates transcription (the on state) of the pertinent structural gene(s). Alternatively, if a protein(s) is not needed, transcription of the protein-coding gene(s) is turned off (the off state).

Frequently, bacterial structural genes that encode proteins for a single metabolic pathway are contiguous. This arrangement is called an operon. Generally, an operon is under the control of a single promoter, and its transcription gives rise to one large mRNA. The placement of a stop codon for

one protein close to the first codon of the next protein within a multigene mRNA generates a set of discrete proteins during translation.

For most of the structural genes of *E. coli*, the promoter region has two DNA binding sites that are specific for RNA polymerase. One of these sites tends to have the sequence $^{\text{TATAAT}}_{\text{ATATTA}}$ (the TATA box, or Pribnow box), and the other is usually $^{\text{TTGAC}}_{\text{AACTG}}$. The TATA box and the TTGAC sequence are located about 10 bp (the -10 region) and 35 bp (the -35 region), respectively, from the site of initiation of transcription (the $+1$ nucleotide [$+1$]) (Fig. 3.19). Generally, the sequence that lies between the TATA box and the $+1$ nucleotide pair plays an essential role in determining whether an operon is transcribed. Depending on the mode of regulation of transcription of an operon, this region is called either an operator or an activating site.

A plethora of elaborate regulatory systems that control the on and off states of various operons have evolved. For example, when a regulatory protein called a repressor binds to an operator region, RNA polymerase cannot move along the DNA, and consequently transcription is blocked (Fig. 3.20). However, a low-molecular-weight compound (effector) can bind to a repressor protein and change its conformation, thus preventing it from binding to its operator region. When an effector-repressor complex fails to

Figure 3.19 Schematic representation of a bacterial transcription unit. **A.** The structural genes (A, B, C, and D) of an operon are under the transcriptional control of an operator (o) and promoter (p) region. RNA polymerase binds to regions that are 10 (-10) and 35 (-35) bp from the site of initiation of transcription ($+1$ [right-angled arrow]). The t denotes a transcription termination signal sequence. After transcription of the operon, the proteins of the operon (α, β, γ, and δ) are produced during translation. **B.** Same as panel A except that binding of RNA polymerase to the promoter region is shown.

A

B

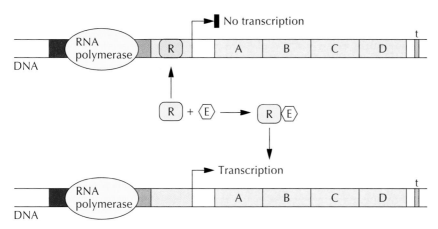

Figure 3.20 Induction of the on state for transcription of a bacterial operon. The repressor protein (R) binds to the operator region and blocks transcription. The binding of an effector molecule (E) to the repressor protein changes the conformation of the repressor protein. The repressor protein–effector (RE) complex cannot bind to the operator region; thus, RNA polymerase can transcribe the operon.

bind to an operator region, the RNA polymerase which binds to the promoter moves along the DNA and the operon is transcribed. As a rule, effector molecules are broken down by cellular activity. When the concentration level of effector molecules is reduced, newly synthesized repressor protein molecules bind to the operator region and the off state is reestablished. Each operator region is specific for its particular operon. Similarly, an effector molecule interacts only with a particular repressor protein.

By way of illustration, if a cell has the enzymatic capability of catabolizing a particular sugar, it is a waste of cellular resources to synthesize the enzymes that break down the sugar if the sugar is not present in the medium. On the other hand, if the sugar is available and is the only carbon source, then the enzymes that are responsible for its cellular utilization are essential. In this case, the sugar acts as an effector, preventing the repressor from binding to the operator region and thus enabling the operon to be transcribed. When the amount of sugar is depleted, the repressor protein binds to the operator region and prevents transcription of the operon.

For other operons, transcription may be the normal state because the repressor protein is inactive. In this case, a specific effector molecule (corepressor) attaches to an inactive repressor and causes a conformational change that enables the repressor-corepressor complex to bind to its corresponding operator region and turns off the transcription of the operon (Fig. 3.21). When the concentration of the corepressor decreases, the on state for the operon resumes because the repressor by itself is unable to bind to the operator.

The regulation of transcription by a repressor protein is called a negatively controlled system. A positively controlled system increases the rate of transcription instead of repressing it. Briefly, a protein (activator protein, or activator) binds to the region between the TATA box and the site of initiation of transcription. The activator-operator combination attracts RNA polymerase to the adjacent promoter region and consequently enhances the transcription of the operon (Fig. 3.22). A bound activator protein does not block the movement of RNA polymerase along the DNA. Rather, it can be viewed as "greasing the wheels" for transcription. Activators are specific

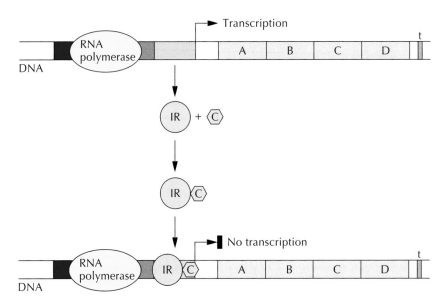

Figure 3.21 Induction of the off state for transcription of a bacterial operon. The binding of a corepressor molecule (C) to an inactive repressor protein (IR) changes the conformation of the repressor protein. The corepressor–repressor protein complex (IR-C) binds to the operator region and blocks transcription of the operon by RNA polymerase.

for particular activator sites. In some instances, an effector molecule converts an active activator to an inactive one and diminishes the rate of transcription of the operon (Fig. 3.22). In other cases, an effector molecule activates an inactive activator. Understanding how the transcription of a bacterial operon is regulated requires detailed molecular studies of mutations that affect a regulatory system and in vitro analyses of the various protein- and DNA-binding sites.

Figure 3.22 Activation and deactivation of a bacterial operon. An activator protein (Act) binds to an activating site and enhances the rate of transcription of the operon. An effector molecule (E) binds to the activator protein. The Act–E complex does not bind to the activating site. The rate of transcription of the operon is diminished when the activating site is not occupied by the activating protein.

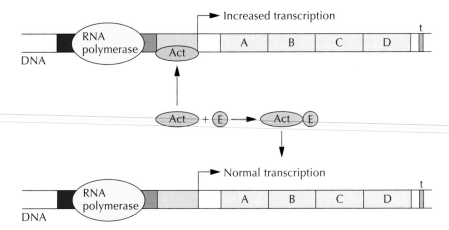

Regulation of mRNA Transcription in Eukaryotes

Most active eukaryotic cells transcribe a common (basal) set of structural genes that maintain routine (household) cellular functions. In some cells, specific structural genes are exclusively transcribed and translated, thus giving the tissue or organ its unique properties. For example, the genes that encode the α and β subunits of adult hemoglobin are expressed only in the cells that develop into red blood cells. The numbers of cell-specific mRNA transcripts range from a few sequences in some cells to dozens of different sequences in others. The ability of cells to turn on (activate) or turn off (repress) transcription of particular structural genes is essential for maintaining cell specificity and conserving cellular energy.

There are a number of diverse, highly specific processes that activate or repress the transcription of various eukaryotic structural genes. In general, the control of transcription in eukaryotes is mediated by proteins that are collectively classified as transcription factors. Many transcription factors bind directly to DNA sequences that are frequently less than 10 bp in length. The naming of these protein-binding sites is idiosyncratic. However, for the most part, they are called boxes, DNA modules, initiator elements, or response elements. Unlike the situation in prokaryotes, operons are almost never found in the genomes of eukaryotes. Consequently, each eukaryotic structural gene has its own set of response elements. Moreover, in addition to DNA-protein interactions, protein-protein associations are important for regulating eukaryotic transcription.

Although eukaryotic structural genes often have their own collection of response elements, a representative eukaryotic structural gene has a promoter sequence (TATA box, or Hogness box) which consists of eight nucleotides that includes a TATA sequence; a CCAAT sequence ("cat" box); and a sequence of repeated GC nucleotides (GC box) that lie about -25, -75, and -90 bp, respectively, from the site of initiation of transcription ($+1$) (Fig. 3.23). The first step in the initiation of transcription of eukaryotic structural genes with a TATA promoter is the binding of transcription factor IID (TFIID, or TATA-binding protein [TBP], which is a complex of at least 14 proteins, to an available TATA sequence. Subsequently, other transcription factors bind to TFIID and the DNA adjacent to the TATA box. Then RNA polymerase II, which is oriented towards the structural gene, binds to the transcription complex. With the aid of additional transcription factors, transcription is initiated at the correct starting point ($+1$ nucleotide) (Fig. 3.24). Clearly, if a TATA sequence is deleted or grossly altered, then transcription of the structural gene cannot occur. Transcription factors that are specific for the CCAAT and GC response elements have been identified, but how these DNA-protein interactions affect the efficiency of transcrip-

Figure 3.23 Promoter and initiator elements of some eukaryotic structural genes. The negative numbers designate the locations of nucleotide pairs in the DNA that lie upstream from the site of initiation ($+1$) of transcription. The right-angled arrow indicates the site of initiation and direction of transcription. The locations of the transcription elements are not drawn to scale.

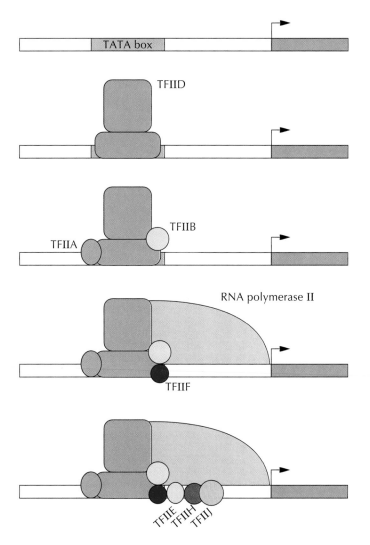

Figure 3.24 Formation of an RNA polymerase II transcription initiation complex at a TATA box. Transcription factor TFIID binds to a TATA box, and, in sequence, other transcription factors and RNA polymerase II bind to form a protein aggregate that is responsible for initiating transcription. The right-angled arrow designates the site of initiation and direction of transcription.

tion when they are 75 or more base pairs from the site of initiation of transcription is not known. In addition, enhancer sequences that increase the rate of transcription of structural genes are located hundreds or even thousands of base pairs from the +1 base pair. Folding, looping, or bending of the chromosomal DNA may bring DNA regions, which in the elongated state are far apart, close to one another. Also, transcription factors that bind to certain enhancers or response elements may form a chain of proteins that create bridges from one DNA site to another.

Some repressed (nonexpressed) structural genes are activated by a cascade of events that is triggered by a specific extracellular signal such as a temperature increase or the presence of a hormone. For example, a hormone that is released into the circulatory system comes into contact with a specific cell type that has a receptor that binds the hormone and facilitates the entry of the hormone into the cell. Once inside the cell, the hormone interacts with

a cellular protein and changes the conformation of the protein. In this altered state, the protein is now able to enter the nucleus, where it binds to an exclusive response element that initiates transcription of the target gene.

Some proteins bind to response elements and prevent transcription. For example, there is a class of about 18 vertebrate genes that are actively transcribed in nerve cells (neurons) and turned off in nonneuronal cells. Each of these neuron-active genes has a 24-bp response element that lies upstream of its +1 bp site. This DNA sequence is called a neuron-restrictive silencer element (NRSE). In nonneuronal cells, a protein called neuron-restrictive silencer factor (NRSF) is synthesized, binds to each NRSE, and prevents transcription of each member of this set of genes. Conversely, NRSF is not produced by neuronal cells, and each gene with an NRSE is transcribed.

On the whole, the regulation of transcription in eukaryotes is complex. A structural gene may have a number of different response elements that can be activated in different cell types by different signals at different times in the life cycle of an organism. Alternatively, some structural genes are under the preferential control of a unique transcription factor. For the off state, specific proteins can interact with certain response elements and prevent transcription. Or, in a more general way, some proteins obstruct transcription by binding to the transcription complex either before initiation or during the elongation process.

MILESTONE

A Structure for Deoxyribose Nucleic Acid
J. D. WATSON and F. H. C. CRICK
Nature **171**:737–738, 1953

The elucidation of the structure of the genetic material (DNA) was undoubtedly one of the most important scientific breakthroughs of the twentieth century. As expressed by Watson and Crick: "We wish to suggest a structure for the salt of deoxyribonucleic acid (D. N. A.). This structure has novel features which are of considerable biological interest. . . . This structure has two helical chains each coiled around the same axis. . . . Both chains follow right handed helices, but . . . the sequences of the atoms in the two chains run in opposite directions. . . . The novel feature of the structure is the manner in which the two chains are held together by the purine and pyrimidine bases. . . . They are joined in pairs, a single base from one chain being hydrogen bonded to a single base from the other chain. . . .

One of the pair must be a purine and the other a pyrimidine for bonding to occur. . . . If adenine forms one member of a pair, on either chain, then . . . the other member must be a thymine; similarly for guanine and cytosine. The sequence of bases on a single chain does not appear to be restricted in any way. . . . It has not escaped our notice that the specific pairing we have postulated immediately suggests a possible copying mechanism for the genetic material." In another article a few months later (*Nature* **171**:964, 1953), Watson and Crick discussed more implications for their model. "The phosphate-sugar backbone of our model is completely regular but any sequence of the pairs of bases can fit into the structure. It follows that in a long molecule many different permutations are possible, and it therefore

seems likely that the precise sequence of bases is the code which carries the genetic information. . . . Our model suggests possible explanations for a number of other phenomena. For example, spontaneous mutations may be due to a base occasionally in one of its less likely tautomeric forms. Again, the pairing between homologous chromosomes at meiosis may depend on pairing between specific bases."

Within a decade of the demonstration of the double-helical nature of DNA with its complementary base pairs, the molecular aspects of DNA replication were known; the cellular processes that are responsible for both the decoding and regulating the synthesis of gene products were understood, and many of the kinds of changes that lead to altered gene products were recognized. From the time of its publication to the present, the scientific impact of the Watson-Crick discovery has been pervasive and, for the most part, inestimable. As one small example, recombinant DNA technology would not exist without knowledge of the structure of DNA.

SUMMARY

A DNA molecule has two polynucleotide strands that form an antiparallel double helix. The monomeric unit of a DNA strand is a nucleotide that consists of an organic base, a deoxyribose sugar, and a phosphate group. The successive nucleotides of a DNA strand are linked by phosphodiester bonds, and the two strands of DNA are held together by specific complementary pairs of bases. The base adenine forms hydrogen bonds only with thymine, and guanine forms bonds only with cytosine. During replication, which is mediated by a number of different proteins, including DNA polymerases, each existing DNA strand acts as a template for the production of a complementary strand. The base pairing requirements ensure that each newly synthesized DNA strand has the correct nucleotide sequence.

Proteins are vital for the maintenance of all biological functions. A protein consists of a specific sequence of amino acids that are linked by peptide bonds. The sequence of amino acids for a protein is encoded in the DNA. The process of decoding genetic information is carried out by RNA molecules, including mRNA, tRNA, and rRNA; various enzymes; and an assortment of protein factors. All RNA is transcribed from DNA. Sequences of DNA in combination with protein factors ensure that transcription is initiated at a precise starting point, that the appropriate strand is used as the template, and that termination occurs at a specified nucleotide site. In eukaryotic organisms, most structural genes consist of coding regions (exons) separated by noncoding segments (introns). Primary transcripts contain both exons and introns. However, a processing system removes the introns and joins the exons, in the proper order, to form a functional mRNA. An mRNA carries the code for the sequence of amino acids of a protein.

Both tRNA and rRNA molecules play important roles in the biosynthesis of proteins (translation). There are over 50 different tRNA molecules in a cell. Each tRNA has two significant features. First, one of the 20 different amino acids that are found in proteins is enzymatically bound to the 3' end of a tRNA. The relationship between a tRNA and its amino acid is absolutely specific. Second, in another location of the tRNA molecule, there is a set of three contiguous nucleotides called the anticodon that enables a tRNA to bind to a complementary set of three nucleotides (codon) of an mRNA during translation. There are two major rRNA molecules. A small rRNA molecule and a large rRNA molecule each combine with different groups of proteins to form a small and a large ribosomal subunit, respectively. The rRNA molecules of prokaryotes are smaller than eukaryotic rRNAs, and the prokaryotic ribosomes have fewer proteins than their eukaryotic counterparts.

Translation in prokaryotes is initiated by the joining of an mRNA with a small ribosome subunit. Then, as a result of codon-to-anticodon base pairing, the initiator tRNA, fMet-tRNA$_F^{Met}$, attaches to the mRNA–small ribosomal subunit complex. The large ribosomal subunit combines with the initiator tRNA–mRNA–small ribosomal subunit complex to form an initiation complex.

Translation in eukaryotes is initiated by the combining of a unique initiator tRNA which carries the amino acid methionine, Met-tRNAMet, with a small ribosomal subunit and then by the threading of an mRNA through the initiator tRNA–small ribosomal subunit complex until the first AUG sequence in the mRNA pairs with the UAC sequence of the anticodon of the

initiator tRNA. The large ribosomal subunit joins the initiator tRNA–small ribosomal subunit–mRNA complex to form a ribosome (initiation complex) that is ready for the translation of the mRNA sequence.

In principle, the elongation and termination phases of translation in prokaryotes and eukaryotes are very similar. After the formation of the initiation complex, the next three nucleotides in the mRNA pair with the anticodon of a tRNA that carries its specific amino acid, e.g., AA_2. The first amino acid, methionine, is cleaved from the initiator tRNA and joined to the second amino acid by a peptide bond. The methionineless initiator tRNA is ejected from the ribosome; the ribosome complex shifts, and the peptidyl-tRNA, i.e., Met-AA_2-tRNAAA2, occupies the site vacated by the ejected initiator tRNA. As a consequence of the shift (translocation), the next codon of the mRNA is available to pair with the appropriate anticodon of the tRNA that carries its specific amino acid, e.g., AA_3. Then AA_2 of the Met-AA_2-tRNAAA2 combination is cleaved from its tRNA and joined to AA_3 with a peptide bond to give Met-AA_2-AA_3-tRNAAA3. The peptidylless tRNAAA2 is ejected from the ribosome; the ribosome complex shifts, and the remaining peptidyl-tRNA, i.e., Met-AA_2-AA_3-tRNAAA3, occupies the site vacated by the previous peptidyl-tRNA. By repeating these steps, a specific sequence of amino acids is formed.

Translation is terminated when one of three codons (stop codons) is encountered on a ribosome. No normally occurring tRNA has an anticodon that is complementary to a stop codon. However, a termination factor(s) recognizes a stop codon, and after this factor attaches to the ribosome, the bond between the last tRNA and the completed amino acid chain is cleaved, causing the tRNA, mRNA, and completed protein to be released. Then the two ribosomal subunits are separated by a ribosome releasing factor.

mRNA synthesis and, consequently, protein synthesis must be regulated because a cell could not survive if all of its structural genes were transcribed and translated simultaneously. In both prokaryotes and eukaryotes, the mRNAs that are essential for maintaining routine cellular functions are synthesized continuously. For the remaining structural genes, elaborate systems have evolved to ensure that transcription occurs only when a particular protein(s) is required and is turned off when the protein(s) is no longer needed. In prokaryotes, transcription is initiated by the binding of RNA polymerase to the TATA and TTGAC sequences of the promoter region of a structural gene or an operon. For some operons, the switching on and off of transcription is mediated by an effector molecule that alters the conformation of a repressor protein and prevents it from blocking transcription. As the cellular concentration of an effector molecule decreases, the repressor protein binds to the DNA region adjacent to the site of initiation of transcription and blocks the movement of the RNA polymerase along the DNA, thereby preventing transcription of the operon. With other operons, a protein (activator protein) binds to the DNA region next to the site of initiation of transcription and enhances the rate of transcription of the operon. The binding of an effector molecule to an activator protein can decrease the rate of transcription. The protein-DNA interactions that regulate transcription are highly specific for a particular structural gene or operon. In eukaryotes, RNA polymerase II, which transcribes structural genes, binds to an array of proteins called transcription factors that attach, in sequence, to a TATA sequence of a promoter region. Other transcription factors that bind to DNA elements of eukaryotic structural genes are responsible for turning on or turning off transcription.

References

Buratowski, S. 1994. The basics of basal transcription by RNA polymerase II. *Cell* **77**:1–3.

Kozak, M. 1991. Structural features in eukaryotic mRNAs that modulate the initiation of translation. *J. Biol. Chem.* **266**:19867–19870.

Lodish, H., D. Baltimore, A. Berk, S. L. Zipursky, P. Matsudaira, and J. Darnell. 1995. *Molecular Cell Biology,* 3rd ed. Scientific American Books, Inc., New York, N.Y.

Nakamura, Y., K. Ito, and L. A. Isaksson. 1996. Emerging understanding of translation termination. *Cell* **87**:147–150.

Schoenherr, C. J., and D. J. Anderson. 1995. The neuron-restrictive silencer factor (NRSF): a coordinate repressor of multiple neuron-specific genes. *Science* **267**:1360–1363.

Tate, W. P., and C. M. Brown. 1992. Translational termination: "stop" for protein synthesis or "pause" for regulation of gene expression. *Biochemistry* **31**:2443–2450.

Tjian, R., and T. Maniatis. 1994. Transcriptional activation: a complex puzzle with few easy pieces. *Cell* **77**:5–8.

REVIEW QUESTIONS

1. Discuss the basic features of DNA replication.

2. Compare and contrast DNA and RNA.

3. Describe the differences and similarities between prokaryotic and eukaryotic structural genes.

4. Describe the elongation phase of translation.

5. Deduce the most likely DNA coding sequence for the following human protein: MAGGTWYQLFPRKMWNDSTLHPFILPMNVAG.

6. Determine the amino acid sequence encoded by the following mRNA: GCGAUCGACGAUGUUUCUAAAAGUAUCUCAUCGAAAU-GAGGGUUCGUAAUAGCGACCCGGGCGG.

7. What is a TATA box?

8. What is an operon? What is the biological significance of an operon?

9. Discuss three different modes of regulation of transcription in prokaryotes.

10. Describe the major DNA elements that are responsible for the transcription of eukaryotic structural genes.

CHAPTER 4

Recombinant DNA Technology

Recombinant DNA technology, which is also called gene cloning or molecular cloning, is an umbrella term that encompasses a number of experimental protocols leading to the transfer of genetic information (deoxyribonucleic acid [DNA]) from one organism to another. There is no single set of methods that can be used to meet this objective; however, a recombinant DNA experiment often follows the following format (Fig. 4.1).

- The DNA (cloned DNA, insert DNA, target DNA, foreign DNA) from a donor organism is extracted, enzymatically cleaved (cut, digested), and joined (ligated) to another DNA entity (a cloning vector) to form a new, recombined DNA molecule (cloning vector–insert DNA construct, DNA construct).
- This cloning vector–insert DNA construct is transferred into and maintained within a host cell. The introduction of DNA into a bacterial host cell is called transformation.
- Those host cells that take up the DNA construct (transformed cells) are identified and selected (separated, isolated) from those that do not.
- If required, a DNA construct can be prepared to ensure that the protein product that is encoded by the cloned DNA sequence is produced by the host cell.

Recombinant DNA technology was developed from discoveries in molecular biology, nucleic acid enzymology, and the molecular genetics of both bacterial viruses and bacterial extrachromosomal DNA elements (plasmids). However, recombinant DNA technology would not exist without the availability of enzymes (restriction enzymes, restriction endonucleases) that recognize specific double-stranded DNA sequences and cleave the DNA in both strands at these sequences.

Restriction Endonucleases

For molecular cloning, both the source DNA that contains the target sequence and the cloning vector must be consistently cut into discrete and reproducible fragments. Subjecting isolated chromosomal DNA either to

45

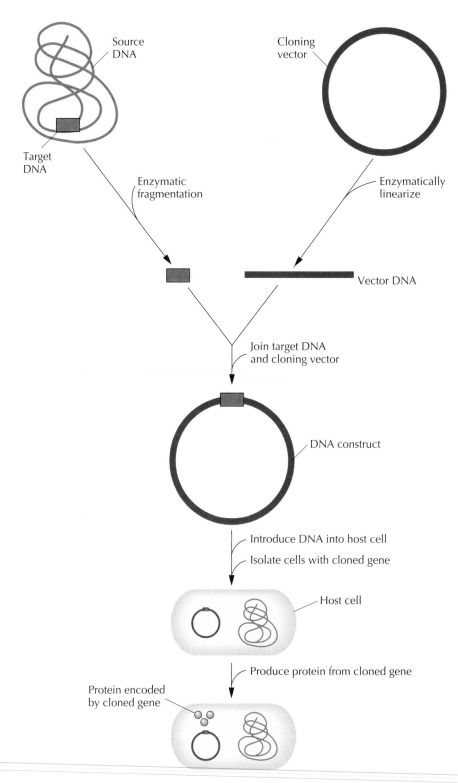

Figure 4.1 Recombinant DNA cloning procedure. DNA from a source organism is cleaved with a restriction endonuclease and inserted into a cloning vector. Then the cloning vector–insert DNA construct is introduced into a target host cell, and those cells that carry the construct are identified and grown. If required, the cloned gene can be expressed in the host cell and its protein can be produced and harvested.

passage through a small-bore needle or to sonication produces double-stranded pieces of DNA that may range from 0.3 to 5 kilobase pairs (kb) in length. Unfortunately, these simple procedures induce breaks randomly, so each time a DNA sample is treated, a different set of fragments is generated. It was only after bacterial enzymes that cut DNA molecules internally at specific base pair sequences were discovered that molecular cloning became feasible. These enzymes are called type II restriction endonucleases.

One of the first of these type II restriction endonucleases to be characterized was from the bacterium *Escherichia coli*, and it was designated *Eco*RI. This enzyme binds to a DNA region with a specific palindromic sequence (the two strands are identical in this region when either is read in the same polarity, i.e., 5′ to 3′) of 6 base pairs (bp) and cuts between the guanine and adenine residues on each strand (Fig. 4.2). It specifically cleaves the internucleotide bond between the oxygen of the 3′ carbon of the sugar of one nucleotide and the phosphate group attached to the 5′ carbon of the sugar of the adjacent nucleotide. The symmetrical staggered cleavage of DNA by *Eco*RI produces two single-stranded, complementary cut ends, each with extensions of four nucleotides. In this case, each single-stranded extension ends in a 5′-phosphate group, and the 3′-hydroxyl group from the opposite strand is recessed.

In addition to *Eco*RI, hundreds of other type II restriction endonucleases have been isolated from various bacteria. The naming protocol for these enzymes is the same as that for *Eco*RI; the genus is the capitalized letter and the first two letters of the species name are in lowercase letters. The strain designation is often omitted from the name, and roman numerals are used to designate the order of characterization of different restriction en-

Figure 4.2 Symmetrical, staggered cleavage of a short fragment of DNA by the type II restriction endonuclease *Eco*RI. The bold arrows show the sites of cleavage in the DNA backbone. S, deoxyribose sugar; P, phosphate group; OH, hydroxyl group; A, adenine; T, thymine; C, cytosine; G, guanine. The *Eco*RI recognition sequence is highlighted by the dashed line.

donucleases from the same organism. For example, *Hpa*I and *Hpa*II are the first and second type II restriction endonucleases that were isolated from *Haemophilus parainfluenzae*.

The palindromic sequences where most type II restriction endonucleases bind and cut a DNA molecule are called recognition sites. Some restriction endonucleases digest (cleave) DNA, leaving 5′-phosphate extensions (protruding ends, sticky ends), some leave 3′-hydroxyl extensions, and some cut the backbone of both strands within a recognition site to produce blunt-ended (flush-ended) DNA molecules (Fig. 4.3). The length of the recognition site for different enzymes can be four, five, six, eight, or more nucleotide pairs (Table 4.1). Because of the frequency with which their recognition sites occur in DNA, restriction endonucleases that cleave within sites of four (four-cutters) and six (six-cutters) nucleotide pairs are used for most of the common molecular cloning protocols.

The importance of the type II restriction endonucleases for gene cloning cannot be overstated. When a DNA sample is treated with one of these enzymes, the same set of fragments is always produced, assuming that all of the recognition sites are cleaved. Moreover, physical maps that designate the linear order of restriction endonuclease sites on a specific piece of DNA can be constructed by treating the DNA molecule singly with different restriction endonucleases and then with combinations of restriction endonucleases. The positions of the cleavage sites can be deduced from an analysis of fragment sizes, which are determined by agarose gel electrophoresis (Box 4.1). For example, the fragment sizes produced by various digestions are shown in Fig. 4.4A. It can be deduced that this piece of DNA has two *Bam*HI

Figure 4.3 Blunt-end cleavage of a short fragment of DNA by the type II restriction endonuclease *Hin*dII. The bold arrows show the sites of cleavage in the DNA backbone. For abbreviations, see the legend to Fig. 4.2. The *Hin*dII recognition sequence is highlighted.

Table 4.1 Recognition sequences of some restriction endonucleases

Enzyme	Recognition site	Type of cut end
*Eco*RI	G $^\downarrow$ A—A—T—T—C C—T—T—A—A $_\uparrow$ G	5′-phosphate extension
*Bam*HI	G $^\downarrow$ G—A—T—C—C C—C—T—A—G $_\uparrow$ G	5′-phosphate extension
*Pst*I	C—T—G—C—A $^\downarrow$ G G $_\uparrow$ A—C—G—T—C	3′-hydroxyl extension
*Sau*3AI	$^\downarrow$ G—A—T—C C—T—A—G $_\uparrow$	5′-phosphate extension
*Pvu*II	C—A—G $^\downarrow$ C—T—G G—T—C $_\uparrow$ G—A—C	Blunt end
*Hpa*I	G—T—T $^\downarrow$ A—A—C C—A—A $_\uparrow$ T—T—G	Blunt end
*Hae*III	G—G $^\downarrow$ C—C C—C $_\uparrow$ G—G	Blunt end
*Not*I	G $^\downarrow$ C—G—G—C—C—G—C C—G—C—C—G—G—C $_\uparrow$ G	5′-phosphate extension

sites and two *Eco*RI sites in a specific order with a specified number of base pairs separating the sites.

A restriction map is formulated by comparing the fragment sizes produced in each single digestion with those from double digestions. In the example shown in Fig. 4.4, the analysis goes as follows. Because each single digestion produces three fragments from a linear DNA molecule, the original piece of DNA must contain two sites for each of the restriction endonucleases. The 300-bp fragment that is produced by the single *Eco*RI digestion remains intact after the double digestion, whereas the 850- and 500-bp *Eco*RI fragments are cleaved. Therefore, the two *Eco*RI sites are 300 bp apart with no intervening *Bam*HI site, and there is a single *Bam*HI site within each of the 850- and 500-bp *Eco*RI fragments. The 950-bp fragment that is produced by the single *Bam*HI digestion is cleaved by *Eco*RI in the double digestion into three pieces (250 + 300 + 400 = 950 bp). Therefore, the two *Bam*HI sites lie 250 and 400 bp to either side of the *Eco*RI sites. Digestion with *Bam*HI cleaves the 850-bp *Eco*RI fragment into 250- and 600-bp fragments, and one of the *Eco*RI sites is 250 bp from a *Bam*HI site, so the 600-bp region must include one of the ends of the original molecule. Using the same logic, we also note that digestion with *Bam*HI cuts the 500-bp *Eco*RI fragment into 100- and 400-bp fragments and that one of the *Eco*RI sites is 400 bp from a *Bam*HI site; therefore, the 100-bp region must include the other end of the original molecule. In the final map (Fig. 4.4B), the assigned locations of the restriction endonuclease sites are consistent with the fragment lengths that were observed in each of the digestion reactions.

Restriction endonuclease cleavage is also used in another way. When two different DNA samples are digested with the same restriction endonuclease that produces a staggered cut (the same 5′ or 3′ extension or sticky end) and then mixed together, new DNA combinations can be formed as a result of base pairing between the extension (overhang) regions (Fig. 4.5).

Restriction enzymes alone are not sufficient for molecular cloning. First, when the extended ends that are created by restriction enzyme (e.g., *Bam*HI) cleavage are aligned, the hydrogen bonds of the four bases that pair are not strong enough to keep two DNA molecules together. A means of re-forming the internucleotide linkage between the 3′-hydroxyl group and the 5′-

BOX 4.1

Gel Electrophoresis

Gel electrophoresis is a commonly used technique for resolving proteins or nucleic acids. In general, a sample of one particular macromolecule (protein, DNA, or RNA) is placed in a well at or near the end of a gel matrix (gel). The composition of an electrophoresis gel is a semisolid, open meshwork of interlinked linear strands. A gel is cast as a thin slab with a number of sample wells. After the wells of a gel are loaded with sample, an electric field is applied across the gel and charged macromolecules of the same size are driven together in the direction of the anode through the gel as discrete invisible bands of material. The distance that a band moves into a gel depends on the mass of its macromolecules and the size of the openings (pore size) of the gel. The smaller macromolecules travel further than the larger ones.

The progress of gel electrophoresis is monitored by observing the migration of a visible dye (tracking dye) through the gel. The tracking dye is a charged, low-molecular-weight compound that is loaded into each sample well at the start of a run. When the tracking dye reaches the end of the gel, the run is terminated. The bands, which are aligned in a lane under each well, are visualized by staining the gel with a dye that is specific for protein, DNA, or RNA. Discrete bands are observed when there is enough material present in a band to bind the dye to make the band visible and when the individual macromolecules of a sam-

ple have distinctly different sizes. Otherwise, a band is not detected. If there is little or no difference among the sizes of the macromolecules in a concentrated sample, a smear of stained material is observed. The intensity of a stained band reflects the frequency of occurrence of a macromolecule in a sample.

The relative molecular mass (molecular weight) of a gel-fractionated macromolecule (band) is determined from a standard curve that is based on a set of macromolecules of known relative molecular mass (size markers) that covers the separation range of the gel system and is run in one or both of the outside lanes (calibrator lanes) of the same gel as the samples. The logarithm of the relative molecular mass of a size marker is related to its relative mobility (R_f) through a gel. The value of R_f is defined as the distance traveled by a band divided by the distance traveled by the tracking dye (ion front). The relationship between the logarithm of the relative molecular mass of each size marker and its R_f value is plotted. Then, with this standard curve, a relative molecular mass is calculated for each band in a lane. The units of relative molecular mass for proteins and double-stranded and single-stranded nucleic acids are daltons, base pairs, and bases, respectively. The size markers are included in the same gel as the samples because the extent of mobility of a macromolecule(s) varies from one electrophoretic run to the next.

Polyacrylamide is the preferred gel system for separating proteins.

Copolymerization of monomeric acrylamide and the cross linker bisacrylamide forms a lattice of cross-linked, linear polyacrylamide strands. The pore size of a polyacrylamide gel is determined by the concentration of acrylamide and the ratio of acrylamide to bisacrylamide. For many applications, a protein sample is treated with the anionic detergent sodium dodecyl sulfate (SDS) before electrophoresis. The SDS binds to proteins and dissociates most multichain proteins. Each SDS-coated protein chain has a similar charge-to-mass ratio. Consequently, during electrophoresis, the separation of the SDS-protein chains is based primarily on size and the effect of conformation is eliminated. SDS-polyacrylamide gel electrophoresis with a 10% polyacrylamide gel resolves proteins that range from 20 to 200 kDa.

Agarose, which is a polysaccharide from seaweed, is used routinely as the gel matrix for the electrophoretic separation of medium-sized nucleic acid molecules. A 1.0% agarose gel will resolve duplex DNA chains that range from about 600 to 20,000 bp. Specialized agarose gel electrophoresis systems are available for fractionating DNA molecules with millions of base pairs, denatured DNA, and denatured RNA. In addition, for specific purposes, polyacrylamide gels are used for separating DNA molecules. For example, DNA chains that are as small as six bases and differ from each other by one nucleotide can be resolved with a 20% polyacrylamide gel.

hosphate group in the backbone at the two broken bond sites (nicks) is required. This problem can be resolved by using the enzyme DNA ligase, usually from bacteriophage T4. This enzyme catalyzes the formation of phosphodiester bonds at the ends of DNA strands that are already held together by the base pairing of two extensions. It also joins blunt ends that come in contact when they both bind to the enzyme (Fig. 4.6).

Second, the ability to join different DNA molecules would not by itself be useful unless the new DNA combination (i.e., recombinant DNA) could be perpetuated in a host cell. One of the DNA molecules must provide the biological information for the cellular maintenance of the recombined

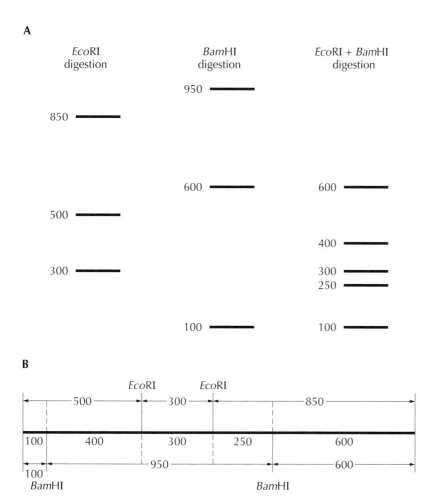

Figure 4.4 Mapping of restriction endonuclease sites. **A.** Restriction endonuclease digestions and electrophoretic separation of fragments. A purified piece of DNA is cut with *Eco*RI and *Bam*HI separately (single digestions) and then with both enzymes together (double digestion). The horizontal lines under each of the digestion conditions represent schematically the locations of the DNA fragments (bands) in the lanes of the gel after electrophoresis and staining of the DNA with ethidium bromide. The numbers denote the length of the digestion products (fragments) in base pairs. **B.** The restriction endonuclease map derived from the digestions and electrophoretic separation shown in **A.**

DNA, and the other must contain the gene that was targeted for cloning. This problem is solved by using cloning vectors.

Third, digestion of the source DNA with a restriction endonuclease produces a mixture of various DNA molecules, and, after ligation with a cloning vector, a number of different DNA constructs are formed. There has to be a way to identify the DNA combination in a host cell that contains the target DNA sequence. Screening procedures that detect host cells carrying a particular cloning vector–DNA insert construct solve this problem.

Plasmid Cloning Vectors

Plasmids are self-replicating, double-stranded, circular DNA molecules that are maintained in bacteria as independent extrachromosomal entities. Virtually all bacterial genera have plasmids. Some plasmids carry informa-

tion for their own transfer from one cell to another (F plasmids), others encode resistance to antibiotics (R plasmids), others carry specific sets of genes for the utilization of unusual metabolites (degradative plasmids), and some have no apparent functional coding genes (cryptic plasmids). Plasmids can range in size from less than 1 to more than 500 kb. Each plasmid has a sequence that functions as an origin of DNA replication; without this site, it cannot replicate in a host cell.

Some plasmids are represented by 10 to 100 copies per host cell; these are called high-copy-number plasmids. Others maintain 1 to 4 copies per cell and are called low-copy-number plasmids. Seldom does the population of plasmids in a bacterium make up more than approximately 0.1 to 5.0% of the total DNA. When two or more types of plasmids cannot coexist in the same host cell, they are said to belong to a single incompatibility group. But

Figure 4.5 Annealing complementary extensions after staggered cleavage with a type II restriction endonuclease. Two different DNA fragments are cut with the restriction endonuclease *Bam*HI, mixed, and annealed. Not all of the possible combinations of annealed DNAs are shown. The four fragments that are generated by the *Bam*HI digestion can anneal to one another to form any of six different DNA molecules. A break in the phosphodiester bond in one strand of duplex DNA is called a nick. The hydrogen bonds of the four base pairs between nicks on opposite strands are not sufficiently strong enough to hold DNA molecules together for long periods in solution. A, C, G, and T represent nucleotides.

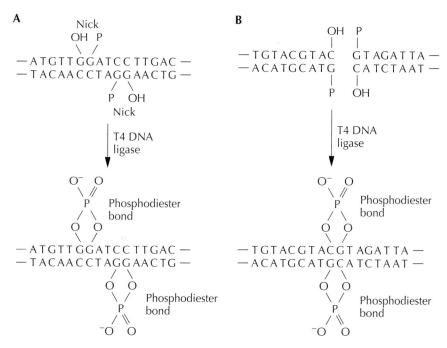

Figure 4.6 Mode of action of T4 DNA ligase. The enzyme T4 DNA ligase forms phosphodiester bonds by joining 5'-phosphate and 3'-hydroxyl groups at nicks in the backbone of double-stranded DNA. **A.** Ligation of sticky-ended DNA. **B.** Ligation of blunt-ended DNA. A, C, G, and T represent nucleotides.

plasmids from different incompatibility groups can be maintained together in the same cell. This coexistence is independent of the copy numbers of the individual plasmids. Some microorganisms have been found to contain as many as 8 to 10 different plasmids. In these instances, each plasmid can carry out different functions and have its own unique copy number, and each belongs to a different incompatibility group. Some plasmids, because of the specificity of their origin of replication, can replicate in only one species of host cell. Other plasmids have less specific origins of replication and can replicate in a number of bacterial species. These plasmids are called narrow- and broad-host-range plasmids, respectively.

As autonomous, self-replicating genetic elements, plasmids have the basic attributes to make them potential vectors for carrying cloned DNA. However, naturally occurring (unmodified, unengineered) plasmids often lack several important features that are required for a high-quality cloning vector. These features are (1) a small size, which is necessary because the efficiency of transfer of exogenous (foreign) DNA into *E. coli* decreases significantly with plasmids that are more than 15 kb long; (2) unique (single) restriction endonuclease recognition sites into which the insert DNA can be cloned; and (3) one or more selectable genetic markers for identifying recipient cells that carry the cloning vector–insert DNA construct. Consequently, plasmid cloning vectors have to be genetically engineered.

Plasmid Cloning Vector pBR322

In the 1980s, one of the best-studied and most often used "general-purpose" plasmid cloning vectors was pBR322. In general, plasmid cloning vectors are designated by a lowercase *p*, which stands for plasmid, and some ab-

breviation that may be descriptive or, as is the case with pBR322, anecdotal. The *BR* of pBR322 recognizes the work of the researchers F. *B*olivar and R. *R*odriguez, who created the plasmid; and *322* is a numerical designation that has relevance to these workers. Plasmid pBR322 contains 4,361 bp. As shown in Fig. 4.7, pBR322 carries two antibiotic resistance genes. One confers resistance to ampicillin (Ampr), and the other confers resistance to tetracycline (Tetr). This plasmid also has unique *Bam*HI, *Hin*dIII, and *Sal*I recognition sites within the Tetr gene; a unique *Pst*I site in the Ampr gene; a unique *Eco*RI site that is not within any coding DNA; and an origin of replication that functions only in *E. coli*. It is maintained at a high copy number in *E. coli* and cannot be readily transferred to other bacteria.

How does pBR322 work as a cloning vector? Purified, closed circular pBR322 molecules are cut with a restriction enzyme that lies within either of the antibiotic resistance genes and cleaves the plasmid DNA only once to create single, linear, sticky-ended DNA molecules. These linear molecules are combined with prepared target DNA from a source organism. This DNA has been cut with the same restriction enzyme, which generates the same sticky ends as those on the plasmid DNA. The DNA mixture is then treated with T4 DNA ligase in the presence of ATP. Under these conditions, a number of different ligated combinations are produced, including the original closed circular plasmid DNA. To reduce the amount of this particular unwanted ligation product, the cleaved plasmid DNA preparation is treated with the enzyme alkaline phosphatase to remove the 5′-phosphate groups from the linearized plasmid DNA. As a consequence, T4 DNA ligase cannot join the ends of the dephosphorylated linear plasmid DNA (Fig. 4.8). However, the two phosphodiester bonds that are formed by T4 DNA ligase after the ligation and circularization of alkaline phosphatase-treated plasmid DNA with restriction endonuclease-digested source DNA, which provides the phosphate groups, are sufficient to hold both molecules together, despite the presence of two nicks (Fig. 4.8). After transformation, these nicks are sealed by the host cell DNA ligase system. In addition, although unwanted, fragments from the source DNA are also joined to each other by T4 DNA ligase.

Transformation and Selection

The next step in a recombinant DNA experiment requires the uptake by *E. coli* of the cloned DNA. The process of introducing purified DNA into a bacterial cell is called transformation. For *E. coli*, one of the methods of transformation requires that the cells be treated with a regimen of high temperature and calcium chloride (CaCl$_2$). But transformation is an inefficient process, with typically no more than one cell in a thousand being transformed. In other words, following transformation, most of the cells have not acquired a new plasmid. Furthermore, a few cells are transformed by recircularized plasmid DNA that escaped dephosphorylation by alkaline phosphatase, others acquire nonplasmid DNA, and a few are transformed by the plasmid–insert DNA construct.

As discussed earlier, extrachromosomal DNA that lacks an origin of replication cannot be maintained within a bacterial cell. Thus, the uptake of nonplasmid DNA is of no consequence in a recombinant DNA experiment. To ensure that a plasmid–cloned DNA construct is perpetuated in its original form, the *E. coli* host cells that are used must lack the genes for restriction endonucleases and are usually not capable of carrying out exchanges between DNA molecules because they have been made recombination negative (RecA$^-$). A plasmid–DNA insert construct is less likely to be degraded

Figure 4.7 Genetic map of the plasmid cloning vector pBR322. Unique *Hin*dIII, *Sal*I, *Bam*HI, and *Pst*I recognition sites are present the genes for tetracycline resistance (Tetr) and ampicillin resistance (Ampr). The unique *Eco*RI site is just outside the tetracycline resistance gene. The origin of replication functions in the bacterium *E. coli*. The complete DNA sequence of pBR322 consists of 4,361 bp.

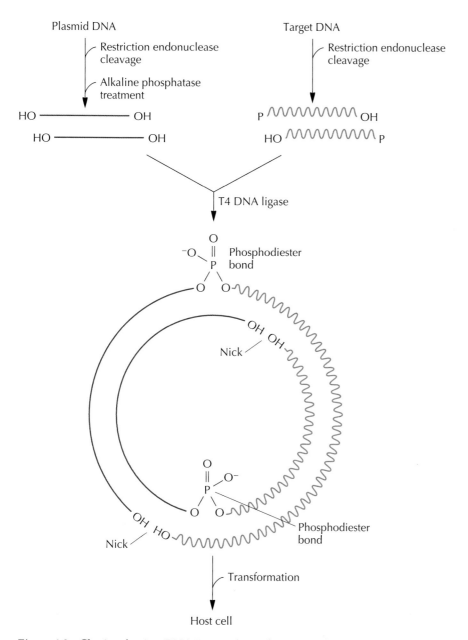

Figure 4.8 Cloning foreign DNA into a plasmid vector. After restriction endonuclease cleavage and alkaline phosphatase treatment, the plasmid DNA is ligated to the restriction endonuclease-digested target DNA, and two of the four nicks are sealed. This molecular configuration is stable, and the two DNA molecules are covalently joined. After introduction into a host cell, ensuing replication cycles produce new complete circular DNA molecules with no nicks.

in a bacterium that does not produce restriction endonucleases. And the insert DNA will not become altered due to homologous recombination events in a recombinant-deficient host cell.

After the transformation step, it is necessary to identify, as easily as possible, those cells that contain plasmid–cloned DNA constructs. In a pBR322 system in which the target DNA was inserted into the *Bam*HI site, this specific identification is accomplished in two steps. First, the cells from the transformation mixture are plated onto medium that contains the antibiotic

ampicillin. Only those cells that have either intact pBR322 or pBR322–cloned DNA constructs, both of which have an intact Ampr gene, can grow under these conditions. The nontransformed cells are sensitive to ampicillin. The *Bam*HI site of pBR322 is within the tetracycline resistance gene (Fig. 4.7), so the insertion of DNA into this gene disrupts the coding sequence and tetracycline resistance is lost. Therefore, cells with these plasmid–cloned DNA constructs are resistant to ampicillin and sensitive to tetracycline. Cells with recircularized pBR322 DNA, however, have an intact Tetr gene and are resistant to both ampicillin and tetracycline.

The second step in the selection scheme distinguishes between these two possibilities. Cells that grow on the ampicillin-containing medium are transferred to a tetracycline-containing medium. Each location of inoculating cells on a tetracycline–agar plate corresponds to the site of a colony on an original ampicillin–agar plate. Cells that form colonies on the tetracycline–agar plates carry recircularized pBR322, because, as noted above, these cells are resistant to both ampicillin and tetracycline. Those cells that do not grow on the tetracycline–agar plates, however, are sensitive to tetracycline and carry pBR322–cloned DNA constructs.

Individual cultures that are sensitive to tetracycline are established from each of the colonies on the ampicillin–agar plates; or, more often, cells from colonies that are ampicillin resistant and tetracycline sensitive are pooled and subcultured together. Later, additional screening procedures can be conducted to identify those cells that carry a specific pBR322–cloned DNA construct. The *Hin*dIII and *Sal*I sites in the tetracycline resistance gene and the *Pst*I site in the ampicillin resistance gene of pBR322 provide alternative cloning locations. When the *Pst*I recognition site is used for cloning, the principle of the selection scheme is the same but the antibiotic sensitivities are reversed; thus, the first set of plates contains tetracycline and the second set contains ampicillin.

Other Plasmid Cloning Vectors

The plasmid pBR322 was a well-conceived cloning vector. It has only a few unique cloning sites, however, and its use requires a time-consuming selection procedure. Thus, it was inevitable that other systems would be developed. For example, the plasmid pUC19 is 2,686 bp long and contains an ampicillin resistance gene, a regulatable segment of the β-galactosidase gene (*lacZ'*) of the lactose operon of *E. coli*, a *lacI* gene that produces a repressor protein that regulates the expression of the *lacZ'* gene, a short sequence with multiple unique cloning sites (i.e., *Eco*RI, *Sac*I, *Kpn*I, *Xma*I, *Sma*I, *Bam*HI, *Xba*I, *Sal*I, *Hinc*II, *Acc*I, *Bsp*MI, *Pst*I, *Sph*I, and *Hin*dIII), and the origin of replication from pBR322 (Fig. 4.9).

The pUC19 selection procedure has the following rationale. When cells carrying unmodified pUC19 are grown in the presence of isopropyl-β-D-thiogalactopyranoside (IPTG), which is an inducer of the *lac* operon, the product of the *lacI* gene cannot bind to the promoter–operator region of the *lacZ'* gene, so the *lacZ'* gene encoded in the plasmid is transcribed and translated. The *lacZ'* protein combines with a protein that is encoded by chromosomal DNA to form an active hybrid β-galactosidase. In pUC19, the multiple cloning sequence is incorporated into the *lacZ'* gene in the plasmid without interfering with the production of the functional hybrid β-galactosidase. Finally, if the substrate 5-bromo-4-chloro-3-indolyl-β-D-galactopyranoside (X-Gal) is present in the medium, it is hydrolyzed by this hybrid β-galactosidase to a blue product. Under these conditions, colonies containing unmodified pUC19 appear blue.

For a pUC19 cloning experiment, DNA from a source organism is cut with one of the restriction endonucleases for which there is a recognition site in the multiple cloning sequence. This source DNA is mixed with pUC19 plasmids that have been treated initially with the same restriction endonuclease and subsequently with alkaline phosphatase. After ligation with T4 DNA ligase, the reaction mixture is introduced by transformation into a host cell which can synthesize that part of β-galactosidase (LacZα) that combines with the product of the *lacZ'* gene to form a functional enzyme. The treated host cells are plated onto medium that contains ampicillin, IPTG, and X-Gal. Nontransformed cells cannot grow in the presence of ampicillin. Cells with recircularized plasmids can grow with ampicillin in the medium; because they can form functional β-galactosidase, they produce blue colonies. In contrast, host cells that carry a plasmid–cloned DNA construct produce white colonies on the same medium. The reason for this is that, usually, DNA inserted into a restriction endonuclease site within the multiple cloning sequence disrupts the correct sequence of DNA codons (reading frame) of the *lacZ'* gene and prevents the production of a functional LacZ' protein, so no active hybrid β-galactosidase is produced. In the absence of β-galactosidase activity, the X-Gal in the medium is not converted to the blue compound by these colonies, so they remain white. The white (positive) colonies subsequently must be screened to identify those that carry a specific target DNA sequence.

In addition to pBR322 and the pUC-series plasmids, many other cloning vectors have been devised. Some vectors have inventive selection systems to identify clones with insert–vector constructs. For example, a vector that is derived from the pUC series carries a gene that, when expressed, encodes a protein that kills the cell. This cell-killing gene (suicide gene) is fused in the correct reading frame to the *lacZ'* gene. A cell with an intact plasmid and no IPTG in the medium does not synthesize the suicide protein. With an insert and IPTG, a nonfunctional suicide protein is produced because the insert, in all likelihood, disrupts the reading frame of the suicide gene. By contrast, cells with a plasmid and no insert in the presence of IPTG synthesize the suicide protein and are killed. Nontransformed cells are sensitive to an antibiotic, whereas recombinant clones have as part of the vector a gene that confers resistance to the antibiotic. In other words, the only surviving cells in the presence of IPTG and antibiotic are those that

Figure 4.9 Genetic map of the plasmid cloning vector pUC19. The multiple cloning sequence contains unique sites for the restriction endonucleases *Eco*RI, *Sac*I, *Kpn*I, *Xma*I, *Sma*I, *Bam*HI, *Xba*I, *Sal*I, *Hinc*II, *Acc*I, *Pst*I, *Bsp*MI, *Sph*I, and *Hin*dIII and is used for the insertion of cloned DNA. The plasmid contains an ampicillin resistance gene, an origin of replication that functions in *E. coli*, and the *lacI* gene, which produces a repressor that blocks the transcription of the *lacZ'* gene in the absence of the inducer IPTG. The complete DNA sequence of pUC19 is 2,686 bp long.

MILESTONE

Cleavage of DNA by R$_I$ Restriction Endonuclease Generates Cohesive Ends

J. E. MERTZ and R. W. DAVIS

Proc. Natl. Acad. Sci. USA **69**:3370–3374, 1972

Recombinant DNA technology requires a vector to carry cloned DNA, the specific joining of vector and cloned (insert) DNA molecules to form a vector–insert DNA construct, the introduction of the vector–insert DNA construct into a host cell, and the identification of host cells that acquired the cloned DNA. Without type II restriction endonucleases, it would be impossible to do recombinant DNA technology routinely. These enzymes facilitate the development of vectors (see, e.g., Bolivar et al., *Gene* **2**:95–113, 1977) and are essential for cloning genes into vectors. In 1968,

M. Meselson and R. Yuan (*Nature* **217**:1110–1114) showed that the capability of a strain of *E. coli* to prevent (restrict) the development of a bacterial virus (bacteriophage) was due to a host cell enzyme that cleaved the DNA of the infecting bacteriophage. The Mertz-Davis study established that the R$_I$ restriction endonuclease, which is now called *Eco*RI, from *E. coli* cut DNA at a specific site and produced complementary extensions. Briefly, after circular DNA was linearized by treatment with *Eco*RI, some of the molecules formed hydrogen-bonded circular DNA molecules which were

converted to covalently closed circular DNA molecules by treating the sample with a DNA ligase. The extensions of all the cut DNA molecules were the same and estimated to be four to six nucleotides long, with the recognition site being six nucleotide pairs. Mertz and Davis concluded that "... any two DNA molecules with R$_I$ sites can be 'recombined' at their restriction sites by the sequential action of R$_I$ endonuclease and DNA ligase to generate hybrid DNA molecules." The discovery of the ability of *Eco*RI to create cohesive ends was one of the most important contributions to the development of recombinant DNA technology because it provided, according to Mertz and Davis, a "simple way ... to generate specifically oriented recombinant DNA molecules *in vitro*."

carry insert DNA. One suicide gene that has been used in this way encodes an enzyme that prevents the rejoining of double-stranded breaks in the chromosomal DNA. These DNA breaks accumulate and are responsible for the death of the cell.

Although a number of vectors have ingenious designs, in principle they all retain the two basic requirements of recombinant DNA technology: a choice of cloning sites and an easy way to identify cells that have plasmid–cloned DNA constructs. It should be noted that unique restriction endonuclease sites have a dual function in recombinant DNA research: they are essential for inserting DNA into a cloning vector, and they allow an inserted DNA sequence to be recovered from the vector. In other words, after a piece of DNA has been cloned into a site, it can be retrieved by cutting a purified plasmid–cloned DNA construct with the original restriction endonuclease because the insertion event creates two recognition sites at the ends of the cloned DNA sequence. Occasionally, the initial recognition site is destroyed. Then the cloned piece of DNA is difficult to retrieve. A recovered DNA fragment can be cloned into cloning vectors for DNA sequencing or vectors that have been specially designed to achieve high levels of expression (transcription and translation) of the cloned gene.

Even though *E. coli*, which is well-known as a laboratory organism, is used for all routine molecular cloning procedures, other bacteria such as *Bacillus subtilis* and *Agrobacterium tumefaciens* are often used as host cells. For many applications, cloning vectors that function in *E. coli* may be provided with a second origin of replication that enables the plasmid to replicate in the alternative host cell. These so-called shuttle cloning vectors allow researchers to conduct the initial cloning steps with *E. coli* and then introduce a fully developed construct into a different host cell. In addition, researchers have constructed a number of plasmid vectors that contain a single broad-host-range origin of DNA replication instead of a narrow-

host-range origin of replication. These vectors can be used with a variety of different microorganisms.

Creating and Screening a Library

Making a Gene Library

The isolation of genes that encode proteins is often the goal of a biotechnology experiment. In prokaryotic organisms, these structural genes each have a continuous coding domain in the genomic DNA; in eukaryotes, however, the coding regions (exons) of structural genes are separated by noncoding regions (introns). Consequently, different cloning strategies have to be used for prokaryotic and eukaryotic genes.

In a prokaryote, the desired sequence (target DNA) is frequently a minuscule portion (0.02%) of the total chromosomal DNA. The problem, then, is how to clone and select the targeted DNA sequence. To do this, the complete DNA of an organism is cut with a restriction endonuclease and each fragment is inserted into a vector. Then, the specific cell line (clone) that carries the target DNA sequence must be identified, isolated, and characterized. The process of subdividing genomic DNA into clonable elements and inserting them into host cells is called creating a library (clone bank, gene bank). A complete library, by definition, contains all of the genomic DNA of the source organism.

One way to create a DNA library is by treating the DNA from a source organism with a four-cutter restriction endonuclease (e.g., *Sau*3AI), which theoretically cleaves the DNA approximately once in every 256 bp. The conditions of the digestion reaction are set to give a partial, not a complete, digestion. In this way, all possible fragment sizes are generated (Fig. 4.10). However, because restriction endonuclease sites are not randomly located, some fragments may be too large to be cloned. In these cases, an incomplete library is available for selection, so it may be difficult or even impossible to find a specific target DNA sequence. This problem can be overcome by forming a library with another restriction endonuclease.

After a library is created, the clone(s) (cell lines) with the target sequence must be identified. Three popular methods of identification are used: DNA hybridization with a labeled DNA probe, followed by radiographic screening for the probe label; immunological screening for the protein product; and screening for protein activity.

Screening by DNA Hybridization

The presence of a target nucleotide sequence in a DNA sample can be determined with a DNA probe. This procedure is called DNA hybridization and depends on the formation of stable base pairs between the probe and the target sequence. DNA hybridization is feasible because double-stranded DNA can be converted into single-stranded DNA by heat or alkali treatment. Heating DNA breaks the hydrogen bonds that hold the bases together (denaturation) but does not affect the phosphodiester bonds of the DNA backbone. If the heated solution is rapidly cooled, the strands remain single standed. However, if the temperature of a heated DNA solution is lowered slowly, the double-stranded, helical conformation of DNA can be reestablished due to the base pairing of complementary nucleotides (renaturation). The process of heating and slowly cooling double-stranded DNA is called annealing. Some of the products of this process contain molecules of hybrid DNA, that is, double-stranded DNA in which the two strands come from different DNA molecules.

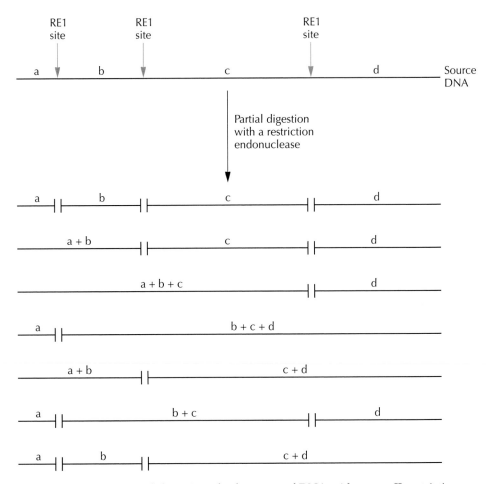

Figure 4.10 Partial digestion of a fragment of DNA with a type II restriction endonuclease. Partial digestions are usually performed by varying either the length of time or the amount of enzyme used for the digestion. In some of the DNA molecules, the restriction endonuclease has cut at all sites (each labeled RE1). In other molecules, fewer cleavages have occurred. The desired outcome is a sample with DNA molecules of all possible lengths.

In general, for a DNA hybridization assay, the target DNA is denatured and the single strands are irreversibly bound to a matrix (e.g., nitrocellulose or nylon). This binding process is often carried out at a high temperature. Then the single strands of a DNA probe, which are labeled with either a radioisotope or another tagging system, are incubated with the bound DNA sample. If the sequence of nucleotides in the DNA probe is complementary to a nucleotide sequence in the sample, then base pairing (i.e., hybridization) occurs (Fig. 4.11). The hybridization can be detected by autoradiography (Box 4.2) or other visualization procedures, depending on the nature of the probe label. If the nucleotide sequence of the probe does not base pair with a DNA sequence in the sample, then no hybridization occurs and the assay gives a negative result. Generally, probes range in length from 100 to more than 1,000 bp, although both larger and smaller probes can be used. Depending on the conditions of the hybridization reaction, stable base pairing requires a match of >80% within a segment of 50 bases.

DNA probes can be labeled in various ways. One strategy, which is called the random primer method, utilizes a mixture of synthetic random

oligonucleotides (oligomers) containing all possible combinations of sequences of six nucleotides that act as primers for DNA synthesis. On the basis of the chance occurrence of complementary sequences, some of the oligomers in the sample will hybridize to complementary sequences on the unlabeled probe DNA template (Fig. 4.12). After the oligomer sample is mixed with the denatured probe template DNA, the four deoxyribonucleotides (deoxyribonucleoside triphosphates [dNTPs]) and a portion of *E. coli* DNA polymerase I called the Klenow fragment are added. The Klenow fragment retains both DNA polymerase and 3′ exonuclease activities but lacks the 5′ exonuclease activity that is normally associated with *E. coli* DNA polymerase I. This 5′ exonuclease activity would degrade some of the

Figure 4.11 DNA hybridization. The source DNA is denatured, and the two strands are kept apart, usually by binding them to a solid matrix such as a nitrocellulose or nylon membrane. Labeled probe DNA (often approximately 100 to 1,000 bp) is denatured and added to the denatured source DNA. Hybridization (base pairing) between the probe and source DNA may occur under these conditions. The membrane is then washed to remove unhybridized probe DNA, and the membranes are assayed. If the probe hybridizes with the source DNA, then it can be detected with an assay that identifies its labeled tag. If the probe does not hybridize, then no label is detected. The asterisks (*) denote the labeled tag (signal) of the probe DNA.

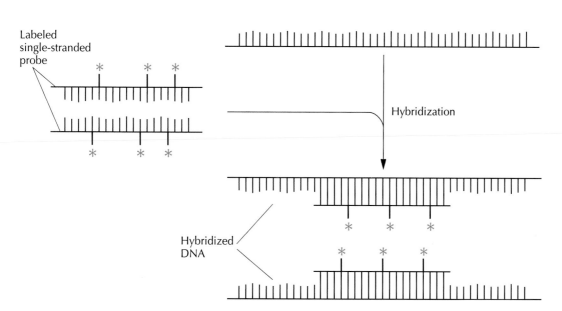

Autoradiography

Autoradiography is used to detect the site of a radiolabeled entity in a cell or sample of fractionated macromolecules. In principle, autoradiography consists of placing a radioactive source next to a radiosensitive photographic film that contains silver bromide. The energy from the decay of the radioisotope hits the photographic emulsion and produces electrons that are trapped by specks of silver bromide crystals in the emulsion. The negatively charged specks attract silver ions, and metallic silver is formed. The grains of metallic silver are visualized by developing the photographic film. Thus, an exposed dark region on a developed film indicates that the underlying material was radiolabeled. Parenthetically, fluorography is the term used for the exposure of light-sensitive photographic film to molecules that directly or indirectly generate light as the source of energy that reduces silver in the photographic emulsion.

Proteins and nucleic acids that are radiolabeled and separated by gel electrophoresis can be visualized by placing an X-ray film on a dried gel and developing the film after a suitable exposure time. All autoradiographic steps are carried out in the dark to avoid inadvertent exposure of the X-ray film by light. A number of autoradiographic techniques have been devised for the quantitative and qualitative analysis of proteins and nucleic acids.

One of the major applications of autoradiography is the detection of the hybridization of a radiolabeled DNA probe to a DNA molecule that has been electrophoretically fractionated. However, DNA molecules in a gel are not accessible to hybridization with a DNA probe. Consequently, the DNA molecules in the gel are transferred by blotting or electrotransfer to a nitrocellulose or nylon membrane. The transfer process retains the same positions on the membrane as the DNA molecules had in the gel. The DNA molecules that are transferred to a

membrane are denatured, bound to the membrane, and hybridized with a radiolabeled DNA probe. Autoradiography of the membrane reveals whether the probe hybridized to a particular DNA band(s).

The transfer of DNA from a gel to a membrane is called Southern blotting (Southern DNA blotting) after Edwin Southern, who devised the original DNA blotting strategy. Northern blotting and Western blotting are methods for the transfer of RNA and protein, respectively, from a gel to a membrane. The terms "Northern" and "Western" have nothing to do with direction and were coined by molecular biologists both to give Edwin Southern further credit for developing the notion of blotting macromolecules from a gel to a membrane and to distinguish which macromolecules were transferred. The designations "Northern" and "Western" are also examples of molecular biology humor.

newly synthesized DNA. With the available 3'-hydroxyl groups of the bound random primers and the strands of the probe as templates, new DNA synthesis occurs (Fig. 4.12). If a radioactive label is used, then one of the dNTPs contains the isotope ^{32}P in the α-position phosphate. Autoradiography can be used later to determine whether the labeled probe sequences hybridize to sequences of a target DNA sample.

For nonisotopic detection of hybridization, biotin (or a similar nonradioactive label) can be attached to one of the four dNTPs that is incorporated during the DNA synthesis step. When a probe with this kind of tag (label) hybridizes to the sample DNA, detection is based on the binding of an intermediary compound (e.g., streptavidin) that carries an appropriate enzyme. Depending on the assay system, the enzyme can be used for the formation of either a chromogenic (colored) molecule that can be visualized directly or a chemiluminescent response that can be detected by autoradiography.

There are at least two possible sources of probes for screening a genomic library. First, cloned DNA from a closely related organism (a heterologous probe) can be used. In this case, the conditions of the hybridization reaction can be adjusted to permit considerable mismatch between the probe and the target DNA to compensate for the natural differences between the two sequences. Second, a probe can be produced by chemical synthesis. The nucleotide sequence of a synthetic probe is based on the

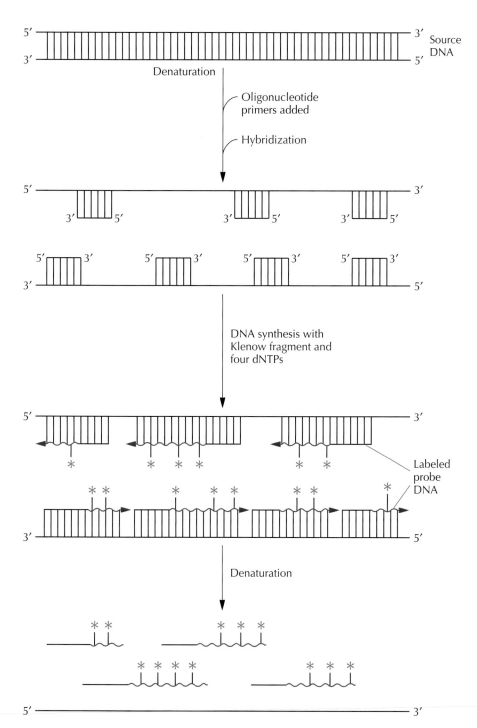

Figure 4.12 Production of labeled probe DNA by the random primer method. The duplex DNA containing the sequence that is to act as the probe is denatured, and an oligonucleotide sample containing all possible sequences of six nucleotides is added. It is a statistical certainty that some of the molecules of the oligonucleotide mixture will hybridize to the unlabeled, denatured probe DNA. In the presence of Klenow fragment and the four dNTPs (one of which is labeled with a tag [*]), the bound oligonucleotides act as primers for DNA synthesis. The synthesized DNA is labeled and can be used as a probe to detect the presence of a complementary DNA sequence in a source DNA sample. In this case, the labeled probe consists of a number of separate DNA molecules that together constitute almost the entire sequence of the original unlabeled template DNA.

probable nucleotide sequence that is deduced from the known amino acid sequence of the protein encoded by the target gene.

Genomic DNA libraries are often screened by plating out the transformed cells on the growth medium of a master plate and then transferring samples of each colony to a solid matrix such as a nitrocellulose or nylon membrane, lysing (breaking) the cells, deproteinizing and denaturing the DNA, and binding the DNA to the matrix. At this stage, a labeled probe is added; if hybridization occurs, signals are observed on an autoradiograph. The colonies from the master plate that correspond to samples containing hybridized DNA are then isolated and cultured (Fig. 4.13). Because most libraries are created from partial digestions, a number of colonies (clones) may give a positive response to the probe. The next task is to determine which clone, if any, contains the complete sequence of the target gene. Preliminary analyses that use the results of gel electrophoresis and restriction

Figure 4.13 Screening a library with a labeled DNA probe (colony hybridization). Cells from the transformation reaction are plated onto solid medium under conditions that permit transformed but not nontransformed cells to grow. (1) From each discrete colony formed on this master plate, a sample is transferred to a solid matrix such as a nitrocellulose or nylon membrane. The pattern of the colonies on the master plate is retained on the matrix. (2) The cells on the matrix are lysed, and the released DNA is denatured, deproteinized, and irreversibly bound to the matrix. (3) A labeled DNA probe is added to the matrix under hybridization conditions. After the nonhybridized probe molecules are washed away, the matrix is processed by autoradiography to determine which cells have bound labeled DNA. (4) A colony on the master plate that corresponds to the region of a positive response on the X-ray film is identified. Cells from the positive colony on the master plate are subcultured because they may carry the desired plasmid–cloned DNA construct.

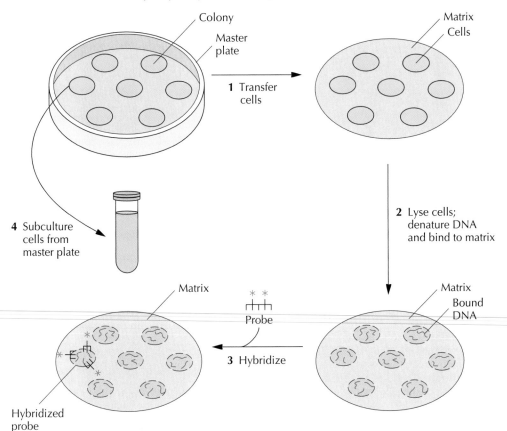

endonuclease mapping reveal the length of each insert and identify those inserts that are the same and those that share overlapping sequences. By using overlapping sequences, it may be possible to join sections of the gene in additional cloning experiments. Alternatively, if an insert in any one of the clones is large enough to include the full gene, then the complete gene can be recognized after DNA sequencing, because it will have start and stop codons and a contiguous set of nucleotides that code for the target protein.

Unfortunately, there is no guarantee that the complete sequence of a target gene will be present in a particular library. If the search for an intact gene fails, then another library can be created with a different restriction endonuclease and screened with either the original probe or probes derived from the first library. Alternatively, as discussed later in this chapter, libraries that contain DNA fragments larger than the average prokaryotic gene can be created to increase the chance that some members of the library will carry a complete version of the target gene.

Screening by Immunological Assay

If a DNA probe is not available, alternative methods can be used to screen a library. For example, if a cloned DNA sequence is transcribed and translated, the presence of the protein, or even part of it, can be determined by an immunological assay. Technically, this procedure has much in common with a DNA hybridization assay. All cell lines (clones) of the library are grown on master plates. A sample of each colony is transferred to a matrix, where the cells are lysed and the released proteins attach to the matrix. The matrix with the bound proteins is treated with an antibody (primary antibody) that specifically binds to the protein encoded by the target gene. Following the interaction of the primary antibody with the target protein (antigen), any unbound antibody is washed away, and the matrix is treated with a second antibody that is specific for the primary antibody. In many assay systems, the secondary antibody has an enzyme, such as alkaline phosphatase, attached to it. After the matrix is washed, a colorless substrate is added. If the secondary antibody has bound to the primary antibody, the colorless substrate is hydrolyzed by the attached enzyme and produces a colored compound that accumulates at the site of the reaction (Fig. 4.14).

The colonies on the master plate that correspond to positive results (colored spots) on the matrix contain either an intact gene or a portion of the gene that is large enough to produce a protein product that is recognized by the primary antibody. After detection by immunoassay of genomic DNA libraries, the positive clones must be characterized further to determine which, if any, carry a complete gene.

Screening by Protein Activity

DNA hybridization and immunological assays work well for many kinds of genes and gene products. If the target gene produces an enzyme that is not normally made by the host cell, a plate assay can be devised to identify members of a library that carry the functional gene encoding that enzyme. For example, the genes for α-amylase, endoglucanase, and β-glucosidase from various organisms have been isolated by plating the genomic library in *E. coli* onto medium supplemented with a specific substrate and then using a selective stain to identify those colonies that are capable of utilizing the substrate.

If the gene that is being sought encodes a product that is essential for the growth of a mutated host cell, then the library can be formed by transformation into these mutant cells. The cells that are able to grow on minimal

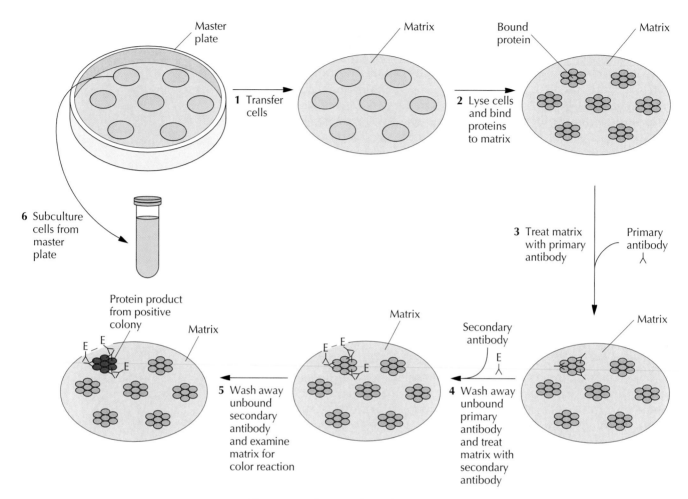

Figure 4.14 Immunological screening of a gene library (colony immunoassay). Cells from the transformation reaction are plated onto solid medium under conditions that permit transformed but not nontransformed cells to grow. (1) From the discrete colonies formed on this master plate, a sample from each colony is transferred to a solid matrix such as a nitrocellulose or nylon membrane. (2) The cells on the matrix are lysed, and their proteins are bound to the matrix. (3) The matrix is treated with a primary antibody that binds only to the targeted protein. (4) Unbound primary antibody is washed away, and the matrix is treated with a secondary antibody that binds only to the primary antibody. (5) Any unbound secondary antibody is washed away, and a colorimetric reaction, which can occur only if the secondary antibody is present, is carried out. (6) A colony on the master plate that corresponds to a positive response on the matrix is identified. Cells from the positive colony on the master plate are subcultured because they may carry the plasmid–insert DNA construct that encodes the protein that binds the primary antibody.

medium in the absence of the required substrate must carry a functional form of the target gene on the cloning vector. Variations of this form of genetic complementation have been used to isolate a variety of important genes, including those for the biosynthesis of antibiotics and the formation of nitrogen-fixing nodules on the roots of certain plants.

Cloning DNA Sequences That Encode Eukaryotic Proteins

Special techniques are required for cloning eukaryotic structural genes. Prokaryotic hosts are not able to remove the introns from transcribed RNA molecules; therefore, this mRNA is not translated correctly in a bacterial

host cell. Moreover, a eukaryotic DNA sequence needs prokaryotic transcriptional and translational control sequences to be properly expressed. A functional eukaryotic messenger RNA (mRNA), which does not have introns, has a G cap at the 5' end and, usually, a string of up to 200 adenine residues [poly(A) tail] at the 3' end.

The poly(A) tail can be used to separate the mRNA fraction of a tissue from the ribosomal RNA (rRNA) and transfer RNA (tRNA). Extracted cellular eukaryotic RNA is passed through a column packed with cellulose beads to which is bound short chains of thymidine residues, each about 15 nucleotides long [oligo(dT), dT_{15}]. The poly(A) tails of the mRNA molecules bind by base pairing to the oligo(dT) chains. The tRNA and rRNA molecules, which lack poly(A) tails, pass through the column. The mRNA is removed (eluted) from the column by treatment with a buffer that breaks the A:T hydrogen bonds, thereby releasing the bound mRNAs.

Before the mRNA molecules can be cloned into a vector, they must be converted to double-stranded DNA. This synthesis is accomplished by using, in succession, two different kinds of nucleic acid polymerases: reverse transcriptase and the Klenow fragment of DNA polymerase I (Fig. 4.15). After the mRNA fraction is purified, short (unbound) sequences of oligo(dT) molecules are added to the sample, along with the enzyme reverse transcriptase and the four dNTPs (dATP, dTTP, dGTP, and dCTP). The oligo(dT) molecules base pair with the poly(A) tail regions and provide an available 3'- hydroxyl group to prime the synthesis of a DNA strand.

Reverse transcriptase, which is produced by certain RNA viruses, uses an RNA strand as a template while directing deoxyribonucleotides into the growing chain. Thus, when an A, G, C, or U residue of the template RNA strand is encountered, the complementary deoxyribonucleotide (i.e., T, C, G, or A) is incorporated into the growing DNA strand. The synthesis of a DNA strand by reverse transcriptase in vitro is often incomplete; however, before synthesis ceases, the DNA strand usually turns back on itself for a few nucleotides (Fig. 4.15) to form a hairpin loop.

The second DNA strand is synthesized by the addition of the Klenow fragment of *E. coli* DNA polymerase I, which uses the first DNA strand as a template and adds deoxyribonucleotides to the growing strand, starting from the end of the hairpin loop. After the reaction is complete, the sample is treated with the enzyme RNase H, which degrades the mRNA molecules, and with S1 nuclease, which opens the hairpin loops and degrades single-stranded DNA extensions. At the end of this procedure, the sample contains a mixture of partial and complete double-stranded complementary DNA (cDNA) copies of the more prevalent mRNAs in the original sample.

This cDNA population can be cloned by blunt-end ligation or other joining mechanisms into a plasmid cloning vector to form a cDNA library. A cDNA library can be screened by either DNA hybridization or immunological assays, as explained earlier, to identify clones that carry a specific plasmid–cDNA construct. For immunological screening, the cDNA must be cloned into a site that puts the cloned sequence under the control of a bacterial promoter, which ensures its transcription. However, with most cloning vectors, there is no guarantee that, after insertion of the cDNA, its nucleotides will have the correct reading frame and will therefore be able to direct the synthesis of the true amino acid sequence of the protein. Nevertheless, positive clones that are isolated by either screening method must be examined further to determine which one(s) carries the complete coding sequence for the target protein.

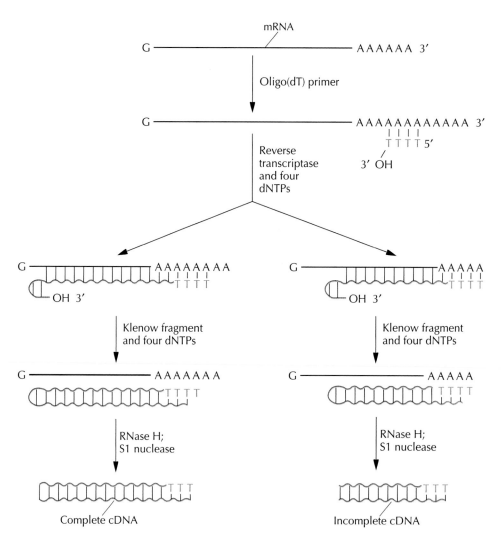

Figure 4.15 Synthesis of cDNA. Oligo(dT) primer is added to a purified mRNA preparation, and reverse transcriptase with the four dNTPs is used for the production of DNA from the RNA template. Reverse transcriptase does not produce full-length cDNA copies of the RNA from every template molecule (lower right). The looping back of the growing DNA strand produced by the reverse transcriptase provides a 3'-hydroxyl group for the Klenow fragment to complete the synthesis of the second strand, using the initial cDNA strand as a template. After synthesis of the second strand is completed, the mRNA is hydrolyzed with the enzyme RNase H and the DNA is treated with S1 nuclease to produce blunt-ended linear molecules without hairpin loops.

Vectors for Cloning Large Pieces of DNA

Bacteriophage λ Vectors

The plasmid-based vectors used for cloning DNA molecules generally carry up to 10 kb of inserted DNA. However, for the formation of a library, it is often helpful to be able to maintain larger pieces of DNA. To meet this end, the *E. coli* virus (bacteriophage, phage) lambda (λ) has been developed as a cloning vehicle.

In its life cycle, bacteriophage λ infects *E. coli*, and, after injection of the viral DNA, two possibilities exist. Bacteriophage λ can enter a lytic cycle which, after 20 minutes, leads to the lysis of the host cell and the release of

about 100 phage particles. Alternatively, the injected bacteriophage λ DNA can be integrated into the *E. coli* chromosome as a prophage and be maintained more or less indefinitely as a benign guest (lysogeny). However, under conditions of nutritional or environmental stress, the integrated bacteriophage λ DNA can be excised and enter a lytic cycle. The bacteriophage λ DNA is about 50 kb in length, of which approximately 20 kb is essential for the integration-excision (I/E) events. For forming genomic libraries, it was reasoned that this 20 kb of DNA could be replaced with 20 kb of cloned DNA. This DNA molecule could then be perpetuated as a "recombinant" bacteriophage λ through compulsory lytic cycles.

To appreciate how bacteriophage λ cloning systems function, some understanding of the molecular aspects of the lytic cycle is necessary. An infective bacteriophage λ consists of a tubular protein tail with a protein tail fiber and a protein head packed with 50 kb of DNA. The production and assembly of the heads and tails and the packaging of DNA are a highly coordinated sequence of events. The DNA within the head of a λ particle is a 50-kb linear molecule, with a 12-base, single-stranded extension at the 5' strand of each end. These extensions are called cohesive (cos) ends, because they contain sequences that are complementary to each other. After the injection through the tail of the λ DNA into *E. coli*, the cos ends base pair to form a circular DNA molecule. During the early phase of the lytic cycle, DNA replication from the circular molecule creates a linear form of λ DNA that is composed of several contiguous lengths of 50-kb units (Fig. 4.16A). Each newly assembled head is filled with one 50-kb unit of DNA; finally, the tail assembly is added, to complete the formation of an infective particle (Fig. 4.16B). The volume of the bacteriophage λ head is sufficient for about 50 kb. If less than 38 kb of DNA is packed into a head, a noninfective bacteriophage particle is produced; in contrast, more than 52 kb of DNA cannot fit into a head. The location of the cos sequences, which are 50 kb apart in the multiple-length linear λ DNA, ensures that each head receives the correct amount of DNA. Located at the opening of the head is an enzyme that recognizes the double-stranded cos sequence and cuts the DNA at this site as the DNA is inserted into the head.

During studies to determine how a bacteriophage λ particle is assembled, an in vitro packaging system was developed. By mixing purified empty heads, bacteriophage λ DNA (50 kb), and tail assemblies, infective particles can be produced in a reaction tube.

One of the many bacteriophage λ cloning vectors that have been devised has two *Bam*HI sites that flank the I/E region. When purified DNA from this bacteriophage is cut with *Bam*HI, three segments are created. The so-called left arm (L region) contains the genetic information for the production of heads and tails, the right arm (R region) carries the genes for DNA replication and cell lysis, and the middle fragment (I/E) has the genes for the integration-excision processes. The objective of this genetic engineering protocol is to replace the middle segment of the λ DNA with cloned DNA that is approximately 20 kb in length (Fig. 4.17). The source DNA is cut with *Bam*HI, and DNA pieces that are 15 to 20 kb in length are isolated. The two DNA samples are combined and incubated with T4 DNA ligase. Then empty bacteriophage heads and tail assemblies are added. Under these conditions, 50-kb units of DNA are packaged into the heads, and infective bacteriophage particles are formed. Other products from the ligation reaction cannot be packaged, because they are either too large (>52 kb) or too small (<38 kb). Recombinant bacteriophage λ can undergo lytic cycles

A

Figure 4.16 Packaging of bacteriophage λ DNA into heads during the lytic cycle. **A.** DNA replication from the circular form of bacteriophage λ creates a linear form that is composed of contiguous, multiple lengths of bacteriophage DNA of approximately 50 kb each. **B.** Each newly assembled head is filled with a 50-kb unit of λ DNA before the tail assembly is attached.

only in an *E. coli* strain that does not allow reconstituted bacteriophage λ with intact I/E regions to grow. Recombinant bacteriophage λ are perpetuated by continual growth on fresh *E. coli*.

Bacteriophage λ libraries can be screened by using either DNA probes or immunological assays. In the plasmid cloning vector systems, bacterial colonies are used for testing; in the bacteriophage vector system, the individual zones of lysis (plaques) that contain bacteriophage are lifted onto a matrix and processed. For DNA hybridization, the bacteriophage proteins are removed and the DNA is denatured and bound to a matrix. For immunological assays, proteins encoded by cloned genes are synthesized during the lytic cycle, so those proteins are transferred with the plaque and subsequently bound to the matrix. On the basis of the sites on the matrix that give positive responses, corresponding plaques on the original plate are subcultured to provide a source of a recombinant bacteriophage, which can be individually cultured on *E. coli*.

Cosmids

Cloning vectors, called cosmids, can carry 40 kb of cloned DNA and can be maintained as plasmids in *E. coli*. Cosmids combine the properties of plasmids and bacteriophage λ vectors. For example, the commonly used cosmid pLFR-5 (approximately 6 kb) has two cos sites (cos ends) from bacterio-

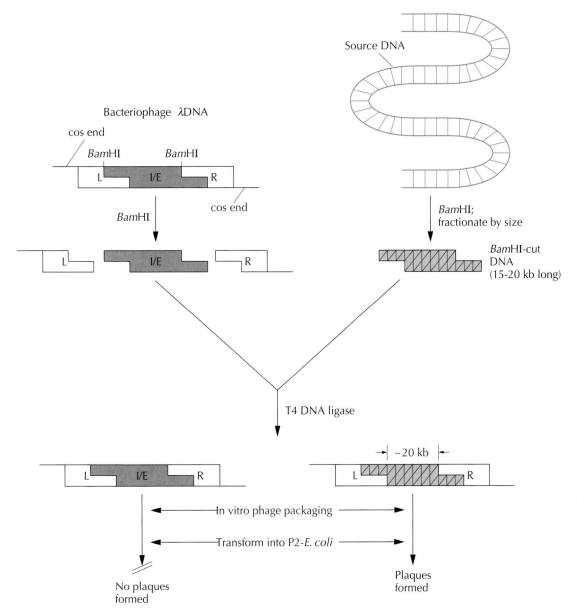

Figure 4.17 Bacteriophage λ cloning system. Bacteriophage λ is engineered to have two *Bam*HI sites that flank the I/E region of the bacteriophage λ genome. The extensions indicate the cos ends of the λ DNA. For cloning, the source DNA is cut with *Bam*HI and fractionated by size to isolate pieces that are 15 to 20 kb long. The bacteriophage λ DNA is also cut with *Bam*HI. The two DNA samples are mixed and treated with T4 DNA ligase. The ligation reaction mixture will contain a number of different DNA molecules, including (1) reconstituted bacteriophage λ and (2) the bacteriophage λ L and R regions with a 20-kb piece of DNA from the source DNA instead of the I/E region. These molecules are packaged into bacteriophage λ heads in vitro, and infective particles are formed after the addition of tail assemblies. After infection of *E. coli* cells that have P2 bacteriophage DNA integrated in their chromosomes, only the molecules with the R and L regions and a cloned ~20-kb piece of DNA can replicate and form infectious bacteriophage λ. In this way, only the bacteriophage λ containing a DNA insert are perpetuated.

phage λ separated by a *Sca*I restriction endonuclease site, a multiple cloning sequence with six unique sites (*Hin*dIII, *Pst*I, *Sal*I, *Bam*HI, *Sma*I, and *Eco*RI), an origin of DNA replication (*ori*), and a tetracycline resistance (Tetr) gene. This cosmid can carry about 40 kb of cloned DNA (Fig. 4.18). For this vector, pieces of DNA that are approximately 40 kb in length are purified by sucrose density gradient centrifugation from a partial restriction endonuclease digestion of the source DNA with *Bam*HI (Fig. 4.18). The pLFR-5 DNA is cleaved initially with *Sca*I and then with *Bam*HI. The two DNA samples are mixed and ligated. Some of the ligated products will have a 40-kb DNA piece inserted between the two fragments that are derived from the digestions of the pLFR-5 DNA. These molecules will be about 50 kb long, with cos sequences that are about 50 kb apart. Consequently, these DNA constructs can be successfully packaged into bacteriophage λ heads in vitro. The bacteriophage λ packaging process accepts only DNA that is about 50 kb in length, so reconstituted pLFR-5 without inserted DNA will not be packaged. After assembly of bacteriophage particles, the DNA is delivered by infection into *E. coli* (Fig. 4.18).

Once inside the host cell, the cos ends, which were cleaved during the in vitro packaging, base pair and enable the linear DNA to circularize. This circular form is stable, so the cloned DNA can be maintained as a plasmid–insert DNA construct because the vector contains a complete set of plasmid functions. Moreover, the tetracycline resistance gene allows colonies that carry the cosmid to grow in the presence of tetracycline; nontransformed cells are sensitive to tetracycline and die.

There are additional cosmid vectors based on bacteriophage λ. As well, other *E. coli* bacteriophages have been used for creating cosmid vectors. For example, the genome of the P1 bacteriophage is 115 kb long. In this case, the P1 cosmid system can carry 85 kb of inserted DNA. The advantages of using cosmids are twofold. First, because the cloning capacity of cosmids is greater than that of plasmids, gene clusters and large genes are easier to clone. Second, for creating a library, a larger insert in the cloning vehicle means that fewer clones have to be screened.

Very-Large-Insert Bacterial Vector Systems

A vector system that carries large inserts (>100 kb) is helpful for the analysis of complex eukaryotic genomes. For example, very-large-insert vector systems are indispensable for the mapping of the human genome and the

Figure 4.18 A cosmid cloning system. The cosmid contains an *E. coli* origin of replication (*ori*) (which allows the cosmid to be maintained as a plasmid in *E. coli*); two intact cos sites closely flanking a unique *Sca*I site; a unique *Bam*HI site near, but outside, one of the cos sites; and a gene for tetracycline resistance (Tetr). The source DNA is cut with *Bam*HI and fractionated by size to isolate molecules that are about 40 kb long. The plasmid DNA is cut with *Sca*I and *Bam*HI. The two DNA samples are mixed and treated with T4 DNA ligase. After ligation, some of the joined DNA molecules will have a 40-kb piece of DNA inserted into the *Bam*HI site of the plasmid; when this happens, the two cos sequences are about 50 kb apart. These molecules are packaged into bacteriophage λ heads in vitro, and infective particles are formed after the addition of tail assemblies. Infective bacteriophage λ deliver a linearized DNA molecule with cos extensions into *E. coli*. After entry into the host cell, the cos ends base pair and the DNA ligase of the host cell seals the nicks. The circular DNA molecule that is created in this way persists as a plasmid in the host cell. In this case, transformed cells can be identified because they are resistant to the antibiotic tetracycline.

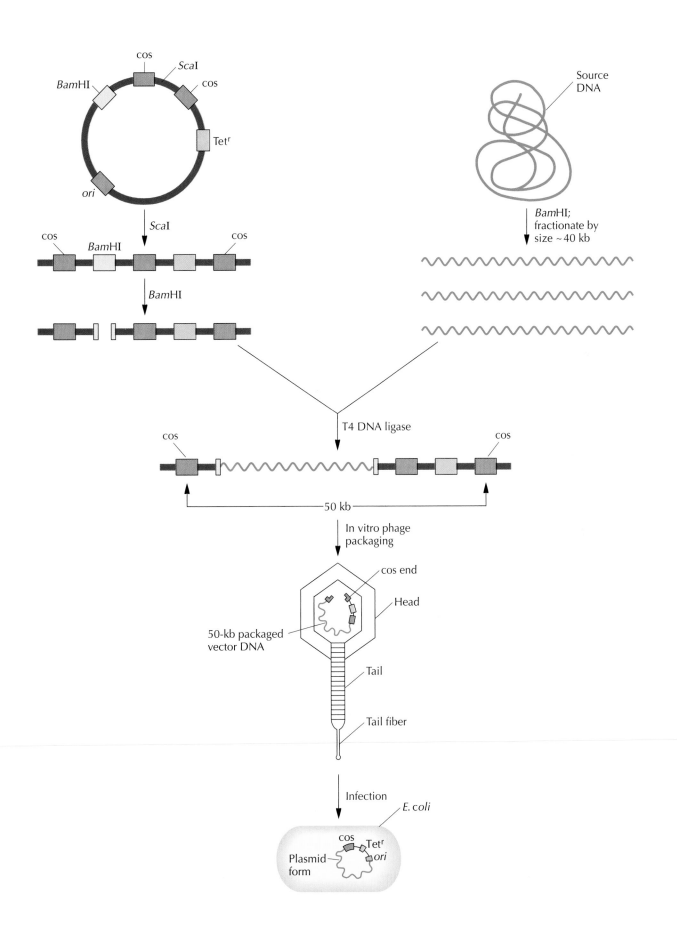

discovery of human genes. In contrast to a small-insert library, a large-insert genomic library is more likely to include all of the genetic material of the organism, there are fewer clones to maintain, and there is a good chance that most or all of a particular gene is present in a single clone. A low-copy-number *E. coli* plasmid vector based on the P1 bacteriophage cloning system has been devised for cloning inserts that are from 100 to 300 kb long. These insert vector constructs are called bacteriophage P1-derived artificial chromosomes. Similarly, the F plasmid (F-factor replicon, sex plasmid, fertility plasmid) of *E. coli*, which is present at 1 to 2 copies per cell, has been used along with the *lacZ'* selection system of the pUC vectors to construct an extremely stable cloning vector that carries DNA inserts from 150 to 300 kb. These F-plasmid-based insert vector constructs are called bacterial artificial chromosomes (BACs).

Genetic Transformation of Prokaryotes

Transferring DNA into *E. coli*

Transformation is the process of introducing free DNA into a bacterial cell. For *E. coli*, which is the main host cell for recombinant DNA research, the uptake of plasmid DNA is usually achieved by treating cells with ice-cold $CaCl_2$ and then exposing them for approximately 90 s to a high temperature (42°C). This method has a maximum transformation frequency, which is defined as the fraction of the cell population that can be transformed, of about 1 transformed cell per 1,000 cells (10^{-3}). For this procedure, the transformation efficiency, which is defined as the number of transformants per microgram of added DNA, is approximately 10^7 to 10^8 transformed colonies per microgram of intact plasmid DNA. Although a 100% transformation frequency would be ideal, selection schemes that enable plasmid-transformed cells to be readily identified overcome the drawback of a low transformation frequency.

The mechanism of action of the $CaCl_2$–heat transformation procedure in *E. coli* is not known. It is assumed that the bacterial cell wall is broken down in localized regions, a condition that allows the plasmid DNA in solution to be taken into the interior of the bacterial cell. Bacterial cells that are able to take up DNA are said to be competent. For *E. coli*, competence must be induced. In some other bacteria, it occurs naturally and sometimes can be enhanced by the use of specific growth media or growth conditions. For bacteria that are refractory (resistant) to chemically induced competence or are not naturally competent, other DNA delivery systems must be used.

Electroporation

The uptake of free DNA can be induced by subjecting bacteria to a high-strength electric field in the presence of DNA. This procedure is called electroporation, a term that is a contraction of the descriptive phrase "electric field-mediated membrane permeabilization." The experimental protocols for electroporation are different for various bacterial species. For *E. coli*, the cells (~50 μL) and DNA are placed in a chamber fitted with electrodes, and a single pulse of approximately 25 microfarads, 2.5 kilovolts, and 200 ohms is administered for about 4.6 milliseconds (ms). This treatment yields transformation efficiencies of 10^9 transformants per microgram of DNA for small plasmids (about 3 kb) and 10^6 for large plasmids (about 136 kb). Similar conditions are used to introduce BAC vector DNA into *E. coli*. Thus, electropo-

ration is an effective way to transform *E. coli* with plasmids containing inserts that are longer than 100 kb. Because it is likely that an appropriate set of electroporation conditions can be found for all bacterial species, this procedure will become the standard way for transforming bacteria.

Very little is known about the mechanism of DNA uptake during electroporation. It has been deduced, along the lines of the explanation of chemically induced transformation, that transient pores are formed in the cell wall as a result of the electroshock and that, after contact with the lipid bilayer of the cell membrane, the DNA is taken into the cell.

Conjugation

For some bacteria, the natural transmission of plasmid DNA from a donor cell to a recipient cell has been used to transfer plasmid–insert DNA constructs to a host cell that is not readily transformed. Some plasmids are genetically equipped to form cell-to-cell junctions through which plasmid DNA is transferred from one cell to another. Effective contact between a donor cell and a recipient cell is due to conjugative functions; the mechanical transfer of the DNA is the consequence of mobilizing functions. Most of the plasmids that are used for recombinant DNA research lack conjugative functions, and therefore the DNA of these plasmids cannot be passed to recipient cells by conjugation. However, some plasmid cloning vectors can be mobilized and transferred if the conjugative functions are supplied by a second plasmid in the same cell. Thus, by introducing a plasmid with conjugative functions into a bacterial cell that carries a mobilizable plasmid cloning vector, it is possible to transfer the plasmid cloning vector to a recipient cell that is difficult to transform by other means.

The standard experimental protocol for this procedure entails mixing three strains together. When the cells are close to each other, the conjugative plasmid, which in this case is also mobilizable, can be self-transferred to the cell with the mobilizable plasmid cloning vector. Then, with the help of the conjugative plasmid, the plasmid cloning vector is transferred to a targeted recipient cell. All possible combinations of plasmid transfer occur among the cells, but the genetic features of the strains and plasmids are designed to select for the targeted recipient cells that receive the cloning vector. For example, one possible selection procedure uses strain A, which carries the conjugative, mobilizable plasmid but cannot grow on minimal growth medium and is sensitive to an antibiotic (say, antibiotic X); strain B, which also cannot grow on minimal growth medium and carries the nonconjugative plasmid cloning vector that has the gene for resistance to antibiotic X; and strain C, the targeted recipient cell, which can grow on minimal growth medium, has no incompatible plasmid, and is sensitive to antibiotic X. After the conjugations are allowed to occur, the cells are grown briefly in complete growth medium in the absence of antibiotic X before being transferred to minimal growth medium with antibiotic X. Under the latter growth conditions, only the targeted recipient cells that have acquired the plasmid cloning vector can grow. Occasionally, the targeted recipient cell may receive both types of plasmids. However, this rare event can be detected by replica plating onto minimal medium and selection for transconjugants that are able to grow in the presence of antibiotic X but unable to grow in the presence of the antibiotic resistance gene (say, antibiotic gene Y) that is contained in the conjugative plasmid from strain A. Because the transfer of plasmid DNA requires conjugation among three bacterial strains, this procedure has been designated tripartite mating.

SUMMARY

Recombinant DNA technology comprises a battery of experimental procedures used to isolate (clone) pieces of DNA that contain specific genes. The success of gene cloning depends on being able to cut DNA molecules reproducibly into fragments of discrete sizes. This precise cleaving of DNA is achieved by using type II restriction endonucleases. These enzymes bind to specific sequences within a DNA molecule and symmetrically cut phosphodiester bonds of each strand at the recognition site.

A typical gene cloning experiment can be summarized as follows. (1) DNA that is isolated from an organism that contains the target gene is cut with a restriction endonuclease. (2) A DNA cloning vector that can be perpetuated in a host cell—usually a plasmid—is cut with the same restriction endonuclease used to digest the source DNA. (3) The two DNA samples are mixed, and the molecules are joined by the action of T4 DNA ligase, an enzyme that forms phosphodiester bonds at the ends of DNA strands. In the ligation reaction, some of the joined DNA molecules contain plasmid DNA and a DNA fragment from the source organism. (4) The DNA molecules that were joined by T4 DNA ligase are used to transform host cells. This transformation produces some cells that carry plasmid–insert DNA constructs. Because the plasmid has a DNA sequence that enables it to be replicated (origin of replication) in the host cell, the entire construct is perpetuated.

Strategies have been devised to select for cells with plasmid–insert DNA constructs. The extent of recircularization of plasmid DNA in the ligation reaction is diminished by treating the cut plasmid DNA with alkaline phosphatase, which removes phosphate groups from the 5′ ends of the linearized DNA so that these ends cannot be joined to each other by T4 DNA ligase. After transformation, cells that have taken up the introduced plasmid are identified (1) by testing for resistance to specific antibiotics or by looking for a specific colorimetric response in growing cells; (2) by immunological or activity assays of protein products encoded by the cloned gene; or (3) by DNA hybridization with a probe that contains a nucleotide sequence present in the target gene.

To increase the chance of cloning a complete gene, the source DNA is incompletely (partially) digested with a restriction endonuclease to produce fragments of various sizes, which are used to construct a genomic library. Vectors based on bacteriophage λ, bacteriophage P1, and the F plasmid have been developed for cloning large pieces of DNA.

To obtain DNA segments that encode eukaryotic proteins, purified mRNA is used as a template for the enzyme reverse transcriptase to synthesize a complementary DNA strand; in turn, this strand acts as a template for DNA polymerase to produce a second complementary DNA strand. After enzymatic treatment, this double-stranded complementary DNA form of an mRNA molecule can be cloned into a vector.

Regardless of the strategy that is used for gene cloning, more laboratory work is required after the initial positive identification of a cloned sequence to determine whether an intact structural gene has been cloned.

References

Berger, S. L., and A. R. Kimmel (ed.). 1987. *Methods in Enzymology*, vol. 152. *Guide to Molecular Cloning Techniques*. Academic Press, London, United Kingdom.

Garfin, D. E. 1995. Electrophoretic methods, p. 53–109. *In* J. A. Glasel and M. P. Deutscher (ed.), *Introduction to Biophysical Methods for Protein and Nucleic Acid Research*. Academic Press, San Diego, Calif.

Grinstead, J., and P. M. Bennett (ed.). 1988. *Methods in Microbiology*, vol. 21. *Plasmid Technology.* Academic Press, London, United Kingdom.

Ioannou, A. P., C. T. Amemiya, J. Garnes, P. M. Kroisel, H. Shizuya, C. Chen, M. A. Batzer, and P. J. de Jong. 1994. A new bacteriophage-derived vector for the propagation of large human DNA fragments. *Nat. Genet.* **6:**84–89.

Kim, U.-J., B. W. Birren, T. Slepak, V. Mancino, C. Boysen, H.-L. Kang, M. I. Simon, and H. Shizuya. 1996. Construction and characterization of a human bacterial artificial chromosome library. *Genomics* **34:**213–218.

Leonardo, E. D., and J. M. Sedivy. 1990. A new vector for cloning large eukaryotic DNA segments in *Escherichia coli. Bio/Technology* **8:**841–844.

Martin, C., L. Bresnick, R.-R. Juo, J. C. Voyta, and I. Bronstein. 1991. Improved chemiluminescent DNA sequencing. *BioTechniques* **11:**110–114.

Old, R. W., and S. B. Primrose. 1985. *Principles of Gene Manipulation*, 3rd ed. Blackwell Scientific Publications, Oxford, United Kingdom.

Sambrook, J., E. F. Fritsch, and T. Maniatis. 1989. *Molecular Cloning: a Laboratory Manual*, 2nd ed. Cold Spring Harbor Laboratory, Cold Spring Harbor, N.Y.

Slightom, J. L., R. F. Drong, and P. P. Chee. 1993. Construction of λ clone banks, p. 121–146. *In* B. R. Glick and J. E. Thompson (ed.), *Methods in Plant Molecular Biology and Biotechnology.* CRC Press, Boca Raton, Fla.

Southern, E. 1975. Detection of specific sequences among DNA fragments separated by gel electrophoresis. *J. Mol. Biol.* **98:**503–507.

Winnacker, E.-L. 1987. *From Genes to Clones: Introduction to Gene Technology.* VCH, New York, N.Y.

REVIEW QUESTIONS

1. What are type II restriction endonucleases, and why are they important for recombinant DNA technology?

2. When circular double-stranded DNA from the plasmid pCEL1 is digested by various restriction endonucleases and combinations thereof, the following bands (with lengths given in kilobases) are observed: *Eco*RI, 6.0; *Bam*HI, 6.0; *Hin*dIII, 6.0; *Hae*II, 3.0, 2.0, and 1.0; *Eco*RI and *Hae*II, 2.0 and 1.0; *Eco*RI and *Hin*dIII, 3.5 and 2.5; *Eco*RI and *Bam*HI, 4.5 and 1.5; *Bam*HI and *Hin*dIII, 5.0 and 1.0; *Bam*HI and *Hae*II, 3.0, 1.5, 1.0, and 0.5; and *Hin*dIII and *Hae*II, 3.0, 1.5, 1.0, and 0.5. Use this information to construct a restriction map of pCEL1.

3. Describe the use of plasmid pBR322 as a cloning vector. What special features does it have?

4. Describe the principal features of the pUC cloning system.

5. A clone bank is often constructed by completely digesting a plasmid vector with restriction enzyme *Bam*HI and ligating it with chromosomal DNA that has been partially digested by restriction enzyme *Sau*3AI.

 a. Why are two different enzymes used in this experiment?

 b. What is a partial digest, and how is it performed?

 c. Why is partial digestion often used for constructing clone banks?

6. For gene cloning experiments, why is cleaved plasmid DNA often treated with alkaline phosphatase prior to the ligation step?

7. Suggest a few different ways for introducing recombinant plasmids into a gram-negative bacterium such as *E. coli.*

8. Outline a few different approaches that could be used to detect a cloned target gene within a library in *E. coli*. What conditions must be satisfied for each type of assay?

9. What is a cDNA library?

10. Why would you use a plasmid, bacteriophage λ, cosmid, or BAC as a cloning vector?

Chemical Synthesis, Sequencing, and Amplification of DNA

Technological advances in any area of science have profound effects on research. New protocols spawn novel experiments, and laboratory procedures that used to be difficult to implement become much easier to perform. The very essence of molecular biotechnology is rooted in a wide range of technical developments, many of which have become commonplace and accessible to both large and small research facilities. For example, it is now standard to chemically synthesize a DNA molecule, sequence another, and amplify yet another with the polymerase chain reaction (PCR). Each of these procedures is derived from information generated by basic studies of the characteristics of DNA and DNA replication. In addition, these experimental methods have become integral components of molecular cloning studies designed to isolate, characterize, manipulate, and transfer specific DNA fragments.

Chemical Synthesis of DNA

The ability to chemically synthesize a strand of DNA with a specific sequence of nucleotides easily, inexpensively, and rapidly has contributed significantly to the methodologies of molecular cloning and DNA characterization. These readily available, chemically synthesized, single-stranded DNA oligonucleotides may be used to assemble whole genes or gene fragments, to amplify specific DNA fragments, to introduce specific mutations into isolated DNAs, as DNA hybridization probes, as adjuncts to gene sequencing, or as linkers to facilitate gene cloning.

Machines that automate the chemical reactions involved in DNA synthesis (DNA synthesizers or "gene machines") have made the synthesis of single-stranded oligonucleotides (≤50 nucleotides) into a more or less routine procedure. Generally, DNA synthesizers consist of a set of valves and pumps that are programmed to introduce, in the correct order, specified nucleotides and the reagents required for the coupling of each consecutive nucleotide to the growing chain. Chemical DNA synthesis does not follow the biological direction of DNA synthesis; rather, in the chemical process each

incoming nucleotide is coupled to the 5'-hydroxyl terminus of the growing chain. All the reactions are carried out in succession in a single reaction column, and both the duration of each reaction and the washing steps are computer controlled.

The Phosphoramidite Method

Currently, the phosphoramidite method of chemical DNA synthesis is the procedure of choice. Before their introduction into the reaction column, the amino groups of the bases adenine, guanine, and cytosine are derivatized by the addition of benzoyl, isobutyryl, and benzoyl groups, respectively, to prevent undesirable side reactions during chain growth. Thymine is not treated because it lacks an amino group. Solid-phase synthesis (i.e., attachment of the growing DNA strand to a solid support) is used so that all the

Figure 5.1 Flowchart for the chemical synthesis of DNA oligonucleotides. After *n* coupling reactions or cycles, a single-stranded piece of DNA of $n + 1$ nucleotides is produced.

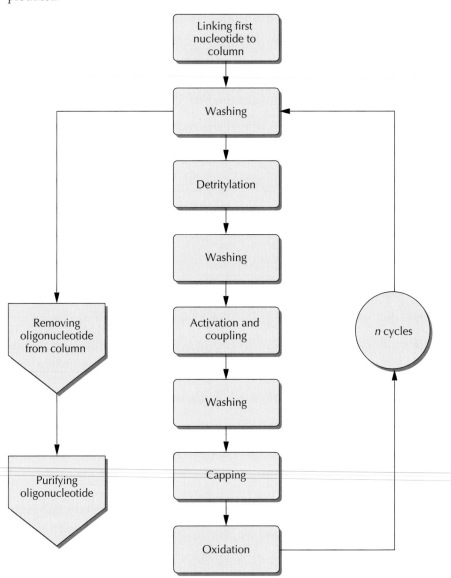

reactions can be conducted in one reaction vessel, the reagents from one re-action step can be readily washed away before the reagents for the next step are added, and the reagents can be used in excess to drive the reactions as close as possible to completion.

The multistep synthesis proceeds as follows (Fig. 5.1). The initial nucle-oside (base and sugar only) is attached to an inert solid support, which is of-ten controlled pore glass (CPG) beads (glass beads with uniformly sized pores). The first nucleoside, which will be the 3′-terminal nucleotide of the synthesized strand, is attached to a spacer molecule by its 3′ terminus, and the spacer molecule is covalently attached to the support material. A dimethoxytrityl (DMT) group has been attached to the 5′ terminus of the initial nucleoside (Fig. 5.2) to prevent the 5′-hydroxyl group from reacting nonspecifically, before the addition of the second nucleotide. Each nu-cleotide that is to be added to the growing chain has a 5′-DMT group and a diisopropylamine group attached to a 3′-phosphite group, which itself is protected by a methyl residue (Fig. 5.3). This molecular configuration is called a phosphoramidite.

After the first nucleoside is attached to the CPG beads, the cycle begins. First, the reaction column is washed extensively with an anhydrous reagent (e.g., acetonitrile) to remove water and any nucleophiles that may be pres-ent. The column is flushed with argon to purge the acetonitrile. Next, the 5′-DMT group is removed (the detritylation step) from the attached nucleo-side by treatment with trichloroacetic acid (TCA) to yield a reactive 5′-hydroxyl group (Fig. 5.4). The reaction column is washed again with ace-tonitrile to remove the TCA and then with argon to flush out the acetoni-trile. The machine is programmed to introduce the next prescribed base (phosphoramidite) and tetrazole simultaneously (the activation and cou-pling steps). The tetrazole activates the phosphoramidite so that its 3′-phos-phite forms a covalent bond with the 5′-hydroxyl group of the initial nucle-oside (Fig. 5.5). Unincorporated phosphoramidite and tetrazole are removed by flushing with argon.

Because not all of the support-bound nucleosides are linked to a phos-phoramidite during the first coupling reaction, the unlinked residues must be prevented from linking to the next nucleotide during the following cy-cle. To do this, acetic anhydride and dimethylaminopyridine are added to

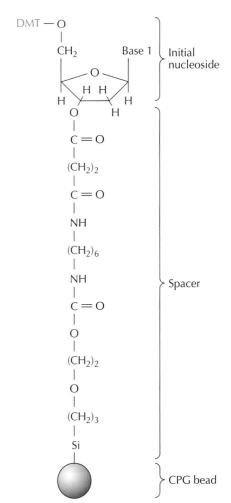

Figure 5.2 Starting complex for the chemical synthesis of a DNA strand. The initial nucleoside has a di-methoxytrityl (DMT) group attached to the 5′-hydroxyl group of the deoxyri-bose moiety and a spacer molecule at-tached to the hydroxyl group of the 3′ carbon of the deoxyribose. The spacer unit is attached to a solid support, which is usually a CPG bead.

Figure 5.3 Structure of a phosphoramidite. Phosphoramidites are available for each of the four bases—A, C, G, and T—that are used for the chemical synthesis of a DNA strand. A diisopropylamine group is attached to the 3′-phosphite group of the nucleoside. A methyl (Me) group protects the 3′-phosphite group and a dimethoxytrityl (DMT) group is on the 5′-hydroxyl group of the deoxyribose sugar.

Figure 5.4 Detritylation. The 5′-dimethoxytrityl (DMT) group is removed by treatment with trichloroacetic acid (TCA). In this example, the detritylation of the first nucleoside is depicted.

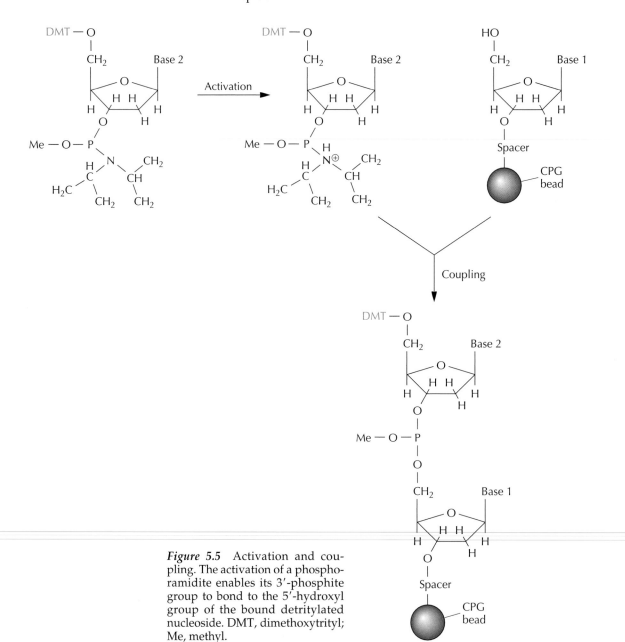

Figure 5.5 Activation and coupling. The activation of a phosphoramidite enables its 3′-phosphite group to bond to the 5′-hydroxyl group of the bound detritylated nucleoside. DMT, dimethoxytrityl; Me, methyl.

Figure 5.6 Capping. The available 5'-hydroxyl groups of unreacted detritylated nucleosides are acetylated to prevent them from participating in the coupling reaction of the next cycle.

acetylate the unreacted 5'-hydroxyl groups (Fig. 5.6). If this capping step is not carried out, then, after a number of cycles, the growing chains will differ in both length and nucleotide sequence.

At this stage of the process, the internucleotide linkage is in the form of a phosphite triester bond, which is unstable and prone to breakage in the presence of either acid or base. Therefore, the phosphite triester is oxidized with an iodine mixture to form the more stable pentavalent phosphate triester (Fig. 5.7). After this oxidation step and a subsequent wash of the reaction column, the cycle of detritylation, phosphoramidite activation, cou-

Figure 5.7 Oxidation. The phosphite triester internucleotide linkage is oxidized to the pentavalent phosphate triester. This reaction stabilizes the phosphodiester bond and makes it less susceptible to cleavage under either acidic or basic conditions. DMT, dimethoxytrityl; Me, methyl.

pling, capping, and oxidation is repeated (Fig. 5.1). This cycling continues with each successive phosphoramidite until the last programmed residue has been added to the growing chain. When the final cycle is completed, the newly synthesized DNA strands are bound to the CPG beads; each phosphate triester contains a methyl group; every guanine, cytosine, and adenine carries its amino-protecting group; and the 5' terminus of the last nucleotide has a DMT group.

The methyl groups are removed by a chemical treatment in the reaction column. The DNA strands are then cleaved from the spacer molecule, leaving a 3'-hydroxyl terminus. The DNA is eluted from the reaction column, and, in succession, the benzoyl and isobutyryl groups are stripped away, and the DNA is detritylated. The 5' terminus of the DNA strand is phosphorylated either by the enzyme T4 polynucleotide kinase and ATP or by a chemical procedure. This reaction can be carried out after detritylation, while the oligonucleotide is still bound to the support.

To achieve a reasonable overall yield, the coupling efficiency should be greater than 98% at each step. The coupling efficiency of each cycle is determined by spectrophotometrically monitoring the amount of trityl units that are released. If, for example, the efficiency is 99% at each cycle in the production of a 20-unit oligonucleotide (20-mer), then 82% (i.e., $0.99^{20} \times 100$) of the product will be 20 nucleotides long. If a 60-mer is synthesized with 99% efficiency at each cycle, then about 55% of the final product will contain 60 nucleotides. However, when the average coupling efficiency is less than 98%, the length of the final oligonucleotide that can be obtained in useful amounts is often limited (Table 5.1). The manufacturers of most commercial DNA synthesizers claim that average coupling efficiencies of approximately 98% are readily attained. But to achieve such high coupling efficiencies, extremely pure reagents and chemicals must be used, a requirement not easy to meet in practice. Generally, researchers observe coupling efficiencies of around 95%, although 99% is occasionally obtainable. Thus, the initial products of most chemical DNA synthesis runs must be subjected to either reverse-phase high-pressure liquid chromatography or gel electrophoresis to purify the true product. This goal is generally achieved, because all of the "failure" sequences are shorter than the target oligonucleotide.

Uses of Synthesized Oligonucleotides

Chemically synthesized oligonucleotides have a number of applications in molecular biotechnology. They can be used as DNA hybridization probes,

Table 5.1 Overall yield of chemically synthesized oligonucleotides for various coupling efficiencies for each cycle

Coupling efficiency (%)	Overall yield of oligonucleotide (%)				
	20-mer	**40-mer**	**60-mer**	**80-mer**	**100-mer**
90	12	1.5	0.18	0.02	0.003
95	36	13	4.6	1.7	0.6
98	67	45	30	20	13
99	82	67	55	45	37
99.5	90	82	74	67	61

to join DNA molecules together in cloning experiments, as primers for DNA sequencing, and for altering the sequences of cloned genes.

1. Single-stranded oligonucleotide probes (20- to 40-mers) can be formulated by deducing the codons from the amino acid sequence of a protein.
2. Double-stranded sequences, which are called linkers, can be created by synthesizing an oligomer, which is a palindromic single-stranded DNA sequence that base pairs (anneals, hybridizes) to itself and contains a restriction endonuclease recognition site to help in the cloning of DNA fragments (Fig. 5.8A and B). One of these short DNA duplexes (6 to 12 base pairs [bp]) can be blunt end ligated to each of the ends of a target DNA molecule (usually a complementary DNA [cDNA]). When this new molecule is cut with the appropriate restriction endonuclease, the single-stranded extensions (sticky ends) enable the target DNA to be cloned into a vector. But, prior to cloning, the digestion mixture is fractionated to remove the excess linker molecules from the now sticky-ended DNA. The vector also is linearized by digestion with the same restriction endonuclease, and the two types of sticky-ended molecules (vectors and target DNAs) are mixed and joined by using T4 DNA ligase. For this procedure to be effective, the target DNA molecule must not have a restriction endonuclease site that is the same as the linker sequence, or it also will be cut.
3. "Adaptor" sequences, which are variations of linker sequences, often contain two or more restriction endonuclease sites (Fig. 5.8C). Adaptors allow cDNAs to be cloned by blunt-end ligation and then later excised from the vector following digestion with a different restriction enzyme. The adaptor shown in Fig. 5.8C is inserted into a *Bam*HI site on the vector before the target DNA (cDNA) is blunt end ligated into the *Sma*I site. The cloned DNA can be excised from the vector by digestion with *Bam*HI. In this example, the vector should be free of *Sma*I sites, and neither the vector nor the cDNA should have *Bam*HI sites.
4. Single-stranded sequences (~17- to 24-mer) can be synthesized for use as primers for DNA sequencing reactions and for various PCR protocols.
5. Single-stranded oligonucleotides can be used for in vitro mutagenesis.
6. It may be necessary to chemically synthesize the DNA sequence that codes for a protein because, although the amino acid sequence of a protein is known, researchers have found it difficult to clone the gene. The nucleotide sequence of the gene can be deduced from the established amino acid sequence and synthesized. In other cases, the original codons of a gene may not be preferred by a host organism and poor levels of translation occur. Consequently, a gene with an effective set of codons (codon optimization) that are better suited to the host organism can be synthesized without changing the true amino acid sequence of the protein.

Synthesis of Genes

If chemically synthesized double-stranded DNA is required for an application such as the synthesis of a gene or a gene fragment, then each com-

Figure 5.8 Typical linker and adaptor sequences. **A.** A 6-mer *Eco*RI linker. **B.** An 8-mer *Eco*RI linker. **C.** A *Bam*HI-*Sma*I adaptor. Both the *Bam*HI cohesive ends and the *Sma*I recognition sequence are indicated.

plementary strand is made separately. The production of short genes (60 to 80 bp) is technically straightforward and can be accomplished by synthesizing the complementary strands and then annealing them. For the production of longer genes (>300 bp), however, special strategies must be invoked, because the coupling efficiency of each cycle during chemical DNA synthesis is never 100%. For example, if a gene contains 999 base pairs and the coupling efficiency is 99% at each cycle, the proportion of full-length single DNA strands after the last cycle is a minuscule 0.004%. To overcome this problem, synthetic (double-stranded) genes are assembled in modular form from (single-stranded) fragments that are from 20 to 100 nucleotides in length.

One method for building a synthetic gene requires the initial production of a set of overlapping, complementary oligonucleotides, each of which is between 20 and 60 nucleotides long. The sequences of the strands are planned so that, after annealing, the two end segments of the gene are aligned to give blunt ends. Each internal section of the gene has complementary 3'- and 5'-terminal extensions that are designed to base pair precisely with an adjacent section (Fig. 5.9). Thus, after the gene is assembled, the only remaining requirement to complete the process is sealing the nicks along the backbones of the two strands with T4 DNA ligase. In addition to

Figure 5.9 Assembly of a synthetic gene from short oligonucleotides. Individual oligonucleotides (20- to 60-mers) are synthesized chemically. Their sequences are designed so that they form a double-stranded array after annealing. T4 DNA ligase is used to seal the nicks and produce an intact version of the gene.

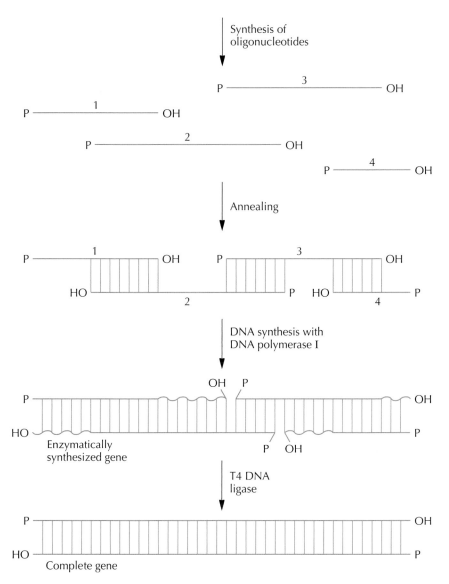

Figure 5.10 Assembly and in vitro enzymatic DNA synthesis of a gene. Individual oligonucleotides are synthesized chemically. Their sequences are designed to enable them to form a molecule with base-paired regions separated by open regions (gaps). The gaps are filled in by in vitro enzymatic DNA synthesis, and the nicks are sealed with T4 DNA ligase.

the protein coding sequence, synthetic genes can be designed with terminal sequences that facilitate insertion into a cloning vector (restriction endonuclease sites) and, if necessary, sequences that contain signals for the proper initiation and termination of transcription and translation.

An alternative way to prepare a full-size gene is to synthesize a specified set of overlapping oligonucleotides (40 to 100 nucleotides). After the 3' and 5' extensions (6 to 10 nucleotides) are annealed, large gaps still remain, but the base-paired regions are both long enough and stable enough to hold the structure together. The duplex is completed and the gaps are filled by enzymatic DNA synthesis with *Escherichia coli* DNA polymerase I. This enzyme uses the 3'-hydroxyl groups as replication initiation points and the

single-stranded regions as templates. After the enzymatic synthesis is completed, the nicks are sealed with T4 DNA ligase (Fig. 5.10).

For larger genes (≥1,000 bp), the complete gene sequence is usually assembled from double-stranded fragments that are each put together by joining four to six overlapping oligonucleotides (20 to 60 bp each). If there is a sufficient amount of the double-stranded fragments after each synthesis and annealing step, they are simply joined to one another. Otherwise, each fragment is cloned into a vector to amplify the amount of DNA available. In both cases, the double-stranded constructs are sequentially linked to one another to form the entire gene sequence. Because it is absolutely essential that a chemically synthesized gene have the correct sequence of nucleotides, each double-stranded fragment, and then the complete sequence, is characterized by DNA sequence analyses.

MILESTONE

DNA Sequencing with Chain-Terminating Inhibitors
F. SANGER, S. NICKLEN, and A. R. COULSON
Proc. Natl. Acad. Sci. USA **74:**5463–5467, 1977

Specific Enzymatic Amplification of DNA In Vitro: the Polymerase Chain Reaction
K. B. MULLIS, F. A. FALOONA, S. J. SCHARF, R. K. SAIKI, G. T. HORN, and H. A. ERLICH
Cold Spring Harbor Symp. Quant. Biol. **51:**263–273, 1986

New techniques are the lifeblood of science. They enable researchers to acquire information that was previously inaccessible and that, in turn, generates insights that stimulate new research and lead to new discoveries. For molecular biotechnology, DNA sequencing and the PCR are powerful procedures that have become laboratory mainstays.

DNA sequencing by enzymatic synthesis with chain elongation inhibitors of DNA polymerase is a rapid, relatively simple, inexpensive, and reliable method. The most definitive form of molecular characterization of a cloned piece of DNA is its sequence. Among other things, the coding content of an insert, potential primer sequences for a PCR, and the presence of a gene mutation can be determined by DNA sequencing. At the time the Sanger (dideoxy) method was published, most DNA sequencing was carried out by the base-specific chemical cleavage method devised by A. M.

Maxam and W. Gilbert (*Proc. Natl. Acad. Sci. USA* **74:**560–564, 1977). Before the development of these techniques, nucleic acid sequencing was more or less limited to RNA molecules. The sequencing of a DNA molecule required transcribing a DNA fragment into RNA with RNA polymerase and then sequencing the RNA product. In general, RNA sequencing entailed treating a radiolabeled RNA molecule with different ribonucleases, chromatographically separating the digestion products, redigesting the separated products, hydrolyzing the products of the second digestion with alkali, chromatographically separating the hydrolysis products, determining the sequence of the oligonucleotides, and constructing the sequence based on overlapping stretches of nucleotides. This approach was time-consuming and tedious. With the advent of the dideoxy method, it became obsolete. Now RNA molecules are sequenced by converting them into DNA molecules. The Sanger method

superseded the Maxam and Gilbert sequencing procedure when the M13 bacteriophage cloning system was developed. The ability to sequence DNA molecules has been, directly and indirectly, responsible for both the dramatic upsurge in studies of the molecular basis of human diseases and the development of diagnostic and therapeutic treatments for these disorders.

The PCR, which is the invention of Kary Mullis (U.S. patent 4,683,202), has had a tremendous impact on many research areas, including molecular biotechnology. The capability of generating large amounts of DNA by amplification from segments of cloned or genomic DNA has facilitated the cloning of DNA versions of rare mRNA molecules, screening gene libraries, diagnostic testing for gene mutations, physical mapping of chromosomes, and a myriad of other applications. The first study using PCR described a diagnostic test for sickle-cell anemia (Saiki et al., *Science* **230:**1350–1354, 1985). The PCR was a unique idea that did not replace any existing technology. Since 1986, more than 10,000 studies have used the PCR in one way or another. Its status as an indispensable method is well established. And, considering the large number of PCR applications that have already been devised, there seems to be no end to its possible uses.

DNA Sequencing Techniques

In molecular terms, the definitive understanding of a DNA molecule comes from determining its nucleotide sequence. The function of a gene can often be deduced from its nucleotide sequence, for example, by comparing the deduced sequence with sequences of genes with known functions. In addition, nucleotide sequence information is essential in molecular cloning studies. Although the nucleotide sequence of a DNA fragment can be determined by using either a chemical procedure that was originally developed by Alan Maxam and Walter Gilbert or an enzymatic procedure developed by Fred Sanger, the enzymatic method, often called the dideoxynucleotide method, is currently the method of choice.

Dideoxynucleotide Procedure for Sequencing DNA

A dideoxynucleotide is a human-made molecule that lacks a hydroxyl group at both the 2' and 3' carbons of the sugar moiety (Fig. 5.11A). In contrast, a (natural) deoxyribonucleotide has a 3'-hydroxyl group on the sugar unit (Fig. 5.11B). During DNA replication, an incoming (natural) nucleoside triphosphate, which is determined by base pairing to the nucleotide of the template strand, is linked by its 5' α-phosphate group to the 3'-hydroxyl group of the last nucleotide of the growing chain (Fig. 5.12). However, if a dideoxynucleotide is incorporated at the end of the growing chain, DNA synthesis stops because a phosphodiester bond cannot be formed with the next incoming nucleotide (Fig. 5.13). The termination of DNA synthesis is the quintessential feature of the dideoxynucleotide DNA sequencing method, although other experimental conditions must be met before a DNA sequence can be determined.

In principle, the first step in the standard laboratory procedure for dideoxynucleotide DNA sequencing entails annealing a synthetic oligonucleotide (17- to 24-mer) to a predetermined segment of a strand of the cloning vector near the insertion site of the cloned DNA. The oligonucleotide acts as a primer sequence by providing a 3'-hydroxyl group for the initiation of DNA synthesis. The primed DNA sample is partitioned into four separate reaction tubes. Each tube contains four deoxyribonucleotides

Figure 5.11 **A.** A dideoxynucleotide. Its 2' and 3' carbons lack hydroxyl groups. **B.** A deoxyribonucleotide.

Figure 5.12 Normal DNA synthesis. An incoming deoxyribonucleotide (deoxyribonucleoside triphosphate [dNTP]) base pairs with the complementary nucleotide of the template strand. The internucleotide linkage occurs between the 3'-hydroxyl group of the last nucleotide of the growing strand and the α-phosphate group of the incoming nucleotide.

(dATP, dCTP, dGTP, and dTTP), one of which is radiolabeled, and one of the four dideoxynucleotides (ddATP, ddCTP, ddGTP, or ddTTP). The concentration of each dideoxynucleotide in each reaction tube is carefully adjusted to ensure that it is incorporated into the mixture of growing chains at every possible site and not just at the first occurrence of the template's complementary nucleotide. (Recall that chain growth stops as soon as a dideoxyribonucleotide is incorporated, so each growing chain will eventually contain one dideoxyribonucleotide at its 3' terminus.) This experimental condition is met in each reaction tube. Consequently, after enzymatic DNA synthesis with DNA polymerase, each reaction tube will contain a unique set of oligonucleotides, each of which includes the primer sequence (Fig. 5.14).

The synthesis reactions are stopped by the addition of formamide, which prevents DNA strands from base pairing, and the DNA molecules are separated by polyacrylamide gel electrophoresis. This separation procedure resolves pieces of DNA that differ in size by as little as a single nucleotide. An autoradiograph of the gel shows only the radiolabeled DNA

Figure 5.13 Blocked DNA synthesis. Chain growth is stopped by the addition of a dideoxynucleotide to the end of the growing strand. The internucleotide linkage between the last nucleotide, which is a dideoxynucleotide, and the next incoming nucleotide cannot be formed because there is no 3'-hydroxyl group on the dideoxynucleotide sugar.

Figure 5.14 Primer extension during DNA synthesis in the presence of dideoxynucleotides. Each of the four reaction tubes contains a unique set of nucleotide extensions attached to the primer because, when a dideoxynucleotide is incorporated in the growing strand, it terminates the synthesis. A few full-length DNA molecules will be synthesized in each reaction tube. dNTP, deoxynucleoside triphosphate.

Contents of reaction tube	Size of primer and extension	Primer and sequence of extension
ddATP + four dNTPs	Primer + 3	Primer – dGdCddA
	Primer + 7	Primer – dGdCdAdTdCdGddA
	Primer + 8	Primer – dGdCdAdTdCdGdAddA
ddCTP + four dNTPs	Primer + 2	Primer – dGddC
	Primer + 5	Primer – dGdCdAdTddC
ddGTP + four dNTPs	Primer + 1	Primer – ddG
	Primer + 6	Primer – dGdCdAdTdCddG
ddTTP + four dNTPs	Primer + 4	Primer – dGdCdAddT
	Primer + 9	Primer – dGdCdAdTdCdGdAdAddT

fragments that were produced during the enzymatic DNA synthesis step, with each of the four lanes in the gel and the autoradiograph corresponding to a reaction tube that contained one of the four dideoxynucleotides.

The sequence of a segment of a cloned piece of DNA is determined by noting the order of the bands, as accurately as possible, from the bottom to the top of the autoradiograph. In the example shown in Fig. 5.15, the first six bases of the sequenced DNA are AGCTGC. In this case, the fastest-migrating band (i.e., the radiolabeled fragment closest to the bottom), which corresponds to the smallest DNA fragment, is in the ddATP lane; the next band is in the ddGTP lane, the next in the ddCTP lane, the next in the ddTTP lane, and so on. About 250 to 350 bands can be resolved clearly on most autoradiographs. Usually, the primer sequence is positioned about 10 to 20 nucleotides away from the insertion site of the cloned DNA so that the researcher can recognize the known sequence at the start of the autoradiograph and thereby identify precisely the first nucleotide of the cloned DNA.

Using Bacteriophage M13 To Obtain Single-Stranded DNA

A number of strategies have been devised to obtain the complete sequence of a cloned piece of DNA. One of the early approaches utilized the *E. coli* bacteriophage M13 as a vector. The M13 system uses the same principle for identifying vectors that have incorporated insert DNA as the pUC series of

Figure 5.15 Simulated autoradiograph of a DNA sequencing gel. Each lane of the gel was loaded with the contents of one of the four reaction tubes (containing either ddATP, ddCTP, ddGTP, or ddTTP). By convention, the bands of the autoradiograph are read from the bottom to the top. In this example, the results of the sequence determination are shown on the right.

vectors. The DNA of an M13 virus particle is a single-stranded circle. When an *E. coli* cell is infected by an M13 virus particle, a double-stranded DNA replicative form is produced first. Single-stranded circular DNA molecules that are packed into progeny M13 bacteriophage particles are synthesized from this double-stranded molecule. Cells that are infected with M13 do not lyse; instead, they continually secrete infective virus particles into the surrounding environment. Each of the new single-stranded M13 DNA molecules is coated with capsid protein as it is released from the cell and passes through the outer membrane of the cell. By removing a nonessential segment of the M13 DNA, it became possible to add cloned DNA without destroying the infectivity of the new virus particles. The M13 system has two advantages: the double-stranded replicative form can be isolated and handled as a plasmid, and the single-stranded phage DNA can be used as a template for DNA sequencing.

Bacteriophage M13 has been used as a combined cloning and DNA sequencing vector system. Usually, target DNA that is not much larger than 500 bp is cloned into a multiple cloning sequence that forms part of a cloned modified *lacZ'* gene on the double-stranded replicative form of M13. The DNA from the ligation reaction is mixed with competent *E. coli* cells. After plating and incubation, both white (colorless) colonies and blue colonies (M13 plaques) develop on medium containing the substrate X-Gal, which turns blue when it is hydrolyzed by hybrid β-galactosidase. The colorless colonies that produce M13 bacteriophage particles have a piece of target DNA disrupting the *lacZ'* gene. In contrast, blue colonies represent infection with reformed M13 bacteriophage that have retained a functional *lacZ'* sequence and do not have inserted DNA. The bacteriophage particles from a white colony are isolated and used as the source of single-stranded DNA for DNA sequencing (Fig. 5.16). A primer oligonucleotide that hybridizes to the vector DNA near the insertion site of the cloned DNA is added to the single-stranded M13–cloned DNA construct, the dideoxynucleotide reactions are carried out, and the sequence of the insert is read from the autoradiographs.

When a large piece of DNA (≤2,000 bp) needs to be sequenced, there are a number of different strategies that may be used. One approach is to clone the DNA fragment into a conventional plasmid cloning vector and create a detailed restriction endonuclease map of the insert. With this information, overlapping segments of 100 to 500 bp of the insert DNA can be identified, and each segment can be subcloned into a different M13 vector and sequenced. The complete sequence of the cloned DNA is reconstructed by determining the sequences of all of the subcloned pieces. To ensure that the correct sequence is obtained and that there is no ambiguity regarding the identity of a single nucleotide, both strands must be sequenced several times. The sequencing of both strands is facilitated by constructing the M13 system so that each fragment of the original DNA insert can be subcloned in two orientations. In one of the orientations the primer initiates the synthesis of one of the strands, and in the other orientation the same primer directs the synthesis of the other strand.

Primer Walking

For very long pieces of DNA (≥5,000 bp), the M13 sequencing system, although feasible, becomes time-consuming because the number of M13 vectors containing overlapping subsequences that have to be created and maintained is very large. To overcome this inconvenience, double-stranded

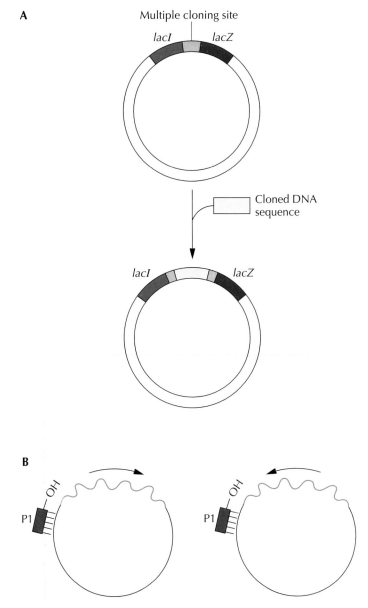

Figure 5.16 Use of bacteriophage M13 as a cloning and DNA sequencing vector. **A.** Cloning DNA into the double-stranded form of M13. **B.** The same primer (P1) can be used to initiate DNA sequencing of complementary DNA strands of a cloned piece of DNA due to the orientation (arrows) of the cloned DNA sequence in the M13 cloning vector.

plasmid DNA sequencing protocols that do not require subcloning steps have been developed. In these protocols, plasmid DNA containing the target DNA is purified and annealed with a synthetic oligonucleotide that base pairs with one strand of the vector near the insertion site of the cloned DNA. The dideoxyribonucleotide sequencing reactions are carried out to determine the identity and order of the first 250 to 350 nucleotides of the insert DNA. On the basis of this analysis, a second oligonucleotide primer designed to hybridize to a region about 300 nucleotides away from the binding site of the initial primer is chemically synthesized and then used to de-

termine the sequence of the next 250 to 350 nucleotides. In a similar manner, a third primer binding site is selected, another oligonucleotide is synthesized, and the sequence of the next 250 to 350 bases is determined (Fig. 5.17). This "primer walking" (primer extension) strategy proceeds until the complete cloned DNA is sequenced.

The complementary strand can be sequenced in the same way, starting with an oligonucleotide primer that hybridizes to this strand near the insertion site of the cloned DNA. However, false priming of DNA synthesis can give erroneous and ambiguous results. This situation arises because a standard-sized primer may bind to more than one region within the target DNA. To prevent this problem, the primers used in this procedure are at least 24 nucleotides long. In addition, stringent annealing conditions that do not permit spurious binding of the primer are maintained. Pieces of DNA that have been cloned into bacteriophage λ (~20 kilobase pairs [kb]) or a cosmid vector (~40 kb) have been sequenced by this technique.

For laboratories that are engaged in sequencing tens of thousands of base pairs of DNA on a regular basis, a number of automated procedures have recently been developed. In many of these protocols, the primers in

Figure 5.17 DNA sequencing by primer walking. 1. DNA sequencing is initiated with a primer (P1) that is complementary to a site on a plasmid near the site of insertion of the cloned DNA. 2. Based on the segment of the cloned DNA that has just been sequenced, a second primer that is complementary to a stretch of about 20 nucleotides near the end of that segment is synthesized. 3. The second primer (P2) is used to sequence the next segment of cloned DNA. 4. Based on the segment of the cloned DNA that has just been sequenced, a third primer that is complementary to a stretch of about 20 nucleotides near the end of that segment is synthesized. 5. The third primer (P3) is used to sequence the next segment of cloned DNA. 6. Based on the segment of the cloned DNA that has just been sequenced, a fourth primer that is complementary to a stretch of about 20 nucleotides near the end of that segment is synthesized. The process of successively synthesizing and using new primers continues until the entire insert is sequenced.

the various reaction tubes are labeled with different fluorescent dye markers that emit distinctive wavelengths when stimulated by a light source. In these automated systems, the contents of all four reaction tubes are pooled and subjected to gel electrophoresis in one lane. The lane is then scanned with a laser beam, and the order of the specific fluorescent responses is recorded by a computer, which collates the data and prints out the nucleotide sequence.

PCR

The PCR is an effective procedure for generating large quantities of a specific DNA sequence in vitro. This amplification, which can be more than a millionfold, is achieved by a three-step cycling process. The essential requirements for the PCR are (1) two synthetic oligonucleotide primers (~20 nucleotides each) that are complementary to regions on opposite strands that flank the target DNA sequence and that, after annealing to the source DNA, have their 3'-hydroxyl ends oriented toward each other; (2) a target sequence in a DNA sample that lies between the pair of primers and can be from 100 to ~35,000 bp in length; (3) a thermostable DNA polymerase that can withstand heating to 95°C or higher; and (4) the four deoxyribonucleotides.

A typical PCR process entails a number of cycles for amplifying a specific DNA sequence; each cycle has three successive steps.

1. *Denaturation.* The first step in the PCR amplification system is the thermal denaturation of the DNA sample by raising the temperature within a reaction tube to 95°C. In addition to the source DNA, this reaction tube contains a vast molar excess of the two oligonucleotide primers, a thermostable DNA polymerase (e.g., *Taq* DNA polymerase, isolated from the bacterium *Thermus aquaticus*), and four deoxyribonucleotides. The temperature is maintained for about 1 minute.
2. *Renaturation.* For the second step, the temperature of the mixture is slowly cooled to ~55°C. During this step, the primers base pair with their complementary sequences in the source DNA.
3. *Synthesis.* In the third step, the temperature is raised to ~75°C, which is optimum for the catalytic functioning of *Taq* DNA polymerase. DNA synthesis is initiated at the 3'-hydroxyl end of each primer (Fig. 5.18).

Each step and the temperature changes required during a PCR cycle are usually carried out in an automated, programmable block heater in which the reaction tubes are immersed. One cycle generally lasts from 3 to 5 minutes.

To understand how the PCR protocol succeeds in amplifying a discrete segment of DNA, it is important to keep in mind the location of each primer sequence and its complementary sequence within the strands that are synthesized during each cycle. During the synthesis phase of the first cycle, the new DNA attached to each primer is extended beyond the endpoint of the sequence that is complementary to the second primer. These new strands form "long templates" that will be used in the second cycle (Fig. 5.18).

During the second cycle, the original DNA strands and the new strands synthesized in the first cycle (long templates) are denatured again and then hybridized with primer sequences. A second round of synthesis again pro-

FIRST CYCLE

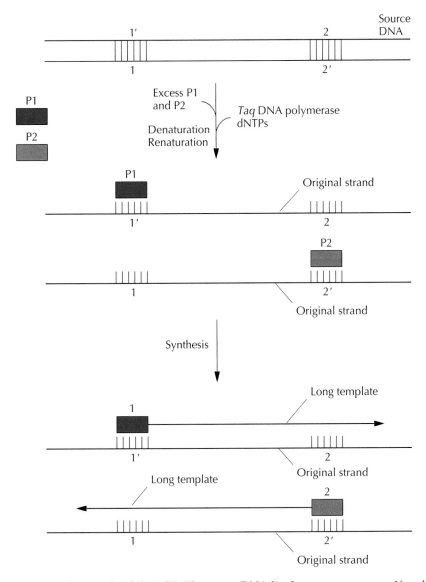

Figure 5.18 First cycle of the PCR. The target DNA lies between sequences 1′ and 2 on one strand and between sequences 1 and 2′ on the complementary strand. Two primers (P1 and P2) are mixed with the sample DNA. The mixture, which also contains *Taq* DNA polymerase and the four deoxyribonucleotides (deoxyribonucleoside triphosphates [dNTPs]), is heated to 95°C to denature the DNA and then slowly cooled to 55°C. The primers, which are present in excess, base pair to the original DNA strand of the sample during the renaturation step. The temperature is raised to about 75°C, and DNA synthesis commences from the 3′-hydroxyl ends of the primer sequences and continues past the regions of the DNA strands that are complementary to the other primer sequence. The products of this reaction are two long strands of DNA, which will serve as DNA templates in the second cycle of the PCR.

duces long templates as well as some DNA strands that have a primer sequence at one end and a sequence complementary to the other primer at the other end ("short templates") (Fig. 5.19).

During the third cycle, short templates, long templates, and original strands all hybridize with primer sequences and are replicated (Fig. 5.20). In subsequent cycles, the short templates preferentially accumulate, and by the 30th cycle these strands are about a million times more abundant than either the original or long template strands (Fig. 5.21).

The PCR has become a pervasive technique. Many PCR procedures will be described in ensuing chapters. One important PCR application is the identification of a pathogenic organism that is associated with a disease in humans, animals, or plants. In these cases, purification of the target DNA is unnecessary and the detection assay can often be carried out with small, crude samples. The nucleotide sequence of the possible pathogenic organism must be known so that a pair of primers that anneal exclusively to sites within the targeted DNA can be synthesized. After PCR cycling, a DNA fragment of a specific size that is equivalent to the lengths of the two primers plus the length of the DNA between them will be amplified only if the target DNA is present in the sample.

Because the PCR technique is so sensitive, exceptionally small amounts of contaminating DNA that can be carried from one run to the next can give false-positive results. This problem can be overcome by implementing simple procedures for treating recycled PCR solutions and for cleaning reaction tubes.

PCR protocols have also been devised to detect naturally occurring mutations, to produce specific mutations in vitro, to assemble whole genes from synthetic oligonucleotides, and for DNA sequencing. For many applications, it is frequently helpful to be able to clone the PCR product. However, direct cloning by blunt-end ligation is not easily achieved, because the *Taq* DNA polymerase has an unusual enzymatic activity that adds an extra adenine nucleotide to the 3' end of a strand. The presence of the single adenine prevents efficient blunt-end ligation. However, if the cloning vector is cut with a restriction endonuclease to give new blunt ends and then is incubated with *Taq* DNA polymerase in the presence of dTTP, one thymidine unit is added to both of its 3' termini. The single-base complementary extensions of the vector and the PCR product are sufficient to enable the two DNA molecules to be joined and ligated.

Cloning cDNA Ends by PCR

PCR protocols have been devised for generating complementary DNAs (cDNAs) from either the 3' or 5' end of a specific messenger RNA (mRNA). The acronym RACE, which is derived from the expression "rapid amplification of cDNA ends," is used to describe these strategies. The terms 3' RACE and 5' RACE are applied when a particular end of an mRNA is amplified. Both RACE systems depend on knowing an internal nucleotide sequence from the coding region of a target mRNA. With this information, gene-specific primers (GSPs) can be produced for cDNA synthesis and cDNA amplification during PCR cycles.

For 3' RACE, the primer for the synthesis of the first cDNA strand is an oligonucleotide with a string of consecutive thymidine residues at its 3' end that is followed by a second primer sequence (Fig. 5.22). The oligo(dT) segment of the primer base pairs with the poly(A) tail of an mRNA, and reverse transcriptase uses the mRNA sequence as a template to produce a cDNA strand. The second cDNA strand is synthesized by adding a GSP, which is

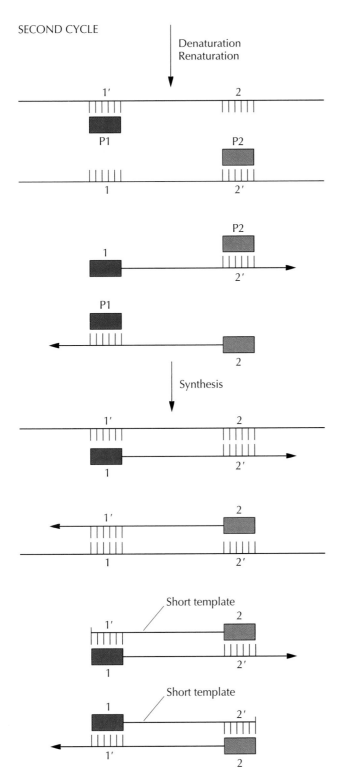

SECOND CYCLE

Figure 5.19 Second cycle of the PCR. The starting material for this step is the mixture of DNA that results from the first PCR cycle (Fig. 5.18). During the denaturation and renaturation steps, the primer sequences hybridize to complementary regions in both the original strands and the long template strands. In vitro enzymatic DNA synthesis produces more long template strands from the original strands and short template strands from the long template strands. A short template has a primer sequence at one end and the complementary sequence to the other primer at the other end.

Figure 5.20 Third cycle of the PCR. During the renaturation step, the primer sequences hybridize to complementary regions of original, long template, and short template strands. In vitro enzymatic DNA synthesis produces more long template strands from the original strands and more short template strands both from the long template strands and from the short template strands.

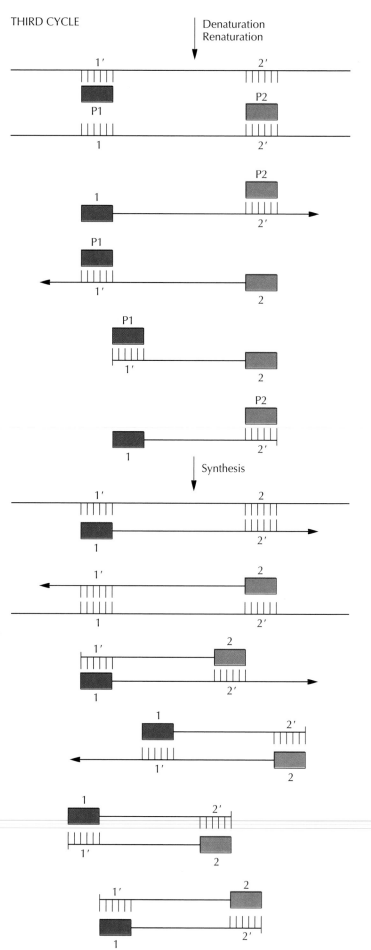

based on the sequence of the mRNA of interest, *Taq* DNA polymerase, and deoxyribonucleotides to a purified preparation of the first cDNA strands. After a limited number of rounds of PCR amplification with these primers, a second pair of primers that bind to sites adjacent to the two initial primers are used for the final PCR cycles. Sets of primers that are arranged close to one another are called nested primers. A second set of primers is required because complete amplification of the target molecule is not achieved when the initial primers are present in the reaction tube. However, amplification is not suppressed when a different set of primers is used. The final PCR product is a cDNA of the 3' end of the target mRNA.

Figure 5.22 Schematic representation of cDNA amplification of the 3' end of an mRNA by PCR. Reverse transcription with a dT oligonucleotide primer produces the first cDNA strand. The second cDNA strand is produced by *Taq* DNA polymerase with a GSP (GSP1). An initial set of PCR amplifications are run with a dT oligonucleotide and GSP1 as the primers. Nested primers GSP2 and P are used for a second set of amplifications to produce a cDNA of the 3' end of the target mRNA.

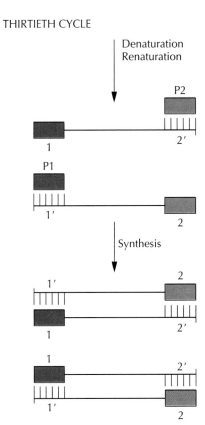

Figure 5.21 Thirtieth cycle of the PCR. By the 30th cycle, the population of DNA molecules in a reaction tube consists almost entirely of short template strands.

For 5' RACE, a GSP is used to produce the first cDNA strand (Fig. 5.23). Then a purified sample of first cDNA strands is treated with the enzyme terminal deoxynucleotidyl transferase and dATP. Terminal deoxynucleotidyl transferase adds deoxyribonucleotides, one at a time and without a template, to the 3' end of a DNA strand. In this case, because there is only dATP in the reaction mixture, a series of adenine residues are linked by phosphodiester linkages to a 3' end of a cDNA strand to form a poly(A) tail. This reaction is called homopolymeric tailing. The dT-containing oligonucleotide that generates the first cDNA strand for 3' RACE primes the synthesis of the second cDNA strand for 5' RACE. After the synthesis of the second cDNA strand, a limited number of PCR cycles are run with the initial primers before a pair of nested primers amplify the 5' end of the target mRNA.

The RACE techniques are useful for a number of reasons. It is usually very difficult to detect a cDNA clone of an mRNA that occurs at a low concentration in a particular tissue. However, the cDNA ends of a known rare mRNA can be generated quickly by the RACE protocol. And, when available, the cDNA ends of either common or rare mRNAs can be used as probes to screen cDNA and genomic libraries. Moreover, since 3'-end partial cDNA clones are more prevalent than full-length clones, the 5' RACE procedure can in some instances provide the missing 5' segment.

Gene Synthesis by PCR

The assembly of a gene by PCR is faster and more economical than filling in overlapping oligonucleotides with DNA polymerase and then sealing the nicks with T4 DNA ligase. One PCR protocol for total gene construction starts with two overlapping oligonucleotides (A and B) that represent sequences from the center of the gene (Fig. 5.24). After annealing, these oligonucleotides have recessed 3'-hydroxyl groups that provide a starting point for DNA synthesis during the elongation phase of a PCR cycle. A double-stranded DNA molecule is formed by a filling-in reaction. Next, two oligonucleotides (C and D) are added to the mixture. Oligonucleotide C overlaps at its 3' end with the 5' end of oligonucleotide A and represents the nucleotide sequence of the gene immediately to the left (upstream) of the oligonucleotide A sequence. Oligonucleotide D overlaps at its 3' end with the 5' end of oligonucleotide B and represents the nucleotide sequence of the gene immediately to the right (downstream) of the oligonucleotide B sequence. After denaturation and renaturation of the mixture, during the first synthesis phase the DNA molecules with extensions are filled in from the 3'-hydroxyl ends. During ensuing PCR cycles, the complementary overlapping regions between strands base pair and elongation from the 3' hydroxyl ends of the overlapped oligonucleotides produces a double-stranded DNA product. The order of the segments of this PCR product is CABD. The overlapped DNA molecule with recessed 5'-phosphate ends cannot be elongated.

For the next set of PCR cycles, two oligonucleotides (E and F) are added to the mixture. Oligonucleotide E overlaps at its 3' end with the 5' end of oligonucleotide C and represents the nucleotide sequence of the gene immediately to the left (upstream) of the oligonucleotide C sequence. Oligonucleotide F overlaps at its 3' end with the 5' end of oligonucleotide D and represents the nucleotide sequence of the gene immediately to the right (downstream) of the oligonucleotide D sequence. After denaturation and renaturation of the mixture, during the first synthesis phase the DNA molecules with extensions are filled in from the 3'-hydroxyl ends. During

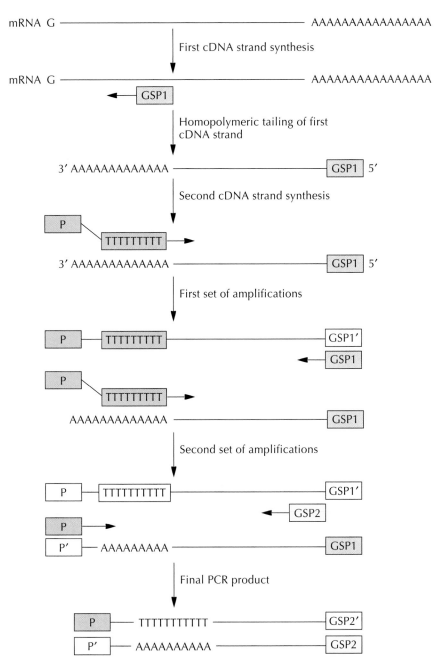

Figure 5.23 Schematic representation of cDNA amplification of the 5′ end of an mRNA by PCR. Reverse transcription with GSP (GSP1) produces the first cDNA strand. Terminal deoxynucleotidyl transferase adds a poly(A) tail to the first cDNA strand. The second cDNA strand is generated with a dT oligonucleotide as the primer. An initial set of PCR amplifications are run with the dT oligonucleotide and GSP1 primers. Nested primers GSP2 and P are used for a second set of amplifications to produce a cDNA of the 5′ end of the target mRNA.

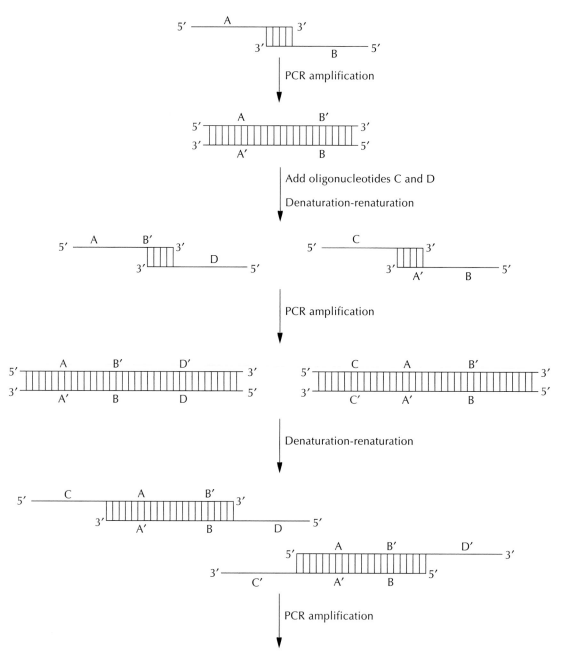

Figure 5.24 Gene synthesis by PCR. Overlapping oligonucleotides (A and B) are filled in from the recessed 3'-hydroxyl ends during the DNA synthesis phase of PCR. Overlapping molecules, which are filled in from the recessed 3'-hydroxyl ends by DNA synthesis, are formed during the denaturation-renaturation phases of the PCR run. A pair of oligonucleotides (C and D) that overlap the ends of the product of the first PCR run are added to a sample, and a second PCR run is initiated. After the second PCR run, a pair of oligonucleotides (E and F) that overlap the ends of the PCR product are added to a sample, and a third PCR run is initiated. The final PCR product is a double-stranded DNA molecule with a specified sequence of nucleotides. The pairs of letters with and without a prime (e.g., A' and A) represent complementary base pairs of a region of a DNA molecule. The sequence of each oligonucleotide is designed to correspond to a segment from a particular DNA strand.

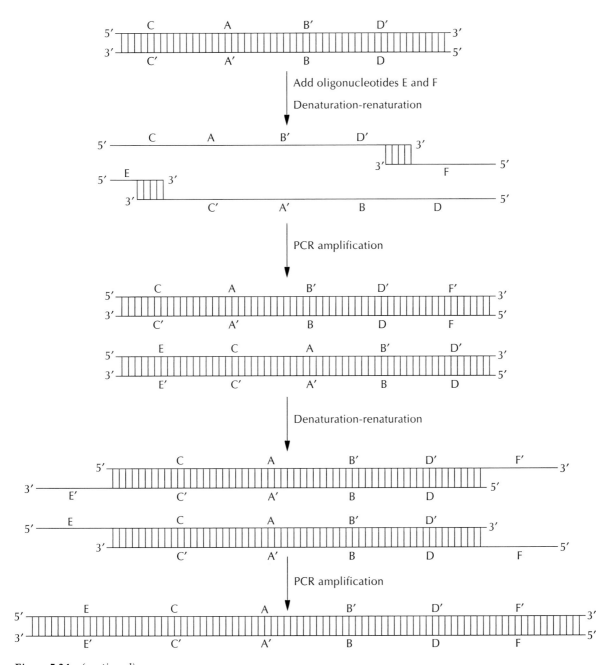

Figure 5.24 *(continued)*

ensuing PCR cycles, the complementary overlapping regions between strands base pair, and elongation from the 3'-hydroxyl ends of the overlapped oligonucleotides produces a double-stranded DNA product. The order of the segments of this PCR product is ECABDF.

Pairs of oligonucleotides—one upstream and the other downstream—are added sequentially and subjected to PCR amplification until the total gene is synthesized. Generally, the oligonucleotides are about 50 nucleotides in length. With this protocol, each PCR run comprises 20 4-minute cycles. Thus, the 10 PCR runs that would be required to synthesize a gene with 1,000 bp can be carried out in 1 day. In addition, as with other meth-

ods for synthesizing genes, the last pair of oligonucleotides (i.e., the 5' and 3' ends of the gene) can be synthesized with supplementary sequences outside the coding region that facilitate the cloning of the gene into a vector.

SUMMARY

In addition to the various gene cloning techniques, there are a number of other procedures that are important to molecular biotechnology. These techniques include the chemical synthesis of DNA, DNA sequencing, and the PCR.

Chemical synthesis of DNA is a stepwise in vitro method that produces single-stranded DNA molecules. A high efficiency of phosphodiester bond formation is essential. Otherwise, at the end of the procedure, the sample contains very few full-length molecules. In vitro-synthesized DNA molecules are often 20 to 30 nucleotides long and seldom more than 80 to 100 nucleotides long. To make double-stranded DNA molecules, the complementary strands are synthesized separately and then mixed together. Chemically synthesized DNA can be used for producing probes to screen gene libraries; linkers and adaptors for cloning genes; oligonucleotides for in vitro mutagenesis studies; and segments of DNA that, when joined, will form a gene that can be cloned.

For biotechnology and other applications, it is necessary to know the complete sequence of a cloned gene. The dideoxynucleotide method that was developed by Sanger and his colleagues is often used for this purpose. This technique relies on the in vitro incorporation of a dideoxynucleotide into a growing DNA strand. This nucleotide terminates synthesis of the strand because it does not have a 3'-hydroxyl group. In a sequencing experiment, four DNA synthesis reactions, each with a different dideoxynucleotide, are carried out simultaneously. The products of these reactions are separated electrophoretically, and the pattern of separation of the synthesized radiolabeled fragments is used to determine the nucleotide sequence of the growing strand.

Bacteriophage M13, which carries cloned pieces of DNA that are up to 500 nucleotides long, is used to generate the template DNA for a sequencing study, because the M13 DNA–DNA insert construct can be readily isolated as single-stranded DNA. Alternatively, double-stranded plasmids with cloned DNA can be used for sequencing studies.

For long pieces of cloned DNA, a primer sequence can be determined from a sequenced segment of the cloned DNA, synthesized chemically, and then used to determine the sequence of the next 250 to 350 nucleotides from the cloned piece of DNA. This "primer walking" method can be used to sequence a large segment of DNA without carrying out the subcloning of DNA fragments that is necessary with the bacteriophage M13 system.

The PCR is a revolutionary procedure that has innumerable uses. This technique enables discrete segments of DNA to be amplified in vitro over a millionfold. To achieve this amplification, two primers that hybridize to segments of DNA on opposite strands are used; the 3'-hydroxyl ends of the primers are oriented toward each other and flank the target sequence. The process entails 30 or more successive cycles. Each cycle consists of denaturation, renaturation, and in vitro DNA synthesis. Because it would be tedious and costly to add DNA polymerase at the end of each cycle, a heat-resistant DNA polymerase that is not denatured at the temperature (95°C) that breaks the hydrogen bonds of the DNA in the sample is used. During

the renaturation step, the primer sequences, which are present in excess, hybridize to the sample DNA in the first cycle and to complementary primer sequences of amplified DNA molecules in subsequent cycles. In the DNA synthesis step, a new DNA strand that is complementary to its template strand grows from the 3' end of the primer. By the 30th cycle, the amplified DNA molecules have one primer sequence at one end and a sequence that is complementary to the second primer at the other end. Among its many applications, PCR can be used to detect the presence of a pathogenic organism in a biological sample, to obtain large amounts for a specific DNA sequence for cloning, to amplify the 5' and 3' ends of a specific mRNA, to synthesize a synthetic gene, or to determine whether a deletion or an addition of DNA has occurred in a gene that causes a genetic disease.

References

Chen, E. Y., and P. H. Seeburg. 1985. Supercoil sequencing: a fast and simple method for sequencing plasmid DNA. *DNA* **4:**165–170.

Climie, S., and D. V. Santi. 1990. Chemical synthesis of the thymidylate synthase gene. *Proc. Natl. Acad. Sci. USA* **87:**633–637.

Di Donato, A., M. de Nigris, N. Russo, S. Di Biase, and G. D'Alessio. 1993. A method for synthesizing genes and cDNAs by the polymerase chain reaction. *Anal. Biochem.* **212:**291–293.

Erlich, H. A., D. Gelfand, and J. J. Sninsky. 1991. Recent advances in the polymerase chain reaction. *Science* **252:**1643–1651.

Fox, D. K., B. Westfall, M. Nathan, A. J. Hughes, Jr., A. Rashtchian, and D. M. Schuster. 1996. Striding new distances with 5' RACE: long 5' RACE of human APC and TSC-2 cDNA. *Focus* **18:**33–37.

Itakura, K., J. J. Rossi, and R. B. Wallace. 1984. Synthesis and use of synthetic oligonucleotides. *Annu. Rev. Biochem.* **53:**323–356.

Mullis, K. B., F. Ferré, and R. A. Gibbs (ed.). 1994. *The Polymerase Chain Reaction.* Birkhäuser, Boston, Mass.

Saiki, R. K., D. H. Gelfand, S. Stoffel, S. Scharf, R. Higuchi, G. T. Horn, K. B. Mullis, and H. A. Erlich. 1988. Primer-directed enzymatic amplification of DNA with a thermostable DNA polymerase. *Science* **239:**487–491.

Sanger, F., S. Nicklen, and A. R. Coulson. 1977. DNA sequencing with chain-terminating inhibitors. *Proc. Natl. Acad. Sci. USA* **74:**5463–5467.

Schuster, D. M., G. W. Buchman, and A. Rashtchian. 1992. A simple and efficient method for amplification of cDNA ends using 5' RACE. *Focus* **14:**46–52.

REVIEW QUESTIONS

1. If your new DNA synthesizer has an average coupling efficiency of 98.5%, what overall synthesis yield would you expect after the synthesis of a 50-mer DNA hybridization probe?

2. Suggest two different strategies for synthesizing a 0.5-kb gene. How would you decide which of these strategies to employ?

3. What is a linker? How is it used?

4. What is a dideoxynucleotide? How is it used to determine the sequence of a DNA molecule?

5. Why is it necessary to make DNA single stranded before determining its sequence?

6. Describe how bacteriophage M13 is used to help determine the sequence of a cloned DNA fragment.

7. A criminal leaves a single hair, containing about 10 to 20 picograms (10^{-12} g) of DNA, at the scene of a crime. To characterize a small portion of this DNA and determine whether or not it matches the same region of the DNA from a tissue sample from your prime suspect, you need about 10 to 100 nanograms (10^{-9} g) of DNA from the crime scene. How would you go about obtaining sufficient DNA to compare the two samples? What information would you need before you could proceed?

8. Discuss the significance of long templates and short templates and their prevalence as the number of cycles increases during the PCR.

9. Discuss how PCR is used to synthesize a gene.

10. Describe how the ends of an mRNA are converted into cDNA.

Manipulation of Gene Expression in Prokaryotes

The primary objective of gene cloning for biotechnological applications is the expression of the cloned gene in a selected host organism. Unfortunately, the insertion of a gene into a cloning vector does not necessarily ensure that it will be successfully expressed. Moreover, for commercial purposes, a high rate of production of the protein encoded by the cloned gene is often required. In response to this need for a high expression rate, many specialized expression vectors have been created by manipulating a number of different genetic elements that control aspects of transcription, translation, protein stability, oxygen limitation, and secretion from the host cell. More specifically, the molecular biological features that have been manipulated to modulate gene expression include (1) the nature of the relevant transcriptional promoter and terminator sequences, (2) the strength of the ribosome binding site, (3) the number of copies of the cloned gene and whether the gene is plasmid borne or integrated into the genome of the host cell, (4) the final cellular location of the synthesized foreign protein, (5) the efficiency of translation in the host organism, and (6) the intrinsic stability of the cloned gene protein within the host cell.

There is no single strategy for achieving maximal expression of every cloned gene. Most cloned genes have distinctive molecular properties that require the investment of considerable time and effort before a specific set of conditions that result in an acceptable level of expression is found.

The level of foreign gene expression achieved also depends on the identity of the host organism. Currently, although a wide range of both prokaryotic and eukaryotic organisms can express foreign genes, most commercially important protein products that are produced by recombinant DNA technology are synthesized in *Escherichia coli*. The extensive use of this organism is understandable in the light of the vast amount of research that has been carried out on its genetics, molecular biology, biochemistry, and physiology. Moreover, many proteins can be produced by *E. coli* rapidly and inexpensively. But other host systems, such as *Bacillus subtilis*, yeast, and animal, plant, and insect cells, are used to express certain cloned genes. Nevertheless, the strategies that have been elaborated for *E. coli* should, in

109

principle, be applicable to all systems, and these strategies are stressed in this chapter.

Gene Expression from Strong and Regulatable Promoters

The minimal requirement for an effective gene expression system is the presence of a strong and regulatable promoter sequence upstream from a cloned gene. A strong promoter is one that has a high affinity for RNA polymerase, with the consequence that the adjacent downstream region is highly (frequently) transcribed. The ability to regulate a promoter enables the cell (and the researcher) to control the extent of transcription in a precise manner. The promoter from the well-studied *lac* (lactose) operon of *E. coli* has been used extensively for expressing cloned genes. However, other promoters have distinctive properties that make them useful for controlling expression, so systems have been designed to isolate promoter regions from a range of organisms. Usually, random DNA fragments are inserted upstream from a gene encoding a product whose activity is readily observed (a reporter gene) that lacks a promoter (Fig. 6.1). Subsequently, if it is found that the reporter gene is expressed, then it is assumed that a functional promoter has been cloned. Most reporter genes, when functional, encode either a product that causes resistance to an antibiotic or an enzyme that can be identified by a simple colorimetric assay.

It might seem that a good way to optimize the expression of a cloned gene would be to insert it into a plasmid under the control of a continuously transcribed strong promoter. However, a high level of continual expression of a cloned gene is often detrimental to the host cell because it creates an energy drain, thereby impairing essential host cell functions. In addition, all or a portion of the plasmid carrying a continually (constitutively) expressed cloned gene may be lost after several division cycles, since cells without a plasmid grow faster and eventually take over the culture. Such plasmid instability is a major problem that may prevent the efficient production of a plasmid-borne gene product on a large scale. To overcome this drawback, it is desirable to control transcription in such a way that a cloned gene is expressed only at a specific stage in the host cell growth cycle and only for a specified duration. This objective is achieved by using a strong regulatable promoter. The plasmids constructed to accomplish this task are called expression vectors.

Regulatable Promoters

The most widely used strong regulatable promoters are those from the *E. coli lac* and *trp* (tryptophan) operons; the *tac* promoter, which is an in vitro construct that includes the −10 region (i.e., 10 nucleotide pairs upstream from the site of initiation of transcription) of the *lac* promoter and the −35 region of the *trp* promoter; the leftward, or p^L, promoter from bacteriophage λ; and the gene 10 promoter from bacteriophage T7. Each of these promoters interacts with repressors, which provide a controllable switch for either turning on or turning off specific transcription of adjacent cloned genes. In addition, each of these promoters is recognized by the major form of the *E. coli* RNA polymerase holoenzyme, which uses a sigma factor (a protein that, when combined with the core proteins of RNA polymerase to form the holoenzyme, directs the binding of the holoenzyme to promoter regions on the DNA) that is present in a much larger quantity than other minor sigma

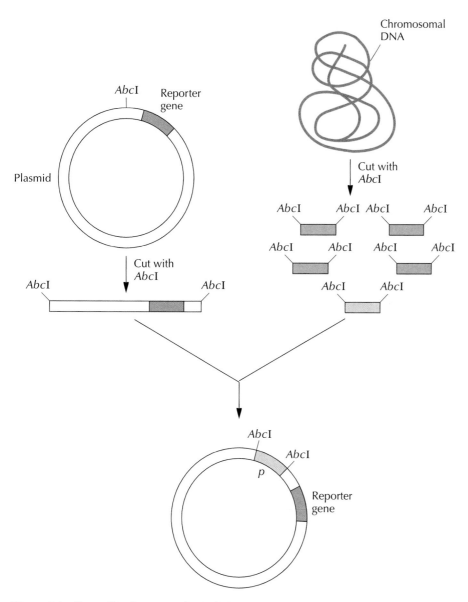

Figure 6.1 Generalized strategy for isolating presumptive promoter sequences. A reporter gene without a promoter is cloned into a plasmid. Chromosomal DNA is cut with a restriction endonuclease (*Abc*I) and cloned into the *Abc*I site on the plasmid. If a promoter sequence (*p*) is present, then the reporter gene will be expressed.

factors. Consequently, transcription is not limited because of a shortage of available sigma factor.

In the absence of lactose in the growth medium, the *E. coli lac* promoter is repressed, i.e., turned off, by the *lac* repressor protein, which prevents the *lac* operon from being transcribed. Induction, or turning on, of the *lac* promoter is achieved by the addition of either lactose or isopropyl-β-D-thiogalactopyranoside (IPTG), a gratuitous inducer, to the medium. Either of these substances prevents the *lac* repressor from binding to the *lac* operator, thereby enabling transcription to occur.

Transcription from the *lac* promoter is also regulated by the binding of the catabolite activator protein (CAP) to the promoter region (Fig. 6.2).

Glucose concentration	Lactose concentration	cAMP concentration	lac promoter-operator region	Level of transcription
Low	Low	High		Low
High	Low	Low		Low
High	High	Low		Low
Low	High	High		High

Figure 6.2 Diagrammatic representation of the effect of the concentrations of glucose, lactose, and cAMP in the growth medium on the level of transcription from the *E. coli lac* promoter. The arrow indicates the direction of transcription. Adapted from Abeles et al., 1992, *Biochemistry*, p. 383, Jones & Bartlett Publishers, Boston, Mass.

When CAP binds to the promoter, it increases the affinity of the promoter for RNA polymerase, thereby increasing transcription of the genes downstream from the promoter. The affinity of CAP for the promoter is enhanced by its association with cyclic AMP (cAMP), whose level is highest when the amount of glucose in the medium is lowest. Thus, when inducer is present and there is no repressor bound to the operator, a high intracellular concentration of cAMP can lead to a high level of transcription of the genes downstream of the *lac* promoter.

In practice, *lacUV5*, a variant of the *lac* promoter that contains an altered nucleotide sequence in the −10 region and is a stronger promoter than the wild-type *lac* promoter, is usually used in plasmid expression vectors. The *tac* promoter is also repressed by the *lac* repressor and derepressed by the addition of lactose or IPTG.

The *trp* promoter is negatively regulated, or turned off, by a tryptophan–*trp* repressor protein complex that binds to the *trp* operator and prevents transcription of the *trp* operon. Derepression, or turning on, of the *trp* promoter is achieved either by removing tryptophan or by adding 3-indoleacrylic acid to the growth medium.

The p^L promoter is controlled by the *c*I repressor protein of bacteriophage λ. In practice, a temperature-sensitive mutant of the *c*I repressor, cI_{857}, is generally used to regulate p^L-directed transcription. Cells carrying the temperature-sensitive *c*I repressor are first grown at 28 to 30°C, a temperature at which the *c*I repressor prevents transcription directed by the p^L promoter. When the cell culture reaches the desired stage of growth, often the mid-log phase, the temperature is shifted to 42°C, at which the thermosensitive *c*I repressor is inactivated and transcription can proceed.

The bacteriophage T7 promoter requires T7 RNA polymerase for transcription. To utilize this promoter, the T7 RNA polymerase gene is inserted in the *E. coli* chromosome on a bacteriophage λ lysogen under the control of the *E. coli lac* promoter. After cells are transformed by a plasmid with a tar-

get gene under the control of the T7 promoter, IPTG is added to the medium. Under these conditions, the T7 RNA polymerase gene is induced, T7 RNA polymerase is produced, and the cloned gene is transcribed and translated. There is often a lag of an hour or more from the time that the T7 RNA polymerase gene is induced until the target gene is transcribed. A series of plasmids called pET vectors have been developed to exploit the strength of the T7 promoter.

The effectiveness of deactivating a repressor protein and thereby activating transcription depends on the ratio of the number of repressor protein molecules to the number of copies of the promoter sequences. If there are too many repressor protein molecules, then it is difficult to induce transcription. Conversely, with too few repressor protein molecules (even when there are more repressor molecules than copies of the promoter), transcription occurs even in the absence of induction. In these cases the promoters are said to be "leaky." Various means have been devised to keep these regulatable systems under complete control. For example, the repressor protein gene and its corresponding promoter may be placed on two different plasmids that maintain different numbers of copies per cell; this arrangement maintains the appropriate ratio between the repressor protein and the promoter. Usually, the repressor gene is placed on a low-copy-number plasmid that maintains about 1 to 8 copies per cell, and the promoter sequence is inserted into a high-copy-number plasmid that maintains about 30 to 100 copies per cell. Alternatively, the repressor protein gene may be carried as a single gene in the chromosomal DNA, an arrangement that keeps repressor protein levels low. In systems that use the *lac* promoter, a mutant form of the *lacI* gene (*lacI*q) produces much higher levels of the *lac* repressor, thereby decreasing the leakiness under noninduced conditions, i.e., transcription of a cloned gene in the absence of inducer.

Increasing Protein Production

Plasmid pPLc2833 was created in an effort to obtain the highest possible level of foreign protein production in a recombinant *E. coli* strain. This plasmid contains a strong promoter, a selectable marker, and a short stretch of DNA with several unique restriction enzyme sites (a multiple cloning site) immediately downstream from the promoter. The effectiveness of this expression vector in directing the synthesis of foreign proteins in *E. coli* can be increased by replacing the origin of DNA replication in plasmid pPLc2833 with that from plasmid pKN402, a substitution that increases the modified plasmid's copy number 5- to 10-fold when the growth temperature is increased to 42°C (Table 6.1). The resultant plasmid, pCP3, contains the p^L promoter and the β-lactamase gene (ampicillin resistance gene) from pPLc2833 and the origin of replication from pKN402 (Fig. 6.3).

Table 6.1 Effect of temperature on the plasmid copy number of three expression vectors

Plasmid	Plasmids/cell at:		p^L present
	28°C	**42°C**	
pKN402	82	521	No
pPLc2833	38	42	Yes
pCP3	60	713	Yes

Adapted from Remaut et al., 1983, *Gene* **22**:103–113.

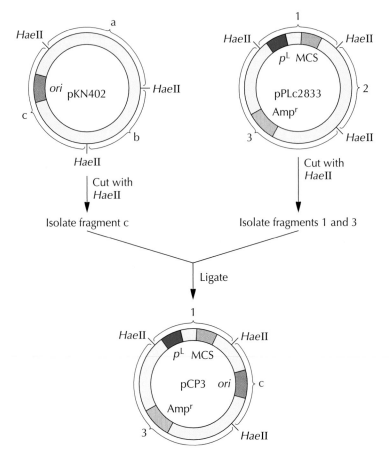

Figure 6.3 Formation of plasmid pCP3. Restriction enzyme *Hae*II fragment c, which contains the temperature-sensitive replication origin (*ori*) of plasmid pKN402, was isolated and ligated to *Hae*II fragments 1 and 3 from plasmid pPLc2833. pPLc2833 *Hae*II fragment 1 contains the p^L promoter and a multiple cloning sequence (MCS). pPLc2833 *Hae*II fragment 3 contains the selectable marker gene for ampicillin resistance (Ampr).

Cells that carry the plasmid pCP3 are first grown at 28°C and then shifted to 42°C. At the lower temperature, the *c*I repressor, which is integrated into the host *E. coli* chromosomal DNA, is functional, the p^L promoter is turned off, and the plasmid copy number is normal (Table 6.1). At the higher temperature, the temperature-sensitive *c*I repressor is inactivated, the p^L promoter is active, and the plasmid copy number increases. These properties make pCP3 a particularly effective expression vector. When the gene for T4 DNA ligase is inserted into the multiple cloning site of pCP3, about 20% of the cellular protein produced at 42°C is T4 DNA ligase. This level of expression is much higher than that for even the most abundant native *E. coli* proteins, such as elongation factor EF-Tu, which have an expression level of about 2%.

Large-Scale Systems

In small culture vessels (1 to 5 liters), induction is readily achieved either by shifting the temperature or by adding a chemical inducer. In pilot plant-sized (20 to 200 liters) and industrial-sized (>200 liters) bioreactors, however, a shift in temperature requires time (30 to 60 minutes) and energy,

both of which can be costly. Similarly, the cost of a chemical inducer, such as IPTG, that is required for the expression of a cloned gene in a large-scale bioreactor can make the overall process uneconomical. To overcome some of the problems associated with the use of the p^L promoter for large-scale fermentations, a two-plasmid system has been developed. The cI repressor was placed under the control of the *trp* promoter and inserted into a low-copy-number plasmid (Fig. 6.4). The use of a low-copy-number plasmid ensures that excess cI repressor molecules are not produced. A second plasmid carries a cloned gene under the control of the p^L promoter. As shown in Fig. 6.4A, the *trp* promoter is turned on in the absence of tryptophan, so the cI repressor protein is synthesized and the p^L promoter is turned off. In contrast, as shown in Fig. 6.4B, the *trp* promoter is turned off in the presence of tryptophan, so the cI repressor protein is not synthesized and the p^L promoter is fully active.

With this two-plasmid system, cells can be grown on an inexpensive medium consisting of molasses and casein hydrolysate, which contains only very small amounts of free tryptophan, and then induced to express the cloned gene by the addition of tryptone to the medium. Tryptone contains enough free tryptophan for efficient induction of transcription. In trial runs of this system, cloned β-galactosidase and citrate synthase genes, after induction by addition of tryptone to the medium, represented 21 and 24% of the cellular protein, respectively. Thus, this system provides a potentially inexpensive means of producing proteins from recombinant microorganisms on a large scale.

Expression in Other Microorganisms

E. coli is not necessarily the microorganism of choice for the expression of all foreign proteins. However, the genetics and molecular biology of most other microorganisms are not as highly developed. Unfortunately, there is no one vector or even promoter-repressor system that gives optimal levels of gene expression in all bacteria, or even in all gram-negative bacteria. Fortunately, many of the approaches and strategies that have been developed for *E. coli* are also useful with a variety of other microorganisms. With this in mind, the ability of various promoters to support transcription in other gram-negative bacteria has been tested. In one study, a set of plasmid expression vectors containing either the *lac*, *tac*, Nm (from the neomycin resistance gene), or S1 (from the ribosomal protein S1 gene of *Rhizobium meliloti*) promoter was constructed. The expression of β-galactosidase under the control of each of these promoters was examined (Table 6.2). The results indicated that (1) all of the promoters were active to some extent in

Table 6.2 β-Galactosidase activity expressed by gram-negative bacteria carrying a plasmid vector with the *E. coli lacZ* gene and a heterologous promoter

Promoter	β-Galactosidase activity (U) in:			
	Escherichia coli	*Rhizobium meliloti*	*Rhizobium leguminosarum*	*Pseudomonas putida*
None	16	110	130	150
Nm	1,400	21,800	13,900	16,300
lac	2,000	9,050	6,250	9,800
tac	11,300	2,850	1,150	2,950
S1	40	3,300	1,200	3,350

Adapted from Labes et al., 1990, *Gene* **89**:37–46.

A No tryptophan in the medium

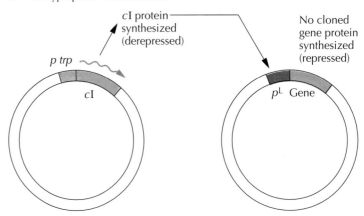

B Tryptophan in the medium

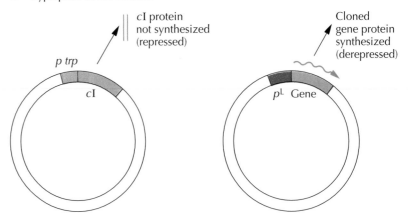

Figure 6.4 Dual plasmid system for controlling the λ p^L promoter by regulating the *c*I repressor with tryptophan. The tryptophan promoter (*p trp*) is inserted next to the *c*I gene on one plasmid, and the p^L promoter is placed adjacent to the cloned gene (Gene) on a second plasmid. The wavy arrows denote transcription. **A.** With no tryptophan in the medium, the *c*I gene is transcribed and translated, and the *c*I repressor protein binds to the p^L promoter, thereby blocking the transcription of the cloned gene. **B.** In the presence of tryptophan, the *c*I gene is repressed, no *c*I product is made, and the cloned gene is transcribed and translated.

each of the bacteria tested, (2) *tac* was the most active promoter in *E. coli* and the least active promoter in the other bacteria, and (3) Nm was the second least active promoter in *E. coli* and the most active promoter in the other bacteria. Clearly, even though gram-negative organisms may utilize similar DNA sequences to promote transcription, the best promoter for use in a particular organism is not necessarily the one that is most efficient in *E. coli*. Nevertheless, depending on the application, known *E. coli* promoters may be satisfactory for regulating the expression of cloned genes in other gram-negative bacteria.

Numerous efforts have been made to construct a "universal" gram-negative bacterial expression vector. To this end, a 70-bp DNA fragment from one of the terminal inverted repeats of transposon 5 (Tn5) was inserted into a multiple cloning site in the low-copy-number broad-host-range plasmid pRK290 to create plasmid pAV10 (Fig. 6.5). The cloned segment of DNA from Tn5 contains two independent but overlapping promoters, each of

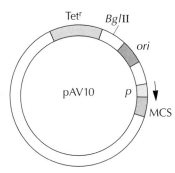

Figure 6.5 Cloning vector pAV10. The arrangement of the tetracycline resistance gene (Tetr), a *Bgl*II restriction endonuclease site, the origin of replication (*ori*), a promoter (*p*), and a multiple cloning sequence (MCS) is shown. Insertion of a cloned gene into the multiple cloning sequence puts it under the control of the inserted Tn*5* promoter (*p*). The arrow shows the direction of transcription. The plasmid is not drawn to scale.

which is necessary for the transcription of a separate essential Tn*5* gene. Because Tn*5* functions efficiently in a wide range of bacterial hosts, Tn*5* promoters should facilitate transcription in the same range of organisms. This supposition was tested by cloning either a chloramphenicol acetyltransferase or a β-galactosidase gene into the multiple cloning site downstream from the cloned Tn*5* promoters. Efficient expression was observed in an *Alcaligenes* sp., *E. coli*, *Enterobacter cloacae*, *Klebsiella pneumoniae*, *Pseudomonas stutzeri*, *Pseudomonas fluorescens*, and *Serratia marcescens*. Thus, it may indeed be possible to use the Tn*5* promoters to direct transcription of foreign genes in a number of different bacterial hosts.

Fusion Proteins

Often, foreign proteins, especially small ones, occur in minute quantities in heterologous host cells. This apparently low level of expression is, in many instances, actually due to degradation of the foreign protein. One way to solve this problem is to covalently attach the cloned gene product to a stable host protein. This combination, which is called a fusion protein, protects the cloned gene product from attack by host cell proteases. Indeed, in a number of studies, cloned gene proteins have been found to be resistant to degradation when they are part of a fusion protein, whereas, as separate intact proteins, they are susceptible to degradation of the protein by proteolytic enzymes (proteolysis).

Fusion proteins are constructed at the DNA level by ligating together the coding regions of the two genes. In its simplest form, a fusion vector system entails the insertion of a target gene or gene segment into the coding region of a cloned host gene. It is essential that the RNA transcribed from the cloned target gene contain the correct base sequence for the production of the cloned gene protein. If the combined DNA has an altered reading frame, i.e., a sequence of successive codons that yields either an incomplete or an incorrect translation product, then a functional version of the protein encoded by the cloned target gene will not be produced. Many strategies have been devised to ensure that a proper reading frame is achieved. Generally, it is necessary to know the precise nucleotide sequence of the two DNA sequences that are to be ligated.

Site of
cleavage of
factor X_a

X_a linker sequence

. . . Thr-Ala-Glu-Gly-Gly-Ser-Ile-Glu-Gly-Arg-Val-His-Leu . . .

Figure 6.6 Proteolytic cleavage of a fusion protein by blood coagulation factor X_a. The factor X_a recognition sequence (X_a linker sequence) lies between the amino acid sequences of two different proteins. A functional cloned gene protein (with Val at its N terminus) is released after cleavage.

Cleavage of Fusion Proteins

Depending on the end use of a cloned gene protein, investigators may find it undesirable to produce a fusion protein as the final product. For example, the presence of the host protein segment makes most fusion proteins unsuitable for clinical use and may affect the biological functioning of the target protein. In addition, fusion proteins require more extensive testing before being approved by regulatory agencies, such as the U.S. Food and Drug Administration. Thus, it is necessary to develop strategies to remove the unwanted amino acid sequence from the target protein. One way to do this is to join the target gene protein to the cloned host gene protein with short stretches of amino acids that are recognized by a specific nonbacterial protease. This addition is made at the DNA level. Oligonucleotide linkers that code for these protease recognition sites can be ligated to the cloned gene before the construct is inserted into a fusion expression vector. For example, an oligonucleotide linker encoding the amino acid sequence Ile-Glu-Gly-Arg can be joined to the cloned gene. Following synthesis and purification of the fusion protein, a blood coagulation factor called X_a can be used to release the cloned gene protein from the host protein because factor X_a is a specific protease that cleaves peptide bonds uniquely on the C-terminal side of the Ile-Glu-Gly-Arg sequence (Fig. 6.6). Moreover, because this peptide sequence occurs rather infrequently in native proteins, this approach can be used to recover many different cloned gene products.

Uses of Fusion Proteins

For some applications, a fusion protein can be a satisfactory end product. For example, an antibody that recognizes a specific portion of a protein is frequently required for research purposes. After a DNA segment encoding a portion of a target gene is cloned into a fusion vector, the fusion protein contains the stabilizing host protein and a protein domain against which antibodies can be generated. The fusion protein can be used as an antigen. The antibodies against the stabilizing host component can be removed by absorption with this protein alone, thus leaving in the antiserum only those antibodies that bind to the targeted protein sequence.

One fusion cloning vector that has been used to generate antibodies against a target protein contains the 5'-terminal segment of the E. coli ompF gene, which encodes an outer membrane protein, and an adjacent portion of the E. coli lacZ (β-galactosidase) gene (Fig. 6.7). The ompF gene segment contributes the signals for the initiation of both transcription and translation as well as for secretion of the fusion protein. Even though the truncated lacZ gene lacks the codons for the first eight amino acids, the shortened protein encoded by this gene fragment is still enzymatically active. This form

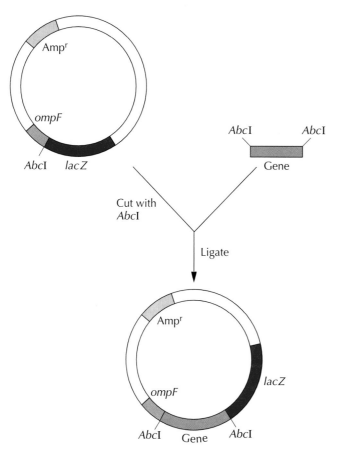

The lac z can't make a final protein ong b/c it's off the reading frame — once you add gene it puts lac z gene in right place

Figure 6.7 Fusion protein cloning vector. The plasmid contains an ampicillin resistance (Amp^r) gene as the selectable marker, a DNA sequence encoding the N-terminal segment of the *E. coli* outer membrane protein (*ompF*), a restriction endonuclease site (*Abc*I) for cloning, and a truncated β-galactosidase gene (*lacZ*). The cloned gene (Gene) is inserted into the *Abc*I site. After transcription and translation, a tribrid protein is produced.

of the enzyme β-galactosidase is able to function with almost any N-terminus-localized peptide fused to it. The *lacZ* gene is cloned on the vector at a location that puts it in an altered reading frame with respect to the *ompF* leader sequence. Therefore, no functional β-galactosidase will be produced. But any cloned target DNA that has both *ompF* and *lacZ* "in frame" will produce a "tribrid" (a three-part hybrid) protein that comprises the OmpF amino acid sequence, the protein encoded by the cloned target gene, and the functional C-terminal portion of β-galactosidase. The tribrid protein can be used either as an antigen to produce antibodies that will cross-react with the cloned gene protein or as a means of isolating small important portions of specific proteins.

In addition to reducing the degradation of polypeptides, a number of fusion proteins have been developed to simplify the purification of recombinant proteins (Table 6.3). For example, a yeast (*Saccharomyces cerevisiae*) plasmid construct that contains the human interleukin-2 gene joined to DNA encoding a marker peptide sequence (Asp-Tyr-Lys-Asp-Asp-Asp-Asp-Lys) has the dual function of reducing the degradation of the expressed interleukin-2 gene product and then enabling the product to be purified. Interleukin-2 is a biological factor that stimulates both T-cell growth

Table 6.3 Some fusion systems used to facilitate the purification of foreign proteins produced in *E. coli*

Fusion partner	Size	Ligand	Elution condition
ZZ	14 kDa	IgG	Low pH
His tail	6–10 aa	Ni^{2+}	Imidazole
Strep-tag	10 aa	Streptavidin	Iminobiotin
PinPoint	13 kDa	Streptavidin	Biotin
MBP	40 kDa	Amylose	Maltose
β-Lactamase	27 kDa	Phenyl-boronate	Borate
GST	25 kDa	Glutathione	Reducing agent
Flag	8 aa	Specific MAb	Low calcium

Adapted from Nygren et al., 1994, *Trends Biotechnol.* **12**:184–188.

Abbreviations: aa, amino acids; kDa, kilodaltons; ZZ, a fragment of *Staphylococcus aureus* protein A; His, histidine; Strep-tag, a peptide with affinity for streptavidin; PinPoint, a protein fragment which is biotinylated in vivo in *E. coli*; MBP, maltose binding protein; GST, glutathione *S*-transferase; Flag, a peptide recognized by enterokinase.

and B-cell antibody synthesis. The eight-amino-acid marker peptide is sold as the Flag peptide. Following expression of this construct in yeast, the secreted fusion protein can be purified in a single step by immunoaffinity chromatography, in which monoclonal antibodies against the marker peptide have been immobilized on a polypropylene support and act as ligands to bind the fusion protein (Fig. 6.8). Because the marker peptide is relatively small, it does not significantly decrease the amount of host cell resources that are available for the production of interleukin-2. This fusion protein has the same biological activity as native interleukin-2. However, to satisfy the government agencies that regulate the use of pharmaceuticals, it is still necessary to remove the marker peptide if the product is to be used for human immunotherapy or other medical purposes. In this system, the marker sequence may be specifically removed by treatment of the fusion protein with bovine intestinal enterokinase.

Many of the proteins that are produced in *E. coli* accumulate in the form of insoluble, intracellular, biologically inactive inclusion bodies. A small amount of biologically active protein can often be recovered from inclusion bodies; however, the extraction procedure requires expensive and time-consuming protein solubilization and refolding procedures. Since the in vivo insolubility of proteins is often a consequence of their incorrect folding, various strategies have been devised to avoid this problem. For example, fusion proteins that contain thioredoxin, an 11.7-kilodalton kDa protein, as the fusion partner remain soluble even when up to 40% of the cellular protein consists of the fusion protein. With this system, the target gene is cloned into a multiple cloning site just downstream from the thioredoxin gene, all under the control of the *p*L promoter on an *E. coli* plasmid vector (Fig. 6.9). The *E. coli* host cell for this system has a *c*I repressor protein construct (a copy of the gene for the *c*I repressor protein under the transcriptional control of the *trp* promoter) integrated into the chromosomal DNA. In the absence of tryptophan (Fig. 6.9A), a sufficient amount of *c*I repressor protein to prevent transcription from the *p*L promoter, and therefore no fusion protein, is synthesized. When tryptophan is added to the growth medium (Fig. 6.9B), the *trp* promoter is turned off by the *trp* repressor, the *c*I repressor protein is not synthesized, and the fusion protein is transcribed from the *p*L promoter on the plasmid and then translated. In this case, the fusion protein (consisting of thioredoxin and the target protein) is synthe-

1 Concentrate secreted protein mixture

2 Prepare immunoaffinity column

3 Add secreted protein mixture
to the column

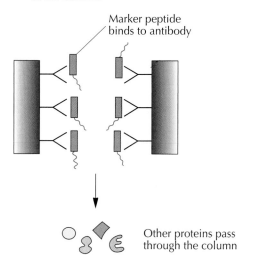

Figure 6.8 Immunoaffinity chromatographic purification of a fusion protein. An antibody that binds to the marker peptide of the fusion protein (anti-marker peptide antibody) is attached to a solid polypropylene support. The secreted proteins are passed through the column containing the bound antibody. The marker peptide portion of the fusion protein is bound to the antibody, and the other proteins pass through. The immunopurified fusion protein can then be selectively eluted from the column.

4 Elute fusion protein

A

B

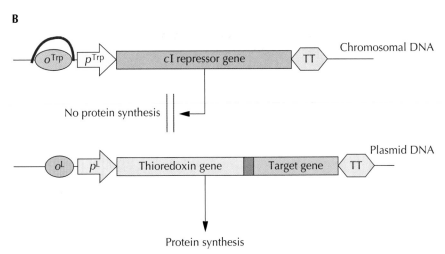

Figure 6.9 Synthesis of a thioredoxin–target protein fusion in the absence (**A**) and in the presence (**B**) of tryptophan in the growth medium. The arrows labeled p^{Trp} and p^L indicate the direction of transcription. Abbreviations and designations: o^{Trp}, the operator region where the *trp* repressor protein binds; o^L, the operator region where the *c*I repressor binds; p^{Trp}, the *trp* promoter; p^L, the leftward promoter from bacteriophage lambda; TT, transcription termination region. The box with the diagonal lines indicates the DNA region that codes for the peptide that acts as the enterokinase cleavage site; the horseshoes indicate the binding of a repressor protein to its operator region.

sized and accumulates preferentially at osmotically sensitive sites called adhesion zones at the inner periphery of the host *E. coli* cytoplasmic membrane. The soluble fusion protein is selectively released by osmotic shock from *E. coli* cells into the growth medium. If required, the native form of the target protein can be released from the fusion protein by treatment with the enzyme enterokinase. Finally, since thioredoxin is stable at temperatures as high as 80°C, in those instances where the target protein is also stable at high temperatures, the fusion protein may be purified by high-temperature incubation, a condition that denatures most of the other cellular proteins.

Surface Display

Specialized fusion protein systems have been devised for screening complementary DNA (cDNA) libraries that contain very large numbers of dif-

ferent clones (sometimes up to 5×10^{10}) for proteins that are encoded by rarely occurring cDNAs. Generally, cDNA molecules are cloned into a surface protein (filament protein, pilus protein) gene of a filamentous bacteriophage such as M13 or a bacterium, and after transcription and translation the fusion protein is incorporated into a surface structure of the bacteriophage or bacterium, where it can be identified by an immunological assay. More specifically, fusions are often made with the protein pIII gene from bacteriophage M13. Protein pIII is normally found at the tip of this tubular-shaped bacteriophage and is responsible for initiating phage infection of *E. coli* by binding to F pili. A plasmid that contains a small piece of M13 DNA which allows the plasmid (phagemid) to be packaged in vitro into M13 phage particles, a protein pIII gene under the control of a regulatable bacterial promoter such as the *E. coli lac* promoter, and a cloning site near the 5′ end of the pIII gene for a cDNA or other coding sequence was constructed. The target protein is fused to bacteriophage M13 protein pIII near its N terminus. After M13 replication in *E. coli* cells, the plaques may be assayed immunologically for the presence of the target protein. Recombinant phagemids isolated from positive plaques can then be used as a source of the target cDNA. This is an extremely powerful selection system that has the capability of finding cDNAs for very rarely expressed but important proteins.

Alternatively, libraries with bacterial surface structures composed of fusion proteins can be screened for clones that carry specific coding sequences. In order to export proteins to the surface of a gram-negative bacterium such as *E. coli*, fusions between the genes for the target protein and for an outer surface protein are created. Bacterial fusion partners that have been used in these types of constructs include *E. coli* outer membrane protein A (OmpA) and peptide-glycan-associated lipoprotein (PAL), as well as *Pseudomonas aeruginosa* outer membrane protein F (OprF). With most bacte-

Figure 6.10 Fusion protein between bacterial outer surface protein and a foreign target protein joined to either the N or C terminus (**A**) or foreign peptides inserted into the surface-exposed loops (**B**). In both cases, the foreign peptide or protein is directed to the bacterial cell's outer surface.

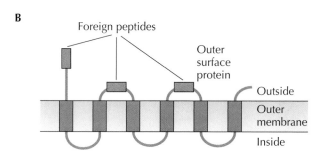

rial surface fusion proteins, the target protein is located at either the N or C terminus of the fusion protein. However, in some instances, short stretches of a target protein can be expressed in the middle of the fusion partner (Fig. 6.10).

In addition to facilitating the screening of large cDNA libraries, surface-displayed proteins can provide an effective means of overexpressing some peptides or proteins. For example, in one study the amino acid-repeating epitope (antigenic determinant) Asn-Ala-Asn-Pro of a protein from the parasite *Plasmodium falciparum*, the causative agent of malaria, was inserted into the regions that encode surface-exposed loops of the major outer membrane protein from *P. aeruginosa* (OprF). Whole bacterial cells expressing this fusion protein reacted positively when challenged with monoclonal antibodies against *P. falciparum*. It may, therefore, be possible to use surface displayed fusion proteins as vaccines (see chapter 11).

Unidirectional Tandem Gene Arrays

Generally, the level of gene expression is proportional to the number of copies of the transcribed gene in the host cell. An obvious inference is that by increasing the plasmid copy number, there will be a concomitant increase in the amount of the cloned gene protein that is synthesized. However, in addition to the cloned gene, plasmids contain other transcribed genes such as those that code for antibiotic resistance. Thus, as the plasmid copy number is increased, more of the energy resources of the cell are diverted to the production of plasmid-encoded proteins and the metabolic activities of the host cell are constrained. Because increasing the plasmid copy

MILESTONE

The *tac* Promoter: a Functional Hybrid Derived from the *trp* and *lac* Promoters

H. A. deBOER, L. J. COMSTOCK, and M. VASSER

Proc. Natl. Acad. Sci. USA **80**:21–25, 1983

DeBoer and his colleagues began their efforts to construct the *tac* promoter with the idea of combining portions of two different strong and regulatable promoters to create an even stronger promoter that would direct very high levels of foreign gene expression. When they undertook their studies, although the DNA sequences of a number of prokaryotic promoters, mostly from *E. coli*, were known, the precise features that enabled a promoter to be efficient at directing transcription were not well understood. It was known that almost all mutations that affected the strength of a prokaryotic promoter were found in either the −10 region or the −35 region, which are approximately 10 or 35

bp upstream of the mRNA transcription start site, respectively. Moreover, only mutations that made an existing promoter more like the consensus sequences for each of these regions, i.e., 5'-TATAAT-3' for the −10 region and 5'-TTGACA-3' for the −35 region, increased the strength of promoter. The consensus sequences had been deduced by comparing the DNA sequences of all known promoters and determining which nucleotides occurred most often. deBoer and his colleagues knew that the *lacUV5* promoter, which is a stronger variant of the *lac* promoter, has a consensus sequence at its −10 but not its −35 region, while the *trp* promoter, which normally controls the transcription of

genes involved in the biosynthesis of tryptophan, has a consensus sequence at its −35 but not its −10 region. They decided to create a fusion promoter which included the −10 region from the *lac* promoter and the −35 region from the *trp* promoter. They tested this new "*tac*" promoter, as they called it, for its ability to direct the synthesis of the enzyme galactose kinase in *E. coli*, and compared it in the same assay system to the *trp* and *lac* promoters. In agreement with the reasoning that had suggested that they make this construction in the first place, the *tac* promoter was found to be approximately 5 times stronger than the *trp* promoter and 10 times stronger than the *lac* promoter. In addition, like the *lac* promoter, the *tac* promoter was repressed by the *lac* repressor and derepressed by IPTG. Thus, this new promoter was not only strong but also regulatable.

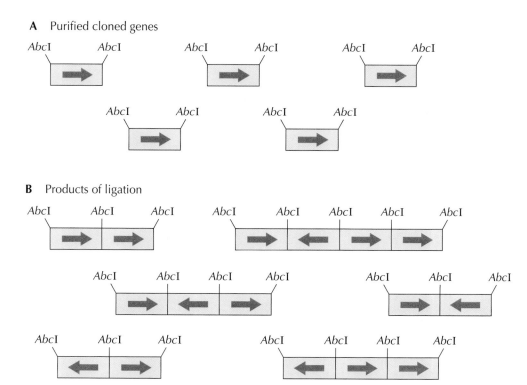

A Purified cloned genes

B Products of ligation

Figure 6.11 Formation of randomly oriented tandem gene arrays. **A.** Cloned gene sequences are excised from the cloning vector with the restriction endonuclease *Abc*I and then separated from the vector DNA. **B.** The conditions of the ligation mixture favor linearization of the DNA. Because the overhang sequences on both ends of the excised genes are the same, they can base pair and be ligated in any orientation. Therefore, the tandem arrays contain randomly oriented DNA sequences. Arrows denote the orientation of the cloned gene.

number is not always effective in maximizing the yield of the cloned gene product, another approach is to clone multiple copies of the gene of interest into a low-copy-number plasmid rather than to clone one copy into a high-copy-number plasmid. The technical problem then becomes the creation of tandem arrays of a gene in which each sequence is in the correct orientation for transcription and translation. Simple end-to-end ligation of DNA results in a random orientation of the arrays of genes; some will be in the correct direction that allows subsequent transcription and translation, but others will be oriented in the opposite direction and cannot be expressed (Fig. 6.11).

The restriction enzyme *Ava*I and its DNA recognition site can be used to overcome this problem. This enzyme recognizes the sequence CTCGGG and cleaves on the 5′ side of the T residue. It is used as follows. A plasmid containing the sequence CTCGGG is cleaved by *Ava*I and treated with DNA polymerase I to fill in the sticky ends. Then an *Eco*RI linker (GAATTC) is ligated at both ends onto the blunt-ended sites. The resultant plasmid contains a segment of DNA that has two *Ava*I sites flanking and overlapping an *Eco*RI site (Fig. 6.12A and B), i.e., the sequence <u>CTCGGG</u>AATT<u>CTCGGG</u>, where the underlined sequences are the *Ava*I recognition sites. The gene, together with translational start and stop signals, is cloned into the *Eco*RI site. The gene is cut out of the plasmid by digestion with *Ava*I (Fig. 6.12C). In this form, the excised DNA fragment will have nonidentical sticky ends. There-

Figure 6.12 Cloning multiple copies of a gene on a single plasmid. **A.** Vector preparation. A unique restriction site on the plasmid vector is digested with *Ava*I, the protruding ends are filled in by *E. coli* DNA polymerase I and the appropriate deoxyribonucleoside triphosphates, and an *Eco*RI linker is inserted by blunt-end ligation. **B.** Insertion of an *Eco*RI linker into the *Ava*I site on the plasmid. **C.** Production of unidirectional tandem gene arrays. Arrows denote the orientation of the cloned gene. The sequences shown are produced by digestion of the cloned gene with *Ava*I. The dots represent ligation sites.

fore, during the subsequent ligation reaction, the DNA segments can be joined together in only one orientation. Then a set of unidirectional tandem gene copies can be inserted into an expression vector. Because there are two possible cloning orientations with respect to the promoter on the expression vector (antisense or its complement) for each tandem array, only 50% of the resulting clones will be able to express the array of cloned genes.

Another approach to cloning multiple copies of a gene involves the use of synthetic directional adaptors, which are short sequences of oligodeoxyribonucleotides added to the ends of linearized plasmid DNA and the gene-containing DNA fragment. During subsequent ligation, the DNA pieces are oriented in only one direction. This procedure is technically easier than the one that utilizes the restriction endonuclease *Ava*I and is not limited by the requirement for the absence of *Ava*I or *Eco*RI sites within the target gene.

In practice, expression of interferon genes is actually found to rise in proportion to the number of tandem gene copies, at least up to four gene copies per plasmid. However, tandem gene copies are sometimes not stable

at the DNA level and, during growth, some or all of the copies may be excised from the plasmid.

Translation Expression Vectors

Putting a cloned gene under the control of a regulatable, strong promoter, although essential, may not be sufficient to maximize the yield of the cloned gene product. Other factors, such as the efficiency of translation and the stability of the newly synthesized cloned gene protein, may also affect the amount of product.

In prokaryotic cells, various messenger RNAs (mRNAs) are not necessarily translated with the same efficiency. In fact, they may be expressed at very different levels (up to several hundredfold). Differences in translational efficiency and in transcriptional regulation enable the cell to have hundreds or even thousands of copies of some proteins and only a few copies of others.

In part, the molecular basis for differential translation is the presence of a translational initiation signal called a ribosome binding site in the transcribed mRNA. A ribosome binding site is a sequence of six to eight nucleotides (e.g., UAAGGAGG) in mRNA that can base pair with a complementary sequence (in this case AUUCCUCC) on the RNA component of the small ribosomal subunit.

Generally, the stronger the binding of the mRNA to the ribosomal RNA, the greater the efficiency of translational initiation. For this reason, many *E. coli* expression vectors have been designed to ensure that the mRNA of a cloned gene contains a strong ribosome binding site. In effect, this means that heterologous prokaryotic and eukaryotic genes can be translated readily in *E. coli*. However, certain other conditions must be satisfied for this approach to function properly. First, the ribosome binding sequence must be located a precise distance from the translational start codon of the cloned gene. (At the RNA level, the translational codon is usually AUG. At the DNA level, the strand that has the ATG sequence is called the coding strand, and the complementary strand, which acts as the template for transcription, is the noncoding strand. By convention, a start codon at the DNA level is designated as ATG.) Second, the DNA sequence that includes the ribosome binding site through the first few codons of the gene of interest must not contain nucleotide sequences that after transcription can fold back, i.e., form intrastrand loops (Fig. 6.13), thereby thwarting the interaction of the mRNA with the ribosome. The local secondary structure of the mRNA, which can either shield or expose the ribosome binding site, determines the extent to which the mRNA can bind to the appropriate sequence on the translation-initiating ribosome. Thus, for each cloned gene, it is important to establish that the ribosome binding site is properly placed and that the secondary structure of the mRNA does not prevent its access to the ribosome.

A number of convenient vector systems that incorporate both transcriptional and translational signals for the expression of cloned eukaryotic genes in *E. coli* have been developed. One such system is called expression vector pKK233-2 and includes the following elements (Fig. 6.14).

- An ampicillin resistance gene as a selectable marker
- The *tac* promoter
- The *lacZ* ribosome binding site

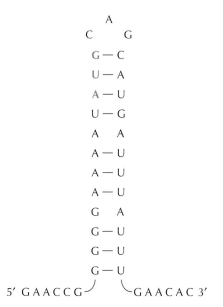

Figure 6.13 Example of secondary structure of the 5′ end of an mRNA that would prevent efficient translation. The ribosome binding site is GGGGG, the initiator codon is AUG (shown in boldface), and the first few codons are CAG-CAU-GAU-UUA-UUU. The mRNA is oriented with its 5′ end to the left and its 3′ end to the right. Note that in addition to the traditional A·U and G·C base pairs in mRNA, G can also base pair to some extent with U.

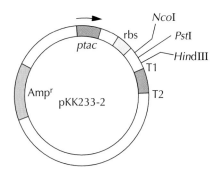

Figure 6.14 The expression vector pKK233-2. The plasmid pKK233-2 codes for the ampicillin resistance gene (Amp^r) as a selectable marker gene, the *tac* promoter (*ptac*), the *lacZ* ribosome binding site (rbs), three restriction endonuclease cloning sites (*Nco*I, *Pst*I, and *Hin*dIII), and two transcription termination sequences (T1 and T2). The arrow indicates the direction of transcription. The plasmid is not drawn to scale.

- An ATG start codon located 8 nucleotides downstream from the ribosome binding site
- The transcription terminators T1 and T2 from bacteriophage λ

The cloned gene is inserted into an *Nco*I, *Pst*I, or *Hin*dIII site which lies between the ribosome binding site and the transcription terminators. If the cloned gene is not in the same reading frame as the ATG start codon, minor adjustments can be made to correct the reading frame. After induction and transcription, the mRNA of a cloned gene is efficiently translated. However, since the nucleotide sequences that encode the amino acids at the N-terminal region of the target protein vary from one gene to another, it is not possible to design a vector that will eliminate the possibility of mRNA foldback in all instances. Therefore, no single optimized translational initiation region can guarantee a high rate of translation initiation for all cloned genes. Consequently, the expression vectors described above are merely starting points for the optimization of translation initiation.

A cellular incompatibility that can interfere with efficient translation occurs when a cloned gene utilizes codons that are rarely used by the host cell. In these cases, the host cell may not produce enough of the transfer RNAs (tRNAs) that recognize those rarely used codons, and the yield of the cloned gene protein is lower than expected. There is little that can be done experimentally to alleviate this problem. However, when the product of the cloned gene is important, it is worthwhile to chemically synthesize a new version of the gene containing codons that are more commonly used by the host cell (codon optimization).

Increasing Protein Stability

Under normal growing conditions, the half-lives of different proteins range from a few minutes to hours. The basis for this differential stability is due both to the extent of disulfide bond formation and to the presence of certain amino acids at the N terminus. For example, when certain amino acids were attached to the N terminus of β-galactosidase, the in vitro survival time of the modified protein varied from approximately 2 minutes to more than 20 hours (Table 6.4). Amino acid additions that extend the intrinsic survival of a protein can be readily incorporated into cloned genes. Often the presence of a single extra amino acid at the N-terminal end is sufficient to stabilize a target protein. Long-lived proteins can accumulate in cells and thereby in-

Table 6.4 Stability of β-galactosidase with certain amino acids added to its N terminus

Amino acid added	Half-life
Met, Ser, Ala	>20 h
Thr, Val, Gly	>20 h
Ile, Glu	>30 min
Tyr, Gln	~10 min
Pro	~7 min
Phe, Leu, Asp, Lys	~3 min
Arg	~2 min

Adapted from Bachmair et al., 1986, *Science* **234**: 179–186.

crease the yield of the product. This phenomenon is seen in both eukaryotes and prokaryotes.

In contrast to the amino acids at the N terminus, which increase the stability of a protein, there are internal amino acid sequences that make a protein more susceptible to proteolytic degradation. These regions of the protein, which are termed PEST sequences, are rich in proline (P), glutamic acid (E), serine (S), and threonine (T). They are often, but not always, flanked by clusters of positively charged amino acids and may act to mark proteins for degradation within the cell. It is possible that the stability of a protein would be enhanced if its PEST regions were altered by genetic manipulation. Such changes, of course, must not alter the function of the target protein.

Overcoming Oxygen Limitation

E. coli and most other microorganisms that are used to express foreign proteins generally require oxygen for growth. Unfortunately, oxygen has only a limited solubility in aqueous media. Thus, as the cell density of a growing culture increases, the cells rapidly deplete the growth medium of dissolved oxygen. Moreover, since oxygen dissolves into the growth medium very slowly, this problem is not alleviated when large amounts of air or oxygen are added to the growth medium, even with high stirring rates. When cells become oxygen limited, exponential growth slows and the culture rapidly enters a stationary phase during which cellular metabolism changes. One consequence of the stationary phase is the production by the host cells of proteases that can degrade the target protein. Modification of the fermentation configuration to optimize the aeration and agitation of cells and addition of chemicals to the growth medium to increase the solubility of oxygen have been tried in an effort to deal with the limited amount of dissolved oxygen. These efforts have met with only limited success.

Use of Protease-Deficient Host Strains

One possible way to stabilize foreign proteins produced in *E. coli* is to develop host strains that are deficient in the production of proteolytic enzymes. However, this is not as simple as it might appear. *E. coli* has at least 25 different proteases, and only a few of these have been studied at the genetic level. Moreover, these proteases are important for the degradation of abnormal or defective proteins, which is a housekeeping function that is necessary for the continued viability of the cells. In one study, strains with mutations in one or more protease genes were constructed. Generally, the strains that were most deficient in overall protease activity grew the slowest. Thus, decreasing protease activity caused cells to be debilitated. However, with *E. coli* strains with mutations in both the gene for the RNA polymerase sigma factor responsible for heat shock protein synthesis (*rpoH*) and the gene for a protease required for cell growth at high temperatures (*degP*), secreted proteins had a 36-fold greater specific activity than when they were produced in wild-type host cells. This apparent increase in activity reflects a decrease in the proteolytic degradation of these secreted proteins.

Bacterial Hemoglobin

Some strains of the *Vitreoscilla* bacterium, a gram-negative obligate aerobe, normally live in oxygen-poor environments such as stagnant ponds. To obtain a sufficient amount of oxygen for their growth and metabolism, these

organisms synthesize a hemoglobin-like molecule that binds oxygen from the environment and increases the level of available oxygen inside cells. When the gene for this protein was cloned and expressed in *E. coli*, the transformants displayed higher levels of protein synthesis of both cellular and recombinant proteins, more efficient proton pumping, a higher ATP production rate, and higher ATP contents, especially at low levels of oxygen in the growth medium, than nontransformed cells. For this strategy to be effective in different host cells, not only must the *Vitreoscilla* sp. hemoglobin gene be efficiently expressed, the host cells must also be able to synthesize the heme portion of the hemoglobin molecule. Once these conditions have been met, this strategy can be used to improve growth as well as foreign gene expression in a range of different industrially important bacteria, including *E. coli, Streptomyces lividans, Corynebacterium glutamicum,* and *Xanthomonas maltophilia.*

DNA Integration into the Host Chromosome

A plasmid imposes a metabolic load on the cell because energy is used for its replication and for the transcription and translation of the proteins that it encodes. For the most part, high-copy number plasmids impose a greater metabolic load than do low-copy-number plasmids. As a consequence, a fraction of the cell population often loses its plasmids during cell growth. Also, cells that lack plasmids generally grow faster than those that retain them, so plasmidless cells eventually dominate the culture. After a number of generations of cell growth, the loss of plasmid-containing cells diminishes the yield of the cloned gene product. Investigators have devised at least two methods of combating this loss. On a laboratory scale, plasmid-containing cells are maintained by growing the cells in the presence of either an antibiotic or an essential metabolite that enables only plasmid-carrying cells to thrive. But the addition of either antibiotics or metabolites to pilot plant or industrial-scale fermentations can be extremely costly. Similarly, for genetically engineered microorganisms that are designed to be released into the environment to remain both effective and environmentally safe, it is essential that the cloned DNA be retained and be neither easily lost nor transferred to other microorganisms. For these reasons, the introduction of cloned DNA directly into the chromosomal DNA of the host organism, a method bypassing the use of plasmids, can overcome the problem of plasmid loss. When DNA is part of the host chromosomal DNA, it is relatively stable and consequently can be maintained for many generations in the absence of selective agents.

When a cloned gene is integrated into the host chromosome, the chromosomal integration site must not be within an essential coding gene. To ensure this, the input DNA sequence must be targeted to a specific nonessential site within the chromosome. In addition, to ensure efficient production of the target protein, the input gene should be under the control of a regulatable promoter.

For integration of DNA into a chromosomal site, the input DNA must share some sequence similarity, usually at least 50 nucleotides, with the chromosomal DNA, and there must be a physical exchange (recombination) between the two DNA molecules. Briefly, a generalized protocol for DNA integration includes the following steps:

1. Identify the desired chromosomal integration site, i.e., a segment of DNA on the host chromosome that can be disrupted without affecting the normal functions of the cell.

2. Isolate and clone part or all of the chromosomal integration site.
3. Ligate a cloned gene and a regulatable promoter either into (Fig. 6.15A) or adjacent to (Fig. 6.15B) the cloned chromosomal integration site.
4. Transfer the chromosomal integration fragment–cloned gene con-

Figure 6.15 Two ways to integrate a cloned gene into a chromosomal site. **A.** On a plasmid, the cloned gene has been inserted in the middle of a cloned segment of DNA (ab) from the host chromosome. Homologous DNA pairing occurs between plasmid-borne DNA regions a and b and host chromosome DNA regions a′ and b′, respectively. A double-crossover event (X—X) results in the integration of the cloned gene. **B.** The cloned gene is inserted adjacent to the cloned DNA (c) from the host chromosome. Homologous DNA pairing occurs between plasmid DNA region c and host chromosome DNA region c′. A single recombination event (X) within the paired c-c′ DNA region results in the integration of the entire plasmid, including the cloned gene.

A

B

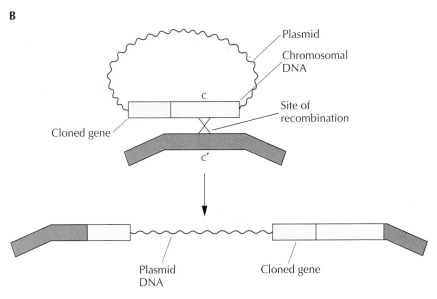

struct into the host cell as part of a plasmid that cannot replicate in the host cell.

5. Select and perpetuate host cells that express the cloned gene. Propagation of the cloned gene can occur only if it has been integrated into the chromosomal DNA of the host cell.

When a host cell is transformed with a nonreplicating plasmid that carries the cloned gene in the middle of a portion of the cloned chromosomal integration site, the DNA on the plasmid can base pair with identical sequences on the host chromosome (Fig. 6.15A). The integration can occur as a result of a host enzyme-catalyzed double crossover. Alternatively, a single crossover incorporates the entire input plasmid into the host chromosome (not shown). Similarly, insertion of the entire input plasmid DNA occurs when the cloned gene is inserted next to the cloned chromosomal integration site (Fig. 6.15B).

The effectiveness of integration of a cloned gene was examined in *B. subtilis*. The investigators constructed an *E. coli* plasmid that contained an α-amylase gene from *Bacillus amyloliquefaciens* which had been inserted into the middle of a DNA fragment from *B. subtilis* but could not replicate in *B. subtilis*. This construct, however, could transform *B. subtilis*. Transformants expressing α-amylase, an enzyme involved in the hydrolysis of starch, were recovered, indicating that the α-amylase gene had been integrated into the *B. subtilis* chromosome and was functioning. The selected transformants were resistant to ampicillin and chloramphenicol. Because both of these antibiotic resistance genes were on the input plasmid, this result indicated that a single recombination event must have occurred and caused the integration of the entire plasmid into the *B. subtilis* chromosomal DNA.

To increase the number of copies of the α-amylase gene that were present on the *B. subtilis* chromosome, the original transformants were grown in the presence of high levels of chloramphenicol. Only cells in which spontaneous DNA duplication of the integrated plasmid had occurred could survive under these conditions. The cells selected for a high level of chloramphenicol resistance were then assayed for α-amylase activity (Table 6.5). With this selection procedure, cells with up to nine copies of the α-amylase gene were identified. The level of enzyme activity expressed from the chromosomally integrated genes far exceeded the activity levels that occurred when the α-amylase gene was present on a multicopy (about 20 to 40 copies per cell) *B. subtilis* plasmid.

In one study, several copies of a foreign gene were inserted into different predetermined sites on the *B. subtilis* chromosome. To do this, a two-step procedure was used for each copy of the foreign gene to be inserted

Table 6.5 α-Amylase gene copy number and activity in *B. subtilis*

Copies/genome	Activity (U/mL of mid-log cells)
2	500
5	2,300
7	3,100
8	3,400
9	4,400
Multicopy plasmid	700

Adapted from Kallio et al., 1987, *Appl. Microbiol. Biotechnol.* **27**:64–71.

(Fig. 6.16). First, a selectable marker gene, such as an antibiotic resistance gene, was inserted into the middle of a defined but nonessential piece of *B. subtilis* chromosomal DNA on a plasmid vector which cannot replicate in *B. subtilis* (Fig. 6.16, step 1). Following transformation of *B. subtilis* with this construct, cells expressing the marker gene were selected. These transformants carry the selectable marker gene integrated into the *B. subtilis* chromosomal DNA. Second, the target gene with its transcriptional and translational control sequences was inserted into the middle of the defined but nonessential piece of *B. subtilis* chromosomal DNA that was introduced into the cell on a nonreplicatable plasmid (Fig. 6.16, step 2). Following transformation of *B. subtilis* with the target gene-plasmid construct, cells that no longer expressed the marker gene were selected after replica plating. These transformants carry the target gene and not the marker gene integrated into the *B. subtilis* chromosomal DNA. To integrate additional copies of the target gene into the host cell chromosomal DNA, the entire procedure was repeated several times with different defined but nonessential regions of *B. subtilis* chromosomal DNA.

Figure 6.16 Insertion of a foreign gene into a unique predetermined site on the *B. subtilis* chromosome. In step 1, a marker gene is integrated into the host cell chromosomal DNA by homologous recombination. In step 2, the marker gene is replaced by the target gene. The process may then be repeated with different nonessential regions of *B. subtilis* chromosomal DNA.

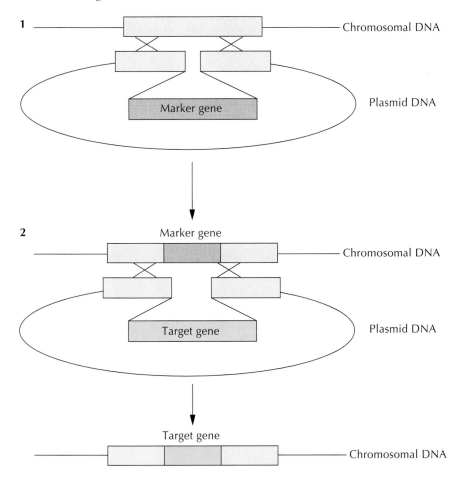

Increasing Secretion

The stability of a cloned gene protein in *E. coli* often depends on its cellular location. For example, recombinant proinsulin is approximately 10 times more stable if it is secreted (exported) into the periplasm (the space between the cytoplasmic and outer membranes) than if it remains in the cytoplasm. Moreover, proteins that are secreted into either the periplasm or the growth medium are easier to purify.

Normally, an amino acid sequence called the signal peptide (signal sequence, leader peptide), located at the N-terminal end of a protein, facilitates its export by enabling the protein to pass through the cell membrane. It is sometimes possible to engineer a protein for secretion by adding the DNA sequence encoding a signal peptide to the cloned gene. However, the presence of a signal peptide sequence does not guarantee a high rate of secretion. In addition, *E. coli* and other gram-negative microorganisms generally cannot secrete proteins into the surrounding medium because of the presence of an outer membrane. There are at least two solutions to this problem. The first one is to use gram-positive prokaryotes or eukaryotic cells, which both lack an outer membrane and can secrete proteins into the medium. The second solution entails the use of genetic engineering to make gram-negative bacteria that can secrete proteins into the medium.

When the fusion of a target gene to a DNA fragment encoding a signal peptide is ineffective in producing a secreted protein product, alternative strategies need to be employed. One approach that was found to be successful for the secretion of the protein interleukin-2 was the fusion of the interleukin-2 gene downstream from the gene for the entire precursor maltose binding protein, rather than just the maltose binding protein signal sequence, with DNA encoding the factor X_a recognition site peptide separating these two genes. When this genetic fusion, on a plasmid vector, was used to transform *E. coli* cells, a large fraction of the fusion protein was found localized in the host cell periplasm. Functional interleukin-2 could then be released from the fusion protein by digestion with factor X_a.

In one study, it was found that the secretion of many heterologous proteins in *E. coli* depended on the translational level of that protein. Foreign proteins that were translated most efficiently were not necessarily secreted to the greatest extent. Sometimes too high a level of translation of a foreign protein can overload the cell's secretion machinery and inhibit the secretion of that protein. Thus, if it is important that a particular protein be secreted, one way to ensure that this occurs most efficiently may be to lower the level of expression of that protein.

Some gram-negative bacteria are capable of secreting a bacteriocidal protein called a bacteriocin into the medium. A cascade mechanism is responsible for this specific secretion. The presence of a bacteriocin release protein activates phospholipase A in the bacterial inner membrane, and consequently both the inner and outer membranes are permeabilized. Some cytoplasmic and periplasmic proteins are released into the culture medium. Thus, by putting the bacteriocin release protein gene onto a plasmid under the control of a strong regulatable promoter, *E. coli* cells can be permeabilized at will. If the *E. coli* cells that carry the bacteriocin release protein gene are transformed with another plasmid carrying a cloned gene that has been fused to a secretion signal peptide sequence, and if it is under the same transcriptional regulatory control as the bacteriocin release protein gene, then the two genes can be induced simultaneously, with the cloned gene protein being secreted into the medium.

In many instances, when foreign proteins engineered for secretion are overproduced in *E. coli*, the precursor form is only partially processed, with about half of the secreted proteins retaining the leader peptide and the other half being fully processed to the mature form. This is probably the result of overloading some of the proteins involved in the secretion process. If this is the case, then it might be possible to increase the ratio of processed to unprocessed proteins by increasing the level of expression of some of the limiting components of the protein secretion pathway. This hypothesis was tested in a series of experiments in which a plasmid containing both the *prlA4* and *secE* genes, which encode major components of the molecular apparatus that physically moves proteins across the membrane, was introduced into *E. coli* host cells. Following this augmentation of the host cell secretory machinery, the fraction of the recombinant protein (the cytokine interleukin-6) that was secreted to the periplasm as the processed or mature form increased from about 50 to more than 90%.

Filamentous fungi that belong to the genus *Aspergillus* are major sources of industrial enzymes, partly because of their ability to secrete large amounts of protein into the growth medium. In one study, a human interferon gene was fused to a secretion signal peptide sequence and put under the control of the *Aspergillus nidulans* glucoamylase promoter sequence, which is inducible by starch. Induction of transformed *A. nidulans* cells by the addition of starch to the medium yielded about 1 mg of secreted human interferon per liter of culture medium, which was equivalent to approximately 5% of the total secreted protein. This study showed that the strategies that have been developed for the modulation of gene expression in *E. coli* can be applied to diverse biological systems.

Metabolic Load

The introduction and expression of foreign DNA in a host organism often change the metabolism of the organism in ways that may impair normal cellular functioning. This phenomenon, which is a multifaceted biological response, is due to a metabolic load (metabolic burden, metabolic drain) that is imposed upon the host by the foreign DNA. A metabolic load can occur as the result of a variety of conditions, including the following.

- Increasing plasmid copy number and/or size requires increasing amounts of cellular energy for plasmid replication and maintenance (Table 6.6).

Table 6.6 Effect of plasmid copy number on host cell growth rate

E. coli **HB101 with plasmid:**	**Plasmid copy number**	**Relative specific growth rate**
None	0	1.00
A	12	0.92
B	24	0.91
C	60	0.87
D	122	0.82
E	408	0.77

Adapted from Seo and Bailey, 1985, *Biotechnol. Bioeng.* **27:**1668–1674.

The different plasmids, designated A, B, C, D and E, encode only β-lactamase and are all the same size. The growth rates were normalized to the growth rate value for *E. coli* HB101 without a plasmid.

- The limited amount of dissolved oxygen in the growth medium is often insufficient for both host cell metabolism and plasmid maintenance and expression.
- Overproduction of foreign proteins may deplete the pools of certain aminoacyl-tRNAs (or even certain amino acids) and/or drain the host cell of its energy (in the form of ATP or GTP).
- When a foreign protein is overexpressed and then exported from the cytoplasm to either the cell membrane or the periplasm, it may "jam" export sites and thereby prevent the proper localization of essential host cell proteins.
- Host cells with unusual metabolic features such as a naturally high rate of respiration, e.g., *Azotobacter* spp., are more likely to be affected by these perturbations than are other host cells.
- The foreign protein per se may interfere with the functioning of the host cell, for example, by converting an important and needed metabolic intermediate into a compound that is irrelevant, or even toxic, to the cell.

A metabolic load can cause a variety of changes to the physiology and functioning of the host cell. One of the most commonly observed changes is a decrease in the rate of cell growth after introduction of foreign DNA. In such cases, plasmid-bearing cells grow more slowly than nontransformed cells (Table 6.6), often resulting in the loss of the recombinant plasmid. Sometimes, a metabolic load may result in plasmid-containing cells losing a portion of the plasmid DNA; in the presence of selective pressure, often all or part of the recombinant gene is deleted from the plasmid. Since cells growing in the presence of a metabolic load generally have a decreased level of energy available for a variety of cellular functions, the cell's energy-intensive metabolic processes such as nitrogen fixation and protein synthesis are invariably adversely affected by a metabolic load. A metabolic load may also lead to changes in the size and shape and increases in the amount of extracellular polysaccharide of a bacterial host cell. This extracellular carbohydrate causes the cells to stick together, making it more difficult to harvest them by cross-flow microfiltration procedures.

A metabolic load from the depletion of nutrients or "building blocks" such as amino acids can result from a high level of expression of a foreign protein that initiates a cellular stress response, including turning on the synthesis of cellular proteases so that the recombinant protein is rapidly degraded. Moreover, depletion of amino acid pools may occur as the consequence of the synthesis of highly expressed antibiotic resistance genes as well as the overexpression of the cloned target protein.

Translational errors occur in growing *E. coli* cells at a rate of about 2×10^{-4} to 2×10^{-3} errors per cell per generation. However, when a particular aminoacyl-tRNA becomes limiting, as is often the case when a foreign protein is overexpressed in *E. coli*, there is an increased probability that an incorrect amino acid will be inserted in place of the limiting amino acid. In addition, translational accuracy, which depends upon the availability of GTP as part of a proofreading mechanism, is likely to be further decreased as a consequence of a metabolic load from foreign protein overexpression. In one study, it was found that the high level expression of mouse epidermal growth factor in *E. coli* caused about 10 times the normal amount of incorrect amino acids to be incorporated into the recombinant protein. This level of errors diminishes the usefulness of the protein as a therapeutic agent,

since (1) the specific activity and stability of the protein may be significantly lower than expected and (2) the presence of incorrect amino acids may cause the protein to be immunogenic in humans.

A well-designed experiment can minimize the impact of the metabolic load, optimize the yield of the recombinant protein, and enhance the stability of the transformed host cell. For example, the extent of the metabolic load can be reduced by using a low-copy-number rather than a high-copy-number plasmid vector. An even better strategy is to avoid the use of plasmid vectors altogether and integrate the introduced foreign DNA directly into the chromosomal DNA of the host organism. In this case, plasmid instability will not be a problem. With an integrated cloned gene without the plasmid vector, the transformed host cell will not waste its resources synthesizing unwanted and unneeded antibiotic resistance marker gene products. Often, one of the major causes of a metabolic load is the high level of synthesis of the products of many antibiotic resistance marker genes carried along with the target gene on the plasmid vector. Chromosome integration is particularly important when a genetically engineered organism is destined to be released directly into the environment. The use of strong but regulatable promoters is also an effective means of reducing metabolic load. In this case, the fermentation is performed in two stages. During the first or growth stage the promoter controlling the transcription of the target gene is "turned off," while during the second or induction stage this promoter is "turned on."

When the codon usage of the foreign gene is different from the codon usage of the host organism, depletion of specific aminoacyl-tRNA pools may be avoided by either completely or partially synthesizing the target gene to better reflect the codon usage of the host organism. However, since this is not a simple procedure, such an approach is likely to be used in only a limited number of instances. This notwithstanding, in one study it was found that levels of the protein streptavidin were 10-fold higher in *E. coli* when expression was directed by a synthetic gene which had a G+C content of 54% than when it was directed by the natural gene with its G+C content of 69%.

Although it may at first seem counterintuitive, one way to increase the amount of foreign protein produced by the fermentation of a recombinant host cell is to accept a modest level of foreign gene expression—perhaps 5% of the total cell protein—and instead focus on attaining a high host cell density. An organism with a 5% foreign protein expression level and a low level of metabolic load that can be grown to a density of 40 g (dry weight) per liter is more efficient than an organism with a 15% expression level for the same protein if a cell density of only 10 g (dry weight) per liter is attained.

It is often difficult to achieve high levels of foreign gene expression and a high cell density at the same time because of the accumulation of harmful waste products, especially acetate, which inhibits both cell growth and protein production. One strategy for reducing acetate accumulation in rich medium without impairing cell growth entails decreasing the glucose uptake rate of the cells by adding methyl α-glucoside, a glucose analog, to the growing cells. The same effect has also been achieved by using an *E. coli* host cell which contained a mutation in *ptsG*, a gene encoding enzyme II in the glucose phosphotransferase system. A comparison of batch cultures of wild-type and *ptsG* mutant *E. coli* in rich medium, with both carrying a plasmid expressing β-galactosidase activity, the wild-type cells attained a density of approximately 10 g (dry weight) per liter while the mutant cells attained more than 15 g (dry weight) per liter. At the same time, the mutant

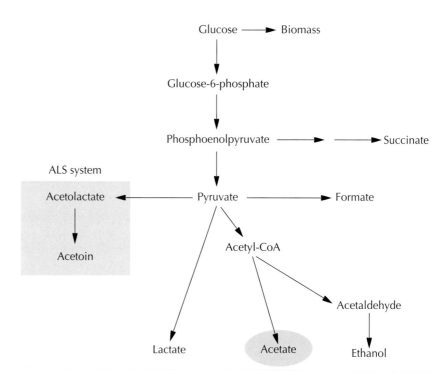

Figure 6.17 Schematic representation of the pathways for glucose metabolism in an *E. coli* strain that has been transformed with a plasmid carrying the genes for acetolactate synthase (ALS).

cells synthesized about 25% more β-galactosidase per gram (dry weight) of cells than the wild-type cells did. Overall, the *ptsG* mutant cells synthesized nearly twice as much β-galactosidase as the wild-type cells.

Since it is much easier and quicker to alter a particular host cell by genetic transformation than by mutagenesis and selection, an alternative means of reducing acetate production in cells was developed. This method includes introducing genes encoding the enzyme acetolactate synthase into *E. coli* host cells. This enzyme catalyzes the formation of acetolactate from pyruvate, thereby decreasing the flux through acetyl coenzyme A to acetate (Fig. 6.17). In practice, the acetolactate synthase genes are introduced into the cell on one plasmid, while the target gene is introduced on a second plasmid from a separate incompatibility group. The cells that were transformed with the acetolactate synthase genes produced much less acetate than the nontransformed cells and instead synthesized acetoin, which is approximately 50-fold less toxic to cells than acetate.

SUMMARY

The production of a protein from a gene requires that the gene be properly transcribed and then that the mRNA be translated. In prokaryotes, a promoter region is necessary for the initiation of transcription of the gene at the correct nucleotide site, and a sequence at the end of the gene (a terminator) is essential for the cessation of transcription. Cloned genes often lack these signals. Consequently, for expression of a cloned gene in a prokaryotic host cell, these signals must be in the correct locations. Moreover, many biotechnology applications require the production of large

amounts of protein, so it is necessary to use a promoter that supports transcription at a high level (strong promoter) and that is compatible with the RNA polymerase of the host cell. But continuous transcription of a cloned gene drains the energy reserves of the host cell, so it is also necessary to use a promoter system whose activity can be regulated either by the addition of a low-molecular-weight compound or by changing the growth temperature.

Efficient protein synthesis from a gene depends on specific sequences in its mRNA, and often, for cloned genes, other manipulations are required to ensure stability of the protein and, if necessary, its secretion. As part of the gene engineering process, a ribosome binding site is placed in the DNA segment that precedes the translation initiation site, which also may need to be added. Finally, a termination codon may be placed at the end of the cloned gene to ensure that translation stops at the correct amino acid. If secretion of the protein is desired, the DNA sequence preceding the cloned gene should include a signal sequence in the same reading frame as the target gene.

Lack of stability of the protein that is encoded by a cloned gene is another complication that is often encountered. For example, the recombinant protein can be degraded by proteolytic enzymes of the host cell. One strategy to overcome this problem is to alter the cloned gene so that it encodes one or more additional amino acids at its N terminus. In this form, the recombinant protein is no longer rapidly degraded. In addition, the amino acids that are added to the recombinant protein can sometimes be used for purifying the fusion protein by, for example, immunoaffinity column chromatography. In these cases, the junction point of a fusion protein is usually designed to be cleaved in vitro either chemically or enzymatically.

Most microorganisms that are used to express foreign proteins require oxygen for growth. However, oxygen has only a limited solubility in water and is rapidly depleted from the growth medium of actively growing cultures, especially when the cultures attain a high cell density. Attempts to deal with the limited amount of dissolved oxygen available for cell growth and maintenance have included (1) utilizing microbial host strains deficient in the production of some proteolytic enzymes and (2) introducing the gene for the *Vitreoscilla* sp. hemoglobin that binds oxygen from the environment, creating a higher level of intracellular oxygen and thereby causing increases in both host and foreign protein synthesis.

By increasing the numbers of copies of the cloned gene, the production of the protein is increased, and under these conditions, the stability of the product is enhanced. However, during scale-up of plasmid-based systems, plasmid–DNA insert constructs can be lost. This plasmid instability is undesirable for commercial systems. To overcome this problem, researchers have developed protocols for integrating a cloned gene into a chromosomal site of the host organism. Under these conditions, the gene is maintained stably as part of the DNA of the host organism.

The introduction and expression of foreign DNA in a host organism often change the metabolism of the organism and thereby impair its normal functioning. This phenomenon is called a metabolic load. A variety of strategies have been developed to minimize the extent of the perturbations caused by a metabolic load and at the same time optimize the yield of the target protein and the stability of the transformed cell.

Expression systems are quite diverse, and investigators tend to develop the set of conditions that optimizes the production of each specific protein

in a particular host cell. Notwithstanding the differences in detail, the same fundamental strategies are used to create a variety of different expression systems.

References

Amann, E., and J. Brosius. 1985. "ATG vectors" for regulated high-level expression of cloned genes in *Escherichia coli. Gene* **40:**183–190.

Aristidou, A. A., K. Y. San, and G. N. Bennett. 1995. Metabolic engineering of *Escherichia coli* to enhance recombinant protein production through acetate reduction. *Biotechnol. Prog.* **11:**475–478.

Bachmair, A., D. Finley, and A. Varshavsky. 1986. In vivo half-life of a protein is a function of its amino-terminal residue. *Science* **234:**179–186.

Bagdasarian, M. M., E. Amann, R. Lurz, B. Ruckert, and M. Bagdasarian. 1983. Activity of the hybrid *trp-lac* (*tac*) promoter of *Escherichia coli* in *Pseudomonas putida.* Construction of broad-host-range, controlled-expression vectors. *Gene* **26:**273–282.

Baker, R. T., and A. Varshavsky. 1991. Inhibition of the N-end rule pathway in living cells. *Proc. Natl. Acad. Sci. USA* **88:**1090–1094.

Chaitan, K., J. E. Curtis, J. DeModena, U. Rinas, and J. E. Bailey. 1990. Expression of intracellular hemoglobin improves protein synthesis in oxygen-limited *Escherichia coli. Bio/Technology* **8:**849–853.

Chou, C. H., G. N. Bennett, and K. Y. San. 1994. Effect of modified glucose uptake using genetic engineering techniques on high-level recombinant protein production in *Escherichia coli* dense cultures. *Biotechnol. Bioeng.* **44:**952–960.

deBoer, H. A., L. J. Comstock, and M. Vasser. 1983. The *tac* promoter: a functional hybrid derived from the *trp* and *lac* promoters. *Proc. Natl. Acad. Sci. USA* **80:**21–25.

Donovan, R. S., C. W. Robinson, and B. R. Glick. 1996. Optimizing inducer and culture conditions for expression of foreign proteins under the control of the *lac* promoter. *J. Ind. Microbiol.* **16:**145–154.

Ernst, J. F. 1988. Codon usage and gene expression. *Trends Biotechnol.* **6:**196–199.

Friesen, J. D., and G. An. 1983. Expression vehicles used in recombinant DNA technology. *Biotechnol. Adv.* **1:**205–227.

Geisow, M. J. 1991. Both bane and blessing—inclusion bodies. *Trends Biotechnol.* **9:**368–369.

Gentz, R., A. Langner, A. C. Y. Chang, S. N. Cohen, and H. Bujard. 1981. Cloning and analysis of strong promoters is made possible by the downstream placement of a RNA termination signal. *Proc. Natl. Acad. Sci. USA* **78:**4936–4940.

Glick, B. R. 1995. Metabolic load and heterologous gene expression. *Biotechnol. Adv.* **13:**247–261.

Glick, B. R., and G. K. Whitney. 1987. Factors affecting the expression of foreign proteins in *Escherichia coli. J. Ind. Microbiol.* **1:**277–282.

Goldstein, M. A., and R. H. Doi. 1995. Prokaryotic promoters in biotechnology, *In* p. 105–128. M. R. El-Gewely (ed.), *Biotechnology Annual Review,* vol. 1. Elsevier Science B.V., Amsterdam, The Netherlands.

Gwynne, D. I., F. P. Buxton, S. A. Williams, S. Garven, and R. W. Davies. 1987. Genetically engineered secretion of active human interferon and a bacterial endoglucanase from *Aspergillus nidulans. Bio/Technology* **5:**713–719.

Halfmann, G., H. Brailly, A. Bernadac, F. A. Montero-Julian, C. Lazdunski, and D. Baty. 1993. Targeting of interleukin-2 to the periplasm of *Escherichia coli. J. Gen. Microbiol.* **139:**2465–2473.

Hartley, J. L., and T. J. Gregori. 1981. Cloning multiple copies of a DNA gene. *Gene* **13:**347–353.

Hockney, R. C. 1994. Recent developments in heterologous protein production in *Escherichia coli. Trends Biotechnol.* **12:**456–463.

Hopp, T. P., K. S. Prickett, V. L. Price, R. T. Libby, C. J. March, D. P. Cerretti, D. L. Urdal, and P. J. Conlon. 1988. A short polypeptide marker sequence useful for re-combinant protein identification and purification. *Bio/Technology* **6**:1204–1210.

Hsiung, H. M., A. Cantrell, J. Luirink, B. Oudega, A. J. Veros, and G. W. Becker. 1989. Use of bacteriocin release protein in *E. coli* for excretion of human growth hormone into the culture medium. *Bio/Technology* **7**:267–271.

Jay, G., G. Khoury, A. K. Seth, and E. Jay. 1981. Construction of a general vector for efficient expression of mammalian proteins in bacteria: use of a synthetic ribosome binding site. *Proc. Natl. Acad. Sci. USA* **78**:5543–5548.

Jespers, L. S., J. H. Messens, A. De Keyser, D. Eeckhout, I. Van Den Brande, Y. G. Gansemans, M. J. Lauwereys, G. P. Vlasuk, and P. E. Stanssens. 1995. Surface expression and ligand-based selection of cDNAs fused to filamentous phage gene VI. *Bio/Technology* **13**:378–382.

Kallio, P., A. Palva, and I. Palva. 1987. Enhancement of α-amylase production by integrating and amplifying the α-amylase gene of *Bacillus amyloliquefaciens* in the genome of *Bacillus subtilis*. *Appl. Microbiol. Biotechnol.* **27**:64–71.

Kiel, J. A. K. W., A. M. ten Berge, P. Borger, and G. Venema. 1995. A general method for the consecutive integration of single copies of a heterologous gene at multiple locations in the *Bacillus subtilis* chromosome by replacement recombination. *Appl. Environ. Microbiol.* **61**:4244–4250.

Kolata, G. 1986. New rule proposed for protein degradation. *Science* **234**:151–153.

Kolowsky, K. S., J. G. K. Williams, and A. A. Szalay. 1984. Length of foreign DNA in chimeric plasmids determines the efficiency of its integration into the chromosome of the cyanobacterium *Synechococcus* R2. *Gene* **27**:289–299.

Kozlowski, M., A. VanBrunschot, G. Nash, and R. W. Davies. 1988. A novel vector allowing the expression of genes in a wide range of gram-negative bacteria. *Gene* **70**:199–204.

Labes, M., A. Puhler, and R. Simon. 1990. A new family of RSF1010-derived expression and *lac*-fusion broad-host-range vectors for gram-negative bacteria. *Gene* **89**:37–46.

Lee, N., J. Cozzitorto, N. Wainwright, and D. Testa. 1984. Cloning with tandem gene systems for high level gene expression. *Nucleic Acids Res.* **12**:6797–6812.

Leemans, R., E. Remaut, and W. Fiers. 1987. Broad-host-range expression vector based on the *p*ᴸ promoter of coliphage 1: regulated synthesis of human interleukin 2 in *Erwinia* and *Serratia* species. *J. Bacteriol.* **169**:1899–1904.

Little, M., F. Breitling, B. Micheel, and S. Dübel. 1994. Surface display of antibodies. *Biotechnol. Adv.* **12**:539–555.

Liu, S. C., D. A. Webster, M. L. Wei, and B. C. Stark. 1996. Genetic engineering to contain the *Vitreoscilla* hemoglobin gene enhances degradation of benzoic acid by *Xanthomonas maltophilia*. *Biotechnol. Bioeng.* **49**:101–105.

Looman, A. C., J. Bodlaender, M. de Gruyter, A. Vogelaar, and P. H. van Knippenberg. 1986. Secondary structure as primary determinant of the efficiency of ribosomal binding sites in *Escherichia coli*. *Nucleic Acids Res.* **14**:5481–5497.

Magnolo, S. K., D. L. Leenutaphong, J. A. DeModena, J. E. Curtis, J. E. Bailey, J. L. Galazzo, and D. E. Hughes. 1991. Actinorhodin production by *Streptomyces coelicolor* and growth of *Streptomyces lividans* are improved by the expression of a bacterial hemoglobin. *Bio/Technology* **9**:473–476.

Meerman, H. J., and G. Georgiou. 1994. Construction and characterization of a set of *E. coli* strains deficient in all known loci affecting the proteolytic stability of secreted recombinant proteins. *Bio/Technology* **12**:1107–1110.

Mieschendahl, M., and B. Müller-Hill. 1985. F'-coded, temperature-sensitive λ *cI*₈₅₇ repressor gene for each construction and regulation of λ promoter-dependent expression systems. *J. Bacteriol.* **164**:1366–1369.

Mieschendahl, M., T. Petris, and U. Hanggi. 1986. A novel prophage independent *trp* regulated λ *p*ᴸ expression system. *Bio/Technology* **4**:802–808.

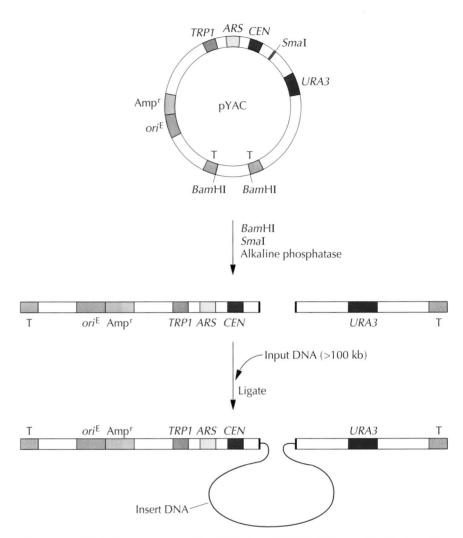

Figure 7.3 YAC cloning system. The YAC plasmid (pYAC) has an *E. coli* selectable marker gene (Amp^r); an origin of replication that functions in *E. coli* (*ori*^E); and yeast DNA sequences, including *URA3*, *CEN*, *TRP1*, and *ARS*. (*CEN* provides centromere function; *ARS* is a yeast autonomous-replicating sequence, which is equivalent to a yeast origin of replication; *URA3* is a gene in the uracil biosynthesis pathway; and *TRP1* is a gene in the tryptophan biosynthesis pathway.) T sites are yeast chromosome telomeric sequences. *Sma*I is the cloning insertion site. pYAC is first treated with *Sma*I, *Bam*HI, and alkaline phosphatase and then ligated with size-fractionated (100 kb) input DNA. The final construct carries cloned DNA and can be stably maintained in a Ura^− Trp^− host yeast cell.

man enzyme superoxide dismutase illustrates the process of heterologous gene expression in *S. cerevisiae*.

Superoxide anion is a by-product of oxygen utilization in aerobic organisms. In humans, this anion helps both to stimulate the inflammatory response of phagocytes and to direct leukocytes to the site of infection. However, an excess of this molecule and its derivatives can cause cellular damage. To minimize these potentially cytotoxic effects, the naturally occurring cytoplasmic enzyme Cu/Zn superoxide dismutase (Cu/Zn-SOD) scavenges the superoxide radical and combines it with a hydrogen ion to form hydrogen peroxide, which in turn is degraded by catalase or peroxidase. Superoxide anion is also produced when blood is allowed to reenter an organ (reperfusion) after it has been

deprived of blood during a surgical procedure. To prevent this source of superoxide anion damage, investigators have proposed that Cu/Zn-SOD be administered to an organ as it is being reperfused. In addition, Cu/Zn-SOD has the potential to act as a therapeutic agent against inflammatory diseases, such as osteoarthritis, rheumatoid arthritis, scleroderma, and ankylosing spondylitis. For both of these uses, an authentic human form of Cu/Zn-SOD is preferred, to avoid any adverse immunological responses that might result from using an enzyme from another species.

Initially, a cDNA for human Cu/Zn-SOD was cloned into an *E. coli* expression system. But the *E. coli* host cells only removed the initiator N-terminal methionine from the Cu/Zn-SOD protein, just as they do for all the proteins they produce. Unfortunately, the next amino acid (alanine) was not acetylated, as it is in human cells. So the researchers cloned the human Cu/Zn-SOD cDNA into a yeast episomal vector in an effort to obtain an authentic enzyme. Yeast cells are not particularly effective at removing introns, so cDNA sequences or chemically synthesized genes must be used for encoding specific gene products. The human Cu/Zn-SOD cDNA–yeast vector (Fig 7.4) contained (1) a yeast gene for leucine biosynthesis (*LEU2*); (2) a 2μm plasmid DNA sequence, which provides a yeast origin of DNA replication and permits this plasmid to replicate in yeast cells; (3) both an *E. coli* selectable marker gene (for ampicillin resistance [Ampr]) and an origin of replication that functions in *E. coli* and allows the routine genetic manipulation involved in the construction of the plasmid to be done conveniently in *E. coli*; and (4) the human Cu/Zn-SOD cDNA cloned between the promoter region of the yeast glyceraldehyde phosphate dehydrogenase gene (*GAPDp*) and a sequence containing the signals for transcription termination and polyadenylation of mRNA from the same gene (*GAPDt*).

A leucine-defective (LEU2$^-$) yeast strain was transformed with this vector, and the cells were plated onto medium that lacked leucine. Only cells with the functional *LEU2* gene, which was supplied by the vector, could grow under these conditions. The *GAPD* promoter is not regulated; rather, it is transcribed continuously during cell growth. Thus, human Cu/Zn-SOD cDNA is synthesized throughout the growth period (constitutively). In this experiment, the yeast cells produced high levels of intracel-

Figure 7.4 *S. cerevisiae* expression vector. The cDNA for human Cu/Zn-SOD has been cloned between the promoter (*GAPDp*) and termination–polyadenylation sequence (*GAPDt*) of the *S. cerevisiae* glyceraldehyde phosphate dehydrogenase gene. The *LEU2* gene, cloned into the middle of the yeast 2μm plasmid DNA, encodes an enzyme of the leucine biosynthesis pathway. The yeast origin of replication is included in the 2μm plasmid sequences. The ampicillin resistance gene (Ampr) and the *E. coli* origin of replication (*ori*E) are derived from plasmid pBR322.

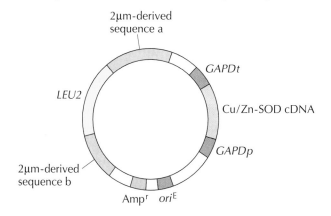

lular Cu/Zn-SOD which, like the authentic protein from human cells, was acetylated on the amino group of the amino terminal alanine residue.

Secretion of Heterologous Proteins by *S. cerevisiae*

In yeast, only secreted proteins are glycosylated, so a secretion system must be used for recombinant proteins that require either O-linked or N-linked sugars for biologic activity. To facilitate protein secretion in yeast, the pre-pro-α-factor, or leader peptide (signal peptide), coding sequence of the yeast mating type factor α1 gene is cloned in front (just upstream) of the cDNA that encodes the desired protein. This gene fusion creates a protein that can be efficiently secreted by yeast.

During the exporting process, disulfide bond formation, proteolytic cleavage, and posttranslational modifications occur; in many cases, an active protein is eventually released to the extracellular environment. The leader peptide enables a protein to pass through the cytoplasmic membrane and be secreted. During this process, the leader peptide is removed by a yeast endoprotease that recognizes the dipeptide Lys-Arg. The Lys-Arg codons are therefore positioned immediately upstream from the cDNA sequence so that, following removal of the leader peptide, the target protein will have the correct amino acid residue at its N terminus.

A properly processed and active form of the protein hirudin was synthesized and secreted by an *S. cerevisiae* strain containing an episomal expression vector that included prepro-α-factor. The gene for hirudin came from an invertebrate, the leech *Hirudo medicinalis*. This protein is a powerful blood anticoagulant that does not have immunological side effects in humans. The availability of large quantities of active hirudin has facilitated the testing of its efficacy in removing venous blood clots and alleviating other thrombotic conditions. In one clinical trial with 12,142 patients, of whom 4,131 had heart disease, recombinant hirudin showed a small advantage as a treatment over heparin. However, because of the higher cost of recombinant hirudin, the slight therapeutic benefit may not warrant its adoption as a standard treatment for acute coronary syndromes.

Additional strategies have been devised to enhance the secretion of recombinant protein by *S. cerevisiae* strains. For example, it was of interest to determine if the overproduction of a naturally occurring secretory system enzyme such as protein disulfide isomerase, which promotes the proper folding of proteins during secretion, would increase the release of a recombinant protein. The constitutive glyceraldehyde phosphate dehydrogenase promoter and a transcription terminator sequence were added to the ends of a cloned yeast protein disulfide isomerase gene, and this construct was integrated into a chromosome site. The modified strain showed a 16-fold increase in protein disulfide isomerase production compared to the wild-type strain. Then an extrachromosomal expression vector carrying the gene for human platelet-derived growth factor B was transformed into the protein disulfide isomerase-overproducing strain. The level of secreted human platelet-derived growth factor B produced by this strain was 10-fold higher than that secreted by a strain that synthesized the normal amount of protein disulfide isomerase. The overproduction of protein disulfide isomerase specifically increases the secretion of proteins with disulfide bonds. The manipulation of other proteins of the yeast secretory system may enhance the release of recombinant proteins with different folding requirements.

Other Yeast Expression Systems

Recombinant protein expression in *S. cerevisiae* has been successful with many proteins. But despite a few high yields, expression generally has been low. In addition, there are other limitations:

- During scale-up, plasmid loss is a frequent occurrence, even when inducible promoters are used.
- The heterologous protein is often hyperglycosylated, containing more than 100 mannose residues in each N-linked oligosaccharide side chain. By contrast, naturally occurring high-mannose oligosaccharide side chains have only 8 to 13 mannose residues. The extra mannose units may alter the product's biological activity or change its immunogenicity.
- In a number of trials, proteins that were designed to be secreted were actually retained within the periplasmic space, a situation that makes purification more difficult.

Therefore, researchers have been examining other yeast species and eukaryotic systems that can act as host cells for the production of heterologous proteins. The attributes that are being sought include the availability of suitable vectors, clonable expression systems with host-specific transcriptional and translational regulatory sequences, the capability for transformation, high protein yields, and the ability of the host organism to grow under industrial conditions. Candidate yeasts that are being considered as alternatives to *S. cerevisiae* are *Kluyveromyces lactis*, which has been used commercially for the production of lactase (*β*-galactosidase); *Schizosaccharomyces pombe*, a yeast that reproduces by fission rather than by budding; *Yarrowia lipolytica*, which uses alkanes as a growth substrate; and *Pichia pastoris* and *Hansenula polymorpha*, which can utilize methanol as the sole source of carbon and energy.

Expression of Hepatitis B Virus Surface Antigen

The methylotrophic yeast *P. pastoris* can be grown easily and economically in large bioreactors. The development of *P. pastoris* as a host organism for the production of heterologous proteins would provide a means for achieving high yields of fully active products. To test whether this approach has merit, an integrating vector system for the production of the hepatitis B virus surface antigen (HBsAg) was developed. First, the HBsAg sequence was cloned between the promoter region of alcohol oxidase gene 1 (*AOX1p*) and the termination and polyadenylation signal region (*AOX1t*) of the same gene (Fig. 7.5). The *AOX1* gene is regulated by methanol in *P. pastoris*. In the presence of methanol, alcohol oxidase can represent as much as 30% of the total cellular protein because of the strength of the *AOX1* promoter. In the absence of methanol, no alcohol oxidase is synthesized.

The complete plasmid (Fig. 7.5) developed for this study has the *AOX1p-HBsAg-AOX1t* unit; an origin of replication that functions in *P. pastoris*; a DNA fragment containing the *E. coli* pBR322 origin of replication and an *E. coli* selectable marker; a segment of DNA that lies downstream from the *AOX1t* sequence (3'-AOX1) and facilitates integration of input DNA into a specific chromosomal site; and a functional histidinol dehydrogenase gene (*HIS4*), which encodes an enzyme that is required for the synthesis of the amino acid histidine. The pBR322 sequence enables the vector to be

Figure 7.5 *P. pastoris* integrating expression vector. The *HBsAg* gene is cloned between the promoter (*AOX1p*) and termination-polyadenylation sequence (*AOX1t*) of the *P. pastoris* alcohol oxidase 1 gene. The *HIS4* gene encodes histidinol dehydrogenase, which is an enzyme in the histidine biosynthesis pathway. An origin of replication from *P. pastoris* (*ori*pp) is included, as are both the ampicillin resistance gene (Amp^r) and an origin of replication (*ori*E) that function in *E. coli*. The segment marked 3'-AOX1 is a piece of DNA from the 3' end of the alcohol oxidase 1 gene of *P. pastoris*. The joined right-angled arrows indicate the DNA region that will be integrated into the *P. pastoris* genome.

maintained in *E. coli* so that cloning steps can be carried out easily and, when required, large amounts of vector DNA can be prepared.

To avoid problems due to plasmid instability, the strategy in this experiment was to integrate the *AOX1p-HBsAg-AOX1t* unit into the genome of *P. pastoris*. To do this, a histidinol dehydrogenase-deficient (HIS4⁻) strain of *P. pastoris* was transformed with the segment of the vector that contained the *AOX1p-HBsAg-AOX1t, HIS4,* and 3′-AOX1 sequences (Fig. 7.5). A double crossover event between the *AOX1p* and 3′-AOX1 sequences of the input DNA and complementary sequences in the chromosomal DNA led to the integration of the *AOX1p-HBsAg-AOX1t* and *HIS4* sequences into the genome, with the concomitant loss of the chromosomal *AOX1* gene (Fig. 7.6). Cells that have an integrated form of the *HIS4* gene can grow in medium lacking histidine. A second mode of selection is slow growth in the presence of methanol, because a second but less efficient gene, *AOX2*, is the only one that is present and functional after the loss of the *AOX1* gene as a result of the double crossover event. Thus, cells that have integrated the input DNA fragment and lost the *AOX1* gene will be histidine independent and grow slowly on methanol.

In the presence of methanol, which activates the *AOX1* promoter, a clone carrying an integrated DNA unit synthesized large quantities of authentic HBsAg, which was retained in the cytoplasm. This protein product was assembled into the same multisubunit complex that occurs in hepatitis B virus-infected human cells and was capable of eliciting neutralizing antibodies against hepatitis B virus. When this clone was grown in 240-liter batch cultures, the yield was equivalent to approximately 9×10^6 doses of

Figure 7.6 Integration of part of an expression vector into the alcohol oxidase 1 gene of *P. pastoris*. The double crossover event occurs within the *AOX1p* and 3′-AOX1 DNA segments (shown at the top). This event results in the integration of the input DNA into the genomic DNA and the loss of most of the alcohol oxidase 1 gene (*AOX1*) from the host chromosome (shown at the bottom). The *HIS4* gene product enables cells with integrated DNA to grow on medium lacking histidine. In the presence of methanol, the *AOX1p* region drives the transcription of the *HBsAg* gene. The *AOX1t* segment provides transcription termination and polyadenylation signals for the *HBsAg* gene.

vaccine. Furthermore, the genetic construct was stable and remained unaltered after 200 hours of culture in the presence of methanol.

Expression of Bovine Lysozyme C2

The ability of *P. pastoris* to secrete a heterologous protein was examined in a system using bovine lysozyme C2 cDNA, which encodes the complete protein and its own leader peptide. Bovine lysozyme is a stomach enzyme that attacks bacterial cell walls, is resistant to proteases, and has a narrow pH range for activity. These properties may enable bovine lysozyme to be used as an additive to animal feed stocks to enhance digestion by cows and other ruminants or to be used with organisms that are given bacteria as a food source.

The vector constructed for this secretion study was identical to the *AOX1p-HBsAg-AOX1t* vector described above, except that the lysozyme cDNA sequence replaced the *HBsAg* coding sequence. In this experiment, the entire plasmid was integrated into a defective genomic copy of the *HIS4* gene of *P. pastoris*. The product of this integration had a bovine lysozyme cDNA expression unit flanked by one functional (*HIS4*) and one defective (*HIS4⁻*) histidinol dehydrogenase gene (Fig. 7.7). The bovine lysozyme precursor was accurately processed by *P. pastoris* and secreted into the medium. The secreted protein had a specific activity equal to that of the au-

Figure 7.7 Integration of an expression vector into the defective chromosomal *HIS4⁻* gene of *P. pastoris*. The input DNA is a plasmid that is first introduced by transformation into a histidine-requiring *P. pastoris* strain. A single crossover within the *HIS4* gene of the plasmid and the *HIS4⁻* gene of the host cell results in the integration of the entire plasmid, which then is flanked by the functional *HIS4* gene and the defective *HIS4⁻* gene. The letters *p*, L, and *t* denote the *AOX1* promoter sequence, bovine lysozyme C2 cDNA, and *AOX1* termination–polyadenylation signal sequence, respectively. The black bar in the *HIS4⁻* gene represents the defective sequence.

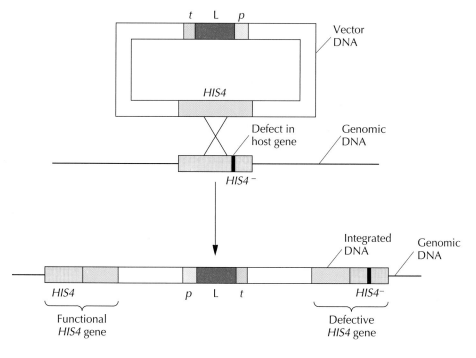

thentic enzyme. With a 200-hour continuous, 10-liter, high-cell-density fermentation system, about 20 g of lysozyme was produced.

Authentic heterologous proteins have been generated in other yeasts. For example, the cDNAs for the α and β globin chains of human hemoglobin A were each cloned between the methanol oxidase promoter (*MOXp*) and transcription terminator (*MOXt*) sequences of *Hansenula polymorpha* and placed in tandem in an expression vector. After 40 generations, an isolate with a fortuitously integrated component from the initial vector was studied further and found to produce functional hemoglobin A that had the correct tetrameric organization of two α and two β globin chains ($\alpha_2\beta_2$). Also, an expression vector for *S. pombe* with mammalian promoters driving both the selectable marker gene and the cloned human gene produced large amounts of recombinant proteins from a number of different human genes.

In sum, yeast expression systems have begun to play an important role in the production of heterologous proteins for research, industrial, and medical applications. However, experience has shown that no one system will be able to produce an authentic protein from every potential gene. For this and other reasons, gene expression systems that use insect and mammalian cells have been developed.

Cultured Insect Cell Expression Systems

Baculoviruses exclusively infect invertebrates, including many insect species. During the infection cycle, two forms of baculovirus are produced. One form consists of single virions that are released by an infected host cell, generally a midgut cell, and are capable of infecting other midgut cells. The second form is made up of a number of virions trapped (occluded) in a protein matrix. The protein of this matrix is called polyhedrin, and a "package" of occluded virions is a polyhedron. Many of these packages are released into the environment after cell lysis and death of a single host organism. A polyhedron protects the virions from being inactivated by environmental agents. Upon subsequent ingestion by a susceptible host, the polyhedrin protein is solubilized and the virions are capable of establishing an infection cycle. During the late stages of the infection cycle, the polyhedrin protein is synthesized in massive quantities. That synthesis is initiated about 36 to 48 hours after infection and continues for 4 to 5 days, until the infected cells lyse and the host organism dies.

The promoter for the polyhedrin gene is exceptionally strong, but the viral reproduction cycle does not depend on the presence of the polyhedrin gene. Consequently, it was reasoned that replacement of the polyhedrin gene with a gene for a heterologous protein, followed by infection of cultured insect cells with the genetically engineered baculovirus, would result in the production of large amounts of a heterologous protein that, because of the similarity of posttranslational modification systems between insects and mammals, would mimic closely, if not precisely, the authentic form of the target protein. On the basis of these premises, investigators developed baculovirus as an expression vector for both mammalian proteins and animal virus proteins.

The specific baculovirus that has been used extensively as an expression vector is the *Autographa californica* multiple nuclear polyhedrosis virus (AcMNPV). *A. californica* (the alfalfa looper) and over 30 other insect species can be infected by AcMNPV. This virus also grows well on many insect cell

Figure 7.8 Organization of the expression unit of a baculovirus (AcMNPV) transfer vector. The gene that is to be expressed will be inserted into the cloning site (cs) that lies between the polyhedrin gene promoter (*Pp*) and polyhedrin gene termination (*Pt*) sequences. AcMNPV DNA upstream from the polyhedrin promoter (5′-AcMNPV DNA) and downstream from the polyhedrin termination sequence (3′-AcMNPV DNA) are included in the vector to provide sequences that will enable the expression unit to be integrated via homologous recombination into AcMNPV DNA in insect cells.

lines. The most commonly used cell line for genetically engineered AcM-NPV is derived from the fall armyworm, *Spodoptera frugiperda*. In these cells, the polyhedrin promoter is exceptionally active, and during infections with wild-type baculovirus, high levels of polyhedrin are synthesized.

Baculovirus Expression Vector System

The first step in the production of a recombinant baculovirus AcMNPV is to create a transfer vector. The transfer vector is an *E. coli*-based plasmid that carries a segment of DNA from AcMNPV (Fig. 7.8) consisting of (1) the polyhedrin promoter region and an adjacent portion of upstream AcMNPV DNA, which provides a region for homologous recombination with AcM-NPV; (2) a cloning site for the input DNA; and (3) the polyhedrin termination and polyadenylation signal regions and an adjacent portion of downstream AcMNPV DNA, which provides a second region for homologous recombination with AcMNPV (Fig. 7.8). The coding region for the polyhedrin gene has been deleted from this block of DNA. A gene of interest is cloned between the polyhedrin promoter and termination sequences, and the construct is propagated in *E. coli*.

Next, insect cells in culture that have been transfected with AcMNPV DNA are transfected with a transfer vector carrying a cloned gene. Within some of the doubly transfected cells, a double crossover event occurs, and the cloned gene with polyhedrin promoter and termination regions then becomes integrated into the AcMNPV DNA (Fig. 7.9) with the concomitant loss of the polyhedrin gene. Virions lacking the polyhedrin gene produce distinctive zones of cell lysis (occlusion-negative plaques) from which recombinant baculovirus can be isolated.

The visual identification of occlusion-negative plaques is tedious and subjective. Consequently, DNA hybridization or a polymerase chain reaction (PCR) assay can be used to detect recombinant baculovirus. Moreover, if the *E. coli lacZ* gene, which encodes β-galactosidase, is put under the control of a baculovirus promoter that is turned on during the early to late stages of the lytic cycle and this construct is made part of the DNA unit that is incorporated into the AcMNPV genome, recombinant plaques turn blue when a chromogenic substrate for β-galactosidase is added to the medium.

When host insect cells in culture are infected with recombinant baculovirus, heterologous protein can be harvested after 4 to 5 days. The baculovirus expression vector system has been used to produce more than 500 different heterologous proteins, of which over 95% had the correct posttranslational modifications (Fig. 7.10).

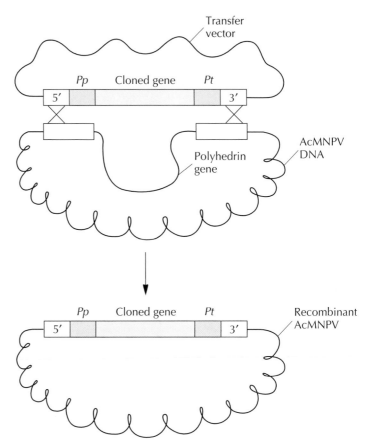

Figure 7.9 Replacement of the polyhedrin gene of AcMNPV with an expression unit from a transfer vector. A double crossover event, indicated by X's upstream (in the 5′ direction [*Pp* and 5′]) and downstream (in the 3′ direction [*Pt* and 3′]), results in the integration of the expression unit into the AcMNPV genome.

Production of Recombinant Baculovirus

The original procedure for producing recombinant baculovirus has been changed in a number of ways. First, the polyhedrin promoter is limited to a certain extent because proteins that are synthesized during the late stages of

Figure 7.10 Some of the recombinant proteins that have been produced by the baculovirus expression vector system. HIV-1, human immunodeficiency virus type 1; HSV, herpes simplex virus.

α-Interferon	G-protein-coupled receptors	Malaria proteins
Adenosine deaminase	HIV-1 envelope protein	Mouse monoclonal antibodies
Anthrax antigen	HSV capsid proteins	Multidrug transporter protein
β-Amyloid precursor protein	Human alkaline phosphatase	Poliovirus proteins
β-Interferon	Human DNA polymerase α	Pseudorabies virus glycoprotein 50
Bovine rhodopsin	Human pancreatic lipase	Rabies virus glycoprotein
Bluetongue virus neutralization antigen	Influenza virus hemagglutinin	Respiratory syncytial virus antigen
Cystic fibrosis transmembrane conductance regulator	Interleukin-2	Simian rotavirus capsid antigen
Dengue virus type 1 antigen	Lassa virus protein	Tissue plasminogen activator
Erythropoietin		

the lytic cycle are often not fully modified. This problem was resolved by substituting one of the strong AcMNPV promoters that are expressed in the early to late stages of the lytic cycle for the polyhedrin promoter. Second, linearization of the AcMNPV genome before transfection into insect cells increases the frequency of recombinant plaques. Cutting the AcMNPV genome at a single site decreases the number of nonrecombinant plaques because linearized baculovirus genomes have reduced infectivity. A double crossover event between a linearized AcMNPV DNA molecule and a circular transfer vector establishes a closed circular DNA molecule that is infective. To ensure consistent linearization from one experiment to the next, a unique *Bsu*361 restriction endonuclease site was incorporated into the polyhedrin gene of the wild-type AcMNPV genome. With the addition of this site, about 30% of the plaques contain recombinant baculovirus, in contrast to <1% when the intact circular AcMNPV genome is used.

The *Bsu*361 linearization system has been developed further to generate a system that produces a very high frequency of recombinant baculovirus plaques. Briefly, the AcMNPV genome was engineered with two *Bsu*361 sites that were placed on either side of the polyhedrin gene (Fig. 7.11). One is in gene 603 (open reading frame 603 [ORF603]), and the other is in a gene (ORF 1629) that is essential for viral replication. When DNA from this modified baculovirus is treated with *Bsu*361 and transfected into insect cells, no viral replication occurs because a segment of the essential gene (ORF1629) is missing. As part of this system, a transfer vector is constructed with the target gene and, if required, a selectable marker gene between intact versions of gene 603 and the essential gene. This transfer vector is transfected into insect cells that had been previously transfected with a linearized AcMNPV genome that has a deletion between the *Bsu*361 sites. A double crossover event both reestablishes a functional form of ORF1629 and incorporates the cloned gene into the AcMNPV genome (Fig. 7.11). With this system, up to 99% of the baculovirus plaques are recombinant.

MILESTONE

Synthesis of Rabbit β-Globin in Cultured Monkey Kidney Cells Following Infection with a SV40 β-Globin Recombinant Genome

R. C. MULLIGAN, B. H. HOWARD, and P. BERG
Nature 277:108–114, 1979

Conceptually, the development of a eukaryotic expression system appears to be a relatively simple matter of assembling the appropriate regulatory sequences, cloning them in the correct order into a vector, and then putting the cloned gene into the precise location that enables it to be expressed. In reality, the development of the first generation of eukaryotic expression vectors was a painstaking process following a trial-and-error approach. Before the Mulligan-Howard-Berg study, a number of genes had been cloned into the mammalian SV40 vectors, but mature, functional mRNAs were never detected after infection of host cells. This problem was overcome by inserting the rabbit cDNA for β-globin into an SV40 gene that had nearly all of its coding region deleted but retained "all the regions implicated in transcriptional initiation and termination, splicing and polyadenylation" Both rabbit β-globin mRNA and protein were synthesized in cells that were transfected with this β-globin cDNA–SV40 construct. Mulligan et al. concluded, "The principal conceptual innovation is the decision to leave intact the regions of the vector implicated in . . . mRNA processing" This study established that an effective eukaryotic expression system could be created by placing the cloned gene under the control of transcription and translation regulatory sequences. Additional studies that pinpointed in detail the structural prerequisites for eukaryotic expression vectors ensued.

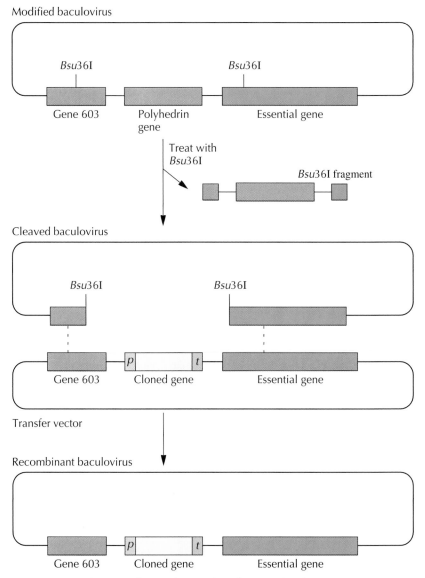

Figure 7.11 Production of recombinant baculovirus. Single *Bsu*36I sites are engineered into gene 603 and a gene (ORF1629) that is essential for baculovirus replication in insect cells (essential gene) of the AcMNPV genome. These genes flank the polyhedrin gene of the AcMNPV genome. After a baculovirus with two engineered *Bsu*36I sites is treated with *Bsu*36I, the segment between the *Bsu*36I sites is deleted. After transfection of an insect cell carrying a *Bsu*36I-treated baculovirus with a transfer vector that has a cloned gene under the control of the promoter (*p*) and terminator (*t*) elements of the polyhedrin gene and the complete sequence of both gene 603 and the essential gene, a double crossover event (dashed lines) generates a recombinant baculovirus with a functional essential gene. This system produces up to 99% recombinant baculoviruses.

Construction of an *E. coli*–Insect Cell Baculovirus Shuttle Vector

A system that makes it possible to carry out all the genetic manipulations for the generation of a baculovirus expression vector in *E. coli* has been developed. In this case, transfection of insect cells is required only for the production of the recombinant protein. Briefly, the system uses a portion of a small *E. coli* plasmid that is flanked by DNA segments that lie outside the 5′ and 3′ ends of the polyhedrin gene and contains a kanamycin resistance

gene, a DNA sequence that acts as an integration site (attachment site) that is inserted into the *lacZ′* gene without impairing its function, and an *E. coli*-based origin of replication. The plasmid is integrated into the AcMNPV genome by a double crossover event, with the concomitant loss of the poly-hedrin gene (Fig. 7.12A). The double crossover between the *E. coli* plasmid and the AcMNPV genome forms a single circular DNA construct that can

Figure 7.12 Construction of a recombinant bacmid. **A.** An *E. coli* plasmid is incorporated into the AcMNPV genome by a double crossover event (dashed lines) between DNA segments (5′ and 3′) that flank the polyhedrin gene to create a shuttle vector that replicates in both *E. coli* and insect cells. The gene for resistance to kanamycin (Kan^r), an attachment site (att) that is inserted in frame in the *lacZ′* sequence, and an *E. coli* origin of replication (*ori*^E) are introduced as part of the plasmid DNA. **B.** The transposition proteins encoded in the helper plasmid facilitate the integration (transposition) of the DNA segment of the donor plasmid that is bounded by two attachment sequences (attR and attL). The gene for resistance to gentamicin (Gm^r) and a cloned gene (gene) that is under the control of the promoter (*p*) and terminator (*t*) elements of the polyhedrin gene are inserted into the attachment site (att) of the bacmid. The helper and donor plasmids carry the genes for resistance to tetracycline (*Tet*^r) and ampicillin (*Amp*^r), respectively. **C.** The recombinant bacmid has a disrupted *lacZ′* gene (*). The right-angled arrow denotes the site of initiation of transcription of the cloned gene after transfection of the recombinant bacmid into an insect cell.

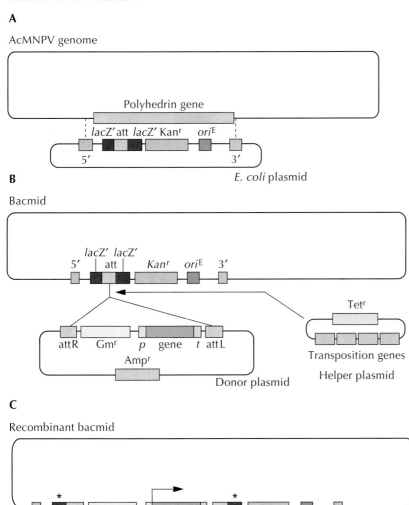

be maintained in *E. coli* as a plasmid and that, after transfection into insect host cells, directs the production of baculoviruses. This *E. coli*–insect cell baculovirus shuttle vector is called a bacmid.

As part of the bacmid system, the target gene is cloned between the polyhedrin promoter and a terminator sequence on another *E. coli* plasmid (donor plasmid). In the donor plasmid, a gentamicin resistance gene and the target gene expression unit are flanked by DNA sequences that bind to the attachment site in the bacmid and an ampicillin resistance gene that lies outside the two attachment sites (Fig. 7.12B). Integration between the corresponding sites of the donor plasmid and the bacmid can occur only in the presence of specific proteins (transposition proteins) which, in this system, are supplied by a third *E. coli* plasmid (helper plasmid) that also carries a gene for resistance to tetracycline (Fig. 7.12B).

Bacterial cells carrying a bacmid are cotransformed with donor and helper plasmids. In some of the doubly transformed bacteria, the DNA segment that is bounded by the two attachment sites is transposed into the attachment site of the bacmid (Fig. 7.12B and C). The integration of the segment of the donor plasmid with the expression unit and the gentamicin resistance gene into the bacmid attachment site destroys the reading frame of the *lacZ'* gene. Consequently, bacteria that carry recombinant (composite) bacmids produce white colonies in the presence of isopropyl-β-D-thiogalactopyranoside (IPTG) and 5-bromo-4-chloro-3-indolyl-β-D-galactopyranoside (X-Gal). Moreover, white colonies that are resistant to kanamycin and sensitive to both ampicillin and tetracycline carry only a recombinant bacmid and no donor and helper plasmids. After all of these manipulations, the integrity of the cloned gene can be confirmed by a PCR assay. Finally, a recombinant bacmid DNA can be transfected into insect cells, where the cloned gene is transcribed and the recombinant protein is produced.

Other protocols have been devised for creating baculovirus expression vectors. In one instance, all of the genetic manipulations of the AcMNPV genome are carried out in yeast cells with a yeast–insect cell shuttle vector before the recombinant baculovirus is introduced into insect cells. In another strategy, an in vitro recombination system that is based on the integration–excision components of bacteriophage P1 is used to generate a cloned gene–AcMNPV genome construct that can be transfected directly into insect cells.

Selective Purification of Recombinant Protein from Insect Cells by Affinity Tagging

Various strategies have been devised for the selective purification of a heterologous protein from cell extracts and mixtures of secreted proteins. In principle, a DNA sequence that encodes a short amino acid sequence that specifically binds to an element, compound, or macromolecule is ligated in the proper reading frame to a cloned gene. The construct is inserted between a promoter and a transcription termination sequence on an expression vector. After transcription and translation of the gene construct in the host cell, the short amino acid sequence, which is part of the recombinant protein, acts as an identification tag (affinity tag). More specifically, after a DNA sequence that encodes six histidine residues (His_6), an in-frame spacer sequence for seven amino acids, and a protease cleavage site of six amino acids is ligated to a cloned gene, the resulting combined (fusion) protein can be isolated by passing all the proteins of a cell extract through a column of nickel–triacetic acid agarose beads. The sequence of six histidines (hexahis-

tidine) binds to the nickel ions (Ni^{2+}). The unbound proteins are eluted from the column, and then the bound fusion protein is selectively removed either by the addition of a competitor compound (e.g., imidazole) that dislodges the hexahistidine tag of the recombinant protein from the nickel ions or by lowering the pH of the elution buffer. The amino acids of the tag are removed from the recombinant protein by digestion of the fusion protein with a proteolytic enzyme (protease) that cleaves only at the engineered cleavage site. The recombinant protein can be isolated from both the amino acid tag and protease by size separation chromatography or by other methods. When a protein is required for some use other than as a therapeutic agent, retention of the hexahistidine tag is acceptable because it seldom alters protein structure or function.

A number of different affinity tags (protein tags) have been developed. For example, glutathione transferase, maltose binding protein, and short amino acid sequences (epitopes) bind to glutathione, maltose, and specific antibodies, respectively. In addition, various cleavage sites that are specific for thrombin, enterokinase, and other proteases are used. A protein tag and associated cleavage site can be placed at either the N or C terminus of the recombinant protein and used for recombinant proteins that are produced by prokaryotic, fungal, insect cell, or mammalian cell expression systems.

Mammalian Cell Expression Vectors

Extrachromosomal mammalian expression vectors are helpful for studying the function and regulation of mammalian genes. In addition, authentic recombinant proteins with the potential to be therapeutic agents can be produced by mammalian cell lines in culture for use in human clinical trials. Although a variety of different mammalian expression vectors have been developed, they all tend to have common features and are very similar in design to other eukaryotic expression vectors.

A representative mammalian expression vector (Fig. 7.13) contains a eukaryotic origin of replication from an animal virus (e.g., Simian virus 40 [SV40]). The promoter sequences that drive both the cloned gene(s) and the selectable marker gene(s) and the transcription termination sequences (polyadenylation signals) for these genes must be eukaryotic and are usu-

Figure 7.13 Generalized mammalian expression vector. The multiple cloning site (mcs) and selectable marker gene (sm) are under the control of eukaryotic promoter (*p*) and polyadenylation (*pa*) sequences. Propagation of the vector in *E. coli* and mammalian cells depends on the origins of replication, *ori*E and *ori*euk, respectively. A marker gene (Amp^r) can be used for selecting transformed *E. coli.*

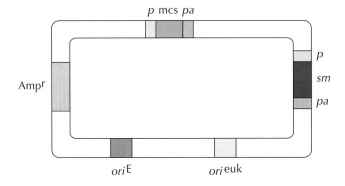

ally taken from either animal viruses (e.g., human cytomegalovirus, SV40, or herpes simplex virus) or mammalian genes (e.g., β-actin, metallothionein, thymidine kinase, or bovine growth hormone). Strong promoters and efficient polyadenylation signals are usually preferred. The sequences that are required for selection and propagation of a mammalian expression vector in *E. coli* are derived from a standard *E. coli* cloning vector (e.g., pBR322).

Selectable Marker Systems for Mammalian Expression Vectors

The bacterial gene (Neo[r]) that encodes neomycin phosphotransferase is often used to select transfected mammalian cells. The toxic agent for this system is the compound G-418 (Geneticin), which kills nontransfected mammalian cells by blocking translation. However, within a transfected cell, G-418 is phosphorylated by neomycin phosphotransferase. The addition of a phosphate group inactivates G-418. Consequently, only those cells that synthesize the Neo[r] gene product survive and proliferate.

Another system for selecting transfected mammalian cells uses the gene that encodes the enzyme dihydrofolate reductase (DHFR). With this protocol, the recipient cells must have a defective *DHFR* gene so they cannot produce any functional DHFR. After transfection of DHFR-deficient cells with a mammalian expression vector carrying an active *DHFR* gene, the chemical compound methotrexate (MTX) is added to the medium. The DHFR-deficient cells that were not transfected cannot grow in the presence of MTX, but the cells that synthesize dihydrofolate reductase survive. After the initial selection for cells with the *DHFR* gene, the concentration of MTX in the medium is increased, which selects for cells with large numbers of copies of the vector. These cells are perpetuated and produce high levels of recombinant protein.

Other dominant marker selection schemes have been developed. For example, the enzyme glutamine synthetase (GS) confers resistance to the cytotoxic effects of the chemical compound methionine sulfoximine (MSX). In the GS-MSX system, a vector carrying a *GS* gene is prepared and introduced into cultured mammalian cells. Selection of those cells that have a high vector copy number is achieved by increasing the concentration of MSX in the medium. In this case, the host cell line need not lack GS, because only multiple copies of the *GS* gene can confer resistance to MSX. This selection scheme gives the GS-MSX system a potential advantage over the DHFR-MTX system.

Many different proteins have been cloned into mammalian expression vectors and expressed in host cells. In some instances, the yield of gene product was enhanced by inserting an intron between the promoter and the cloned gene. The reason for the improved production of recombinant protein is not clear. In some instances, the primary transcript from a cloned gene may contain intron splice sites by chance (cryptic sites) and part of the coding region of the cloned gene may be spliced out, whereas when an intron is present, splicing at the cryptic sites is less likely to occur.

A high level of production of a recombinant protein has been achieved by coordinating the expression of a selectable marker and a cloned gene. For example, a *DHFR* gene is inserted upstream and close to a cloned gene, with the two genes under the control of a single promoter and polyadenylation signal and the *DHFR* gene flanked by intron-removing sequences. DHFR and the recombinant protein are translated from the primary transcript and the spliced mRNA, respectively (Fig. 7.14).

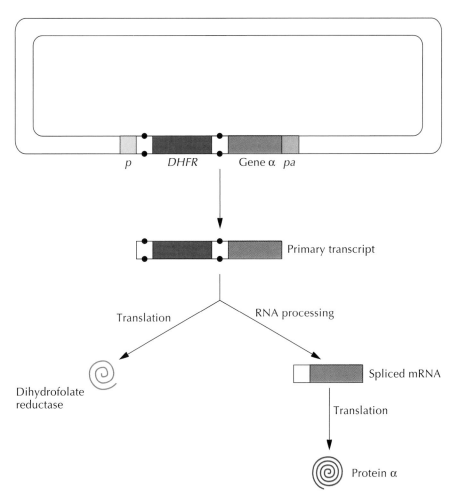

Figure 7.14 Coordinated expression of DHFR and a recombinant protein. A *DHFR* gene is cloned between intron donor and acceptor splice sites (dots) and upstream from a cloned gene (gene α). Both the *DHFR* and cloned genes are under the control of eukaryotic promoter (*p*) and polyadenylation (*pa*) sequences. Dihydrofolate reductase and the heterologous protein (protein α) are translated from the unspliced (primary) and processed (spliced) transcripts, respectively.

Expression of Two Cloned Genes in One Mammalian Cell

The active form of some commercially important proteins consists of two different protein chains. For example, human thyroid-stimulating hormone is a two-chain protein (heterodimer), and hemoglobin is a tetramer with two copies of each subunit (i.e., $\alpha_2\beta_2$). It is possible to clone the gene or cDNA for each subunit of a multimeric protein, synthesize and purify each subunit separately, and then mix the chains together in a test tube. Unfortunately, relatively few multichain proteins are properly assembled in vitro. By contrast, in vivo assembly of dimeric and tetrameric proteins is quite efficient. Consequently, various strategies have been devised for the production of two different recombinant proteins within the same cell.

Two mammalian expression vectors, each with the gene or cDNA for one of the subunits and a different selectable marker gene, are cotransfected into host cells (Fig. 7.15). The transfected cells are treated with both

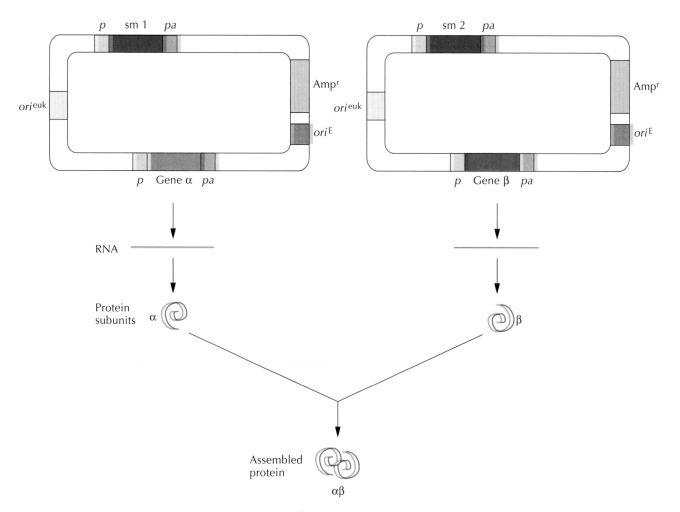

Figure 7.15 Two-vector expression system. The cloned genes (gene α and gene β) encode subunits of a protein dimer ($\alpha\beta$). After cotransfection, both subunits (α and β) are synthesized and assembled into a functional protein dimer. Both vectors carry origins of replication for *E. coli* (*ori*E) and mammalian cells (*ori*euk), a marker gene (Ampr) for selecting transformed *E. coli*; and eukaryotic promoter (*p*) and polyadenylation (*pa*) sequences that control a selectable marker gene (sm) and each of the cloned genes (gene α and gene β).

selecting agents, and the cells that survive carry both vectors. Two-vector systems have been used successfully for the production of authentic dimeric and tetrameric recombinant proteins. However, loss of one of the two vectors in doubly transfected cells is common. Moreover, the two vectors are not always maintained with the same copy number, so one subunit may be overproduced relative to the other and yields of the final product may be reduced. To overcome these problems, single vectors that carry two cloned genes have been developed. In some instances, the two genes are placed under the control of independent promoters and polyadenylation signals (double cassette vectors) (Fig. 7.16). Alternatively, to ensure that equal amounts of the recombinant proteins are synthesized, vectors (dicistronic vectors) have been constructed with the two cloned genes separated from each other by a DNA sequence that contains an internal ribosomal entry site (IRES). Parenthetically, IRESs are found in

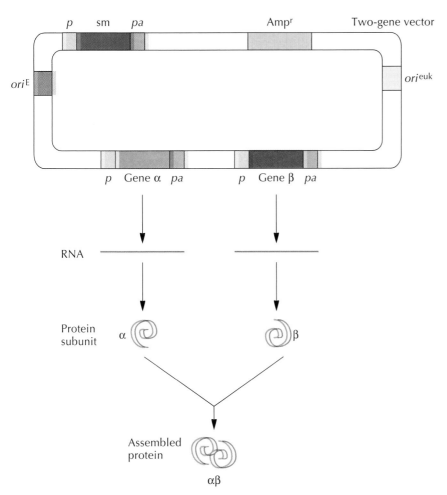

Figure 7.16 Two-gene expression vector. The cloned genes (gene α and gene β) encode subunits of a protein dimer ($\alpha\beta$). Each cloned gene is inserted into a vector as part of its own transcription unit under the control of a eukaryotic promoter (p) and polyadenylation sequence (pa). Each subunit is translated from a separate mRNA, and a functional protein dimer ($\alpha\beta$) is assembled. Each vector carries origins of replication for *E. coli* (ori^{E}) and mammalian cells (ori^{euk}), a marker gene (Ampr) for selecting transformed *E. coli*, and a selectable marker gene (sm) that is under the control of a eukaryotic promoter (p) and sequence polyadenylation (pa).

mammalian virus genomes and after transcription allow simultaneous translation of different proteins from a polycistronic mRNA molecule. Transcription of the "gene–IRES–gene" construct is controlled by one promoter and polyadenylation signal. A single "two-gene" (dicistronic) transcript is synthesized, and translation proceeds from the 5' end of the mRNA to produce one of the chains and internally from the IRES element to produce the second chain (Fig. 7.17).

In sum, mammalian expression vectors are as versatile and effective as the vectors for other eukaryotic expression systems when authentic recombinant proteins are required for research purposes and clinical trials. However, industrial production of a recombinant protein with engineered mammalian cells is costly. Consequently, less expensive expression systems are favored unless authenticity of an important recombinant protein can be obtained only with mammalian cells.

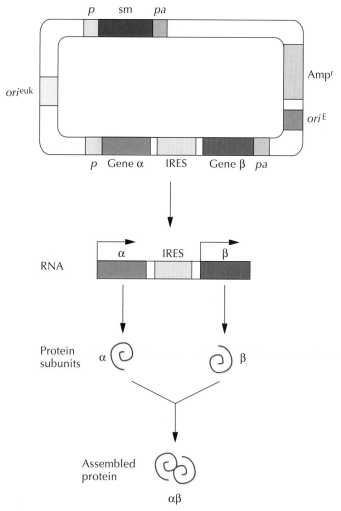

Figure 7.17 Dicistronic expression vector. The cloned genes (gene α and gene β) encode subunits of a protein dimer (αβ). Each cloned gene is inserted into a vector, and they are separated by a DNA sequence that after transcription, as part of an mRNA, acts as an IRES. The two cloned genes are under the control of a eukaryotic promoter (p) and sequence polyadenylation (pa). Translation of the mRNA occurs from the 5′ end and internally (right-angled arrows). Both subunits (α and β) are synthesized and assembled into a functional protein dimer. The vector carries origins of replication for *E. coli* (*ori*E) and mammalian cells (*ori*euk), a selectable marker (Ampr) for selecting transformed *E. coli*, and a selectable marker gene (sm) that is under the control of a eukaryotic promoter (p) and polyadenylation sequence (pa).

SUMMARY

Prokaryotic expression systems work well for the production of a number of proteins; however, some proteins require specific posttranslational modifications such as glycosylation, phosphorylation, and acetylation. Bacteria do not perform these modifications. To resolve this problem, cloned genes have been expressed in eukaryotic cells. Eukaryotic expression vectors are engineered with the same objectives as those applied to the development of prokaryotic expression vectors.

The yeast *S. cerevisiae*, which is well characterized genetically and can be grown in large fermenters, has been used for the production of a number

of different proteins from cloned genes. *S. cerevisiae* vectors that support secretion of heterologous proteins have been devised to simplify the purification process. Numerous authentic proteins have been produced in *S. cerevisiae*. However, with *S. cerevisiae* as the host cell, a number of recombinant proteins are not properly modified and protein yields are often low. Consequently, other yeast systems have been developed for the production of recombinant proteins.

In the search for other nonfungal eukaryotic expression systems that produce biologically active proteins, attention has focused on baculovirus expression vectors. The baculovirus AcMNPV, which infects insect cells, has been developed as a eukaryotic expression vector, with promising results. In the original method, a transfer vector that carried the cloned gene flanked by AcMNPV-specific sequences was transfected into an insect cell that had been infected with AcMNPV. A double crossover event between the transfer vector and the AcMNPV genome inserted the cloned gene into the AcMNPV genome and put it under the control of a strong promoter that is active during the late stage of the lytic cycle. A heterologous protein is produced by infecting insect cells with recombinant baculovirus.

The frequency of occurrence of recombinant baculovirus has been increased from <1% with the original protocol to ~99%. AcMNPV DNA is treated with a restriction endonuclease which cuts the DNA at two specific sites and removes a section of the genome that includes a portion of a gene essential for the lytic cycle. Insect cells are then transfected with the treated viral DNA. These transfected cells are subsequently transfected with a transfer vector, and, in some cells, a double crossover event creates a recircularized AcMNPV genome with the incorporation of the cloned gene and reconstitution of a functional form of the essential gene. As a result, within the limits of the technique, virtually all of the virulent baculoviruses are recombinant.

A further improvement to the baculovirus expression system has been the development of an *E. coli*–insect cell shuttle vector (bacmid) so that all the genetic manipulations can be carried out with *E. coli* as the host cell. Insect cells are transfected with a recombinant bacmid only when the heterologous protein is required. About 95% of all the heterologous proteins that have been synthesized with the baculovirus expression system have the appropriate posttranslational modifications.

In general, extrachromosomal mammalian expression vectors are used to produce heterologous proteins for research purposes and clinical trials. Mammalian expression vectors are shuttle vectors with origins of replication from an animal virus and an *E. coli*-based plasmid. The elements that regulate the transcription units originate from animal viruses or mammalian genes. Dominant selectable marker gene systems are used for selecting transfected cells. And, with some of these systems, such as the DHFR-MTX system, the level of the cytotoxic compound selects for cells with a high vector copy number, increasing the yield of a heterologous protein.

Mammalian expression systems have been devised for the in vivo assembly of proteins that consist of two different subunits. The production and assembly of two-chain and four-chain proteins were achieved by cotransfection of host cells with two vectors that each carry a gene for one of the subunits. Alternative approaches include using one vector that carries the two genes as separate transcription units (double-cassette vector) or forming a single transcription unit with the two genes (dicistronic vector),

with one of the proteins being translated from the 5' end of the mRNA and the other one from an IRES.

References

Cockett, M. I., C. R. Bebbington, and G. T. Yarranton. 1990. High level expression of tissue inhibitor of metalloproteinases in Chinese hamster ovary cells using glutamine synthetase gene amplification. *Bio/Technology* **8**:662–667.

Cole, E. S., K. Lee, K. Lauziere, C. Kelton, S. Chappel, B. Weintraub, D. Ferrara, P. Peterson, R. Bernasconi, T. Edmunds, S. Richards, L. Dickrell, J. M. Kleeman, J. H. McPherson, and B. M. Pratt. 1993. Recombinant human thyroid stimulating hormone: development of a biotechnology product for detection of metastatic lesions of thyroid carcinoma. *Bio/Technology* **11**:1014–1024.

Cregg, J. M., J. F. Tschopp, C. Stillman, R. Siegel, M. Akong, W. S. Craig, R. G. Buckholz, K. R. Madden, P. A. Kellaris, G. R. Davis, B. L. Smiley, J. Cruze, R. Torregrossa, G. Velicelebi, and G. P. Thill. 1987. High-level expression and efficient assembly of hepatitis B surface antigen in the methylotrophic yeast, *Pichia pastoris. Bio/Technology* **5**:479–485.

Davies, A. H. 1994. Current methods for manipulating baculovirus. *Bio/Technology* **12**:47–50.

Digan, M. E., S. V. Lair, R. A. Brierley, R. S. Siegel, M. E. Williams, S. B. Ellis, P. A. Kellaris, S. A. Provow, W. S. Craig, G. Velicelebi, M. M. Harpold, and G. P. Thill. 1989. Continuous production of a novel lysozyme via secretion from the yeast, *Pichia pastoris. Bio/Technology* **7**:160–164.

Dirks, W., M. Wirth, and H. Hauser. 1993. Dicistronic transcription units for gene expression in mammalian cells. *Gene* **128**:247–249.

Gellissen, G., Z. A. Janowicz, U. Weydemann, K. Melber, A. W. M. Strasser, and C. P. Hollenberg. 1992. High-level expression of foreign genes in *Hansenula polymorpha. Biotechnol. Adv.* **10**:179–189.

Giga-Hama, Y., H. Tohda, H. Okada, M. K. Owada, H. Okayama, and H. Kumagai. 1994. High-level expression of human lipocortin I in the fission yeast *Schizosaccharomyces pombe* using a novel expression vector. *Bio/Technology* **12**:400–404.

Gilbert, S. C., H. van Urk, A. J. Greenfield, M. J. McAvoy, K. A. Denton, D. Coghlan, G. D. Jones, and D. J. Mead. 1994. Increase in copy number of an integrated vector during continuous culture of *Hansenula polymorpha* expressing functional human hemoglobin. *Yeast* **10**:1569–1580.

The Global Use of Strategies to Open Occluded Arteries (GUSTO) IIb Investigators. 1996. A comparison of recombinant hirudin with heparin for the treatment of acute coronary syndromes. *N. Engl. J. Med.* **335**:775–782.

Hallewell, R. A., R. Mills, P. Tekamp-Olson, R. Blacher, S. Rosenberg, F. Otting, F. R. Masiarz, and C. J. Scandella. 1987. Amino terminal acetylation of authentic human Cu,Zn superoxide dismutase produced in yeast. *Bio/Technology* **5**:363–366.

Kidd, I. M., and V. C. Emery. 1993. The use of baculoviruses as expression vectors. *Appl. Biochem. Biotechnol.* **42**:137–159.

Kitts, P. A., and R. D. Possee. 1993. A method for producing recombinant baculovirus expression vectors at high frequency. *BioTechniques* **14**:810–817.

Laroche, Y., V. Strome, J. DeMeutter, J. Messens, and M. Lauwereys. 1994. High-level secretion and very efficient isotopic labeling of tick anticoagulant peptide (TAP) expressed in the methylotrophic yeast, *Pichia pastoris. Bio/Technology* **12**:1119–1124.

Loison, G., A. Findeli, S. Bernard, M. Nguyen-Juilleret, M. Marquet, N. Riehl-Bellon, D. Carvallo, L. Guerra-Santos, S. W. Brown, M. Courtney, C. Roitsch, and Y. Lemoine. 1988. Expression and secretion in *S. cerevisiae* of biologically active leech hirudin. *Bio/Technology* **6**:72–77.

Lucas, B. K., L. M. Giere, R. A. DeMarco, A. Shen, V. Chisholm, and C. W. Crowley. 1996. High-level production of recombinant proteins in CHO cells using a dicistronic DHFR intron expression vector. *Nucleic Acids Res.* **24:**1774–1779.

Lucknow, V. A., S. C. Lee, G. F. Barry, and P. O. Olins. 1993. Efficient generation of infectious recombinant baculoviruses by site-specific transposon-mediated insertion of foreign genes into a baculovirus genome propagated in *Escherichia coli. J. Virol.* **67:**4566–4579.

Peng, S., M. Sommerfelt, J. Logan, Z. Huang, T. Jilling, K. Kirk, E. Hunter, and E. Sorscher. 1993. One-step affinity isolation of recombinant protein using the baculovirus/insect cell expression system. *Protein Expr. Purif.* **4:**95–100.

Robinson, A. S., V. Hines, and K. D. Wittrup. 1994. Protein disulfide isomerase overexpression increases secretion of foreign proteins in *Saccharomyces cerevisiae. Bio/Technology* **12:**381–384.

Romanos, M. A., C. A. Scorer, and J. J. Clare. 1992. Foreign gene expression in yeast: a review. *Yeast* **8:**423–488.

Vozza, L. A., L. Wittmer, D. R. Higgins, T. J. Purcell, M. Bergseid, L. A. Collins-Racie, E. R. LaVallie, and J. P. Hoeffler. 1996. Production of a recombinant bovine enterokinase catalytic subunit in the methylotrophic yeast, *Pichia pastoris. Bio/Technology* **14:**77–81.

REVIEW QUESTIONS

1. Why would a eukaryotic cell rather than a prokaryotic cell be used to express a target protein?

2. What are the advantages and disadvantages of the different classes of yeast vectors for producing a biotechnology product?

3. Discuss the salient features of a *P. pastoris* high-expression integrating vector system.

4. What are baculoviruses? Describe the strategy of the original baculovirus expression system and how this protocol has been improved.

5. What is a bacmid? How is it used?

6. What is an affinity tag? How is it used?

7. Describe the main features of an extrachromosomal mammalian cell expression vector.

8. Describe at least two selectable marker systems that are used with mammalian expression vectors.

9. Describe the various strategies for expressing two recombinant proteins in one mammalian cell.

10. What criteria are used to decide if a particular recombinant protein should be expressed in a yeast, baculovirus, or mammalian cell system?

Directed Mutagenesis and Protein Engineering

It is possible with recombinant DNA technology to isolate the gene for any protein that exists in nature, to express it in a specific host organism, and to produce a purified product that can be used commercially. However, the physical and chemical properties of these "naturally occurring" proteins are often not well suited for an industrial application. In some instances, it is possible to obtain a protein that is better suited for a particular task by using a gene from an organism that grows in an unusual, often extreme environment. For example, when an enzyme such as α-amylase was required to function at a high temperature, the gene for that enzyme was isolated from *Bacillus stearothermophilus*, an organism whose natural niche is a 90°C hot spring. In this case, the cloned gene was used to produce α-amylase that is stable at the high temperatures used in the industrial production of alcohol from starch. In addition to isolating natural genes that encode proteins with useful properties, conventional mutagenesis and selection schemes can be used in an attempt to create and perpetuate a mutant form of a gene that encodes a protein with the desired properties. However, the number of mutant proteins—each with a different amino acid change—that are possible after the alteration of individual nucleotides within a structural gene by conventional mutagenesis is extremely large. In practice, the mutagenesis–selection strategy rarely results in any significant beneficial changes to the targeted protein because most amino acid changes decrease the activity of an enzyme.

By using a set of techniques that specifically change amino acids encoded by a cloned gene, proteins with properties that are better suited than naturally occurring counterparts can be created for therapeutic and industrial applications. For example:

- By altering both the Michaelis constant (K_m), which reflects the "tightness" of substrate binding to the enzyme and the maximal rate of conversion of the substrate into product under defined conditions (V_{max}) for an enzyme-catalyzed reaction to improve the overall catalytic efficiency (V_{max}/K_m) of the reaction, where V_{max} equals the

total amount of enzyme present (E_0) times the catalytic rate constant (k_{cat})

- By changing the thermal tolerance or pH stability, or both, of a protein, enabling the altered protein to be used under conditions that would inactivate the native version
- By modifying the reactivity of an enzyme in nonaqueous solvents so that chemical reactions can be catalyzed under nonphysiological conditions
- By changing an enzyme so that a cofactor is no longer required for certain continuous industrial production processes in which the cofactor must be supplied on a regular basis
- By modifying the substrate binding site of an enzyme to increase its specificity, thereby decreasing the extent of undesirable side reactions
- By increasing the resistance of a protein to cellular proteases, which would simplify purification and increase the recoverable yield
- By altering the allosteric regulation of an enzyme to diminish the impact of metabolite feedback inhibition and increase the product yield

Directed Mutagenesis Procedures

It is not a simple matter to produce a new protein with specified predetermined properties. However, it is quite feasible to modify the existing properties of known proteins. Theoretically, these changes can be carried out at either the protein or the gene level. However, chemical modifications of proteins generally are harsh, nonspecific, and required repeatedly, for each batch of protein, so it is preferable to manipulate the DNA sequence of a cloned gene to create an altered protein with novel properties. Unfortunately, it is not always possible to know in advance which individual amino acids or short sequences of amino acids contribute to a particular physical, kinetic, or chemical property. For example, a particular property of a protein may be the consequence of two or more amino acid residues that are far apart from each other in the linear sequence but are juxtaposed as a result of the folding of the protein. In this case, two or more amino acid residues may have to be changed to produce a protein with the desired properties. In the not too distant future, computer programs may be able to make accurate predictions of protein function on the basis of deduced amino acid sequences, thereby simplifying the task of producing a protein with specific predetermined properties. At present, although it is relatively straightforward to introduce new coding information into cloned genes, large numbers of novel proteins must be assayed to determine whether a particular property has been created.

The process for generating amino acid coding changes at the DNA level is called directed mutagenesis. Determining which amino acids of a protein should be changed to attain a specific property is easier if the three-dimensional structure of the protein has been well characterized by X-ray crystallographic analysis and other analytical procedures. But for most proteins, such detailed information is lacking, so directed mutagenesis becomes a trial-and-error strategy in which changes are made to those nucleotides that are most likely to yield a particular change in a protein property. And then, of course, the protein encoded by each mutated gene has to be tested to ascertain whether the mutagenesis process has indeed generated the desired change.

A large number of experimental approaches have been devised for the directed mutagenesis of cloned genes. Some of the methods are designed to mutate specific nucleotides of a cloned gene. Others introduce mutations randomly over a short segment of the cloned gene, thereby creating a panel of mutated proteins among which one may have the desired activity.

Oligonucleotide-Directed Mutagenesis with M13 DNA

Oligonucleotide-directed mutagenesis (site-specific mutagenesis) is a straightforward method for producing defined point mutations in a cloned gene (Fig. 8.1). For this procedure, the investigator must know (1) the pre-

Figure 8.1 Oligonucleotide-directed mutagenesis. Single-stranded bacteriophage M13 (M13 + strand), carrying a cloned gene, is annealed with a complementary synthetic oligonucleotide containing one mismatched base, i.e., one base that is not complementary to its counterpart in the target DNA. With the oligonucleotide as the primer, DNA synthesis is catalyzed by the Klenow fragment of *E. coli* DNA polymerase I; the cloned gene and the M13 vector are the template. Synthesis continues until the entire strand is copied. The newly synthesized DNA strand is circularized by T4 DNA ligase. The ligation reaction mixture is used to transform *E. coli*. Both the target DNA with its original sequence and the mutated sequence are present in the progeny M13 phage. dNTPs, deoxynucleoside triphosphates.

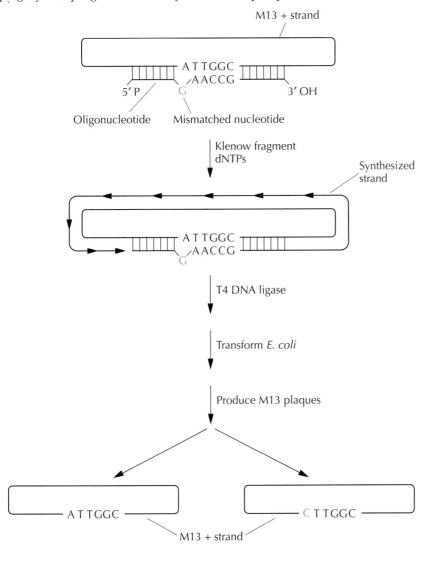

cise nucleotide sequence in the region of DNA that encodes the mRNA codon that is to be changed and (2) the amino acid changes that are being introduced. In one commonly used method, the cloned gene is inserted into the double-stranded form of an M13 bacteriophage vector. The single-stranded form (M13 + strand) of the recombinant vector is isolated and mixed with a synthetic oligonucleotide. The oligonucleotide has, except for one nucleotide, the sequence exactly complementary to a segment of the cloned gene. The nucleotide difference (i.e., mismatch) coincides precisely with the nucleotide of the messenger RNA (mRNA) codon that is targeted for change. In Fig. 8.1, the sequence ATT, which encodes the isoleucine codon AUU, is to be changed to CTT, which encodes the leucine codon CUU. The oligonucleotide will hybridize to the complementary region of the cloned gene if it is added in an amount much in excess of that of the M13 DNA, if the mismatch is near the middle of the oligonucleotide, and if the mixing is done at a low temperature in the presence of a high concentration of salt. The 3′ end of the hybridized oligonucleotide acts as a primer site for the initiation of DNA synthesis that uses the intact M13 strand as the template. The replication, which uses the four deoxyribonucleoside triphosphates, is catalyzed by the Klenow fragment of *Escherichia coli* DNA polymerase I. T4 DNA ligase is added to ensure that the last nucleotide of the synthesized strand is joined to the 5′ end of the primer. But in vitro DNA synthesis is often incomplete, so partially double-stranded M13 molecules must be removed from the mixture by sucrose gradient centrifugation.

The complete double-stranded M13 molecules, now containing the mismatched nucleotides, are introduced into *E. coli* cells by transformation. Infected cells produce M13 virus particles, which eventually lyse the cells and form plaques. Theoretically, because plasmid DNA is replicated semiconservatively, half of the phage that are formed carry the wild-type sequence and the other half contain the mutated sequence that has the specified nucleotide change. Phage produced in the initial transformation step are propagated in *E. coli*, and particles that contain only the mutated gene are identified by DNA hybridization under highly stringent conditions. The original oligonucleotide is the probe in these hybridization experiments.

After the double-stranded form of M13 is isolated, the mutated gene is excised by digestion with restriction enzymes and then spliced onto an *E. coli* plasmid expression vector. For further study, the altered protein is expressed in and purified from the *E. coli* cells.

However, in actuality, when oligonucleotide-directed mutagenesis is used, the expected 50% of the M13 viruses carrying the mutated form of the target gene are not recovered. Rather, for a variety of technical reasons, only 1 to 5% of the plaques actually contain phage carrying the mutated gene. Consequently, the oligonucleotide-directed mutagenesis method has been modified in a number of different ways to enrich for the number of mutant phage plaques that can be obtained. For example, one approach is to introduce the M13 viral vector carrying the gene that is to be mutagenized into an *E. coli* strain that has two defective enzymes of DNA metabolism (Fig. 8.2). One is a defective form of dUTPase (*dut*). Cells without a functional dUTPase have an elevated intracellular level of dUTP, which in turn causes a few dUTP residues to be incorporated into DNA during replication instead of dTTP. The other enzyme is a defective uracil *N*-glycosylase (*ung*). In the absence of functional uracil *N*-glycosylase, the dUTP residues that were spuriously incorporated into DNA cannot be removed. The single-stranded M13 DNA that is produced by this *E. coli* strain has approximately

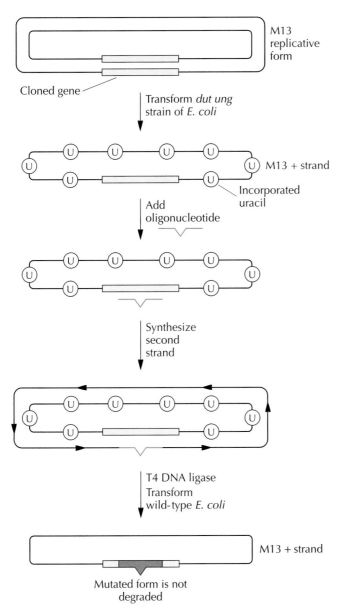

Figure 8.2 Enrichment of mutated M13 by passage of the parental DNA through a *dut ung* strain of *E. coli*. The target DNA is cloned into the double-stranded replicative form of bacteriophage M13, which is then used to transform a *dut ung* strain of *E. coli*. The *dut* mutation causes the intracellular level of dUTP to be elevated; the high level of nucleotide leads to the incorporation of a few dUTP residues (U). The *ung* mutation prevents the removal of any incorporated uracil residues. Following in vitro oligonucleotide-directed mutagenesis, the double-stranded M13 vector with the target DNA is introduced into wild-type *E. coli*. The wild-type *ung* gene product (uracil *N*-glycosylase) removes any uracil residues from the parental strand, so a significant portion of the parental strand is degraded. The mutated strand remains intact because it does not contain uracil. It serves as a template for DNA replication, thereby enriching the yield of M13 bacteriophage carrying the mutated gene.

1% of its thymidines replaced by uridines. A mismatched oligonucleotide primer is hybridized to this substituted M13 DNA, and the second strand of M13 is prepared by in vitro synthesis and ligation. The double-stranded bacteriophage DNA is next introduced into an *E. coli* strain that contains a

functional *ung* gene. The active uracil *N*-glycosylase in the host cells then removes uridine residues from the transforming M13 DNA (Fig. 8.2). As a result, the original M13 template strand is degraded and only the mutated strand, which contains no dUTP, is replicated. In this way, the yield of M13 bacteriophage carrying a gene with a site-specific mutation is significantly increased.

Oligonucleotide-Directed Mutagenesis with Plasmid DNA

The major drawback to bacteriophage M13 oligonucleotide-directed mutagenesis is the large number of time-consuming steps that need to be performed before the mutated form of the target gene is eventually isolated. As an alternative to the M13 system, a number of approaches which allow oligonucleotide-directed mutagenesis to be performed with plasmid rather than M13 DNA have been developed. With this approach, the need to subclone a target gene from a plasmid into M13 and then, after mutagenesis, clone it back into a plasmid is avoided. In one of these methods, the target DNA is inserted into a multiple cloning site on a plasmid vector that contains a functional tetracycline resistance gene and a nonfunctional ampicillin resistance gene as the result of a single nucleotide substitution in the middle of the gene (Fig. 8.3). Following transformation of the vector carrying the target DNA into *E. coli* host cells, the double-stranded plasmid DNA is denatured by treatment with alkaline to form single-stranded circular DNA molecules. Three different oligonucleotides that anneal to one of the single-stranded circular DNA molecules are added to the sample of denatured plasmid DNA. One oligonucleotide is designed specifically to alter the target DNA, another is designed to correct the substituted nucleotide in the ampicillin resistance gene, and the third is designed to change a single nucleotide in the tetracycline resistance gene so that this gene will become nonfunctional. The four deoxyribonucleoside triphosphates and T4 DNA polymerase, which has the same activity as the Klenow fragment of *E. coli* DNA polymerase I, are added to the reaction mixture, and the 3' ends of the annealed oligonucleotides act as primers for DNA synthesis with the intact circular DNA molecule as the template. The nicks in the synthesized strand are sealed by T4 DNA ligase. After synthesis and ligation are complete, the reaction mixture is used to transform *E. coli* cells. Transformants are selected for ampicillin resistance and tetracycline sensitivity. With this procedure, about 90% of the selected transformants have the specified mutation in the target gene. In the remaining transformants, the target gene is unchanged because the oligonucleotide did not anneal to the target gene or the oligonucleotide was bypassed during DNA synthesis. The cells with the specified mutation in the target gene are identified by DNA hybridization. All of the plasmids, strains, enzymes, oligonucleotides, other than the one needed to alter the target gene, and buffers for this procedure are sold as a kit, facilitating the widespread use of this procedure.

PCR-Amplified Oligonucleotide-Directed Mutagenesis

Researchers continually try to develop simpler and faster protocols. For site-specific mutagenesis, the polymerase chain reaction (PCR) can be exploited both to enrich the mutated gene and to avoid using the bacteriophage M13 system. In one of the PCR-based mutagenesis protocols, the targeted gene is cloned onto a plasmid vector (Fig. 8.4). This plasmid is then dispensed into two reaction tubes. Two specific PCR oligonucleotide primers are added to each reaction tube; one primer is completely comple-

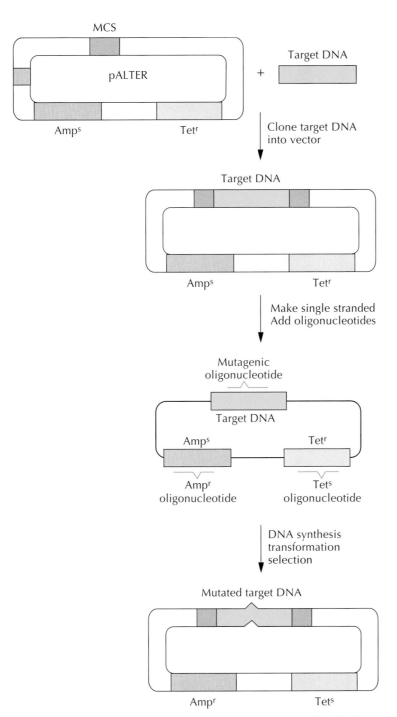

Figure 8.3 Oligonucleotide-directed mutagenesis with plasmid DNA. The target DNA is inserted into the multiple cloning site on the vector, pALTER. Plasmid DNA isolated from the *E. coli* cells is alkaline denatured before the mutagenic oligonucleotide, the ampicillin resistance (Ampr) oligonucleotide, and the tetracycline sensitivity (Tets) oligonucleotide are annealed. The oligonucleotides act as primers for DNA synthesis by T4 DNA polymerase with the original strand as the template. The gaps between the synthesized pieces of DNA are sealed by T4 DNA ligase. The reaction mixture is used to transform *E. coli* host cells, and cells that are Ampr and Tets are selected.

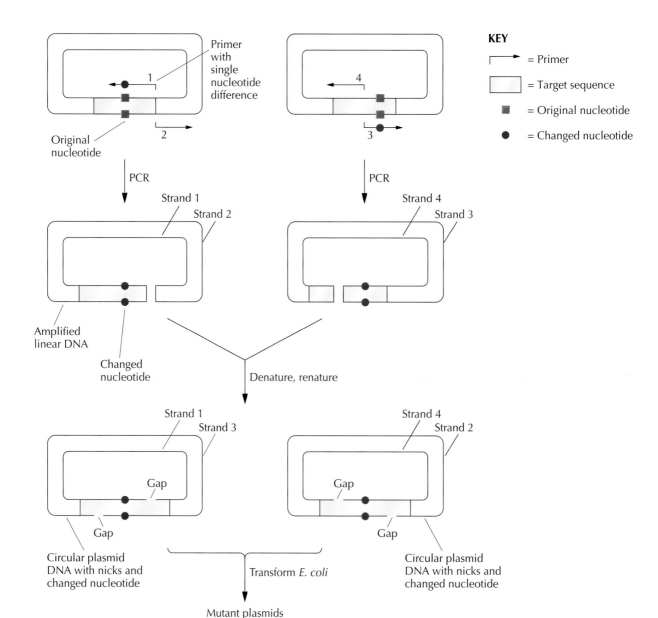

Figure 8.4 Oligonucleotide-directed mutagenesis by PCR. This procedure is performed in two separate reaction tubes, which contain the same double-stranded plasmid DNA but different sets of PCR primers. Both primers 1 and 3 contain the desired base change but are complementary to different strands of the plasmid DNA. Primers 2 and 4, respectively, anneal to DNA strands that are complementary to the strands to which primer 1 and primer 3 anneal. The pairs of primers are positioned at different sites, and each pair of primers is aligned end to end. After PCR amplification of the plasmid DNA, linear molecules are produced in each reaction tube. The linear PCR products from each reaction tube are mixed, denatured, and then renatured. As a result, in addition to the two original linear PCR-amplified molecules, two circular plasmid DNAs, each with two offset nicks (gaps), are produced. When the circular molecules are introduced into *E. coli* cells, the nicks are repaired by the host cell enzymes and the plasmid can then be propagated indefinitely. The linearized DNA molecules are not maintained in *E. coli*.

mentary to a sequence within or adjacent to the cloned gene, and the other is complementary to a different segment of the cloned gene, except for one nucleotide, which has been targeted for change. In the two reaction tubes, the primers with the one-nucleotide mismatch (primers 1 and 3) anneal to opposite strands, so both nucleotides of a specific base pair are targeted. The second primer pair (primers 2 and 4) in each reaction tube is completely complementary to a short DNA segment. The positioning of the hybridization regions of the primers in the two reaction tubes is such that after PCR cycling, the amplified DNA molecules have different ends. Following PCR amplification, the contents of the two reaction tubes are mixed, denatured, and renatured. As a result of the different positions of the ends of the amplified DNA in the two original reaction tubes, a strand from one reaction tube hybridizes with its complementary strand from the other reaction tube to form circular DNA molecules with two offset nicks. These nicks are repaired in vivo after they are introduced by transformation into *E. coli*. If two complementary strands from the same reaction tube happen to hybridize with one another, then linear DNA molecules are formed. The circular forms but not the linear ones are stably maintained and inherited in *E. coli*. Thus, not only do the circularized DNA molecules become established as plasmids, but all of these molecules carry the site-specific mutated nucleotide base pair. In sum, this procedure introduces a specified point mutation into a cloned gene without the need to insert the cloned gene into M13, to use enrichment procedures such as the *dut ung* system, or to subclone the mutated gene from M13 onto an expression plasmid vector.

Random Mutagenesis with Degenerate Oligonucleotide Primers

Unfortunately, investigators seldom know which specific nucleotide changes are needed to be introduced into a cloned gene to modify the properties of the target protein. Consequently, they must use methods that gen-

MILESTONE

Oligonucleotide-Directed Mutagenesis Using M13-Derived Vectors: an Efficient and General Procedure for the Production of Point Mutations in Any Fragment of DNA

M. J. ZOLLER and M. SMITH
Nucleic Acids Res. **10**:6487–6500, 1982

The technique of oligonucleotide-directed mutagenesis (site-specific mutagenesis) was developed mainly in the laboratory of Michael Smith as incremental changes to the technique of "marker rescue." In the marker rescue procedure, a mutation in a bacteriophage genomic DNA is corrected after the mutant DNA is annealed to a fragment of complementary wild-type DNA. Subsequently, it was demonstrated that a chemically synthesized oligonucleotide annealed to bacteriophage genomic DNA could produce a specific mutation. Unfortunately, these and other early procedures for oligonucleotide-directed mutagenesis required specialized skills and initially could only be performed in a few research laboratories. However, the procedure described by Zoller and Smith made it a relatively straightforward matter for thousands of laboratories throughout the world to specifically and rapidly alter the DNA sequence of any cloned gene. The key to the success of the protocol developed by Zoller and Smith lay in the use of *E. coli* bacteriophage M13. It was possible to clone foreign DNA into the double-stranded form of the virus, add an oligonucleotide with a specified change to the single-stranded form to produce a mutated DNA copy, and then recover the mutated double-stranded form in a relatively high yield. Since it was originally described, this procedure has been enhanced, simplified, and optimized and has been used by a large number of researchers to specifically modify thousands of different genes.

erate all the possible amino acid changes at one particular site. For example, oligonucleotide primers can be synthesized with any of the four nucleotides at defined positions. This pattern of sequence degeneracy is generally achieved by programming an automated DNA synthesis reaction to add a low level, usually a few percent, of each of the three alternative nucleotides each time a particular nucleotide is added to the chain (Fig. 8.5). In this way, the oligonucleotide primer preparation contains a heterogeneous set of DNA sequences that will generate a series of mutations that are clustered in a defined portion of the target gene.

This approach has two advantages. (1) Detailed information regarding the role of particular amino acid residues in the functioning of the protein is not required. (2) Unexpected mutants encoding proteins with a range of interesting and useful properties may be generated because the introduced changes are not limited to one amino acid. Of course, should none of the mutants yield a protein with the properties that are being sought, then it may be necessary to repeat the entire procedure with a set of degenerate primers that is complementary to a different region of the gene.

In practice, partially degenerate oligonucleotides may be incorporated into a target gene by a variety of different procedures. One strategy entails inserting a target gene into a plasmid between two unique restriction endonuclease sites and using PCR, in separate reactions, to amplify overlapping fragments (Fig. 8.6). The primer pair that is used to amplify the left fragment consists of mismatched oligonucleotides (synthesized to contain degenerate oligonucleotides) that bind to the lower strand of the target DNA and a reg-

Figure 8.5 Chemical synthesis of oligonucleotide primers with any of the four nucleotides at defined positions. In this case, the flask with G phosphoramidite consists of a mixture of nucleotides such as 94% G, 2% A, 2% C, and 2% T, leading to a mixture of oligonucleotides that may have A, C, or T at the sites where G is the specified nucleotide.

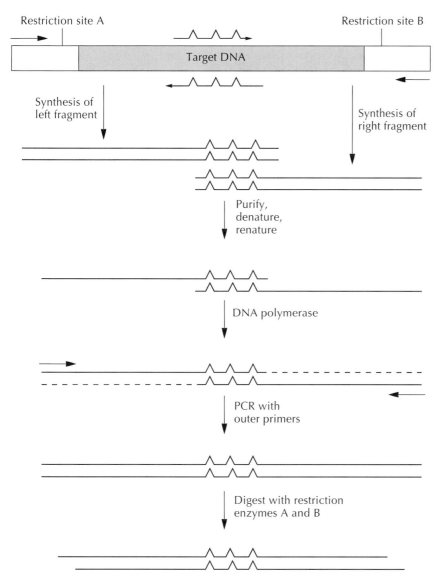

Figure 8.6 Random mutagenesis of a target DNA by using degenerate oligonucleotides and PCR. The left and right portions of the target DNA are amplified separately by PCR. The primer pairs are shown by horizontal arrows. A mutation-producing oligonucleotide is shown as a line with three spikes; each spike denotes a position that contains a nucleotide that is not found in the native gene. The amplified fragments are purified, denatured to make them single stranded, and then reannealed. Complementary regions of overlap are formed between complementary mutation-producing oligonucleotides. The single-stranded regions are made double stranded with DNA polymerase, and then the entire fragment is amplified by PCR. The resultant product is digested with restriction endonucleases A and B and then cloned into a vector that has been digested with the same enzymes.

ular, completely complementary primer that hybridizes to a region of the upper strand that flanks the left unique restriction endonuclease site. For the right fragment, the PCR primers are the mismatched oligonucleotides (synthesized to contain degenerate oligonucleotides) that bind to the upper strand of the target DNA and a primer that is complementary to a region of the lower strand that lies outside the second (right) unique restriction endo-

nuclease site. After PCR amplification, the products are purified and combined. Denaturation and reannealing of the DNA in the mixture produce some DNA molecules that overlap in the target region. DNA polymerase is then used to form complete double-stranded DNA molecules. These molecules are amplified by PCR with a pair of primers that bind to opposite ends of the DNA molecule. The amplified DNA is then treated with the two restriction enzymes for which there are unique sites at the ends of the fragment, and the DNA is then cloned into a suitable plasmid vector. This procedure results in the production of altered genes in which the overlapping fragments used in the original PCR amplification contain randomly mutagenized DNA.

Random Mutagenesis with Nucleotide Analogs

In addition to M13-based methods for introducing a series of mutations into a cloned gene, a number of other procedures that use plasmid DNA have been devised. Figure 8.7 shows one of these procedures. A cloned gene is inserted into a plasmid cloning site next to two closely placed restriction sites. These sites are specifically chosen and arranged so that, after a double digestion, both a 3′ and a 5′ recessed end are produced. On the end of the cut DNA that is next to the cloned gene, the 3′ end is recessed, whereas on the end that is next to the plasmid DNA, the 3′ end protrudes.

The *E. coli* enzyme exonuclease III (ExoIII) degrades DNA exclusively from recessed 3′ ends. It cannot degrade from 3′ protruding ends or from either 5′ end. When ExoIII is added to the digestion mixture, the recessed 3′ strand is degraded one nucleotide at a time. After a specified time, the ExoIII reaction is terminated and the gap is filled in with the Klenow fragment of DNA polymerase I. A deoxyribonucleotide mixture that contains the four normal deoxyribonucleotides and one deoxyribonucleotide analog supplies the nucleotides for this synthesis. *E. coli* cells are transformed with repaired plasmids which contain the nucleotide analog at one or more sites within the target gene. During subsequent plasmid replication, the nucleotide analog will direct the incorporation of a nucleotide that is different from the one present in the wild-type DNA into the complementary strand of the cloned gene.

In addition to the procedure described above, several other approaches have been used to introduce random changes into specific regions of a DNA molecule. For example, random mutagenesis can be achieved by a variant of the oligonucleotide-directed mutagenesis procedure that uses M13 DNA. In this case, DNA synthesis is primed by an oligonucleotide mixture in which the nucleotide sequence of some positions has been varied at random. The result of this method or its variations is the generation of libraries that contain clones with many different mutated sites. The drawback with producing a large number of random mutations within a cloned gene by any method is that each clone must be assayed to discover which one produces a protein with the desired properties. This sort of testing is not a simple task, but often it is the only way to find proteins with novel features. Once a mutant of potential interest has been identified, the sequence of the cloned gene can be determined to ascertain which site(s) has been mutated.

Protein Engineering

About 20 of the many thousands of enzymes that have been studied and characterized biochemically account for over 90% of the enzymes that are currently being used industrially. Table 8.1 lists some of the most important

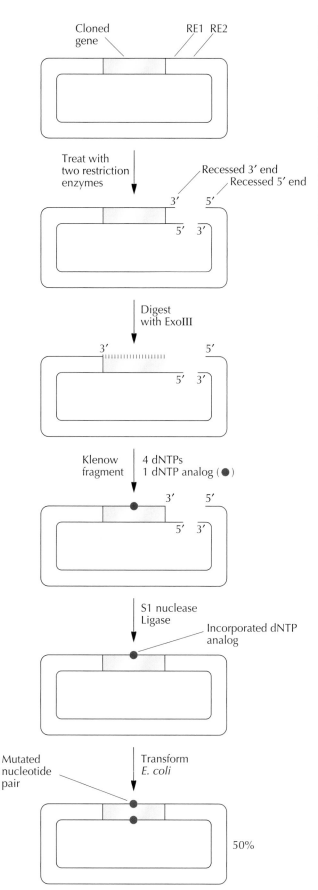

Figure 8.7 Random mutagenesis of a cloned gene. The vector carrying the cloned gene is digested with restriction enzymes 1 and 2 (RE1 and RE2). The double digestion generates one 3' and one 5' recessed end (and one 3' and one 5' protruding end). The vector is then treated with ExoIII, which digests only the strand with the 3' recessed end and, as a result, removes a portion of one strand of the cleaved DNA. The missing nucleotides (gapped DNA) are filled in by DNA synthesis catalyzed by the Klenow fragment of *E. coli* DNA polymerase I. The four deoxynucleotide triphosphates (dNTPs) or, at a low frequency, an analog of one of them is incorporated into the vector during this repair. After S1 nuclease treatment to generate blunt ends and T4 DNA ligase-catalyzed blunt-end ligation, the DNA is introduced into *E. coli*. During subsequent replication, the vectors that contain nucleotide analogs will direct the incorporation into the complementary strand of a nucleotide that differs from the original nucleotide, thereby creating a mutated gene.

Table 8.1 Some industrial enzymes and their commercial uses

Enzyme	Industrial use(s)
α-Amylase	Beer making, alcohol production
Aminoacylase	Preparation of L-amino acids
Bromelain	Meat tenderizer, juice clarification
Catalase	Antioxidant in prepared foods
Cellulase	Alcohol and glucose production
Ficin	Meat tenderizer, juice clarification
Glucoamylase	Beer making, alcohol production
Glucose isomerase	Manufacture of high-fructose syrups
Glucose oxidase	Antioxidant in prepared foods
Invertase	Sucrose inversion
Lactase	Whey utilization, lactose hydrolysis
Lipase	Cheese making, preparation of flavorings
Papain	Meat tenderizer, juice clarification
Pectinase	Clarifying fruit juices, alcohol production
Protease	Detergent, alcohol production
Rennet	Cheese making

enzymes and their primary uses. A major reason why additional enzymes are not used in industrial processes is that an activity that evolved under natural conditions usually is not well suited for a highly specialized in vitro function. Most enzymes are easily denatured by exposure to the conditions, such as high temperature and the presence of organic solvents, that are used in many industrial processes. Although thermotolerant enzymes can be isolated from thermophilic microorganisms, these organisms often lack the particular enzyme that is required for use in industrial processes. However, with the availability of directed mutagenesis and gene cloning, these constraints are no longer significant.

Adding Disulfide Bonds

The thermostability of a protein can be increased by creating a molecule that will not readily unfold at elevated temperatures. In addition, these thermostable enzymes are often resistant to denaturation by organic solvents and nonphysiological conditions such as extremes of pH. The addition of disulfide bonds can significantly increase the stability of a protein. The problem is whether extra disulfide bonds perturb the normal functioning of a protein. In one study, six variants of the enzyme T4 lysozyme were constructed by oligonucleotide-directed mutagenesis, which introduced new internal disulfide bonds. Specifically, two, four, or six amino acid residues at a time were changed to cysteine, thereby generating one, two, or three disulfide bonds, respectively (Table 8.2).

The amino acid residues that were targeted to become cysteine residues were spatially close to each other in the active enzyme. This proximity ensured that the overall conformation of the molecule would remain essentially unaffected by the formation of the new disulfide linkages. The targeted amino acids were not involved in the active site of the enzyme, which is the portion of the enzyme molecule that is probably most sensitive to small changes in conformation. The newly introduced cysteines created disulfide bonds between positions 3 and 97, 9 and 164, and 21 and 142 of the

Table 8.2 Properties of T4 lysozyme and six engineered variants

Enzyme	Amino acid at position:							No. of -S-S-	% Activity	T_m (°C)
	3	**9**	**21**	**54**	**97**	**142**	**164**			
wt	Ile	Ile	Thr	Cys	Cys	Thr	Leu	0	100	41.9
pwt	Ile	Ile	Thr	Thr	Ala	Thr	Leu	0	100	41.9
A	Cys	Ile	Thr	Thr	Cys	Thr	Leu	1	96	46.7
B	Ile	Cys	Thr	Thr	Ala	Thr	Cys	1	106	48.3
C	Ile	Ile	Cys	Thr	Ala	Cys	Leu	1	0	52.9
D	Cys	Cys	Thr	Thr	Cys	Thr	Cys	2	95	57.6
E	Ile	Cys	Cys	Thr	Ala	Cys	Cys	2	0	58.9
F	Cys	Cys	Cys	Thr	Cys	Cys	Cys	3	0	65.5

Adapted from Matsumura et al., *Nature* **342**:291–293, 1989.

wt, wild-type T4 lysozyme; pwt, pseudo-wild-type enzyme; A through F, six engineered cysteine variants; -S-S-, disulfide bonds; T_m, "melting" temperature (a measure of thermostability).

enzyme (the numbers denote the amino acid position in the polypeptide, starting from the N terminus).

After mutagenesis, each mutated gene was identified and expressed in *E. coli*. The engineered enzymes were purified, and the enzymatic activity and thermostability of each were determined (Table 8.2). The thermostability of a protein is often defined as the temperature at which the overall structure of the protein is 50% denatured; the state of denaturation can be assessed by monitoring the circular dichroism of the protein in solution. The wild-type (native) form of the enzyme T4 lysozyme has two free cysteine residues, neither of which is involved in a disulfide bond. In the so-called pseudo-wild-type enzyme, these cysteines were changed by oligonucleotide-directed mutagenesis to Thr and Ala without altering either the activity or the thermostability of the enzyme. Consequently, the pseudo-wild-type sequence provided a standard for comparing variants with potentially thermostabilizing disulfide bonds and also prevented spurious disulfide bonding between the mutated cysteine residues and the naturally occurring ones. The constructed lysozyme derivatives had from one to three disulfide bonds.

The results of this experiment indicate that the thermal stability of the enzyme increases as a result of the presence of disulfide bonds, with the most thermostable variant being the one with the largest number of disulfide bonds, and that some variants (C, E, and F), which are more thermostable than the wild type or pseudo-wild type, have lost their enzymatic activity. The loss of enzymatic activity in three of the variants probably reflects a distortion of the peptide backbone of the molecule containing a disulfide linkage between residues 21 and 142. Often, the engineering of a protein is a trial-and-error process. Hence, the precise amino acid changes that yield the "best" variant are not always obvious. However, from this experiment it is clear that increasing disulfide bonds to enhance protein stability is feasible.

In a similar study, the development of a temperature-stable mutant of the enzyme xylanase from *Bacillus circulans* was undertaken. During the making of paper, wood pulp is chemically bleached to remove the hemicellulose. Unfortunately, this step results in the creation of large amounts of potentially toxic effluent. From an environmental perspective, treatment of wood pulp with xylanase is preferred to bleaching. Treatment of wood pulp

with this enzyme lowers the amount of bleaching chemical that would otherwise be required as a part of this process. However, the stage of the process at which the enzyme is added follows hot alkali treatment of the pulp. While it is possible to lower the pH of this material by adding acid, current industry practice is directed toward using less water to cool the pulp, so any xylanase used in this process must be able to function efficiently at relatively high temperatures.

Computer modeling of the three-dimensional structure of xylanase was used to predict sites along the polypeptide chain where one, two, or three disulfide bridges could be introduced in order to stabilize the enzyme without disrupting its catalytic activity. All of the eight derivatives of the *B. circulans* xylanase that were generated showed an increase in thermostability compared to the wild type, and three of the eight mutants were as active as the wild type at 60°C. Moreover, one mutant, in which a disulfide bridge between the N- and C-terminal ends of the enzyme was introduced, was nearly twice as active as the wild type at 60°C and retained more than 85% of its activity after a 2-hour incubation at 60°C, whereas the wild type lost all of its activity after only 30 minutes at this temperature. The success of this study in improving the thermostability of an enzyme while retaining or increasing its activity indicates that this is probably a generally useful strategy that may be used to improve the thermostability of a variety of enzymes, provided that sufficiently detailed X-ray crystallographic information is available. The success of the protein engineering notwithstanding, it remains to be determined whether a thermostable xylanase can be effectively incorporated into a commercial process for the manufacture of paper.

Changing Asparagine to Other Amino Acids

When proteins are exposed to high temperatures, asparagine and glutamine residues may undergo deamidation, a reaction that releases ammonia. With the loss of the amide moiety, these amino acids become aspartic acid and glutamic acid, respectively, resulting in localized changes in the folding of the peptide chain that may lead to a loss of activity.

In one study, the effect of changing some asparagine residues in the *Saccharomyces cerevisiae* (yeast) enzyme triosephosphate isomerase was examined. This enzyme consists of two identical subunits; each subunit has two asparagine residues that may contribute to the thermosensitivity of the enzyme because they are located at the subunit interface. By oligonucleotide-directed mutagenesis, the asparagine residues at positions 14 and 78 were

Table 8.3 Stability at 100°C of the yeast enzyme triosephosphate isomerase and its engineered derivatives

Enzyme	Amino acid at position:		Half-life (min)
	14	78	
Wild type	Asn	Asn	13
Variant A	Asn	Thr	17
Variant B	Asn	Ile	16
Variant C	Thr	Ile	25
Variant D	Asp	Asn	11

Adapted from Ahern et al., *Proc. Natl. Acad. Sci. USA* **84:**675–679, 1987.

Enzyme stability is expressed as the half-life, or rate of enzyme inactivation, at 100°C. A longer half-life indicates a more stable enzyme.

targeted for change (Table 8.3). Converting either of these asparagine residues to threonine or isoleucine enhanced the thermostability of the enzyme, whereas, as predicted, changing one of the asparagine residues to aspartic acid reduced thermostability. When both asparagine residues were changed to aspartic acid residues, the resulting enzyme was unstable even at ambient temperature and had low enzymatic activity (not shown in Table 8.3).

The engineered proteins were also tested for their sensitivity to proteolytic digestion. A positive correlation between thermostability and resistance to proteolysis was observed. On the basis of these results, it should be possible to generate thermostable forms of other enzymes by mutating nonessential asparagine codons.

Reducing the Number of Free Sulfhydryl Residues

Occasionally, an expressed foreign protein is less active than expected. Protein engineering can be used to increase this activity. For example, when the complementary DNA (cDNA) for human β-interferon (IFNβ) was cloned and expressed in *E. coli*, the protein product showed only a disappointing 10% of the antiviral activity of the authentic, glycosylated form. And although reasonable amounts of IFNβ were synthesized, most of the IFNβ was found to exist as dimers and higher oligomers that were inactive.

Analysis of the DNA sequence of the IFNβ gene revealed that the encoded protein has three cysteine residues, so one or more of these residues could be involved in the intermolecular disulfide bonding that produced the dimers and oligomers in *E. coli* but not in human cells. It was reasoned that conversion of one or more of the cysteine codons into serine codons might result in an IFNβ derivative that would not form oligomers. Serine was chosen to replace cysteine because the structures of the two amino acids are identical, except that serine contains oxygen instead of sulfur and as a result cannot form disulfide bonds.

When this study was undertaken, detailed information about the structure of β-interferon was lacking, so the researchers were forced to rely on data derived from related proteins. In other words, they did not know which of the three cysteine residues contributed to intermolecular disulfide bond formation. However, the locations of the cysteine residues that form the internal disulfide bonds in IFNα, a structurally similar molecule, were known, so a comparison of the primary amino acid sequence of this molecule with that of IFNβ was made (Fig. 8.8). Alignment of the amino acid sequences of these two protein molecules indicated that Cys-31 and Cys-141 in IFNβ are located in positions similar to those of Cys-29 and Cys-138 in IFNα. Because Cys-29 and Cys-138 in IFNα are involved in the formation of an intramolecular disulfide bond, it seemed reasonable to assume that Cys-17 of IFNβ was not involved in intramolecular disulfide bonding and was therefore a good candidate for modification.

This deduction proved to be correct. No multimeric complexes were formed when a Ser-17 variant of IFNβ (Ser-17 IFNβ) was expressed in *E. coli*. Moreover, the Ser-17 IFNβ had a specific activity similar to that of authentic, native IFNβ and was more stable during long-term storage than the native form.

Increasing Enzymatic Activity

In addition to stabilizing an enzyme by directed mutagenesis, it may be feasible to modify its catalytic function. Currently, detailed information about

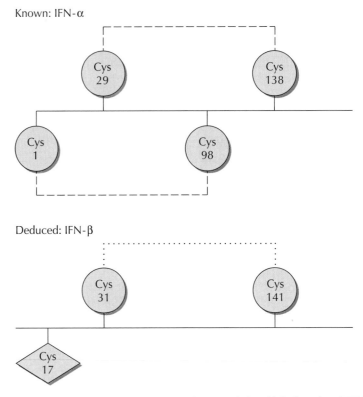

Figure 8.8 Positions of the cysteine residues and disulfide bonds of IFNα and IFNβ. The known intramolecular disulfide bond in IFNα is indicated by a dashed line, and the deduced intramolecular disulfide bond in IFNβ is indicated by a dotted line.

the geometry of the active site of a well-characterized enzyme is required in order to alter enzymatic activity in a meaningful way. With such data, researchers are able to deduce which specific changes might be necessary to modulate the substrate-binding specificity of an enzyme.

The enzyme tyrosyl-tRNA synthetase from *B. stearothermophilus* has been modified with respect to substrate binding. These results are illustrative of what can be done in this area of research. Tyrosyl-tRNA synthetase catalyzes the aminoacylation of a transfer RNA (tRNA) that specifically accepts tyrosine (i.e., tRNATyr) in a two-step process:

(1) Tyr + ATP → Tyr-A + PP$_i$
(2) Tyr-A + tRNATyr → Tyr-tRNATyr + AMP

In step 1, tyrosine (Tyr) is activated by ATP to yield enzyme-bound tyrosyl adenylate (Tyr-A), with the concomitant formation of pyrophosphate (PP$_i$). In step 2, the tyrosyl adenylate is hydrolyzed by the free 3′-hydroxyl of the incoming tRNA molecule, so the tyrosine moiety becomes attached to the tRNA and AMP is released. Both of these reactions take place while the substrates are bound to tyrosyl-tRNA synthetase.

The three-dimensional structure of tyrosyl-tRNA synthetase from *B. stearothermophilus* had already been determined, and the active site had been mapped. With the aid of computer graphics, predictions were made about the effects of changing one or more amino acid residues of the active site on the interaction of the enzyme with the reaction substrates. To test

whether these predictions were correct, the gene for tyrosyl-tRNA synthetase was specifically modified by oligonucleotide-directed mutagenesis. A threonine residue at position 51, i.e., Thr-51, was replaced by either an alanine or a proline residue. In the native enzyme, the hydroxyl group of Thr-51 forms a long hydrogen bond with the ring oxygen of the ribose moiety of tyrosine adenylate. It was deduced that the removal of this weak hydrogen bond might improve the affinity of the enzyme for ATP.

The resultant enzyme variants were characterized by determining their kinetic constants, and some of the observed changes were more dramatic than anticipated (Table 8.4). For the Ala-51 variant, the binding affinity (K_m) of the enzyme for ATP increased approximately twofold, without any significant change in the catalytic rate constant (k_{cat}). By contrast, the enzyme that contained a proline in position 51 bound ATP more than 100-fold more tightly than did the native enzyme. The catalytic efficiency (k_{cat}/K_m) of the aminoacylation reaction was increased with both of the variants. The result obtained with the Pro-51 variant was unexpected because, theoretically, the addition of a proline residue should distort at least locally the α-helical polypeptide backbone found in this region. Thus, one might expect that this conformational change would reduce substrate binding.

This study demonstrates that even though it may be difficult to predict the precise effect that a particular amino acid change will have on the reaction kinetics, this approach can correctly identify those amino acid side chains that might be altered to improve the kinetic behavior of an enzyme. Moreover, as a consequence of these kinds of experiments, it is clear that the affinity of an enzyme for its substrate as well as the catalytic efficiency of the enzyme-catalyzed reaction can be improved by in vitro manipulation of a cloned gene.

Modifying Metal Cofactor Requirements

Subtilisins are a group of serine proteases that are secreted into growth medium by gram-positive bacteria and are widely used as biodegradable cleaning agents in laundry detergents. All subtilisins bind tightly [affinity constant (K_a) = 10^7 M] to one or more molecules of calcium per molecule of enzyme. Calcium binding stabilizes the enzyme. Unfortunately, since subtilisins are used in industrial settings where there are a large number of metal-chelating agents that can bind calcium, these enzymes are rapidly inactivated under these conditions. To circumvent this problem, it is initially necessary to abolish completely the ability of a subtilisin to bind calcium and then to attempt to increase the stability of this modified enzyme.

The starting point for the development of a modified subtilisin was the isolated subtilisin BPN' gene from *Bacillus amyloliquefaciens*. Prior to this

Table 8.4 Aminoacylation activity of native (Thr-51) and modified (Ala-51 and Pro-51) tyrosyl-tRNA synthetases

Enzyme	k_{cat} (s^{-1})	K_m (mM)	k_{cat}/K_m (s^{-1} M^{-1})
Thr-51	4.7	2.5	1,860
Ala-51	4.0	1.2	3,200
Pro-51	1.8	0.019	95,800

Adapted from Wilkinson et al., *Nature* **307**:187–188, 1984.

The units for K_m, the binding constant of the enzyme for ATP, are millimolar units (mM); the units for k_{cat}, the catalytic rate constant, are reciprocal seconds (s^{-1}); and the units for k_{cat}/K_m, the catalytic efficiency, are s^{-1} M^{-1}.

work, the subtilisin BPN' protein had been well characterized and its high-resolution X-ray crystallographic structure had been determined. Oligonucleotide-directed mutagenesis was used to construct a mutant form of the gene for this enzyme by deleting the nucleotides encoding the portion of the protein, amino acid residues 75 to 83, that is responsible for binding to calcium. The protein with this stretch of amino acids removed does not bind calcium and surprisingly retains an overall conformation that is similar to the native form.

The next steps in the development of a stable subtilisin from one that lacked a calcium-binding domain entailed determining which sites might be modified and which amino acid residues should be placed in these sites. The researchers assumed that any amino acids that had previously interacted with the calcium binding loop in the native form of the enzyme would not be the best amino acids to have in those positions. In total, 10 amino acids were considered to be candidates for possible modification. Moreover, since it was not known a priori which amino acid residues might best contribute to stabilizing the enzyme molecule, random mutagenesis was used for each of these sites.

Amino acids selected for mutagenesis came from four separate regions of the protein: the N terminus (residues 2 to 5), the omega loop (residues 36 to 44), an α-helical region (residues 63 to 85) and a β-pleated region (residues 202 to 220). To identify the best amino acid at a particular position, mutant clones were grown in the wells of microtiter plates, heated to 65°C for 1 hour, allowed to cool, and then assayed for subtilisin activity. It was necessary to express the active calcium-free subtilisin in *B. subtilis* because it was found to be lethal when expressed in *E. coli*.

After the initial screening, stabilizing mutations were identified at 7 of the 10 positions that were examined (Table 8.5). When these stabilizing mutations were combined into a single gene, the enzyme that was produced had kinetic properties that were very similar to those of the native form of subtilisin. Moreover, the modified form of subtilisin was nearly 10 times more stable than the native form of the enzyme in the absence of calcium and, surprisingly, about 50% more stable than the native enzyme in the presence of calcium. Although this work was somewhat time-consuming

Table 8.5 Effect of random mutations of selected amino acid residues on the stability of subtilisin BPN' lacking a calcium-binding domain

Region of protein	Amino acid residue	Stabilizing mutation	Fold increase in half-life
N terminus	2	Gln→Lys	2.0
	3	Ser→Cys	17.0
	4	None found	Nil
	5	Pro→Ser	1.2
Omega loop	41	Asp→Ala	1.5
	43	Lys→Asn	1.2
α-Helix	73	Ala→Leu	2.6
	74	None found	Nil
β-Pleat structure	206	Gln→Cys	17.0
	214	None found	Nil

Adapted from Strausberg et al., *Bio/Technology* **13:**669–673, 1995.

The mutations at positions 3 and 206 to Cys occur in the same clone and provide such a high level of stability because of the formation of a disulfide bridge between these residues.

and painstaking, it demonstrates that complex properties of enzymes that involve a large number of different amino acid residues can be genetically engineered.

Modifying Enzyme Specificity

Although protein engineering studies using oligonucleotide-directed mutagenesis have focused on modifying and enhancing existing properties of specialized enzymes, it is conceivable that an enzyme could be redesigned to produce an entirely unique catalytic entity. For example, new site-specific endonucleases have been created from the relatively nonspecific enzyme *Fok*I endonuclease.

To date, more than 2,500 restriction-modification enzymes have been discovered from a large number of different organisms. Since many of these enzymes recognize the same DNA sequence, there are only about 200 different recognition sites, with the vast majority of the known enzymes recognizing DNA sequences that are 4 to 6 base pairs (bp) in length. Restriction enzymes that recognize 4 to 6 bp cut DNA relatively frequently and are not as useful for producing large DNA fragments as are restriction endonucleases that recognize DNA sequences that are 8 bp or longer (i.e., rare cutters). Since the discovery of new restriction enzymes is tedious, time-consuming, and unlikely to yield many new enzymes that recognize DNA sequences that are 8 bp or longer, an alternative means of obtaining new rare cutters is to develop these enzymes by protein engineering.

The class of proteins that contain unique structural domains that each bind a molecule of Zn^{2+} are called zinc finger proteins. These proteins bind to DNA in a sequence-specific manner by inserting a protein α-helical region into the major groove of the DNA double helix. For example, a mouse protein called Zif268 has three separate zinc finger domains, and each zinc finger interacts with a specific DNA triplet codon. Moreover, since the zinc fingers bind to the DNA independently of one another, they can be linked together in a peptide so that they will bind to a predetermined site on the DNA. Thus, it is possible to engineer nucleases that can cut DNA at specific sites by fusing the several zinc finger-encoding sequences with the portion of the gene for the nonspecific nuclease *Fok*I from the bacterium *Flavobacterium okeanokoites*. To test this idea, a hybrid gene that included six consecutive histidine residues at the N terminus to facilitate purification of the fusion protein, three zinc fingers, a $(Gly_4Ser)_3$ linker peptide to confer flexibility to the fusion protein, and the portion of the *Fok*I gene that encodes the nuclease activity was constructed (Fig. 8.9). After purification of the expressed protein, the N-terminal histidine residues may be removed by treatment with thrombin.

Bacteria that produce restriction enzymes protect their own DNA from being cleaved by these restriction enzymes by synthesizing enzymes that bind to and methylate the same sequence on the DNA as the restriction enzyme. However, the host genome is not protected from digestion by the synthetic *Fok*I hybrid restriction endonuclease. Therefore, during cell growth the expression of the hybrid enzyme was prevented by placing it under the control of the bacteriophage T7 expression system.

Figure 8.9 Gene construct for a zinc finger–*Fok*I hybrid restriction endonuclease.

Table 8.6 Stability and activity of various modified versions of tPA

tPA variant	Modification(s)	Stability in plasma	Fibrin binding	Activity in plasma	Activity vs clots
1	Thr(103)→Asn	10	0.34	0.68	0.56
2	LysHisArgArg(296–299)→AlaAlaAlaAla	0.85	0.93	0.13	1.01
3	Thr(103)→Asn, LysHisArgArg(296–299)→AlaAlaAlaAla	5.3	0.33	0.13	0.65
4	Thr(103)→Asn, Asn(117)→Gln	3.4	1.0	1.13	1.17
5	LysHisArgArg(296–299)→AlaAlaAlaAla, Asn(117)→Gln	1.2	1.33	0.16	1.38
6	Thr(103)→Asn, LysHisArgArg(296–299)→AlaAlaAlaAla, Asn(117)→Gln	8.3	0.87	0.06	0.85

Adapted from Peña et al., *Proc. Natl. Acad. Sci. USA* **91**:3670–3674, 1994.

All of the values shown are normalized to the wild type. Plasma stability is the reciprocal of the time it takes for plasma clearance; larger numbers indicate a more stable derivative. Fibrin specificity is reflected by a high activity versus clots and a low activity in plasma.

Two *Fok*I hybrid restriction endonucleases, each designed to cleave bacteriophage λ DNA at different single sites, were produced. One cleaved bacteriophage λ DNA at its target site, and the other cleaved λ at the expected site and to a lesser extent at two other sites. The latter result is not entirely surprising, since zinc fingers recognize triplet codons by interacting primarily with two of the three bases. Thus, although these hybrid enzymes are not yet ready for routine laboratory use, the strategy of using zinc finger motifs and a nuclease domain to create unique restriction endonucleases appears promising.

Increasing Enzyme Stability and Specificity

The enzyme tissue plasminogen activator (tPA) is a multidomain serine protease that is medically useful for the dissolution of blood clots. However, since tPA is rapidly cleared from the circulation, it must be administered by infusion. Therefore, to be effective with this form of delivery, high initial concentrations of tPA must be used. Unfortunately, under these conditions, tPA can cause nonspecific internal bleeding. Thus, a long-lived tPA that has an increased specificity for fibrin in blood clots and is not prone to induce nonspecific bleeding would be desirable. It was found that these three properties could be separately introduced by directed mutagenesis into the gene for the native form of tPA. First, changing Thr-103 to Asn causes the enzyme to persist in rabbit plasma approximately 10 times as long as the native form. Second, changing amino acids 296 to 299 from Lys-His-Arg-Arg to Ala-Ala-Ala-Ala produces an enzyme that is much more fibrin specific than the native form. Third, changing Asn-117 to Gln causes the enzyme to retain the level of fibrinolytic activity found in the native form. Moreover, combining these three mutations in a single construct allows all three activities to be expressed simultaneously (Table 8.6). Additional work is needed to determine whether a modified form of tPA is an acceptable replacement for native tPA.

SUMMARY

The proper functioning of a protein is due to its conformation, which is a consequence of its amino acid sequence and subunit structure. Certain amino acids in a protein chain play important roles in determining the specificity, thermostability, and other properties of a protein. Changing even a single nucleotide of the gene encoding a target protein can result in

the incorporation of an amino acid that can either disrupt the normal activity or enhance a specific property of a protein. With the emergence of recombinant DNA technology, it has become possible to alter or replace nucleotides of a cloned gene and produce proteins with specific amino acids at defined sites. This procedure is called directed mutagenesis, and it can be performed in various ways. Commonly, the target gene is first subcloned onto bacteriophage M13 DNA. The single-stranded form of this bacteriophage is then copied by using an oligonucleotide primer that is designed to introduce a specified nucleotide into the target gene. *E. coli* cells are transformed with the double-stranded M13, and some of the M13 bacteriophage progeny carry a cloned gene variant that contains the directed mutation. These bacteriophages are identified, the altered gene is subcloned into an expression vector, and the expressed protein is tested for activity. There are also plasmid- and PCR-based strategies for introducing these kinds of changes into cloned genes. In many instances, the amino acid change(s) that might enhance a particular property of a target protein is not known a priori. In these cases, random mutagenesis rather than oligonucleotide-directed mutagenesis is preferred.

The choice of which amino acid to change is often based on knowledge of the role of a particular amino acid in the functional protein. This knowledge comes from genetic studies or X-ray crystallographic data of the three-dimensional organization of the protein. Specific sites or regions can be altered to improve the thermostability, pH tolerance, specificity, allosteric regulation, cofactor requirements, and other properties of enzymes that are used in industrial processes. For example, thermostability has been enhanced by changing amino acids at two sites in the enzyme triosephosphate isomerase. Not only is this approach helpful in engineering new properties for existing proteins, but it may eventually be used to design unique enzymes.

References

Ahern, T. J., J. I. Casal, G. A. Petsko, and A. M. Klibanov. 1987. Control of oligomeric enzyme thermostability by protein engineering. *Proc. Natl. Acad. Sci. USA* **84:**675–679.

Brange, J., U. Ribel, J. F. Hansen, G. Dodson, M. T. Hansen, S. Havelund, S. G. Melberg, F. Norris, K. Norris, L. Snel, A. R. Sorensen, and H. O. Voigt. 1988. Monomeric insulins obtained by protein engineering and their medical implications. *Nature* **333:**679–682.

Chen, K., and F. H. Arnold. 1991. Enzyme engineering for nonaqueous solvents: random mutagenesis to enhance activity of subtilisin E in polar organic media. *Bio/Technology* **9:**1073–1077.

Deng, W. P., and J. A. Nickoloff. 1992. Site-directed mutagenesis of virtually any plasmid by eliminating a unique site. *Anal. Biochem.* **200:**81–88.

Geisselsoder, J., F. Witney, and P. Yuckenberg. 1987. Efficient site-directed in vitro mutagenesis. *Bio Techniques* **5:**786–791.

Herlitze, S., and M. Koenen. 1990. A general and rapid mutagenesis method using polymerase chain reaction. *Gene* **91:**143–147.

Hermes, J. D., S. M. Parekh, S. C. Blacklow, H. Köster, and J. R. Knowles. 1989. A reliable method for random mutagenesis: the generation of mutant libraries using spiked oligodeoxyribonucleotide primers. *Gene* **84:**143–151.

Keyt, B. A., N. F. Paoni, C. J. Refino, L. Berleau, H. Nguyen, A. Chow, J. Lai, L. Peña, C. Pater, J. Ogez, T. Etcheverry, D. Botstein, and W. F. Bennett. 1994. A faster-acting and more potent form of tissue plasminogen activator. *Proc. Natl. Acad. Sci. USA* **91:**3670–3674.

Kim, Y.-G., J. Cha, and S. Chandrasegaran. 1996. Hybrid restriction enzymes: zinc finger fusions to *Fok*I cleavage domain. *Proc. Natl. Acad. Sci. USA* **93:**1156–1160.

Kirchhoff, F., and R. C. Desrosiers. 1995. Random mutagenesis of short target DNA sequences via PCR with degenerate oligonucleotides. *Methods Mol. Biol.* **57:**323–333.

Landt, O., H.-P. Grunert, and U. Hahn. 1990. A general method for rapid site-directed mutagenesis using the polymerase chain reaction. *Gene* **96:**125–128.

Mark, D. F., S. D. Lu, A. A. Creasey, R. Yamamoto, and L. S. Lin. 1984. Site-specific mutagenesis of the human fibroblast interferon gene. *Proc. Natl. Acad. Sci. USA* **81:**5662–5666.

Matsumura, M., G. Signor, and B. W. Mathews. 1989. Substantial increase of protein stability by multiple disulfide bonds. *Nature* **342:**291–293.

Nosoh, Y., and T. Sekiguchi. 1990. Protein engineering for thermostability. *Trends Biotechnol.* **8:**16–20.

Piechocki, M. P., and R. N. Hines. 1994. Oligonucleotide design and optimized protocol for site-directed mutagenesis. *Bio Techniques* **16:**702–707.

Shiraishi, H., and Y. Shimura. 1988. A rapid and efficient method for targeted random mutagenesis. *Gene* **64:**313–319.

Smith, M. 1985. In vitro mutagenesis. *Annu. Rev. Genet.* **19:**423–462.

Strausberg, S. L., P. A. Alexander, D. T. Gallagher, G. L. Gilliland, B. L. Barnett, and P. N. Bryan. 1995. Directed evolution of a subtilisin with calcium-independent stability. *Bio/Technology* **13:**669–673.

Wakarchuk, W. W., W. L. Sung, R. L. Campbell, A. Cunningham, D. C. Watson, and M. Yaguchi. 1994. Thermostabilization of the *Bacillus circulans* xylanase by the introduction of disulfide bonds. *Protein Eng.* **7:**1379–1386.

Wetzel, R., L. J. Perry, and C. Veilleux. 1991. Mutations in human interferon gamma affecting inclusion body formation identified by a general immunochemical screen. *Bio/Technology* **9:**731–737.

Wilkinson, A. J., A. R. Fersht, D. M. Blow, P. Carter, and G. Winter. 1984. A large increase in enzyme-substrate affinity by protein engineering. *Nature* **307:**187–188.

Zoller, M. J., and M. Smith. 1982. Oligonucleotide-directed mutagenesis using M13-derived vectors: an efficient and general procedure for the production of point mutations in any fragment of DNA. *Nucleic Acids Res.* **10:**6487–6500.

REVIEW QUESTIONS

1. What physical and chemical properties of enzymes are targets for enhancement by directed mutagenesis?

2. You have cloned a bacterial gene that is expressed in *E. coli*, and now you want to alter its activity. However, as a consequence of technical problems with the original M13 protocol, only a very small fraction of mutagenized clones of this gene actually carry the modified gene; the vast majority of the clones contain the unaltered form of the gene. How would you perform site-specific mutagenesis so that a much larger proportion of the clones have the desired mutation?

3. You have isolated a gene for an enzyme that is expressed in *E. coli*. Describe how you would alter the catalytic activity of the enzyme. Assume that you know the DNA sequence of the gene but do not know anything about which regions of the enzyme are important for catalytic activity.

4. Discuss the advantages and disadvantages of oligonucleotide-directed mutagenesis with either bacteriophage M13 or PCR.

5. Describe a strategy for oligonucleotide-directed mutagenesis with plasmid DNA.

6. How can degenerate oligonucleotides be used to generate randomly mutagenized DNA?

7. Describe a strategy for increasing the stability of a protein that has (a) no cysteine residues and (b) an odd number of cysteine residues.

8. How does replacing asparagine with another amino acid residue affect the stability of a protein?

9. How can the cofactor requirements of an enzyme be altered?

10. Describe how you would go about changing the catalytic activity or substrate specificity of an enzyme whose gene you have isolated. Why would you want to do this?

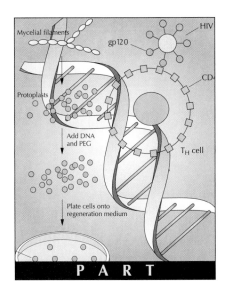

Molecular Biotechnology of Microbial Systems

WITH THE ADVENT of recombinant DNA technology, many of the properties of microorganisms that might be useful in a variety of applications can be more readily exploited. In Part II, we examine some of the uses for genetically engineered microbial systems.

Currently, bacteria are being genetically manipulated to act as biological factories for the production of pharmaceutical proteins, restriction endonucleases, chemical compounds, amino acids, antibiotics, and biopolymers. In some applications, a cloned gene has been introduced into a bacterial host cell to create a novel biosynthetic pathway that produces a unique metabolite. Genes from pathogenic organisms have been cloned and used as diagnostic probes for the diagnosis of disease in livestock and humans. In other instances, isolated genes have been used to produce safer and more effective vaccines.

Genetic manipulation of microbial systems also entails enhancing the natural ability of certain bacterial species that carry out specific biological processes. For example, researchers have developed bacterial strains that can degrade noxious environmental pollutants, improve the growth of plant crops, degrade cellulosic biomass into utilizable low molecular weight carbon compounds, and kill specific insect pests.

It is often assumed that growth of large quantities of microbes is a routine procedure. Successful large-scale production of recombinant proteins, however, requires that many different factors be controlled during both the growth phase and the purification process to ensure that high yields of a pure product are obtained.

Molecular Diagnostics

The success of modern medicine and agriculture often depends on the ability of workers in these fields to detect the presence of specific viruses, bacteria, fungi, parasites, proteins, and small molecules in humans, animals, plants, water, and soil. For example, the prevention, control, or treatment of infectious disease is generally facilitated by the early and accurate identification of the causative pathogenic organism. Many of these detection procedures require the growth in culture of the potential pathogen and then the analysis of a spectrum of physiological properties that facilitate its identification. Although tests of this type are effective and reasonably specific, they are often slow and costly. These constraints apply to the identification of both bacterial and parasitic (Table 9.1) organisms. In addition, if the pathogenic organism does not grow well or cannot be cultivated at all, the opportunity to detect the disease-causing organism is severely limited. For example, *Chlamydia trachomatis*, an obligately intracellular parasite, causes a sexually transmitted disease prevalent in North America and Europe. Clinical diagnosis of chlamydial infection is difficult, because long-term cell culture is required. Frequently, false-negative results (i.e., the diagnosis of the absence of the organism is erroneous) are obtained, and consequently adequate treatment procedures are not implemented. Certainly, if growth were required for detection, then at best only a few of all known pathogenic organisms could ever be routinely identified. To overcome this major constraint, molecular diagnostic procedures using either immunological or DNA detection methodologies have been devised.

In general, any useful detection strategy must be specific, sensitive, and simple. Specificity means that the assay must yield a positive response for only the target organism or molecule. Sensitivity means that the diagnostic test must identify very small amounts of the target organism or molecule, even in the presence of other potentially interfering organisms or substances. Simplicity is required for the test to be run efficiently, effectively, and inexpensively on a routine basis.

It is estimated that worldwide sales of immunodiagnostics accounted for approximately $3.4 billion (U.S.) in 1993, and this figure is expected to

Table 9.1 A comparison of some of the methods used to diagnose parasite infection

Method	Advantages	Disadvantages
Microscopic examination	Simple Direct detection of parasite Differentiates morphologically distinct organisms	Slow, laborious, and tedious Low sensitivity Cannot discriminate between similar organisms Requires a high skill level
In vitro culture and mouse inoculation	Detects only viable parasites Measures virulence and infectivity	Slow and expensive Different strains show a range of responses Parasite may lose its viability in the specimen Uses animals
Detection of antibodies in serum	Simple and fast Automatable Can be used to screen a large number of samples	Not always specific Does not distinguish between active and latent infections
DNA hybridization and PCR	Fast, sensitive, and specific Detects parasite directly Can distinguish different species Independent of previous infections Parasites need not be viable Automatable	Expensive and multistep Does not distinguish between live and dead organisms Possible false positives and false negatives

Adapted from Weiss, *Clin. Microbiol. Rev.* **8:**113–130, 1995.

continue to increase by 5 to 10% for the next 10 to 15 years. DNA-based diagnostic procedures had a market of around $80 million in 1994 and are expected to be worth $600 million in 2000 and $2 billion in 2004. In this chapter, the principles behind some of these molecular diagnostic procedures and the use of these procedures for a variety of different applications will be discussed.

Immunological Diagnostic Procedures

Many immunological detection systems are sensitive, specific, and simple. They can be used for a wide range of applications, including drug testing, assessment and monitoring of various cancers, detection of specific metabolites, and pathogen identification and monitoring. However, there are limitations. If the target is a protein, then the use of antibodies requires that the genes contributing to the presence of the target site must be expressed and that the target site must not be masked or blocked in any way that would prevent the binding of the antibody.

In principle, traditional diagnostic procedures for infectious agents relied on either a discrete set of traits that is characteristic of the pathogenic agent or, preferably, one unique, readily distinguishable feature. The clinical microbiologist searches for the smallest number of biological characteristics that can, with complete certainty, reveal the presence and precise identity of the pathogenic agent. For example, some infectious agents produce distinctive biochemical components. The problem is how to determine when the identifying component is present in a biological sample. Often, such a marker molecule can be identified directly in a specialized biochemical assay that is very specific for the marker molecule. But this approach would lead to a proliferation of highly individualized detection systems for different pathogenic organisms. A standardized method of identifying any key marker molecule, regardless of its chemical nature, is preferred. Because antibodies bind with high specificity to discrete target sites (antigens), assays based solely on identifying specific antibody–antigen complexes

have abolished the need to devise a unique identification procedure for each particular marker molecule.

Enzyme-Linked Immunosorbent Assay

There are a number of different ways to determine whether an antibody has bound to its target antigen. The enzyme-linked immunosorbent assay (ELISA) is one method, and it is frequently used for diagnostic detection. A generalized ELISA protocol has the following steps (Fig. 9.1).

1. Bind the sample being tested for the presence of a specific molecule or organism to a solid support such as a plastic microtiter plate, which usually contains 96 sample wells (Fig. 9.1A).
2. Add a marker-specific antibody (primary antibody) to the bound material, and then wash the support to remove unbound primary antibody (Fig. 9.1B).
3. Add a second antibody (secondary antibody) that binds specifically to the primary antibody and not to the target molecule (Fig. 9.1C). Bound (conjugated) to the secondary antibody is an enzyme such as alkaline phosphatase, peroxidase, or urease, which can catalyze a reaction that converts a colorless substrate into a colored product. Wash the mixture to remove any unbound secondary antibody–enzyme conjugate.
4. Add the colorless substrate (Fig. 9.1D).
5. Observe or measure the amount of colored product.

If the primary antibody does not bind to a target site in the sample, the first washing step removes it. Consequently, the secondary antibody–enzyme conjugate has nothing to bind to and is removed during the second washing step, and the final mixture remains colorless. Conversely, if the target site is present in the sample, then the primary antibody binds to it, the secondary antibody binds to the primary antibody, and the attached enzyme catalyzes the reaction to form an easily detectable colored product.

The principal feature of an ELISA system is the specific binding of the primary antibody to the target site. If the target molecule is, for example, a protein, then a purified preparation of this protein is generally used to generate the antibodies that will be used to detect the target. The resulting antibody mixture, which is found in the serum (antiserum) of an inoculated animal (usually a rabbit), contains a number of different antibodies that would each bind to a different antigenic determinant (epitope) on the target molecule. Such a mixture of antibodies is called a polyclonal preparation. For certain diagnostic assays, the use of polyclonal antibodies has two drawbacks: (1) the amounts of the different antibodies within a polyclonal preparation may vary from one batch to the next, and (2) polyclonal antibodies cannot be used to distinguish between two similar targets, e.g., when the difference between the pathogenic form (target) and the nonpathogenic one (nontarget) is a single determinant. However, these problems can be overcome, because it is now possible to generate an antibody preparation that is directed against a single antigenic determinant, namely, a monoclonal antibody.

Monoclonal Antibodies

In mammals, a complex set of cellular systems has evolved to protect the body from toxic substances and invasion by infectious agents. As part of the defensive response, cells of the lymphatic system can be induced to produce

A Bind sample to support

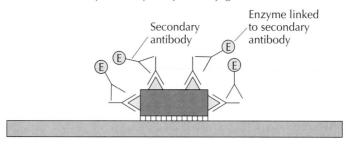

B Add primary antibody; wash

C Add secondary antibody–enzyme conjugate; wash

D Add substrate

Figure 9.1 Generalized ELISA protocol for detecting a target antigen. The enzyme (E) is conjugated to the secondary antibody.

specific proteins (antibodies) that bind to foreign substances (antigens) and—with the help of other immune system proteins, including the complement system—neutralize their biological impact. In response to an immunological challenge, each antibody-producing cell synthesizes and secretes a single antibody that recognizes with high affinity a discrete region (epitope, antigenic determinant) of the immunizing antigen. Because an antigen generally has several different epitopes, normally several cells of the immune system each produce a different antibody against one of the many epitopes of the antigen. Such a set of antibodies, all of which react with the same antigen, is designated as a polyclonal antibody.

Early in this century, although the polyclonal nature of antibodies was not appreciated, it was realized that antibody specificity could be used to

prevent infections. Later, antibodies were used as diagnostic agents to determine the presence of toxic substances in clinical samples. Unfortunately, the effectiveness of a polyclonal antibody preparation varies from batch to batch because, in some immunizations, certain antigenic determinants of a particular antigen are strong stimulators of antibody-producing cells, whereas at other times the immune system responds more actively to different epitopes of the same antigen. This variation can affect the ability of different preparations to neutralize antigens because different epitopes have different potencies (stimulating abilities). Hence, one batch of polyclonal antibody may have a low level of antibody molecules directed against a major epitope and not be as effective as a previous antibody preparation.

Consequently, a fundamental objective for the applied use of antibodies (e.g., as diagnostic agents or as components of therapeutic agents) was to discover how to create a cell line that could be grown in culture and would produce a single type of antibody molecule (monoclonal antibody) with a high affinity for a specific target antigen. Such a cell line would provide a consistent and continual source of identical antibody molecules. Unfortunately, the B lymphocytes (B cells) which synthesize antibodies do not reproduce in culture. However, it was envisioned that a hybrid cell type could be created to solve this problem. This hybrid would have the B-cell genetic components for producing antibodies and the cell division functions of a compatible cell type to enable the cells to grow in culture. It was known that normal B lymphocytes sometimes become cancer cells (myelomas) that acquire the ability to grow in culture while retaining many of the attributes of B cells. Thus, myeloma cells, especially those that did not produce antibody molecules, became candidates for fusion with antibody-producing B cells. In the mid-1970s, these ideas became reality.

Formation and Selection of Hybrid Cells

The initial step leading to the production of a hybrid cell line that produces a single antibody entails the inoculation of mice with an antigen. After several inoculations over a period of a few weeks, the animals are tested to determine whether they have developed an immune response. If they have, they are killed and their spleens are removed, washed, minced, and gently agitated to release individual cells, some of which are antibody-producing B cells. The splenic cell suspension is mixed with a suspension of myeloma cells that are genetically defective for the enzyme hypoxanthine-guanine phosphoribosyltransferase (HGPRT$^-$). The combined cell suspensions are mixed with 35% polyethylene glycol for a few minutes and then transferred to a growth medium containing hypoxanthine, aminopterin, and thymidine (HAT medium).

The polyethylene glycol treatment facilitates fusion between cells. Nevertheless, the fusion events are rare and random. There will be myeloma cells, spleen cells, myeloma–spleen fusion cells, myeloma–myeloma fusion cells, and spleen–spleen fusion cells in the mixture. The HAT medium, however, allows only the myeloma–spleen fusion cells to grow, because none of the other cell types can proliferate in this medium. Unfused spleen cells and spleen–spleen fusion cells cannot grow in any culture medium. The HGPRT$^-$ myeloma and the myeloma–myeloma fusion cells cannot use hypoxanthine as a precursor for the biosynthesis of the purines guanine and adenine, which are, of course, essential for nucleic acid synthesis. But they have a second, naturally occurring pathway for purine biosynthesis that

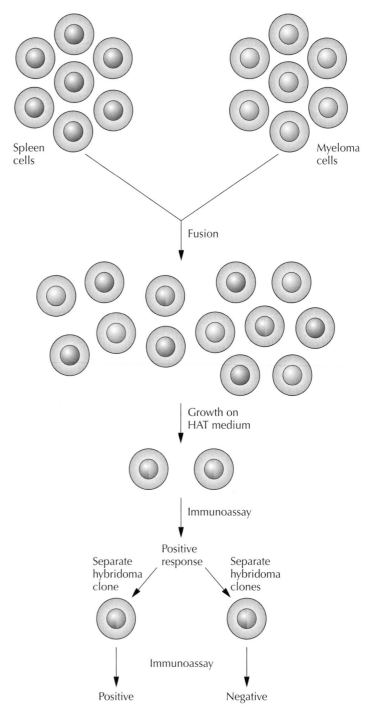

Figure 9.2 Screening for the production of a monoclonal antibody. Spleen cells from a mouse that was immunized with a specific antigen are isolated and fused in culture with myeloma cells that do not produce antibody chains. Fused cells are selected for the ability to grow on HAT medium, which contains hypoxanthine, aminopterin, and thymidine. Cells that produce a specific antibody to the immunizing antigen (hybridomas) are identified by an immunoassay and individually subcultured. A hybridoma, which grows in culture and secretes a single type of antibody molecules, is the source of a monoclonal antibody.

utilizes the enzyme dihydrofolate reductase. Therefore, aminopterin is included in the medium because it inhibits dihydrofolate reductase activity. Hence, HGPRT$^-$ myeloma and myeloma–myeloma fusion cells are unable to synthesize purines in HAT medium, so they die.

The spleen–myeloma fusion cells survive in HAT medium because the spleen cell contributes a functional HGPRT, which can utilize the exogenous hypoxanthine in the medium even though purine production by means of dihydrofolate reductase is blocked by aminopterin, and because the cell division functions of the myeloma cell are active. Thymidine is provided to overcome the block in pyrimidine production that is caused by the inhibition of dihydrofolate reductase by aminopterin. About 10 to 14 days after the fusion treatment, only spleen–myeloma fusion cells have survived and grown in the HAT medium. These cells are then distributed into the wells of plastic microtiter plates and grown on complete culture medium without HAT.

Identification of Specific Antibody-Producing Hybrid Cell Lines

The next task is to identify those hybrid cells that produce antibody against the immunizing antigen. One common screening procedure uses the culture medium, which contains secreted antibodies. The medium is collected from the wells that have growing cells and is added to the well of another microtiter plate that has been precoated with the target antigen. If the culture medium contains an antibody (primary antibody) that recognizes an epitope of the antigen, it will bind to the antigen and not be washed away during a subsequent washing step. A second antibody (secondary antibody) that is specific for mouse antibodies is added to the wells of the test plate. It will bind to any primary antibody that is bound to the antigen.

Before its use in the immunoassay, the second antibody is conjugated to an enzyme that can convert a colorless substrate to a colored compound. The presence of color in one of the test wells denotes that the original culture medium contained an antibody that was specific for the antigen (Fig. 9.2). If the culture medium does not contain an antibody that binds to the antigen, then the first wash will remove all the antibodies. Therefore, when the secondary antibody is added, it has nothing to bind to and is removed by the second washing step. In a well where such a sequence of events occurs, the substrate solution remains colorless.

Those wells of the original microtiter plate whose media give a positive (color) response in the immunoassay may contain a mixture of cell fusions. These cells are therefore diluted with culture medium and seeded into fresh wells to establish cell lines from single cells (clones). After the clones have been cultured, their media are tested again to determine which cell lines (hybridomas) produce monoclonal antibody molecules that recognize the target antigen. If more than one specific hybridoma is isolated, further tests are conducted to determine whether the different clones produce antibody against the same antigenic determinant. Each monoclonal antibody-producing clone can be maintained, more or less indefinitely, in culture. In addition, samples can be frozen in liquid nitrogen to provide a source of cells for future use.

Because a monoclonal antibody binds to a single discrete site, the specificity of an ELISA protocol can be considerably enhanced by using a monoclonal rather than a polyclonal antibody. Many monoclonal antibodies have been developed for use as immunodiagnostic agents for a variety of different compounds and pathogenic organisms (Fig. 9.3). As an alternative to the iso-

POLYPEPTIDE HORMONES
 Chorionic gonadotropin
 Growth hormone
 Luteinizing hormone
 Follicle-stimulating
 hormone
 Thyroid-stimulating hormone
 Prolactin
TUMOR MARKERS
 Carcinoembryonic antigen
 Prostate-specific antigen
 Interleukin-2 receptor
 Epidermal growth factor receptor
CYTOKINES
 Interleukins 1—8
 Colony-stimulating factor
DRUG MONITORING
 Theophylline
 Gentamicin
 Cyclosporin
MISCELLANEOUS TARGETS
 Thyroxine
 Vitamin B$_{12}$
 Ferritin
 Fibrin degradation products
 Tau protein
INFECTIOUS DISEASES
 Chlamydia
 Herpes
 Rubella
 Hepatitis B
 Legionella
 HIV

Figure 9.3 Targets for diagnostic monoclonal antibodies. HIV, human immunodeficiency virus.

lation and synthesis of monoclonal antibodies in hybridoma cells in culture, monoclonal antibodies and parts of antibodies (Fv fragments) directed against a target antigen may be selected and produced in *E. coli* (see Chapter 10).

DNA Diagnostic Systems

The genetic material of an organism contains the essential information that contributes to its various features and characteristics. For example, bacterial pathogenicity may be due to the presence of a specific gene or set of genes. Similarly, alteration of a gene may cause an inherited genetic disease in humans. In theory, the sequence of nucleotides that contributes to a particular biological characteristic is a distinctive signature that, if detectable, can be used as a definitive diagnostic determinant.

Nucleic acid hybridization is the basis for rapid and reliable assays. The physical basis of these systems is precise nucleotide base pairing and hydrogen bonding between one string of nucleotides and a complementary nucleotide sequence. A general laboratory nucleic acid hybridization scheme is as follows.

1. Bind single-stranded DNA (the target) to a membrane support.
2. Add single-stranded labeled DNA (the probe) under appropriate conditions of temperature and ionic strength to promote base pairing between the probe and the target DNAs.
3. Wash the support to remove excess unbound labeled probe DNA.
4. Detect the hybrid sequences that form between the probe and target DNA.

A nucleic acid hybridization diagnostic test has three critical elements: probe DNA, target DNA, and signal detection. This type of detection system can be both extremely specific and highly sensitive.

Hybridization Probes

To be effective, a nucleic acid hybridization probe must have a high degree of specificity. In other words, the probe must hybridize exclusively to the selected target nucleic acid sequence. False positives (i.e., responses in the absence of the target sequence) and false negatives severely undermine the utility of a diagnostic procedure. Probes can be specific at different organismic levels. They can distinguish between two or more species, determine particular strains within a given species, or identify differences between genes. Depending on the requirements of the test protocol, probes can be DNA or RNA; long (>100 nucleotides) or short (<50 nucleotides); and chemically synthesized, cloned intact genes, or isolated regions of a gene.

Sequences that might make effective probes can be isolated in a number of ways. For example, the DNA from a pathogenic organism can be cut with a restriction endonuclease and cloned into a plasmid vector. Recombinant plasmids are screened with the genomic DNA from both pathogenic and nonpathogenic strains. Those plasmids that contain sequences that hybridize only to the pathogenic strain form the basis for species-specific probes. Additional hybridization tests with DNA from a wide range of organisms are then conducted to ensure that the candidate probe sequences do not cross-hybridize. Each potential probe is also tested under simulated sample conditions, including the presence of mixed cultures, to determine its level of sensitivity.

The ability to perform nucleic acid probe diagnostic assays directly on

available samples without either additional culturing or time-consuming extraction procedures is extremely desirable, especially with clinical specimens. Researchers have successfully used probes that hybridize to target DNA from fecal samples, urine, blood, throat washings, and tissue samples without extensive DNA purification. If a target sequence is rare in the working sample, the polymerase chain reaction (PCR) can be used to amplify it.

Diagnosis of Malaria

An example of a diagnostic protocol that utilizes a DNA probe as a means of detection is the procedure developed for the detection of *Plasmodium falciparum*. This parasite causes malaria, a disease that affects about one-third of the world's population. The parasite infects and destroys red blood cells, leading to fever and, in severe cases, damage to the brain, kidneys, and other organs. Sensitive, simple, and inexpensive methods are required to identify the source(s) of the parasite in various localities, to assess the progress of eradication programs, and to facilitate early treatment. Currently, malarial infections are diagnosed by microscopic examination of blood smears, an effective but labor-intensive and time-consuming process. Although immunological procedures for *Plasmodium* detection, such as ELISAs, are rapid and amenable to automation, they do not discriminate between present and past infections, because they are designed simply to detect anti-*Plasmodium* antibodies in the blood of affected individuals.

A DNA diagnostic procedure that selectively measures only current infections, i.e., the presence of DNA-containing organisms, was developed by using highly repeated DNA from *P. falciparum*. First, a genomic library of the parasite DNA was screened with labeled parasite DNA. The most intensely labeled hybridizing colonies were selected because they were expected to contain repetitive DNA. The DNA from each of the selected colonies was then tested for its ability to hybridize with DNA from several other *Plasmodium* species that do not cause malaria. The DNA sequence that was chosen as a specific probe hybridized with *P. falciparum* but not with *P. vivax*, *P. cynomolgi*, or human DNA. This probe can detect as little as 10 pg of purified *P. falciparum* DNA or 1 ng of *P. falciparum* DNA in blood.

More than 100 different DNA diagnostic probes have been isolated and characterized for the detection of various pathogenic strains of bacteria, viruses, and parasites. For example, probes have been developed for the diagnosis of human bacterial infections caused by *Legionella pneumophila* (respiratory failure), *Salmonella typhi* (food poisoning), *Campylobacter hyointestinalis* (gastritis), and enterotoxigenic *Escherichia coli* (gastroenteritis). Clearly, this is just the "tip of the iceberg," because, in principle, nearly all pathogenic organisms can be detected by this procedure.

Detection of *Trypanosoma cruzi*

The protozoan parasite *Trypanosoma cruzi* is the causative agent of Chagas' disease. In this disease, the parasites invade the liver, spleen, lymph nodes, and central nervous system, where they multiply and destroy the parasitized cells. The parasite *T. cruzi* is quite common in Latin America. It is spread by insects and is responsible for approximately 50,000 deaths per year. Diagnosis of acute Chagas' disease is usually made by microscopic examination of a fresh blood sample. Alternatively, a test that takes a longer time but ensures that the parasite has not been overlooked entails feeding a patient's blood to uninfected insects and then examining with a microscope the contents of the insects' intestines for parasites about 30 to 40 days later.

Both of these tests are laborious, time-consuming, and costly. Chagas' disease can also be diagnosed by immunological tests; however, these tests are notorious for producing false-positive responses. As a possible alternative to these less than satisfactory procedures, several PCR-based assays have been developed. At present, PCR assays for Chagas' disease are used as adjuncts to the traditional diagnostic procedures that are currently in widespread use.

In one of the PCR procedures, a 188-base-pair (bp) DNA fragment that is present in multiple copies in the *T. cruzi* genome but is absent from the genomic DNA of several related parasites is the target sequence. The presence of the amplified 188-bp DNA fragment is readily detected by polyacrylamide gel electrophoresis. In general, with minor variations in the methodology, such as the primer sequences, PCR can facilitate detection of a wide range of bacteria, viruses, and parasites.

Nonradioactive Hybridization Procedures

In most research laboratories, nucleic acid hybridization is routinely detected by labeling the probe with a radioactive isotope, commonly phosphorus-32. High specific activity ensures an excellent signal-to-noise ratio. In a standard detection system, a radiolabeled probe is mixed with target DNA that is bound to a membrane support. After the support is washed free of nonhybridized probe DNA, the presence of radioactivity is determined by laying the membrane on X-ray film (autoradiography).

However, phosphorus-32 is short-lived, is potentially dangerous, and requires special laboratory equipment for handling and safe disposal, so nonradioactive systems for signaling hybrid DNA formation have also been

MILESTONE

Detection of Sickle Cell β^S-Globin Allele by Hybridization with Synthetic Oligonucleotides

B. J. CONNER, A. A. REYES, C. MORIN, K. ITAKURA,
R. L. TEPLITZ, and R. B. WALLACE
Proc. Natl. Acad. Sci. USA **80**:278–282, 1983

The development of techniques in the early 1980s for rapid, efficient, and inexpensive chemical synthesis of DNA oligonucleotides opened up the possibility of using radiolabeled oligonucleotides as hybridization probes for the detection of a variety of DNAs, including mutations in human genes. At the time, relatively few human genes or even complementary DNAs (cDNAs) had been isolated. Thus, it was not always easy to find a homologous hybridization probe for a particular human gene. Moreover, even if the relevant cDNAs had been isolated, it was not possible to discriminate by DNA hybridization between wild-type and mutant human

genes which had only a single-base-pair difference when the hybridization probe was large ($\geq$100 bp). However, Conner et al. synthesized specific oligonucleotides that could recognize either the wild-type or the mutant DNA. Moreover, oligonucleotides complementary to the sequences of both the wild type and the mutant were synthesized so that it could be determined whether a person was heterozygous as well as whether one was homozygous. In the case of sickle-cell anemia, two 19-base-long oligonucleotides were used: one complementary to the normal β-globin gene (β^A) and one complementary to the sickle-cell gene (β^S). DNA from normal indi-

viduals ($\beta^A\beta^A$) hybridized only with the β^A probe, DNA from individuals with sickle-cell anemia ($\beta^S\beta^S$) hybridized only with the β^S probe, and DNA from heterozygous individuals ($\beta^A\beta^S$) hybridized with both probes. This model system was the first demonstration of the feasibility of determining genotypes by DNA hybridization and opened up the possibility of detecting a range of different human genetic disorders by hybridization with oligonucleotide probes. This possibility has been realized with the development of a large number of DNA-based gene mutation tests. Although it has been largely supplanted by newer techniques such as PCR and OLA, it was important in establishing that single DNA base pair mutations could be accurately and easily detected.

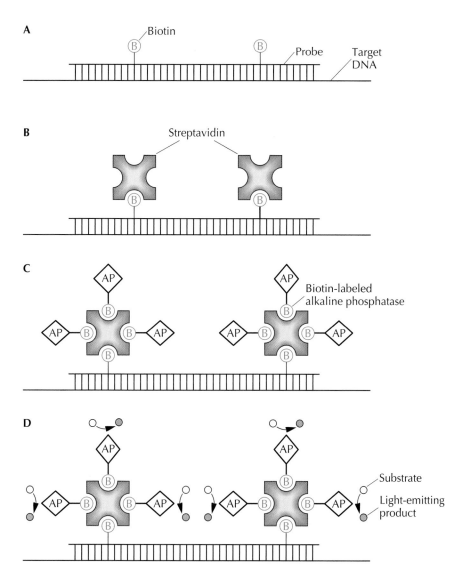

Figure 9.4 Chemiluminescent detection of target DNA. B, biotin; AP, alkaline phosphatase. **A.** A biotin-labeled probe is bound to the target DNA. **B.** Streptavidin is bound to the biotin molecules. **C.** Biotin-labeled alkaline phosphatase binds to streptavidin. **D.** Alkaline phosphatase converts the substrate into a light-emitting product.

developed. The nonradioactive detection systems achieve signal amplification by enzymatic conversion of either chromogenic or chemiluminescent substrates. Chromogenic substrates change color and chemiluminescent substrates give off light when they are converted into a specific product by an appropriate enzyme. The signal is detected, in most of these systems, by incorporating biotin-labeled nucleotides into the DNA probe and following a more or less standard procedure:

1. The biotin-labeled probe is hybridized to the target DNA (Fig. 9.4A).
2. Either avidin, a chicken egg white protein, or streptavidin, a bacterial analog of avidin, is added (Fig. 9.4B).
3. A biotin-labeled enzyme such as alkaline phosphatase or peroxidase is added (Fig. 9.4C).

4. Depending on which biotin-labeled enzyme is used in the previous step, either a chromogenic or chemiluminescent substrate is added and either the color change or the light produced as a consequence of the conversion of substrate into product is measured (Fig. 9.4D).

Alternatively, following hybridization with a biotin-labeled probe (step 2 above), a streptavidin–enzyme complex with an available biotin binding site can be added directly.

Both avidin and streptavidin bind very tightly [diffusion constant (K_d) = 10^{-15}] to biotin; in addition, each of these proteins has four separate biotin-binding sites, so a single molecule of avidin or streptavidin can bind both a biotin-labeled enzyme and a biotin-labeled probe. Enzymatic activity is not impaired by biotin labeling or binding to streptavidin. In chromogenic detection systems, the action of the enzyme on the substrate creates a colored insoluble dye that remains at the site of the hybrid DNA. In chemiluminescent systems, enzymatic alteration of the substrate generates a product that emits light at the site of the hybrid DNA.

Nonradioactive systems have other advantages: biotin-labeled DNA is stable for at least 1 year at room temperature, devices detecting chemiluminescence are as sensitive as those that detect radioactive signals, and detection of the emitted light with either X-ray film or a luminometer or scoring of a color change can be completed within a few hours. It is likely that the use of chemiluminescence, which is more sensitive than chromogenic dyes, will soon become the detection signal system of choice for many nucleic acid probe-based diagnostic assays. For PCR-based assays, the amplification product can be labeled by a fluorescent dye that is bound to the 5' end of each primer. A fluorescent compound emits light of a longer wavelength

Figure 9.5 Use of fluorescent dyes that are attached to primers for detecting amplified PCR products. The primers are marked as P1 and P2.

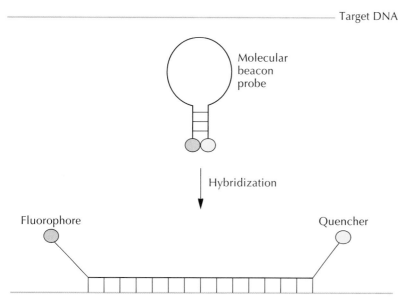

Target DNA

Molecular
beacon
probe

Hybridization

Fluorophore

Quencher

Target DNA–probe hybrid

Figure 9.6 Hybridization of a molecular beacon probe to target DNA. The un-paired single-stranded portion of the molecular beacon base pairs with its complementary sequence on the target DNA, and the stem structure becomes destabilized, physically separating the fluorophore and quencher moieties. As a consequence of the binding of the molecular beacon probe to the target DNA, the fluorescence of the fluorophore is not quenched. The appearance of fluorescence indicates that the probe has hybridized to its target sequence. Adapted from Tyagi and Kramer, *Nat. Biotechnol.* **14:**303–308, 1996.

after it absorbs light of a shorter wavelength. Fluorescein, which appears green under certain wavelengths of light, and rhodamine, which appears red, are often used for this purpose. After PCR amplification of a target DNA with fluorescent-labeled primers, the primers are separated from the amplification product and the presence of the label is detected (Fig. 9.5). If the target DNA is not present in the sample, then no fluorescent product will be observed.

A novel nonradioactive method for detecting specific sequences of nucleic acids involves using "molecular beacon" probes (Fig. 9.6). A typical molecular beacon probe is 25 nucleotides long. The middle 15 nucleotides are complementary to the target DNA and do not base pair with one another, and the five nucleotides at each end are complementary to each other. A fluorescent chromophore (fluorophore) is coupled to the 5′ end, and a nonfluorescent chromophore (quencher) that can absorb the energy emitted by the fluorophore before it fluoresces is coupled to the 3′ end. In solution at room temperature, the conformation of the molecular beacon ensures that the fluorophore and quencher are close to one another, and the fluorophore is quenched, i.e., it does not fluoresce. On the other hand, when the 15 middle nucleotides of the molecular beacon probe hybridize to a target DNA or RNA sequence, the fluorophore and quencher are separated from one another and the fluorophore is not quenched, i.e., it fluoresces. With this procedure, care must be taken to maintain the reaction mixture at near-ambient temperatures, since high temperatures can also cause the nucleotides in the hairpin stem to become unpaired, with the result that the molecule fluoresces. For this procedure to be effective, all 15 nucleotides in

the molecular beacon probe must be perfectly complementary to the target DNA or RNA.

DNA Fingerprinting

DNA fingerprinting (DNA typing) is a technique that is used to characterize biological samples for legal proceedings. This technique has been used to establish both that some suspects committed a crime and that some individuals had been wrongly convicted of crimes that they did not commit. For DNA typing, after a biological sample (e.g. blood, semen, skin, or hair) is collected, a portion is analyzed to ensure that there is a sufficient amount of intact (undegraded) DNA for the intended analyses. If there is enough undegraded DNA, it is digested with a restriction enzyme, and the fragments are separated on an agarose gel and transferred by blotting onto a nylon membrane. The membrane is hybridized sequentially with four or five separate radiolabeled probes that each recognize a distinct DNA sequence. After each hybridization reaction, the bands where the probe has bound to the digested DNA sample are visualized by autoradiography and the banding pattern for each sample is noted (Fig. 9.7). Before the next probe is used, the first probe is completely removed (stripped) from the membrane. Since each hybridization and autoradiography step can take up to 10 to 14 days, the entire process may take many weeks and even several months. A commonly used set of probes for this type of analysis consists of human minisatellite DNAs. These sequences occur throughout the human genome and consist of tandemly repeated sequences. The lengths of the repeats range from 9 to 40 bp, and the number of repeats in the minisatellites ranges from about 10 to 30. The same minisatellite DNA sequence can have a different length in different individuals. This variability is due to either a gain or a loss of tandem repeats, probably during DNA replication. These changes do not have any biological effect because minisatellite DNA does not code for any proteins. Unrelated individuals generally have minisatellites that differ in length. On the other hand, a child will inherit one minisatellite DNA sequence from one parent and another one from the other parent.

A minisatellite DNA fingerprint represents the repertoire of the lengths of some of these sequences in an individual. Because of the extensive variability in human minisatellite DNA sequences, the chance of finding two individuals in the population with the same DNA fingerprint is about one in 10^5 to 10^8. In other words, an individual's DNA banding pattern based on minisatellite DNA sequences is almost as unique as his or her fingerprints. In addition to the forensic applications of DNA fingerprinting, this technique may be used to help determine the paternity of a child. The DNA fingerprint banding pattern of the child should be a composite of the patterns of its mother and father.

If the amount of DNA in a forensic sample is quite small but the DNA is relatively undegraded, it may be possible to amplify small portions of the minisatellite DNA by PCR. Then DNA sequence analysis, a procedure which is even more sensitive than a comparison of length differences of tandem repeats, can be used to characterize the DNA.

Random Amplified Polymorphic DNA

Not only are DNA banding patterns important for forensic analyses, but they are also useful in distinguishing among different plant cultivars. One way that this can be achieved is using random amplified polymorphic DNA (RAPD) markers. In this procedure, arbitrary oligonucleotide primers, usu-

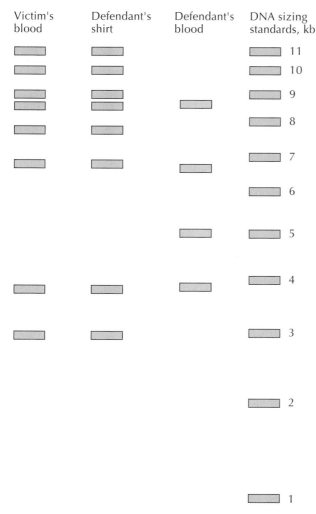

Figure 9.7 Southern blot of a forensic DNA sample. The DNA samples from the victim, the defendant's shirt, and the defendant were treated with the same restriction enzyme. Here, the banding pattern of the DNA extracted from the blood on the defendant's shirt is identical to the victim's DNA banding pattern and different from the defendant's pattern. The sizes of the DNA molecules in these bands are estimated by comparison with the positions of the sizing standards.

ally 9 to 10 bp long, that do not contain any palindromic sequences and have a G+C content of 50 to 80% are individually added to a sample of plant chromosomal DNA. Each PCR is initiated by the addition of a single primer, so the same oligonucleotide must be able to bind to opposite strands on the target DNA. Although the sequence of each primer is known, it is not known which oligonucleotide, if any, will be effective in priming the PCR. Whenever a primer can hybridize to both strands of the target DNA in the proper orientation and the two sites are about 100 to 3,000 bp from each other, then the intervening DNA region will be amplified and a characteristically sized DNA fragment which can be visualized following gel electrophoresis will be synthesized. The number of DNA fragments in a sample that are amplified is dependent on the primer and the genomic DNA used. Each time that the same primer is used with the same target DNA, the amplified products will be the same. A single nucleotide substitution in a primer will result in a complete change in the RAPD pattern. Thus, the RAPD fingerprints of different

plant cultivars can be compared when the same set of oligonucleotide primers is used. To fingerprint the DNA of two very similar plant strains or cultivars, it is often necessary to perform the RAPD procedure with a number of different arbitrary primers of known sequence until differences are revealed (Fig. 9.8). Like other molecular markers, RAPDs can be used to characterize whole genomes, individual chromosomes, or specific genes.

In comparison with other procedures for characterizing complex DNA, the RAPD procedure has a number of advantages. These include the following: (1) the same (universal) set of oligonucleotide primers can be used for all plant species; (2) no genomic libraries, radioactivity, Southern transfers, or DNA hybridization reactions are required, so that with this procedure a large number of samples may be easily and rapidly characterized; and (3) the process can be automated. Moreover, with conventional PCR analysis it is necessary to know ahead of time the sequence of a specific gene or gene segment that is the target for amplification. On the other hand, amplification in RAPD analysis occurs anywhere in a genome that contains two sequences complementary to the primer that are within the length limits of the PCR.

With this technology, scientists were able to distinguish six inbred maize lines from each other, and maize hybrids were shown to have the complement of PCR products of their parental inbred lines. RAPD markers have also been used to screen different strains of the fungus *Leptosphaeria maculans*, which is the causal agent of blackleg disease in crucifers. Differences between avirulent (nondisease-causing) and virulent (disease-causing) strains could be distinguished on the basis of specific RAPD markers.

Figure 9.8 Ethidium bromide-stained bands following polyacrylamide gel electrophoresis of PCR-amplified plant DNA. Three separate random primers were used to amplify fragments from each of the two cultivars. Cultivars 1 and 2 show identical patterns of bands with primers A and C. However, they have different patterns when primer B is used; hence, primer B can be used to distinguish between cultivars 1 and 2.

Molecular Diagnosis of Genetic Disease

The ability to diagnose the occurrence of specific inherited diseases in humans at the genetic level makes it possible for individuals to discover whether they or their offspring are at risk. DNA analysis can be used for the identification of carriers of hereditary disorders, for prenatal diagnosis of serious genetic conditions, and for early diagnosis before the onset of symptoms.

Tests at the DNA level are definitive for determining the existence of specific genetic mutations. Previously, genetic testing relied almost exclusively on biochemical assays that scored either the presence or absence of a gene product. A DNA-based test does not, however, require expression for detection of the mutant gene, thereby making it theoretically possible to develop screening assays for all single-gene diseases.

Sickle-Cell Anemia

Sickle-cell anemia is a genetic disease that is the result of a single nucleotide change in the codon for the sixth amino acid of the β chain of the hemoglobin molecule. In individuals homozygous for the defect (S/S), the shape of the red blood cells is irregular (sickle shaped) because the conformation of the hemoglobin molecule is distorted by a single amino acid change from valine to glutamic acid. The biological ramifications of this genetic alteration are severe anemia and progressive damage to the heart, lungs, brain, joints, and major organ systems. The anemia is caused by the inability of the mutated hemoglobin to carry sufficient oxygen. The life expectancy for S/S homozygotes is quite short. Heterozygous individuals (A/S) (i.e., genetic carriers) have normal-shaped red blood cells and no symptoms unless subjected to extreme conditions, such as high altitude or extremes of temperature, that lower the oxygen supply. If both parents are heterozygous (A/S), then there is a 25% chance that a child of theirs will have sickle-cell anemia, i.e., will be an S/S homozygote. The sickle-cell anemia gene occurs with a high frequency among black Africans and their descendants and in Hispanic populations. Carrier screening for the sickle-cell anemia gene is routinely conducted in the United States so that those individuals who are at risk for transmitting this gene to their offspring can be identified. One of the test systems is described below.

The single nucleotide change in the β-globin gene that causes sickle-cell anemia by chance abolishes a CvnI restriction endonuclease site. This restriction enzyme recognizes the sequence CCTNAGG and cleaves the DNA between the C and the T. (The letter N indicates that any one of the four nucleotides can occupy this position.) In the normal gene the DNA sequence is CCTGAGG, whereas in the sickle-cell anemia gene the sequence is CCTGTGG. This difference forms the basis for a DNA diagnostic assay (Fig. 9.9).

After two primer sequences that flank the CvnI site are added, a small amount of sample DNA can be amplified by PCR (Fig. 9.9A). The amplified DNA is digested with CvnI, and the cleavage products are separated by gel electrophoresis and visualized by ethidium bromide staining of the DNA in the gel. If the CvnI site is present, then a specific set of DNA fragments is observed (Fig. 9.9C). A different profile of DNA fragments occurs if the CvnI site is absent. By this procedure, the genetic makeup of a tested person can be determined quickly, directly, and easily. Moreover, because of the fortuitous loss of the CvnI site, this assay functions without the need for a target-probe hybridization reaction.

using this avoids the need for hybridization

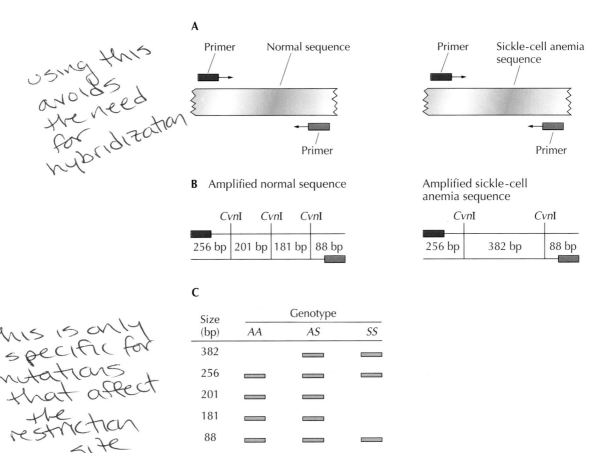

this is only specific for mutations that affect the restriction site

Figure 9.9 Detection of the sickle-cell anemia gene at the DNA level. **A.** PCR amplification of the portion of the β-globin gene containing the *Cvn*I recognition site that is altered in the mutant gene. **B.** *Cvn*I digestion of the PCR products generated as shown in panel A. The normal (wild-type) gene has three *Cvn*I sites between the PCR primers, and the mutant gene has two. **C.** Size distribution of fragments following gel electrophoresis of *Cvn*I-digested PCR-amplified β-globin DNA. *AA*, homozygous condition for the normal β-globin gene; *AS*, heterozygous condition; *SS*, homozygous condition for the sickle-cell anemia β-globin gene.

The PCR/OLA Procedure

Obviously, not all genetic changes that produce defective genes affect existing restriction endonuclease sites. Therefore, other strategies for detecting single nucleotide changes are required. One of these procedures combines PCR with an oligonucleotide ligation assay (OLA); not surprisingly, it is called PCR/OLA.

Let us assume that in a normal gene at a specific site (say, nucleotide number 106) the nucleotide pair is A·T; in the mutant form the nucleotide pair at this site is G·C. Knowledge of the sequence of nucleotides on both sides of position 106 enables two short (20-nucleotide) adjacent nucleotide sequences (oligonucleotides) that are complementary to one of the two native DNA strands to be synthesized (Fig. 9.10). The essential feature of this pair of oligonucleotides is that one of them (probe X) has as its last base at the 3′ end the nucleotide that is complementary to the nucleotide at position 106 of the normal sequence. The other oligonucleotide (probe Y) starts at its 5′ end with a nucleotide that is complementary to the nucleotide immediately adjacent to position 106. When these two probes are hybridized with

A Synthesize a pair of oligonucleotide probes

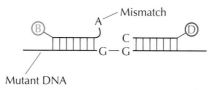

B Hybridize probes to PCR-amplified DNA

C Add ligase to hybridized DNA

D Bind probes to streptavidin; wash

E Add antidigoxigenin antibody–alkaline phosphatase conjugate; wash; add substrate

Figure 9.10 PCR/OLA procedure. B, biotin; D, digoxigenin; AP, alkaline phosphatase; SA, streptavidin; A, adenine; C, cytosine; G, guanine; T, thymine.

target DNA (which has been amplified by PCR) containing the normal sequence, the nucleotide at the 3' end of probe X base pairs with the target DNA, and probe Y is aligned so that its 5' end lies next to the 3' end of probe X. The addition of DNA ligase to the reaction covalently joins probe X and probe Y. By contrast, when these two probes are hybridized to mutant DNA in which the nucleotide at position 106 is altered, the nucleotide at the 3' end of probe X is mismatched and is not able to pair with nucleotide 106 in the target DNA sequence; probe Y, however, is perfectly aligned. In this case, DNA ligase cannot join probe X and probe Y because of the single nucleotide misalignment.

Other oligonucleotides (probes) also can be chemically synthesized to give a perfect base pair match when nucleotide 106 is mutated. Obviously, with this second set of probes, ligation occurs when they are hybridized to target DNA that contains the mutant nucleotide, whereas with normal target DNA, the single nucleotide pair mismatch prevents the ligation of the probes. In short, PCR/OLA is designed to distinguish between two possibilities: ligation and no ligation of two input probes.

To determine whether ligation has occurred between two indicator probes, probe X is labeled at its 5' end with biotin and probe Y is labeled at its 3' end with digoxigenin. The low-molecular-weight compound digoxigenin serves as an antibody binding indicator. After the hybridization and ligation steps are carried out, the DNA is denatured to release the hybridized probes, and the entire mixture is transferred to a small plastic well that has been coated with streptavidin. The well is then washed to remove unbound material, so only the biotin-labeled probe DNA remains bound. Next, antidigoxigenin antibodies, which have been previously coupled to alkaline phosphatase, are added to the well. After an additional washing step to remove unbound conjugated antidigoxigenin antibodies, a colorless chromogenic substrate is added. The appearance of color in the well indicates that antidigoxigenin antibodies have bound to digoxigenin and that the digoxigenin-labeled probe was ligated to the biotinylated probe. If no color appears, then no ligation occurred.

With two pairs of probes, it is possible to ascertain the genetic makeup of any tested individual. For example, heterozygous individuals yield positive results with both pairs of probes. The DNA from people with two copies of the normal gene gives a positive response only with the set of probes that contains the nucleotide complementary to the nucleotide at the normal site. Finally, DNA from individuals with two altered gene copies will give a positive response only with the set of probes that is designed to detect the mutant site. To minimize the amount of the original sample DNA that is required for the assay, the segment of the target DNA sequence that contains the nucleotide site to be tested is amplified by PCR before the hybridization reaction.

Overall, the PCR/OLA system is rapid, sensitive, and highly specific. It has even been automated with a robotic workstation to carry out the steps of the assay procedures. Under these conditions, as many as 1,200 ligation reactions can be conducted per day.

The ligase chain reaction assay is a simpler, albeit less sensitive variant of the PCR/OLA system. Sample DNA is mixed with an excess amount of a pair of OLA indicator probes (as described above) in the presence of a heat-resistant DNA ligase. After an initial ligation reaction at 65°C, the temperature is raised to 94°C to denature the probe–target DNA hybrid and then lowered to 65°C to allow hybridization of the free nonligated OLA in-

dicator probes to the target DNA. The cycle is repeated 20 times. If the OLA indicator probes match the target DNA perfectly, then ligation will occur at 65°C during each cycle, and after 20 cycles enough ligation product (i.e., probe X joined to probe Y) will accumulate to be observed by either gel electrophoresis or an ELISA detection system. If no ligation occurs because of a mismatch, then no joined probe product will be produced or detected.

Genotyping with Fluorescence-Labeled PCR Primers

PCR primers labeled with different fluorescent dyes can be used in the development of nonradioactive color-based detection systems. To distinguish between mutant and wild-type DNA, PCR is performed with two different primers. One is exactly complementary to the wild-type DNA and is labeled at its 5′ end with rhodamine (red). The other is complementary to the mutant DNA and is labeled at its 5′ end with fluorescein (green) (Fig. 9.11). In both cases, amplification is programmed by a third unlabeled primer that is complementary to the opposite strand. Since PCR amplification can occur only when the primer is exactly complementary to the target DNA, the presence of these three primers in the same reaction mixture will result in the amplification of either the wild-type or the mutant DNA or both, depending on which target DNAs are initially present to act as PCR templates. If an individual is homozygous for the wild-type DNA, after PCR and removal of unincorporated primer, the reaction mixture will fluoresce red; if he or she is homozygous for the mutant DNA, the reaction mixture will fluoresce green; and if he or she has both mutant and wild-type DNA (i.e., is heterozygous), the reaction mixture will fluoresce yellow. This assay can be automated and can be adapted for any single nucleotide target site of any gene that has been sequenced.

Mutation at Different Sites within One Gene

Not all genetic diseases are due to a single specific nucleotide change within a gene. In most cases, a variety of different intragenic sites can mutate, and each mutation can cause the same form of the genetic disease. For example, β-thalassemia is a genetic disease that is due to a loss in the activity of β-globin. Heterozygotes (carriers) tend to have only a mild form of anemia. In contrast, people who are homozygous for one of eight or more possible mutations must receive regular blood transfusions and other treatment to survive. Because a mutation at any one of eight or more different sites within the β-globin gene can cause β-thalassemia, testing for a change at only one particular nucleotide site is insufficient; at least eight separate tests are necessary. Although this is feasible, it is costly.

Therefore, a PCR/hybridization strategy that uses one reaction assay system has been devised to screen for mutant nucleotide sites at different locations within a single gene. A set of sequence-specific oligonucleotide probes (each 20 nucleotides in length) is synthesized. Each oligonucleotide can form a perfect match with a segment of the target gene that corresponds to the site of a known mutation. A thymidine homopolymer [poly(dT)] tail approximately 400 nucleotides long is added to the 3′ end of each of the probes. The tail enables the probe DNA to be physically bound to a predesignated discrete spot on a nylon filter while ensuring that the rest of the probe is accessible for hybridization (Fig. 9.12). Segments of the sample (test) DNA that span each of the possible mutant sites within the gene are amplified simultaneously by PCR. In each case, one of each of the pairs of primers is labeled at the 5′ end with biotin. The amplified target DNA is

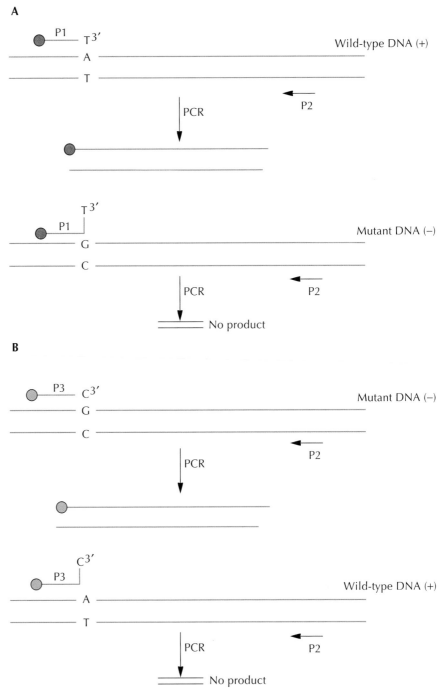

Figure 9.11 Detection of a single-base mutation with fluorescence-labeled PCR primers. **A.** Primers P1 and P2 amplify DNA from the wild-type sequence. The same primers cannot amplify DNA from the mutant sequence because primer P1 mismatches with this DNA. Primer P1 is labeled at its 5′ end with rhodamine. Primer P2 is unlabeled. **B.** Primers P3 and P2 amplify DNA from the mutant but not the wild-type sequence. Primer P3 is labeled at its 5′ end with fluorescein. Primer P2 is unlabeled. The plus and minus signs denote wild-type and mutant sites, respectively. The genotypes +/+, +/−, and −/− produce PCR products that contain rhodamine only, rhodamine and fluorescein, and fluorescein only and that fluoresce red, yellow, and green, respectively.

A Bind sequence-specific probes to a membrane

Membrane

B Hybridize with PCR-amplified biotinylated
segments of target gene

C Add streptavidin–alkaline phosphatase conjugate

D Add substrate

Figure 9.12 Procedure for the detection of mutations at different sites within one
gene. B, biotin; TTTTT, poly(dT) tail; SA, streptavidin; AP, alkaline phosphatase.

then hybridized to the filter-bound probes under conditions that allow only
perfect matches to hybridize. Streptavidin with attached alkaline phos-
phatase is added during the hybridization reaction (horseradish peroxidase
or urease can also be used). After hybridization, the filter is washed and a
colorless substrate is added. A colored spot on the filter appears wherever
there is a perfect nucleotide match between an amplified target DNA seg-
ment and one of the specific oligonucleotide probes. One filter can be "spot-
ted" with a number of different sequence-specific oligonucleotide probes.
Thus, with a single filter assay, one of a number of different mutant sites can
be identified.

Prospects

The development of molecular diagnostic assays is a growing and dynamic field. Although the technical details of various tests may differ, the general principles have been established. At present, PCR has contributed significantly to overcoming the problem of the limited availability of target DNA. The use of PCR for probe systems has eliminated concern about the sensitivity of the detection signal, with the result that nonradioactive chromogenic, chemiluminescent, or fluorescent systems can be used reliably for certain assays. Moreover, in a number of tests, PCR treatment and electrophoretic analysis are sufficient to determine the presence of either a genetic mutation or an infectious agent in the targeted sample. Undoubtedly, many novel DNA-based systems will be created for the diagnosis of most, if not all, of the common genetic, infectious, and malignant diseases.

SUMMARY

To be effective, a diagnostic test must be (1) specific for the target molecule; (2) sensitive enough to detect minute levels of the target; and (3) technically simple, with unequivocal results that can be obtained readily. There are two classes of molecular diagnostic techniques. One category relies on the specificity of an antibody for a particular antigen. The other uses nucleic acid hybridization or PCR to detect a specific nucleic acid sequence.

A common assay that uses antibodies is the ELISA. In this procedure, (1) a sample is bound to a solid support, (2) a primary antibody specific for the target antigen is added and binds to the target antigen, (3) a secondary antibody–enzyme conjugate that binds to the primary antibody is added, and (4) a colorless substrate that is transformed into a colored compound by the enzyme in the conjugate is added. The appearance of a color response in an ELISA denotes the presence of the target molecule in the sample.

ELISAs have been used for detecting various proteins, identifying viruses and bacteria, and determining the presence of low-molecular-weight compounds in a wide range of biological samples.

To increase the specificity of the primary antibody and to ensure the reliability of the antibody preparation, monoclonal antibodies are often used for diagnostic ELISAs. More recently, to overcome the problem of the high cost of monoclonal antibodies, antibody fragments have been produced in *E. coli* by using a cloning strategy that yields a wide range of Fv combinations (combinatorial library).

Nucleic acid hybridization can be a highly sensitive and specific method for detecting the presence of a nucleic acid sequence in a biological sample. This method has been used to develop diagnostic assays for disease-causing organisms in a clinical setting and other organisms in the environment.

Because a nucleic acid detection assay is directed toward a known DNA sequence, primers for PCR can be synthesized and then used to amplify the target sequence. The detection assay can be run in a nonradioactive system, such as the biotin–streptavidin–chemiluminescence protocol, or the amplified PCR product can be scored by gel electrophoresis. Also, a PCR product can be labeled with a fluorescent dye that is attached to the 5′ ends of the primer.

One way to characterize forensic samples is by DNA fingerprinting. In

this technique, human minisatellite DNA, which does not encode any proteins and is highly variable in sequence, is usually used as a hybridization probe. The extensive variability of different human minisatellite DNA sequences means that each human being produces a unique set of hybridized DNA bands.

To characterize plant DNA, a set of arbitrary oligonucleotide primers can be used to amplify random segments of the plant DNA by PCR, and, after electrophoresis, to produce a specific set of DNA bands. This procedure is called the RAPD procedure. Any particular set of primer sequences will produce a unique collection of amplified DNA fragments that is characteristic of the genomic DNA of a particular plant cultivar.

DNA diagnostic assays can also be used to detect the presence of a single nucleotide change in a particular gene. One of these methods distinguishes between the ligation and the absence of ligation of two oligonucleotides. A single nucleotide mismatch at the junction of the hybridized oligonucleotides prevents ligation. In general, the use of PCR increases the resolution of nucleic acid diagnostic tests and should also decrease the overall costs of these assays.

References

Barany, F. 1991. Single-nucleotide genetic disease detection using cloned thermostable ligase. *Proc. 1991 Miami Bio/Technol. Winter Symp.* **1**:88.

Barker, R. H., L. Suebsaeng, W. Rooney, G. C. Alecrim, H. V. Dourado, and D. F. Wirth. 1986. Specific DNA probe for the diagnosis of *Plasmodium falciparum* malaria. *Science* **231**:1434–1436.

Bugawan, T. L., R. K. Saiki, C. H. Levenson, R. M. Watson, and H. A. Erlich. 1988. The use of non-radioactive oligonucleotide probes to analyze enzymatically amplified DNA for prenatal diagnosis and forensic HLA typing. *Bio/Technology* **6**:943–947.

Carlson, D. P., C. Superko, J. Mackey, M. E. Gaskill, and P. Hansen. 1990. Chemiluminescent detection of nucleic acid hybridization. *Focus* **12**:9–12.

Caskey, C. T. 1987. Disease diagnosis by recombinant DNA methods. *Science* **236**:1223–1229.

Chehab, F. F., and Y. W. Kan. 1989. Detection of specific DNA sequences by fluorescence amplification: a color complementation assay. *Proc. Natl. Acad. Sci. USA* **86**:9178–9182.

Debenham, P. G. 1992. Probing identity: the changing face of DNA fingerprinting. *Trends Biotechnol.* **10**:96–102.

Erlich, H. A., D. Gelfand, and J. J. Sninsky. 1991. Recent advances in the polymerase chain reaction. *Science* **252**:1643–1651.

Gillam, I. C. 1987. Non-radioactive probes for specific DNA sequences. *Trends Biotechnol.* **5**:332–334.

Hartskeerl, R. A., M. Y. L. De Wit, and P. R. Klatser. 1989. Polymerase chain reaction for the detection of *Mycobacterium leprae*. *J. Gen. Microbiol.* **135**:2357–2364.

Jeffreys, A. J., A. MacLeod, K. Tamaki, D. L. Neil, and D. G. Monckton. 1991. Minisatellite repeat coding as a digital approach to DNA typing. *Nature* **354**:204–209.

Kingsbury, D. T. 1987. DNA probes in the diagnosis of genetic and infectious diseases. *Trends Biotechnol.* **5**:107–111.

Klevan, L., and G. Gebeyehu. 1990. Biotinylated nucleotides for labeling and detecting DNA. *Methods Enzymol.* **184**:561–577.

Kohler, G., and C. Milstein. 1975. Continuous cultures of fused cells secreting antibody of predefined specificity. *Nature* **256**:495–497.

Kuppuswamy, M. N., J. W. Hoffmann, C. K. Kasper, S. G. Spitzer, S. L. Groce, and S. P. Bajaj. 1991. Single nucleotide primer extension to detect genetic diseases: experimental application to hemophilia B (factor IX) and cystic fibrosis genes. *Proc. Natl. Acad. Sci. USA* **88**:1143–1147.

Mathews, J. A., and L. J. Kricka. 1988. Analytical strategies for the use of DNA probes. *Anal. Biochem.* **169**:1–25.

Nickerson, D. A., R. Kaiser, S. Lappin, J. Stewart, L. Hood, and U. Landegren. 1990. Automated DNA diagnostic using an ELISA-based oligonucleotide ligation assay. *Proc. Natl. Acad. Sci. USA* **87**:8923–8927.

Persing, D. H., T. F. Smith, F. C. Tenover, and T. J. White (ed.). 1993. *Diagnostic Molecular Microbiology: Principles and Applications.* American Society for Microbiology, Washington, D.C.

Plikaytis, B. B., R. H. Gelber, and T. M. Shinnick. 1990. Rapid and sensitive detection of *Mycobacterium leprae* using a nested-primer gene amplification assay. *J. Clin. Microbiol.* **28**:1913–1917.

Pollard-Knight, D., A. C. Simmonds, A. P. Schaap, H. Akhavan, and M. A. W. Brady. 1990. Nonradioactive DNA detection on Southern blots by enzymatically triggered chemiluminescence. *Anal. Biochem.* **185**:353–358.

Rafalski, J. A., and S. V. Tingey. 1993. Genetic diagnostics in plant breeding: RAPDs, microsatellites and machines. *Trends Genet.* **9**:275–279.

Saiki, R. K., P. S. Walsh, C. H. Levenson, and H. A. Erlich. 1989. Genetic analysis of amplified DNA with immobilized sequence-specific oligonucleotide probes. *Proc. Natl. Acad. Sci. USA* **86**:6230–6234.

Sayler, G. S., and A. C. Layton. 1990. Environmental application of nucleic acid hybridization. *Annu. Rev. Microbiol.* **44**:625–648.

Tyagi, S., and F. R. Kramer. 1996. Molecular beacons: probes that fluoresce upon hybridization. *Nat. Biotechnol.* **14**:303–308.

Waldmann, T. A. 1991. Monoclonal antibodies in diagnosis and therapy. *Science* **252**:1657–1662.

Weiss, J. B. 1995. DNA probes and PCR for diagnosis of parasitic infections. *Clin. Microbiol. Rev.* **8**:113–130.

White, T. J., N. Arnheim, and H. A. Erlich. 1989. The polymerase chain reaction. *Trends Genet.* **5**:185–188.

White, T. J., R. Madej, and D. H. Persing. 1992. The polymerase chain reaction: clinical applications. *Adv. Clin. Chem.* **29**:161–196.

Winter, G., and C. Milstein. 1991. Man-made antibodies. *Nature* **349**:293–299.

Yu, K. F., A. Van Deynze, and K. P. Pauls. 1993. Random amplified polymorphic DNA (RAPD) analysis, p. 287–301. *In* B.R. Glick and J.E. Thompson (ed.), *Methods in Plant Molecular Biology and Biotechnology.* CRC Press, Boca Raton, Fla.

REVIEW QUESTIONS

1. Briefly describe how the change in the human β-globin gene that gives rise to sickle-cell anemia can be detected by using PCR.

2. Describe and discuss the PCR/OLA detection protocol.

3. What is an ELISA? How does it work?

4. Describe three types of nonradioactive DNA labels. What are the advantages of nonradioactive detection procedures?

5. You have been given the task of developing a simple, sensitive, and reproducible diagnostic procedure for a double-stranded DNA virus that is devastating a local cattle population. Because effective treatment of this dis-

ect diagnosis, you need to be able to detect
 that are present in infected animals before
3riefly explain how you would proceed and
ar course of action.

s meant by sensitivity, specificity, and sim-

rently diagnosed? How might the existing

d to facilitate the detection of specific DNA

probe, and how does it work?

, and how is it used to characterize traces of

re, and how can it be used to characterize

CHAPTER 10

Microbial Production of Therapeutic Agents

Prior to the development of recombinant DNA technology, most human protein pharmaceuticals were available in only limited quantities, they were costly to produce, and, in a number of cases, their biological mode of action was not well characterized. When recombinant DNA technology was first developed, it was heralded as a means of producing a whole range of possible human therapeutic agents in sufficient quantities for both efficacy testing and eventual human use. This forecast has turned out to be true. Today, the "genes" (mostly cDNAs) for over 400 different proteins that are potential human therapeutic agents have been cloned. Most of these sequences have been expressed in host cells and are currently being tested for the treatment of various human diseases (Table 10.1). Although more than 30 of these biotechnology drugs have been approved for use in the United States (Table 10.2), it will be several years before many of these proteins are commercially available, because medical products must first be tested rigorously in animals and then undergo thorough human trials before being approved for general use. However, the monetary incentive for pharmaceutical companies is considerable. It has been estimated that the annual global market for human pharmaceuticals is about $150 billion and growing. The worldwide market for recombinant protein drugs is growing at an annual rate of approximately 12 to 14%, and it has been estimated that it will be worth around $20 billion by the year 2000.

The development of treatments, preventive procedures, and cures for human diseases has been the outstanding contribution of medicine and science to human well-being in the 20th century. This process, however, is a continuous one. So-called old diseases (e.g., tuberculosis) can reappear if preventive measures are relaxed or if resistant organisms arise. The idea of using antibodies as therapeutic agents is appealing, and it seems probable that specific antibodies could be used to attack toxins, bacteria, viruses, or even cancer cells. An antibody may be viewed as a target-seeking missile or a magic bullet that either can directly neutralize an offending agent or, if equipped with a warhead or poison arrow, can destroy a specific target cell. Unfortunately, despite their theoretical promise, the use of antibodies for

227

Table 10.1 Some human proteins that have been produced by recombinant DNA technology for treating various disorders

Protein	Disorder(s)
α_1-Antitrypsin	Emphysema
Adrenocorticotropic hormone	Rheumatic diseases
B-cell growth factors	Immune disorders
Bactericidal/permeability-increasing protein	Infections
Brain-derived neurotrophic factor	Amyotrophic lateral sclerosis (Lou Gehrig's disease)
Calcitonin	Osteomalacia
Colony-stimulating factors	Cancer
Chorionic gonadatropin	Female infertility
Endorphins and enkephalins	Pain
Epidermal growth factor	Burns
Erythropoietin	Anemia, kidney disorders
Factor VIII	Hemophilia
Factor IX	Hemophilia
Growth hormone	Growth defects
Growth hormone-releasing factor	Growth defects
Hemoglobin	Anemia
Insulin	Diabetes
Insulin-like growth factor	Diabetes, renal failure
Interferons (α, β, γ)	Viral diseases, cancer, multiple sclerosis
Interleukins	Cancer, immune disorders
Interleukin-1 receptor	Asthma, rheumatoid arthritis
Lymphotoxin	Cancer
Macrophage-activating factor	Cancer
Nerve growth factor	Nerve damage
Platelet-derived growth factor	Atherosclerosis
Relaxin	Birthing
Serum albumin	Insufficient plasma proteins
Somatomedin C	Growth defects
Thyroid-stimulating hormone	Thyroid cancer
Tissue plasminogen activator	Blood clots
Tumor necrosis factor	Cancer
Urogastrone	Ulcers
Urokinase	Blood clots

preventing or treating diseases and other conditions has been limited. However, recently, the development of recombinant DNA technology and the means of producing monoclonal antibodies has combined with a rapidly growing understanding of the molecular structure and function of immunoglobulin molecules to reactivate interest in using specific antibodies to treat various diseases.

Pharmaceuticals

Isolation of Interferon cDNAs

A number of different strategies have been used to isolate either the genes or cDNAs for human proteins. In some cases, the target protein is isolated and a portion of the amino acid sequence is determined. From this infor-

Table 10.2 Some recombinant proteins that have been approved for human use by the U.S. Food and Drug Administration

Compound	Company	Disorder
Antihemophilic factor	Miles, Baxter Healthcare, Genetics Institute	Hemophilia A
DNase I	Genentech	Cystic fibrosis
Erythropoietin	Amgen and Ortho Biotech	Anemia, kidney disease
Glucocerebrosidase	Genzyme	Gaucher disease
Growth hormone	Genentech	Growth hormone deficiency in children
Insulin	Eli Lilly	Diabetes
IFN-α_{2a}	Hoffmann-La Roche	Hairy cell leukemia, Kaposi sarcoma
IFN-α_{2b}	Schering-Plough	Hairy cell leukemia, genital warts, Kaposi sarcoma, hepatitis B and C
IFN-α_{n3}	Interferon Sciences	Genital warts
IFN-β_{1b}	Berlex Laboratories and Chiron	Relapsing multiple sclerosis
IFN-γ_{1b}	Genentech	Chronic granulomatous disease
Interleukin-2	Chiron	Renal cell carcinoma
Somatotropin	Eli Lilly	Growth hormone deficiency
Tissue plasminogen activator	Genentech	Acute myocardial infarction, acute massive pulmonary embolism

mation, a DNA coding sequence is deduced. The appropriate oligonucleotide is synthesized and used as a DNA hybridization probe to isolate the gene or cDNA from either a genomic or cDNA library. Alternatively, antibodies are raised against the purified protein and used to screen a gene expression library. For human proteins that are synthesized primarily in a single tissue, a cDNA library from the mRNA of this tissue will be enriched for the target DNA sequence. For example, the major protein synthesized by the islets of Langerhans of the pancreas is insulin; 70% of the mRNA fraction isolated from these cells encodes insulin.

A simple cDNA enrichment procedure does not work well for human proteins that are found in very low concentrations or whose site of synthesis is not known. Consequently, it may be necessary to devise additional experimental approaches. The human interferon (IFN) proteins, which include the α-, β-, and γ-interferons (IFN-α, IFN-β, and IFN-γ) are naturally occurring proteins; each one might have a different therapeutic use (Table 10.3). When their cDNAs were initially isolated, a novel scheme had to be devised to overcome the scarcity of both the mRNAs and proteins. The procedure used to isolate IFN cDNAs included the following steps.

1. Size-fractionated mRNA was isolated from human leukocytes, reverse transcribed, and inserted into the *Pst*I site of plasmid pBR322.
2. The approximately 6,000 clones that were produced following transformation of *Escherichia coli* were divided into 12 pools of 512 clones each. Pools of clones rather than individual clones were tested to speed the identification process.
3. Each pool was hybridized to a crude interferon mRNA preparation.
4. The hybridized input mRNA was isolated from the cloned

Table 10.3 Possible human therapeutic uses for some of the different types of human interferon

Interferon	Disorder
α_{2a}	Hepatitis C, hairy cell leukemia
α_{2b}	Bladder cancer, head and neck cancer, malignant melanoma, multiple myeloma, non-Hodgkin's lymphoma, renal cell carcinoma, Crohn disease, HIV infection
α_{n3}	AIDS, cervical dysplasia, papillomavirus infections, chronic hepatitis C infection, genital warts
β_{1a}	Multiple sclerosis
β_{1b}	Chronic progressive multiple sclerosis
γ_{1b}	Renal cell carcinoma, chronic granulomatous disease

DNA–mRNA hybrids and translated in a cell-free protein synthesis system.

5. Each translation mixture was then assayed for IFN antiviral activity. The pools that showed IFN activity contained a clone with a cDNA that hybridized to IFN mRNA.
6. Positive pools were divided into eight subgroups of 64 clones each and retested. This subgrouping process was repeated until a clone with the complete cDNA for a human IFN was identified.

Subsequently, whenever large quantities of the IFN are required, the IFN cDNAs can be subcloned into an *E. coli* expression vector and expressed at high levels.

Engineering Human Interferons

After the isolation of the first IFN gene in the early 1980s, it was determined that there are a number of different IFNs. On the basis of chemical and biological properties, the IFNs can be classified, as noted earlier, into three different groups: IFN-α, IFN-β, and IFN-γ. The proteins IFN-α and IFN-β are synthesized in cells that have been exposed to viruses or viral RNA; IFN-γ is synthesized in response to cell growth-stimulating agents. IFN-α is encoded by a family of at least 15 nonallelic genes, whereas IFN-β and IFN-γ are each encoded by a single gene. The IFN-α subtypes have different specificities. For example, the antiviral activities of IFN-α_2 and IFN-α_1 are approximately the same when assessed with a virus-challenged bovine cell line, but IFN-α_2 is seven times more effective than IFN-α_1 when human cells are treated with virus. IFN-α_2 is 30 times less effective than IFN-α_1 when mouse cells are used in this assay.

Several groups of workers have attempted to form engineered IFNs with combined properties by taking advantage of the fact that different members of the IFN-α gene family vary in the extent and specificity of their antiviral activity. Theoretically, this can be achieved by splicing a portion of one IFN-α gene with a DNA sequence from a different IFN-α gene to create a hybrid protein that exhibits novel properties, i.e., properties different from those of either of the contributing genes. In one study, hybrid genes from IFN-α_2 and IFN-α_3 were constructed in an effort to create proteins with novel IFN activities (Fig. 10.1).

Comparison of the sequences of the two IFN-α cDNAs indicated that they have common restriction sites at positions 60, 92, and 150. Digestion of both cDNAs at the common restriction sites and ligation of the DNA frag-

Figure 10.1 Structure of the IFNα₂ and IFNα₃ genes and four hybrid genes. Comparison of the sequences of the IFNα₂ and IFNα₃ genes shows shared restriction enzyme sites (RE1, RE2, and RE3). Digestion of the genes at the indicated restriction sites and ligation of the resulting fragments generate a number of different hybrid IFN genes, of which four possibilities are shown.

ments yielded a number of hybrid derivatives of the original genes (Fig. 10.1). These hybrids were expressed in *E. coli*, and the resultant proteins were purified and examined for various biological functions. When tested for the extent of protection of mammalian cells in culture, some of the hybrid IFNs were found to have greater activity than the parental molecules. In addition, many of the hybrid IFNs induced test cells to synthesize (2'-5') oligoisoadenylate synthetase. This enzyme generates (2'-5')-linked oligonucleotides, which, in turn, activate a latent cellular endoribonuclease that cleaves viral mRNA. Other hybrid IFNs had an antiproliferative activity against various human cancers that was greater than that of the parental molecules. The creation of these hybrid interferons demonstrates that new potential therapeutic molecules can be constructed by combining functional domains from related genes.

Engineering Human Growth Hormone

The strategy of designing proteins by either functional domain shuffling or directed mutagenesis can be used to augment or constrain the mode of action of a protein. For example, native human growth hormone (hGH) binds to both growth hormone and prolactin receptors on a number of different cell types. To avoid unwanted side effects during therapy, it is desirable that hGH bind only to growth hormone receptors. Because the segment of the growth hormone molecule that binds to the growth hormone receptor overlaps but is not identical to the portion of the molecule that binds to the prolactin receptor, it should be possible to selectively decrease the binding to the prolactin receptor.

Site-specific mutagenesis of the cloned hGH cDNA was used to change some of the amino acid side chains that act as ligands for Zn^{2+} (i.e., His-18,

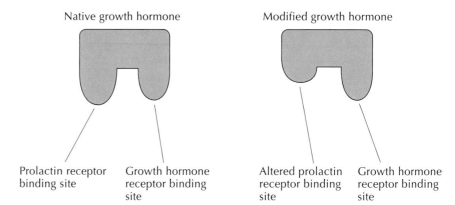

Figure 10.2 Schematic representation of native and modified hGH. Oligonucleotide-directed mutagenesis was used to alter hGH so that it no longer bound to the prolactin receptor but retained its specificity for the growth hormone receptor.

His-21, and Glu-174), an ion required for the high-affinity binding of hGH to the prolactin receptor (Fig. 10.2). As hoped, these modifications yielded hGH derivatives that bind to the growth hormone receptor but not to the prolactin receptor. Although this work is interesting, it is still uncertain whether these derivatives will be commercialized.

Optimizing Gene Expression

After designing a protein, it is important that the expression of the gene be optimized. Initially, researchers determine whether an authentic product can be synthesized in sufficient quantities in either a prokaryotic or a eukaryotic expression system. It is preferable to use prokaryotic organisms, because scale-up costs tend to be lower and yields tend to be higher in such systems. Unfortunately, not all organisms are equally effective in synthesizing a functional form of a heterologous protein; therefore, comparative assessments must be conducted.

In a study of the expression of the gene for human interleukin-3 in different host cells, the "best" organism was *Bacillus licheniformis* (Table 10.4). Although somewhat higher levels of expression were achieved with one of the *E. coli* constructs, the product was a 20-kilodalton (kDa) fusion protein of interleukin-3 and a portion of *E. coli* β-galactosidase and not the 15-kDa mature authentic protein. Such a fusion protein is not usually acceptable as a therapeutic agent. The yeasts *Kluyveromyces lactis* and *Saccharomyces cere-*

Table 10.4 Levels of interleukin-3 synthesis achieved in different host systems

Host system	Promoter	Expression level (units)	Protein form
B. licheniformis	Amylase	300	15 kDa (mature)
E. coli	*lacZ*	500	20 kDa (fusion)
E. coli	*lacZ*	20	15 kDa (mature)
Human cells	Metallothionein	2	20–40 kDa
K. lactis	Lactase	20	20–100 kDa
S. cerevisiae	Mating factor α	20	20–100 kDa

Adapted from van Leen et al., *Bio/Technology* **9**:47–52, 1991.

In each case, one of the strongest available promoters for that system was used.

visiae and the human cells were able to glycosylate interleukin-3, although the level of expression in these cell types was relatively low. Glycosylation, which is not essential for the activity of interleukin-3, resulted in an apparent size heterogeneity.

Enzymes

DNase I

Cystic fibrosis is the most common fatal hereditary disease among Caucasians, with approximately 30,000 diagnosed cases in the United States and another 23,000 cases in Canada and Europe. Individuals with cystic fibrosis are highly susceptible to bacterial infections in their lungs. Antibiotic treatment of patients who have these recurring infections eventually leads to the selection of antibiotic-resistant bacteria. The presence of bacteria, some alive and some lysed, causes the accumulation of a thick mucus in the lungs of these patients, making breathing very difficult. A portion of the thick mucus in the lungs is the result of the DNA that is released when bacterial cells are lysed. To address this problem, scientists at the U.S. biotechnology company Genentech isolated and expressed the gene for the enzyme DNase I, which can hydrolyze long polymeric DNA chains into much shorter oligonucleotides. They then delivered the purified enzyme in an aerosol to the lungs of patients with cystic fibrosis. The DNase I treatment acted to decrease the viscosity of the mucus in the lungs and made it easier for these patients to breathe. While this treatment is not a cure for cystic fibrosis, it nevertheless relieves the most severe symptom of the disease in most patients. This enzyme has recently been approved for use by the U.S. Food and Drug Administration; it is predicted to have sales of approximately $100 million (U.S.) per annum by around the year 2000.

Alginate Lyase

Alginate is a polysaccharide polymer that is produced by a wide range of seaweeds and both soil and marine bacteria. Alginate is composed of chains of the sugars β-D-mannuronate and α-L-guluronate. The properties of a particular alginate depend on the relative amounts and distribution of these two saccharides. For example, stretches of α-L-guluronate residues form both interchain and intrachain cross-links by binding calcium ions; the β-D-mannuronate residues bind other metal ions. The cross-linked alginate polymer forms an elastic gel. In general, the physical properties of an alginate polymer is related to its viscosity, which is in turn directly proportional to its molecular size.

The excretion of alginate by mucoid strains of *Pseudomonas aeruginosa* significantly contributes to the viscosity of the mucus found in the lungs of patients with cystic fibrosis. Therefore, either instead of or in addition to the DNase I treatment mentioned above, depolymerization of the alginate would help to clear the lungs in these patients. The enzyme alginate lyase can liquefy viscous bacterial alginate. Therefore, it is reasonable to consider this protein as a possible treatment for alleviating the blocked airways of cystic fibrosis patients.

An alginate lyase gene has been isolated from a *Flavobacterium* sp., a gram-negative soil bacterium that is a strong producer of this enzyme activity. A *Flavobacterium* clone bank was constructed in *E. coli* and screened for alginate lyase-producing clones by plating the entire clone bank onto solid medium containing alginate. Following growth, colonies that pro-

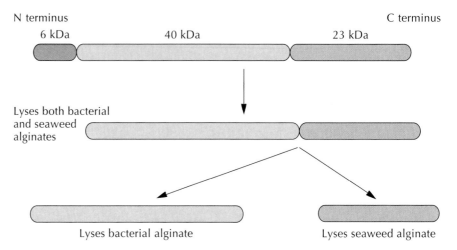

N terminus · C terminus

6 kDa · 40 kDa · 23 kDa

Lyses both bacterial and seaweed alginates

Lyses bacterial alginate · Lyses seaweed alginate

Figure 10.3 Schematic representation of the processing of the recombinant *Flavobacterium* alginate lyase protein precursor in *E. coli*. A 6-kDa peptide is removed from the N terminus of the 69-kDa precursor to yield a 63-kDa protein that can depolymerize alginate from both seaweed and bacteria. A second cleavage event converts the 63-kDa protein to a 23-kDa protein that is active against seaweed alginate and a 40-kDa protein that hydrolyzes bacterial alginate.

duced alginate lyase formed a halo around the colony when calcium was added to the plate. In the presence of calcium, all of the alginate in the medium, except in the immediate vicinity of an alginate lyase-positive clone, becomes cross-linked and opaque. The hydrolyzed alginate chains were no longer able to form cross-links, so that the medium in the vicinity of an alginate lyase-positive clone was transparent. Analysis of a cloned DNA fragment from one of the positive colonies revealed the presence of an open reading frame encoding a polypeptide with a molecular mass of approximately 69,000 Da. Detailed biochemical and genetic studies indicated that this polypeptide is probably a precursor of the three alginate lyases produced by the *Flavobacterium* sp. (Fig. 10.3). After the 69,000-Da precursor is produced, a proteolytic enzyme cleaves off an N-terminal peptide of about 6,000 Da. This 63,000-Da protein can lyse both bacterial and seaweed alginates. Cleavage of the 63,000-Da protein yields a 23,000-Da enzyme that can depolymerize seaweed alginate and a 40,000-Da enzyme that is effective against bacterial alginate. To produce large amounts of the 40,000-Da enzyme, the DNA corresponding to this enzyme was amplified by the polymerase chain reaction (PCR) and then inserted into a *Bacillus subtilis* plasmid vector fused to a *B. subtilis* α-amylase leader peptide to direct the secretion of this protein, under the transcriptional control of a penicillinase gene expression system (Fig. 10.4). Transformation of *B. subtilis* cells with

Figure 10.4 DNA encoding the 40,000-Da alginate lyase. The leader peptide from a *B. subtilis* α-amylase gene is fused to the N terminus of the alginate lyase coding sequence. The construct is under the transcriptional control of a *B. subtilis* penicillinase gene expression system.

Penicillinase promoter · α-Amylase leader peptide · 40,000-Da alginate lyase

this construct yielded colonies with large halos on solid medium containing alginate when calcium was added. When these transformants were grown in liquid medium, the recombinant alginate lyase was secreted into the culture broth. Further tests showed that this enzyme efficiently liquefied alginates that were produced by mucoid strains of *P. aeruginosa* isolated from the lungs of patients with cystic fibrosis. Additional studies are necessary to determine whether recombinant alginate lyase is a candidate for testing its efficacy in human clinical trials.

Monoclonal Antibodies as Therapeutic Agents

About 100 years ago, horses were inoculated with the bacterium *Corynebacterium diphtheriae*, which causes diphtheria in humans. The resulting crude horse antiserum was used to treat this often fatal childhood disease. (In those days, mortality sometimes reached 45%.) *C. diphtheriae* infects the throat or tonsils and produces an exotoxin that is lethal to human cells. This exotoxin enters the bloodstream and damages organs that are distant from the primary site of infection. However, the administration of horse antiserum containing antibodies against the exotoxin provides passive immunity, protecting the patient from a fatal outcome when the antiserum is given within the first few days of the onset of infection.

Unfortunately, this kind of antibody therapy carries considerable risks and is not widely used today. Patients often develop their own antibodies against the foreign proteins of either whole or partially purified therapeutic antiserum. After a second treatment, the sensitized patient may go into anaphylactic shock and die.

Today, with the advent of hybridoma methodology, antibodies are once again seen as potential therapeutic agents. This technique can be used to maintain a continuous supply of pure monospecific antibody. However, the problems of cross-reactivity leading to an immune response and anaphylaxis have not been overcome, because the recipient might still produce antibodies to a monoclonal antibody which carries mouse (murine) determinants. Therefore, the present goal is to design and produce human monoclonal antibodies with both specific immunotherapeutic properties and lowered potential for immunogenicity.

Structure and Function of Antibodies

An antibody molecule (immunoglobulin) consists of two identical "light" (L) protein chains and two identical "heavy" (H) protein chains, all held together by both hydrogen bonding and precisely localized disulfide linkages. The N-terminal regions of the L and H chains together form the antigen recognition site of each antibody. Antibody genes can be readily manipulated because the various functions of an antibody molecule are confined to discrete domains (regions) (Fig. 10.5). The sites that recognize and bind antigens consist of three complementarity-determining regions (CDRs) that lie within the variable (V_H and V_L) regions at the N-terminal ends of the two H and two L chains. The CDRs are the part of an antibody molecule with the greatest variability in amino acid sequence. In addition to the variable regions (V_H and V_L), each L chain contains one constant region, or domain (C_L), and each H chain has three constant regions, or domains (C_{H1}, C_{H2}, and C_{H3}). When antibodies are digested with the proteolytic enzyme papain, three fragments are released: two identical (Fab)

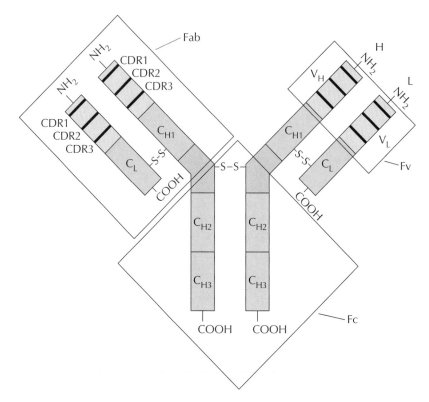

Figure 10.5 Structure of an antibody molecule. The H and L chains contain variable (V_L and V_H) and constant (C_L, C_{H1}, C_{H2}, and C_{H3}) domains with their CDRs (CDR1, CDR2, and CDR3). The Fv, Fab, and Fc portions of an antibody molecule are delineated. The N-terminal (NH_2) and C-terminal (COOH) ends of each polypeptide chain are indicated.

fragments, each of which contains an intact L chain linked by a disulfide bond to the V_L and C_{H1} regions of the H chain; and one Fc fragment, which consists of two H chain fragments, each containing the C_{H2} and C_{H3} domains and joined by a disulfide bond. The Fab fragment retains the antigen binding activity. In fact, the N-terminal half of the Fab fragment, which is called the Fv fragment, contains all of the antigen binding activity of the intact antibody molecule (Fig. 10.5). The amino acid sequence of this portion of the antibody varies considerably from one molecule to another.

In an intact antibody molecule, the Fc portion elicits several immunological responses after antigen–antibody binding occurs:

- The complement cascade is activated. The components of this system break down cell membranes, activate phagocytes, and generate signals to mobilize other components of the immunological response system.
- Antibody-dependent cell-mediated cytotoxicity (ADCC), which is the result of the binding of the Fc portion of the antibody to an Fc receptor of an ADCC effector cell, is produced. The bound effector cell releases substances that lyse the foreign cell to which the Fab portion of the antibody molecule is bound.
- After the Fab region binds to a soluble antigen, the Fc portion of an antibody can be bound to Fc receptors of phagocytic cells, which engulf and destroy the antibody–antigen complex.

Preventing Rejection of Transplanted Organs

In the 1970s, passive immunization was reconsidered as a way of preventing immunological rejection of a transplanted organ. The rationale was to administer to patients a specific antibody that would bind to certain lymphocytes and diminish the immune response directed against the transplanted organ.

The mouse monoclonal antibody OKT3 was the first to be approved by the U.S. Food and Drug Administration for use as an immunosuppressive agent after organ transplantation in humans. Lymphocytes that differentiate in the thymus are called T cells. Various members of the T-cell population act as immunological helper and effector cells and are responsible for organ rejection. The OKT3 monoclonal antibody binds to a cell surface receptor called CD3, which is present on all T cells. This action prevents a full immunological response and spares the transplanted organ from rejection. Immunosuppression by this means was reasonably effective, although, as anticipated, there were some side effects including fever and rash formation.

Chemically Linked Monoclonal Antibodies

Drugs that are very effective when tested in vitro—usually in cell culture—are often much less potent in vivo. This apparent loss of potency is due to the drug not reaching its targeted site at a concentration sufficient to be effective. Increasing the dose of a drug is not the answer to this problem, because high drug concentrations often have deleterious side effects. Moreover, to avoid unwanted side effects, many drugs are administered at lower than optimal levels, thereby decreasing their efficacy. A number of different strategies may be used to enhance the delivery of a drug to its target site. (1) Drugs may be encapsulated in liposomes, particles in which the drug is surrounded by a specific lipid surface, that can be targeted to certain organs. (2) Certain toxin genes may be incorporated into tumor-infiltrating lymphocytes. These cells can deliver the incorporated toxin directly to the site of a tumor. (3) Some drugs can be coupled to monoclonal antibodies that are specific for proteins found only on the surface of certain cells, e.g., tumor cells (Fig. 10.6). (4) Inert forms of some drugs, called prodrugs, may be used. These prodrugs can be enzymatically converted to the active form of the drug. To ensure that the drug is released only in the vicinity of the target cells, the enzyme is coupled to a monoclonal antibody directed against specific cell surface antigens (Fig. 10.6).

For this type of therapeutic system to be effective, the monoclonal antibody or single-chain antibody that is complexed with the prodrug-converting enzyme must be available in quantity in a relatively pure form; bind to a protein that is highly specific to the target cell; be stable under physiological conditions but cleared rapidly from circulation; and, when necessary, be able to penetrate into tumor masses so that all of the cells can be exposed to the drug. With this approach, only specifically targeted cells are exposed to the drug, permitting the use of a much lower level of the drug than if it were administered directly. Since mouse monoclonal antibodies would generate an immune response when used in this manner, it is essential to use human or humanized antibody fragments.

The majority of the natural deaths in North America and Europe are the consequence of blockage of a cerebral or coronary artery by a blood clot (thrombus). A thrombus consists of a network of fibrin, a blood-clotting

Figure 10.6 Schematic representation of monclonal antibody-based drug delivery system. **A.** The drug is coupled to a monoclonal antibody. **B.** An enzyme that converts an inactive prodrug to an active drug is attached to a monoclonal antibody. The active drug is formed only in the immediate vicinity of the target cells. In both cases, the monoclonal antibody binds to one specific protein on the surface of the target cell.

A

B

Figure 10.7 Activation of plasminogen to plasmin and breakdown of the two substrates (fibrinogen and fibrin) by plasmin in the blood system.

agent, which is formed in response to a defect in the wall of a blood vessel. Under natural conditions, plasmin, which is a serine protease and is produced by the activation of plasminogen by plasminogen activator (Fig. 10.7), degrades the fibrin in a blood clot and causes the clot to dissolve. In many instances, however, arterial blockage is caused when this biological system does not remove blood clots efficiently. Thus, it was reasoned that plasminogen activators could be used as therapeutic agents to induce higher levels of plasmin, which, in turn, would efficiently remove the arterial thrombi, thereby reducing their impact on brain and heart arteries.

Plasmin can also degrade fibrinogen (Fig. 10.7), which is the precursor of fibrin; therefore, this therapy can lead to excessive internal bleeding as a result of the depletion of fibrinogen levels. Consequently, thrombolytic agents that can degrade only the fibrin present in blood clots have been developed. Researchers reasoned that an antibody covalently coupled to a plasminogen activator could be used both to target the fibrin and, after binding, to cause an increase in the localized levels of plasmin by converting plasminogen to plasmin (Fig. 10.8). To test this concept, tissue-specific plasminogen activator was chemically coupled to an antifibrin monoclonal antibody. Under trial conditions in model systems, the complex bound to blood clots and caused their lysis without causing significant fibrinogen breakdown. Other antibody–plasminogen activator conjugates have been created and have also been found to cause a localized formation of plasmin that enhanced the lysis of blood clots.

Human Monoclonal Antibodies

Although the initial studies of immunotherapeutic agents were promising, there are drawbacks to chemical coupling and the use of a nonhuman monoclonal antibody. The yields of chemically linked molecules are low, and the coupling occurs at random sites. The chemical procedure may also inactivate the enzymatic activity of the plasminogen activator or, for other therapies, the moieties that are used. Finally, if the therapy requires multiple treatments, the antibody component should be from a human source to prevent immunological cross-reactivity and sensitization of the patient.

It is very difficult to create specific non-cross-reactive antibodies because of problems associated with obtaining human monoclonal antibodies by conventional hybridoma techniques.

Figure 10.8 Structure of an immunotherapeutic thrombolytic agent. Antifibrin antibody, a monoclonal antibody that is specific for the fibrin found in blood clots, is coupled to plasminogen activator (PA). After the complex binds to the fibrin of a blood clot, the plasminogen activator causes plasmin to accumulate in the vicinity of the clot. The plasmin then degrades the clot.

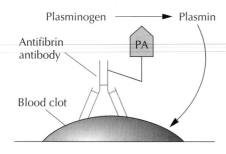

- The human chromosomes of fused human lymphocyte–mouse myeloma cells are unstable, and so cells producing a human monoclonal antibody are extremely rare.
- No effective human myeloma cell line that can replace the mouse myeloma cell line in the procedure has been discovered.
- For ethical reasons, it is not acceptable to immunize humans against various antigens.

Therefore, other approaches for obtaining human monoclonal antibodies have to be devised.

In one scheme, human B lymphocytes that were actively producing antibody were isolated from subjects and treated with a fluorescence-labeled antigen. Fluorescence-activated cell sorting was then used to enrich the cell sample for specific antibody-producing B lymphocytes. Because B cells do not grow well in culture, they were transformed with Epstein-Barr virus, which allows them to grow more readily in culture. Some of these transformed B-cell clones were found to secrete human monoclonal antibody that interacted with the selecting antigen. Unfortunately, with this strategy, the yields tend to be small and the monoclonal antibodies have weak antigen binding affinities. In addition, there is a low probability that a nonimmunized individual will have antibody-secreting cells that recognize the selecting antigen.

It may also be possible to introduce cells of the human immune system into a mutant mouse strain that lacks, for the most part, its own natural immunological cell repertoire. After the transplantation of human immune system stem cells, such a mouse, with severe combined immunodeficiency (*scid* mouse), acquires a human cell immune system and, in response to challenge by antigen, can produce human antibodies.

MILESTONE

Synthesis in *E. coli* of a Polypeptide with Human Leukocyte Interferon Activity
S. NAGATA, H. TAIRA, A. HALL, L. JOHNSRUD, M. STREULI, J. ECSÖDI, W. BOLL, K. CANTELL, and C. WEISSMANN
Nature **284**:316–320, 1980

The late 1970s and early 1980s were a time of tremendous excitement in molecular biotechnology. The promise of this new technology was being touted to both the public and large institutional investors. One of the products of the new biotechnology that captured the imagination of a large number of people was interferon, which at the time was seen by many as a possible miracle cure for a wide range of virally caused diseases and cancer. Thus, the isolation of a human interferon cDNA and its subsequent expression in *Escherichia coli* was reported in newspapers and magazines around the world.

Several features of interferon made it particularly difficult to isolate a cDNA encoding this polypeptide. First, although interferon had been purified more than 80,000-fold, only minuscule amounts were available, so that at the time researchers did not even have an accurate estimate of its molecular mass. Second, unlike many other proteins, interferon did not have a chemical or biological activity that was easy to monitor: its activity was measured by the reduction in the cytopathic effect of an animal virus on cells in culture, which was an extremely complex and time-consuming process. Third, unlike insulin, researchers had no idea if there was a particular type of human cell that could produce high levels of interferon and therefore serve as a source of mRNA that was enriched in interferon mRNA. These problems notwithstanding, a cDNA encoding interferon was eventually isolated and characterized. Since that time, researchers have discovered that there are several different types of interferons. Unfortunately, interferon is not the panacea that was dreamed of by both investors and the press. However, the genes for several interferons have been isolated, and clinical trials have shown that they are effective treatments for a variety of viral diseases.

Other workers are attempting to introduce human immunoglobulin genes into the germ line of mice to create transgenic mice that could produce a human immunoglobulin after immunization with a particular antigen. Standard hybridoma methodology could be used to isolate cells that secreted a specific monoclonal antibody from the transgenic animals; then these positive cell lines could be screened to determine which lines produced antibody that was encoded by the human immunoglobulin transgenes. In fact, a transgenic mouse that expresses the native form of both H and L human immunoglobulin genes has recently been reported.

The humanization of the immune system of *scid* mice and the generation of transgenic mouse lines are both laborious ways to produce human monoclonal antibodies. Consequently, researchers have been examining genetic engineering strategies to create both human therapeutic antibodies and effective dual-function proteins that have the ability to bind a target and then destroy it.

Hybrid Human–Mouse Monoclonal Antibodies

The modular nature of antibody functions has made it possible to convert a mouse monoclonal antibody into one that has some human segments but still retains its original antigen binding specificity. This hybrid antibody is called a chimeric antibody, or a "humanized" antibody. The first portion of a mouse monoclonal antibody that was targeted for replacement with a human sequence was the mouse Fc fragment. It was chosen because the mouse Fc fragment functions poorly as an effector of immunological responses in humans; it is also the most likely fragment to elicit the production of human antibodies. To diminish immunogenicity and to introduce human Fc effector capabilities, the DNA coding sequences for the Fv regions of both the L and the H chains of a human immunoglobulin were substituted for the FvDNA sequences for the L and H chains from a specific mouse monoclonal antibody (Fig. 10.9). This replacement of Fv coding regions can be accomplished either by using oligonucleotides and in vitro DNA replication or by using subcloned segments. The DNA constructs for both chimeric chains were cloned into an expression vector and transfected into cultured B lymphocytes from which the chimeric antibody was collected.

When a chimeric antibody that contained the binding site from a mouse monoclonal antibody directed against the surface of human colon cancer cells was tested in patients with colorectal cancer, it remained in the blood system about six times longer than the complete mouse monoclonal antibody did, thereby extending the period of effectiveness. Only 1 patient of the 10 developed a mild antibody response to the chimeric antibody. However, in this trial, no antitumor effects were observed, an outcome that may be due to the low dose levels used or to the advanced stage of the cancer in the subjects. With in vitro studies, this kind of chimeric antibody had high effector activity, and so there is still hope that these molecules may, in other cases, aid in the destruction of targeted tumor cells.

The "humanizing" of mouse and rat monoclonal antibodies has been taken one step further than the formation of chimeric molecules described above by substituting into human antibodies only the CDRs of the rodent monoclonal antibodies (Fig. 10.10). Because these "reshaped" human antibodies have antigen binding affinities similar to those of the original rodent monoclonal antibodies, they may be effective therapeutic agents.

Humanizing rodent monoclonal antibodies may be performed as follows. Starting with a rodent hybridoma cell line, cDNAs for the L and H

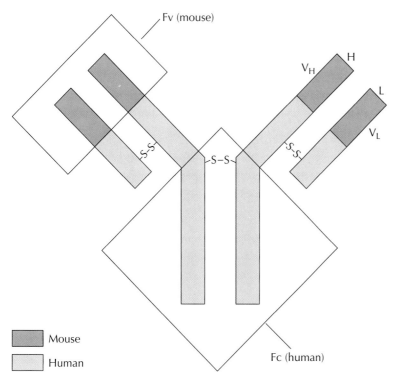

Figure 10.9 Genetically engineered humanized antibody. The V_L and V_H DNA regions from the immunoglobulin L and H genes that encode part of a mouse monoclonal antibody were substituted for the V_L and V_H DNA regions of a human immunoglobulin molecule. The product of the constructed gene is a humanized (chimeric) immunoglobulin with the antigen binding specificity of the mouse monoclonal antibody and both lowered immunogenicity in humans and human Fc effector capabilities.

Figure 10.10 Genetically engineered humanized antibody. The CDRs (CDR1, CDR2, and CDR3) from the genes for H and L immunoglobulin chains of a mouse monoclonal antibody replace the CDRs of the genes for a human antibody. The product of this constructed gene is an immunoglobulin with the antigen binding specificity of the mouse monoclonal antibody (pink) and all the other properties of a human antibody molecule (light blue).

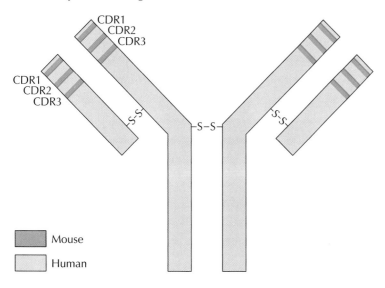

chains can be isolated. The variable regions of these cDNAs can be amplified by PCR. The oligonucleotide primers that are used for this amplification are complementary to the sequences at the 5′ and 3′ ends of the DNA encoding the variable regions, where there is a high degree of nucleotide sequence conservation from one antibody gene to another. From the nucleotide sequences of the cDNAs for the light and heavy regions (V_L and V_H), it is possible to delineate the limits of the CDRs. It is usually straightforward to determine where the CDRs begin and end, since these regions are highly variable in sequence while the sequences of the framework regions are relatively conserved. On the basis of the sequences of the DNAs encoding the rodent CDRs, six pairs of oligonucleotide PCR primers are synthesized. Each pair of primers is designed to initiate the synthesis of the DNA for one of the six rodent CDRs—three from the L chain and three from the H chain. In addition, each primer includes an extra 12 nucleotides at its 5′ end, complementary to the flanking regions within the human framework DNA into which the DNA for the rodent CDRs is targeted (Fig. 10.11). Oligonucleotide-directed mutagenesis is then used to replace, one at a time, the complete DNA sequence for each of the human CDRs with the amplified DNA for the rodent CDRs. In this way, it is necessary to carry out six cycles of oligonucleotide-directed mutagenesis, one to replace each CDR. This procedure, in effect, "grafts" the rodent CDRs onto the human antibody framework. The humanized variable-region cDNAs are then cloned into expression vectors, which are then introduced into appropriate host cells, usually either *E. coli* or mammalian cells, for the production of antibodies.

This procedure assumes that only the CDRs and not the framework regions contribute to the ability of an antibody to bind to an antigen. However, if the humanized antibody does not bind sufficiently tightly to the target antigen, it may be necessary to modify some of the amino acids in the framework region by oligonucleotide-directed mutagenesis.

To date, with this procedure, more than 50 different monoclonal antibodies have been humanized. While this technology is clearly effective and widely

Figure 10.11 PCR amplification of CDR1 from a rodent monoclonal antibody L chain cDNA. The PCR primers P1 and P2 contain oligonucleotides complementary to the rodent CDR1 DNA. In addition, P1 and P2 each contain 12 nucleotides at their 5′ end that is complementary to the framework regions of human monoclonal L chain cDNAs. Using six separate pairs of oligonucleotide primers—three for the V_L region and three for the V_H region—each of the rodent CDRs is separately amplified by PCR. Then, by oligonucleotide-directed mutagenesis, the amplified rodent CDRs are spliced into human antibody genes in place of the resident CDRs. This grafting is made possible by the presence of DNA complementary to the human framework regions on the amplified rodent CDR DNAs.

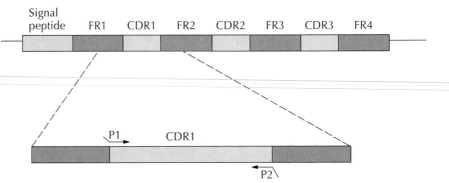

applicable, it is nevertheless time-consuming and expensive. Probably, in the future, phage display combinatorial libraries that are constructed from mRNA from human B cells of nonimmunized donors will be the preferred way to produce human antibodies and antibody fragments (as described below).

Production of Antibodies in *E. coli*

Hybridoma cells, like most other animal cells in culture, grow relatively slowly, do not attain high cell densities, and require a complex and expensive growth medium. The cost of their production is an impediment to the more widespread use of monoclonal antibodies as therapeutic agents. To circumvent this problem, attempts have been made to genetically engineer

Figure 10.12 Procedure to create in *E. coli* a combinatorial library of the V_L and V_H regions of antibody chains.

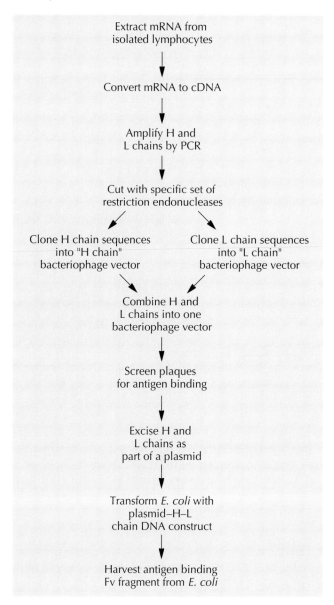

Extract mRNA from
isolated lymphocytes

Convert mRNA to cDNA

Amplify H and
L chains by PCR

Cut with specific set of
restriction endonucleases

Clone H chain sequences
into "H chain"
bacteriophage vector

Clone L chain sequences
into "L chain"
bacteriophage vector

Combine H and
L chains into one
bacteriophage vector

Screen plaques
for antigen binding

Excise H and
L chains as
part of a plasmid

Transform *E. coli* with
plasmid–H–L
chain DNA construct

Harvest antigen binding
Fv fragment from *E. coli*

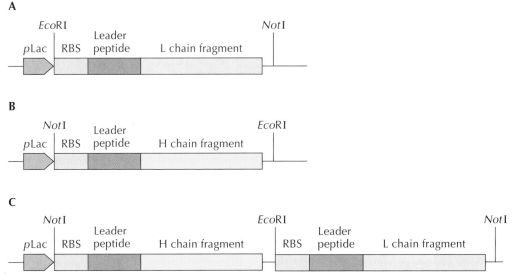

Figure 10.13 DNA constructs of an Fv combinatorial gene library cloned into bacteriophage λ DNA. **A and B.** Portions of the cDNAs of the L **(A)** and H **(B)** chains are separately cloned into bacteriophage λ vectors. **C.** Each of these libraries is digested with *Eco*RI, and then the fragments from the H chain library are ligated to the fragments from the L chain library, thereby creating a combinatorial library that contains all possible combinations of L and H chain fragments expressed on the same vector. Abbreviations: *p*Lac, the *E. coli lac* promoter; RBS, ribosome binding site.

bacteria, plants, and animals to act as "bioreactors" for the production of monoclonal antibodies. For effective delivery and function of some immunotherapeutic agents, only the antigen binding region of an antibody (the Fab or Fv fragment) is required; in other words, the Fc portion of an antibody is dispensable.

An elaborate series of manipulations makes it possible to select, as well as produce, functional antibodies in *E. coli* (Fig. 10.12).

1. cDNA is synthesized from mRNA isolated from either mouse or human antibody-producing cells (B lymphocytes).
2. The H and L chain sequences in the cDNA preparation are amplified separately by the PCR procedure.
3. The amplified cDNA preparation is treated with a specific set of restriction endonucleases and cloned into a bacteriophage λ vector. The cDNA sequences of the H and L chains each have distinctive restriction endonuclease recognition sites, an arrangement that facilitates the directional cloning of each sequence into a separate bacteriophage λ vector. At this stage of the process, many different H and L chain sequences are cloned (Fig. 10.13A and B).
4. The cDNAs of one H and one L chain are cloned into a single "combinatorial" vector, thereby enabling the bacteriophage to coexpress both chains, thus forming an assembled antibody Fv fragment (Fig. 10.13C).

The H and L chains are expressed during the lytic cycle of bacteriophage λ, so that the library of combinatorial bacteriophage clones can be screened for the presence of antigen binding activity.

The step in which L and H chain cDNAs are combined on one vector creates a vast array of diverse antibody genes, some of which will encode unique target binding sites whose isolation would never have been possible by standard hybridoma procedures. The mammalian antibody repertoire has the potential to produce approximately 10^6 to 10^8 different antibodies. A phage library contains approximately this number of clones, so that one combinatorial library can be expected to produce as many different antibodies (Fv molecules) as any mammal. In addition, once an initial combinatorial library has been constructed, it is possible to shuffle the L and H chains to obtain Fv molecules that recognize unusual epitopes. And even greater variation may be achieved by random mutagenesis of the DNAs in the combinatorial library. Because millions of bacteriophage plaques can be screened in a relatively short period, the identification of Fv molecules with the desired specificity takes only about 7 to 14 days. By contrast, months are usually required to screen a few hundred hybridoma cell lines.

Bacteriophage λ vectors are not particularly useful for the production of large quantities of protein. To overcome this drawback, the bacteriophage λ vector was engineered so that the H and L chain DNA sequences are inserted into a site that is flanked by plasmid DNA sequences. This plasmid DNA, containing an H and L chain DNA combination, can be excised from the bacteriophage λ vector and transformed into *E. coli* (Fig. 10.12). As part of a plasmid, large amounts of Fv fragments can be produced within the *E. coli* cells and harvested for either diagnostic or therapeutic purposes.

As an alternative to the use of bacteriophage λ, filamentous bacteriophages such as M13 or fd have been used for the production of combinatorial libraries (Fig. 10.14). In these cases, the antibody fragment is synthesized as part of a fusion protein that is located on the surface of the bacteriophage. A combinatorial library of antibody fragments displayed on the surface of a filamentous bacteriophage can be screened by an enzyme-linked immunosorbent assay-like system. Briefly, samples (aliquots) of the library are added to the wells of a multiwell plate that are coated with the target antigen. The wells are rinsed to remove any unbound bacteriophage. Next, an antibody that binds to the bacteriophage coat protein and is conjugated with an enzyme is added to each well. The wells are rinsed to remove any unbound antibody–enzyme complex. The phage particles bound to the target antigen will be recognized by the antibody–enzyme complex. A chromogenic substrate that is cleaved by the bound enzyme is then used to determine which wells contain a phage carrying antibodies to the target antibody. This approach makes it easier than plaque assays with bacteriophage λ to select and subsequently purify a bacteriophage producing an antibody fragment that binds to a specific antigen. Once a desired antibody fragment-producing bacteriophage has been isolated, the DNA can be isolated and subcloned into an expression vector. Variants of the selected antibody that have an increased affinity for the target antigen can be produced by replacing portions of the DNA for the V_L and V_H regions or through random mutagenesis of the antibody genes.

In addition to producing Fv fragments, researchers have attempted to determine whether a single protein chain consisting of only V_L and V_H domains would form a functional antigen binding molecule. Computer simulations of the three-dimensional structure of a potential single-chain antibody showed that the V_L and V_H domains have to be separated by a linker peptide to assume the correct conformation for antigen binding. On the basis of these design constraints, DNA constructs of V_L and V_H sequences

Figure 10.14 Use of the filamentous bacteriophage M13 to produce a combinatorial library synthesizing Fv antibody fragments. The cDNAs for the V_L and V_H regions are amplified by PCR and then ligated to the DNA for a short linker peptide. The DNAs encoding the single-chain antibody combinatorial library are cloned into the genome of bacteriophage M13 fused to M13 gene 3, which encodes a phage surface protein. There are three copies of the phage gene 3 protein per M13 bacteriophage, so that in recombinant M13 carrying the single-chain antibody combinatorial library, there are three copies of the gene 3 protein–single-chain antibody fusion protein.

from a cDNA of a cloned monoclonal antibody were each ligated to a chemically synthesized DNA linker fragment in the order V_L-linker-V_H. After expression in *E. coli*, the single-chain protein was purified, and both its affinity and specificity were found to be equivalent to those of the original intact monoclonal antibody. Thus, functional single-chain antibodies can be readily produced in *E. coli*.

Single-chain antibodies may be used for a variety of therapeutic and diagnostic applications in which Fc effector functions are not required and when a small size (single-chain antibodies have a molecular mass of approximately 27 kDa compared to approximately 150 kDa for immunoglobulin G molecules) is an advantage. In addition, a protein-coding sequence can be linked to a single-chain antibody sequence to create a dual-function

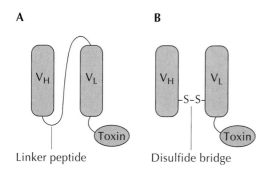

A

B

Linker peptide

Disulfide bridge

Figure 10.15 Schematic representation of a single-chain Fv immunotoxin **(A)** and a disulfide-stabilized Fv immunotoxin **(B)**.

molecule that can bind to a specific target and carry with it a specific activity.

In one study, instead of linking the V_H and V_L chains with a short peptide, amino acids in the framework region were modified to include a disulfide linkage between these two peptides. The effectiveness of this disulfide-stabilized Fv molecule coupled to a toxic peptide intended to destroy a cancer cell was compared with that of a single-chain Fv molecule coupled to the same toxin (Fig. 10.15). The disulfide-stabilized and single-chain Fv immunotoxins had the same activity and specificity, and the former was severalfold more stable than the latter. This suggests that disulfide-stabilized Fv molecules may be more useful than single-chain Fv molecules in some therapeutic applications.

HIV Therapeutic Agents

With acquired immunodeficiency syndrome (AIDS), which is caused by human immunodeficiency virus (HIV), scientists have so far been unable to produce an effective vaccine. While the development of an HIV vaccine has been proceeding, researchers have also been seeking other ways to keep the infection under control.

The target cell for HIV is the helper T lymphocyte (T_H cell) of the immune system. During the normal immunological response to pathogens, the T_H cells bind degradation products of specific antigens and release factors that stimulate other immune cells to participate in the immune response. The T_H cells play a pivotal role in this process, but HIV infection disrupts their critical function. Once HIV enters a T_H cell, it is sheltered from immunological attack and proceeds to destroy T_H cells in several different ways:

- An infected cell can be lysed by the production and release of HIV particles.
- An infected cell can act as a production plant for the synthesis and release of an HIV glycoprotein (gp120) that leads to the destruction of other T_H cells and other cells of the T-lymphocyte family.
- An infected cell can fuse with other T_H cells to form a syncytium that incapacitates individual T_H-cell functions.

The primary consequence of HIV infection is that the immune system is not able to provide protection against common bacterial and viral diseases; these eventually cause death despite antibiotic and other therapies.

The first step in the infection of T_H cells by HIV is an interaction between gp120, a 120-kDa HIV envelope glycoprotein, and the cell surface re-

A

B

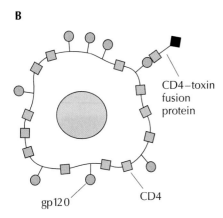

Figure 10.16 HIV infection and treatment. **A.** Initial binding of HIV to the T$_H$ cell is the result of contact between the viral protein gp120 and the CD4 surface protein of the T$_H$ cell. **B.** The HIV-infected cell has gp120 on its surface. Free CD4–toxin fusion protein can bind to the gp120 on the surface of infected cells. Once inside an HIV-infected cell, the toxin portion of the CD4–toxin molecule can kill it.

ceptor of T$_H$ cells called CD4 (Fig. 10.16A). Antibodies to CD4 block HIV infection of T$_H$ cells in vitro. Also, in the presence of excess free CD4 protein, HIV infection is diminished in vitro. Although both of these treatments block HIV infection, neither destroys the virus. One approach to achieving both protection of T$_H$ cells and killing of HIV has been to engineer a fusion protein that consists of a portion of the CD4 protein and the Fc portion of an immunoglobulin gene. The fusion protein that is produced is called a CD4 immunoadhesin. This molecule possesses gp120 binding and HIV-blocking properties resulting from the CD4 component, the ability to bind to antibody receptor cells, and the long half-life in plasma resulting from the immunoglobulin portion of the molecule. After the binding of the CD4 immunoadhesin to either free virus or HIV-infected cells, an ADCC response is elicited, which leads to the destruction of the virus or HIV-infected cells.

Another approach to controlling HIV infection entails creating a cell-targeted system for specifically killing HIV-infected cells. For example, the CD4 coding sequence was ligated to the coding sequence for the intracellular toxin *Pseudomonas* exotoxin A to form a gene for a single fusion protein with these combined functions (Fig. 10.16B). *Pseudomonas* exotoxin A is a 66-kDa protein containing three separate domains: domain I is responsible for cell binding, domain II is responsible for translocation of the protein into the cell, and domain III is responsible for ADP-ribosylation. The CD4–*Pseudomonas* exotoxin A fusion protein includes most of the CD4 sequence in place of domain I (Fig. 10.17), and so it retains the cytotoxic activity of the *Pseudomonas* exotoxin as well as the gp120 binding activity of CD4. Because HIV-infected cells have gp120 on their surfaces, the CD4 protein domain of the fusion protein binds exclusively to these infected cells. After this initial binding, because of a domain provided by the *Pseudomonas* exotoxin A protein, the fusion protein enters the HIV-infected cell. Then, the exotoxin portion of the fusion protein inactivates the protein synthesis elongation factor EF-2. The loss of EF-2 activity prevents further protein synthesis, eventually killing the cell. In other words, the CD4 domain "fingers" the HIV-infected cells and the exotoxin acts as the "hired killer."

After expression in *E. coli*, this fusion protein formed an insoluble inclusion body in the cytoplasm. It was solubilized with guanidine hydrochloride and isolated by rapid dilution and anion-exchange chromatog-

Figure 10.17 Genetic construction of a CD4–*Pseudomonas* exotoxin A fusion protein. The T7 promoter is from the *E. coli* bacteriophage T7.

raphy. The solubilized fusion protein functions well in a test cell culture system. However, because the *Pseudomonas* portion of the fusion protein might provoke an immune response in humans, it may be necessary to administer it together with a known immunosuppressant such as cyclosporin. It is important to bear in mind that although the experimental protocol outlined above may eventually become an effective means of treating HIV infections, this approach is still at an early stage of development.

With this type of immunotherapy, dose levels should be low and potency should be high so that immunological side effects are minimal. Moreover, this therapeutic strategy should be effective for various types of cancer as well as HIV-infected cells and could replace, for some cancers, the need for chemotherapy. Cytotoxic proteins other than the *Pseudomonas* exotoxin, such as diphtheria toxin and the plant toxin ricin, may also be targeted to infected cells. However, even with the best results, it will take a number of years before recombinant immunotoxins are used routinely as therapeutic agents.

SUMMARY

A large number of proteins that have potential as pharmaceutical agents have been synthesized from cloned genes in bacteria. Because most of these proteins are from eukaryotic organisms, the strategy for the isolation of a target gene often involves isolating mRNA enriched in the messenger of interest, synthesizing a cDNA library, and subcloning the selected target cDNA into an appropriate expression vector. In some instances, novel and useful variants of these proteins can be constructed either by shuffling functional domains of related genes or by directed replacement of functional domains of the cloned gene.

In some instances, enzymes may be used as therapeutic agents. For example, both recombinant DNase I and alginate lyase have been used in an aerosol form to decrease the viscosity of the mucus found in the lungs of patients with cystic fibrosis.

The development of recombinant DNA and monoclonal antibody technologies, combined with a growing understanding of the molecular structure and function of immunoglobulin molecules, has created interest in using specific antibodies as therapeutic agents to treat various diseases. Antibody genes can be readily manipulated because the various functions of an antibody molecule are confined to discrete domains.

Drugs or enzymes may be coupled to monoclonal antibodies or Fv fragments that are specific for proteins found only on the surface of certain cells, e.g., tumor cells. Inert forms of the drugs, called prodrugs, may be used. These antibody–drug or antibody–enzyme constructs can then be used as therapeutic agents. However, if the therapy requires multiple treatments, the antibody component should be from a human source to prevent immunological cross-reactivity and sensitization of the patient. To do this, rodent monoclonal antibodies can be "humanized" by substituting into human antibodies only the CDRs of the rodent monoclonal antibodies. In addition, it has recently become possible to produce and select human monoclonal antibodies in *E. coli*.

It is possible, by using genetic manipulation, to produce a unique type of therapeutic agent consisting of a hybrid between a protein that binds to a specific cell type—such as a cell infected with HIV—and a toxin molecule.

This approach is still at an early stage of development but promises to be an effective means of treating a variety of diseases.

References

Barbas, C. F., III, and D. R. Burton. 1996. Selection and evolution of high-affinity human anti-viral antibodies. *Trends Biotechnol.* **14:**230–234.

Bird, R. E., and B. W. Walker. 1991. Single chain antibody variable regions. *Trends Biotechnol.* **9:**132–137.

Brinkmann, U., L. H. Pai, D. J. FitzGerald, M. Willingham, and I. Pastan. 1991. B3(Fv)-PE38KDEL, a single-chain immunotoxin that causes complete regression of a human carcinoma in mice. *Proc. Natl. Acad. Sci. USA* **88:**8616–8620.

Brüggemann, M., H. M. Caskey, C. Teale, H. Waldmann, G. T. Williams, M. A. Surani, and M. S. Neuberger. 1989. A repertoire of monoclonal antibodies with human heavy chains from transgenic mice. *Proc. Natl. Acad. Sci. USA* **86:**6709–6713.

Bryn, R. A., J. Mordenti, C. Lucas, D. Smith, S. A. Marsters, J. S. Johnson, P. Cossum, S. M. Chamow, F. M. Wurm, T. Gregory, J. E. Groopman, and D. J. Capon. 1990. Biological properties of a CD4 immunoadhesin. *Nature* **344:**667–670.

Buchner, J., and R. Rudolph. 1991. Renaturation, purification and characterization of recombinant Fab-fragments produced in *E. coli. Bio/Technology* **9:**157–162.

Burton, D. R. 1991. Human and mouse monoclonal antibodies by repertoire cloning. *Trends Biotechnol.* **9:**169–175.

Capon, D. J., S. M. Chamow, J. Mordenti, S. A. Marsters, T. Gregory, H. Mitsuya, R. A. Byrn, C. Lucas, F. M. Wurm, J. E. Groopman, S. Broder, and D. H. Smith. 1989. Designing CD4 immunoadhesins for AIDS therapy. *Nature* **337:**525–530.

Chamow, S. M., and A. Ashkenazi. 1996. Immunoadhesins: principles and applications. *Trends Biotechnol.* **14:**52–60.

Chaudhary, V. K., T. Mizukami, T. R. Fuerst, D. J. FitzGerald, B. Moss, I. Pastan, and E. A. Berger. 1988. Selective killing of HIV-infected cells by recombinant human CD4-*Pseudomonas* exotoxin hybrid protein. *Nature* **335:**369–372.

Chester, K. A., and R. E. Hawkins. 1995. Clinical issues in antibody design. *Trends Biotechnol.* **13:**294–300.

Chiswell, D. J., and J. McCafferty. 1992. Phage antibodies: will new 'coliclonal' antibodies replace monoclonal antibodies? *Trends Biotechnol.* **10:**80–84.

Collet, T. A., P. Roben, R. O'Kennedy, C. F. Barbas III, D. R. Burton, and R. A. Lerner. 1992. A binary plasmid system for shuffling combinatorial antibody libraries. *Proc. Natl. Acad. Sci. USA* **89:**10026–10030.

Cunningham, B. C., and J. A. Wells. 1991. Rational design of receptor-specific variants of human growth hormone. *Proc. Natl. Acad. Sci. USA* **88:**3407–3411.

Davis, G. T., W. D. Bedzyk, E. W. Voss, and T. W. Jacobs. 1991. Single chain antibody (SCA) encoding genes: one-step construction and expression in eukaryotic cells. *Bio/Technology* **9:**165–169.

Dewerchin, M., and D. Collen. 1991. Enhancement of the thrombolytic potency of plasminogen activators by conjugation with clot-specific monoclonal antibodies. *Bioconjugate Chem.* **2:**293–300.

Gram, H., L. A. Marconi, C. F. Barbas III, T. A. Collet, R. A. Lerner, and A. S. Kang. 1992. *In vitro* selection and affinity maturation of antibodies from a naive combinatorial immunoglobulin library. *Proc. Natl. Acad. Sci. USA* **89:**3576–3580.

Harris, W. J. 1994. Humanizing monoclonal antibodies for *in vivo* use. *Animal Cell Biotechnol.* **6:**259–279.

Hodgson, J. 1991. Making monoclonals in microbes. *Bio/Technology* **9:**421–425.

Huennekens, F. M. 1994. Tumor targeting: activation of prodrugs by enzyme-monoclonal antibody conjugates. *Trends Biotechnol.* **12:**234–239.

Huse, W. D., L. Sastry, S. A. Iverson, A. S. Kang, M. Alting-Mees, D. R. Burton, S. J. Benkovic, and R. A. Lerner. 1989. Generation of a large combinatorial library of the immunoglobulin repertoire in phage lambda. *Science* **246:**1275–1281.

Johnson, I. S. 1983. Human insulin from recombinant DNA technology. *Science* **219:**632–637.

Little, M., F. Breitling, S. Dübel, P. Fuchs, and M. Braunagel. 1995. Human antibody libraries in *Escherichia coli. J. Biotechnol.* **41:**187–195.

LoBuglio, A. G., R. H. Wheeler, J. Trang, A. Haynes, K. Rogers, E. B. Harvey, L. Sun, J. Ghrayeb, and M. B. Khazaeli. 1989. Mouse/human chimeric monoclonal antibody in man: kinetics and immune response. *Proc. Natl. Acad. Sci. USA* **86:**4220–4224.

Marks, J. D., A. D. Griffiths, M. Malmqvist, T. P. Clackson, J. M. Bye, and G. Winter. 1992. By-passing immunization: building high affinity antibodies by chain shuffling. *Bio/Technology* **10:**779–783.

Meyer, F., A. Hinnen, A. Meister, M. G. Grutter, and S. Alkan. December 1989. Hybrid interferons. U.S. patent 4,885,166.

Mullinax, R. L., E. A. Gross, J. R. Amberg, B. N. Hay, H. H. Hogrefe, M. M. Kubitz, A. Greener, M. Alting-Mees, D. Ardourel, J. M. Short, J. A. Sorge, and B. Shopes. 1990. Identification of human antibody fragment clones specific for tetanus toxoid in a bacteriophage λ immunoexpression library. *Proc. Natl. Acad. Sci. USA* **87:**8095–8099.

Murata, K., T. Inose, T. Hisano, S. Abe, Y. Yonemoto, T. Yamashita, M. Takagi, K. Sakaguchi, A. Kimura, and T. Imanaka. 1993. Bacterial alginate lyase: enzymology, genetics and application. *J. Ferment. Bioeng.* **76:**427–437.

Nagata, S., H. Taira, A. Hall, L. Johnsrud, M. Streuli, J. Ecsodi, W. Boll, K. Cantell, and C. Weissmann. 1980. Synthesis in *E. coli* of a polypeptide with human leukocyte interferon activity. *Nature* **284:**316–320.

Pastan, I., and D. FitzGerald. 1991. Recombinant toxins for cancer treatment. *Science* **254:**1173–1177.

Plückthun, A. 1991. Antibody engineering: advances from the use of *E. coli* expression systems. *Bio/Technology* **9:**545–551.

Primrose, S. B. 1986. The application of genetically engineered microorganisms in the production of drugs. *J. Appl. Bacteriol.* **61:**99–116.

Queen, C., W. P. Schneider, H. E. Selick, P. W. Payne, N. F. Landolf, J. F. Duncan, N. M. Avdalovic, M. Levitt, R. P. Junghans, and T. A. Waldmann. 1989. A humanized antibody that binds to the interleukin 2 receptor. *Proc. Natl. Acad. Sci. USA* **86:**10029–10033.

Reiter, Y., U. Brinkmann, K. O. Webber, S.-H. Jung, and I. Pastan. 1994. Engineering interchain disulfide bonds into conserved framework regions of Fv fragments: improved biochemical characteristics of recombinant immunotoxins containing disulfide-stabilized Fv. *Protein Eng.* **7:**697–704.

Riechmann, L., M. Clark, H. Waldmann, and G. Winter. 1988. Reshaping human antibodies for therapy. *Nature* **332:**323–327.

Taniguchi, T., Y. Fujii-Kuriyama, and M. Muramatsu. 1980. Molecular cloning of human interferon cDNA. *Proc. Natl. Acad. Sci. USA* **77:**4003–4006.

van Leen, R. W., J. G. Bakhuis, R. F. W. C. van Beckhoven, H. Burger, L. C. J. Dorssers, R. W. J. Hommes, P. J. Lemson, B. Noordam, N. L. M. Persoon, and G. Wagemaker. 1991. Production of human interleukin-3 using industrial microorganisms. *Bio/Technology* **9:**47–52.

Vaughan, T. J., A. J. Williams, K. Pritchard, J. K. Osbourn, A. R. Pope, J. C. Earnshaw, J. McCafferty, R. A. Hodits, J. Wilton, and K. S. Johnson. 1996. Human antibodies with sub-nanomolar affinities isolated from a large non-immunized phage display library. *Nat. Biotechnol.* **14:**309–314.

Waldmann, T. A. 1991. Monoclonal antibodies in diagnosis and therapy. *Science* **252:**1657–1662.

Winter, G., and C. Milstein. 1991. Man-made antibodies. *Nature* **349:**293–299.

REVIEW QUESTIONS

1. You have been given the task of cloning and expressing a DNA sequence encoding human interferon. You do not have a DNA hybridization probe for human interferon, although you have isolated a human cell line that can be induced to synthesize interferon approximately 100-fold over background levels. Explain your strategy for cloning and expressing this DNA.

2. What is the Fc portion of an antibody molecule? The Fab portion? The Fv portion? The CDR portion?

3. How are antibody light and heavy chains coordinately synthesized in *E. coli*?

4. Why would DNase I and alginate lyase be useful in the treatment of cystic fibrosis?

5. How can the production of alginate lyase from a cloned gene be detected in *E. coli* transformants?

6. What is a combinatorial cDNA library?

7. How is bacteriophage M13 used to select Fv fragments that bind to specific target antigens?

8. What are disulfide-stabilized and single-chain Fv molecules?

9. How can enzymes coupled to monoclonal antibodies or Fv fragments be used as therapeutic agents?

10. How are mouse monoclonal antibodies "humanized"? Discuss the reasons for creating humanized monoclonal antibodies.

11. Describe a protocol for producing a therapeutic agent that targets and kills a specific cell type.

Vaccines

Vaccination protects a recipient from pathogenic agents by establishing an immunological resistance to infection. An injected or oral vaccine induces the host to generate antibodies against the disease-causing organism; therefore, during future exposures, the infectious agent is inactivated (neutralized or killed), its proliferation is prevented, and the disease state is not established.

Just over 200 years ago—in 1796—Dr. Edward Jenner experimentally tested the folklore-based notion that human infection with a mild cattle disease called cowpox would protect infected individuals against the human disease smallpox. Smallpox is an extremely virulent disease with a high death rate; if one survives, permanent disfigurement, mental derangement, and blindness often follow. Jenner inoculated James Phipps, an 8-year-old, with exudate from a cowpox pustule. In two separate trials after the initial vaccination, the boy was fully protected against human smallpox. This country doctor had discovered the principle of vaccination.

Communicable diseases such as tuberculosis, smallpox, cholera, typhus, bubonic plague, and poliomyelitis have in the past been a scourge for humankind. With the advent of vaccination, antibiotics, and effective public health measures, these epidemic diseases have, for the most part, been brought under control. Occasionally, however, protective measures become ineffective, and devastating new outbreaks occur. In 1991, a cholera epidemic struck Peru, producing, over the next 3 years, approximately 1 million infections and several thousand deaths. Also, for many current human and animal diseases, there are no vaccines. Today, over 2 billion humans suffer from diseases that, theoretically, could be curtailed by vaccination. In addition, new diseases such as acquired immune deficiency syndrome (AIDS), for which a vaccine might be useful, continue to emerge.

Modern vaccines typically consist of either a killed (inactivated) or a live, nonvirulent (attenuated) form of an infectious agent. Traditionally, the infectious agent is grown in culture, purified, and either inactivated or attenuated without, of course, losing the ability to evoke an immune response that is effective against the virulent form of the infectious organism.

Notwithstanding the considerable success that has been achieved in creating effective vaccines against diseases such as German measles, diphtheria, whooping cough, tetanus, smallpox, and poliomyelitis, there are a number of limitations to the current mode of vaccine production:

- Not all infectious agents can be grown in culture, and so no vaccines have been developed for a number of diseases.
- Production of animal and human viruses requires animal cell culture, which is expensive.
- Both the yield and rate of production of animal and human viruses in culture are often quite low, thereby making vaccine production costly.
- Extensive safety precautions are necessary to ensure that laboratory and production personnel are not exposed to a pathogenic agent.
- Batches of vaccine may not be killed or may be insufficiently attenuated during the production process, thereby introducing virulent organisms into the vaccine and inadvertently spreading the disease.
- Attenuated strains may revert, a possibility that requires continual testing to ensure that the reacquisition of virulence has not occurred.
- Not all diseases (e.g., AIDS) are preventable through the use of traditional vaccines.
- Most current vaccines have a limited shelf life and often require refrigeration to maintain potency. This requirement creates storage problems in countries with large, unelectrified rural areas.

Within the last decade, recombinant DNA technology has provided a means of creating a new generation of vaccines that overcome the drawbacks of traditional vaccines. The availability of gene cloning has enabled researchers to contemplate various novel strategies for vaccine development:

- Virulence genes could be deleted from an infectious agent that retains the ability to stimulate an immunological response. In this case, the genetically engineered agent could be used as a live vaccine without concern about reversion to virulence, because it is impossible for a whole gene to be reacquired spontaneously during growth in pure culture.
- Live nonpathogenic carrier systems that carry discrete antigenic determinants of an unrelated pathogenic agent can be created. In this form, the carrier system facilitates the induction of a strong immunological response directed against the pathogenic agent.
- For infectious agents that cannot be maintained in culture, the genes for the proteins that have critical antigenic determinants can be isolated, cloned, and expressed in an alternative host system such as *Escherichia coli* or a mammalian cell line. These cloned gene proteins can be formulated into a "subunit" vaccine (see the next section).
- There are some infectious agents that do not damage host cells directly; instead, the disease condition results when the host immune system attacks its own (infected) cells. For these diseases, it may be possible to create a targeted cell-specific killing system. Although not a true vaccine, this type of system attacks only infected cells, thereby removing the source of the adverse immunological response. In these cases, the gene for a fusion protein is constructed. First, one part of

Table 11.1 Human disease agents for which recombinant vaccines are currently being developed

Pathogenic agent	Disease
Viruses	
Varicella-zoster virus	Chicken pox
Cytomegalovirus	Infection in infants and immunocompromised patients
Dengue virus	Hemorrhagic fever
Hepatitis A virus	High fever, liver damage
Hepatitis B virus	Long-term liver damage
Herpes simplex virus type 2	Genital ulcers
Influenza A and B viruses	Acute respiratory disease
Japanese encephalitis virus	Encephalitis
Parainfluenza virus	Inflammation of the upper respiratory tract
Rabies virus	Encephalitis
Respiratory syncytial virus	Upper and lower respiratory tract lesions
Rotavirus	Acute infantile gastroenteritis
Yellow fever virus	Lesions of heart, kidney, and liver
Human immunodeficiency virus	AIDS
Bacteria	
Vibrio cholerae	Cholera
E. coli enterotoxin strains	Diarrheal disease
Neisseria gonorrhoeae	Gonorrhea
Haemophilus influenzae	Meningitis, septicemic conditions
Mycobacterium leprae	Leprosy
Neisseria meningitidis	Meningitis
Bordetella pertussis	Whooping cough
Shigella strains	Dysentery
Streptococcus group A	Scarlet fever, rheumatic fever, throat infection
Streptococcus group B	Sepsis, urogenital tract infection
Streptococcus pneumoniae	Pneumonia, meningitis
Clostridium tetani	Tetanus
Mycobacterium tuberculosis	Tuberculosis
Salmonella typhi	Typhoid fever
Parasites	
Onchocerca volvulus	River blindness
Leishmania spp.	Internal and external lesions
Plasmodium spp.	Malaria
Schistosoma mansoni	Schistosomiasis
Trypanosoma spp.	Sleeping sickness
Wuchereria bancrofti	Filariasis

this fusion protein binds to an infected cell. Then, the other part kills the infected cell.

Because of less stringent regulatory requirements, the first vaccines that were produced by recombinant DNA techniques were for animal diseases such as foot-and-mouth disease, rabies, scours, and a diarrheal disease of piglets. In addition, many more animal vaccines are currently being developed. For human diseases, a large number of recombinant vaccines should be available by the year 2000 (Table 11.1).

Subunit Vaccines

Vaccines generally consist of either killed or attenuated forms of the whole pathogenic agent. The antibodies elicited by these vaccines initiate an immune response to inactivate (neutralize) pathogenic organisms by binding

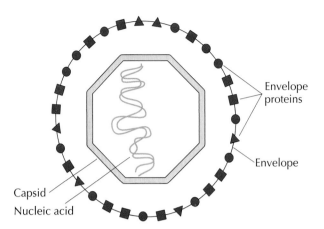

Figure 11.1 Schematic representation of an animal virus. Viruses generally consist of a relatively small nucleic acid genome (3 to 200 kb of either double- or single-stranded DNA or RNA) within a viral protein capsid that is sometimes, depending on the virus, surrounded by a protein-containing viral envelope (membrane).

to proteins on the outer surface of the agent. So, do vaccines need to contain the whole organism, or will specific portions of pathogenic organisms suffice? For disease-causing viruses, it has been shown that purified outer surface viral proteins, either capsid or envelope proteins (Fig. 11.1), are alone sufficient for eliciting neutralizing antibodies in the host organism. Vaccines that use components of a pathogenic organism rather than the whole organism are called "subunit" vaccines; recombinant DNA technology is very well suited for developing new subunit vaccines.

There are advantages and disadvantages to the use of subunit vaccines. On the positive side, using purified protein(s) as an immunogen ensures that the preparation is stable and safe, is precisely defined chemically, and is free of extraneous proteins and nucleic acids that can initiate undesirable side effects in the host organism. On the negative side, purification of a specific protein can be costly and, in certain instances, an isolated protein may not have the same conformation as it does in situ (i.e., within the viral capsid or envelope), with the result that its antigenicity is altered. Obviously, the decision to produce a subunit vaccine depends on an assessment of several biological and economic factors.

Herpes Simplex Virus

Herpes simplex virus (HSV) has been implicated as a cancer-causing (oncogenic) agent—in addition to its more common roles in causing sexually transmitted disease, severe eye infections, and encephalitis—so that prevention of HSV infection by vaccination with either killed or attenuated virus may put the recipient at risk for cancer. Thus, protection against HSV would be best achieved by a subunit vaccine, which would not be oncogenic.

The primary requirement for creating any subunit vaccine is identification of the component(s) of the infectious agent that elicits antibodies that react against the intact form of the infectious agent. The HSV type 1 (HSV-1) envelope glycoprotein D (gD) is such a component, because, after injection into mice, it elicits antibodies that neutralize intact HSV. The HSV-1 gD gene was isolated and then cloned into a mammalian expression vector and expressed in Chinese hamster ovary (CHO) cells, which, unlike the *Esche-*

richia coli system, allows foreign proteins to be properly glycosylated. The complete sequence of the gD gene encodes a protein that normally is bound to the mammalian host cell membrane (Fig. 11.2A). However, a membrane-bound protein is more difficult to purify than a soluble protein, and so the gD gene was modified by removing the nucleotides encoding the C-terminal transmembrane binding domain (Fig. 11.2B). The modified gene was then transformed into CHO cells, where the product was glycosylated and secreted into the external medium since it could not be incorporated into the cell membrane. In laboratory trials testing its neutralizing ability, the modified form of gD was effective against both HSV-1 and HSV-2.

Foot-and-Mouth Disease

Foot-and-mouth disease virus (FMDV) has a devastating impact on cattle and swine and is extremely virulent, but it has been possible to keep the negative effects of this virus to a minimum by using formalin-killed FMDV preparations as a vaccine. Approximately 1 billion doses of this killed-virus vaccine are used worldwide each year.

Research on FMDV found that the major antigenic determinant inducing neutralizing antibody is the capsid viral protein 1 (VP1). Although purified VP1 is a much less potent antigen than intact viral particles, it can still induce neutralizing antibodies by itself and therefore can protect animals from infection by FMDV. Thus, the gene for VP1 became a target for cloning.

Figure 11.2 **A.** Location in the cytoplasmic membrane of HSV-1 gD with the transmembrane domain. **B.** Extracellular location of a soluble gD without the transmembrane domain.

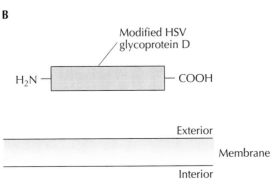

The genome of FMDV is composed of single-stranded RNA. Therefore, it was necessary first to synthesize a double-stranded cDNA of the entire genome (approximately 8,000 nucleotides long). This cDNA was then digested with restriction enzymes, and the fragments were cloned in an *E. coli* expression vector. The product of the VP1 coding sequence was identified immunologically as part of a fusion protein under the control of the p^L promoter-*c*I repressor system. The fusion protein was 396 amino acids long and consisted of a portion of the carrier protein (i.e., the bacteriophage MS2 replicase protein) as well as the entire sequence of the FMDV VP1 protein. The fusion protein containing the VP1 protein fragment was able to generate neutralizing antibodies to FMDV.

A fusion protein, however, faces more government regulatory hurdles than intact VP1 would. Therefore, the complete VP1 sequence will probably have to be subcloned onto a different expression vector. Nevertheless, a subunit vaccine for foot-and-mouth disease should soon be ready for preclinical trials.

Tuberculosis

Tuberculosis, one of the most important infectious diseases worldwide, is caused by the bacterium *Mycobacterium tuberculosis*. This bacterium can form lesions in any tissue or organ, which leads to cell death. The lungs are most commonly affected. In addition, patients suffer fever and loss of body weight, and without treatment tuberculosis is often fatal. It is estimated that approximately 2 billion people are currently infected with this organism and that nearly 3 million deaths a year result from these infections. Over the past 50 years, antibiotics have been used to treat patients infected with *M. tuberculosis*. However, numerous multidrug-resistant strains of *M. tuberculosis* are now prevalent. Consequently, a bacterial disease that was thought to have been defeated has become a serious public health problem in many parts of the world.

Currently, in some countries, bacillus Calmette-Guérin (BCG), a strain of *Mycobacterium bovis*, is used as a vaccine against tuberculosis. However, the efficacy of this approach has been seriously questioned. First, live BCG cells can cause serious disease in immunocompromised individuals, such as AIDS patients, who are susceptible to tuberculosis. Second, individuals treated with BCG cells respond positively to a common diagnostic procedure for the presence of the tuberculosis-causing bacterium, so that it becomes impossible to distinguish between individuals infected with *M. tuberculosis* and those inoculated with BCG cells. For these reasons, the BCG strain is not approved for use in a number of countries including the United States.

In an initial attempt to determine whether a safer and more effective subunit vaccine against tuberculosis might be developed, one group of researchers decided to first examine the immunoprotection that was provided by purified *M. tuberculosis* extracellular proteins. Following growth of the bacterium in liquid culture, 6 of the most abundant of the approximately 100 secreted proteins were purified. Each of these proteins was used separately and then in combination with the others to immunize guinea pigs. The immunized animals were then challenged with an aerosol containing approximately 200 cells of live *M. tuberculosis*—a large dose for these animals. The animals were observed for 9 to 10 weeks before their lungs and spleen were examined for the presence of disease-causing organisms. In these experiments, some of the purified protein combinations provided ap-

proximately the same level of protection against weight loss, death, and infection of lungs and spleen as did the live BCG vaccine. A safe and efficacious subunit vaccine for the prevention of tuberculosis in humans can be developed once it is determined whether the protective *M. tuberculosis* proteins produced by recombinant DNA technology are as effective as secreted proteins.

Peptide Vaccines

The question arises whether a small discrete portion (domain) of a protein can act as an effective subunit vaccine and induce the production of neutralizing antibodies. Intuitively, one would expect that only the portions, or domains, of a protein that are accessible to antibody binding, that is, those on the exterior surface of the virus, would be immunologically essential and that the inaccessible regions inside the virus particle may be ignored if they do not contribute to the conformation of the immunogenic domain (Fig. 11.3). If this argument has validity, it is possible that short peptides that mimic epitopes (antigenic determinants) will be immunogenic and could be used as vaccines (peptide vaccines).

Chemically synthesized domains of FMDV VP1 were tested as potential peptide vaccines. Peptides corresponding to amino acids 141 to 160, 151 to 160, and 200 to 213, which are located near the C-terminal end of VP1, and amino acids 9 to 24, 17 to 32, and 25 to 41, which are located near the N-terminal end of VP1, were each bound to a separate inert carrier protein (keyhole limpet hemocyanin)—very small peptides are usually rapidly degraded unless they are bound to the surface of a larger carrier molecule—and injected into guinea pigs (Fig. 11.4). A single inoculation with peptide 141 to 160 elicited sufficient antibody to protect the animal against subsequent challenges with FMDV. By contrast, inoculation with complete VP1 or peptides 9 to 24, 17 to 32, or 25 to 41 yielded a lower level of neutralizing antibodies.

In an additional experiment, a longer peptide consisting of amino acids 141 to 158 joined to amino acids 200 to 213 by two proline residues elicited high levels of neutralizing antibodies in guinea pigs, even when it was injected without any carrier protein. This "two-peptide" molecule was more effective than either of the single peptides alone and prevented FMDV proliferation in cattle as well as in guinea pigs.

Although these results were promising, the amount (dose) of peptide material that had to be used to elicit an immunological response was

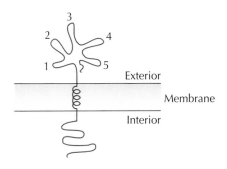

Figure 11.3 Generalized envelope-bound protein with external epitopes (1 to 5) that might elicit an immune response.

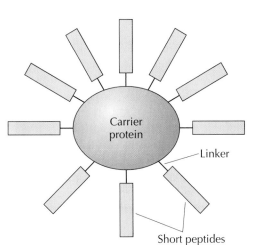

Figure 11.4 Structure of a peptide vaccine composed of identical short peptides bound to a carrier protein.

approximately 1,000 times the amount of inactivated FMDV. To overcome this problem, DNA encoding FMDV VP1 peptide 142 to 160 was linked to the gene encoding a highly immunogenic carrier molecule, hepatitis B core protein (HBcAg). When the gene for this fusion protein was expressed in either *E. coli* or animal cells in culture, the protein molecules self-assembled into stable "27-nm particles," with the FMDV VP1 peptide located on the outer surface of the particle. These particles are highly immunogenic in laboratory animals. Therefore, HBcAg may be an effective carrier molecule for such short synthetic peptides. A comparison of the immunogenicity, in guinea pigs, of a variety of FMDV peptide vaccines, all of which contained the VP1 peptide 142-to-160 sequence, revealed that a fusion protein containing HBcAg and FMDV VP1 amino acids 142 to 160 was approximately 1/10 as immunogenic as inactivated FMDV particles, 35 times more immunogenic than a fusion protein containing *E. coli* β-galactosidase and FMDV VP1 amino acids 137 to 162, and 500 times more immunogenic than the free synthetic peptide composed of amino acids 142 to 160. Because synthetic peptides fused to HBcAg do not seem to interfere with the assembly of the 27-nm hepatitis B virus-like particles and because these particles are nearly as immunogenic as the intact virus from which the synthetic peptide was derived, this approach may become a general method for the delivery of peptide vaccines.

Nevertheless, there are certain limitations to using short peptides as vaccines:

- To be effective, an epitope must consist of a short stretch of contiguous amino acids, which does not always occur naturally.
- The peptide must be able to assume the same conformation that the epitope has in the intact viral particle.
- A single epitope may not be sufficiently immunogenic.

Obviously, more research needs to be done, but in the future, synthetic peptide vaccines could become highly specific, relatively inexpensive, safe, and effective alternatives to traditional vaccines.

Genetic Immunization

A novel strategy that elicits an antibody response without the introduction of an antigen has been developed. In this case, the gene encoding an antigenic protein is incorporated into cells in a target animal, where the antigen is synthesized. In the initial experiments, a cloned gene encoding an antigen under the transcriptional control of an animal viral promoter on an *E. coli* plasmid was used to coat gold microprojectiles, which were introduced into cells in the ears of mice by using a pressurized delivery (biolistic) system. Subsequently, other workers found that cloned cDNAs could be introduced into mouse cells by injecting large amounts of the plasmid carrying the target DNA into the muscles of the test animal. At present, direct injection into muscles requires 3 to 4 orders of magnitude more DNA than the biolistic delivery system does. In one experiment, in more than 75% of the trials, the input gene was taken up into the mouse cells and directed the synthesis of the protein antigen, which in turn activated in the mouse an immune response that led to the production of antibodies against the target protein. The unique feature of this protocol is that the costly and time-consuming procedure of either purifying an antigen or creating a recombinant vaccine vehicle is bypassed. Moreover, proteins produced by this procedure are more likely to be correctly posttranslationally modified than are proteins that are

produced by different host organisms. "Genetic immunization," as this method has been called, may hold promise for the vaccination of domestic animals.

The feasibility of genetic immunization has been examined in detail. In one series of experiments, mice were injected in the quadriceps of both legs with an *E. coli* plasmid carrying the cDNA for influenza A virus nucleoprotein under the transcriptional control of either a Rous sarcoma virus or a cytomegalovirus promoter. Although the expression of the nucleoprotein was too low to detect, nucleoprotein-specific antibodies were observed in the blood of the test mice 2 weeks after the initial injection. In comparison to control mice, nucleoprotein-injected mice were significantly protected against the lethal effects of influenza virus infection (Fig. 11.5). Moreover, nucleoprotein-injected mice were also protected against a different strain of influenza virus. This cross-protection is in sharp contrast to traditional influenza virus vaccines, which are directed against surface antigens of the virus, so that each vaccine is specific to a single strain of influenza virus. Moreover, traditional vaccines work only as long as the antigens on the surface of the virus do not change. Unfortunately, the genes for the surface antigens mutate at a high rate, which creates significant differences among strains. By contrast, core pieces of the virus such as the nucleoprotein are relatively invariant and can activate the immune system by a mechanism that is different from the way in which surface antigens do so.

An advantage of DNA immunization, besides bypassing the need for purified protein antigens, is that it can trigger a response only against the protein encoded on the plasmid and not against the plasmid itself. Thus, the same vector can be used to deliver different proteins to an individual or the administration of the same gene can be repeated a number of times.

At this time, the fate of the introduced DNA is not known. It could have the undesirable effect of integrating into the genome of the host, possibly interrupting an important gene and leading to cancer. However, this risk is currently considered to be extremely low. It is more than likely that the

Figure 11.5 Survival of DNA-immunized mice. Injected mice were immunized with DNA that contained influenza A virus nucleoprotein under the control of the Rous sarcoma virus promoter on an *E. coli* plasmid. The control mice were injected with plasmid DNA only. The *x* axis represents the number of days after the animals were challenged with the live influenza virus.

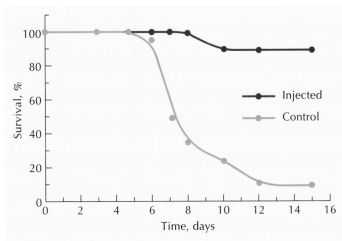

DNA will exist for a short period as a nonreplicating extrachromosomal element before it is degraded. To date, genetic immunization has been used to induce immune responses in animals, but not in humans, against a number of pathogenic organisms including influenza A virus, human immunodeficiency virus type 1, bovine herpesvirus, rabies virus, *Plasmodium* species (which cause malaria), and hepatitis B virus.

A modified strain of the bacterium *Shigella flexneri* has been developed to facilitate the delivery into animal cells of DNA for genetic immunization. *Shigella* can enter animal epithelial cells, escaping the phagocytic vacuole, and can direct plasmid DNA to the cytoplasm of the host cell, where, if the introduced gene(s) contains a eukaryotic promoter, it can be transcribed and translated. *Shigella* is normally a pathogenic organism and would not be an acceptable DNA delivery system in this form. Therefore, to use *Shigella*, it was first necessary to construct a nonpathogenic version of the wild-type organism by making a deletion mutation in the *Shigella asd* gene, which encodes the enzyme aspartate β-semialdehyde dehydrogenase. This enzyme is normally involved in the synthesis of the bacterial cell wall constituent diaminopimelic acid; therefore, the mutant cannot grow unless diaminopimelic acid is added to the growth medium. *Shigella* strains with the *asd* mutation can invade animal epithelial cells and deliver their plasmid DNA; however, once present, the *Shigella* cells are unable to proliferate.

Determination of the safety of using *Shigella* as a vector for the delivery of DNA to animal cells must await the results of human trials; however, the

MILESTONE

Protection against Foot-and-Mouth Disease by Immunization with a Chemically Synthesized Peptide Predicted from the Viral Nucleotide Sequence

J. L. BITTLE, R. A. HOUGHTEN, H. ALEXANDER, T. M. SHINNICK, J. G. SUTCLIFFE, R. A. LERNER, D. J. ROWLANDS, and F. BROWN
Nature **298**:30–33, 1982

Since Jenner developed the first vaccine over 200 years ago, most human vaccines against viral diseases have included partially killed or attenuated preparations of the disease-causing or a similar nonpathogenic version of the virus. While this approach has been undeniably effective and prevents the spread of a number of viral diseases, it is clearly limited. For example, not all viruses can be grown in culture, which precludes the development of vaccines against these viruses; production of traditional vaccines is expensive and potentially dangerous; and not all viral diseases are preventable through the use of these traditional vaccines. With the advent of molecular biotechnology, a number of researchers began to examine the

possibility of developing safer, less expensive, and more effective vaccines that would not have the limitations of traditional vaccines. Since vaccines immunize individuals by priming their immune systems, it was thought that for some viruses short synthetic peptides might elicit the same antibody response as the antigenic determinants normally found on the external surface of the virus. These peptides were designed with the identical linear sequence of amino acids that made up the viral antigenic determinant in the first place. Of course, this approach could be expected to work only when the amino acids of an antigenic region were contiguous.

Bittle et al. isolated and characterized the viral RNA for foot-and-mouth

disease virus and then determined the sequence of viral protein 1 (VP1), the major antigenic protein of the whole virus. Based on other experiments, they reasoned that the major antigenic determinants on VP1 would probably be found at either the N- or C-terminal end of the protein. They then chemically synthesized a series of peptides based on the amino acid sequence of the N and C termini, chemically linked these peptides to carrier proteins, and then used them to inoculate rabbits. When rabbits and guinea pigs were inoculated with peptides from the C-terminal end of VP1, these animals synthesized antibodies that protected them against disease from whole foot-and-mouth disease virus. This work established the principle that a protein domain(s) is sufficient to induce antibodies that can neutralize intact virus particles, and hence a new type of vaccine, which after the initial stages of development does not utilize or depend upon the disease-causing virus, is possible.

results of experiments with guinea pigs are promising. The greatest perceived advantage of this approach is that with the *Shigella* system, DNA for vaccination may be delivered orally, greatly simplifying the delivery of a variety of vaccines.

Attenuated Vaccines

In some instances, genetic manipulation may be used to construct modified organisms (bacteria or viruses) that are used as live recombinant vaccines. These vaccines are either nonpathogenic organisms that have been engineered to carry and express antigenic determinants from a target pathogenic agent or engineered strains of pathogenic organisms in which the virulence genes have been modified or deleted. In these instances, as a part of a bacterium or a virus, the important antigenic determinants are presented to the immune system with a conformation that is very similar to the form of the antigen in the disease-causing organism. By contrast, purified antigen alone often lacks the native conformation and elicits a weak immunological response.

Cholera

If it is possible to develop a live vaccine, it is usually advantageous to do so, because live vaccines are generally much more effective than killed or subunit vaccines. The major requirement for a live vaccine is that no virulent forms be present in the inoculation material. With this objective in mind, a live cholera vaccine has been developed. Cholera is a fast-acting intestinal disease characterized by fever, dehydration, abdominal pain, and diarrhea. It is transmitted by drinking water contaminated with fecal matter. In developing countries, the threat of cholera is a real and significant health concern whenever water purification and sewage disposal systems are inadequate.

The bacterium *Vibrio cholerae*, the causative agent of cholera, colonizes the small intestine and secretes large amounts of a hexameric enterotoxin, which is the actual pathogenic agent. This protein consists of one subunit, the A subunit, that has ADP-ribosylation activity and stimulates adenylate cyclase, and five identical B subunits that bind specifically to an intestinal mucosal cell receptor. The A subunit has two functional domains: the A_1 peptide, which contains the toxic activity, and the A_2 peptide, which joins the A subunit to the B subunits. The cholera vaccine that is currently used consists of phenol-killed *V. cholerae*. This vaccine generates only moderate protection, which normally lasts from about 3 to 6 months. However, other possible types of cholera vaccines have been examined.

Previous studies had indicated that a subunit vaccine consisting of inactivated hexameric cholera enterotoxin was not effective in providing immunity against *V. cholerae*. Since *V. cholerae* colonizes the surface of the intestinal mucosa, it is thought that an effective cholera vaccine should probably be directed to this structure and should therefore be administered orally. With this in mind, a strain of *V. cholerae* in which part of the coding sequence for the A_1 peptide was deleted was created. This strain cannot produce enterotoxin; therefore, it is nonpathogenic and is a good candidate for a live vaccine.

Specifically, in this experiment, a tetracycline resistance gene was incorporated into the A_1 peptide DNA sequence on the *V. cholerae* chromosome. This insertion inactivated the A_1 peptide activity and also made the

strain resistant to tetracycline. Although the A_1 peptide sequence was disrupted, this strain is not acceptable as a vaccine because the inserted tetracycline resistance gene can excise spontaneously, thereby restoring enterotoxin activity. Consequently, it was necessary to engineer a strain carrying a defective A_1 peptide sequence that could not revert (Fig. 11.6).

1. A plasmid containing the cloned DNA segment for the A_1 peptide was digested with the restriction enzymes *Cla*I and *Xba*I, each of which cut only within the A_1 peptide-coding sequence of the insert.

Figure 11.6 Strategy for deleting part of the cholera toxin A_1 peptide DNA sequence from a strain of *V. cholerae*.

2. To recircularize the plasmid, an *Xba*I linker was added to the *Cla*I site and then cut with *Xba*I.

3. T4 DNA ligase was used to join the plasmid at the *Xba*I sites, thereby deleting a 550-base-pair (bp) segment from the middle of the A_1 peptide-coding region. This deletion removed 183 of the 194 amino acids of the A_1 peptide.

4. Then, by conjugation, the plasmid containing the deleted A_1 peptide-coding sequence was transferred into the *V. cholerae* strain carrying the tetracycline resistance gene within its A_1 peptide DNA sequence.

5. Recombination can occur between the remaining A_1 coding sequence on the plasmid and the Tet^r gene-disrupted A_1 peptide gene on the chromosome. Such a double-crossover event replaces the chromosomal A_1 peptide-coding sequence with the homologous segment on the plasmid carrying the deletion.

6. After growth for a number of generations, the extrachromosomal plasmid, which is unstable in *V. cholerae,* was spontaneously lost.

7. Cells with an integrated defective A_1 peptide were selected on the basis of their tetracycline sensitivity. The desired cells no longer had the tetracycline resistance gene but carried the deleted A_1 peptide sequence.

A stable strain with a deleted A_1 peptide sequence was selected in this way. This strain did not produce active enterotoxin but nevertheless retained all the other biochemical features of the pathogenic form of *V. cholerae*. This strain is currently being evaluated in clinical trials to test its effectiveness as a cholera vaccine. The results so far have been equivocal. While this vaccine conferred nearly 90% protection against diarrheal disease in volunteers, it induced side effects in some of those who were tested. It is possible that this strain will require modification at an another chromosomal locus before it can be used as a vaccine.

Salmonella Species

Other attempts to engineer nonpathogenic strains of pathogenic bacteria that could be used as live vaccines have involved deletions in chromosomal regions that code for independent and essential functions. At least two deletions are preferred, because the probability that both sets of functions can be simultaneously reacquired is very small. It is assumed that a "doubly deleted" strain would have a limited ability to proliferate after its injection as a vaccine, thereby curtailing its pathogenicity while allowing it to stimulate an immunological response.

Strains of the *Salmonella* genus cause enteric fever, infant death, typhoid fever, and food poisoning. Therefore, an effective vaccine against these organisms is needed. Deletions in the *aro* genes, which encode enzymes involved in the biosynthesis of aromatic compounds, and in the *pur* genes, which encode enzymes involved in purine metabolism, have been used to attenuate various *Salmonella* strains. These doubly deleted strains, which can be grown on a complete and enriched medium, generally establish only low-level infections; overall, their virulence is reduced by 10^6-fold or more. These attenuated *Salmonella* strains have been shown to be effective oral vaccines in mice, sheep, cattle, chickens, and, more recently, in humans.

Leishmania Species

Although the human immune system can respond to infections by protozoan parasites of the genus *Leishmania*, it has been difficult to develop an

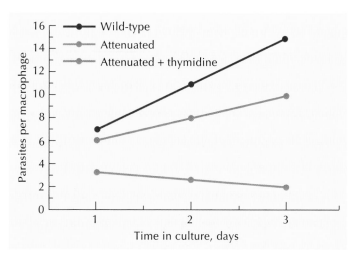

Figure 11.7 Proliferation of wild-type and attenuated *L. major* in mouse macrophages. At time zero, macrophages were infected with the same amount of stationary-phase *L. major*. The wild-type parasite and the attenuated parasite in the presence of thymidine were able to proliferate under these conditions, while the attenuated strain did not proliferate without thymidine in the medium. Adapted from Titus et al., *Proc. Natl. Acad. Sci. USA* **92**:10267–10268, 1995.

effective vaccine against these organisms. Attenuated strains of *Leishmania* are sometimes effective as vaccines; however, they often revert to virulence. Also, the attenuated parasite can persist for long periods in an infected but apparently asymptomatic individual. Such individuals can act as reservoirs for the parasite, which can be transferred to other people by an intermediate host. To overcome these problems, an attenuated strain of *Leishmania* that is unable to revert to virulence can be created by targeted deletion of an essential metabolic gene such as the one encoding dihydrofolate reductase-thymidylate synthase. In one of these attenuated strains, *Leishmania major*

Figure 11.8 Immunity to virulent *L. major* induced in BALB/c mice inoculated with attenuated *L. major*. At time zero, mice that were previously inoculated with attenuated *L. major* were challenged with virulent *L. major*, and the sizes of the parasite-induced lesions were measured at various times. Control mice were not vaccinated with attenuated *L. major*. Adapted from Titus et al., *Proc. Natl. Acad. Sci. USA* **92**:10267–10271, 1995.

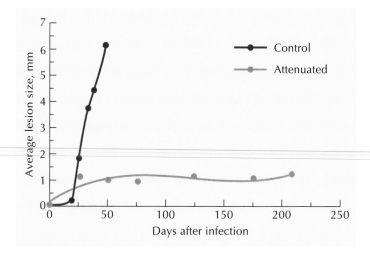

E10-5A3, the two dihydrofolate reductase-thymidylate synthase genes that are present in wild-type strains are replaced with the genes encoding resistance to the antibiotics G-418 and hygromycin. For growth in culture, it is necessary to add thymidine to the medium used to propagate the attenuated, but not the wild-type, strain. In addition, unlike the wild type, the attenuated strain is unable to replicate in macrophages in tissue culture unless thymidine is added to the growth medium (Fig. 11.7). Importantly, the attenuated strain survives for only a few days when inoculated into mice; in that time, it does not cause any disease. This period is sufficient to induce substantial immunity against leishmania lesions in BALB/c mice after administration of the wild-type parasite (Fig. 11.8). Since the attenuated parasite did not establish a persistent infection or cause disease, even in the most susceptible strains of mice tested, it is considered to be a strong candidate vaccine. Following additional experiments with animals, it should be possible to test whether this attenuated parasite is effective as a vaccine in humans.

Vector Vaccines

Vaccines Directed against Viruses

Vaccinia virus has, in the form of a live vaccine, led to the eradication of smallpox globally. Vaccinia virus is a member of the poxvirus family. This completely sequenced virus has a double-stranded DNA genome that contains 187 kilobase pairs (kb) and encodes approximately 200 different proteins. Vaccinia virus DNA replicates within the cytoplasm of infected cells. Cytoplasmic, rather than nuclear, replication is possible because vaccinia virus DNA contains genes for DNA polymerase, RNA polymerase, and the enzymes to cap, methylate, and polyadenylate mRNA. Thus, if a foreign gene is inserted into the vaccinia virus genome under the control of a vaccinia virus promoter, it will be expressed independently of host regulatory and enzymatic functions. In addition to humans, the virus can infect many other vertebrates and invertebrates.

Vaccinia virus has a broad host range, is well characterized at the molecular level, is stable for years after lyophilization (freeze-drying), and is a usually benign virus. Therefore, it is a strong candidate as a vector vaccine. The important features of a vector vaccine are the delivery and expression of cloned genes encoding antigens that elicit neutralizing antibodies against pathogenic agents. However, the vaccinia virus genome is very large and lacks unique restriction sites. Therefore, it is not possible to insert additional DNA directly into the viral genome. Of necessity, the genes for specific antigens must be introduced into the viral genome by in vivo homologous recombination:

1. The DNA sequence coding for a specific antigen, such as HBcAg, is inserted into a plasmid vector immediately downstream of a cloned vaccinia virus promoter and in the middle of a nonessential vaccinia virus gene, such as the gene for the enzyme thymidine kinase (Fig. 11.9A).
2. This plasmid is used to transform thymidine kinase-negative animal cells in culture, usually chicken embryo fibroblasts, that have previously been infected with wild-type vaccinia virus, which produces a functional thymidine kinase.
3. Recombination between DNA sequences that flank the promoter

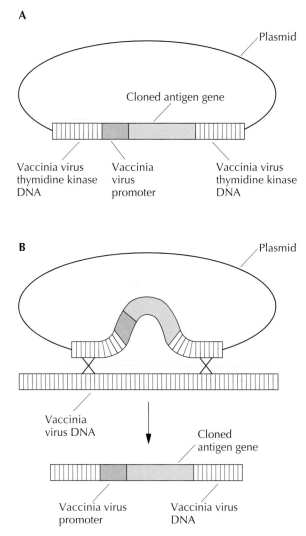

Figure 11.9 Method for the integration into vaccinia virus of a gene whose protein product, generally a viral antigen, elicits an immunological response. **A.** Structure of a plasmid carrying a cloned expressible antigen gene. **B.** A double-crossover event results in the integration of the antigen gene into vaccinia virus DNA.

and the neutralizing antigen gene on the plasmid and the homologous sequences on the viral genome results in the incorporation of the cloned gene into the viral DNA (Fig. 11.9B). Although the recombination event is rare, the absence of thymidine kinase activity in the host cells and the disruption of the thymidine kinase gene in the recombined virus render the host cells resistant to the otherwise toxic effects of bromodeoxyuridine. This selection system can be used to enrich for cell lines that carry a recombinant vaccinia virus.

4. The final selection is made by using a DNA hybridization probe for the antigen gene.

Since thymidine kinase-negative mutants of vaccinia virus arise spontaneously at a relatively high frequency of about 1 virus particle in 10^3 to 10^4, scientists often cotransfer a selectable marker with the target gene. This makes it much easier to distinguish a spontaneous thymidine kinase mutant from a mutant deliberately generated by homologous recombination.

The *neo* gene, which codes for the enzyme neomycin phosophotransferase II and confers resistance to the kanamycin analog G-418, is often used as the selectable marker. This gene, unlike some other selectable markers, is quite stable once it is inserted into the vaccinia virus genome.

To avoid disrupting any vaccinia virus genes or the necessity of screening for selectable markers, a novel system has been devised in which every recombinant virus that can form a plaque will contain and express the target gene. Wild-type vaccinia virus contains a gene, *vp37*, that is responsible for the formation of plaques when the virus is grown on an animal cell monolayer (Fig. 11.10A). Deleting the *vp37* gene and replacing it with an *E. coli* marker gene (Fig. 11.10B) creates a vaccinia virus mutant that does not form plaques after 2 to 3 days of growth in cell culture. Target genes are introduced into the mutant vaccinia virus by homologous recombination with a transfer vector that carries the *vp37* gene as well as the target gene (Fig. 11.10C). If homologous recombination between the non-plaque-forming mutant and the transfer vector occurs, the viruses that can form plaques have acquired the *vp37* gene. As well, the target gene is inserted into the vaccinia virus genome and the marker gene is lost. Since the *vp37* gene has been deleted in the mutant vaccinia virus, it is impossible for this mutation to revert to the wild type. Therefore, every virus that forms a plaque carries the desired construct. This procedure is simple and straightforward, is ap-

Figure 11.10 **A.** Portion of a wild-type vaccinia virus genome. **B.** Portion of a mutant vaccinia virus genome. **C.** Portion of a vaccinia virus transfer vector. "Left flank" and "right flank" refer to the DNA sequences that immediately precede and follow the *vp37* gene in the wild-type vaccinia virus genome. The native *vp37* promoter is part of the *vp37* gene sequence (not shown). MCS is a multiple cloning site with seven unique restriction enzyme sites. *p7.5* is a strong early/late vaccinia virus promoter. The target gene is inserted into the multiple cloning site. Subsequently, homologous recombination between the transfer vector **(C)** and the genomic DNA of the mutant virus **(B)** results in the replacement of the *E. coli* marker gene with the *vp37* gene together with a target gene.

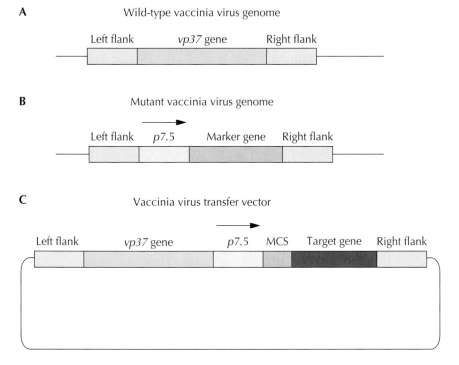

plicable to the cloning and expression of any target gene, does not require any extra marker genes, and does not disrupt any vaccinia virus genes.

A number of antigen genes have been successfully inserted into the vaccinia virus genome and subsequently expressed in animal cells in culture. These antigens include rabies virus G protein, hepatitis B surface antigen, Sindbis virus surface proteins, influenza virus NP and HA proteins, vesicular stomatitis virus N and G proteins, and herpes simplex virus glycoproteins. Several of these recombinant vaccinia virus vehicles have been shown to be effective vaccines. A recombinant vaccinia virus that expresses the herpes simplex virus type 1 glycoprotein D gene prevents herpes infections in mice. Another recombinant vaccinia virus, which expresses the rabies virus surface antigen gene, was able to elicit neutralizing antibodies in foxes, which are major carriers of rabies in Europe.

The use of vector vaccines constructed from vaccinia virus also offers the possibility of vaccinating individuals against several different diseases with one treatment. This may be achieved by using a recombinant vaccinia virus carrying cloned genes encoding a number of different antigens.

The timing of the production of a foreign protein whose gene is carried in a vaccinia virus depends on whether a vaccinia virus promoter functions during the early or late phase of the infection cycle, and the strength of the promoter determines the amount of an antigen that is produced. For the most part, late promoters for an 11-kilodalton (kDa) protein (p11) and the cowpox virus A-type inclusion protein (pCAE) have been used to achieve high levels of expression. When several different foreign proteins are inserted onto one vaccinia virus, each is placed under the control of a different vaccinia virus promoter to avert the possibility of homologous recombination between different portions of the virus that might cause the cloned genes to be lost.

A live recombinant viral vaccine has several advantages over killed virus or subunit vaccines. First, the virus can express the authentic antigen(s) in a manner that closely resembles a natural infection. Second, the virus can replicate within the host, thereby amplifying the amount of antigen which activates the release of antibodies from B cells (humoral response) and stimulating the production of T cells (cell-mediated immune response). Third, the insertion of antigen genes into one or more sites on the vaccinia virus genome further reduces its virulence.

A disadvantage of using a live recombinant viral vaccine is that vaccination of an immunosuppressed host, such as an individual with AIDS, can lead to a serious viral infection. One way to avoid this problem may be to insert the gene encoding human interleukin-2 into the viral vector. The presence of the interleukin-2 gene would enable the recipient to limit immunologically the proliferation of the viral vector by enhancing the response of the T cells of the immune system and thereby decreasing the possibility of an unwanted infection.

If the proliferation of vaccinia virus has deleterious effects in certain patients, it would be helpful to kill or inhibit the virus after vaccination. One approach to meet this end is to create an interferon-sensitive vaccinia virus—wild-type vaccinia virus is relatively resistant to interferon—whose proliferation could be curtailed. Such a virus vector would be amenable to drug intervention if complications from vaccination of individuals with vaccinia virus vectors arise.

The basis of the resistance to interferon of vaccinia virus was not known until a vaccinia virus open reading frame (K3L) was found to code for a

10.5-kDa protein that has an amino acid sequence that is very similar to a portion of the 36.1-kDa host cell eukaryotic initiation factor elF-2α. The N-terminal regions of both of these proteins contain 87 amino acids that are nearly identical. Moreover, this overlap region contains a serine residue, amino acid 51, which in elF-2α is normally phosphorylated by interferon-activated P1 kinase. When this serine residue in elF-2α is phosphorylated in interferon-treated cells, protein synthesis is inhibited. Thus, vaccinia virus may avoid inhibition by interferon because the K3L protein acts as a competitive inhibitor of elF-2α phosphorylation. Therefore, deletion of all or a portion of the K3L gene from vaccinia virus should make this virus sensitive to interferon. A K3L-minus mutant of vaccinia virus was constructed by polymerase chain reaction (PCR) mutagenesis of the K3L gene carried on a plasmid followed by homologous recombination to replace the wild-type K3L sequence with the modified version. When the wild-type and mutant versions of vaccinia virus were tested for their sensitivity to interferon, the mutant was 10 to 15 times more sensitive to interferon than was the wild-type version (Fig. 11.11). Reinsertion of the wild-type K3L sequence into the mutant virus restored the level of interferon sensitivity found in the wild type. This indicates that K3L is indeed involved in the interferon resistance phenotype of vaccinia virus. This work is an important step in the development of safer vaccinia virus vectors. Moreover, other interferon-resistant viruses may contain sequences comparable to K3L and therefore may be amenable to the construction of interferon-sensitive deletion mutants.

While much of the work on the development of live viral vaccines has been done with vaccinia virus, other viruses, such as adenovirus, poliovirus, and varicella-zoster virus, are also being tested as potential vaccine vehicles. For example, while the use of live attenuated poliovirus as a vaccine vector is still at a very early stage of development, its attraction is that it can be delivered orally rather than being injected. Such a mucosal vaccine—a vaccine directed to receptors in the lungs or gastrointestinal tract—might be useful against a range of diseases including cholera, typhoid fever, influenza, pneumonia, mononucleosis, rabies, AIDS, and

Figure 11.11 Sensitivity of wild-type (K3L$^+$) and mutant (K3L$^-$) vaccinia virus to interferon. Mouse L929 cells were pretreated with a mixture of mouse alpha and beta interferon for 24 hours before the virus was added. Plates that did not have interferon added had approximately 3×10^7 viral plaques per plate. Adapted from Paoletti and Tartaglia, U.S. patent 5,378,657, 1995.

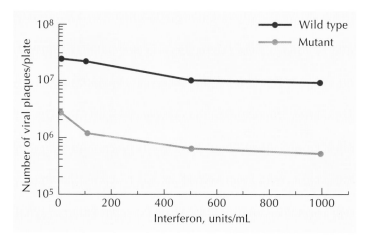

Lyme disease. However, the safety and efficacy of any apparently benign virus as a gene delivery and expression system must be firmly established before clinical trials are undertaken. For example, notwithstanding the routine use of vaccinia virus, it causes complications at a frequency of approximately 3.0×10^{-6} in human recipients. For human use of this recombinant viral vaccine, it may be advisable to delete the genes that contribute to this virulence.

Vaccines Directed against Bacteria

Since the discovery and subsequent widespread dissemination of antibiotics such as penicillin and streptomycin, only a modest level of research has been directed toward the development of vaccines for diseases caused by bacteria. However, there are good reasons for developing bacterial vaccines:

- Not all bacterial diseases are readily treated with antibiotics.
- The use of antibiotics over the last 40 years has resulted in the proliferation of bacterial strains that are resistant to several antibiotics.
- Refrigeration facilities for the storage of antibiotics are not commonly available in many tropical countries.
- It is often difficult to ensure that individuals receiving antibiotic therapy undergo the full course of treatment.

Given the need to produce vaccines that will be effective against bacterial diseases, the question is, What strategies are likely to be most effective? In instances in which the disease-causing bacterium does not grow well in culture, the development of an attenuated strain is not feasible. For these bacteria, alternative approaches must be used. For example, *Rickettsia rickettsii*, a gram-negative obligately intracellular bacterium that causes Rocky Mountain spotted fever, does not grow in culture. In this case, a cloned 155-kDa protein that is a major surface antigen of *R. rickettsii* was used as a subunit vaccine and was found to protect immunized mice against infection by disease-causing bacteria.

Bacteria as Antigen Delivery Systems

Antigens that are located on the outer surface of a bacterial cell are more likely to be immunogenic than are those in the cytoplasm. Thus, one strategy is to place a neutralizing antigen from a pathogenic bacterium on the surface of a live nonpathogenic bacterium. Flagella are made up of filaments of a single protein, flagellin. Under a microscope, they appear as conspicuous thread-like structures on the outer surface of some bacteria. If the flagella of a nonpathogenic organism could be made to carry a specific epitope from a pathogenic bacterium, protective immunogenicity might be easily achieved.

This strategy was used to engineer a cholera vaccine. A synthetic oligonucleotide specifying an epitope of the cholera toxin B subunit was inserted into a portion of the *Salmonella* flagellin gene that varies considerably from one strain to another (hypervariable segment). The construct was then introduced into a flagellin-negative strain of *Salmonella*. The epitope, which consisted of amino acid residues 50 to 64 of the cholera toxin B subunit, was shown previously to elicit antibodies directed against intact cholera toxin. The chimeric flagellin functioned normally. Furthermore, the epitope was present at the flagellum surface. Immunization of mice by intraperitoneal injections of approximately 5×10^6 live or formalin-killed "flagella-engineered" bacteria elicited high levels of antibodies directed against both the

peptide (i.e., amino acids 50 to 64) and the intact cholera toxin molecule. With this system, it may be possible to insert two or three different epitopes into a single *Salmonella* flagellin gene, thereby creating a multivalent bacterial vaccine.

Attenuated *Salmonella* strains can be administered orally and can potentially deliver a range of bacterial, viral, and parasite antigens to the immune system. For this purpose, the choice of promoter that drives the transcription of the foreign antigen is important. If too strong a promoter is used, the metabolic load might constrain bacterial proliferation. However, unlike a closed system such as a fermentation vessel, shifting the temperature or adding specific metabolites to induce foreign gene expression is not possible when the bacterial vector is added to a host animal. On the other hand, promoters that respond to environmental signals may provide effective means of controlling the expression of the foreign antigen gene. For example, the *E. coli nirB* promoter is regulated by both nitrite and the oxygen tension of the environment, becoming active under anaerobic conditions. In one series of experiments, the *nirB* promoter was used to direct the expression in an attenuated strain of *Salmonella* of the nontoxic immunogenic fragment C of tetanus toxin. It is estimated that more than 1 million deaths per year in the developing world are the result of *Clostridium tetani* infections. When the engineered *Salmonella* strain was grown aerobically in culture, tetanus toxin fragment C was not synthesized; however, following oral administration of this bacterium to test mice, fragment C was produced and the animals generated antibodies against this peptide. Thus, the *Salmonella* strain that was engineered with tetanus toxin fragment C under the control of the *nirB* promoter may be useful as a live oral tetanus vaccine. However, additional studies are required to determine if this strategy is effective in humans.

SUMMARY

Traditionally, vaccines are either inactivated or attenuated infectious agents (bacteria or viruses) that are injected into an antibody-producing organism to produce immunity. There are a number of drawbacks to these vaccines. For example, not all pathogenic organisms can be grown to the large volumes needed to make a vaccine; there are safety concerns when large volumes of pathogenic organisms are being handled; attenuated strains may revert to the infectious state; inactivation may be incomplete; and shelf life is often dependent on refrigeration.

Recombinant DNA technology can be used in various ways to create reliable vaccines. Immunologically active, noninfectious agents can be produced by deleting the genes that cause virulence; with this deletion, a live vaccine would never be able to revert to the infectious form. It is possible to clone the gene(s) that encodes the major antigenic determinant(s) from a pathogenic organism into the genome of a benign carrier organism (usually a virus), which can be used as a vaccine without concern that pathogenic particles are present in the vaccine. Finally, the genes or segments of genes that encode the major antigenic determinants of pathogenic organisms can be cloned into expression vectors, and large amounts of the product can be harvested, purified, and used as a vaccine. With the last strategy, complete genes produce subunit vaccines and cloned domains of the major antigenic determinants produce peptide vaccines. Peptide vaccines may also be produced by chemical peptide synthesis.

References

Andio, R., D. Silvera, S. D. Suggett, P. L. Achacoso, C. J. Miller, D. Baltimore, and M. B. Feinberg. 1994. Engineering poliovirus as a vaccine vector for the expression of diverse antigens. *Science* **265:**1448–1451.

Bittle, J. L., R. A. Houghten, H. Alexander, T. M. Shinnick, J. G. Sutcliffe, R. A. Lerner, D. J. Rowlands, and F. Brown. 1982. Protection against foot-and-mouth disease by immunization with a chemically synthesized peptide predicted from the viral nucleotide sequence. *Nature* **298:**30–33.

Blancou, J., M. P. Kieny, R. Lathe, J. P. Lecoq, P. P. Pastoret, J. P. Soulebot, and P. Desmettre. 1986. Oral vaccination of the fox against rabies using a live recombinant vaccinia vaccine. *Nature* **322:**373–375.

Blasco, R., and B. Moss. 1995. Selection of recombinant vaccinia viruses on the basis of plaque formation. *Gene* **149:**157–162.

Boothroyd, J. C., P. E. Highfield, G. A. M. Cross, D. J. Rowlands, P. A. Lowe, F. Brown, and T. J. R. Harris. 1981. Molecular cloning of foot and mouth disease virus genome and nucleotide sequences in the structural protein genes. *Nature* **290:**800–802.

Brown, F. 1984. Synthetic viral vaccines. *Annu. Rev. Microbiol.* **38:**221–235.

Brown, F. 1985. Peptides as the next generation of foot-and-mouth disease vaccines. *Bio/Technology* **3:**445–448.

Burnette, W. N. 1990. The advent of recombinant pertussis vaccines. *Bio/Technology* **8:**1003–1005.

Charles, I., and G. Dougan. 1990. Gene expression and the development of live enteric vaccines. *Trends Biotechnol.* **8:**117–121.

Chatfield, S. N., I. G. Charles, A. J. Makoff, M. D. Oxer, G. Dougan, D. Pickard, D. Slater, and N. F. Fairweather. 1992. Use of the *nir*B promoter to direct the stable expression of heterologous antigens in *Salmonella* oral vaccine strains: development of a single-dose oral tetanus vaccine. *Bio/Technology* **10:**888–892.

Chow, M., R. Yabrov, J. Bittle, J. Hogle, and D. Baltimore. 1985. Synthetic peptides from four separate regions of the poliovirus type 1 capsid protein VP1 induce neutralizing antibodies. *Proc. Natl. Acad. Sci. USA* **82:**910–914.

Clarke, B. E., S. E. Newton, A. R. Carroll, M. J. Francis, G. Appleyard, A. D. Syred, P. E. Highfield, D. J. Rowlands, and F. Brown. 1987. Improved immunogenicity of a peptide epitope after fusion to hepatitis B core protein. *Nature* **330:**381–384.

Cohen, J. 1993. Naked DNA points way to vaccines. *Science* **259:**1691–1692.

Cremer, K. J., M. Mackett, C. Wohlenberg, A. L. Notkins, and B. Moss. 1985. Vaccinia virus recombinant expressing herpes simplex virus type 1 glycoprotein D prevents latent herpes in mice. *Science* **228:**737–740.

DiMarchi, R., G. Brooke, C. Gale, V. Cracknell, T. Doel, and N. Mowat. 1986. Protection of cattle against foot-and-mouth disease by a synthetic peptide. *Science* **232:**639–641.

Ferguson, M. 1991. Progress towards rabies control. *Trends Biotechnol.* **9:**7–11.

Finkelstein, A., and R. F. Silva. 1989. Live recombinant vaccines for poultry. *Trends Biotechnol.* **7:**273–277.

Flexner, C., A. Hugin, and B. Moss. 1987. Prevention of vaccinia virus infection in immunodeficient mice by vector-directed IL-2 expression. *Nature* **330:**259–262.

Graham, F. L. 1990. Adenoviruses as expression vectors and recombinant vaccines. *Trends Biotechnol.* **8:**85–87.

Horwitz, M. A., B.-W. E. Lee, B. J. Dillon, and G. Harth. 1995. Protective immunity against tuberculosis induced by vaccination with major extracellular proteins of *Mycobacterium tuberculosis. Proc. Natl. Acad. Sci. USA* **92:**1530–1534.

Jones, T. R., and S. L. Hoffman. 1994. Malarial vaccine development. *Clin. Microbiol. Rev.* **7:**303–310.

Kaper, J. B., H. Lockman, M. M. Baldini, and M. M. Levine. 1984. A recombinant live oral cholera vaccine. *Bio/Technology* **2:**345–349.

Kaper, J. B., H. Lockman, M. M. Baldini, and M. M. Levine. 1984. Recombinant nontoxinogenic *Vibrio cholerae* strains as attenuated cholera vaccine candidates. *Nature* **308:**655–658.

Kaper, J. B., J. G. Morris, Jr., and M. M. Levine. 1995. Cholera. *Clin. Microbiol. Rev.* **8:**48–86.

Kaslow, D. C., S. N. Isaacs, I. A. Quakyi, R. W. Gwadz, B. Moss, and D. B. Keister. 1991. Induction of *Plasmodium falciparum* transmission-blocking antibodies by recombinant vaccinia virus. *Science* **252:**1310–1313.

Kupper, H., W. Keller, C. Kurz, S. Forss, H. Schaller, R. Franze, K. Strohmaier, O. Marquardt, V. G. Zaslavsky, and P. H. Hofschneider. 1981. Cloning of cDNA of major antigen of foot and mouth disease virus and expression in *E. coli. Nature* **289:**555–559.

Lasky, L. A., D. Dowbenko, C. C. Simonsen, and P. W. Berman. 1984. Protection of mice from lethal herpes simplex virus infection by vaccination with a secreted form of cloned glycoprotein D. *Bio/Technology* **2:**527–532.

Lowe, R. S., P. M. Keller, B. J. Keech, A. J. Davison, Y. Whang, A. J. Morgan, E. Kieff, and R. W. Ellis. 1987. Varicella-zoster virus as a live vector for the expression of foreign genes. *Proc. Natl. Acad. Sci. USA* **84:**3896–3900.

Mekalanos, J. J., and J. C. Sadoff. 1994. Cholera vaccines: fighting an ancient scourge. *Science* **265:**1387–1389.

Michel, M.-L., H. L. Davis, M. Schleef, M. Mancini, P. Tiollais, and R. G. Whalen. 1995. DNA-mediated immunization to the hepatitis B surface antigen in mice: aspects of the humoral response mimic hepatitis B viral infection in humans. *Proc. Natl. Acad. Sci. USA* **92:**5307–5311.

Miner, J. N., and D. E. Hruby. 1990. Vaccinia virus: a versatile tool for molecular biologists. *Trends Biotechnol.* **8:**20–25.

Moss, B. 1991. Vaccinia virus: a tool for research and vaccine development. *Science* **252:**1662–1667.

Nabel, G. J., and P. L. Felgner. 1993. Direct gene transfer for immunotherapy and immunization. *Trends Biotechnol.* **11:**211–215.

Newton, S. M. C., C. O. Jacob, and B. A. D. Stocker. 1989. Immunoresponse to cholera toxin epitope inserted in *Salmonella* flagellin. *Science* **244:**70–72.

Nussenzweig, R. S., and C. A. Long. 1994. Malaria vaccines: multiple targets. *Science* **265:**1381–1383.

Oehen, S., H. Hengartner, and R. M. Zinkernagel. 1991. Vaccination for disease. *Science* **251:**195–197.

Paoletti, E., and J. Tartaglia. January 1995. Interferon sensitive recombinant poxvirus vaccine. U.S. patent 5,378,457.

Ratafia, M. 1987. Worldwide opportunities in genetically engineered vaccines. *Bio/Technology* **5:**1154–1158.

Sizemore, D. R., A. A. Branstrom, and J. C. Sadoff. 1995. Attenuated Shigella as a DNA delivery vehicle for DNA-mediated immunization. *Science* **270:**299–302.

Stover, C. K., V. F. de la Cruz, T. R. Fuerst, J. E. Burlein, L. A. Benson, L. T. Bennett, G. P. Bansal, J. F. Young, M. H. Lee, G. F. Hatfull, S. B. Snapper, R. G. Barletta, W. R. Jacobs, Jr., and B. R. Bloom. 1991. New use of BCG for recombinant vaccines. *Nature* **351:**456–460.

Tang, D.-C., M. DeVit, and S. A. Johnston. 1992. Genetic immunization is a simple method for eliciting an immune response. *Nature* **356:**152–154.

Tartaglia, J., and E. Paoletti. 1988. Recombinant vaccinia virus vaccines. *Trends Biotechnol.* **6:**43–46.

Titus, R. G., J. G. Gueiros-Filho, L. A. R. De Freitas, and S. M. Beverley. 1995. Development of a safe live *Leishmania* vaccine line by gene replacement. *Proc. Natl. Acad. Sci. USA* **92:**10267–10271.

Ulmer, J. B., J. J. Donnelly, S. . Parker, G. H. Rhodes, P. L. Felgner, V. J. Dwarki, S. H. Gromkowski, R. R. Deck, C. M. DeWitt, A. Friedman, L. A. Hawe, K. R. Leander, D. Martinez, H. C. Perry, J. W. Shiver, D. L. Montgomery, and M. A. Liu. 1993. Heterologous protection against influenza by injection of DNA encoding a viral protein. *Science* **259:**1745–1749.

REVIEW QUESTIONS

1. Briefly describe a protocol for developing a vaccine against a toxin-producing bacteria.

2. What factors limit the usefulness of conventional vaccines?

3. As part of your work with an international animal health organization, you are given the task of developing a vaccine against a bovine virus that is the cause of tens of thousands of cattle deaths around the world. The viral genome consists of a 10-kb linear piece of single-stranded RNA with a poly(A) tail that encodes eight different genes. The virus does not have a viral envelope, and the major antigenic determinant is the capsid protein viral protein 2. Outline an experimental strategy to develop a vaccine against this virus.

4. Discuss the development of peptide vaccines that are directed against viruses.

5. What is vaccinia virus, and how can it be used to produce unique live recombinant vaccines?

6. You have isolated the RNA virus that causes rabies in skunks and raccoons. How would you use this purified virus to develop a recombinant vaccine to protect these animals against rabies?

7. As an employee of the World Health Organization, you have to decide on the best strategy for eradicating rabies in wild animal populations. Assuming that you must choose between a peptide and a vaccinia virus-based vaccine, select one approach and justify its use.

8. Describe how a filamentous bacterium could be used as a vaccine.

9. Discuss the advantages of a live recombinant viral vaccine over both killed and subunit vaccines.

10. Discuss some of the different strategies that have been used to produce a vaccine against cholera.

Synthesis of Commercial Products by Recombinant Microorganisms

To date, molecular biotechnology research has focused largely on the production of a range of different proteins. However, recombinant DNA techniques can also be used to enhance the production of a range of low-molecular-weight compounds such as vitamins, amino acids, and antibiotics.

With efficient expression systems, it is a relatively straightforward matter to clone and express a particular target protein. The expressed protein is either the final product, e.g., a restriction enzyme, or a catalyst for a particular chemical reaction, e.g., for antibiotic biosynthesis. Sometimes, a novel catalytic activity is introduced into a microorganism by genetic manipulation and used to produce in vivo low-molecular-weight compounds including vitamins, amino acids, dyes, antibiotics, and precursors of biopolymers. In these cases, the host microorganism is engineered to become a factory for the production of useful metabolites.

Restriction Endonucleases

Recombinant DNA technology would not be possible without a ready supply of a range of different restriction endonucleases. Currently, more than 300 different restriction endonucleases are commercially available. These enzymes are synthesized by a wide range of microorganisms, including species that are aerobic, anaerobic, photosynthetic, diazotrophic, mesophilic, thermophilic, psychrophilic, and either slow or fast growing. For each of these organisms, a detailed fermentation protocol—specifying the temperature, pH, medium composition, and oxygen tension—has to be developed and optimized to achieve the maximum yield of the desired enzyme. To avoid having to maintain a large number of different microorganisms, stock a very wide range of medium components, design several different fermenters, and spend an inordinate amount of time developing optimal growth conditions for a large number of different organisms, investigators often clone restriction endonuclease genes into *Escherichia coli*. Exclusive use of *E. coli* allows bioengineers to standardize the production

conditions. In addition, *E. coli* cells grow rapidly to high cell densities and can be engineered to significantly overexpress the target restriction enzyme.

Although the technology for isolating and expressing foreign genes in *E. coli* and some other host organisms is fairly well established, it should be remembered that the host organism is a living entity that can be dramatically affected by the production or presence of a heterologous protein. For example, overexpression of a heterologous protein may drain the host organism of important metabolic resources and, as a result, adversely affect its growth. In addition, the presence of a heterologous protein may be lethal to the host. For example, restriction endonucleases digest DNA at sites that are present on all DNA molecules. As a result, an organism that expresses a cloned restriction endonuclease gene is likely to have its own DNA degraded unless a protection mechanism is present.

Microorganisms that make restriction endonucleases have evolved a self-protection system. Methylation of one or more of the bases of the DNA within the recognition sequence prevents the homologous restriction endonuclease from cutting the DNA at this site. Gram-negative microorganisms have an added mode of protection that entails localizing the restriction endonuclease within the periplasmic space. This compartmentalization physically separates the restriction endonuclease from the DNA while ensuring that the methylation (modification) enzyme has ready access to the chromosomal DNA. In addition to protecting the cell from the toxic effects of the restriction enzyme, this segregation provides a cellular defense against attack by any foreign DNA that might enter the periplasm, such as a DNA virus.

One way to circumvent the problem of host DNA degradation by heterologous restriction endonucleases is to clone and express the genes for both the restriction enzyme and its cognate modification enzyme in the recipient organism. Cloning both of these genes into the same organism is technically complex unless both the target restriction endonuclease and the target methylation genes are close to each other on the chromosome. In addition, to prevent the digestion of the host DNA by the restriction endonuclease, it is imperative that after transformation, the methylation activity be expressed prior to the production of the restriction endonuclease.

The strategy that was used to isolate the gene for the restriction enzyme *Pst*I from *Providencia stuartii*, a gram-negative bacterium, shows how one restriction endonuclease gene was cloned into *E. coli* (Fig. 12.1).

1. The chromosomal DNA from *P. stuartii* was digested with *Hind*III and ligated into the *Hind*III site on plasmid pBR322.
2. Following the introduction of the *P. stuartii* clone bank (library) into *E. coli* HB101 cells, transformants were grown in liquid medium before being infected with bacteriophage λ. (When a restriction enzyme gene is expressed, the host cells become resistant to the lytic action of DNA bacteriophages such as λ because the restriction endonuclease extensively degrades the infecting bacteriophage DNA.)
3. Transformants that were resistant to lysis by λ were grown, and samples were osmotically shocked to release the periplasmic proteins, which were assayed for *Pst*I restriction enzyme activity.
4. Positive clones were assayed for *Pst*I methylase activity.

One positive clone from this experiment contained, within a 4.0-kilobase

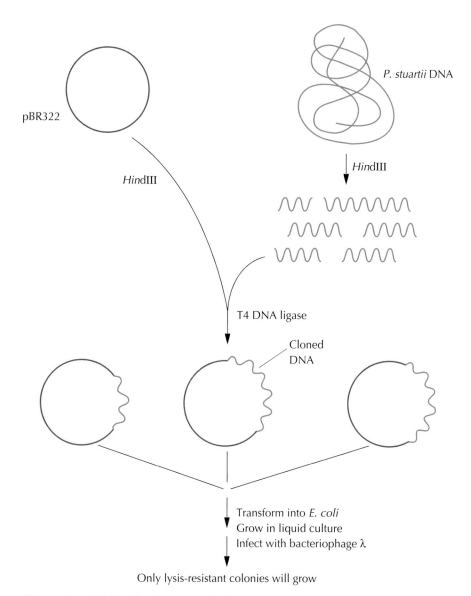

Figure 12.1 Method for cloning and selecting the gene for the restriction enzyme *Pst*I. The *P. stuartii* chromosomal DNA is digested with *Hin*dIII and ligated into the *Hin*dIII site of plasmid pBR322. Transformants are grown in liquid medium before being infected with bacteriophage λ. The resistance of some transformants to lysis by λ is due to the presence and expression of a cloned *Pst*I gene.

(kb) DNA fragment, an intact *Pst*I restriction endonuclease and methylation operon, including the *P. stuartii* promoter. In this construct, the natural temporal order of synthesis—the methylation enzyme preceding the restriction endonuclease—was maintained. The level of the *Pst*I restriction enzyme expressed in *E. coli* was approximately 10-fold greater than that in *P. stuartii*. As expected, *Pst*I was localized in the periplasm and the methylation enzyme in the cytoplasm. Production of *Pst*I by using this *E. coli* clone is simpler and more efficient than production by using *P. stuartii*.

Another strategy also has been used to isolate the genes for restriction and modification (methylation) enzyme systems. It consists of the following steps.

1. A clone bank is made from the DNA of a donor organism that has a previously identified restriction endonuclease. The plasmid vector should have at least one recognition site for the target restriction endonuclease.
2. The clone bank is introduced into *E. coli* by transformation.
3. Plasmid DNA is isolated from transformed cells that have been grown in liquid media under conditions that select for the presence of the plasmid.
4. The plasmid DNA preparation is treated with the target restriction endonuclease.
5. *E. coli* cells are transformed with the restriction endonuclease-treated plasmid DNA preparation.

The rationale for this procedure is that the clones that carry and express the target modification enzyme will produce plasmid DNA that is resistant to digestion by the target restriction endonuclease because their DNA will be methylated at the recognition sites. For example, following transformation of *E. coli* by a pBR322–*Hin*dIII clone bank of *Desulfovibrio desulfuricans* DNA, plasmid DNA was isolated and digested with the restriction enzyme *Dde*I. Plasmids that encode and express the *Dde*I modification enzyme should not be digested by *Dde*I because the eight *Dde*I recognition sites of pBR322 are methylated. After the *Dde*I treatment, the remaining plasmid mixture is used to transform *E. coli*. Only intact circular plasmids yield transformants, and these carry the gene for a functional *Dde*I modification enzyme. All other plasmids are degraded by the restriction endonuclease. The resulting transformants must then be assayed for the *Dde*I restriction enzyme activity to determine which clones have the genes for both the modification enzyme and the restriction endonuclease. This strategy is effective for any restriction enzyme gene that is physically close to its modification enzyme gene and is cloned into a plasmid vector that has at least one recognition site for the target enzyme.

Small Biological Molecules

With recombinant DNA technology, it is possible to modify metabolic pathways of organisms either by introducing new genes or by altering existing ones. The goal is to create an organism with a novel enzymatic activity that can convert an existing substrate into a commercial compound that with current technology can be produced only by a combination of chemical treatments and fermentation steps.

Synthesis of L-Ascorbic Acid

L-Ascorbic acid (vitamin C) is currently synthesized commercially by an expensive process that includes one microbial fermentation step and a number of chemical steps starting with D-glucose (Fig. 12.2). The last step in this process is the acid-catalyzed conversion of 2-keto-L-gulonic acid (2-KLG) to L-ascorbic acid. Biochemical studies of the metabolic pathways of a number of different microorganisms have shown that it may be possible to synthesize 2-KLG by a different pathway. For example, some bacteria (*Acetobacter*, *Gluconobacter*, and *Erwinia*) can convert glucose to 2,5-diketo-D-gluconic acid (2,5-DKG), and others (*Corynebacterium*, *Brevibacterium*, and *Arthrobacter*) have the enzyme 2,5-DKG reductase, which converts 2,5-DKG to 2-KLG.

The current procedure for synthesizing ascorbic acid could be im-

Figure 12.2 Commercial synthesis of L-ascorbic acid. Except for the microbial conversion of D-sorbitol to L-sorbose, all the steps are chemical reactions. The microbial conversion is carried out by *Acetobacter suboxydans,* which produces the enzyme sorbitol dehydrogenase.

proved by producing 2-KLG from glucose by cofermentation with suitable organisms. Unfortunately, cocultivation has problems of its own. For example, the two fermenting organisms might have different temperature and pH optima. The medium requirements and growth rates also might differ in such a way that the fermentation conditions are optimal for one organism and suboptimal for the other. This situation leads to the eventual "washout" (depletion or loss) of one of the organisms. Some of these incompatibilities may be overcome by utilizing a tandem fermentation process in which the two organisms are cultivated in succession (Fig. 12.3). Of course, this approach requires two separate fermentations rather than one; and if the organisms have different growth requirements, it is difficult to run the process on a continuous basis. Therefore, the best way to convert glucose into 2-KLG would be to engineer a single microorganism that carried all of the required enzymes. The conversion of D-glucose to 2,5-DKG by *Erwinia herbicola* includes several enzymatic steps, whereas the transformation of 2,5-DKG to 2-KLG by a *Corynebacterium* sp. requires only one. Consequently, the simplest strategy for constructing a single organism that is able to convert D-glucose to 2-KLG is to isolate the 2,5-DKG reductase gene from the *Corynebacterium* sp. and express it in *Erwinia herbicola.*

The first step in cloning the 2,5-DKG reductase gene from the *Corynebacterium* sp. involved purifying the enzyme and determining the sequence of the first 40 amino acids from the N-terminal end of the molecule. On the basis of the known amino acid sequence, two 43-nucleotide DNA hybridization probes, each corresponding to a different portion of the pro-

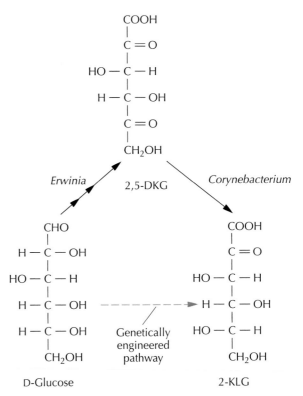

Figure 12.3 Biological production of 2-KLG. *Erwinia* has a set of three enzymes that synthesize 2,5-DKG from D-glucose. *Corynebacterium* has an enzyme that converts 2,5-DKG to 2-KLG. Thus, 2-KLG, which is the immediate precursor of L-ascorbic acid, can be produced from D-glucose either by cocultivating these two microorganisms or by genetically engineering *Erwinia* to express the enzyme from *Corynebacterium*, which converts 2,5-DKG to 2-KLG.

tein molecule, were synthesized. Because 71% of the nucleotides in the *Corynebacterium* sp. are either G or C, the probes were designed to include, where possible, a G or C in the third position of all codons, thereby minimizing the extent of the mismatch between the probe and the target DNA.

A *Corynebacterium* DNA clone bank was screened with these probes. Any clones that hybridized with only one of the probes was discarded. It was assumed that any DNA that interacted with only one probe was probably not the target DNA. A clone that contained the 2,5-DKG reductase gene was isolated and sequenced. The DNA sequences that were upstream of the ATG start signal were deleted and replaced with transcriptional and translational signals that function in *E. coli*, because the regulatory sequences from gram-positive microorganisms such as *Corynebacterium* spp. are not efficiently utilized by *E. coli*. This construct expressed 2,5-DKG reductase activity in *E. coli* and subsequently was subcloned onto a broad-host-range vector, which was used to transform *Erwinia herbicola*.

The transformed *Erwinia* cells were able to convert D-glucose directly to 2-KLG. The endogenous *Erwinia* enzymes, localized in the inner membrane of the bacterium, converted glucose to 2,5-DKG, and the cloned 2,5-DKG reductase, localized in the cytoplasm, catalyzed the conversion of 2,5-DKG to 2-KLG (Fig. 12.4). Thus, by genetic manipulation, the metabolic capabilities of two very dissimilar microorganisms were combined into one organism, which was able to produce the end product of the engineered metabolic

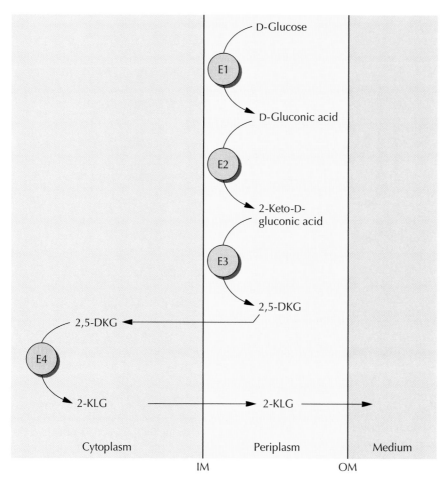

Figure 12.4 Conversion of D-glucose to 2-KLG by recombinant *Erwinia herbicola*. The cellular locations of all of the component enzymes are indicated by the letter E and are numbered consecutively. The major intermediates in the pathway are named. IM and OM denote the inner and outer membranes, respectively.

pathway. An organism of this sort should be useful as a source of 2-KLG for the production of L-ascorbic acid, thereby replacing the first three steps of the currently used process (Fig. 12.2).

The commercial utility of the cloned 2,5-DKG reductase gene might be improved by replacing certain of the amino acids of the enzyme to create mutants with increased catalytic activity and enhanced thermal stability. When the 2,5-DKG reductase gene was isolated, the amino acid residues that contributed to the active site of this enzyme were not known. However, from the primary amino acid sequence of the enzyme, computer modeling predicted an enzyme structure that has an eight-stranded α/β barrel (Fig. 12.5). This structure consists of eight twisted parallel β-strands arranged close together, surrounded by eight α-helices that are joined to the β-strands through loops of various lengths. This folding pattern had previously been observed for 17 other enzymes whose crystal structures were known. By analogy to what was known about the structures of these other proteins, three of the loops that might be involved in substrate binding were identified (Fig. 12.5). By using oligonucleotide-directed mutagenesis, 12 different mutants, each with a single amino acid change in one of these loops,

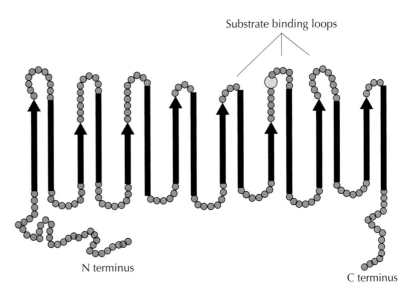

Substrate binding loops

N terminus

C terminus

Figure 12.5 Predicted structure of 2,5-DKG reductase. The solid arrows indicate β-stranded regions; the solid bars are α-helical regions; and the circles are amino acid residues either at the N or C terminus or involved in loops connecting the β-strands to the α-helices. The three loops that may be involved in substrate binding are indicated. Amino acid residue 192 is shown as a stippled circle.

were constructed. Of the 12 mutants, 11 produced enzymes with a lower 2,5-DKG reductase specific activity than that of the native form of the enzyme. The 12th mutant, in which amino acid residue 192 was changed from glutamine to arginine, had approximately twice the activity of the native enzyme. Kinetic studies revealed that this increase in activity resulted from a 1.8-fold increase in the maximal rate of the enzyme-catalyzed reaction, V_{max}, and a 25% decrease in the Michaelis constant, K_m, of the enzyme-catalyzed reaction. Oligonucleotide-directed mutagenesis of the glycine residues at positions 55 and 57 to alanine produced an enzyme that was more temperature stable than the native form. Future work on 2,5-DKG reductase is likely to focus on combining these amino acid changes into a single version of the enzyme.

Microbial Synthesis of Indigo

A number of bacteria, most notably *Pseudomonas* species, have the ability to use a variety of organic compounds such as naphthalene, toluene, xylene, and phenol as their sole carbon source. In many instances, the genes encoding the enzymes for the degradation of these organic compounds are located on large naturally occurring plasmids (approximately 50 to 200 kb). Research on these bacteria often requires detailed genetic and biochemical studies so that the genes encoding enzymes catalyzing important steps in the pathway can be targeted for modification. Occasionally, despite the original purpose of a study, an unexpected but useful discovery is made. For example, plasmid NAH7 has two separate and distinct operons that allow pseudomonads that contain this plasmid to grow on naphthalene as the sole carbon source. As a first step toward characterizing these genes, NAH7 plasmid DNA was digested with *Hin*dIII and the fragments were ligated with linear *Hin*dIII-digested plasmid pBR322. This clone bank was introduced into *E. coli* cells, and transformants were selected on the basis of their resistance to ampicillin and sensitivity to tetracycline. All transformants

were then tested for the production of nonvolatile metabolites that might result from the hydrolysis of radiolabeled naphthalene.

During the course of characterizing one transformant that had a 10.5-kb insert and could convert naphthalene to salicylic acid, it was observed that minimal growth medium that contained tryptophan turned blue. A thorough analysis of the blue color revealed that the transformed *E. coli* cells were synthesizing the dye indigo. This synthesis was achieved in four steps (Fig. 12.6):

1. conversion of tryptophan in the growth medium to indole by the enzyme tryptophanase, which is synthesized by the *E. coli* host cell;
2. oxidation of indole to *cis*-indole-2,3-dihydrodiol by naphthalene dioxygenase, which is encoded by the DNA that was cloned from the NAH7 plasmid;
3. spontaneous elimination of water; and
4. air oxidation to form indigo.

Thus, the combination of enzymes from two different pathways and two different organisms results in the synthesis of an unexpected compound, the dye indigo. In addition, introduction of the gene for the enzyme xylene oxidase, which is encoded in the TOL plasmid, can convert tryptophan to indoxyl, which then spontaneously oxidizes to indigo (Fig. 12.6).

Indigo, a commercially important blue pigment that is used to dye both cotton and wool, was originally isolated from plants but is currently synthesized chemically. At present, approximately 13×10^6 kg of indigo, worth

Figure 12.6 Indigo biosynthesis from tryptophan in genetically engineered *E. coli*. Tryptophanase is an *E. coli* enzyme. In pathway A, the naphthalene dioxygenase is derived from the NAH plasmid; in pathway B, the xylene oxidase is from the TOL plasmid. *E. coli* transformants that synthesize indigo contain either pathway A or B but not both pathways.

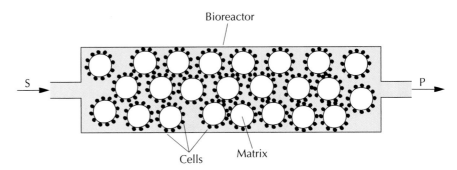

Figure 12.7 Schematic representation of a bioreactor that might be suitable for the production of indigo from recombinant *E. coli* cells. The cells are immobilized either chemically or physically to particles of a stable matrix. The substrate, tryptophan (S), is added at one end, and the product, indigo (P), is removed continuously from the other end. The rate of flow through the column would be limited by the rate of conversion of substrate into product.

around $200 million, is produced every year. Indigo, the coloring agent in blue jeans, is the largest-selling dye in the world. The ability to produce indigo from bacteria opens the possibility of developing an efficient and economical commercial microbial process for its production. This process would avoid the use of hazardous compounds such as aniline, formaldehyde, and cyanide, which are needed in the chemical synthesis of indigo. The conditions for large-scale growth of the *E. coli* strain capable of synthesizing indigo, including temperature, pH, and the amount of tryptophan that must be added to the medium to give maximum yields, are being tested. Although this system has not yet been commercialized, a microbial process for the synthesis of indigo might consist of a bioreactor in which the recombinant *E. coli* is chemically immobilized to a solid matrix (e.g., cellulose or silica gel). This unit could be run continuously by adding tryptophan to one end and removing indigo from the other end (Fig. 12.7).

Synthesis of Amino Acids

Amino acids are used extensively in the food industry as flavor enhancers, antioxidants, and nutritional supplements; in agriculture as feed additives; in medicine in infusion solutions for postoperative treatment; and in the chemical industry as starting materials for the manufacture of polymers and cosmetics (Table 12.1). It is estimated that more than 800,000 tons of amino acids, worth more than $5 billion, are produced worldwide annually. L-Glutamic acid, which is used in the manufacture of the flavor enhancer monosodium glutamate, makes up more than half of the total volume.

For the most part, amino acids are commercially produced either by extraction from protein hydrolysates or as fermentation products of either *Corynebacterium* or *Brevibacterium* spp., which are both nonsporulating gram-positive soil bacteria. Traditionally, the productivity of these organisms has been improved by mutagenesis and subsequent screening for strains that overproduce certain amino acids. However, this way of developing new strains is slow and inefficient. By using detailed biochemical information about the enzymes that are involved in the biosynthesis of various commercially important amino acids, it may be more expeditious to isolate and manipulate the specific genes encoding the key components of a particular pathway. However, this type of genetic engineering may not be a simple matter. For example, the pathway(s) leading to the biosynthesis of certain amino acids may contain a number of different enzymes, each of

Table 12.1 Commercial applications of amino acids

Amino acid	Application(s)
Alanine	Flavor enhancer
Arginine	Therapy for liver diseases
Aspartic acid	Flavor enhancer; sweetener synthesis
Asparagine	Diuretic
Cysteine	Bread production; therapy for bronchitis; antioxidant
Glutamic acid	Flavor enhancer
Glutamine	Therapy for ulcers
Glycine	Sweetener synthesis
Histidine	Therapy for ulcers; antioxidant
Isoleucine	Intravenous solutions
Leucine	Intravenous solutions
Lysine	Feed additive; food additive
Methionine	Feed additive
Phenylalanine	Infusions; sweetener synthesis
Proline	Intravenous solutions
Serine	Cosmetics
Threonine	Feed additive
Tryptophan	Intravenous solutions; antioxidant
Tyrosine	Intravenous solutions; precursor for L-DOPA
Valine	Intravenous solutions

which may be either activated or inhibited by a number of metabolites present in the cell, which makes it difficult to know which enzyme to manipulate in order to enhance the yield of the end product. Moreover, at this time, such detailed biochemical information and genetic manipulations of these organisms are at the formative stage and both expression vectors and transformation protocols are being developed for gram-positive organisms such as *Corynebacterium* and *Brevibacterium* spp.

Because most broad-host-range plasmid vectors replicate only in gram-negative organisms, it is necessary to construct expression vectors that are specifically suited for *Corynebacterium* and *Brevibacterium* spp. Such cloning vehicles might take the form of *E. coli–Corynebacterium* shuttle vectors. The *E. coli* portion of the plasmid could encode resistance to the antibiotic tetracycline, chloramphenicol, or kanamycin. Because both *E. coli* and *Corynebacterium* spp. are susceptible to these antibiotics, they could be used as selectable markers in both organisms.

An efficient transformation protocol for *C. glutamicum*, the species of *Corynebacterium* often used in these experiments, is still needed. Many *C. glutamicum* genes are not efficiently expressed in *E. coli*. Therefore, for selection schemes that depend on gene expression (e.g., complementation), the entire clone bank should be transformed into *C. glutamicum*. Unfortunately, the transformation frequency is very low when DNA is introduced into *C. glutamicum* by either direct transformation or electroporation. However, effective transformation of *C. glutamicum* is achieved when foreign DNA is introduced by conjugation or after the formation of cells in which the cell wall has been removed with lysozyme (protoplasts). Transformation of protoplasts is made possible by adding polyethylene glycol to facilitate the uptake of exogenous plasmid DNA.

Some preliminary progress has been made in increasing the amino acid

output of *C. glutamicum*. For example, the synthesis of the essential amino acid tryptophan by *C. glutamicum* was enhanced by introducing into wild-type *C. glutamicum* cells a second copy of the gene encoding anthranilate synthetase, which is the rate-limiting enzyme in the normal tryptophan biosynthetic pathway (Fig. 12.8). The following protocol describes one way to isolate the anthranilate synthetase gene.

1. A library of *Brevibacterium flavum* chromosomal DNA was cloned into a *C. glutamicum–E. coli* shuttle vector and introduced into a mutant strain of *C. glutamicum* that produced no active anthranilic acid synthetase.
2. The mutant was unable to grow on minimal medium unless anthranilic acid was added; therefore, transformants were selected by their ability to grow in the absence of anthranilic acid.
3. The vector carrying the anthranilic acid synthetase gene was then transferred to a wild-type strain of *C. glutamicum*.

The amount of tryptophan produced in the mutant and wild-type *C. glu-*

Figure 12.8 Simplified pathway and regulation of tryptophan biosynthesis in *C. glutamicum*. The abbreviations DS, ANS, and PRT represent the enzymes 3-deoxy-D-arabino-heptulosonate 7-phosphate synthase, anthranilate synthase, and anthranilate phosphoribosyltransferase, respectively. The solid lines represent the synthetic pathways. The dashed lines denote feedback inhibition. Indole is produced in a side reaction and is converted into tryptophan by the action of tryptophan synthase β.

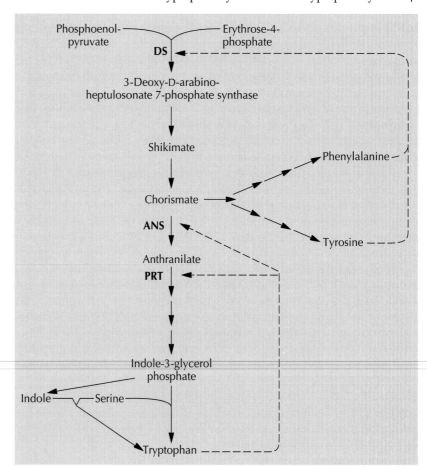

Table 12.2 Production of tryptophan under standard growth
conditions by certain strains of *C. glutamicum*

Strain	Tryptophan concentration (mg mL^{-1})
Mutant	0.00
Mutant with vector	0.34
Wild type	0.48
Wild type with vector	1.12

Adapted from Ozaki et al., U.S. patent 4,874,698, 1989.

tamicum strains—one without and one with the vector carrying the cloned
anthranilic acid synthetase gene—was measured (Table 12.2). The cloned
gene did indeed restore most of the capacity of the mutant to synthesize
tryptophan. However, the effect of adding this gene to wild-type *C. glutam-
icum* was much more dramatic, with the synthesis of tryptophan being in-
creased by approximately 130%. This level of overproduction reflects more
efficient utilization of available precursor material. Thus, by cloning an ad-
ditional gene of an amino acid biosynthesis pathway into an organism, it
was possible to generate much more of the end product. An even higher
level of tryptophan production was achieved when modified genes for the
three key enzymes, 3-deoxy-D-arabino-heptulosonate 7-phosphate syn-
thase, anthranilate synthase, and anthranilate phosphoribosyltransferase,
were introduced into *C. glutamicum* cells. The genes encoding these en-
zymes were mutagenized to render them insensitive to inhibition by the
end product (feedback inhibition).

An alternative to producing amino acids in *Corynebacterium* and *Bre-
vibacterium* spp. is to produce them in *E. coli*, where both metabolic path-
ways and procedures for genetic manipulation are much better character-
ized. Its relative ease of manipulation makes *E. coli* an attractive host
organism for this sort of metabolic engineering.

Antibiotics

Since the discovery of penicillin in the late 1920s, more than 6,000 antibiotics
with different specificities and a variety of modes of action have been iso-
lated from various microorganisms. The universal use of antibiotics to treat
bacterial diseases has resulted in an enormous improvement in human
health and has undoubtedly saved millions of lives. The vast majority of the
most important antibiotics have been isolated from the gram-positive soil
bacterium *Streptomyces*, although fungi and other gram-positive and gram-
negative bacteria are also sources of antibiotics. Worldwide, over 100,000
tons of antibiotics are produced per year, with annual gross sales of about
$5 billion, including over $100 million for antibiotics used in animal feed
and as animal growth promoters.

An estimated 100 to 200 new antibiotics are discovered each year, pri-
marily through labor-intensive research programs in which many thou-
sands of different microorganisms are screened to find those that produce
unique antibiotics. However, the high costs of development and clinical
testing mean that only the compounds that show significant therapeutic
and economic promise are marketed: only 1 to 2% of newly discovered an-
tibiotics are added annually to the disease-fighting arsenal. Recombinant

DNA technology can have a positive impact on this endeavor in two ways. First, this technology can be used to develop new, structurally unique antibiotics with increased activities against selected targets and decreased side effects. Second, genetic manipulation can be used to enhance yields and hence lower the cost of production.

For the genetic manipulation of *Streptomyces*, which is the major organism used for the production of antibiotics, it is essential that it can be transformed and that the transformed cells can be readily selected. However, unlike *E. coli*, *Streptomyces* strains do not exist as individual cells but exist as extended aggregates of mycelial filaments; therefore, the cell wall must be removed to release individual cells that no longer have cell walls (protoplasts) prior to DNA transformation (Fig. 12.9). Without this step, it would not be possible to distinguish transformed from nontransformed cells, because visible colonies on a solid medium would each have started from an aggregate rather than from an individual cell, so that colonies that grew in the presence of a selective antibiotic would contain a mixture of transformed and nontransformed cells. However, as a consequence of protoplast formation prior to transformation, all colonies that grow in the presence of a selective antibiotic contain only transformed cells. The uptake of plasmid DNA into *Streptomyces* protoplasts is enhanced by polyethylene glycol. Following transformation, the protoplasts are first plated onto a solid medium to enable the cell walls to regenerate and then overlaid with a selective medium that often contains either neomycin or thiostrepton, both of which act as selection agents for transformed cells.

Cloning Antibiotic Biosynthesis Genes

The biosynthesis of an antibiotic may include 10 to 30 separate enzyme-catalyzed steps, so that cloning all the genes for the synthesis of a particular antibiotic is not an easy task. One strategy for isolating the complete set of antibiotic biosynthesis genes consists of transforming one or more mutant strains that are unable to synthesize the antibiotic with DNA from a clone bank constructed from wild-type chromosomal DNA. Following the introduction of the clone bank into mutant cells, transformants are screened for their ability to produce the antibiotic. Then the plasmid DNA from the clone that supplies a functional gene and gene product, i.e., provides the missing function of (complements) a mutant strain, is used as a DNA hybridization probe to screen another clone bank of wild-type chromosomal DNA and isolate clones with regions that overlap the probe sequence. In this way, DNA segments that are adjacent to and usually bigger than the initial complementing DNA can be identified and cloned and a complete gene cluster can be reconstructed. If the antibiotic biosynthesis genes are clustered at a single site on the chromosomal DNA, genes adjacent to the complementing gene are also likely to be involved in the biosynthesis of the target antibiotic. However, if the antibiotic biosynthesis genes are scattered in small clusters at different chromosomal locations, it is necessary to have at least one mutant per gene cluster to obtain a DNA clone with which to identify the rest of the genes in that cluster.

This approach has been successfully used to isolate some of the genes for the biosynthesis of the antibiotic undecylprodigiosin from *Streptomyces coelicolor* A3 (Fig. 12.10). In this case, the complementation assay is simple and entails scoring the color of the colonies. Colonies of wild-type organisms are red because of the presence of the antibiotic, and mutant colonies are cream. Complementation produces a red colony.

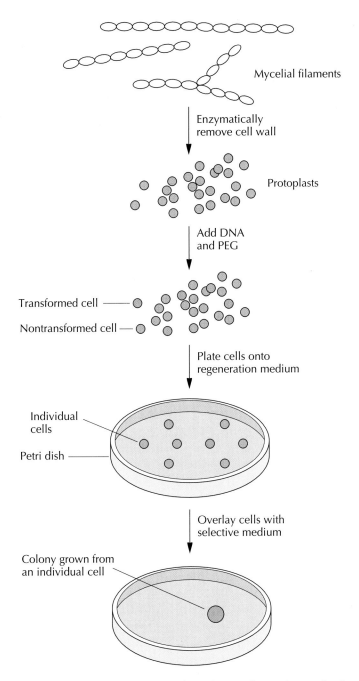

Figure 12.9 Schematic representation of DNA transformation and selection of transformants of *Streptomyces* strains. The pink circles represent transformed cells, and the green circles represent nontransformed cells. PEG, polyethylene glycol.

In addition to cloning antibiotic biosynthesis genes by complementation, more direct strategies can be employed. One or more of the key enzymes in a biosynthetic pathway can be identified through either genetic or biochemical studies and then purified. The N-terminal amino acid sequence of the enzyme can then be determined, and with this information, oligodeoxyribonucleotide probes for the gene can be prepared. This approach has been used to isolate the gene for isopenicillin N synthetase from *Penicillium chrysogenum*. This enzyme catalyzes the oxidative condensation

Figure 12.10 Chemical structure of the antibiotic undecylprodigiosin.

of the compound δ-(L-α-aminodipyl)-L-cysteinyl-D-valine to isopenicillin N, a key intermediate in the biosynthesis of penicillins, cephalosporins, and cephamycins (Fig. 12.11).

Synthesis of Novel Antibiotics

New antibiotics with unique properties and specificities may be produced by genetic manipulation of the genes involved in the biosynthesis of existing antibiotics. In one of the first experiments in which a novel antibiotic was produced, researchers began by examining the consequences of placing two slightly different antibiotic production pathways into one organism.

Figure 12.11 Biosynthetic pathway for penicillins and cephalosporins in *P. chrysogenum.* Isopenicillin N synthetase catalyzes the synthesis of isopenicillin N from δ-(L-α-aminoadipyl)-L-cysteinyl-D-valine. Isopenicillin N is a precursor in the synthesis of penicillin G, penicillin N, and cephalosporin C.

Figure 12.12 Structures of various isochromanequinone antibiotics produced in *Streptomyces* species. Wild-type *S. coelicolor* and plasmid pIJ2303 code for actinorhodine; *Streptomyces* sp. produces medermycin; and *S. violaceoruber* produces both granaticin and dihydrogranaticin. The hybrid antibiotics produced are mederrhodine A and dihydrogranatirhodine.

A *Streptomyces* plasmid (pIJ2303) carrying a 32.5-kb fragment of *S. coelicolor* chromosomal DNA contains all of the genes encoding the enzymes responsible for the biosynthesis of the antibiotic actinorhodine, starting from acetate. This antibiotic is a member of the family of antibiotics called isochromanequinones (Fig. 12.12). The intact plasmid and various subclones carrying portions of the 32.5-kb DNA fragment (e.g., pIJ2315) were introduced into either *Streptomyces* sp. strain AM-7161, which produces the related antibiotic medermycin, or *S. violaceoruber* B1140 or Tü22, both of which produce the related antibiotics granaticin and dihydrogranaticin.

Each of the antibiotics actinorhodine, medermycin, granaticin, and di-

Table 12.3 Antibiotics produced by various *Streptomyces* strains and those transformed with plasmids pIJ2303 and pIJ2315

Strain/plasmid	Color of culture		Antibiotic(s)
	Acidic	**Alkaline**	
S. coelicolor	Red	Blue	Actinorhodine
Streptomyces sp.	Yellow	Brown	Medermycin
Streptomyces sp./pIJ2303	Red	Blue	Medermycin, actinorhodine
Streptomyces sp./pIJ2315	Red	Purple	Mederrhodine A, medermycin
S. violaceoruber B1140	Red	Blue-purple	Granaticin, dihydrogranaticin
S. violaceoruber B1140/pIJ2303	Red	Blue-purple	Granaticin, dihydrogranaticin, actinorhodine
S. violaceoruber Tü22	Red	Blue-purple	Granaticin, dihydrogranaticin
S. violaceoruber Tü22/pIJ2303	Red	Blue-purple	Dihydrogranatirhodine, actinorhodine

Adapted from Hopwood et al., *Nature* **314**:642–644, 1985.

hydrogranaticin (Fig. 12.12) functions as an acid-base indicator, conferring on a growing culture a characteristic color that depends on the pH of the medium (Table 12.3). The pH (and color), in turn, depends on the compound(s) being synthesized. Mutants of the *S. coelicolor* parental strain that are unable to produce actinorhodine are colorless. The appearance of new colors in some cases following the transformation of *Streptomyces* sp. strain AM-7161 or *S. violaceoruber* B1140 or Tü22 with a plasmid carrying either all or some of the genes encoding the enzymes that synthesize actinorhodine suggests that a novel antibiotic has been produced (Fig. 12.12; Table 12.3). *Streptomyces* sp. strain AM-7161 and *S. violaceoruber* B1140 transformants containing pIJ2303 produce the antibiotics encoded by both the chromosomal and plasmid DNAs. However, when *S. violaceoruber* Tü22 is transformed with pIJ2303, a new antibiotic—dihydrogranatirhodine—is synthesized along with actinorhodine. When *Streptomyces* sp. strain AM-7161 is transformed with pIJ2315, a second new antibiotic—mederrhodine A—is produced.

These new antibiotics represent minor structural variants of the preexisting antibiotics actinorhodine, medermycin, granaticin, and hydrogranaticin and probably arise when an intermediate compound in one biosynthetic pathway acts as a substrate for an enzyme from the other pathway. As the biochemistry of various antibiotic biosynthetic pathways becomes better understood, it will be possible to design new unique and specific antibiotics by genetic manipulation of the genes encoding the relevant enzymes.

Engineering Polyketide Antibiotics

The term "polyketide" defines a class of antibiotics that are synthesized through the successive enzymatic condensation of small carboxylic acids such as acetate, propionate and butyrate. While various polyketides are produced by plants and fungi, the majority are produced by actinomycetes

as secondary metabolites. To create new polyketide antibiotics, the functioning of the enzymes that synthesize these antibiotics must be understood before the genes encoding these enzymes can be manipulated.

Polyketide antibiotics are synthesized by a mechanism analogous to that used for the synthesis of long-chain fatty acids. Each condensation cycle results in the formation, on a growing carbon chain, of a β-keto group. Polyketide synthesis consists of a number of steps which are each repeated several times, including ketoreduction, dehydration, and enoylreduction of the β-group of the growing polyketide chain. There are two classes of polyketide synthases, the enzyme complexes responsible for the synthesis of polyketide antibiotics (Fig. 12.13). The synthases that catalyze the biosynthesis of aromatic polyketides comprise one class; they consist of polypeptides with an active site in which successive reactions can take place. Generally, for aromatic polyketide synthases, the active site is on a single polypeptide (Fig. 12.13A). The second class includes synthases which are assemblies of several polypeptides that have separate and distinct active sites for every catalyzed step in polyketide biosynthesis (Fig. 12.13B). These enzymes have a number of different domains (regions A through E in Fig. 12.13), where each domain has a separate enzymatic activity and active site catalyzing a particular step in the process. The complete synthesis of a polyketide antibiotic generally requires the participation of several of these multifunctional enzymes; together, they make up the subunits of the polyketide synthase.

If each of the enzymatic activities that is catalyzed by a domain on a multifunctional polyketide synthase subunit catalyzes only a single biochemical step in the pathway, the loss of any one activity should affect only a single step in the synthesis. Moreover, alteration of a catalytic domain whose function has been established should allow predictable changes to

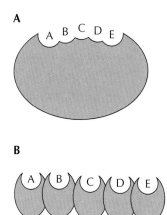

Figure 12.13 Schematic representation of polyketide synthase for aromatic polyketides where the active site is on a single polypeptide **(A)** and for those which are assemblies of polypeptides with separate and distinct active sites **(B).** Both types of enzymes have different domains (regions A through E), and each domain has a separate enzymatic activity.

MILESTONE

Production of 2-Keto-L-Gulonate, an Intermediate in L-Ascorbate Synthesis, by a Genetically Modified *Erwinia herbicola*
S. ANDERSON, C. B. MARKS, R. LAZARUS, J. MILLER,
K. STAFFORD, J. SEYMOUR, D. LIGHT, W. RASTETTER, and D. ESTELL
Science **230**:144–149, 1985

Theoretically, prokaryotes that produce specific metabolites can be genetically manipulated in two different ways. First, the activity or amount of one or more of the enzymes in a pathway encoding the synthesis of a particular metabolite can be modified so that the amount of metabolite that the bacterium produces is increased. Second, foreign genes can be introduced that produce enzymes that can use an endogenous metabolite as a substrate for the production of another metabolite not nor-

mally produced by the host bacterium. While these sorts of manipulations are easy in theory, it is not necessarily easy to isolate and manipulate the necessary genes or to establish the appropriate conditions to make this work.

To create a bacterium that synthesized 2-keto-L-gulonic acid, which is the immediate precursor of vitamin C when this vitamin is commercially synthesized, Anderson et al. isolated the *Corynebacterium* gene encoding the enzyme that converts 2,5-diketo-D-gluconic acid to 2-keto-L-gulonic acid and

transferred it to an *Erwinia* sp., a bacterium that synthesizes 2,5-diketo-D-gluconic acid from D-glucose. The isolation of this gene was difficult because the enzyme had not been previously studied to any great extent. Therefore, before the gene could be isolated, the protein had to be purified and partially sequenced so that DNA hybridization probes based on the amino acid sequence of the protein could be designed.

This work is an early example of what some workers have come to call metabolic engineering, which entails taking the genetic information for part of a metabolic pathway from one organism and transferring it into another organism to create a novel metabolic pathway.

the structure of the synthesized antibiotic. For example, a detailed knowledge of the genetics and biochemistry of the components involved in the synthesis of the antibiotic erythromycin allowed researchers to alter the biosynthetic genes in a predetermined manner and produce predictably altered derivatives of erythromycin. Erythromycin is synthesized by *Saccharopolyspora erythraea*, and the entire 56-kb DNA segment that contains the *ery* gene cluster has been sequenced. The erythromycin polyketide synthase was altered in two different ways: either (1) a DNA region that encoded β-ketoreductase activity was deleted or (2) a DNA region encoding enoylreductase was mutated. With the β-ketoreductase deletion, the erythromycin intermediates that accumulated had a carbonyl moiety rather than a hydroxyl group at the C-5 carbon of the ring (Fig. 12.14). Similarly, with the enoylreductase mutation, a carbon-carbon double bond was introduced at positions C-6 and C-7 of the ring (Fig. 12.14). These experiments suggest that once the cluster of genes encoding the biosynthesis of a particular polyketide antibiotic has been isolated and characterized, it should be possible by altering specific DNA fragments to modify the structure of the antibiotic in a predictable manner. Moreover, it should eventually be possible to cut and splice DNA fragments and thereby shift polyketide synthase domains around to create novel polyketide antibiotics.

The gene clusters for aromatic polyketides all contain a set of three genes encoding a so-called minimal polyketide synthase. These include a ketosynthase (with an acyltransferase domain), a chain length factor, and an acyl carrier protein. The minimal polyketide synthase is responsible for the synthesis of the aromatic polyketide backbone; modifications to the structure are catalyzed by other enzymes acting in concert with the minimal polyketide synthase. The genes encoding these proteins are generally organized into a single cluster (Fig. 12.15). Each minimal polyketide synthase gene cluster encodes the synthesis of a particular antibiotic. By interchanging genes between clusters, two new aromatic polyketide antibiotics were created (Fig. 12.16). This experiment demonstrates the potential of using genetic manipulation to design and produce novel aromatic polyketide antibiotics. This approach promises to dramatically accelerate the process of new antibiotic discovery.

Improving Antibiotic Production

In addition to being a means of developing new antibiotics, genetic engineering can be used to enhance the yield and rate of production of known antibiotics. The large-scale production of antibiotics by *Streptomyces* spp. is often limited by the amount of oxygen available to the cells. The low solubility of oxygen in aqueous media, combined with the highly dense nature of filamentous *Streptomyces* cultures, often results in an oxygen-depleted culture medium, a condition that causes poor cell growth and reduced antibiotic yield. To overcome this problem, it may be possible to improve the design of the bioreactors that are used to grow antibiotic-producing *Streptomyces* cultures and to develop, by genetic manipulation, *Streptomyces* strains that are better able to utilize the available oxygen. These two approaches are not mutually exclusive.

One strategy used by some aerobic microorganisms to cope with oxygen-poor environments is to synthesize a hemoglobin-like molecule that can sequester oxygen from the medium and then deliver it to the cells. For example, the aerobic bacterium *Vitreoscilla* sp. produces a homodimeric

Figure 12.14 Altered erythromycin derivatives produced through genetic manipulation. **A.** A mutation in an enoylreductase gene caused a carbon-carbon double bond to be introduced at positions C-6 and C-7 of the ring (highlighted). **B.** A deletion in a β-ketoreductase gene caused the erythromycin to have a carbonyl moiety rather than a hydroxyl group at the C-5 carbon of the ring (highlighted). Adapted from Katz and Donadio, *Annu. Rev. Microbiol.* **47:**875–912, 1993.

heme protein that is functionally similar to eukaryotic hemoglobin. The gene for the *Vitreoscilla* hemoglobin was isolated and subcloned onto a *Streptomyces* plasmid vector. Following expression of the *Vitreoscilla* hemoglobin gene in *S. coelicolor*, the *Vitreoscilla* hemoglobin represented approximately 0.1% of the total cellular protein, even though the expression was controlled by the native *Vitreoscilla* hemoglobin gene promoter rather than by a *Streptomyces* promoter. When both transformed and nontransformed *S. coelicolor* cultures were grown in the presence of a low level of dissolved oxygen (i.e., approximately 5% saturation), the transformed cells with a functional *Vitreoscilla* hemoglobin produced 10 times more actinorhodine per gram (dry weight) of cells and had greater cell densities than did the nontransformed cells. The expression of the *Vitreoscilla* hemoglobin

Figure 12.15 Gene clusters for the biosynthesis of the aromatic polyketide antibiotics actinorhodine (*act*), tetracenomycin (*tcm*), frenolicin (*fren*), and griseusin (*gris*). Each cluster contains genes encoding a minimal polyketide synthase (PKS), which is responsible for the synthesis of the polyketide backbone. The enzymes encoded by the other genes act to modify the nascent polyketide chain. Each gene is shown pointed in the direction in which it is transcribed.

Figure 12.16 Theoretical biosynthetic pathway for the production of the rationally designed polyketides SEK43 and SEK26. Abbreviations: *act*, actinorhodine; *tcm*, tetracenomycin; *fren*, frenolicin; *gris*, griseusin; min PKS, minimal polyketide synthase; KR, β-ketoreductase; ARO, aromatase; CYC, cyclase.

gene in oxygen-starved microbial cells may provide them with a general mechanism for obtaining sufficient oxygen to allow proliferation under otherwise limiting conditions.

The compound 7-aminocephalosporanic acid (7ACA) is synthesized chemically from the antibiotic cephalosporin C (Fig. 12.11) and is used as the starting material for the chemical synthesis of a number of cephem-type antibiotics (cephalosporins). Cephalosporins have few toxic effects and protect against many different bacteria. Unfortunately, there is no known or-

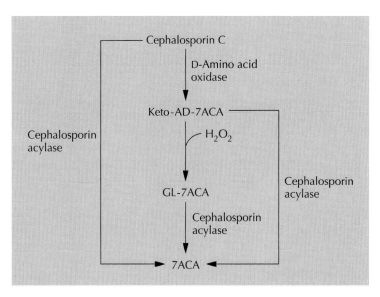

Figure 12.17 Genetically engineered biosynthetic pathway for the synthesis of 7-aminocephalosporanic acid (7ACA) from cephalosporin C. The D-amino acid oxidase gene is from the fungus *F. solani*, and the cephalosporin acylase gene is from the bacterium *P. diminuta*.

ganism that can synthesize 7ACA. However, a novel 7ACA biosynthetic pathway has been constructed in the fungus *Acremonium chrysogenum*, which normally synthesizes only cephalosporin C. The genes involved in this novel engineered pathway consist of a cDNA that encodes D-amino acid oxidase and comes from the fungus *Fusarium solani* and genomic DNA that encodes cephalosporin acylase and comes from *Pseudomonas diminuta*. Both of these genes were subcloned onto an *A. chrysogenum* plasmid expression vector under the control of an *A. chrysogenum* promoter. In the first step of this new pathway, cephalosporin C is converted into the compound 7-β-(5-carboxy-5-oxopentanamido)cephalosporanic acid (keto-AD-7ACA) by D-amino acid oxidase (Fig. 12.17). Some of this product reacts with the hydrogen peroxide that is a by-product of the reaction to form 7-β-(4-carboxybutanamido)cephalosporanic acid (GL-7ACA) (Fig. 12.17). Cephalosporin C, keto-AD-7ACA, and GL-7ACA can each be hydrolyzed by cephalosporin acylase to form 7ACA. However, in the absence of the D-amino acid oxidase step, only 5% of the cephalosporin C is converted to 7ACA; therefore, both enzymes are essential for high yields of 7ACA. This work demonstrates the feasibility of producing 7ACA biologically.

Biopolymers

Biopolymers are large, multiunit macromolecules obtained from living organisms. Some of these polymers have physical and chemical properties that are useful to the food-processing, manufacturing, and pharmaceutical industries. Biopolymers can be derived from microorganisms, plants, and animals. The ability to genetically engineer organisms has stimulated researchers to design new biopolymers, replace synthetic polymers with biological equivalents, modify existing biopolymers to enhance their physical

and structural characteristics, and find ways to increase the yields and decrease the costs of existing industrial processes.

Engineering *Xanthomonas campestris* for Inexpensive Xanthan Gum Production

Xanthomonas campestris is a gram-negative obligatory aerobic soil bacterium that produces the commercially important biopolymer xanthan gum, a high-molecular-weight exopolysaccharide, as a by-product of its metabolism. This polymer has a cellulosic backbone made up of a straight-chain polymer of glucose units. Each of its trisaccharide side chains includes one glucuronic acid and two mannose residues, which are attached to every second glucose residue of the backbone (Fig. 12.18). Xanthan gum has a high viscosity, is stable in extreme physical and chemical environments, and exhibits physical and chemical properties similar to those of a plastic. In particular, its physical properties make it useful as a stabilizing, emulsifying, thickening, or suspending agent. For successful commercial production of xanthan gum, *X. campestris* should be grown on an inexpensive and plentiful carbon source. Wild-type *X. campestris* can efficiently utilize glucose, sucrose, and starch, but not lactose, as a carbon source.

Whey is a waste by-product of the cheese-making process; it consists of water (94 to 95%), lactose (3.5 to 4%), and small amounts of protein, minerals, and low-molecular-weight organic compounds. Enormous quantities of whey are generated by the dairy industry, and its disposal is a major problem. Releasing whey into rivers and lakes can deplete the amount of available oxygen, thereby killing many of the aquatic organisms. Transporting whey to landfill sites is exceptionally expensive, and potential groundwater contamination by the discarded whey is a major concern. Finally, the

Figure 12.18 Structure of xanthan gum. The repeating unit, designated n, forms a chain of glucose units. The trisaccharide is attached to alternate glucose residues of the repeating chain.

costs of removing the solid component of whey are prohibitive. Consequently, many schemes have been devised to use whey in creative ways.

Theoretically, whey could be used as a carbon source for growing industrially important microorganisms. With this in mind, *X. campestris* was genetically engineered to grow on whey. The *E. coli lacZY* genes, which encode the enzymes β-galactosidase and lactose permease, were cloned onto a broad-host-range plasmid under the transcriptional control of an *X. campestris* bacteriophage promoter. This construct was introduced into *E. coli* and then transferred from *E. coli* to *X. campestris* by tripartite mating. Transformants maintained the plasmid, expressed the enzymes β-galactosidase and lactose permease at high levels, utilized lactose as the sole carbon source, and produced high levels of xanthan gum with glucose, lactose, or whey as the carbon source (Table 12.4). Recall that wild-type *X. campestris* produces high levels of xanthan gum only when grown on glucose (Table 12.4). With further work, this system may well be able to convert a nuisance waste product into the substrate for the production of an economically valuable biopolymer.

Isolation of Melanin Biosynthesis Genes

Melanins are a large diverse family of light-absorbing biopolymers that are synthesized by animals, plants, bacteria, and fungi. It has been suggested that these pigments might be useful as topical sunscreens, sunlight-protective coatings for plastics, or additives for cosmetic products. Currently, melanins are obtained in small quantities either by extraction from natural sources or by chemical synthesis. However, recombinant DNA technology may make it possible to produce a range of melanins with different physical properties inexpensively and on a large scale.

Biochemically, melanins are irregular, somewhat random polymers that are composed of indoles, benzthiazoles, and amino acids. The first step in their synthesis, which is catalyzed by the copper-containing monooxygenase tyrosinase, is the oxidation of tyrosine to dihydroxyphenylalanine quinone. The final stages of the polymerization of melanin are nonenzymatic, and, depending on the chemical nature of the nonquinone components that are incorporated into the polymeric structure, the end product can be black, brown, yellow, red, or violet.

The genes involved in melanin biosynthesis in the bacterium *Streptomyces antibioticus* have been isolated and analyzed. They consist of two open reading frames (ORFs), one encoding tyrosinase (molecular weight, 30,600) and one (ORF438) encoding a protein of unknown function with a molecular weight of approximately 14,800. To test whether both of these genes are required for melanin production, they were first subcloned onto an *E. coli*

Table 12.4 Production of xanthan gum by wild-type and transformed *X. campestris*

X. campestris	Amount of xanthan gum produced (μg/mL) with:		
	0.4% Glucose	**0.4% Lactose**	**10% Whey**
Wild type	3,530	245	224
Transformant	3,711	3,608	4,241

Adapted from Fu and Tseng, *Appl. Environ. Microbiol.* **56:**919–923, 1990.

The amount of product is expressed as micrograms per milliliter of culture grown on a minimal medium with either 0.4% glucose or 0.4% lactose added or on diluted whey (10%) which contains approximately 0.4% lactose. The transformant carries the *E. coli lacZY* genes on a plasmid.

Figure 12.19 *E. coli* expression plasmids carrying melanin biosynthesis genes. Plasmid pBGC619 contains the tyrosinase gene. Plasmid pBGC620.3 contains an ORF (ORF438) for melanin synthesis and the tyrosinase gene. Transcription of the cloned genes is under the control of the *E. coli* bacteriophage T7 promoter (P_{T7}). RBS1 and RBS2 denote two different ribosome binding sites. The plasmids both carry genes that confer resistance to ampicillin (Ampr).

expression vector, where one construct contained only the tyrosinase gene and another carried both the tyrosinase and the ORF438 genes (Fig. 12.19). The vector with the tyrosinase gene directs the synthesis of a higher level of tyrosinase than does the vector containing both the tyrosinase and the ORF438 genes. However, the level of tyrosinase was irrelevant, because it turned out that melanin biosynthesis required the products of both genes. The protein encoded by ORF438 may act as a copper donor to apotyrosinase, the inactive precursor form of tyrosinase. Apotyrosinases are activated by acquiring copper ions. Under natural conditions, after dihydroxyphenylalanine quinone is produced by tyrosinase, a variety of low-molecular-weight compounds (nonquinones) can be incorporated into the final polymer. The chemical and physical nature, including the color, of the melanin that is formed after cloning of the key genes into *E. coli* may be manipulated to some extent—to form melanins with different properties—by the addition of different amounts of specific low-molecular-weight compounds to the medium.

Synthesis of an Animal Adhesive Biopolymer in Microbial Cells

Other researchers are trying to inexpensively produce an adhesive protein biopolymer originally isolated from a mussel. The biopolymer produced by the blue mussel, *Mytilus edulis*, is an exceptionally strong, waterproof, adhesive protein, called byssal adhesive, that enables the mussel to attach very tightly to a variety of surfaces. Following its secretion, the byssal adhesive becomes highly cross-linked (randomly) and consequently cannot be sequenced. Without this information, it is impossible to deduce potential nucleic acid sequences that might be used for the synthesis of DNA hybridization probes. However, it was possible to isolate an intracellular precursor form of the adhesive protein (called the 130-kilodalton [kDa] precursor protein), which can be analyzed biochemically. It was found that the 130-kDa precursor protein is rich in serine, threonine, lysine, proline, and tyrosine; 60 to 70% of the amino acids contain a hydroxyl group. Most of the proline residues are hydroxylated to either 3- or 4-hydroxyproline (Hyp), and the majority of the tyrosines are hydroxylated to 3,4-dihydroxyphenylalanine (DOPA). Amino acid sequence analysis of the precursor protein further revealed that it is composed largely of repeating units that consist of

a decapeptide with the sequence Ala-Lys-(Pro or Hyp)-Ser-(Tyr or DOPA)-Hyp-Hyp-Thr-DOPA-Lys.

The cDNA for the 130-kDa precursor adhesive protein was isolated from a cDNA library that was constructed with mRNA isolated from the gland that actively secretes the byssal adhesive. Both the adhesive protein and the cDNA have unusual features that might make cloning, expression, and production of a functional adhesive protein difficult. First, the highly repetitive nature of the adhesive protein cDNA could make it unstable as a result of homologous recombination and subsequent loss of portions of the cloned sequence. Second, proline, lysine, and tyrosine represent about 70% of the amino acids of the protein; therefore, very high levels of synthesis may not be achievable because the intracellular aminoacyl-tRNA pools might be limiting.

When either complete or partial cDNAs for the adhesive protein were cloned onto yeast expression vectors and introduced into yeast cells, active novel forms of the adhesive protein, ranging from 20 to 100 kDa, were synthesized and represented a significant fraction (2 to 5%) of the total cell protein. Thus, there were no problems concerning either the stability of the cloned cDNA or the production of moderate amounts of the adhesive protein. Considerably higher expression levels were attained when a chemically synthesized adhesive protein gene sequence was used (Fig. 12.20). In this case, repeating DNA units that encode the consensus decapeptide repeat of the adhesive protein were used to construct a 600-base-pair (bp) synthetic gene that encoded a protein with a molecular mass of approximately 25 kDa. The 30-bp repeat, the fundamental building block of the synthetic gene, consisted of codons optimized for *E. coli* expression. The synthetic gene was expressed at very high levels by using the T7 promoter. Notwithstanding the level of expression of the adhesive protein, most microorganisms are limited in their ability to hydroxylate amino acids posttranslationally, so that the final protein might be underhydroxylated. In fact, a number of tyrosine residues of the protein were not converted to DOPA, an outcome potentially limiting the ability of the protein to form cross-links. This deficiency was overcome by creating an in vitro hydroxylation system that used a bacterial tyrosinase in the presence of ascorbic acid to hydroxylate the tyrosine residues to DOPA (Fig. 12.21). Ascorbic acid was included in the reaction mixture to prevent the premature oxidation of the DOPA residues to *o*-quinone. Oxidation must be controlled because it leads to cross-linking of the adhesive protein subunits. Like many other adhesives or glues, the protein adhesive must not be activated (cross-linked) prior to its actual use.

When the precursor form of the adhesive protein is oxidized, the cross-linked protein can bind to a variety of surfaces, including polystyrene, glass, hydrogel, and collagen. Moreover, the "strength" and specificity of

Figure 12.20 Synthetic oligonucleotide used in the assembly of a gene for the bioadhesive protein produced by the mussel *M. edulis*. Two synthetic oligonucleotides are designed to base pair and form a DNA module with complementary extensions, which base pair to form a linear DNA molecule with a repeating module. The modules are joined by the enzyme T4 DNA ligase. The amino acid sequence encoded by the DNA module is shown.

```
5′ CCA ACC TAC AAA GCT AAG CCG TCT TAT CCG 3′
       3′ TTT CGA TTC GGC AGA ATA GGC GGT TGG ATG 5′
   Pro – Thr – Tyr – Lys – Ala – Lys – Pro – Ser – Tyr – Pro – Pro – Thr – Tyr
```

Figure 12.21 Pathway for in vitro posttranslational hydroxylation of some of the tyrosine residues in the *M. edulis* adhesive protein. Tyrosine is converted to DOPA by the action of the enzyme tyrosinase and then can be oxidized to *o*-quinone by either catechol oxidase or tyrosinase. The oxidation of DOPA to *o*-quinone can be prevented by the addition of ascorbic acid.

the final adhesive can be manipulated by adding other proteins to the adhesive protein mixture before oxidation and cross-linking. By varying the kinds and amounts of the accessory proteins, adhesives with unique properties can be created. It is anticipated that biopolymeric adhesives will be used extensively in both medicine and dentistry.

Microbial Synthesis of Rubber

Natural rubber, *cis*-1,4-polyisoprene, is an extensively used biopolymer that is obtained from a large number of different plants. The biosynthesis of rubber starts from simple sugars and requires approximately 17 enzyme-catalyzed steps, with the final step being the polymerization of isopentenyl pyrophosphate onto an allylic pyrophosphate. The last step is catalyzed by the enzyme rubber polymerase.

Studies have been undertaken to determine whether rubber can be synthesized by genetically engineered microorganisms. As an initial step in this direction, a cDNA library was constructed by using mRNA from the rubber-producing plant *Hevea brasiliensis*. This library was then screened with a short synthetic DNA hybridization probe whose sequence was based on the amino acid sequence of a portion of the rubber polymerase enzyme. Antibodies directed against the purified enzyme were used to prove unequivocally that the cloned cDNA expressed rubber polymerase. This cDNA clone can now be used, possibly in concert with other genes in the rubber synthesis pathway, in an attempt to produce natural rubber in a microbial system. Alternatively, it can be used as a source of rubber polymerase to develop an in vitro catalytic system. In either case, research that might lead to a new synthetic route for the production of rubber is under way.

Microbial Production of Polyhydroxyalkanoates

Polyhydroxyalkanoates are a class of biodegradable polymers that are produced by a number of different microorganisms, most notably *Alcaligenes eutrophus*, and used as an intracellular carbon and energy storage material. These compounds have thermoplastic or elastic properties, depending on

the polymer composition, and are being considered for use in the synthesis of biodegradable plastics, such as packing material. It has been estimated that the current worldwide market for biodegradable plastics is around $1.3 billion per annum.

Poly(3-hydroxybutyric acid) is the most thoroughly studied and characterized polyhydroxyalkanoate. Both the polymer and the *A. eutrophus* genes that encode its synthesis have been characterized. Poly(3-hydroxybutyric acid), its copolymer [poly(3-hydroxybutyrate-co-3-hydroxyvalerate)], and another polyhydroxyalkanoate [poly(3-hydroxyvaleric acid)] are produced commercially in the United Kingdom by the fermentation of *A. eutrophus*.

Although it is possible to produce poly(3-hydroxybutyric acid) as a by-product of the growth of *A. eutrophus*, this organism grows relatively slowly and utilizes only a limited number of carbon sources for growth, making production costs relatively expensive. On the other hand, when the genes for the biosynthesis of this polymer were transferred to *E. coli*, the resultant transformant grew rapidly to a high cell density and accumulated large amounts (up to 95% of the dry cell weight) of poly(3-hydroxybutyric acid). Poly(3-hydroxybutyric acid) is synthesized from acetyl coenzyme A (acetyl-GA) in three steps catalyzed by three enzymes (Fig. 12.22). The operon containing these genes was cloned into a plasmid as part of a 5.2-kb insert. Unfortunately, plasmids expressing the poly(3-hydroxybutyric acid) operon in *E. coli* were unstable. In the absence of selective pressure such as the addition of antibiotics to the growth medium, about half of the *E. coli* cells lost the plasmid after approximately 50 generations. Plasmid loss of this magnitude, while not a major problem in small-scale batch cultures, becomes more of a problem with large-scale or continuous cultures (see Chapter 16). This problem was overcome by inserting the *parB* genetic locus from another plasmid onto plasmids carrying the poly(3-hydroxybutyric acid) operon. This gene mediates plasmid stabilization by postsegregational killing of plasmid-free cells. The modified plasmids were quite stable even though the poly(3-hydroxybutyric acid) was produced constitutively. *E. coli* transformants synthesizing poly(3-hydroxybutyric acid) produced very little acetate, which can be deleterious for cell growth, presumably because all of the excess acetyl-CoA of the cell was converted to poly(3-hydroxybutyric acid) rather than acetate. An added benefit of producing poly(3-hydroxy-

Figure 12.22 The synthesis of poly(3-hydroxybutyric acid) from acetyl-CoA. The enzyme that catalyzes each of the reactions is shown to the right of the arrow.

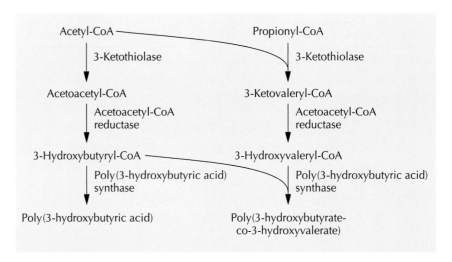

Figure 12.23 Model of the microbial synthesis of the copolymer poly(3-hydroxybutyrate-co-3-hydroxyvalerate).

butyric acid) in *E. coli* instead of *A. eutrophus* is that when the poly(3-hydroxybutyric acid) is recovered by extraction with an alkaline hypochlorite solution, the polymer is degraded to a much lesser extent than when it is produced in *A. eutrophus*. This is probably because most of the poly(3-hydroxybutyric acid) in *E. coli* is produced in a crystalline state while in *A. eutrophus* it is in an amorphous state. Nevertheless, the polymers extracted from the two organisms had identical polymer properties.

The copolymer poly(3-hydroxybutyrate-co-3-hydroxyvalerate) has properties that are similar to those of polypropylene. Consequently, there is considerable commercial interest in the biological production of poly(3-hydroxybutyrate-co-3-hydroxyvalerate). However, *E. coli* strains that expressed the three polymer biosynthetic genes synthesized only poly(3-hydroxybutyric acid) and not the copolymer. This limitation can be overcome with *E. coli* cells that are mutated at both the *fadR* and *atoC* loci. *fadR* is a negative regulator of fatty acid biosynthesis, and *atoC* is a positive regulator of fatty acid uptake. The role of *fadR* is unclear, but the gene product from the *atoC* mutation turns on the synthesis of the proteins encoded by *atoA* and *atoD*, whose gene products facilitate the uptake of propionate from the growth medium into the cell. The propionate is converted to propionyl-CoA and then condensed with acetyl-CoA to form 3-ketovaleryl-CoA, which can be converted into 3-hydroxyvaleryl-CoA before its incorporation into the copolymer (Fig. 12.23).

SUMMARY

In addition to using bacteria as factories for the production of proteins such as restriction enzymes, it is possible to modify metabolic pathways of organisms, either by introducing new genes or by altering existing ones. In this way, various organisms can be genetically engineered for the production of a range of low-molecular-weight compounds such as L-ascorbic acid, indigo, amino acids, antibiotics, and the monomeric subunits of various biopolymers. Here, the strategy is to insert one or more specific enzymes into the host organism by transformation with a vector-cloned gene construct. The inserted enzymes either create a new pathway or augment a

preexisting pathway for the synthesis of a specific compound. Several studies have shown that the creation of such unusual enzymatic pathways is technically feasible. This approach should facilitate the development of new and more efficient synthetic routes for a variety of important compounds.

References

Anderson, S., C. B. Marks, R. Lazarus, J. Miller, K. Stafford, J. Seymour, D. Light, W. Rastetter, and D. Estell. 1985. Production of 2-keto-L-gulonate, an intermediate in L-ascorbate synthesis by a genetically modified *Erwinia herbicola. Science* **230:**144–149.

Bailey, J. E. 1991. Toward a science of metabolic engineering. *Science* **252:**1668–1675.

Berry, A. 1996. Improving production of aromatic compounds in *Escherichia coli* by metabolic engineering. *Trends Biotechnol.* **14:**250–256.

Brooks, J. E., P. D. Nathan, D. Landry, L. A. Sznyter, P. Waite-Rees, C. L. Ives, L. S. Moran, B. E. Slatko, and J. E. Benner. 1991. Characterization of the cloned *Bam*HI restriction modification system: Its nucleotide sequence, properties of the methylase, and expression in heterologous hosts. *Nucleic Acids Res.* **19:**841–850.

Cohen, G., D. Shiffman, M. Mevarech, and Y. Aharonowitz. 1990. Microbial isopenicillin N synthase genes: structure, function, diversity and evolution. *Trends Biotechnol.* **8:**105–111.

della-Cioppa, G., S. J. Garger, G. G. Sverlow, T. H. Turpen, and L. K. Grill. 1990. Melanin production in *Escherichia coli* from a cloned tyrosinase gene. *Bio/Technology* **8:**634–638.

Ensley, B. D., B. J. Ratzkin, T. D. Osslund, M. J. Simon, L. P. Wackett, and D. T. Gibson. 1983. Expression of naphthalene oxidation genes in *Escherichia coli* results in the biosynthesis of indigo. *Science* **222:**167–169.

Flores, N., J. Xiao, A. Berry, F. Bolivar, and F. Valle. 1996. Pathway engineering for the production of aromatic compounds in *Escherichia coli. Nat. Biotechnol.* **14:**620–623.

Floss, H. G. 1987. Hybrid antibiotics—the contribution of the new gene combinations. *Trends Biotechnol.* **5:**111–115.

Fu, J.-F., and Y.-H. Tseng. 1990. Construction of lactose-utilizing *Xanthomonas campestris* and production of xanthan gum from whey. *Appl. Environ. Microbiol.* **56:**919–923.

Hahn, S. K., Y. K. Chang, and S. Y. Lee. 1995. Recovery and characterization of poly(3-hydroxybutyric acid) synthesized in *Alcaligenes eutrophus* and recombinant *Escherichia coli. Appl. Environ. Microbiol.* **61:**34–39.

Hillemann, D., A. Puhler, and W. Wolleben. 1991. Gene disruption and gene replacement in *Streptomyces* via single stranded DNA transformation of integration vectors. *Nucleic Acids Res.* **19:**727–731.

Hopwood, D. A., M. J. Bibb, C. J. Bruton, K. F. Chater, J. S. Feitelson, and J. A. Gil. 1983. Cloning *Streptomyces* genes for antibiotic production. *Trends Biotechnol.* **1:**42–48.

Hopwood, D. A., F. Malpartida, H. M. Kieser, H. Ikeda, J. Duncan, I. Fujii, B. A. M. Rudd, H. G. Floss, and S. Omura. 1985. Production of hybrid antibiotics by genetic engineering. *Nature* **314:**642–644.

Howard, K. A., C. Card, J. S. Benner, H. L. Callahan, R. Maunus, K. Silber, G. Wilson, and J. E. Brooks. 1986. Cloning the *Dde*I restriction-modification system using a two-step method. *Nucleic Acids Res.* **14:**7939–7951.

Hutchinson, C. R., H. Decker, K. Madduri, S. L. Otten, and L. Tang. 1993. Genetic control of polyketide biosynthesis in the genus *Streptomyces. Antonie Leeuwenhoek* **64:**165–176.

Hutchinson, C. R. 1994. Drug synthesis by genetically engineered microorganisms. *Bio/Technology* **12:**375–380.

Hutchinson, C. R., and I. Fujii. 1995. Polyketide synthase gene manipulation: a

structure-function approach in engineering novel antibiotics. *Annu. Rev. Microbiol.* **49**:201–238.

Ikeda, M., K. Nakanishi, K. Kino, and R. Katsumata. 1994. Fermentative production of tryptophan by a stable recombinant strain of *Corynebacterium glutamicum* with a modified serine-biosynthetic pathway. *Biosci. Biotechnol. Biochem.* **58**:674–678.

Ishida, M., K. Miwa, S. Nakamori, and K. Sano. December 1989. Process for producing L-tryptophan. U.S. patent 4,885,245.

Isogai, T., M. Fukagawa, I. Aramori, M. Iwami, H. Kojo, T. Ono, Y. Ueda, M. Kohsaka, and H. Imanaka. 1991. Construction of a 7-aminocephalosporanic acid (7ACA) biosynthetic operon and direct production of 7ACA in *Acremonium chrysogenum*. *Bio/Technology* **9**:188–191.

Katz, L., and S. Donadio. 1993. Polyketide synthesis: prospects for hybrid antibiotics. *Annu. Rev. Microbiol.* **47**:875–912.

Kleinkauf, H., and H. von Dohren. 1990. Antibiotics—cloning of biosynthetic pathways. *FEBS Lett.* **268**:405–407.

Krämer, R. 1996. Genetic and physiological approaches for the production of amino acids. *J. Biotechnol.* **45**:1–21.

Lazarus, R. A., M. Hurle, S. Anderson, and D. B. Powers. December 1994. Enzymes for the production of 2-keto-L-gulonic acid. U.S. patent 5,376,544.

Lee, S. Y., H. N. Chang, and Y. K. Chang. 1994. Production of poly(β-hydroxybutyric acid) by recombinant *Escherichia coli*. *Ann. N. Y. Acad. Sci.* **721**:43–53.

Lee, S. Y., K. S. Yim, H. N. Chang, and Y. K. Chang. 1994. Construction of plasmids, estimation of plasmid stability, and use of stable plasmids for the production of poly(β-hydroxybutyric acid) by recombinant *Escherichia coli*. *J. Biotechnol.* **32**:203–211.

Lee, S. Y., and H. N. Chang. 1995. Production of poly(3-hydroxybutyric acid) by recombinant *Escherichia coli* strains: genetic and fermentation studies. *Can. J. Microbiol.* **41**:207–219.

Magnolo, S. K., D. L. Leenutaphong, J. A. DeModena, J. E. Curtis, J. E. Bailey, J. L. Galazzo, and D. E. Hughes. 1991. Actinorhodin production by *Streptomyces coelicolor* and growth of *Streptomyces lividans* are improved by the expression of a bacterial hemoglobin. *Bio/Technology* **9**:473–476.

Martin, J. F. 1987. Cloning of genes involved in penicillin and cephalosporin biosynthesis. *Trends Biotechnol.* **5**:306–308.

McDaniel, R., S. Ebert-Khosla, D. A. Hopwood, and C. Khosla. 1995. Rational design of aromatic polyketide natural products by recombinant assembly of enzymatic subunits. *Nature* **375**:549–554.

Mermod, N., S. Harayama, and K. N. Timmis. 1986. New route to bacterial production of indigo. *Bio/Technology* **4**:321–324.

Ozaki, A., R. Katsumata, and T. Oka. October 1989. Process for producing tryptophan. U.S. patent 4,874,698.

Piekarowicz, A., R. Yuan, and D. C. Stein. 1991. A new method for the rapid identification of genes encoding restriction and modification enzymes. *Nucleic Acids Res.* **19**:1831–1835.

Rhie, H. G., and D. Dennis. 1995. Role of *fadR* and *atoC*(Con) mutations in poly(3-hydroxybutyrate-co-3-hydroxyvalerate) synthesis in recombinant *pha+ Escherichia coli*. *Appl. Environ. Microbiol.* **61**:2487–2492.

Salerno, A. J., and I. Goldberg. 1993. Cloning, expression, and characterization of a synthetic analog to the bioadhesive precursor protein of the sea mussel *Mytilus edulis*. *Appl. Microbiol. Biotechnol.* **39**:221–226.

Schwarzer, A., and A. Puhler. 1991. Manipulation of *Corynebacterium glutamicum* by gene disruption and replacement. *Bio/Technology* **9**:84–87.

Sikora, L. A. January 1991. DNA fragment encoding a rubber polymerase and its use. U.S. patent 4,983,729.

Stachelhaus, T., A. Schneider, and M. A. Marahiel. 1995. Rational design of peptide antibiotics by targeted replacement of bacterial and fungal domains. *Science* **269:**69–72.

Strausberg, R. L., and R. P. Link. 1990. Protein-based medical adhesives. *Trends Biotechnol.* **8:**53–57.

Terasawa, M., M. Fukushima, Y. Kurusu, and H. Yukawa. 1990. L-Tryptophan production by the application of high expressed tryptophanase in *Escherichia coli. Process Biochem. Int.* **25:**172–175.

Walder, R. Y., J. L. Hartley, J. E. Donelson, and J. A. Walder. 1981. Cloning and expression of the *Pst*I restriction-modification system in *Escherichia coli. Proc. Natl. Acad. Sci. USA* **78:**1503–1507.

Weber, J. M., J. O. Leung, S. J. Swanson, K. B. Idler, and J. B. McAlpine. 1991. An erythromycin derivative produced by targeted gene disruption in *Saccharopolyspora erythracea. Science* **252:**114–117.

Yim, K. S., S. Y. Lee, and H. N. Chang. 1996. Synthesis of poly(3-hydroxybutyate-co-3-hydroxyvalerate) by recombinant *Escherichia coli. Biotechnol. Bioeng.* **49:**495–503.

REVIEW QUESTIONS

1. Describe a strategy for isolating the gene for the restriction endonuclease *Eco*RI.

2. Outline a strategy for cloning the gene for 2,5-diketo-D-gluconic acid reductase from *Corynebacterium* into *Erwinia*. Why would anyone want to do this?

3. Suggest a strategy for improving the commercial utility of a cloned 2,5-diketo-D-gluconic acid reductase gene.

4. How can indigo be produced in *E. coli*?

5. Outline a strategy for increasing the production of the amino acid tryptophan by *Corynebacterium glutamicum*.

6. Suggest a strategy for isolating some of the genes that are involved in the biosynthesis of the antibiotic undecylprodigiosin, which is normally synthesized by *Streptomyces coelicolor*.

7. Why is it difficult to genetically transform various *Streptomyces* species? How can this difficulty be overcome?

8. Suggest a simple strategy for increasing the yield of an antibiotic by the genetic manipulation of the *Streptomyces* strain that produces the antibiotic.

9. Suggest an approach for producing modified versions of nonaromatic polyketide antibiotics such as erythromycin.

10. How can an adhesive protein biopolymer that is normally produced by the blue mussel, *Mytilus edulis*, be synthesized in *E. coli*?

11. Suggest a scheme for producing poly(3-hydroxybutyric acid) in *E. coli*.

12. What is whey? How can it be used to produce industrially important compounds?

Bioremediation and Biomass Utilization

For centuries, humans believed that atmospheric, terrestrial, and aquatic systems were sufficient to absorb and break down wastes from population centers, industry, and farming. We now know that this is not true. Today, there are two fundamental problems. First, how do we dispose of the large quantities of wastes that are continually being produced? Second, how do we remove the toxic compounds that have been accumulating at dump sites, in the soil, and in water systems over the last few decades? Governments have tried to meet the challenge of environmental contamination by instituting antipollution regulations, but these rules often remain unenforced. Governments have also encouraged the three R's—reduce, reuse, and recycle.

Researchers are currently testing a number of technological strategies, including biotechnological schemes, to deal with large-scale wastes such as lignocellulosics and toxic substances that persist in ecosystems.

The term "bioremediation" has been introduced to describe the process of using biological agents to remove toxic wastes from the environment. "Biomass" is the term used to describe the materials that are produced by the food and agricultural industries (e.g., starch and lignocellulosics) and were discarded as waste in the past. Biomass is now being considered as a source material for the production of a variety of economically important products.

Microbial Degradation of Xenobiotics

The problem of toxic waste disposal is enormous. Worldwide production in 1985 of just one chemical that is released into the environment—pentachlorophenol—was more than 50,000 tons. Incineration and chemical treatment have been used to break down toxic chemicals, but these methods are costly and often create new environmental difficulties. With the discovery in the mid-1960s of a number of soil microorganisms that are capable of degrading xenobiotic ("unnatural," synthetic; from xenos, meaning "foreign") chemicals such as herbicides, pesticides, refrigerants, solvents,

and other organic compounds, the notion that microbial degradation might provide an economic and effective means of disposing of toxic chemical wastes gained credence.

Members of the genus *Pseudomonas* are the most predominant group of soil microorganisms that degrade xenobiotic compounds. Biochemical assays have shown that various *Pseudomonas* strains can break down and, as a consequence, detoxify more than 100 different organic compounds. In many cases, one strain can use several different related compounds as the sole carbon source.

The biodegradation of complex organic molecules generally requires the concerted efforts of several different enzymes. The genes that code for the enzymes of these biodegradative pathways are sometimes located in the chromosomal DNA, although they are more often found on large (approximately 50- to 200-kilobase [kb]) plasmids (Table 13.1). In some organisms, the genes that contribute to the degradative pathway are found on both chromosomal and plasmid DNA.

Degradative bacteria, in most cases, enzymatically convert xenobiotic, nonhalogenated aromatic compounds to either catechol (Fig. 13.1) or protocatechuate (Fig. 13.2). Then, through a series of oxidative cleavage reactions, catechol and protocatechuate are processed to yield either acetyl coenzyme A (acetyl-CoA) and succinate (Fig. 13.3) or pyruvate and acetaldehyde (Fig. 13.4), compounds that are readily metabolized by almost all organisms. Halogenated aromatic compounds, which are the main components of most pesticides and herbicides, are converted to catechol, protocatechuate, hydroquinones, or the corresponding halogenated derivatives by the same enzymes that degrade the nonhalogenated compounds. However, for the halogenated compounds, the rate of degradation is inversely related to the number of halogen atoms that are initially present on the target compound. Dehalogenation—the removal of a halogen substituent from an organic compound—is the critical requirement for detoxification and often occurs by a nonselective dioxygenase reaction that replaces the halogen on a benzene

Table 13.1 *Pseudomonas* plasmids, their degradative pathways, and their sizes

Name of plasmid	Compound(s) degraded	Plasmid size (kb)
SAL	Salicylate	60
SAL	Salicylate	72
SAL	Salicylate	83
TOL	Xylene and toluene	113
pJP1	2,4-D	87
pJP2	2,4-D	54
pJP3	2,4-D	78
CAM	Camphor	225
XYL	Xylene	15
pAC31	3,5-Dichlorobenzoate	108
pAC25	3-Chlorobenzoate	102
pWWO	Xylene and toluene	176
NAH	Naphthalene	69
XYL-K	Xylene and Toluene	135

Adapted from Cork and Krueger, *Adv. Appl. Microbiol.* **36**:1–66, 1991.

Plasmids with the same name encode a similar degradative pathway even though they have different sizes and were described in different laboratories. 2,4-D, 2,4-dichlorophenoxyacetic acid.

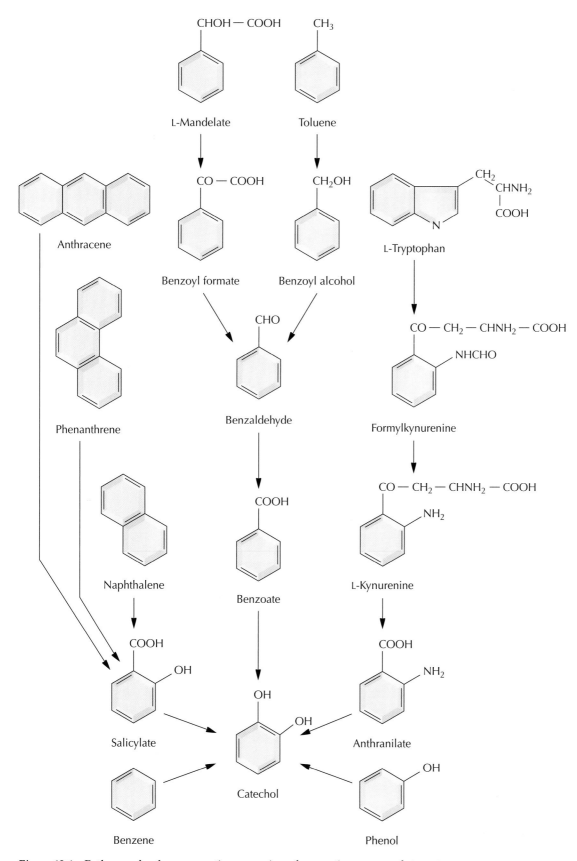

Figure 13.1 Pathways for the enzymatic conversion of aromatic compounds to catechol by degradative bacteria.

Figure 13.2 Pathways for the enzymatic conversion of aromatic compounds to protocatechuate by degradative bacteria.

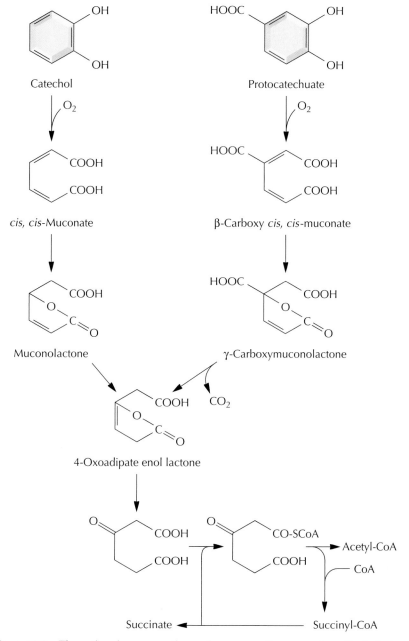

Figure 13.3 The ortho-cleavage pathway for enzymatic conversion of catechol and protocatechuate to acetyl-CoA and succinate.

Figure 13.4 The meta-cleavage pathway for the enzymatic conversion of catechol and protocatechuate to pyruvate and acetaldehyde.

ring with a hydroxyl group. This step may occur either during or after the biodegradation of the original halogenated compound.

Genetic Engineering of Biodegradative Pathways

Despite the ability of many naturally occurring microorganisms to degrade a number of different xenobiotic chemicals, there are limitations to the biological treatment of these waste materials. For example: (1) no single microorganism can degrade all organic wastes; (2) high concentrations of

some organic compounds can inhibit the activity or growth of degradative microorganisms; (3) most contaminated sites contain mixtures of chemicals, and an organism that can degrade one or more of the components of the mixture may be inhibited by other components; (4) many nonpolar compounds adsorb onto particulate matter in soils or sediments and become less available to degradative microorganisms; and (5) microbial biodegradation of organic compounds is often quite slow. One way to address some of these problems is to transfer by conjugation into a recipient strain plasmids that carry genes for different degradative pathways (Fig. 13.5). If two resident plasmids contain homologous regions of DNA, recombination can occur and a single, larger "fusion" plasmid with combined functions can be created. Alternatively, if two plasmids do not contain homologous regions and, in addition, belong to different incompatibility groups, they can coexist within a single bacterium.

Figure 13.5 Schematic representation of the development of a bacterial strain that can degrade camphor, octane, xylene, and naphthalene. Strain A, which contains a CAM (camphor-degrading) plasmid, is mated with strain B, which carries an OCT (octane-degrading) plasmid. Following plasmid transfer and homologous recombination between the two plasmids, strain E carries a CAM and OCT biodegradative fusion plasmid. Strain C, which contains a XYL (xylene-degrading) plasmid, is mated with strain D, which contains a NAH (naphthalene-degrading) plasmid, to form strain F, which carries both of these plasmids. Finally, strain E and strain F are mated to yield strain G, which carries the CAM/OCT fusion plasmid, the XYL plasmid, and the NAH plasmid.

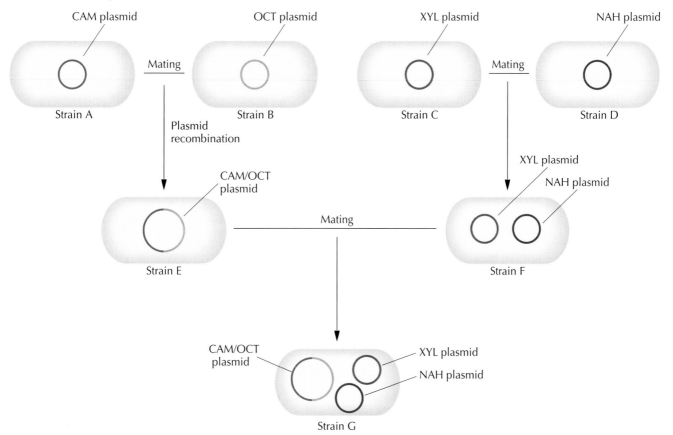

Manipulation by Transfer of Plasmids

Bacterial strains with expanded degradative capabilities were first created in the 1970s by Chakrabarty and coworkers. They used different plasmids to construct a bacterial strain that degraded a number of the hydrocarbon components found in petroleum (Fig. 13.5). This strain has been called a "superbug" because of its increased metabolic capabilities. The CAM (camphor-degrading) plasmid was transferred by conjugation into a strain carrying the OCT (octane-degrading) plasmid. These two plasmids were incompatible and could not be maintained in the same cell as separate plasmids. However, when recombination occurred between the two plasmids, the resulting single plasmid was perpetuated and carried both CAM and OCT degradative activities. The NAH (naphthalene-degrading) plasmid was transferred by conjugation into a strain carrying the XYL (xylene-degrading) plasmid. The NAH and XYL plasmids were compatible and could therefore coexist within the same host cell. Finally, the CAM/OCT fusion plasmid was transferred by conjugation into the strain carrying the NAH and XYL plasmids. The final result of these manipulations was the generation of a strain that grew better on crude oil than did any of the single-plasmid strains either alone or in combination.

Although this particular multiple degradative strain has not been used to clean up oil spills, it has played a critical role in the development of the biotechnology industry. The inventor of this "superbug" was granted a U.S. patent describing its construction and use. This was the first patent ever granted for a genetically engineered microorganism and represented a watershed court decision, because it implied that biotechnology companies could protect their inventions in the same way as the chemical and pharmaceutical industries had in the past.

Most of the degradative bacteria that have been genetically manipulated by plasmid transfer are mesophiles, organisms that grow well only at temperatures between 20 and 40°C. However, rivers, lakes, and oceans that are polluted generally have temperatures that range from 0 to 20°C. To test whether bacteria with enhanced degradative abilities could be created for cold environments, a TOL (toluene-degrading) plasmid from a mesophilic *Pseudomonas putida* strain was transferred by conjugation into a psychrophile (an organism with a low temperature optimum) that was able to degrade salicylate (but not toluene) and use it as a sole carbon source at temperatures as low as 0°C. The transformed strain carried the introduced TOL plasmid and its own SAL (salicylate-degrading) plasmid and was able to use either salicylate or toluene as its sole carbon source at 0°C (Table 13.2). The wild-type (nontransformed) psychrophilic strain was unable to grow at any temperature when toluene (or toluate) was the only carbon source (not shown). This work is preliminary, and additional research is required before an effective psychrophilic degradative bacterium will be available for use in the environment.

Manipulation by Gene Alteration

Bringing together different intact degradative pathways by conjugation is only one way to create bacteria with novel properties. It may also be possible to extend the degradative capability of a strain by altering the genes of the degradative pathway. The feasibility of this approach was examined for the toluene–xylene-degrading pathway of plasmid pWWO. This plasmid encodes a "meta-cleavage" pathway involving 12 different genes and enables pseudomonads carrying the plasmid to utilize various alkylbenzoates as carbon sources (Fig. 13.6). The genes in the toluene–xylene pathway of

Table 13.2 Generation time of wild-type (nontransformed) and transformed psychrotrophic strains of *P. putida* on salicylate or toluate as the sole carbon source at various temperatures

Temp (°C)	Generation time (h) for:		
	Wild-type + salicylate	Transformant + salicylate	Transformant + toluate
37	No growth	No growth	No growth
30	2.2	2.5	2.0
25	2.1	3.2	1.3
20	2.6	3.8	1.9
15	3.2	4.2	2.9
10	6.3	5.6	3.3
5	13.9	12.9	12.2
0	18.6	18.1	24.4

Adapted from Kolenc et al., *Appl. Environ. Microbiol.* **54:**638–641, 1988.

The wild-type strain is unable to utilize toluate for growth at any temperature because it lacks the enzymes to metabolize this compound.

Figure 13.6 The meta-cleavage pathway and the *xyl* operon of the toluene- and xylene-degrading plasmid pWWO. Transcription of the *xyl* operon is controlled by the P_m promoter, which is regulated by the XylS gene product, which must be activated by one of the initial pathway substrates. The genes from *xylX* to *xylH* are under the control of the P_m promoter. The *xylS* gene, which is not part of this operon, is constitutively expressed. Some of the primary substrates are benzoate, where R and R′ = H; 3-methylbenzoate, where R = H and R′ = CH₃; 3-ethylbenzoate, where R = H and R′ = CH₂CH₃; and 4-methylbenzoate, where R = CH₃ and R′ = H. The *xylXYZ* genes encode toluene dioxygenase; *xylL* encodes dihydroxycyclohexadiene carboxylate dehydrogenase; *xylE* encodes catechol 2,3-dioxygenase; *xylF* encodes hydroxymuconic semialdehyde hydrolase; *xylG* encodes hydroxymuconic semialdehyde dehydrogenase; *xylH* encodes 4-oxalocrotonate tautomerase; *xylI* encodes 4-oxalocrotonate decarboxylase; *xylJ* encodes 2-oxopent-4-enoatehydratase; and *xylK* encodes 2-oxo-4-hydroxypentonate aldolase.

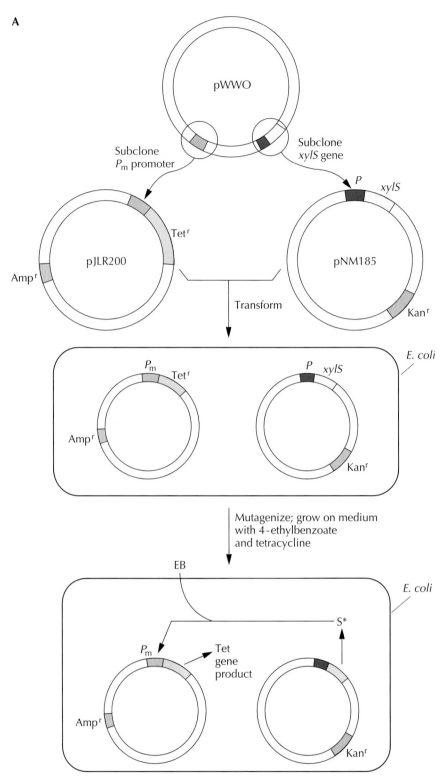

Figure 13.7 **A.** Protocol used to create a XylS protein that can be activated by 4-ethylbenzoate. The P_m promoter is cloned onto plasmid pBR322 and replaces the tetracycline resistance gene promoter to form plasmid pJLR200; the *xylS* gene and its promoter are spliced onto a broad-host-range plasmid containing a kanamycin resistance gene. *E. coli* is transformed with both of these plasmids. Transformants are selected by their resistance to both ampicillin and kanamycin and then chemically mutagenized with ethyl methanesulfonate. The cells with a mutation (S*) in the *xylS* gene that allows the XylS protein to be activated by 4-ethylbenzoate (EB) are the only ones that can grow on medium that contains both 4-ethylbenzoate and tetracycline, because only these cells are resistant to tetracycline. *(continued)*

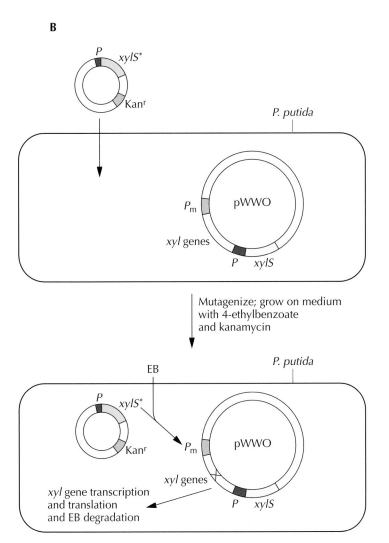

B

Figure 13.7 (*Continued*) **B.** Protocol used to create a modified catechol 2,3-dioxygenase that is not inhibited by 4-ethylcatechol. A *P. putida* strain carrying pWWO is transformed with a broad-host-range plasmid carrying the mutated *xylS** gene, whose product can activate the P_m promoter. Transformants are chemically mutagenized and then grown on minimal medium that contains 4-ethylbenzoate and kanamycin. Any cell that can grow on this medium carries a mutated catechol 2,3-dioxygenase gene. This mutation is indicated by X in the middle of the *xyl* gene cluster.

pWWO are part of a single operon, called the *xyl* operon, under the control of the P_m promoter. Transcription from the P_m promoter is positively regulated by the *xylS* gene product, which is activated by most of the initial substrates (such as benzoate and 3-methylbenzoate) of the pathway (Fig. 13.6). Detailed biochemical and genetic analyses showed that bacteria carrying pWWO could degrade 4-ethylbenzoate to 4-ethylcatechol (which accumulated in the medium) but no further. 4-Ethylcatechol prevented its own degradation by inactivating one of the most important enzymes in the biodegradative pathway, catechol 2,3-dioxygenase, the product of the *xylE* gene. In addition, 4-ethylbenzoate, unlike most other alkylbenzoates, does not activate the XylS protein; consequently, transcription of the operon from the P_m promoter did not occur to any significant extent when 4-ethyl-

benzoate was the only substrate. Thus, there are two major problems with the naturally occurring meta-cleavage pathway system: (1) how to overcome the inactivation of an important enzyme in the degradative process by 4-ethylbenzoate and (2) how to induce transcription of the genes of this pathway with 4-ethylbenzoate as the inducer.

To find a mutant that could solve the second problem, a tetracycline resistance gene was placed under the control of the P_m promoter on one plasmid, which also carried an ampicillin resistance gene. The *xylS* gene was cloned onto another plasmid carrying a kanamycin resistance gene. Transformants carrying both of these plasmids were selected on the basis of resistance to both ampicillin and kanamycin (Fig. 13.7A). The *Escherichia coli* cells carrying both of these plasmids were treated with the mutagen ethyl methanesulfonate and plated onto a medium containing both tetracycline and 4-ethylbenzoate. The only cells that could grow on this medium carried an altered XylS protein (S*) that could interact with 4-ethylbenzoate and cause the tetracycline resistance gene to be transcribed. Thus, the degradative pathway involving the mutated *xylS* gene can be induced by 4-ethylbenzoate. To address the catechol 2,3-dioxygenase inactivation problem, the mutated *xylS* gene was subcloned onto a broad-host-range plasmid carrying a kanamycin resistance gene and introduced into *P. putida* cells carrying pWWO (Fig. 13.7B). The transformed cells were plated, at a high cell density, onto a minimal medium containing 4-ethylbenzoate as the sole carbon source, kanamycin to select for the presence of the plasmid, and ethyl methanesulfonate. Cells that were able to grow on this medium produced an altered form of the enzyme catechol 2,3-dioxygenase that was not inhibited by 4-ethylcatechol. Additional analysis confirmed that the catechol 2,3-dioxygenase gene on pWWO had been mutated and that both mutant versions of the *xylS* gene and the catechol 2,3-dioxygenase gene were required for the degradation of 4-ethylbenzoate.

An important aspect of this work is the fact that the two genes that were altered, i.e., those encoding XylS and catechol 2,3-dioxygenase, are the major determinants of the range of compounds that can be degraded by this pathway. Consequently, the strategy that was used to engineer the degradation of 4-ethylbenzoate can be used for the breakdown of other compounds. In this vein, other mutations of this pathway such as a mutant that overproduces XylS may increase transcription from the P_m promoter, which in turn would lead to an increased rate of substrate degradation. It is also possible to selectively modify the P_m promoter and make it a stronger promoter while retaining its ability to interact with XylS. In sum, the work with 4-ethylbenzoate demonstrates that by combining recombinant DNA technology, conventional mutagenesis, and the appropriate selection protocols, novel properties can be added to a degradative pathway.

The compound trichloroethylene is widely used as a solvent and a degreasing agent, with the result that it is one of the most common contaminants of soil and groundwater. Trichloroethylene persists in the environment for years and is a suspected carcinogen. In addition, anaerobic soil bacteria can reductively dehalogenate it to produce vinyl chloride, which is even more toxic.

Studies showed that some of the strains of *P. putida* that could degrade aromatic compounds such as toluene could also degrade trichloroethylene. Genetic studies established that the complete meta-cleavage degradative pathway was not necessary to completely detoxify trichloroethylene. In fact, only the enzyme toluene dioxygenase, which normally catalyzes the oxidation of toluene to *cis*-toluene dihydrodiol, was required.

Four genes (Fig. 13.8A) are involved in the production of a functional toluene dioxygenase. These genes were isolated and expressed in *E. coli* under the control of the strong and inducible *tac* promoter. When *E. coli* cells carrying these genes were induced by the addition of isopropyl-β-D-thiogalactopyranoside (IPTG), an inducer of the *tac* promoter, trichloroethylene was efficiently broken down to harmless compounds. Although the initial rates of trichloroethylene degradation were lower with *E. coli* than with the original *P. putida* strain, the *E. coli* cells maintained these rates for longer than *P. putida* did. This difference may reflect the fact that *E. coli* is not as susceptible to damage from trichoroethylene as *P. putida* is.

In a variation of this experiment, a hybrid *Pseudomonas* strain with elements of two separate degradative pathways was constructed. Bacterial strains that can degrade the compound biphenyl use the enzyme biphenyl dioxygenase, which includes a two-subunit terminal dioxygenase, ferredoxin, and ferredoxin reductase. Biphenyl dioxygenase is quite similar in both structure and function to the enzyme toluene dioxygenase. Despite the similarities of their enzymes, biphenyl-utilizing pseudomonads cannot grow on toluene and toluene-utilizing strains cannot grow on biphenyl. However, when the *bphA1* gene (coding for the large subunit of biphenyl dioxygenase) from *P. putida* KF715 was replaced by homologous recombination with the *todC1* gene (coding for the large subunit of the toluene dioxygenase) from *P. putida* F1, the resultant strain was able to degrade trichloroethylene (Table 13.3). In fact, the engineered strain grew well on a range of aromatic compounds and also was very efficient at degrading

Figure 13.8 A cloned toluene dioxygenase operon under the control of the *tac* promoter in *E. coli*. **A.** Toluene dioxygenase activity is due to the products of four genes (*todA*, *todB*, *todC1*, and *todC2*). *todA* encodes a flavoprotein that accepts electrons from reduced nicotinamide adenine dinucleotide (NADH) and transfers them to a ferredoxin encoded by *todB*, which reduces the terminal dioxygenase that is encoded by *todC1* and *todC2*. These genes are equivalent to the genes *xylXYZ* shown in Fig. 13.7. **B.** Toluene is converted to *cis*-toluene dihydrodiol by the concerted enzymatic activities of the Tod proteins.

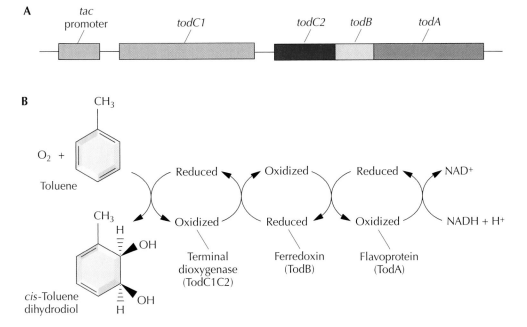

Table 13.3 Growth of parental and engineered *Pseudomonas* strains on various aromatic compounds

Strain	Growth on:				
	Biphenyl	**Diphenyl-methane**	**Toluene**	**Benzene**	**Trichloro-ethylene**
P. putida KF715	+ + +	+ + +	−	−	−
P. putida F1	−	−	+ + +	+ + +	+
P. putida KF715-D5	+ +	+	+ + +	+ + +	+ + +

Adapted from Suyama et al., *J. Bacteriol.* **178**:4039–4046, 1996.

In *P. putida* KF715-D5, the *bphA1* gene from *P. putida* KF715 is replaced with the *todC1* gene from *P. putida* F1. Symbols: + + +, good growth; + +, moderate growth; +, poor growth; −, very poor or no growth.

trichloroethylene, demonstrating that it is possible to rationally engineer bacterial strains that can degrade a number of different compounds.

Utilization of Starch and Sugars

Starch, the major food reserve polysaccharide in plants, consists of a mixture of linear homopolymers (amylose) and branched homopolymers (amylopectin) of D-glucose. Amylose is made up of linear chains of 1×10^2 to 4×10^5 D-glucose residues linked by α-1,4-bonds (Fig. 13.9A). Amylopectin consists of short linear chains of approximately 17 to 23 glucose units that are linked by α-1,4-bonds and joined by 1,6-linkages and some 1,3-linkages to form a highly branched structure that contains 1×10^4 to 4×10^7 glucose residues (Fig. 13.9B). The degree of branching and the ratio of amylose to amylopectin vary with the source and age of the starch.

Commercial Production of Fructose and Alcohol

The major use of starch is in the food and brewing industries, where it is usually hydrolyzed to low-molecular-weight components before being converted into other compounds, especially fructose and alcohol. The most important enzymes for breaking down and transforming starch are α-amylase, glucoamylase, and glucose isomerase. Together, the use of these three enzymes accounts for approximately 30% of the cost of all enzymes currently used for industrial processes.

Fructose and alcohol are produced commercially from starch by multistep processes that include both enzymatic and nonenzymatic reactions as follows (Fig. 13.10).

1. The procedure begins with the gelatinization of milled grain (often corn, which is approximately 40% starch). This treatment—steam cooking under pressure—exposes the surface area of the starch, thereby making it more readily available for subsequent enzymatic hydrolysis. The product of this process has a gel-like consistency.

2. The gelatinized starch is cooled to 50 to 60°C, and α-amylase is added. In this liquefaction step, the gel-like starch is enzymatically digested by the hydrolysis of the available α-1,4-linkages to form low-molecular-weight polysaccharides. A high temperature is used because enzyme hydrolysis is more rapid at high temperatures and the enzyme does not efficiently penetrate the gelatinized starch at lower temperatures.

Figure 13.9 **A.** Pathway for the enzymatic hydrolysis of amylose. **B.** Pathway for the enzymatic hydrolysis of amylopectin. The blue circles represent D-glucose residues.

3. The final step for the release of glucose includes the addition of glucoamylase and results in the saccharification—complete hydrolysis—of the remaining polysaccharides, including both linear and cross-linked molecules.

The end product of these treatments is glucose, which may either be converted into alcohol as a result of fermentation by yeast cells or be transformed into fructose by the enzyme glucose isomerase. The success of the latter enzymatic conversion has led to the replacement of sucrose by fructose, which is much cheaper than sucrose, as a sweetener for prepared foods and beverages in North America. The source of starch for commercial fructose production is generally corn, and the final product is called high-fruc-

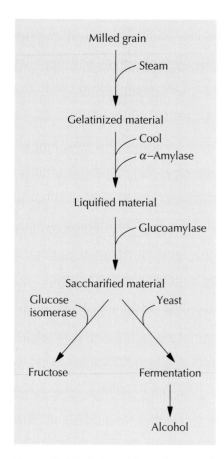

Milled grain

↓ — Steam

Gelatinized material

↓ — Cool
— α–Amylase

Liquified material

↓ — Glucoamylase

Saccharified material

Glucose isomerase ↙ ↘ Yeast

Fructose Fermentation

↓

Alcohol

Figure 13.10 Industrial production of fructose and alcohol from starch.

tose corn syrup or high-fructose syrup, although it contains about half fructose and half glucose.

The enzyme α-amylase randomly hydrolyzes α-1,4-linkages in both amylose and amylopectin chains, yielding a mixture of glucose, maltose (two glucose molecules joined by an α-1,4-linkage), maltotriose (three glucose molecules joined by α-1,4-linkages), and a series of α-limit dextrins, which are the portions of the amylopectin chains that contain cross-links (Fig. 13.9). Although α-amylase can be isolated from a variety of microorganisms, it is commonly obtained from *Bacillus amyloliquefaciens* for industrial purposes.

For some applications, the enzyme β-amylase is used in addition to or in place of α-amylase to digest starch. By hydrolyzing alternate α-1,4-linkages from the ends of amylose and amylopectin chains, β-amylase cleavage yields maltose residues and various β-limit dextrins.

The enzyme glucoamylase hydrolyzes α-1,3-, α-1,4-, and α-1,6-linkages; however, because it is less efficient than α-amylase in cleaving α-1,4-linkages, it is usually used in conjunction with α-amylase. The major role of glucoamylase is digestion of the cross-links of amylopectin, which results in its complete breakdown to glucose. Glucoamylase and other enzymes are used to reduce the carbohydrate (limit dextrin) content of normal beers to produce the so-called light and dry varieties. Although glucoamylase digestion is usually performed prior to the onset of the fermentation, the two steps may be combined. A number of organisms can produce glucoamylase, but it is usually obtained from the fungus *Aspergillus niger*.

Improving Fructose and Alcohol Production

The enzymes that are used in the production of alcohol or fructose from milled grain are major components of the overall cost of the process. These enzymes are often used only once and then discarded. Thus, innovative approaches to the inexpensive large-scale production of these enzymes could be a factor in lowering the cost of alcohol or fructose production. There are several ways to achieve this end.

- Each of the enzymes could be overproduced in a fast-growing recombinant microorganism that utilizes an inexpensive substrate. This alternative would cost less than enzyme production from native organisms.

- Variants of α-amylase (either naturally occurring or genetically manipulated) could be used that function efficiently at 80 to 90°C and allow the liquefaction step to be performed at this temperature. This alternative would speed the hydrolysis of gelatinized starch while decreasing the amount of energy required to cool the gelatinized starch to a temperature suitable for starch hydrolysis.

- The α-amylase and glucoamylase genes could be altered so that each enzyme would have the same temperature and pH optimum, thereby enabling the liquefaction and saccharification steps to be performed under the same conditions.

- An enzyme that could efficiently degrade raw starch could be found or engineered, obviating the need for the gelatinization step and saving a large amount of energy.

- A fermentation organism that can synthesize and secrete glucoamylase could be developed, eliminating the need to add it during fermentation.

A considerable amount of research has been initiated to determine whether these possibilities are feasible.

Genes that code for α-amylase have been isolated from a number of different organisms, including *B. amyloliquefaciens* and the high-temperature-tolerant bacterium *Bacillus stearothermophilus*. Briefly, chromosomal DNA was isolated, partially digested with the restriction enzyme *Sau*3AI, and then ligated to *Bam*HI-digested pUB110 DNA. This plasmid has a unique *Bam*HI site and carries a kanamycin resistance gene. The clone bank was transformed into *Bacillus subtilis*, which does not have α-amylase activity, and transformants were selected for resistance to kanamycin. All transformants were tested for the production and secretion of α-amylase as follows. After the transformants had formed colonies at 65°C on solid medium containing starch, the plates were exposed to iodine vapor. The colonies producing α-amylase were surrounded by a distinctive halo or clear zone, indicating that the starch in the immediate vicinity of these cells had been hydrolyzed. A positive starch–iodine test signifies that the α-amylase gene is transcribed from its own promoter, because the vector does not carry a promoter, and that a secretion signal is present, because the substrate is too large to enter the cell. The availability of α-amylase genes from varied sources will enable researchers to carry out specific genetic modifications that suit the needs of specific industrial processes.

With the aim of bypassing the saccharification step during the production of alcohol from starch, researchers isolated a full-length glucoamylase cDNA from the fungus *Aspergillus awamori* and cloned it into a *Saccharomyces cerevisiae* plasmid under the control of the promoter and transcription terminator regulatory signals from the yeast enolase (*ENO1*) gene. A laboratory strain of *S. cerevisiae* that was transformed with the plasmid carrying the glucoamylase cDNA was able to express this activity and to ferment soluble starch to alcohol, thus demonstrating that this approach is feasible.

MILESTONE

Microorganisms Having Multiple Degradative Energy-Generating Plasmids and Preparation Thereof

A. M. CHAKRABARTY
U.S. patent 4,259,444, 1981

Before the advent of recombinant DNA technology, one of the ways that DNA could be moved from one microorganism to another was by conjugation, in which entire plasmids were transferred from one organism to another. Chakrabarty transferred degradative plasmids—plasmids encoding all of the enzymes involved in the biodegradation of a particular compound—from one bacterium to another until he had constructed bacterial

strains that contained several degradative plasmids. Each degradative pathway encoded by the genes on a plasmid degraded a different organic molecule. Starting with four separate bacteria, Chakrabarty constructed a single bacterium that contained the pathways to degrade camphor, octane, salicylate, and naphthalene. While this work was scientifically interesting and innovative in and of itself—especially since it was conducted in the early

1970s, before the development of most of the techniques of recombinant DNA that are now taken for granted—the key issue for the biotechnology industry was that this invention was awarded a U.S. patent in March 1981, nearly 9 years after the application was first filed. Following the landmark decision by the U.S. Supreme Court in Chakrabarty's favor, it was ruled that genetically engineered microorganisms were inventions that could be patented in the same way that any other invention could be patented. As much as any key scientific experiment, this court decision and the subsequent award of the patent has become an essential component in boosting the development of the biotechnology industry.

Unfortunately, this laboratory strain of *S. cerevisiae* has a number of properties that make it ill suited for use in a commercial process, including an inability to tolerate high levels of alcohol, inefficient expression of the glucoamylase cDNA, and loss of plasmids unless special conditions are used for their maintenance (selective pressure). These problems, however, are not insurmountable. First, the level of glucoamylase expression was increased approximately fivefold by deleting a 175-base-pair (bp) negative regulatory region from the *ENO1* promoter on the plasmid. Second, the plasmid was modified by deleting its yeast origin of replication and adding a segment of DNA that is homologous to a yeast chromosomal site, thereby converting it to an integrating vector. With this form of the plasmid, the complete glucoamylase construct was incorporated into a chromosomal site and stably maintained. Third, another *S. cerevisiae* strain (brewer's yeast) that tolerates high levels of alcohol was used as the host cell. The integrating vector was used to transform this yeast strain.

As a result of these modifications, the investigators created two novel yeast strains that performed better than a natural amylolytic (starch-hydrolyzing) yeast, *Saccharomyces diastaticus*, which is closely related to *S. cerevisiae* and can hydrolyze and ferment soluble starch (Table 13.4). The performance of the brewer's strain, which contained the integrated glucoamylase gene, was superior to the performance of the laboratory strain with the same gene on a multicopy plasmid, a result possibly reflecting plasmid instability and loss of the introduced glycoamylase gene. Neither the laboratory strain nor the brewer's strain of *S. cerevisiae* was able to utilize soluble starch unless they were transformed with the cloned glucoamylase gene. In both the plasmid and integrated form, the *A. awamori* glucoamylase cDNA was under the control of *ENO1* regulatory signals from which the 175-bp negative regulatory region had been removed. The plasmid was maintained under selective pressure.

In an effort to produce large amounts of glucoamylase, several copies of the glucoamylase gene were integrated into the chromosomal DNA of the fungus *A. niger*. Surprisingly, there was no correlation between the number of copies of the glucoamylase gene in the chromosomal DNA and the amount of measured enzyme activity. On the other hand, the level of enzyme activity was strongly dependent upon the sites where the genes were inserted. Thus, merely increasing the gene copy number is not sufficient to produce a larger amount of active enzyme.

The enzyme glucose isomerase should really be called xylose/glucose isomerase, because it primarily catalyzes the conversion of the five-carbon sugar D-xylose to D-xylulose, with the conversion of D-glucose to D-fructose being a secondary or side reaction. Kinetically, xylose/glucose isomerase

Table 13.4 Fermentation of soluble starch (25%, wt/vol) by various yeast strains

Strain	Carbohydrate utilized (%)	Ethanol produced (g/L)	Ethanol yield (g/g of substrate)
Laboratory	5	<0.1	0
Laboratory + plasmid	68	75.6	0.41
Brewer's	<1	3.1	0
Brewer's + integrated gene	93	118.2	0.48
S. diastaticus	43	44.2	0.38

Adapted from Cole et al., *Bio/Technology* **6**:417–421, 1988.

has a lower k_{cat} (catalytic rate constant) and a higher K_m (binding constant) for glucose than for xylose, which means that xylose is bound more tightly to the enzyme than is glucose and that xylose is converted more rapidly to xylulose than glucose is converted to fructose.

Xylose/glucose isomerases are intracellular enzymes and, as such, do not yield the same quantities or purity of product as do the extracellular, or secreted, enzymes that are used in many industrial processes. Most industrial enzymes are used without any extensive purification. An extracellular enzyme preparation generally contains many fewer proteins than an intracellular extract does. In addition, the preparation of an intracellular protein extract requires separation of the cells from the growth medium, mechanical disruption of the cells, and removal of cell debris following disruption. These factors lead to higher production costs for xylose/glucose isomerase than for many other industrial enzymes. One way to overcome this cost problem is to use a batch of xylose/glucose isomerase more than once. This recycling can be achieved by immobilizing the enzyme on a solid support, which both stabilizes the enzyme and facilitates its reuse.

The isomerization of glucose to fructose is a reversible reaction, and the final fructose content is dependent on the reaction temperature. The higher the temperature, the greater the fructose content in the final product. Most commercial processes use conversion temperatures of around 60°C. As a consequence, increasing the temperature optimum for the enzymatic activity and thermostability of xylose/glucose isomerase is one way to make it more efficient.

The thermophilic bacterium *Thermus thermophilus* produces a xylose/glucose isomerase that not only is active at 95°C but also is very stable at high temperatures. Therefore, this enzyme is a good candidate for use in industrial processes. However, unfortunately, wild-type *T. thermophilus* does not produce large amounts of this enzyme. To circumvent this problem, the *T. thermophilus* xylose/glucose isomerase gene was isolated and expressed in *E. coli* and *Bacillus brevis* under the control of various promoters and ribosome binding sites (Table 13.5). One of the constructs (the last one listed in Table 13.5) overproduced xylose/glucose isomerase more than 1,000-fold relative to the amount found in the original organism. Therefore, with this construct, high yields of a thermostable xylose/glucose isomerase can be produced for the industrial synthesis of fructose from glucose.

Table 13.5 Amounts of *T. thermophilus* xylose/glucose isomerase in different bacteria

Species	Plasmid copy number	Promoter source	Source of ribosome binding site	Enzyme activity (units/L)
T. thermophilus	None	T. thermophilus	T. thermophilus	20
E. coli	200	E. coli lac	T. thermophilus	190
E. coli	20	E. coli tac	T. thermophilus	1,790
E. coli	20	E. coli tac	E. coli	3,260
E. coli	20	Phage T7 f10	E. coli	7,050
B. brevis	20	B. brevis cwp	T. thermophilus	1,400
B. brevis	20	B. brevis cwp	B. brevis	25,000

Adapted from Dekker et al., *Appl. Microbiol. Biotechnol.* **36:**727–732, 1992.

The first row presents data for the original enzyme-producing strain. All the other strains are transformants carrying the *T. thermophilus* xylose/glucose isomerase gene on a multicopy plasmid.

Table 13.6 Catalytic efficiency of wild-type and mutant xylose/glucose isomerases from *C. thermosulfurogenes*

Amino acid changes	Catalytic efficiency (k_{cat}/K_m) (min^{-1} mM^{-1}) toward:	
	Glucose	**Xylose**
None (wild type)	5.8	97.2
Trp-139 → Phe	15	13.6
Val-186 → Thr	9.7	55.4
Trp-139 → Phe/Val-186 → Thr	32.9	21.6

Adapted from Meng et al., *Proc. Natl. Acad. Sci. USA* **88:**4015–4019, 1991.

In addition, the substrate specificity of the enzyme can be enhanced. In one series of experiments, site-directed mutagenesis was used to change the nucleotides encoding either one or two amino acids of the xylose/glucose isomerase from the thermophilic organism *Clostridium thermosulfurogenes*. The targeted sites were selected for modification because of other evidence indicating that the corresponding amino acids were involved in substrate binding. Changing the tryptophan at amino acid residue 139 to phenylalanine or the valine at amino acid residue 186 to threonine produces a 1.7- to 2.6-fold increase in the catalytic efficiency (k_{cat}/K_m) of the enzyme toward glucose (Table 13.6). Moreover, these changes cause a two- to sevenfold reduction in the k_{cat}/K_m values of the enzyme toward xylose. When an enzyme has the two amino acid changes, the k_{cat}/K_m value for glucose increases by 5.7-fold and the k_{cat}/K_m value for xylose decreases by 4.5-fold. The double amino acid modification changes an enzyme that was initially 17 times more reactive with xylose than with glucose to one that is now 1.5 times more reactive with glucose than with xylose. The shift in specificity that has been achieved, together with the thermostability of this xylose/glucose isomerase, should make it attractive for use in the industrial conversion of glucose to fructose.

Zymomonas mobilis

Although industrial fermentations to produce alcohol are performed almost exclusively with *S. cerevisiae*, the bacterium *Zymomonas mobilis* is a potentially useful organism for the commercial production of alcohol. *Zymomonas* is a gram-negative, rod-shaped organism that can ferment glucose, fructose, and sucrose, thus producing relatively high yields of alcohol (Table 13.7). This high yield of ethanol is probably related to the fact that *Zymomonas* does not proliferate extensively (i.e., produce biomass) during fermentation. Since *Zymomonas* uses less substrate for biomass formation, more is available for ethanol production. In this regard, yeast produces 2 mol of ATP per mol of glucose, whereas *Zymomonas* uses a different pathway and produces only 1 mol of ATP per mol of glucose. Historically, *Zymomonas* has been used in the tropical regions of the world as a fermentative agent for the production of alcoholic beverages. For example, it is the fermentative agent in a popular Mexican drink called pulque, which is a milky viscous beverage containing 3 to 5% alcohol and is made from the sap of the agave plant.

Zymomonas produces alcohol at a much higher rate than does *S. cerevisiae*, even though these organisms are similar in other features (Table

Table 13.7 Comparison of *Z. mobilis* and *S. cerevisiae* as alcohol producers

Attribute	Value for:	
	Z. mobilis	Yeast
Conversion of sugar to ethanol (%)	96	96
Maximum ethanol concentration (%)	12	12
Ethanol productivity rate (g g^{-1} h^{-1})	5.67	0.67
Volumetric ethanol productivity rate (g L^{-1} h^{-1})	200	29
Sugar tolerance (%)	>40	>40
pH range for ethanol production	3.5–7.5	2–6.5
Optimum temperature (°C)	25–30	30–38

Adapted from Buchholz et al., *Trends Biotechnol.* **5:**199–204, 1987.

The ethanol productivity rate was measured under batch fermentation conditions. The volumetric ethanol productivity rate was measured during continuous culture. Both strains yielded the same maximum ethanol concentration (12%) and had the same sugar tolerance (>40%).

13.7). However, there are biological and technical constraints that prevent *Zymomonas* from being used more widely for alcohol production. First, it can use only a limited number of carbon substrates for the production of alcohol. Second, broad-host-range cloning vectors and, as a consequence, foreign genes are difficult to maintain in this species. Third, *Zymomonas* is naturally resistant to many of the more commonly used antibiotics; this precludes the use of the standard antibiotic resistance marker systems for cloning experiments.

Despite these difficulties, scientists have successfully introduced and expressed a number of foreign genes in *Zymomonas*. Many of these experiments have focused on expanding the range of substrates that *Zymomonas* can utilize. For example, genes encoding enzymes that hydrolyze lactose, starch, cellulose, xylose, and cellobiose have all been introduced into *Zymomonas*. Transformants were able to express all of these genes to some extent; however, in most of these cases, the transformed bacterium was unable to utilize the novel substrate as the sole carbon source. It is not a straightforward matter to engineer additional metabolic capabilities into *Zymomonas*, although in at least one instance *Zymomonas* has been used to convert waste materials that contain xylose into alcohol.

In early studies directed toward developing strains of *Z. mobilis* capable of growth and ethanol production with xylose as a substrate, this bacterium was transformed with genes encoding the xylose utilization enzymes glucose/xylose isomerase and xylulokinase. However, these transformants were limited by their inability to further metabolize the pentoses (xylulose-5-phosphate, ribulose-5-phosphate, and ribose-5-phosphate) that formed following the assimilation of xylose. To remedy this situation, *Z. mobilis* was transformed with a plasmid carrying two synthetic operons, one with the two xylose assimilation genes and one with two pentose metabolism genes (Fig. 13.11). The pentose metabolism genes encoded the enzymes transketolase and transaldolase. The xylose assimilation genes were placed under the transcriptional control of a strong constitutive promoter from the *Z. mobilis* gene for glyceraldehyde-3-phosphate dehydrogenase. The pentose metabolism genes were placed under the control of the *Z. mobilis* enolase promoter. These two synthetic operons were cloned onto an *E. coli*–*Z. mobilis* shuttle vector, which was then used to transform *Z. mobilis*. As expected, these transformants assimilated xylose and converted the pentoses that formed to fructose-6-phosphate and glyceraldehyde-3-phosphate,

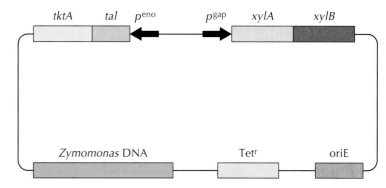

Figure 13.11 *Zymomonas–E. coli* shuttle vector carrying one operon with genes encoding enzymes used for xylose assimilation (*xylA* and *xylB*) and another with genes encoding enzymes involved in pentose metabolism (*tktA* and *tal*). The arrows indicate the direction of transcription from each promoter. Abbreviations: p^{eno}, enolase promoter; p^{gap}, glyceraldehyde-3-phosphate dehydrogenase promoter; *xylA*, xylose isomerase gene; *xylB*, xylulokinase gene; *tktA*, transketolase gene; *tal*, transaldolase gene; Tetr, tetracycline resistance gene; oriE, *E. coli* origin of replication. The *Zymomonas* DNA contains a *Zymomonas* origin of replication.

metabolites that were readily converted to ethanol by the Entner-Doudoroff pathway of *Z. mobilis*. Moreover, the transformants grew efficiently on both glucose and xylose, as well as on glucose–xylose mixtures, and converted xylose to ethanol at high yield. This work demonstrates the feasibility of metabolically engineering *Z. mobilis* as an ethanol producer by using xylose, a waste material produced as a by-product of industrial processes such as pulp and paper making, as a carbon source.

Silage Fermentation

Crops such as grasses, corn, and alfalfa need to be preserved so that they can be used as animal feed many months after the crop is harvested. Traditionally, these crops are preserved by naturally occurring lactic acid bacteria that use the crop as a fermentation substrate to produce lactic and acetic acids. The resulting low pH restricts the growth and metabolic activity of other microorganisms and ensures that the crop is preserved. This preservation strategy is called the making of silage. Often, the numbers of lactic acid bacteria that are found on a fresh crop are quite small, so that a bacterial inoculum (typically *Lactobacillus plantarum*) must be added. Unfortunately, these bacterial inoculants are not especially effective when the amount of water-soluble carbohydrates in the fresh crop is insufficient to support both bacterial growth and lactic acid production.

To develop a bacterium that might be effective in silage fermentation, an α-amylase gene from a strain of *Lactobacillus amylovorus* that does not support silage fermentation was spliced into the chromosomal gene for conjugated acid bile hydrolase (*cbh*) in a strain of *L. plantarum* (Fig. 13.12). The *cbh* gene encodes an enzyme that is active when the bacterium is located in an animal's intestine. Therefore, the gene is dispensable when the bacterium is grown on silage. This work is an important first step in the development of *L. plantarum* strains that are more effective in the fermentation of silage from crops such as alfalfa, which contain a high level of starch.

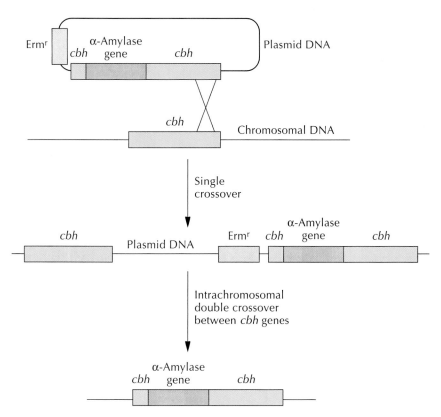

Figure 13.12 Chromosomal integration of an α-amylase gene into *L. plantarum*. The α-amylase gene is cloned into the *L. amylovorus cbh* gene on an *E. coli–L. plantarum* shuttle plasmid, which is then used to transform *L. plantarum*. Erythromycin-resistant and α-amylase-positive clones of *L. plantarum* result from a single crossover between the chromosomal and plasmid DNA at the *cbh* locus. After the growth of transformed *L. plantarum* for approximately 30 generations in the absence of selective pressure, intrachromosomal recombination resulted in the excision of the erythromycin resistance (Erm^r) gene, the chromosomal copy of the *cbh* gene, and the plasmid DNA.

Utilization of Cellulose

The polymers lignin, hemicellulose, and cellulose combine in various proportions to form a "lignocellulosic" structural support system for nearly all terrestrial plants (Table 13.8). This material constitutes a vast biomass that is often a waste product of agriculture, timber processing, and other human activity and needs to be disposed of in a safe and efficient manner or used as a resource. From this perspective, lignocellulosic materials have been grouped into three classes.

- Primary cellulosics include plants that are harvested specifically for cellulosic content, structural use, or feed value, e.g., cotton, timber, and hay.
- Agricultural waste cellulosics are the plant materials that remain after harvesting and processing, e.g., straw, corn stovers, rice hulls, sugar cane bagasse, animal manures, and timber residues.
- Municipal waste cellulosics encompass waste paper and other discarded paper products.

Table 13.8 Typical composition of various lignocellulosic materials

Raw material	Amount (%) of:		
	Lignin	**Cellulose**	**Hemicellulose**
Pine wood	27.8	44.0	26.0
Birch wood	19.5	40.0	39.0
Sugar cane bagasse	18.9	33.4	30.0
Rice straw	12.5	32.1	24.0
Cotton	None	80–95	5–20

Adapted from Brown, *Philos. Trans. R. Soc. Lond. B* **300**:305–322, 1983.

Components of Lignocellulose

Lignin is a three-dimensional, globular, irregular, insoluble, high-molecular-weight (>10,000) polymer made up of phenylpropane subunits with no chains of regular repeating units or any bonds that are easily hydrolyzed either enzymatically or chemically (Fig. 13.13). The lignin polymer molecule has many different types of chemical linkages between aromatic phenyl-

Figure 13.13 Representative lignin structure, showing some of the various possible linkages between the phenylpropane (a C_6 aromatic group attached to a C_3 alkyl chain) units. The phenylpropane units are linked in an unorganized, nonrepeating fashion.

propane units. The physical and chemical characteristics of lignin are generally attributed to the last step in lignin biosynthesis, the nonenzymatic free-radical-based joining of the phenylpropane units in a more or less random fashion. In plants, lignin is chemically bonded to hemicellulose and wraps around fibers composed of cellulose. Lignin is responsible for the rigidity of plants and for their resistance to mechanical stress and microbial attack.

Hemicelluloses are short-chain, heterogeneous polymers that contain both hexoses (six-carbon sugars, such as glucose, mannose, and galactose) and pentoses (five-carbon sugars, such as xylose and arabinose). The three major types of hemicelluloses are xylans, which have a backbone of poly-β-1,4-xylan, with side links to arabinose, glucuronic acid, and arabinoglucuronic acid; mannans, which are composed of glucomannans and galactomannans; and arabinogalactans. The origin of the lignocellulosic material usually defines the nature of the hemicelluloses. For example, xylan hemicellulose is particularly common in hardwoods, and glucomannans are characteristic of softwoods.

Cellulose, which is the simplest of the components found in lignocellulosic material, is the most abundant polymer in the biosphere. It is composed of long chains of D-glucose molecules linked in a β-1,4 configuration (Fig. 13.14). Both cellulose and starch can be hydrolyzed to glucose, but their structures are very different. Starch is an energy storage molecule in which the glucose residues are linked in a manner that prevents a tightly ordered arrangement of the polymer chains. This open mesh-like structure is easily penetrated by water; as a result, starch is both water soluble and readily hydrolyzable by amylases and glucoamylases. By contrast, cellulose is a plant-supporting structural molecule. The glucose chains in cellulose are arranged in a manner that permits them to pack together in a crystal-like structure that is impervious to water. Consequently, the cellulose polymer is both insoluble and resistant to hydrolysis.

Nevertheless, cellulose is still a form of stored glucose, so it is the component of lignocellulosics that has the most potential for conversion into a variety of useful compounds, such as alcohol. However, before cellulose can be utilized, it must be released from its complex with lignin and hemicellulose. For most lignocellulosic materials, this separation requires treatment with either a strong acid or a strong base or the use of high temperature and pressure. Regardless of how the cellulose is separated from the lignocellulose complex, the energy that is necessary to achieve this adds significantly to the cost of the final product.

The annual production of lignocellulosic materials is huge, and effective ways of enzymatically degrading cellulose and, to a lesser extent, hemi-

Figure 13.14 Structure of a portion of a cellulose chain. Glucose residues are joined head to tail by β-1,4-linkages.

cellulose are being sought. Chemical and enzymatic methods for the selective degradation of lignin also are being investigated, with less success.

Isolation of Prokaryotic Cellulase Genes

A wide range of bacteria and fungi are naturally capable of degrading cellulose through the concerted action of several enzymes collectively referred to as cellulase. In a number of microorganisms, cellulase activity is found as part of a multiprotein complex that is called a cellulosome. This complex lies on the external surface of the cell and contains several copies of each of the following enzymes:

- endoglucanase, which hydrolyzes β-1,4-linkages between adjacent glucose molecules within the amorphous (loosely packed) regions of the cellulose polymer, thereby breaking the chain in the middle (Fig. 13.15);
- exoglucanase, which degrades the nicked cellulose chains from their nonreducing ends and produces glucose, cellobiose (i.e., two glucose units), and cellotriose (i.e., three glucose units);
- cellobiohydrolase, which is often found in cellulolytic fungi and is a type of exoglucanase that removes units of 10 or more glucose residues from the nonreducing ends of the cellulose molecule; and
- β-glucosidase, or cellobiase, which converts cellobiose and cellotriose to glucose (Fig. 13.15).

The breakdown of cellulose by cellulolytic organisms is slow and often incomplete. Therefore, genetic engineering strategies have been used to create organisms that have a more effective cellulase activity. For this purpose, genes coding for the individual enzymatic functions of cellulase activity have been isolated from both prokaryotic and eukaryotic organisms.

Prokaryotic endoglucanase genes have been cloned by the following simple yet effective identification technique.

1. A clone bank of DNA from a cellulolytic prokaryote is constructed in *E. coli*, and the host cells are grown overnight on solid medium containing a selective antibiotic.
2. The colonies are then overlaid with agar containing carboxymethyl cellulose (CMC), a soluble derivative of cellulose, and the petri plates are incubated at 37°C for several additional hours. During this time, the carboxymethyl cellulose molecules that are present in the immediate vicinity of a colony that both synthesizes and secretes an endoglucanase will be partially digested. Transformants that synthesize but do not secrete the cloned endoglucanase will not be able to degrade the substrate, which cannot enter the cell because of its large size.
3. The digested regions of the CMC are visualized by flooding the petri plate with a solution of the dye Congo red, which is not toxic to the bacteria, and then with a solution of sodium chloride. Congo red selectively binds to high-molecular-weight cellulose chains and gives a red color; conversely, it binds weakly to low-molecular-weight polysaccharides and produces a yellow hue. The sodium chloride treatment stabilizes the colors of the stain. If a bacterial colony produces a secreted endoglucanase, it will be surrounded by a yellow halo; the background, where the CMC has not been degraded, will be red.

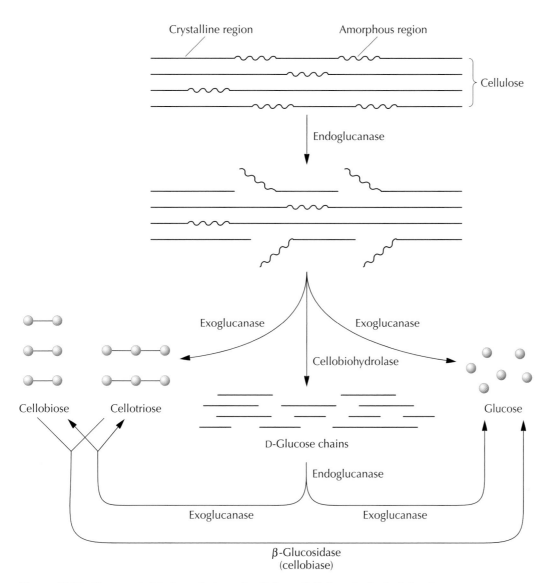

Figure 13.15 Enzymatic biodegradation of cellulose. Cellulose hydrolysis begins with the cleavage of β-1,4-linkages within the accessible amorphous regions of the cellulose chains by endoglucanase(s). This reaction is followed by the removal of oligosaccharides from the reducing ends of the partially cleaved cellulose chains by exoglucanase(s) and cellobiohydrolase(s). The degradation of cellulose is completed when the cellobiose and cellotriose are converted to glucose by β-glucosidase.

This technique has been successfully used in the isolation of endoglucanase genes from *Streptomyces, Clostridium, Thermoanaerobacter, Thermomonospora, Erwinia, Pseudomonas, Cellvibrio, Ruminococcus, Cellulomonas, Fibrobacter,* and *Bacillus* species.

There is no convenient plate assay for detecting cells with a cloned exoglucanase gene, and so immunological screening has been used to pick out the recombinant clones that express exoglucanase. Although this approach requires specific antibodies directed against the target protein, the protein does not have to be secreted to be detected. Recombinant cells can be lysed in situ (e.g., by exposure to chloroform vapor) before the cytoplasmic proteins are transferred to a nylon or nitrocellulose membrane for subsequent

immunological testing. In these tests, replica plates are used to ensure that viable cells are available for further propagation and use.

Prokaryotic β-glucosidase genes have been isolated by transforming a clone bank from a β-glucosidase-producing microorganism into *E. coli* and then selecting for transformants that can grow on minimal media with cellobiose as the sole carbon source. Alternatively, clones that express β-glucosidase activity can be detected with a chromogenic substrate such as 5-bromo-4-chloro-3-indolyl-β-D-glucopyranoside in the plating medium or by using MacConkey-cellobiose agar. In both cases, β-glucosidase-positive colonies turn red.

Isolation of Eukaryotic Cellulase Genes

The strategy of DNA hybridization screening of either cDNA or genomic clone banks with a heterologous probe has not been particularly effective for isolating cellulase genes, because the sequences of cellulolytic enzymes from diverse sources are not very similar. To isolate the mRNAs that encode cellulolytic enzymes from fungi or plants, a novel protocol had to be implemented. Unfortunately, these mRNAs usually constitute only a small fraction of the total mRNA population. Therefore, it is often necessary to enrich for the target mRNA or cDNA and to eliminate cDNA clones that do not carry the target sequences. To meet these ends, the technique of "differential hybridization" has been used for the isolation of a number of different eukaryotic cellulase genes as follows (Fig. 13.16).

1. mRNA is isolated both from cells grown without cellulose (i.e., noninduced cells) and from cells grown in the presence of cellulose or cellulose derivatives to enhance the synthesis of cellulase enzymes (i.e., induced cells).

2. Each mRNA population is fractionated on a sucrose gradient, and each fraction is translated in a cell-free system (either rabbit reticulocytes or wheat germ) to determine which one(s) contains an mRNA for a cellulase enzyme, using molecular mass and an immunological assay as criteria. A cellulose-induced protein(s) is identified, after separation of the cell-free translation products on a polyacrylamide gel, by the presence of unique bands that appear from the induced cells but not from the noninduced cells.

3. The mRNA fractions from the induced cells that direct the synthesis of cellulase enzymes and the comparable fractions of the sucrose gradient from the noninduced cells are used separately to program the synthesis of cDNA.

4. The cDNA sample from the induced-cell population is cloned into either a plasmid or a bacteriophage vector, introduced into *E. coli*, replica plated, and then separately screened with radiolabeled cDNA from both the induced and noninduced fractions as hybridization probes. Clones that hybridize only with the cDNA from the induced cells and not with cDNA from the noninduced cells potentially carry cellulase genes and are characterized further.

5. To prove conclusively that the positive cDNA clones encode cellulase enzymes, the DNA from these clones is hybridized with total mRNA from induced cells, the selected mRNAs are translated in vitro in a cell-free protein synthesis system, and the products are detected with antibodies to the enzymes of the cellulase complex.

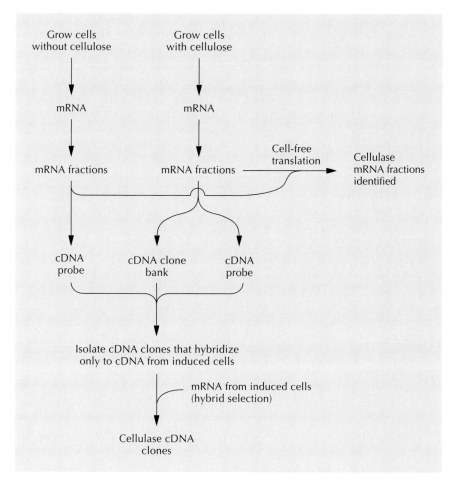

Figure 13.16 Differential hybridization for isolating eukaryotic cellulase cDNA clones.

6. The sequence of each of the positive cDNA clones is determined so that the clones that encode the same protein can be distinguished from those that encode different members of the cellulase complex. In principle, this scheme can be used for the isolation of any induced eukaryotic gene(s).

Manipulation of Cellulase Genes

In the future, there will be a variety of uses for cloned cellulase genes. In some cases, the cellulose binding domain encoded in cellulase genes may facilitate the purification of recombinant proteins. In other instances, the cellulolytic activity can be expressed in organisms that will convert waste cellulose into a commercial product such as alcohol.

The amino acid chain of a cellulase enzyme typically is divided into three separate domains: a catalytic region; a hinge region that is often rich in proline, serine, or threonine residues; and a cellulose binding region. The catalytic and binding domains act independently. This separation of functions can be exploited by cloning the sequence that encodes the cellulose binding domain as part of a fusion gene, where the other portion of the gene encodes a commercial protein. After expression of this fusion protein, it can be purified by passing a crude extract through a column packed with cellu-

lose. Under these conditions, only the fusion protein will bind to the cellulose. Then the fusion protein, in homogeneous form, can be eluted from the column. The commercial protein can be retrieved by removal by proteolytic cleavage of the cellulose binding domain. This system is similar in principle to immunoaffinity chromatography, except that it should be less expensive than using antibodies.

For convenience, most cellulase genes are initially cloned and expressed in *E. coli*, but other useful microorganisms might be developed by the introduction of cellulase genes. For example, *S. cerevisiae* and *Z. mobilis*, which both efficiently convert simple sugars such as glucose into alcohol, have been used as hosts for the expression of cellulase genes. The idea behind these studies was to test whether the presence of cellulase activity would enable these organisms to convert cellulose directly to alcohol.

In one series of experiments, endoglucanase and exoglucanase genes from the bacterium *Cellulomonas fimi* were each put under the control of an *S. cerevisiae* promoter and signal peptide sequence, subcloned onto the same plasmid vector, and introduced into *S. cerevisiae*. Some transformants secreted about 70% of each activity into the growth medium and were able to degrade the cellulose in filter paper and pretreated wood chips. The rate and extent of hydrolysis of both of these substrates were increased by adding β-glucosidase to the mixture, thereby decreasing the amount of cellobiose that accumulates and preventing end product (feedback) inhibition of the endoglucanase and exoglucanase activities by cellobiose. The role of β-glucosidase in the cellulolytic process has been examined in more detail. Cellobiose acts as a feedback inhibitor of cellulose hydrolysis, and glucose inhibits cellobiose cleavage. These two regulatory mechanisms may prevent complete enzymatic breakdown of cellulose. A β-glucosidase gene was isolated from the fungus *Trichoderma reesei*, cloned onto a multiple-copy plasmid, and reintroduced into *T. reesei*. The transformant strain overproduced β-glucosidase activity 5.5-fold, and it degraded Avicel, a cellulose derivative, 33% faster than the nontransformed strain did. These results support the observation that the presence of β-glucosidase enhances the enzymatic utilization of cellulose and suggest that a useful strategy for genetically engineering more effective cellulolytic microorganisms may involve cloning of β-glucosidase genes into existing cellulolytic organisms.

In addition to converting cellulosic wastes into useful materials, endoglucanase genes may have some novel uses. For example, when a wine-making yeast was transformed with an endoglucanase gene under the control of the constitutively expressed yeast actin gene promoter, the wine that was produced had an increased fruity aroma. This improvement is attributed to an increase in the amounts of at least 12 different volatile compounds including ethyl proprionate, 2-butanol, isoamyl acetate, isoamyl alcohol, and isobutyric acid. This type of genetic modification should decrease the variability that is often found in fermentations. Instead, it opens the possibility of genetically engineering yeast strains that yield wines with particular desirable characteristics.

The participation of cellulase enzymes in an industrial process for the bioconversion of waste paper to alcohol has been examined. Waste paper was partially digested by the addition of cellulase enzymes at 45°C; then the released glucose was fermented by *S. cerevisiae* at 37°C without removing the added cellulase enzymes. On the basis of small-scale results, researchers calculated that yields of 400 liters of ethanol per ton of waste paper were obtainable by this approach. If all 100 million tons of waste paper generated

annually in North America were converted into ethanol and used as fuel, approximately 16% of the gasoline that is currently being used in North America could be saved.

Production of Single-Cell Protein

Single-cell protein (SCP) is the term that refers to a monoculture of microbial cells or the total protein extracted from pure cell cultures, either of which can be used as a human or animal protein supplement. SCP that is deemed suitable for human consumption is defined as food grade; when it is acceptable for animal but not human consumption, it is said to be feed grade. The use of microbial biomass as a food source deserves serious consideration because of the insufficient world food supply and the high protein content of most microorganisms—approximately 60 to 80% of the dry cell weight. In addition, the high levels of methionine, lysine, vitamins, and essential minerals often make SCP more nutritious than some plant and animal foods. However, there are drawbacks to the widespread use of SCP.

- The high nucleic acid content of SCP may be hazardous to the health of individuals with certain disorders.
- The potential presence of toxic substances that are either absorbed from the growth substrate (e.g., heavy metals) or produced by the microorganism (e.g., mycotoxins) requires costly quality control analyses.
- The slow digestion of microbial cells in the digestive tract may lead to indigestion or allergic reactions in some individuals.
- SCP is more expensive than other sources of protein, such as soy meal.

A variety of different microorganisms including bacteria, yeasts, fungi, algae, and actinomycetes and a range of different substrate materials have been used to produce SCP (Table 13.9). The first deliberate production of SCP in modern times took place in Germany during World War I, when *S. cerevisiae*, grown on molasses as a carbon source and ammonium salts as a nitrogen source, was used as a thickening agent in soups and sausages.

Until 1973, petroleum was considered to be a cheap and plentiful resource. As a consequence, several major oil companies initiated projects to produce SCP with petroleum or refinery by-products as a growth medium. However, interest in these projects flagged as the price of oil increased. In the 1970s, Imperial Chemical Industries (ICI) in England successfully developed a continuous methanol fermentation process for the commercial production of SCP (which was called Pruteen) from the bacterium *Methylophilus methylotrophus*. This bacterium can utilize methane as a substrate for

Table 13.9 Substrates and organisms used for SCP production

Raw material	Microorganism	Type of organism
Carbon dioxide	*Spirulina maxima*	Cyanobacterium
Whey (lactose)	*Kluyveromyces fragilis*	Yeast
Petroleum alkanes	*Candida lipolytica*	Yeast
Cellulosic wastes	*Chaetomium cellulolyticum*	Fungus
Methane (methanol)	*Methylophilus methylotrophus*	Bacterium

growth, although in practice, the methane is converted into methanol, which is used as the principal substrate. In 1979, ICI constructed a plant capable of producing up to 50,000 tons of SCP per year. This facility included the world's largest continuous airlift fermenter. However, despite a very considerable investment by ICI of more than $200 million as well as the considerable technical achievements from both engineering and biological perspectives, by 1987 little or no SCP was being produced in this plant because SCP production from *M. methylotrophus* had become uneconomical.

There has recently been renewed interest in SCP as a means of utilizing various waste materials (e.g., cellulosics and whey). Some of these schemes use naturally occurring microorganisms, whereas others envision the use of genetically manipulated microorganisms. However, regardless of the nature of the microorganism, economic considerations are likely to be the major determinant of success or failure of SCP commercialization. Perhaps an economical process could be developed to produce SCP from the by-products of a waste treatment process. To facilitate such an eventuality, scientists need to study the growth kinetics, metabolism, genetic manipulation, palatability, and safety of a number of different organisms.

A novel approach to providing cattle with protein enriched in essential amino acids has been devised. Since most of the protein in the diets of dairy cattle comes not from the feed but from the microorganisms in the rumen, it seemed reasonable that the diet could be enriched for essential amino acids by genetically engineering rumen bacteria. Supplementation of cattle feed with proteins that are rich in essential amino acids is expensive and not especially effective, because the proteins and amino acids are degraded by the rumen bacteria before the cattle can utilize them. As a first step, a protein with a large percentage of methionine, threonine, lysine, and leucine residues and a stable structure was designed. This artificial protein has 100 amino acids, of which 57 are essential amino acids, arranged in an α-helical bundle structure. A gene for this protein was synthesized by using a total of 14 overlapping oligonucleotides, ligated to the maltose binding protein gene, and expressed in *E. coli* as a fusion protein under the transcriptional control of the *tac* promoter. When induced, the fusion protein accounted for approximately 12% of the intracellular protein. The fusion protein is highly enriched in essential amino acids. Whether this protein will be expressed in high yield when it is introduced into a rumen bacterium remains to be determined. If this approach is eventually successful, it will represent a unique system for providing cattle with essential amino acids on a continual basis.

SUMMARY

Bioremediation is the term that is applied to the use of microorganisms to clear the environment of contaminating substances. Many members of the bacterial genus *Pseudomonas* carry plasmids that encode enzymes capable of degrading aromatic and halogenated organic compounds. In most cases, a single plasmid carries the genes encoding enzymes for a specific degradative pathway. By combining plasmids from different pseudomonad strains within a single host, it is possible to create an organism with multiple degradation capabilities. In addition, by genetic manipulation, it is possible to extend the range of substrates degraded by a particular enzymatic pathway.

Raw biological material is called biomass and is often used as a starting material in industrial processes. The use of milled grain for the production of alcohol or fructose requires a number of enzymatic steps. The enzymes that are used in these processes are often used only once and then discarded. To enhance enzymatic conversions and decrease costs, investigators are cloning, characterizing, and testing bacterial genes encoding enzymes that are thermostable, highly efficient catalytically, or tolerant to alcohol.

To improve the commercial production of alcohol, some workers have genetically transformed the bacterium *Zymomonas mobilis* with genes that allow it to utilize a broad range of compounds as carbon sources. Enzymes that degrade starch can also be used to facilitate the ability of microorganisms such as *Lactobacillus plantarum* to ferment silage.

Often, as the consequence of processing biological material, large amounts of lignocellulose remain. This material generally has been treated as a waste product. However, there is now interest in using lignocellulose as a resource for carbon-containing compounds, especially glucose, that can be used in other processes. Retrieving glucose from lignocellulose is not an easy matter. Lignocellulose is a complex of lignin, hemicellulose, and cellulose; without pretreatment, it is usually refractory to enzymatic degradation. Much recent research has focused on characterizing the mechanism of breakdown of cellulose to glucose. The genes for endoglucanases, exoglucanases, and β-glucosidases from a variety of organisms have been cloned and characterized, but to date there has been little success in formulating a set of enzymes that efficiently degrades cellulose in vitro on a large scale.

A number of different types of biomass such as whey, petroleum alkanes, cellulosic wastes, and methane have been used as substrates for the growth of microorganisms in culture. It was hoped that these pure cell cultures, which are called single-cell protein (SCP), could be used as food supplements for humans and animals. Unfortunately, due to high costs, lack of palatability and, occasionally, toxicity, few of the SCP formulations have been commercially successful. However, it may still be possible in some instances to devise a low-cost SCP food supplement by genetic manipulation.

References

Barnett, C. C. 1991. Cloning and amplification of the gene encoding an extracellular β-glucosidase from *Trichoderma reesei*: evidence for improved rates of saccharification of cellulosic substrates. *Bio/Technology* **9**:562–567.

Beauregard, M., C. Dupont, R. M. Teather, and M. A. Hefford. 1995. Design, expression, and initial characterization of MB1, a *de novo* protein enriched in essential amino acids. *Bio/Technology* **13**:974–981.

Béguin, P. 1990. Molecular biology of cellulose degradation. *Annu. Rev. Microbiol.* **44**:219–248.

Brown, D. E. 1983. Lignocellulose hydrolysis. *Philos. Trans. R. Soc. Lond. B* **300**:305–322.

Buchholz, S. E., M. M. Dooley, and D. E. Eveleigh. 1987. *Zymomonas*—an alcoholic enigma. *Trends Biotechnol.* **5**:199–204.

Buchholz, S. E., and D. E. Eveleigh. 1990. Genetic modification of *Zymomonas mobilis*. *Biotechnol. Adv.* **8**:547–581.

Chakrabarty, A. M. March 1981. Microorganisms having multiple compatible degradative energy-generating plasmids and preparation thereof. U.S. patent 4,259,444.

Cole, G. E., P. C. McCabe, D. Inlow, D. H. Gelfand, A. Ben-Bassat, and M. A. Innis. 1988. Stable expression of *Aspergillus awamori* glucoamylase in distiller's yeast. *Bio/Technology* **6**:417–421.

Cork, D. J., and J. P. Krueger. 1991. Microbial transformation of herbicides and pesticides. *Adv. Appl. Microbiol.* **36**:1–66.

Dekker, K., A. Sugiura, H. Yamagata, K. Sakaguchi, and S. Udaka. 1992. Efficient production of thermostable *Thermus thermophilus* xylose isomerase in *Escherichia coli* and *Bacillus brevis*. *Appl. Microbiol. Biotechnol.* **36**:727–732.

Eveleigh, D. E. 1987. Cellulase: a perspective. *Philos. Trans. R. Soc. Lond. B* **321**:435–447.

Fitzsimons, A., P. Hols, J. Jore, R. J. Leer, M. O'Connell, and J. Delcour. 1994. Development of an amylolytic *Lactobacillus plantarum* silage strain expressing the *Lactobacillus amylovorus* α-amylase gene. *Appl. Environ. Microbiol.* **60**:3529–3535.

Ghosal, D., I.-S. You, D. K. Chatterjee, and A. M. Chakrabarty. 1985. Microbial degradation of halogenated compounds. *Science* **228**:135–142.

Glick, B. R., and J. J. Pasternak. 1989. Isolation, characterization and manipulation of cellulase genes. *Biotechnol. Adv.* **7**:361–386.

Innis, M. A., M. J. Holland, P. C. McCabe, G. E. Cole, V. P. Wittman, R. Tal, K. W. K. Watt, D. H. Gelfand, J. P. Holland, and J. H. Meade. 1985. Expression, glycosylation and secretion of an *Aspergillus* glucoamylase by *Saccharomyces cerevisiae*. *Science* **228**:21–26.

Kallio, P., A. Palva, and I. Palva. 1987. Enhancement of α-amylase production by integrating and amplifying the α-amylase gene of *Bacillus amyloliquefaciens* in the genome of *Bacillus subtilis*. *Appl. Microbiol. Biotechnol.* **27**:64–71.

Kennedy, J. F., V. M. Cabalda, and C. A. White. 1988. Enzymic starch utilization and genetic engineering. *Trends Biotechnol.* **6**:184–189.

Knowles, J., P. Lehtovaara, and T. Teeri. 1987. Cellulase families and their genes. *Trends Biotechnol.* **5**:255–261.

Kolenc, R. J., W. E. Inniss, B. R. Glick, C. W. Robinson, and C. I. Mayfield. 1988. Transfer and expression of mesophilic plasmid-mediated degradative capacity in a psychrotrophic bacterium. *Appl. Environ. Microbiol.* **54**:638–641.

Kumar, V., S. Ramakrishnan, T. T. Teeri, J. K. C. Knowles, and B. S. Hartley. 1992. *Saccharomyces cerevisiae* cells secreting an *Aspergillus niger* β-galactosidase grow on whey permeate. *Bio/Technology* **10**:82–85.

Lamed, R., J. Naimark, E. Morgenstern, and E. A. Bayer. 1987. Specialized cell surface structures in cellulolytic bacteria. *J. Bacteriol.* **169**:3792–3800.

Lynd, L. R., J. H. Cushman, R. J. Nichols, and C. E. Wyman. 1991. Fuel ethanol from cellulosic biomass. *Science* **251**:1318–1323.

Meng, M., C. Lee, M. Bagdasarian, and J. G. Zeikus. 1991. Switching substrate preference of thermophilic xylose isomerase from D-xylose to D-glucose by redesigning the substrate binding pocket. *Proc. Natl. Acad. Sci. USA* **88**:4015–4019.

Pérez-González, J. A., R. González, A. Querol, J. Sendra, and D. Ramón. 1993. Construction of a recombinant wine yeast strain expressing β-(1,4)-endoglucanase and its use in microvinification processes. *Appl. Environ. Microbiol.* **59**:2801–2806.

Quax, W. J., N. T. Mrabet, R. G. M. Luiten, P. W. Schuurhuizen, P. Stanssens, and I. Lasters. 1991. Enhancing the thermostability of glucose isomerase by protein engineering. *Bio/Technology* **9**:738–742.

Ramos, J. L., A. Wasserfallen, K. Rose, and K. N. Timmis. 1987. Redesigning metabolic routes: manipulation of TOL plasmid pathway for catabolism of alkylbenzoates. *Science* **235**:593–596.

Suyama, A., R. Iwakiri, N. Kimura, A. Nishi, K. Nakamura, and K. Furukawa. 1996. Engineering hybrid pseudomonads capable of utilizing a wide range of aromatic hydrocarbons and of efficient degradation of trichloroethylene. *J. Bacteriol.* **178**:4039–4046.

Timmis, K. N., R. J. Steffan, and R. Unterman. 1994. Designing microorganisms for the treatment of toxic wastes. *Annu. Rev. Microbiol.* **48:**525–557.

Verdoes, J. C., A. D. van Diepeningen, P. J. Punt, A. J. M. Debets, A. H. Stouthamer, and C. A. M. J. J. van den Hondel. 1994. Evaluation of molecular and genetic approaches to generate glucoamylase overproducing strains of *Aspergillus niger. J. Biotechnol.* **36:**165–175.

Wayman, M., S. Chen, and K. Doan. 1992. Bioconversion of waste paper to ethanol. *Process Biochem.* **27:**239–245.

Windass, J. D., M. J. Worsey, E. M. Pioli, D. Pioli, P. T. Barth, K. T. Atherton, E. C. Dart, D. Byrom, K. Powell, and P. J. Senior. 1980. Improved conversion of methanol to single-cell protein by *Methylophilus methylotrophus. Nature* **287:**396–401.

Winter, R. B., K.-M. Yen, and B. D. Ensley. 1989. Efficient degradation of trichloroethylene by a recombinant *Escherichia coli. Bio/Technology* **7:**282–285.

Witholt, B., M.-J. de Smet, J. Kingma, J. B. van Beilen, M. Kok, R. G. Lageveen, and G. Eggink. 1990. Bioconversions of aliphatic compounds by *Pseudomonas oleovorans* in multiphase bioreactors: background and economic potential. *Trends Biotechnol.* **8:**46–52.

Wong, W. K. R., C. Curry, R. S. Parekh, S. R. Parekh, M. Wayman, R. W. Davies, D. G. Kilburn, and N. Skipper. 1988. Wood hydrolysis by *Cellulomonas fimi* endoglucanase and exoglucanase coexpressed as secreted enzymes in *Saccharomyces cerevisiae. Bio/Technology* **6:**713–719.

Zhang, M., C. Eddy, K. Deanda, M. Finkelstein, and S. Picataggio. 1995. Metabolic engineering of a pentose metabolism pathway in ethanologenic *Zymomonas mobilis. Science* **267:**240–243.

Zylstra, G. J., L. P. Wackett, and D. T. Gibson. 1989. Trichloroethylene degradation by *Escherichia coli* containing the cloned *Pseudomonas putida* F1 toluene dioxygenase genes. *Appl. Environ. Microbiol.* **55:**3162–3166.

REVIEW QUESTIONS

1. How would you genetically engineer a bacterium to degrade trichloroethylene?

2. Outline a protocol that you would use to clone fungal cellulase genes.

3. Delineate the role of α-amylase and glucoamylase in the industrial production of alcohol. How might genetic manipulation of the genes encoding these enzymes be used to improve this process?

4. What is glucose isomerase? Why is it important? How and why would you modify the gene encoding this enzyme?

5. Elaborate some of the advantages and disadvantages of using *Zymomonas mobilis* instead of *Saccharomyces cerevisiae* for alcohol production. How would you improve the industrial performance of *Z. mobilis*?

6. How can *Z. mobilis* be engineered to produce ethanol from xylose?

7. Starting with a *Pseudomonas* strain that can utilize phenol as its sole carbon source at 0°C, a *Pseudomonas* strain that can degrade anthracene to catechol at 35°C, and a *Pseudomonas* strain that can degrade *p*-toluene to protocatechuate at 35°C, suggest a strategy for developing a strain that can utilize phenol, anthracene, or *p*-toluene as its sole carbon source at 0°C.

8. Explain how a *Pseudomonas* strain that carries plasmid pWWO and does

not normally degrade 4-ethylbenzoate can be genetically manipulated to perform the hydrolysis of this compound.

9. Suggest schemes for the isolation of prokaryotic endoglucanase and β-glucosidase genes.

10. What is a "superbug"?

11. How can *Lactobacillus plantarum* be manipulated to improve its ability to ferment silage?

12. How might cattle nutrition be improved by manipulating rumen bacteria?

Plant Growth-Promoting Bacteria

Under natural environmental conditions, successful plant development and high crop yields depend on the genetic constitution of the crop species, the availability of nutrients, the presence of certain beneficial microorganisms, and the absence of pathogenic ones (called phytopathogens, from *phyto*, meaning plant) in the surrounding soil. Beneficial indigenous soil bacteria and fungi include some species that act directly by providing a plant growth-enhancing product and other species that act indirectly. The latter organisms inhibit the growth of pathogenic soil microorganisms, thereby preventing them from hindering plant growth.

In general, microbial plant growth-promoting mechanisms that affect plants directly include fixing atmospheric nitrogen and supplying it to the plant, increasing the availability of and/or sequestering iron and phosphorus from the soil and providing these minerals in a form that can be used by the plant, and synthesizing phytohormones that trigger plant cell proliferation. The indirect stimulation of plant growth occurs when a beneficial microbial strain prevents the growth of a phytopathogenic soil microorganism that could interfere with normal plant growth and development. This action is called antibiosis and can be due either to the depletion of a scarce resource by the beneficial microbial strain or to the production and release by the beneficial strain of a compound that impedes the growth of the phytopathogenic strain.

Most of the recent genetic research directed at creating microbial strains with augmented plant growth-promoting activity has focused on the following four areas of study:

- The molecular basis of nitrogen fixation. This topic has been investigated thoroughly to determine whether it is possible to increase the level of microbial nitrogen fixation and consequently lessen the current dependency on chemical fertilizers for crop plants.

- Root nodule formation by symbiotic bacteria. This process has been studied with the aim of producing genetically engineered bacteria that can outcompete naturally occurring symbiotic bacteria.

- The microbial synthesis of iron-sequestering compounds (siderophores). These reactions are being characterized in the hope that it might be possible to produce beneficial strains that prevent the growth of phytopathogenic microorganisms.
- The pathways of microbial phytohormone production. These pathways are being studied in an effort to develop strains that will produce and release specified levels of selected phytohormones that could stimulate plant proliferation.

Current research in this area mainly deals with plant growth-promoting bacteria rather than fungi. This is at least partly due to the fact that scientists have been unable to grow beneficial fungi in culture, so not only is it difficult to manipulate them in the laboratory, it is also impossible to obtain large amounts of these organisms for inoculation.

Nitrogen Fixation

Nitrogen gas (N_2), which makes up approximately 80% (by volume) of the air that we breathe, cannot be used directly by plants or animals to synthesize essential biological nitrogen-containing compounds such as amino acids and nucleotides. Rather, it must first be converted (fixed) into ammonia. This conversion requires a high input of energy because the triple bond of N_2 ($N \equiv N$) is extremely stable. The energy for the biological fixation of nitrogen comes from the hydrolysis of large amounts of adenosine triphosphate (ATP). Similarly, the chemical (industrial) conversion of N_2 to ammonia uses a considerable amount of energy in the form of high temperature and pressure.

More than 100 million tons of fixed nitrogen are needed annually to sustain global food production. Synthetic (chemically produced) fertilizers account for about half of this nitrogen supply, and most of the remainder is derived from nitrogen-fixing (diazotrophic) bacteria such as *Rhizobium, Frankia, Azospirillum, Azotobacter,* and cyanobacteria. No eukaryote is known to fix nitrogen.

Chemical fertilizers have helped considerably in increasing crop yields, but their continual use has led to pollution problems as a result of runoff and depletion of the nutrient reserves in the soil. Moreover, their cost has been rising steadily. These factors have provided an incentive for developing alternative sources of fixed nitrogen, including the development of diazotrophic microorganisms as "bacterial fertilizers."

A wide range of bacteria can fix nitrogen, and a number of these have potential as crop fertilizers. However, until a bacterial fertilizer has been shown conclusively to be as effective as a chemical formulation, there will be a reluctance to change current practices. For example, soybeans, which constitute the second largest crop in the United States in terms of both cash value and total acres planted, form a beneficial symbiotic relationship with the bacterium *Bradyrhizobium japonicum*. In this symbiosis, the bacteria provide the plant with fixed nitrogen and, in turn, receive from the plant photosynthetically fixed carbon. When plants are inoculated with specific strains of *B. japonicum*, the final yield of plant material can be increased by 25 to 50% and the inoculated plants no longer require the addition of chemically fixed nitrogen. Although approximately 50% of the world's soybean crop is produced in just a few locales in the United States and agricultural practices tend to be similar throughout these locales, at present only a small

fraction of this crop is treated with *B. japonicum*. Most of these farmers continue to depend on the naturally occurring strains of *B. japonicum* in the soil and chemical fertilizers.

Bacterial fertilization as an effective agricultural practice has a dubious reputation. During the 1950s in the Soviet Union, more than 10 million hectares (about 39,000 square miles) of farmland were treated with a diazotrophic bacterial mixture that consisted primarily of *Azotobacter chroococcum* and *Bacillus megaterium*. In these experiments, about 60% of the time, yields of various crops were increased by 10 to 20%. However, these field trials were poorly designed and not replicable, so many researchers were skeptical about the validity of the work and tended to discount the use of bacterial inoculants as fertilizing agents on a large scale. Today, however, economic pressures, environmental concerns, and new research technologies have encouraged researchers to reexamine the usefulness of bacterial fertilization.

The most important microorganisms that are presently used agriculturally belong to two related genera, *Rhizobium* and *Bradyrhizobium*. These bacteria are gram negative, flagellated, and rod shaped, and they form symbiotic relationships with legumes. Each *Rhizobium* and *Bradyrhizobium* species is specific for a limited number of plants and will not interact with plants other than its natural hosts (Table 14.1).

As part of their life cycle, rhizobial bacteria invade plant root cells and initiate a complex series of developmental changes that leads to the formation of a root nodule. Inside the root nodule, the bacteria proliferate and persist in a form that has no cell wall. The bacteria within the nodules fix atmospheric nitrogen by means of the enzyme nitrogenase. The structural and biochemical interactions between a symbiotic rhizobacterium and a host plant are quite intricate and mutually beneficial. Inside a nodule, nitrogenase is protected from the toxic effects of atmospheric oxygen in two ways. First, oxygen does not readily diffuse into a nodule. Second, the oxygen content within a nodule is regulated by the protein leghemoglobin. The heme moiety of this oxygen-binding protein is synthesized by the bacterium, and the globin portion of the molecule is encoded by a plant gene. The plant also provides the bacterium with photosynthetically fixed carbon,

Table 14.1 Plant specificity of various *Rhizobium* and *Bradyrhizobium* species

Bacterial species	Host plant(s)
Bradyrhizobium japonicum	Soybean
Rhizobium meliloti	Alfalfa
Rhizobium leguminosarum bv. *trifolii*	Clover
Rhizobium leguminosarum bv. *viciae*	Pea, broad bean
Rhizobium leguminosarum bv. *phaseoli*	Kidney bean, mung bean
Rhizobium loti	Lotus
Rhizobium huakuii	*Astragalus sinicus*
Rhizobium ciceri	Chickpea
Rhizobium tropici	*Leucaena* spp., *Macroptilium* spp.
Rhizobium galegae	*Galega officinalis*, *G. orientalis*
Rhizobium fredii	Soybean
Rhizobium sp. strain NGR234	Tropical legumes
Rhizobium elti	Kidney bean, mung bean
Bradyrhizobium elkanii	Soybean

which the bacterium requires for growth. For its part, the plant benefits from this symbiotic relationship by receiving fixed nitrogen from the bacterium.

Nitrogenase

The renewed interest in diazotrophs as biological fertilizers overlapped the development of techniques for gene isolation and manipulation and provided the impetus for studying the biochemical and molecular biological aspects of nitrogen fixation. Initially, scientists believed that these studies would lead to the development of improved nitrogen-fixing organisms that would enhance crop yields. Some researchers even went so far as to suggest that bacterial genes for nitrogen fixation might be introduced directly into plants to enable them to fix their own nitrogen. Although this overly optimistic prediction has not materialized, a detailed understanding of the process of nitrogen fixation has emerged. And with this understanding, the possibility of improving the nitrogen-fixing activity of some diazotrophs by genetic manipulation is a little closer to becoming a reality.

Components of Nitrogenase

All known nitrogenases have two oxygen-sensitive components. Component I is a complex of two α-protein subunits (approximately 50,000 daltons each), two β-protein subunits (approximately 60,000 daltons each), 24 molecules of iron, 2 molecules of molybdenum, and an iron–molybdenum cofactor, often called FeMoCo (Fig. 14.1). Component II has two α-protein subunits (approximately 32,000 daltons each), which are not the same as the α-protein subunits of component I, and an as yet undetermined number of iron molecules. The catalysis of nitrogen to ammonia requires the combination of components I and II, a complex of magnesium and ATP, and a source of reducing equivalents (reaction 1; the upward-pointing arrow indicates a gas).

$$N_2 + 8H^+ + 8e^- + 16MgATP \rightarrow 2NH_3 + H_2\uparrow + 16MgADP + 16P_i \quad (1)$$

In addition to fixing nitrogen, the nitrogenase can reduce the gas acetylene to ethylene (reaction 2).

$$H-C\equiv C-H + 2H^+ \rightarrow H_2{=}C{=}C{=}H_2 \quad (2)$$

The measurement by gas chromatography of ethylene production as a function of time provides a convenient assay for nitrogenase activity. This assay can be performed with intact cells in solution (Fig. 14.2), bacteria associated with plant roots, crude cell extracts, or highly purified enzyme prepara-

Figure 14.1 Proposed structure of the iron–molybdenum cofactor bound to a molecule of dinitrogen (N$_2$).

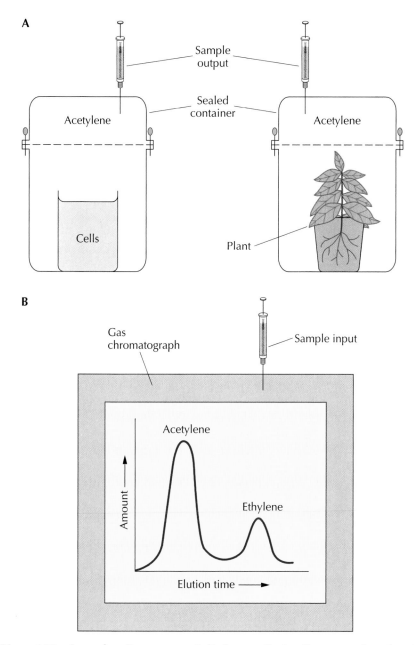

Figure 14.2 Assay for nitrogenase activity by monitoring the conversion of acetylene to ethylene. **A.** The source of the nitrogenase enzyme, for example, bacteria growing in culture, bacteria associated with plant roots, or a purified enzyme preparation (not shown) is placed in a sealed container under an atmosphere of acetylene. **B.** Samples are periodically withdrawn from the sealed container, and the levels of acetylene and ethylene are measured by gas chromatography. The extent of nitrogenase activity is proportional to the amount of ethylene produced.

tions. Component I catalyzes the actual reduction of N_2, and component II donates electrons to component I. Both components are extremely sensitive to oxygen and can be rapidly and irreversibly inactivated when the oxygen concentration is too high. In addition to components I and II, the activity of a complete, functional nitrogenase depends on 15 to 20 additional accessory proteins. The roles of some of the accessory proteins have been delineated

and include the transfer of electrons to component II and the biosynthesis of the iron–molybdenum cofactor that is a part of component I.

Genetic Engineering of the Nitrogenase Gene Cluster

Nitrogen fixation is a very complicated process requiring the concerted action of a large number of different proteins. Therefore, it was not realistic to expect either that an intact single DNA fragment containing all the genetic information for nitrogen fixation could be readily cloned from a diazotrophic microorganism and transferred into a nondiazotrophic organism or that a recipient organism could maintain the physiological conditions needed for nitrogenase activity. Consequently, the most direct way to isolate the genes involved in nitrogen fixation (*nif* genes) was to identify and characterize those clones of a wild-type library that restore nitrogen fixation to various mutants of the original organism. This process is called genetic complementation.

The first *nif* genes identified by complementation were isolated from clone banks of the diazotroph *Klebsiella pneumoniae*. This well-studied enteric organism is found in soil and water and in the human intestine. The isolation protocol comprises the following steps (Fig. 14.3).

1. *K. pneumoniae* cells are treated with a dose of a mutagenic agent that allows approximately 0.1 to 1.0% of the cells to survive. Some of the mutagenized cells are able to grow on a minimal medium containing a source of fixed nitrogen such as NH_4Cl but do not grow in the absence of fixed nitrogen. They are likely to have a mutation in a *nif* gene and are designated Nif$^-$.
2. A clone bank that consists of chromosomal DNA from wild-type (Nif$^+$) *K. pneumoniae* cells is constructed into a broad-host-range plasmid expression vector and maintained in *Escherichia coli*.
3. The Nif$^-$ *K. pneumoniae* cells are conjugated with the *E. coli* cells that carry the clone bank on a plasmid shuttle vector.
4. The transformed *K. pneumoniae* cells are selected for the acquisition of the Nif$^+$ phenotype by plating onto a minimal medium that does not contain a source of fixed nitrogen. The only cells that grow under these conditions are Nif$^-$ *K. pneumoniae* cells containing a plasmid encoding the protein that is either missing or nonfunctional in the Nif$^-$ mutant.

The DNA fragment in the plasmid that complements the Nif$^-$ chromosomal mutation contains a *nif* gene that can be characterized more thoroughly and used to isolate other *nif* genes.

Two other approaches have been used to isolate other genes that are involved in the nitrogen fixation process. First, the *K. pneumoniae* clone bank has been used to complement a series of independently derived Nif$^-$ mutants, increasing the likelihood that in each case a different *nif* gene will be isolated. Second, isolated *nif* genes have been used as DNA hybridization probes, which have then been used to screen a *K. pneumoniae* chromosomal DNA clone bank that carries large (7- to 10-kilobase-pair [kb]) inserts. The premise behind the latter scheme is based on the observation that in prokaryotic organisms many of the genes involved in one pathway tend to be clustered and are often arranged in operons. Thus, DNA hybridization enables investigators to identify clones containing additional *nif* genes that are adjacent to the sequence initially isolated.

As the result of a considerable amount of research, the entire set of *nif*

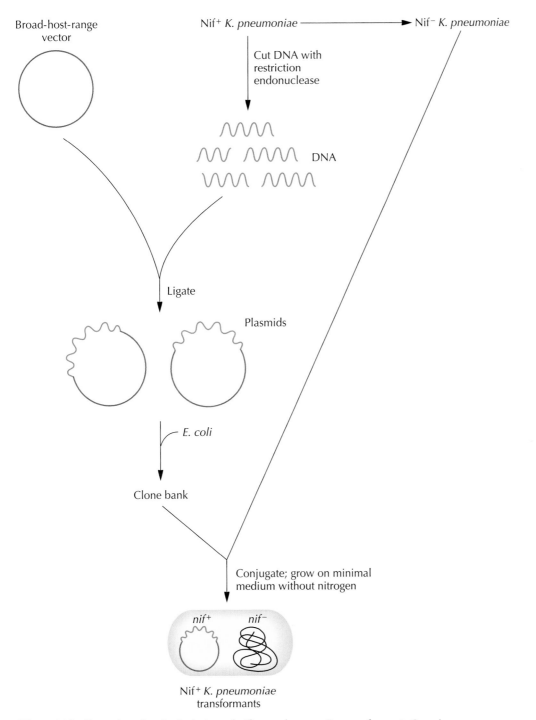

Figure 14.3 Procedure for the isolation of *nif* genes by genetic complementation. A clone bank that was constructed with Nif⁺ *K. pneumoniae* DNA is used to complement a Nif⁻ *K. pneumoniae* strain. Transformants are selected for growth on minimal medium that does not contain fixed nitrogen.

genes from *K. pneumoniae* has been isolated and characterized. These genes are arranged in a single cluster that occupies approximately 24 kb of the bacterial genome (Fig. 14.4). The cluster contains seven separate operons that together encode 20 distinct proteins (Table 14.2). All of the *nif* genes must be transcribed and translated in a concerted fashion (under the regulatory control of the *nifA* and *nifL* genes) to produce a functional nitrogenase. The NifA protein is a positive regulatory factor. It turns on the transcription of all of the *nif* operons except its own by binding to a specific DNA sequence (5′-TGT-N_{10}-ACA-3′) that is part of each promoter of each *nif* operon. There is a site on the DNA approximately 80 to 150 nucleotides upstream of each transcriptional start site where the NifA protein binds. The DNA-bound NifA protein then interacts with a specific transcription initiation protein called σ^{54} before transcription from the *nif* promoter is initiated. The NifL protein is a negative regulatory factor. In the presence of

Figure 14.4 Arrangement of the genes in the *K. pneumoniae nif* gene cluster and some of the functions that they encode. The *nif* genes are represented by italic uppercase letters; each red arrow below sets of these letters denotes a specific *nif* operon and the direction of its transcription. The arrows pointing up and away from the gene designations show how some of the various gene products participate in nitrogen fixation. F, flavodoxin molecule; F.O., pyruvate:flavodoxin oxidoreductase; FeMoCo, the iron–molybdenum cofactor; Acetyl-CoA, acetyl coenzyme A.

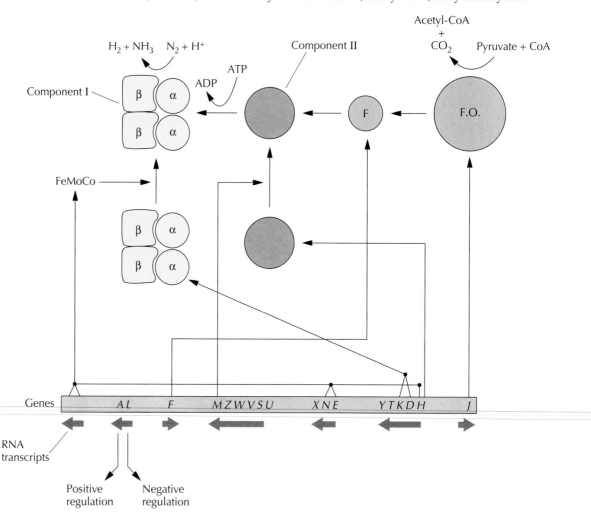

Table 14.2 *K. pneumoniae* genes involved in nitrogen fixation
and the functions of the proteins that they encode

nif gene	Function
D	Nitrogenase component I α subunit
K	Nitrogenase component I β subunit
H	Nitrogenase component II
F	Flavodoxin
J	Pyruvate:flavodoxin oxidoreductase
Q, B, N, E, V	FeMoCo synthesis
M	Processing of dinitrogenase reductase
A	Positive activator
L	Negative regulator
S	Maturation of component I
W, Z, T, Y, U, X	Other, less well defined functions

either oxygen or fixed nitrogen, it acts as an antagonist of the NifA protein and, as a result, turns off the transcription of all other *nif* genes.

The bacterium *K. pneumoniae* does not make a major contribution to the overall global biological fixation of nitrogen. Therefore, to genetically engineer nitrogen fixation in soil bacteria that are more important in promoting plant growth, other *nif* genes have been cloned and characterized. To do this, the *nif* genes from *K. pneumoniae* have been used as hybridization probes to isolate *nif* genes from clone banks of other diazotrophic microorganisms. Most diazotrophic organisms have a similar array of genes encoding their nitrogen-fixing apparatus, and the DNA sequences of these genes do not vary much from one organism to another.

On the basis of information from molecular genetic studies, it may be possible to increase the amount of nitrogen fixed by diazotrophic organisms by manipulating the *nifA* and *nifL* genes. After researchers genetically engineered extra copies of the *nifA* gene into a *Rhizobium meliloti* strain, alfalfa plants that were inoculated by this transformant grew larger and produced more biomass than plants that were treated with the nontransformed strain. Similarly, it may be possible to engineer the *nifL* gene so that the NifL protein (the negative regulator) is less sensitive to the presence of fixed nitrogen. With this kind of deregulation, an organism would fix more nitrogen for its plant partner. However, the available evidence indicates that not all nitrogen-fixing organisms have a NifL protein (in some organisms, the essential regions of NifL may be a part of NifA) so this sort of manipulation may be limited to only certain bacterial strains. Moreover, increasing the amount of nitrogen that an organism can fix also increases the amount of energy, usually in the form of fixed carbon, that is needed to power its metabolism. Consequently, an engineered microorganism may lose its effectiveness as a plant growth-promoting agent because of a diminished growth rate.

Due to the complexity of nitrogen fixation by microorganisms, it is clear that the simple addition of one or two *nif* genes will not confer on a nondiazotrophic recipient cell the ability to fix nitrogen. Moreover, genetic modification of plants with the entire 24-kb *nif* gene cluster would not be effective because the normal level of oxygen in the host cell would inactivate nitrogenase. And, if this level were reduced, the host plant cell would probably die. In addition, the engineering of nitrogen fixation in plant cells re-

quires resolving major, if not insurmountable, transcriptional, translational, and regulatory problems. For example, it is difficult to conceive how the regulation of nitrogen fixation could be achieved, since there are no plant promoters that respond to the NifA protein. Consequently, *nif* genes would not be turned on in such a transgenic plant. All of the *nif* genes would also have to be under the control of separate promoters to be able to respond to the level of fixed nitrogen in the cell, because plant cells cannot process multigene transcripts. The introduction of a functional nitrogen fixation capability into plants is therefore extremely unlikely.

Hydrogenase

An undesirable side reaction of nitrogen fixation is the reduction of H^+ to H_2 (hydrogen gas) by nitrogenase. Energy in the form of ATP is wasted on the production of hydrogen, which is eventually lost to the atmosphere. Because of this side reaction, only 40 to 60% of the electron flux through the nitrogenase system is transferred to N_2, thereby significantly lowering the overall efficiency of the nitrogen-fixing process. Theoretically, if H_2 could be recycled to H^+, the extent of energy loss could be diminished and the nitrogen-fixing process would become more efficient. It is probably impossible to prevent this side reaction directly because it is a consequence of the chemistry of the active site of the nitrogenase and hence blocking the side reaction by altering nitrogenase would concomitantly inhibit nitrogenase activity.

Hydrogen Metabolism

In the mid-1970s, it was discovered that some strains of *B. japonicum* could use hydrogen as an energy source for growth under microaerophilic (low-oxygen-concentration) conditions. These strains have an enzyme called hydrogenase that is able to take up H_2 from the atmosphere and convert it into H^+ (Fig. 14.5). Experiments were undertaken to test whether the presence of hydrogenase in *B. japonicum* had an impact on growth of soybean plants. Plants inoculated with strains that produce hydrogenase (Hup^+) had more biomass and nitrogen than plants that were treated with non-hydrogenase-producing (Hup^-) strains, despite higher levels of nitrogenase activity in the Hup^- strains (Table 14.3). From this and other similar experiments, it

Figure 14.5 Recycling of the hydrogen gas that is produced as a by-product of nitrogen fixation. Hydrogen is generated by nitrogenase at the expense of ATP, but, by using this pathway, the hydrogen can be recaptured by hydrogenase.

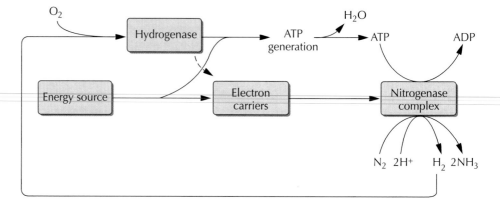

Table 14.3 Relative enzyme activities and growth-stimulating performance of a parental Hup⁺ *B. japonicum* strain (SR) and three Hup⁻ mutants (SR1, SR2, and SR3)

B. japonicum strain	Relative nitrogenase activity	Relative hydrogenase activity	Relative plant dry weight	Relative nitrogen content
SR	1.00	1.00	1.00	1.00
SR1	1.27	0.01	0.81	0.93
SR2	1.13	0.01	0.74	0.91
SR3	1.23	0.01	0.65	0.85

Adapted from Albrecht et al., *Science* **203**:1255–1257, 1979.

Nitrogenase activity was assessed by monitoring the amount of acetylene that was reduced to ethylene as a function of time. Hydrogenase activity was measured by means of a hydrogen electrode. Plant dry weight included the weights of both the leaf material and root material. The nitrogen content was calculated as the fraction of the dry weight of the plant that is nitrogen. All values have been normalized relative to the parental strain.

was concluded that the presence of a hydrogen uptake system in a symbiotic diazotroph such as *B. japonicum* improves its ability to stimulate plant growth, presumably by binding and then recycling the hydrogen gas that is formed inside the nodule by the action of nitrogenase (Fig. 14.5). Within the nodule, the contribution of atmospheric hydrogen is negligible.

Although it is clearly beneficial to the plant to obtain its nitrogen from a symbiotic diazotroph that has a hydrogen uptake system, this trait is not common in naturally occurring rhizobial strains. In one study, it was found that the majority of naturally occurring *Rhizobium* and *Bradyrhizobium* strains examined were Hup⁻ (Table 14.4). In that study, the data were based on a small number of strains for each species except *B. japonicum*, for which over 1,400 strains were assayed. The conclusion that can be drawn from this work is that commercial Hup⁻ rhizobial strains are prime candidates for transformation to a Hup⁺ phenotype once the genetic nature of the hydrogenase system is better understood and the appropriate genes are available.

Genetic Engineering of Hydrogenase Genes

Although a considerable amount of effort has been directed over the past 20 years or so toward studying hydrogenases from both diazotrophic and non-diazotrophic microorganisms, an in-depth understanding of the structure and function of these enzymes remains to be elucidated. Many organisms have more than one hydrogenase, and many hydrogenases consist of more than a single polypeptide chain. Some hydrogenases are active only in the

Table 14.4 Percentages of native *Rhizobium* and *Bradyrhizobium* strains that have functional hydrogen uptake systems (Hup⁺)

Bacterium	Hup⁺ strains (%)
Rhizobium leguminosarum bv. *leguminosarum*	9.3
Rhizobium meliloti	21
Rhizobium leguminosarum bv. *trifolii*	0
Rhizobium leguminosarum bv. *phaseoli*	0
Bradyrhizobium japonicum	21
Bradyrhizobium sp.	91

Adapted from Evans et al., *Annu. Rev. Microbiol.* **41**:335–361, 1987.

uptake of hydrogen from the atmosphere, whereas others, depending on the conditions, can also synthesize hydrogen. One result of this complexity is that the conversion of a Hup⁻ strain of *Rhizobium* into a Hup⁺ strain may not be readily achieved by the introduction of just any hydrogenase gene. Rather, the introduced gene(s) must encode all of the enzyme's subunits and must be able to interact with the appropriate electron transport molecule within the host organism.

The most common strategy for isolating hydrogenase genes has been genetic complementation. The first hydrogenase gene to be isolated (the gene for an *E. coli* membrane-bound hydrogenase) was selected by complementation of an *E. coli* mutant that no longer expressed this activity with a clone bank of wild-type *E. coli* DNA in plasmid pBR322. The mutant, which was previously shown to have a defective membrane hydrogenase, although the activity of its cytoplasmic hydrogenase was unaltered, was unable to proliferate on a minimal medium containing formate. Those transformants that were able to grow on this medium were assayed for the presence of hydrogenase activity. A transformant that restored hydrogenase activity to the wild-type level contained a plasmid that encoded a protein of approximately 60,000 daltons, the reported size of one of the subunits of the *E. coli* membrane hydrogenase. Additional work in other laboratories has shown that a number of genes are involved in the *E. coli* hydrogenase system.

After the work with *E. coli*, workers reported isolating hydrogenase (*hup*) genes from *B. japonicum* by using a clone bank of wild-type DNA constructed in the broad-host-range cosmid vector pLAFR1 to complement *B. japonicum* Hup⁻ mutants. The presence of a hydrogenase that takes up hydrogen from the atmosphere in the transformed Hup⁻ mutants was indicated by the ability of the active enzyme to reduce the dye methylene blue in a hydrogen atmosphere. More detailed studies of the *B. japonicum hup* genes showed that they were organized into at least two, and possibly three, transcriptional units covering approximately 15 kb of the genome. Subsequent work on the *hup* genes from *Rhizobium leguminosarum* has indicated that these genes are similar in both DNA sequence and gene organization to the *hup* genes from *B. japonicum*. Thus, the isolated *hup* genes from *B. japonicum* may be used as DNA hybridization probes to select homologous genes from a clone bank of *R. leguminosarum*.

Following the isolation of *R. leguminosarum hup* genes, and despite the complexity of this system, it has been possible to transfer the uptake hydrogenase system from a Hup⁺ strain of *R. leguminosarum* to a Hup⁻ strain (Table 14.5). When a strain of *R. leguminosarum* that had been transformed to Hup⁺ was used to inoculate legumes, the plants that were nodulated by the Hup⁺ transformants grew larger and contained more nitrogen than the plants inoculated by the Hup⁻ parental strain (Table 14.5).

Although hydrogenase genes have not received as much attention as *nif* genes, this experiment is one of the most convincing demonstrations of the use of genetic manipulation to improve the ability of diazotrophs to stimulate plant growth. It now remains to be seen whether the acquisition of *hup* genes by other diazotrophs (nonsymbiotic as well as symbiotic) will likewise be effective in improving their ability to stimulate plant growth.

In addition to increasing the efficiency of nitrogen fixation, hydrogenase has many other potential applications. For example, purified hydrogenase may be used for converting and storing solar energy, regenerating cofactors that are used in industrial enzymatic reactions, producing specialty chemicals that require H_2 as a reductant, removing tritium from the aque-

Table 14.5 Plant growth and nitrogen assimilation after the introduction of *hup* genes into a Hup⁻ strain of *R. leguminosarum*

Bacterial phenotype	Relative plant dry weight	Relative nitrogen amount	Relative leaf area	Relative nitrogen concentration
Hup⁻	1.00	1.00	1.00	1.00
Hup⁺	1.35	1.52	1.53	1.15

Adapted from Brewin and Johnston, U.S. patent 4,567,146, January 1986.
The data have been normalized relative to the Hup⁻ parental strain.

ous effluents of nuclear power plants, generating H_2 from organic wastes, and driving hydrogen–oxygen fuel cells. Although more than a dozen different hydrogenase genes have been isolated and characterized, none of them has been used for the large-scale production of hydrogenase.

Nodulation

Competition among Nodulating Organisms

A major goal of agricultural biotechnology research is the development, by genetic manipulation, of *Rhizobium* strains that can increase plant productivity more than naturally occurring strains can. Many commercial inoculant strains that have been developed by mutation and selection to be superior nitrogen fixers are not very effective at establishing nodules on host plant roots when placed in competition with *Rhizobium* strains that are already present in the soil. Conversely, although many of the strains that are indigenous to the soil are highly successful in establishing nodules in competitive situations, they are not especially efficient at nitrogen fixation. Therefore, to make use of the commercial inoculant strains, either the nodulation capability of these strains must be enhanced or indigenous rhizobial strains must be inhibited.

Studies were undertaken to determine the genetic basis of this "competitiveness" with the aim of adding these genes to the strains that are used as inoculants. The nature of the competitive advantage of soil rhizobial species is not known, but it was reasoned that the indigenous bacteria might be more efficient at nodulation and, as a consequence, might prevent an inoculated strain from becoming established.

Genetic Engineering of Nodulation Genes

The absence of any specific information about the biochemical or genetic basis of nodulation meant that a strategy had to be devised for the identification of nodulation (*nod*) genes. Therefore, once again, genetic complementation was used. Nodulation-defective (Nod⁻) mutants of *R. meliloti* were transformed with a clone bank of wild-type chromosomal DNA from *R. meliloti*, and those colonies that had acquired the ability to nodulate alfalfa roots were isolated (Fig. 14.6). More specifically, the steps of the procedure were as follows.

1. A clone bank of wild-type (Nod⁺) *R. meliloti* was constructed by partial digestion of *R. meliloti* DNA with *Eco*RI and insertion into the unique *Eco*RI site of the broad-host-range cosmid pLAFR1, which can carry up to 40 kb of cloned DNA.

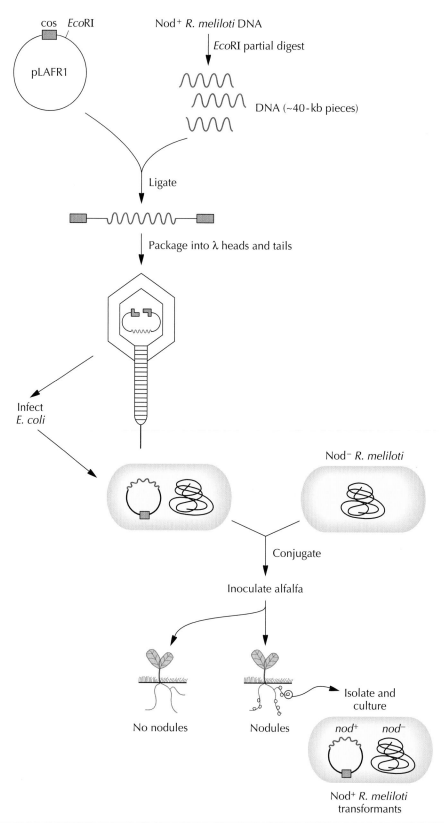

cos *Eco*RI

pLAFR1

Nod⁺ *R. meliloti* DNA

*Eco*RI partial digest

DNA (~40-kb pieces)

Ligate

Package into λ heads and tails

Infect
E. coli

Nod⁻ *R. meliloti*

Conjugate

Inoculate alfalfa

No nodules

Nodules

Isolate and
culture

nod⁺ *nod*⁻

Nod⁺ *R. meliloti*
transformants

Figure 14.6 Procedure for the isolation of *R. meliloti* nodulation genes. The DNA from wild-type *R. meliloti* is cloned into the broad-host-range cosmid pLAFR1, packaged into bacteriophage λ, and introduced into *E. coli*. The clone bank is then transferred from *E. coli* to a Nod⁻ *R. meliloti* strain by conjugation. Alfalfa plants are inoculated with transformed Nod⁻ *R. meliloti*; those plants that develop root nodules have been infected with Nod⁺ *R. meliloti* cells that presumably carry a complementing nodulation gene inserted into the cosmid vector. The transformed Nod⁺ *R. meliloti* cells can be isolated directly from the root nodules.

2. The clone bank was packaged into bacteriophage λ, introduced into *E. coli*, and subsequently transferred to Nod⁻ mutants of *R. meliloti* by conjugation. The vector carries a tetracycline resistance gene that can be used as a selectable marker in both *E. coli* and *R. meliloti*.

3. After conjugation, suspensions of 200 to 300 transformed *R. meliloti* cells were tested for their ability to nodulate sterile alfalfa plants, with the expectation that only those transformants that carried and expressed a gene that complemented the nodulation defect in the *R. meliloti* host would produce nodules.

4. The bacteria that formed nodules on the test plants were recovered from within the nodules and used to retrieve the vector carrying the complementing gene. The specific portion of the large insert DNA that carried the complementing gene was then subcloned and analyzed further.

5. Once a single nodulation gene was identified, it was used as a DNA hybridization probe to identify adjacent regions of *R. meliloti* chromosomal DNA in a genomic library.

With considerable effort, the complete repertoire of nodulation genes from *R. meliloti* has been characterized. Detailed biochemical and genetic studies have revealed that nodulation and its regulation are complex processes that require the functioning of a large number of genes (approximately 20; see Table 14.6). Some of the nodulation genes are highly conserved (common) among nodulating microorganisms, and others are

Table 14.6 Some *Rhizobium* nodulation gene products and their probable functions

Nod protein	Probable function
NodA	Common, cytoplasmic membrane, with NodB stimulates cell division
NodB	Common, cytoplasmic membrane, with NodA stimulates cell division
NodC	Common, outer membrane, chitin synthase
NodD	Common, positive transcriptional regulator, constitutive
NodE	Cytoplasmic membrane, β-ketoacyl synthase
NodF	Cytoplasmic, acyl carrier protein
NodG	Host specific gene, dehydrogenase
NodH	Host specific gene, sulfotransferase
NodIJ	Common, cytoplasmic membrane, capsular polysaccharide secretion
NodK	Affects onset of nodulation in some bradyrhizobia
NodL	Cytoplasmic membrane, acetyltransferase
NodM	D-Glucosamine synthetase
NodN	Unknown
NodO	Secreted, hemolysin
NodP	With NodQ, ATP sulfurylase
NodQ	With NodP, ATP sulfurylase
NodS	Methyltransferase
NodT	Outer membrane, secretion
NodU	Unknown
NodX	Cultivar specificity

Where biochemical or genetic evidence for the function of a particular protein is lacking, a possible function is assigned based on homology of the amino acid sequence to a protein of known sequence. Different rhizobial strains contain different subsets of these proteins. "Common" genes perform the same function in all species of rhizobia.

species specific. The *nod* genes may be grouped into three separate classes: common genes, host-specific genes, and the regulatory *nodD* gene. Thus, for example, the *nodABC* genes are common to all *Rhizobium* species and are structurally interchangeable. In most species, the *nodABC* genes are found on a single operon.

A number of events are now known to occur during nodulation. First, the *nodD* gene product, which is constitutively expressed, recognizes and binds to a flavonoid molecule, which is excreted by the roots of the potential host plants. Flavonoids are a class of plant phenolic molecules with a basic structure that consists of 15 carbons arranged as two aromatic rings connected by a three-carbon bridge. They perform a number of different functions for the plant, such as pigmentation and defense against fungi or insects. The binding of flavonoids to the NodD gene product is one of the major determinants of rhizobial host specificity because each rhizobial species recognizes only a limited number of flavonoid structures and each plant species produces its own specific set of flavonoid molecules. Some strains, such as *R. leguminosarum* biovar (bv.) *trifolii*, have a very narrow host range, responding to only a few kinds of flavonoids, while others such

MILESTONE

Cloning of *Rhizobium meliloti* Nodulation Genes by Direct Complementation of Nod⁻ Mutants

S. R. LONG, W. J. BUIKEMA, and F. M. AUSUBEL
Nature **298**:485–488, 1982

The isolation of a gene, when there is no heterologous probe or information about the gene, can be a daunting experience. In these instances, it is often necessary to devise an innovative selection scheme. The selection scheme may be based on immunological detection of the target protein, determination of the activity of the target protein, DNA hybridization with an oligonucleotide probe whose sequence is deduced from the partial amino acid sequence of the purified target protein, or mutant complementation. Often, after a gene that encodes a particular function has been isolated, other similar genes from different organisms can be isolated by using the first gene as a heterologous DNA hybridization probe. This approach depends on nucleotide sequence similarity between the probe and the target gene. This strategy generally works well for genes that are evolutionarily conserved, such as those that encode proteins involved in the process of nitrogen fixation, and

poorly for many other genes, including those that encode proteins that degrade cellulose.

When Long and her coworkers set out to isolate and characterize the nodulation genes from *Rhizobium meliloti*, nodulation genes had never been isolated. Very little was known about how many genes were involved in this process or what products they might encode. Before isolating the nodulation genes, these workers had isolated and characterized several nodulation-defective mutants of *R. meliloti*. However, these studies had not given them any clue as to the functions of these genes. Therefore, they set about to select nodulation genes that could complement their nodulation-defective *R. meliloti* mutants. This work was further complicated by the fact that, at that time, clone banks were almost always constructed and maintained in *E. coli*. Therefore, to facilitate the assembly of the *R. meliloti* clone bank and its subsequent transfer to nodulation-defective mutant strains

of *R. meliloti*, these researchers first constructed a broad-host-range cosmid vector in which large DNA fragments with an average size of approximately 23 kb could be stably maintained in a number of different gram-negative bacteria. The large size of the DNA on the cosmid increased the likelihood that other genes involved in nodulation would be on the same DNA fragment as the complementing DNA sequence. After the cosmids had been transferred by conjugation to *R. meliloti*, the transformed bacteria were tested for their ability to nodulate alfalfa plants. In previous experiments, these researchers had found that even one nodulation-proficient bacterium in the presence of 10^4 defective ones could successfully nodulate alfalfa plants. Bacteria with cosmids carrying a complementing gene were isolated directly from the nodules.

In summary, these researchers were the first to isolate nodulation genes. They devised a clever and effective selection scheme. The broad-host-range cosmid vector that they developed has since been used many times by a large number of other researchers.

as *Rhizobium* sp. strain NGR234 have a very broad host range and respond to a much larger number of different flavonoids.

The binding of a flavonoid molecule activates the NodD gene product, presumably causing it to undergo a conformational change, and enables the flavonoid–NodD complex to attach to a nodulation promoter element called a *nod* box. This promoter element is located upstream from all the nodulation genes except the *nodD* gene, and it turns on the transcription of these genes.

The *nodABC* genes encode proteins that cause the plant root hair tips to swell and curl, an effect that is recognized as the initial step in the infection of the plant root by the bacterium. Together, the plant and bacteria synthesize an oligosaccharide factor that is modified by the NodH gene product and perhaps also by the NodQ and NodP gene products. This factor, called NodRm-1 (Fig. 14.7), elicits in the plant a host-specific response that includes root curling and deformation.

Finally, depending on the *Rhizobium* or *Bradyrhizobium* strain, up to approximately 20 additional *nod* gene products are synthesized. These proteins, together with some plant-encoded proteins, contribute to the formation of the nodule.

More research will be necessary to elucidate the role of each of the known *nod* genes; furthermore, new *nod* genes may yet be discovered. For example, DNA sequencing and computer analysis revealed that in a slow-growing variant of a *Bradyrhizobium* sp., the region of the DNA between the *nodD* and *nodABC* genes contained an open reading frame; the fast-growing form lacks this sequence. This open reading frame was designated *nodK*. When plants were inoculated with a *Bradyrhizobium* sp. that had a mutagenized *nodK* gene (NodK⁻) and were compared with those treated with the wild-type strain, the onset of nodulation in the NodK⁻ strain-treated plants was 5 days earlier, the nodulation number was doubled, and there was a 120% enhancement of plant yield.

Figure 14.7 Proposed structure of an oligosaccharide factor, NodRm-1. This compound elicits a host-specific plant response that includes root curling and deformation.

To date, no simple genetic means has been devised to use *nod* genes to enable inoculated strains of *Rhizobium* to outcompete indigenous strains. However, host specificity can be altered by the transfer of a *nodD* gene from a broad-specificity rhizobial strain to one with a narrow specificity. It has become clear that the process of nodulation is quite complicated and that considerable additional effort will be required before it is possible to further enhance the competitiveness of rhizobial strains by genetic engineering.

Biocontrol of Pathogens

Plant growth-promoting bacteria can affect plant growth either directly or indirectly. The direct promotion of plant growth usually entails providing the plant either with a compound that is synthesized by the bacterium, such as fixed nitrogen, or with a plant hormone. As well, these bacteria can facilitate the uptake by the plant of certain nutrients such as iron or phosphorus from the environment. The indirect promotion of plant growth occurs when plant growth-promoting bacteria lessen or prevent the deleterious effects of one or more phytopathogenic organisms, either fungi or bacteria. Phytopathogens can reduce crop yields from 25 to 100%. This is an enormous loss of productivity. Phytopathogen damage to crops is generally dealt with by the use of chemical agents, although other treatments have also been employed. For most bacterial diseases, plants may be symptomless for prolonged periods before changes in environmental conditions which favor the proliferation of the bacteria cause a rapid outbreak of the disease. Under these conditions, severe damage can occur and destroy an entire crop. These field epidemics are difficult and costly to control.

Many of the chemicals that are used to control phytopathogens are hazardous to animals and humans and both persist and accumulate in natural ecosystems. It is therefore desirable to replace these chemical agents with biological control agents that are more "friendly" to the environment. One biological approach for the control of phytopathogenic agents is the development of transgenic plants that are resistant to one or more pathogenic agents (this is discussed in Chapter 18). Alternatively, scientists have begun to use plant growth-promoting bacteria as biocontrol agents to suppress or prevent phytopathogen damage. Plant growth-promoting bacteria can produce a variety of substances that can be used to limit the damage to plants by phytopathogens. These substances include siderophores, antibiotics, other small molecules, and a variety of enzymes.

The use of plant growth-promoting bacteria to limit phytopathogen-caused damage is still at an early stage of development. And, while this approach appears to have tremendous potential, most of the studies to date have been carried out under controlled laboratory, growth chamber, or greenhouse conditions. However, the ultimate utility of a strategy based on a particular mechanism can be assessed only under field conditions.

Siderophores

Iron is one of the most abundant minerals on Earth and is an essential requirement for living organisms. However, iron in the soil is unavailable for direct assimilation by microorganisms because ferric ion, or Fe(III), which is the predominant form in nature, is only sparingly soluble, i.e., its solubility is about 10^{-18} M at pH 7.4. This amount of soluble iron is much too low to support microbial growth. Consequently, to survive in this environment, soil microorganisms synthesize and secrete low-molecular-mass (~400 to

1,000-dalton) iron-binding molecules known as siderophores (Fig. 14.8). Siderophores bind Fe(III) with a very high affinity (dissociation constant $[K_d] = 10^{-20}$ to 10^{-50}) and transport it back to the microbial cell, where it is taken up by a cellular receptor. Once inside a cell, the iron is released and is then available to support microbial growth.

Plant growth-promoting bacteria can prevent the proliferation of fungal phytopathogens by producing siderophores that bind most of the Fe (III) in the area around the plant root (the rhizosphere). The resulting lack of iron prevents any fungal pathogens from proliferating in this immediate vicinity. Fungal phytopathogens also synthesize siderophores, but these generally have a lower affinity for iron than do the siderophores produced by plant growth-promoting bacteria. In effect, the plant growth-promoting bacteria outcompete fungal phytopathogens for available iron.

Unlike microbial phytopathogens, plants are not generally harmed by the localized depletion of iron in the soil caused by plant growth-promoting bacteria. Most plants can grow at much lower iron concentrations than microorganisms. In addition, some studies have shown that iron that has been sequestered by bacterial siderophores may be taken up from the soil by the plant, to the plant's benefit.

Because the sequestering of iron by a siderophore-producing bacterium can simultaneously prevent the proliferation of a number of different phytopathogenic microorganisms, siderophore genes are being examined to determine whether they can be used to create more effective biocontrol inoculants.

Many plant growth-promoting, fluorescent pseudomonads excrete a siderophore that consists of a linear hexapeptide of alternating L- and D-amino acids linked to a fluorescent chromophore (Fig. 14.8). One such siderophore, called pseudobactin, has been estimated to bind to Fe(III) with an affinity constant of approximately 10^{25} L mol^{-1}. All fluorescent pseudomonads produce structurally related siderophores that differ mainly in the number and configuration of the amino acids in the peptide chain.

Investigators have also begun to examine the synthesis and regulation of pseudobactin in the plant growth-promoting bacterium *Pseudomonas putida* WCS358. Mutagenesis was used to generate a set of 28 mutants that were defective for siderophore production. Two criteria were used for identifying the siderophore-deficient mutants: (1) lack of fluorescence under UV light and (2) inability to grow in the presence of bipyridyl, a molecule that

Figure 14.8 Structure of the siderophore called pseudobactin, which was isolated from *Pseudomonas* strain B10. One molecule of Fe(III) is bound to the siderophore.

sequesters most of the iron in the growth medium. When most of the iron is unavailable, only a cell that produces siderophores can grow.

A clone bank of *P. putida* WCS358 DNA was constructed in the broad-host-range cosmid vector pLAFR1 and was introduced by conjugation into each of the 28 siderophore mutants. All of the resultant transformants were tested by complementation for restoration of fluorescence and/or the ability to grow in the presence of bipyridyl. Thirteen separate complementing cosmid clones, with an average insert size of 26 kb, were identified. After detailed analyses, these clones were found to represent at least five separate gene clusters.

One of these gene clusters has been studied further. It has a minimum length of 33.5 kb and contains five transcriptional units with at least seven separate genes. Thus, like nitrogen fixation and nodulation, siderophore biosynthesis is a complex process. Since each siderophore is encoded by a number of different genes, genetically engineering bacteria to produce modified siderophores is not a simple matter. However, there may be other ways to improve the effectiveness of plant growth-promoting bacteria as biocontrol agents. For example, it may be possible to extend the range of iron–siderophore complexes that one bacterial strain can utilize so that a genetically altered plant growth-promoting biocontrol bacterial strain could take up and use siderophores synthesized by other soil microorganisms, thereby giving it a competitive advantage. This was done by cloning the genes for iron–siderophore receptors from one plant growth control bacterium and introducing them into other strains.

Antibiotics

One of the most effective mechanisms by which a plant growth-promoting bacterium can prevent phytopathogen proliferation is the synthesis of antibiotics. For example, some of the antibiotics synthesized by biocontrol pseudomonads include agrocin 84, agrocin 434, 2,4-diacetylphloroglucinol, herbicolin, oomycin, phenazines, pyoluteorin, and pyrrolnitrin.

The biocontrol activity of a plant growth-promoting bacterium may be improved by providing it with genes that encode the biosynthesis of antibiotics that are normally produced by other bacteria. In this way, the range of phytopathogens that a single biocontrol bacterium can suppress can be extended. Moreover, by limiting the growth of other soil microorganisms, antibiotic-secreting plant growth-promoting bacteria should facilitate their own proliferation, since they will have fewer competitors for limited nutritional resources. In addition, genetic manipulation can be used to increase the amount of antibiotic that a bacterium synthesizes.

Since the production of a number of antifungal metabolites that are produced by pseudomonads appears to be controlled by a protein that acts as a global transcriptional regulator, it should be possible to enhance antibiotic production by modifying this global regulation. For example, antibiotic production was enhanced after *Pseudomonas fluorescens* CHA0 was transformed with a vector carrying the gene encoding the housekeeping RNA polymerase sigma factor σ^{70}. The modified strain was more effective at protecting cucumber plants against a root disease caused by the fungus *Pythium ultimum* (Table 14.7). In another study, inactivation of the *pqq* genes, which are involved in the biosynthesis of pyrroloquinoline quinone, a cofactor of different hydrogenases, in *P. fluorescens* CHA0 stimulated the production of the antibiotic pyoluteorin. While the precise mechanism of this stimulation is unclear, it is possible that this mutation enhances the flux

Table 14.7 Effect of additional copies of the *rpoD* gene from *Pseudomonas fluorescens* CHA0 on the ability of the bacterium to prevent damage to cucumber roots caused by the pathogenic fungus *Pythium ultimum*

Plant growth-promoting bacterium added	Average root fresh weight (mg)	
	Without *P. ultimum*	With *P. ultimum*
None	382	44
P. fluorescens CHA0	386	177
P. fluorescens CHA0 with vector	365	146
P. fluorescens CHA0 with vector and *rpoD* gene	371	335

Adapted from Schnider et al., *J. Bacteriol.* **177:**5387–5392, 1995.

In the absence of the plant growth-promoting bacterium, *P. fluorescens* CHA0, the pathogen, *P. ultimum*, dramatically inhibits root growth. The presence of the *rpoD* gene on the plasmid vector enhanced the activity of the plant growth-promoting bacterium. Plants were grown for 2 weeks before their roots were measured.

of metabolites from other metabolic pathways into the pyoluteorin biosynthesis pathway.

At the present time, there is only one commercially available genetically engineered biocontrol bacterium. A modified strain of *Agrobacterium radiobacter* K84 has been marketed in Australia since 1989 as a means of controlling crown gall disease, which is caused by the bacterium *Agrobacterium tumefaciens*. This disease affects almond trees and stone fruit trees, such as peach trees. The antibiotic agrocin 84, which is produced by *A. radiobacter*, is toxic to *A. tumefaciens*. However, agrocin 84-resistant strains of *A. tumefaciens* can develop if the plasmid carrying the genes for the biosynthesis of agrocin 84 is accidentally transferred from *A. radiobacter*. To avoid this possibility, the region of DNA responsible for plasmid transfer was removed from the agrocin 84 plasmid, pAgK84 (Fig. 14.9). As a result of this deletion, the *A. radiobacter* strain retains the capacity to act as a biocontrol agent, since it can no longer transfer the modified agrocin plasmid to pathogenic agrobacteria.

Enzymes

Some plant growth-promoting bacteria can produce enzymes, including chitinase, β-1,3-glucanase, protease, and lipase, that can degrade fungal cell walls and thereby lyse the fungal cells. In one study, the incidence of plant disease caused by the phytopathogenic fungi *Rhizoctonia solani*, *Sclerotium rolfsii*, and *Pythium ultimum* was reduced by using a β-1,3-glucanase-producing strain of *Pseudomonas cepacia* which was able to degrade fungal mycelia. In another study, the antifungal activity of three different strains of the plant growth-promoting bacterium *Enterobacter agglomerans* was attributed to a complex of four separate polypeptides that act together to degrade the chitin in fungal cell walls. When tested, these bacteria significantly decreased the damage to cotton plants following infection with *Rhizoctonia solani*. Moreover, Tn5 mutants of *E. agglomerans* that were deficient in chitinase activity were unable to protect plants against damage caused by the fungal pathogen.

Many of the bacterial enzymes that can lyse fungal cells, including chitinases and β-glucanases, are encoded by a single gene. It should therefore be straightforward to isolate these genes and transfer them to plant growth-promoting bacteria, constructing strains that, for example, produce both antibiotics and fungus-degrading enzymes. In one series of experiments, a

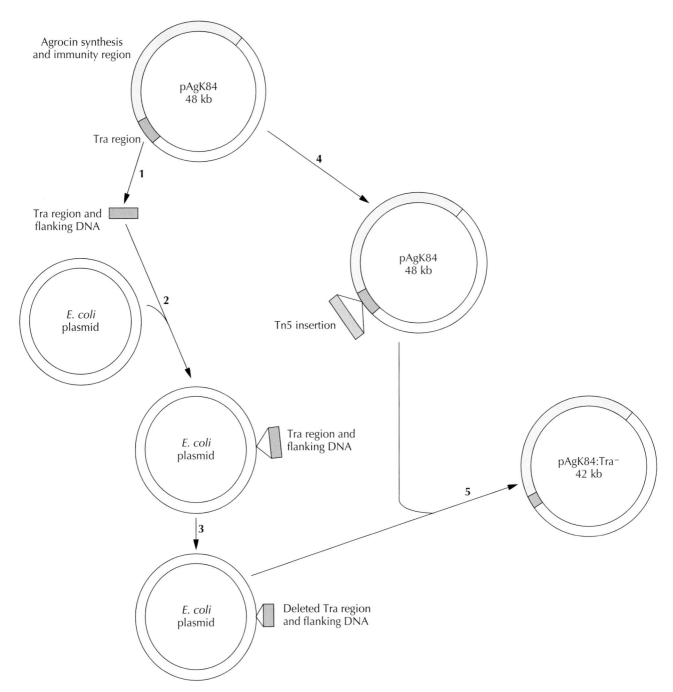

Figure 14.9 Construction of a transfer-deficient (Tra⁻) derivative of plasmid pAgK84 from *A. radiobacter*, which encodes both synthesis of and immunity to the antibiotic agrocin 84. Based on a knowledge of the restriction enzyme map of pAgK84, a DNA fragment containing the transfer (Tra) region together with some of the flanking DNA is isolated (**1**) and spliced into an *E. coli* plasmid vector (**2**). By restriction enzyme digestion, approximately 80% of the Tra region and some of the flanking DNA (a total of about 6 kb) is deleted from the cloned DNA containing the Tra sequence (**3**). Homologous recombination of the *E. coli* plasmid containing the deleted Tra region with plasmid pAgK84 in which transposon Tn5, which carries a kanamycin resistance gene, has been inserted into the Tra region (**4**) results in some derivatives of pAgK84 in which a portion of the Tra region of the plasmid has been deleted (**5**). The resultant Tra⁻ mutant of pAgK84 can no longer be conjugationally transferred to other agrobacteria, although it is still able to synthesize and provide immunity to agrocin 84. None of the DNA fragments is shown to scale.

chitinase gene was isolated from the bacterium *Serratia marcescens* and then transferred into *Trichoderma harzianum* and *R. meliloti* cells. In both cases, the transformed microorganisms expressed the chitinase and subsequently displayed increased antifungal activity. When the *S. marcescens* chitinase gene was introduced into a strain of *P. fluorescens* that promotes plant growth, the transformant stably expressed and secreted active chitinase and effectively controlled the phytopathogen *Rhizoctonia solani*.

Ice Nucleation and Antifreeze Proteins

One of the ways in which some pathogenic leaf bacteria such as *Pseudomonas syringae* damage plants is by synthesizing ice nucleation proteins. These proteins, which are produced at low temperatures, are present on the surface of the bacterium and act as sites that facilitate the formation of ice crystals at freezing temperatures. As the ice crystals grow, they can pierce the plant cells and irreparably damage the plant. The bacteria benefit from this damage by gaining direct access to the nutrients from the lysed plant cells. In the absence of ice nucleation proteins on the leaf surface, a brief overnight frost would not necessarily damage the plant, since the water in a plant cell must usually be several degrees below the freezing point before ice crystals begin to form (i.e., it must be supercooled). One way to prevent freezing damage caused by *P. syringae* to susceptible crops such as strawberries is by spraying the plants, prior to the frost, with a mutant form of the ice-nucleating bacterium. Such a mutant, which may be constructed by either recombinant DNA manipulation or conventional mutagenesis and selection, lacks the ability to produce the ice nucleation protein, and therefore ice crystals are not formed on the leaf surface. If a sufficient number of these "ice-minus" mutant bacteria are sprayed onto a susceptible plant, the mutant will displace the wild-type (ice-plus) bacteria, thereby preventing ice nucleation.

An important facet of the effectiveness of a biocontrol plant growth-promoting bacterium is its ability to persist and proliferate in the natural environment. In areas such as Canada, Scandinavia, and the northern United States, these organisms must survive long, cold winters and then grow at cool soil temperatures in the spring (~5 to 10°C). While microorganisms have a variety of adaptive strategies for thriving under adverse conditions, it may be possible to engineer organisms that are better able to deal with cold temperatures. It was recently discovered that some soil bacteria (that also happen to be capable of plant growth promotion) can both grow at 5°C and secrete antifreeze proteins into the surrounding medium when grown at low temperatures. An antifreeze protein regulates the formation of ice crystals inside the bacterium and thereby provides protection for the bacterium in the soil. In the presence of bacterial antifreeze protein, ice crystals still form, but their size is limited. In the absence of antifreeze proteins, ice crystals can grow to a large size and eventually puncture the bacterial cell wall and membrane, causing cell lysis. After the genes for bacterial antifreeze proteins have been isolated, it should be possible to transfer them to certain strains of plant growth-promoting bacteria with the idea that the transformed bacteria will be better able to persist and proliferate at cold temperatures. Thus, these bacteria should be more effective because they will be able to thrive under adverse conditions. Although there is currently no evidence to link antifreeze activity with the mechanism that bacteria use to function at low temperatures, it will be interesting to examine experi-

mentally whether antifreeze protein activity is a part of the adaptive strategy used by some bacteria for cold tolerance.

Growth Promotion by Free-Living Bacteria

Plant growth-promoting bacteria can directly stimulate plant growth in several different ways. They can (1) fix atmospheric nitrogen that is used by the plant, (2) synthesize siderophores that solubilize and sequester iron from the soil and provide it to plant cells, (3) synthesize several different phytohormones that enhance various stages of plant growth, (4) solubilize minerals such as phosphorus that are used by the plant, and (5) synthesize enzymes that can modulate plant hormone levels. A particular plant growth-promoting bacterium may have one or more of these mechanisms.

In general, nitrogen fixation is probably only a minor component of the benefit that a plant growth-promoting bacterium provides to a plant. In fact, not all plant growth-promoting bacteria are diazotrophic, and many of those that are fix only limited amounts of nitrogen.

A number of plants use bacterial iron–siderophore complexes to obtain iron from the soil. Without this mechanism, plant growth in most soils would be severely limited. However, while bacterial siderophores undoubtedly contribute to the nutrition, and hence to the growth, of plants, in most instances this effect is probably small.

There is some controversy regarding the mechanism that plant growth-promoting bacteria employ in the uptake of minerals such as phosphorus. On one hand, it is argued that the increased mineral uptake in plants treated with plant growth-promoting bacteria is an indirect by-product of that interaction and actually reflects a better-developed root system and an overall healthier plant. On the other hand, experiments with *Azospirillum* have suggested that this organism specifically enhances mineral uptake, possibly by producing and secreting organic acids that can solubilize and bind some minerals.

The mechanism most commonly invoked to explain the various effects of plant growth-promoting bacteria on plants is the production of phytohormones. Most of the research in this area has focused on the role of a class of phytohormones called auxins. The most common and best-characterized auxin is indole-3-acetic acid (IAA), which stimulates in plants both rapid responses, such as increases in cell elongation, and long-term effects, such as increases in cell division and differentiation. Plants as well as plant growth-promoting bacteria can synthesize auxin. This makes it difficult for researchers to distinguish between plant responses that result from bacterial auxin synthesis and those that result from plant auxin synthesis. This problem notwithstanding, there is considerable evidence to suggest that plant growth-promoting bacteria facilitate plant growth by altering the hormonal balance within a plant.

It has recently been discovered that many plant growth-promoting bacteria contain an enzyme that can modulate levels of the plant hormone ethylene. This enzyme 1-aminocyclopropane-1-carboxylate (ACC) deaminase hydrolyzes ACC, which is the immediate biosynthetic precursor of ethylene in plants. One model that explains the role of this enzyme is as follows. The bacterium binds to seed coats or plant roots and then sequesters and hydrolyzes ACC, thereby lowering the level of ethylene in the developing plant. In many plants, ethylene stimulates germination and breaks the dormancy of the seeds; however, if the level of ethylene is too high after germ-

Figure 14.10 Schematic representation of how a plant growth-promoting bacterium bound to either a seed or a plant root lowers the ethylene concentration and thereby prevents ethylene inhibition of root elongation. The arrows indicate chemical or physical steps in the mechanism, and the symbol ⊥ indicates inhibition of root elongation by ethylene. IAA is synthesized and secreted by a plant growth-promoting bacterium that is bound to the surface of either the seed or the root of a developing plant. After being taken up by the plant, together with the IAA from the plant, the bacterial IAA can either stimulate plant cell proliferation and elongation or the activity of the enzyme ACC synthase, which converts **S**-adenosylmethionine (AdoMet) to ACC. A significant portion of the ACC is exuded from plant roots or seeds along with other small molecules normally present in seed or root exudates, taken up by the bacterium, and hydrolyzed by the enzyme ACC deaminase to ammonia and α-ketobutyrate (α-KB). This decreases the amount of ACC outside the plant. To maintain the equilibrium between internal and external ACC, the plant exudes more ACC. Consequently, the concentration of ethylene in the plant is lowered. Adapted from Glick et al., *J. Theor. Biol.*, in press.

ination, root elongation is inhibited. Thus, the ACC deaminase that is provided by a plant growth-promoting bacterium prevents the inhibition of root elongation, and consequently the plant produces longer roots during early development, resulting in a healthier and larger plant. In addition, many plant growth-promoting bacteria synthesize IAA, and the excess IAA that is not used to promote plant cell elongation or proliferation stimulates the enzyme ACC synthase in the plant, thereby increasing the ethylene concentration. When ACC deaminase activity is present, it may prevent the buildup of ACC, even in the presence of high levels of IAA, so that the ethylene level does not become elevated to the point where plant growth is impaired (Fig. 14.10). As a better understanding of the mechanisms used by plant growth-promoting bacteria emerges, it will become possible to genetically engineer improved organisms that can stimulate the growth of a wide range of plants in a variety of different environments.

SUMMARY

Many soil microorganisms have the ability to stimulate the growth of plants. With an eye on diminishing the dependency on chemical fertilizers, the molecular mechanisms of how bacteria promote plant growth have been examined. Plant growth promotion may be achieved directly by the ability of the bacteria to fix nitrogen, sequester iron, facilitate phosphorus uptake, or produce phytohormones that trigger responses in a growing plant. Some soil bacteria can also stimulate plant growth indirectly by inhibiting the growth of phytopathogenic microorganisms.

Of the plant growth-promoting bacteria that have been studied in detail and used in agricultural practice, much of the research has focused on members of the genera *Rhizobium* and *Bradyrhizobium*. These organisms form a complex, obligatory symbiotic relationship with specific plants.

The molecular basis of nitrogen fixation has been examined extensively with *K. pneumoniae*, which serves as a model system for the symbiotic rhizobial bacteria. Nitrogenase, the nitrogen-fixing enzyme, has been characterized in detail. Molecular genetic studies have established that bacterial nitrogen fixation is a complex process that requires seven operons that are coordinately regulated, with a total of 20 different proteins. This complexity has, to date, frustrated attempts by genetic engineering to create plants that can fix nitrogen and has prevented transfer of the ability to fix nitrogen to other bacteria.

As part of the action of nitrogenase, hydrogen gas (H_2) is generated at the expense of ATP. In some *Rhizobium* strains, the enzyme hydrogenase is present. This enzyme is able to recycle H_2 in vivo to $H+$, an activity that increases the efficiency of nitrogen fixation. When strains are defective in hydrogenase activity, the ability to fix nitrogen and promote plant growth is diminished. With this in mind, attempts are being made to put cloned hydrogenase genes into strains of rhizobial bacteria that form symbiotic relationships with crop plants. Initial studies indicate that genetic engineering of the hydrogenase gene can produce rhizobial strains with enhanced growth-promoting properties.

Part of the interaction between symbiotic rhizobial strains and plants is the formation of nodules on the roots of plants. These nodules are the sites of bacterial nitrogen fixation. It has been reasoned that enhancing nodulation by genetic engineering will enable inoculated rhizobial strains to be more effective competitors for sites on the roots of target plants than indigenous strains. Studies to date have shown that the genetic basis of nodulation is complex, involving a number of different genes. At present, there is no simple way to manipulate this process genetically.

The indirect promotion of plant growth occurs when plant growth-promoting bacteria decrease or prevent the damage that is caused by either fungal or bacterial phytopathogens. Some of the substances produced by plant growth-promoting bacteria help to limit the damage to plants by phytopathogens; these include siderophores, antibiotics, other small molecules, and various enzymes. Other products produced by plant growth-promoting bacteria, including phytohormones and ACC deaminase, can directly stimulate plant growth. In the future, manipulation of the genes for these compounds may be used to create more effective plant growth-promoting bacteria.

References

Adams, M. W. W., L. E. Mortenson, and J. S. Chen. 1981. Hydrogenase. *Biochim. Biophys. Acta* **594:**105–176.

Albrecht, S. L., R. J. Maier, F. J. Hanus, S. A. Russell, D. W. Emerich, and H. J. Evans. 1979. Hydrogenase in *Rhizobium japonicum* increases nitrogen fixation by nodulated soybeans. *Science* **203:**1255–1257.

Arp, D. J. 1990. H_2 cycling in N_2 fixation: past, present, and future outlook, p. 67–76. *In* P.M. Gresshoff, L.E. Roth, G. Stacey, and W.E. Newton (ed.), *Nitrogen Fixation: Achievements and Objectives.* Chapman & Hall, New York, N.Y.

Bar-Ness, E., Y. Chen, Y. Hadar, H. Marschner, and V. Römheld 1991. Siderophores of *Pseudomonas putida* as an iron source for dicot and monocot plants, p. 271–281. *In* Y. Chen and Y. Hadar (ed.), *Iron Nutrition and Interactions in Plants.* Kluwer Academic Publishers, Dordrecht, The Netherlands.

Cantrell, M. A., R. A. Haugland, and H. J. Evans. 1983. Construction of a *Rhizobium japonicum* gene bank and use in isolation of a hydrogen uptake gene. *Proc. Natl. Acad. Sci. USA* **80:**181–185.

Chet, I., and J. Inbar. 1994. Biological control of fungal pathogens. *Appl. Biochem. Biotechnol.* **48:**37–43.

Crosa, J. H. 1989. Genetics and molecular biology of siderophore-mediated iron transport in bacteria. *Microbiol. Rev.* **53:**517–530.

Evans, H. J., A. R. Harker, H. Papen, S. A. Russell, F. J. Hanus, and M. Zuber. 1987. Physiology, biochemistry, and genetics of the uptake hydrogenase in *Rhizobia. Annu. Rev. Microbiol.* **41:**335–361.

Fischer, H.-M. 1994. Genetic regulation of nitrogen fixation in Rhizobia. *Microbiol. Rev.* **58:**352–386.

Glick, B. R. 1995. The enhancement of plant growth by free-living bacteria. *Can. J. Microbiol.* **41:**109–117.

Glick, B. R., and Y. Bashan. 1997. Genetic manipulation of plant growth-promoting bacteria to enhance biocontrol of phytopathogens. *Biotechnol. Adv.* **15:**353–378.

Glick, B. R., C. B. Jacobson, M. M. K. Schwarze, and J. J. Pasternak. 1994. 1-Aminocyclopropane-1-carboxylic acid deaminase mutants of the plant growth-promoting rhizobacterium *Pseudomonas putida* GR12-2 do not stimulate canola root elongation. *Can. J. Microbiol.* **40:**911–915.

Glick, B. R., D. M. Penrose, and J. Li. A model for the lowering of plant ethylene concentrations by plant growth-promoting bacteria. *J. Theor. Biol.,* in press.

Glick, B. R., J. Zeisler, A. M. Banaszuk, J. D. Friesen, and W. G. Martin. 1981. The identification and partial characterization of a plasmid containing the gene for the membrane-associated hydrogenase from *E. coli. Gene* **15:**201–206.

Gresshoff, P. M., L. E. Roth, G. Stacey, and W. E. Newton (ed.). 1990. *Nitrogen Fixation: Achievements and Objectives.* Chapman & Hall, New York, N.Y.

Hennecke, H. 1990. Nitrogen fixation genes involved in the *Bradyrhizobium japonicum*-soybean symbiosis. *FEBS Lett.* **268:**422–426.

Higashi, S. 1993. (*Brady*)*Rhizobium*-plant communications involved in infection and nodulation. *J. Plant Res.* **106:**201–211.

Jones, D. A., M. H. Ryder, B. G. Clare, S. K. Farrand, and A. Kerr. 1988. Construction of a Tra− deletion mutant of pAgK84 to safeguard the biological control of crown gall. *Mol. Gen. Genet.* **212:**207–214.

Kloepper, J. W., R. Lifshitz, and M. N. Schroth. 1988. *Pseudomonas* inoculants to benefit plant production. *ISI Atlas Sci. Anim. Plant Sci.* 60–64.

Layva, A., J. M. Palacios, T. Mozo, and T. Ruiz-Argüeso. 1987. Cloning and characterization of hydrogen uptake genes from *Rhizobium leguminosarum. J. Bacteriol.* **169:**4929–4934.

Lerouge, P., P. Roche, C. Faucher, F. Maillet, G. Truchet, J. C. Promé, and J. Dénarié. 1990. Symbiotic host-specificity of *Rhizobium meliloti* is determined by a sulphated and acylated oligosaccharide signal. *Nature* **344**:781–784.

Long, S. R., W. J. Buikema, and F. M. Ausubel. 1982. Cloning of *Rhizobium meliloti* nodulation genes by direct complementation of Nod− mutants. *Nature* **298**:485–488.

Lynch, J. M. 1990. Beneficial interactions between microorganisms and roots. *Biotechnol. Adv.* **8**:335–346.

Marugg, J. D., M. van Spanje, W. P. M. Hoekstra, B. Schippers, and P. J. Weisbeek. 1985. Isolation and analysis of genes involved in siderophore biosynthesis in plant-growth-stimulating *Pseudomonas putida* WCS358. *J. Bacteriol.* **164**:563–570.

Marugg, J. D., H. B. Nielander, A. J. G. Horrevoets, I. van Megen, I. van Genderen, and P. J. Weisbeek. 1988. Genetic organization and transcriptional analysis of a major gene cluster involved in siderophore biosynthesis in *Pseudomonas putida* WCS358. *J. Bacteriol.* **170**:1812–1819.

Morris, R. O. 1986. Genes specifying auxin and cytokinin biosynthesis in phytopathogens. *Annu. Rev. Plant Physiol.* **37**:509–538.

Mylona, P., K. Pawlowski, and T. Bisseling. 1995. Symbiotic nitrogen fixation. *Plant Cell* **7**:869–885.

Nap, J.-P., and T. Bisseling. 1990. Developmental biology of a plant-prokaryote symbiosis: the legume root nodule. *Science* **250**:948–954.

Neilands, J. B., and S. A. Leong. 1986. Siderophores in relation to plant growth and disease. *Annu. Rev. Plant Physiol.* **37**:187–208.

O'Sullivan, D. J., and F. O'Gara. 1992. Traits of fluorescent *Pseudomonas* spp. involved in suppression of plant root pathogens. *Microbiol. Rev.* **56**:662–676.

Paau, A. S. 1991. Improvement of *Rhizobium* inoculants by mutation, genetic engineering, and formulation. *Biotechnol. Adv.* **9**:173–184.

Patten, C. L., and B. R. Glick. 1996. Bacterial biosynthesis of indole-3-acetic acid. *Can. J. Microbiol.* **42**:207–220.

Peters, J. W., K. Fisher, and D. R. Dean. 1995. Nitrogenase structure and function: a biochemical-genetic perspective. *Annu. Rev. Microbiol.* **49**:335–366.

Rossen, L., E. O. Davis, and A. W. B. Johnston. 1987. Plant-induced expression of *Rhizobium* genes involved in host specificity and early stages of nodulation. *Trends Biol. Sci.* **12**:430–433.

Schnider, U., C. Keel, C. Blumer, J. Troxler, G. Défago, and D. Haas. 1995. Amplification of the housekeeping sigma factor in *Pseudomonas fluorescens* CHA0 enhances antibiotic production and improves biocontrol abilities. *J. Bacteriol.* **177**:5387–5392.

Spaink, H. P., C. A. Wijffelman, E. Pees, R. J. H. Okker, and B. J. J. Lugtenberg. 1987. *Rhizobium* nodulation gene *nodD* as a determinant of host specificity. *Nature* **328**:337–340.

Sprent, J. I. 1986. Benefits of *Rhizobium* to agriculture. *Trends Biotechnol.* **4**:124–129.

Stacey, G. 1995. *Bradyrhizobium japonicum* nodulation genetics. *FEMS Microbiol. Lett.* **127**:1–9.

Sun, X., M. Griffith, J. J. Pasternak, and B. R. Glick. 1995. Low temperature growth, freezing survival and production of antifreeze protein by the plant growth-promoting rhizobacterium *Pseudomonas putida* GR12-2. *Can. J. Microbiol.* **41**:776–784.

van Rhijn, P., and J. Vanderleyden. 1995. The *Rhizobium*-plant symbiosis. *Microbiol. Rev.* **59**:124–142.

REVIEW QUESTIONS

1. Starting with a strain of *Rhizobium japonicum* that can fix nitrogen and form a symbiotic relationship with soybean roots, and assuming that you do not have a DNA hybridization probe for *nod* genes, outline a scheme that

you would use for isolating the cluster of nodulation genes from this organism.

2. What is hydrogenase? How could it be used to improve the yield of alfalfa?

3. Suggest a strategy that you might employ to isolate all of the genes involved in nitrogen fixation from *Azotobacter vinelandii*, assuming that you do not have *nif* genes from other microorganisms to use as hybridization probes.

4. What would you expect the effect of mutagenizing either *nifA* or *nifL* to be on the amount of nitrogen that an organism can fix?

5. Discuss whether it is possible to genetically engineer plants to fix their own nitrogen.

6. Suggest a scheme for isolating hydrogenase genes.

7. What are siderophores? How could genetic manipulation of siderophore genes enable bacteria to enhance plant growth?

8. Suggest a scheme for isolating siderophore biosynthesis genes.

9. What are the advantages of microbial fertilizers over chemical fertilizers?

10. How can *Agrobacterium radiobacter* be engineered to be a more effective biocontrol agent?

11. Which enzymes secreted by plant growth-promoting bacteria contribute to their ability to act as biocontrol agents? How do these enzymes work?

Microbial Insecticides

Of all classes of organisms, insects have the largest number of described species ($>7.5 \times 10^5$). Insects negatively affect humans in a variety of ways: they cause massive crop damage, and they act as vectors of both human and animal diseases. During the 1940s, a number of chemical insecticides were developed as a means of controlling the proliferation of noxious insect populations. One of these was the chlorinated hydrocarbon DDT (dichlorodiphenyltrichloroethane), which had originally been synthesized in the 1870s but was not recognized as an insecticide until the late 1930s. DDT proved to be exceptionally effective in killing and controlling many species of pests. Chlorinated hydrocarbons like DDT function by attacking the nervous system and muscle tissue of insects. Other chlorinated hydrocarbons such as dieldrin, aldrin, chlordane, lindane, and toxophene have since been synthesized and applied on a massive scale.

Another class of chemical insecticides is called organophosphates and includes malathion, parathion, and diazinon. The first generation of organophosphates was developed as chemical warfare agents. Now they are used to control insect populations by inhibiting the enzyme acetylcholinesterase, which hydrolyzes the nerve transmitter acetylcholine. These insecticides disrupt the functioning of motor neurons and neurons in the brain of the insect.

By the early 1960s, over 100 million acres of U.S. agricultural land was being treated annually with chemical insecticides. At about that time, researchers realized that chlorinated hydrocarbon insecticides (to a large extent) and organophosphate insecticides (to a lesser extent) had dramatic and immediate side effects and long-term and indirect effects on animals, ecosystems, and humans. Chlorinated hydrocarbons, exemplified by DDT, were found to persist in the environment for 15 to 20 years and to accumulate in increasing concentrations through food chains. This bioaccumulation in fatty tissues had a significant biological impact on many organisms. For example, in North America, many species of birds, including peregrine falcons, sparrow hawks, bald eagles, brown pelicans, and double-crested cormorants, were severely depopulated.

During the 1950s, as the targeted insect pest populations became increasingly resistant to treatment with many chemical insecticides, higher concentrations of the insecticides had to be applied to control the pest species. In addition, chemical insecticides were found to lack specificity; consequently, beneficial insects were being killed along with those that were considered to be pests. In fact, in some instances, the natural enemies of the insect pest species were killed off more efficiently than the target organisms, with the bizarre result that pesticide treatment led to higher levels of the insect pest after the application of the pesticide.

Given all the drawbacks associated with the use of chemical insecticides, alternative means of controlling harmful insects have been sought over the past 20 years. Using insecticides that are produced naturally by either microorganisms or plants was an obvious choice. On the positive side, these compounds are usually highly specific for a target insect species, biodegradable, and slow to select for resistance. But, on the negative side, their low potency and high cost of production limit their use for a variety of applications. These negative attributes might be overcome by recombinant DNA technology. Researchers can now manipulate genes that encode insect pathogenic agents to increase the effectiveness of microbial insecticides. In particular, the insecticidal activities of the bacterium *Bacillus thuringiensis* and insect baculoviruses systems appear to be safe, specific, and effective.

The worldwide market for pesticides is enormous: currently more than $20 billion per year and growing rapidly. Although biopesticides, mostly *B. thuringiensis*, make up only about 1% of this total, it is expected that most of the growth in this field in the future will involve biopesticides.

Insecticidal Toxin of *Bacillus thuringiensis*

Mode of Action and Use

A microbial insecticide can be an organism that either produces a toxic substance that kills an insect species or has the ability to fatally infect a specific target insect. The most studied, most effective, and most often utilized microbial insecticides are the toxins synthesized by *B. thuringiensis*. This bacterium comprises a number of different strains and subspecies (subsp.), each of which produces a different toxin that can kill certain specific insects (Table 15.1). For example, *B. thuringiensis* subsp. *kurstaki* is toxic to lepidopteran larvae, including moth, butterfly, and skipper larvae, cabbage worms, and spruce budworms. *B. thuringiensis* subsp. *israelensis* kills diptera such as mosquitoes and blackflies. *B. thuringiensis* subsp. *tenebrionis* (also known as *san diego*) is effective against coleoptera (beetles) such as the potato beetle and the boll weevil. There are other *B. thuringiensis* strains, and each is toxic to a distinct set of insects.

The insecticidal activity (toxin) of *B. thuringiensis* subsp. *kurstaki* and other strains is contained within a very large structure called the parasporal crystal, which is synthesized during bacterial sporulation. No significant role for the bacterium has been attributed to the parasporal crystal structure. The parasporal crystal comprises approximately 20 to 30% of the dry weight of the sporulated culture and usually consists mainly of protein (~95%) and a small amount of carbohydrate (~5%). The crystal is an aggregate of protein that can generally be dissociated by mild alkali treatment into subunits. The subunits can be further dissociated in vitro by treatment with β-mercaptoethanol, which reduces disulfide linkages (Fig. 15.1).

The insecticidal toxins from the several thousand *B. thuringiensis* strains

Table 15.1 Some properties of the insecticidal toxins from various strains of *B. thuringiensis*

B. thuringiensis strain or subspecies	**Toxin class**	**Protoxin size (kDa)**	**Target insects**	**Serotype**
berliner	CryI	130–140	Lepidoptera	1
kurstaki KTO, HD-1	CryI	130–140	Lepidoptera	3
entomocidus 6.01	CryI	130–140	Lepidoptera	6
aizawai 7.29	CryI	130–140	Lepidoptera	7
aizawai IC 1	CryI	135	Lepidoptera, Diptera	7
kurstaki HD-1	CryII	71	Lepidoptera, Diptera	3
tenebrionis (*san diego*)	CryIII	66–73	Coleoptera	8
morrisoni PG14	CryIV	125–145	Diptera	8
israelensis	CryIV	68	Diptera	14

Adapted from Lereclus et al., p. 37–69, *in* Entwistle et al., (ed.), Bacillus thuringiensis, *an Environmental Biopesticide: Theory and Practice*, 1993.

that have been isolated and characterized, based on the insecticidal activity of the toxin, can be grouped into four major classes: CryI, CryII, CryIII, and CryIV. The CryI proteins are generally toxic to lepidoptera, CryII to both lepidoptera and diptera, CryIII to coleoptera, and CryIV to diptera. These proteins are further arranged into subclasses (A, B, C,...) and subgroups (a, b, c,...) according to the DNA sequence of the toxin gene. For example, the *cryI* gene class contains six subclasses (i.e., *cryIA* to *F*) and the *cryIA* gene subclass has three subgroups [i.e., *cryIA(a)* to *(c)*]. In addition, based on immunological differences, about 30 different serotypes of *B. thuringiensis* have been reported (for example, see Table 15.1). A separate serotype is due to a particular set of antigenic differences (epitopes) on the surface of one *B. thuringiensis* strain in contrast to another.

The parasporal crystal does not usually contain the active form of the insecticide; rather, once the crystal has been solubilized, the protein that is released is a protoxin, a precursor of the active toxin. The protoxin of the CryI toxin group has a molecular mass of approximately 130 kilodaltons (kDa) (Fig. 15.1). When a parasporal crystal is ingested by a target insect, the protoxin is activated within its gut by the combination of alkaline pH (7.5 to 8.0) and specific digestive proteases, which converts the protoxin into an active toxin with a molecular mass of approximately 68 kDa (Fig. 15.1). In its active form, the toxic protein inserts itself into the membrane of the gut epithelial cells of the insect and creates an ion channel through which there is believed to be an excessive loss of cellular ATP (Fig. 15.2). About 15 minutes after this ion channel forms, cellular metabolism ceases, and the insect stops feeding, becomes dehydrated, and eventually dies. Because the conversion of the protoxin to the active toxin requires *both* alkaline pH and the presence of specific proteases, it is unlikely that nontarget species such as humans and farm animals will be affected.

The mode of action of *B. thuringiensis* toxins imposes certain constraints on its application. To kill an insect pest, *B. thuringiensis* must be ingested. Contact of the bacterium or the insecticidal toxin with the surface of an insect has no effect on the target organism. Ingestion, in part, limits the susceptibility of nontarget insects and other animals to this insecticide. *B. thuringiensis* is generally applied by spraying, so it is usually formulated with insect attractants to increase the probability that the target insect will ingest the toxin. But insects that bore into plants or attack plant roots are less likely to ingest a *B. thuringiensis* toxin that has been sprayed on a host plant,

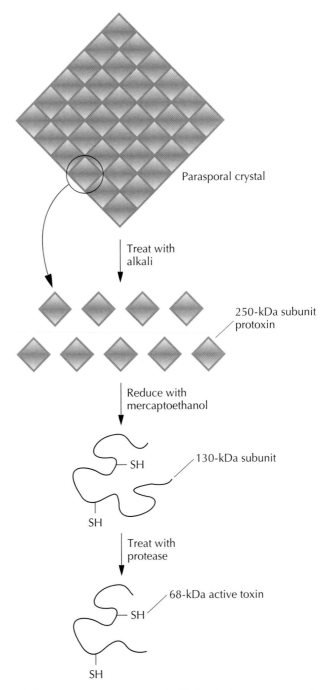

Parasporal crystal

Treat with
alkali

250-kDa subunit
protoxin

Reduce with
mercaptoethanol

130-kDa subunit

Treat with
protease

68-kDa active toxin

Figure 15.1 Schematic representation of a *B. thuringiensis* parasporal crystal composed of CryI protoxin protein. Each 250-kDa protein subunit of the parasporal crystal contains two 130-kDa polypeptides. (Molecular masses determined by polyacrylamide gel electrophoresis are approximations and do not always provide exact multiples.) Conversion of the 130-kDa protoxin into an active 68-kDa toxin requires the combination of a slightly alkaline pH (7.5 to 8) and the action of a specific protease(s), both of which are found in the insect gut.

so other strategies have been devised to control such pests. One approach is to create transgenic plants that carry and express a *B. thuringiensis* toxin gene and consequently are protected from infestation throughout the growing season.

A second limiting feature of the action of the *B. thuringiensis* toxin is that it can kill a susceptible insect only during a specific developmental stage. Therefore, the toxin must be applied when the pest population is at a particular stage in its life cycle.

B. thuringiensis subsp. *kurstaki* was first discovered in 1901, although its commercial potential was largely ignored until 1951. Within the last decade, however, *B. thuringiensis* subsp. *kurstaki* has become the major means of controlling the spruce budworm in Canada. In 1979 approximately 1% of the forests in Canada that were treated with an insecticide to combat the spruce budworm (about 2 million hectares, or 8,000 square miles) were sprayed with *B. thuringiensis* subsp. *kurstaki*. The remainder were sprayed with chemical insecticides. By 1986, the use of *B. thuringiensis* subsp. *kurstaki* had increased dramatically: it was used to treat approximately 74% of the forests sprayed in that year for spruce budworm. In other countries, *B. thuringiensis* subsp. *kurstaki* has been used against tent caterpillars, gypsy moths, cabbage worms, cabbage loopers, and tobacco hornworms. The major impediment to even more widespread application of *B. thuringiensis* subsp. *kurstaki* is that it costs from 1.5 to 3 times as much as chemical insecticides.

For the biological control (biocontrol) of insect pests, *B. thuringiensis* subsp. *kurstaki* is applied by spraying approximately 1.3×10^8 to 2.6×10^8 spores per square foot of the target area. Administration of the spores is timed to coincide with the peak of the larval population of the target organism because the parasporal crystals, being sensitive to sunlight, are short-lived in the environment. In simulated conditions, sunlight degrades over 60% of the tryptophan residues of the parasporal crystal within a 24-hour period. Depending on the amount of sunlight present, parasporal crystals may persist in the environment for as little as a day or as long as a month. The lack of persistence in the natural environment of the insecticidal protoxin means that normally selection of resistant insects is highly unlikely.

However, when *B. thuringiensis* subsp. *kurstaki* is used as an insecticide in a controlled environment where there is no sunlight (e.g., when stored grain is being protected against insect predation), resistant target insects have been found to develop within a few generations. One form of this in-

Figure 15.2 Insertion of the *B. thuringiensis* toxin into the membrane of an insect gut epithelial cell. The toxin forms an ion channel between the cell cytoplasm and the external environment.

herited resistance is due to an alteration in a midgut membrane protein that normally acts as a receptor for the *B. thuringiensis* subsp. *kurstaki* toxin. Resistance probably develops because the protoxin persists under these conditions and acts as a selection agent. The lesson here is that the simplest way to avoid selecting for insects that are resistant to *B. thuringiensis* subsp. *kurstaki* is to limit the use of this bacterium to field applications. However, extensive annual use even under natural conditions may result in a level of persistence high enough to allow selection. Certainly, as larger quantities of *B. thuringiensis* are used over a wider and wider geographical area, the probability that resistant strains of insects will be selected will increase.

Toxin Gene Isolation

To develop *B. thuringiensis*-based insecticides that have greater potencies and broader host ranges, it is necessary as a first step to isolate and characterize the protoxin gene(s). The first aspect of this problem is to determine whether the toxin genes are located on a plasmid or on the chromosomal DNA. To test for plasmid-borne toxin genes, the source *B. thuringiensis* strain can be conjugated with a strain that lacks insecticidal activity. If the latter strain acquires the ability to synthesize the insecticidal toxin, then the toxin gene(s) is most likely present on a plasmid because the transfer of chromosomal DNA during conjugation is a rare event.

The procedure for isolating a protoxin-encoding DNA sequence is a familiar one. *B. thuringiensis* cells are grown in laboratory culture and lysed. The total cellular DNA is isolated and separated into plasmid and chromosomal DNA fractions by cesium chloride (CsCl) gradient centrifugation. When the protoxin gene is part of the genome, a clone bank is constructed from the chromosomal DNA. When the toxin gene(s) is plasmid encoded, the plasmid DNA can be further fractionated by sucrose gradient centrifugation, which separates different plasmids according to their size and enriches for the DNA that serves as the starting material for the isolation of a protoxin gene(s) (Fig. 15.3).

B. thuringiensis subsp. *kurstaki* contains an insecticidal protoxin gene on one of seven different plasmids that are approximately 2.0, 7.4, 7.8, 8.2, 14.4, 45, and 71 kilobase pairs (kb) in length. To determine which *B. thuringiensis* subsp. *kurstaki* plasmid carries the protoxin gene, following sucrose gradient centrifugation, the plasmid DNA sample is divided into three fractions that respectively contain the small (2.0-kb), medium (7.4-, 7.8-, 8.2-, and 14.4-kb), and large (45- and 71-kb) plasmids. The fraction with the small plasmid (2.0 kb) is discarded because this plasmid is too small to encode a protein equivalent to the 130-kDa protoxin. A protein of this size requires at least 4.0 kb of coding DNA. The medium and large plasmid fractions are each partially digested with the restriction enzyme *Sau*3AI and then ligated into the *Bam*HI site of plasmid pBR322. In the original experiments, these clone banks were transformed into *E. coli* and then screened immunologically by the following procedure.

1. Colonies are transferred from agar plates to a nitrocellulose membrane.
2. The transferred colonies are partially lysed with organic solvents.
3. All available sites on the membrane to which primary and secondary antibodies could bind are blocked by treating the membrane with bovine serum albumin.
4. The bovine serum albumin-treated membranes are treated with rabbit antiserum that contains antibodies against the insecticidal toxin.

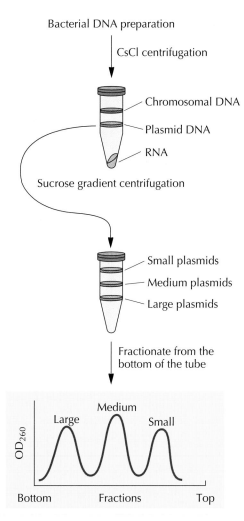

Bacterial DNA preparation

CsCl centrifugation

Chromosomal DNA

Plasmid DNA

RNA

Sucrose gradient centrifugation

Small plasmids

Medium plasmids

Large plasmids

Fractionate from the bottom of the tube

Figure 15.3 Procedure for the isolation and partial enrichment of plasmid DNA fractions from a microorganism with a number of different plasmids, one of which encodes an insecticidal protoxin.

5. The membranes are washed to remove unbound antibodies and then treated with [125]I-labeled *Staphylococcus aureus* protein A, which binds to the Fc portion of the bound antibodies.

6. Spots on the membrane corresponding to colonies that actively synthesize the insecticidal toxin are visualized by autoradiography.

When the isolated protoxin gene was used as a DNA hybridization probe, the 71-kb plasmid of *B. thuringiensis* subsp. *kurstaki* was found to encode this sequence. Similar cloning and screening procedures have been used to isolate *B. thuringiensis* toxin genes located either on plasmid DNA or, less often, on chromosomal DNA of other strains.

Engineering of *B. thuringiensis* Toxin Genes

Once the isolation and sequencing of a toxin gene were accomplished, the complete amino acid sequence was deduced. Comparisons of amino acid sequences from other toxin proteins showed that a common toxic domain

exists in some of the strains. Moreover, a subcloned segment of the complete coding sequence produced a truncated protein that retained full insecticidal activity. Thus, an intact toxin gene, a portion of one, or a chemically synthesized coding sequence can be used for further genetic manipulation.

Under normal conditions, most *B. thuringiensis* protoxin proteins are synthesized only during the sporulation phase of growth. In other words, only a portion of the growth cycle of the organism is devoted to parasporal crystal production. It might therefore be advantageous in terms of increased yield and decreased production time to have the toxin gene transcribed and translated during vegetative growth. Furthermore, production of the insecticidal toxin during vegetative growth would permit the toxin to be synthesized by a *continuous* fermentation process, potentially significantly decreasing the cost of producing the toxin. Continuous fermentations are carried out with smaller-scale—and therefore less expensive—bioreactors and downstream processing equipment than conventional batch fermentations. (See Chapter 16 for additional details.)

During the sporulation of *B. thuringiensis*, a specific transcription initiation factor (sigma factor) interacts with the promoters of genes that are active only within this phase of the bacterial life cycle. This factor turns on the transcription of the messenger RNAs (mRNAs) that are unique to sporulation. In fact, when a *B. thuringiensis* toxin gene with its sporulation-specific promoter was cloned and expressed in *Bacillus subtilis*, *Bacillus megaterium*, or *B. thuringiensis*, gene transcription occurred only during sporulation. Thus, to express a *B. thuringiensis* insecticidal toxin during vegetative growth, it is necessary to place the toxin-producing gene(s) under the control of a promoter that is active during vegetative growth.

When a DNA fragment containing a toxin gene that lacked its native promoter was cloned into a plasmid under the control of a continuously active, constitutive promoter from a tetracycline resistance gene that had been originally isolated from a *Bacillus cereus* plasmid and reintroduced into *B. thuringiensis*, active toxin protein was produced continuously throughout the growth cycle, including both the vegetative and sporulation phases (Fig. 15.4). In addition, when the construct was used to transform a sporulation-defective mutant of *B. thuringiensis*, toxin synthesis occurred in the absence of sporulation. Under these conditions, toxin synthesis is more efficient than in wild-type cells; i.e., the final amount of protein obtained is larger in the transformed cells, and the amounts of time and substrate that are required to produce it are smaller.

A refinement of this system might entail integration of this vegetatively expressed toxin gene into the chromosomal DNA of the sporulation-defective *B. thuringiensis* host. This manipulation would ensure that the insecticidal toxin gene is not lost due to plasmid instability during a continuous fermentation process.

Unlike most other *B. thuringiensis* toxin protein (*cry*) genes, the expression of *cryIIIA* is normally controlled by a vegetative promoter rather than by a sporulation-specific promoter. The *cryIIIA* gene encodes a toxin that is directed against coleopteran larvae. When a mutant strain of *B. thuringiensis* that was unable to form spores was transformed with a plasmid carrying a cloned *cryIIIA* gene, the insecticidal toxin was both overproduced and stabilized in comparison to when this protein was produced in the wild-type strain. This result suggests that other *cry* genes that are normally expressed only during sporulation could be placed under the control of the *cryIIIA*

Figure 15.4 Procedure for subcloning the *B. thuringiensis* subsp. *kurstaki* insecticidal toxin gene so that it is expressed constitutively under the control of the promoter of the tetracycline resistance gene (*p*Tet). The isolated *B. thuringiensis* toxin gene is removed from its promoter by digestion of the isolated DNA fragment with restriction enzymes RE1 and RE2. It is spliced by T4 DNA ligase into the plasmid vector downstream from *p*Tet in place of the tetracycline resistance gene, which has been removed by digestion with restriction enzymes RE1 and RE2.

promoter and overproduced by merely expressing these constructs in a sporulation-defective *B. thuringiensis* mutant.

Because many crops are attacked by more than one insect species, it would be advantageous, if feasible, to create microbial insecticides that are each effective against a broad spectrum of target insects. Such a broad-specificity compound could be obtained by (1) transferring the gene for a particular toxin (e.g., one against diptera) into a *B. thuringiensis* strain that normally synthesizes a different species-specific toxin (e.g., one against coleoptera) or (2) by fusing portions of two different species-specific toxin genes to one another so that a unique dual-acting toxin (hybrid toxin) is produced.

In one experiment designed to test whether the spectrum of target insect pests could be widened, the insecticidal toxin genes from *B. thuringiensis* subspecies *aizawai* and *tenebrionis* were cloned into shuttle vectors that could be maintained in both *B. thuringiensis* and *Escherichia coli.* These genetic constructs were then introduced by electroporation into *B. thuringien-*

MILESTONE

Cloning and Expression of the *Bacillus thuringiensis* Crystal Protein Gene in *Escherichia coli*

H. E. SCHNEPF and H. R. WHITELEY
Proc. Natl. Acad. Sci. USA **78**:2893–2897, 1981

Although it had been well known for a long time that the parasporal crystal that is produced by *B. thuringiensis* contained insecticidal activity, it took scientists many years before the conditions for solubilizing the crystal were discovered and the insecticidal protein toxin was isolated in a pure form. Moreover, although some *B. thuringiensis* mutants defective in the synthesis of the parasporal crystal were known, at the time protocols for the genetic transformation of *B. thuringiensis* were not well developed. Therefore, when Schnepf

and Whiteley decided to isolate *B. thuringiensis* insecticidal toxin protein genes, they were limited to screening *E. coli* transformants carrying *B. thuringiensis* DNA either immunologically, using antibodies directed against the whole crystal, or by the insecticidal activity of extracts of the transformants. Since evidence at the time suggested that the insecticidal toxin was probably plasmid encoded, clone banks were constructed from fractionated plasmid preparations, which significantly enriched the clone banks for the presence of insecticidal toxin

genes. Moreover, despite some concerns that antibodies against the whole (glycosylated) crystal protein might not interact with (nonglycosylated) crystal protein subunits in solution, this did not turn out to be a problem. Finally, extracts of the particulate fraction of *E. coli* transformants carrying the *B. thuringiensis* insecticidal toxin gene were found to be toxic to susceptible insects. This first cloning of a *B. thuringiensis* insecticidal toxin gene made it clear to workers in this field that these genes could be isolated in a straightforward manner and provided an impetus for increased activity both in the search for new strains of *B. thuringiensis* and in studies of the biochemistry of the insecticidal toxin.

sis subspecies *aizawai*, *kurstaki*, *israelensis*, and *tenebrionis*, and all the transformed strains were tested for their toxicity toward three different larvae.

In each case the toxicity of the endogenous host toxin gene(s) was maintained, and in most cases the introduced toxin gene also expressed the toxic specificity of its source (Table 15.2). In addition, and surprisingly, when the *B. thuringiensis* subsp. *tenebrionis* toxin gene was introduced into *B. thuringiensis* subsp. *israelensis*, the resultant transformant was somewhat toxic to *Pieris brassicae* (the cabbage white butterfly), against which neither of the gene products alone has insecticidal activity.

In many instances, researchers have found that introduced plasmid vectors carrying isolated *cry* genes are unstable in *B. thuringiensis*. Often, in the absence of selective pressure, all or a portion of these plasmids are lost. Interestingly, this occurs despite the fact that in nature most *cry* genes are located on plasmids. This instability provides a strong argument for the integration of a cloned *cry* gene(s) into the chromosomal DNA of the host. The problem of plasmid instability of introduced genes was overcome by integrating cloned *cry* genes into the chromosomal DNA of the host cell. One group of researchers attempted to broaden the insect specificity of a strain of *B. thuringiensis* subsp. *kurstaki* which normally encodes five different insecticidal toxin genes, *cryIA(a)*, *cryIA(b)*, *cryIA(c)*, *cryIIA*, and *cryIIB*. While the products of these *cry* genes are active against a variety of lepidopteran species, they are not effective against *Spodoptera* spp. Therefore, a *cryIC* gene which is normally found only in *B. thuringiensis* subsp. *aizawai* and subsp. *entomocidus* was introduced into the chromosomal DNA of the *B. thuringiensis* subsp. *kurstaki* host strain. The transformed *B. thuringiensis* subsp. *kurstaki* strain showed a sixfold increase in its ability to kill *Spodoptera exigua* (beet armyworm) larvae.

As noted above, fusion of the coding portions of the active regions of

Table 15.2 Toxicity of naturally occurring and transformed subspecies of *B. thuringiensis* against the insects *Pieris brassicae* (cabbage white butterfly), *Aedes aegypti* (mosquito), and *Phaedon cochleariae* (beetle)

Source of toxin		Toxicity to:		
Host DNA	**Introduced DNA**	*Pieris*	*Aedes*	*Phaedon*
aizawai	None	+ +	+	−
israelensis	None	−	+ +	−
israelensis	*aizawai*	+ +	+ +	−
israelensis	*tenebrionis*	+	+ +	+ +
kurstaki	None	+ +	+	−
kurstaki	*tenebrionis*	+ +	+	+ +
tenebrionis	None	−	−	+ +
tenebrionis	*aizawai*	+ +	+	+

Adapted from Crickmore et al., *Biochem. J.* **270**:133–136, 1990.

In these experiments, the toxicity was graded as follows: + +, 0 to 5% of the leaf was consumed (for *Phaedon* and *Pieris*) or 100% mortality occurred within 1 hour (*Aedes*); +, 5 to 50% of the leaf was consumed (*Phaedon* and *Pieris*) or 50 to 100% mortality occurred within 24 hours (*Aedes*); −, >50% of the leaf was consumed (*Phaedon* and *Pieris*) or no mortality occurred within 24 hours (*Aedes*). The test plant was either cabbage leaf (for *Pieris*) or turnip leaf (for *Phaedon*).

two different toxin genes is another way of generating a novel protein with extended toxicity. This idea has been examined in laboratory experiments. When a series of lepidopteran-specific hybrid toxins were constructed, some of them were more toxic than the products of either of the contributing genes were by themselves, and in one case a hybrid protein had acquired a totally new biological activity.

The toxic moiety of many Cry proteins is composed of three separate domains. Domain I, at the N-terminal region of the toxin, is involved in the specific binding of the toxin to protein receptors that are found on the surface of insect midgut epithelial cells. Domain III, which is located at the C-terminal end of the toxin molecule, is thought to contain the actual toxin portion of this molecule. Generally, resistance to *B. thuringiensis* insecticidal toxins is the consequence of a mutation(s) that alters an insect midgut receptor protein(s) so that it no longer binds to the Cry protein. However, if a toxin gene was engineered so that the toxin bound to other midgut cell surface proteins, then resistance might be less likely to arise.

The insecticidal proteins CryIC and CryIE are both toxic to lepidoptera but have different species specificities. CryIC is active against *S. exigua*, *Mamestra brassicae*, and *Manduca sexta*, while CryIE is active only against *M. sexta*. In one experiment, hybrid CryIC–CryIE proteins were constructed and tested for their toxicity to different insect species as well as for their ability to bind to different receptors (Fig. 15.5). The hybrid toxin G27, which contained domain III from CryIC, was toxic to *S. exigua* larvae even though it bound to the CryIE receptor but not to the CryIC receptor. Conversely, the hybrid toxin F26 was not toxic to *S. exigua* larvae even though it bound to the CryIC receptor. Since the CryIC and G27 proteins bind to different insect midgut receptors although both are toxic to *S. exigua*, either simultaneous or alternating treatments of *S. exigua* with these two *B. thuringiensis* insecticidal toxins might limit the development of strains that are resistant to these toxins. Resistance to both CryIC and G27 would require mutations in two separate midgut proteins.

		Toxic to *S. exigua*	Competes for CryIC site	Competes for CryIE site
CryIC	Domain I / Domain II / Domain III	+	+	−
CryIE	Domain I / Domain II / Domain III	−	−	+
G27	Domain I / Domain II / Domain III	+	−	+
F26	Domain I / Domain II / Domain III	−	+	−

Figure 15.5 Toxicity and binding specificity of CryIC, CryIE, and hybrid toxins G27 and F26. A toxin–receptor binding assay was used to determine binding specificities. Unlabeled CryIC or CryIE was added to a complex of the *S. exigua* midgut receptor protein and radiolabeled toxin, and the extent of binding of radiolabeled toxin was determined. Adapted from Bosch et al., *Bio/Technology* **12:**915–918, 1994.

The *B. thuringiensis* subsp. *israelensis* insecticidal protein is highly toxic when ingested by mosquito larvae. However, the parasporal crystal of this species sinks rapidly after it is sprayed on water, which effectively removes it from the feeding area of mosquito larvae and dramatically decreases its efficacy as a mosquitocide. To overcome this shortcoming, the insecticidal toxin gene might be introduced into an organism that is a common food source for mosquito larvae. Good candidate organisms for this purpose include *Synechocystis* and *Synechococcus* spp., which are photosynthetic cyanobacteria that are a food source for the larvae and proliferate near the water surface, where there is sufficient light for their growth and where mosquito larvae are normally found.

Another organism with the potential to be a host for the expression of foreign insecticidal toxin genes is *Caulobacter crescentus*, an aquatic bacterium that generally is widely distributed throughout aquatic environments where mosquito larvae feed. The toxin gene from *B. thuringiensis* subsp. *israelensis* was introduced into and expressed in these organisms. In laboratory trials, the insecticidal toxin that was produced by either transformed cyanobacteria or *C. crescentus* was toxic to mosquito larvae. However, in field trials, transformed cyanobacteria or *C. crescentus* expressing *B. thuringiensis* insecticidal toxin genes had poor viability and the cloned genes were expressed at a low level.

A possible alternative host for the expression of *cry* genes that produce a protein toxic to mosquitoes is *Asticcacaulis excentricus*, a gram-negative aerobe that is found in aqueous environments near the surface of the water. In a series of experiments, *A. excentricus* was transformed with a broad-host-range plasmid vector that carried the genes for mosquitocidal toxin proteins produced by a strain of *Bacillus sphaericus* (a bacterium similar to *B. thuringiensis*) under the control of the *tac1* promoter, which is a variant of the *tac* promoter. This transformant produced insecticidal toxin proteins of 51 and 42 kDa and was almost as toxic to *Anopheles* and *Culex* mosquito larvae as the naturally occurring high toxicity strains of *B. sphaericus* were.

However, unlike *B. sphaericus*, *A. excentricus* does not sink when it is sprayed onto ponds infested with mosquito larvae. Moreover, *A. excentricus* is inexpensive to produce, as it can be grown on much simpler media than either *B. sphaericus* or *B. thuringiensis*. It does not have a high level of protease activity, so the insecticidal toxin will not readily be degraded. It is well adapted to environments such as those near the surface of standing water that are exposed to relatively high levels of ultraviolet light. Thus, *A. excentricus* cells should not be as sensitive to inactivation by ultraviolet light as either *B. sphaericus* or *B. thuringiensis*. However, to use a genetically engineered strain of *A. excentricus* to control mosquito populations in the environment, it will be necessary to integrate the insecticidal toxin genes into the chromosomal DNA without any antibiotic resistance genes.

Insects that attack the roots of plants are not affected by *B. thuringiensis*-based insecticides that are sprayed onto leaves and shoots. However, it is possible to introduce the toxin gene from a *B. thuringiensis* strain into a bacterial species that colonizes the region adjacent to plant roots (the rhizosphere). The engineered bacteria could be introduced into the soil, where they would synthesize the insecticidal toxin and release it into the area immediately surrounding the plant roots, thereby conferring protection against root-attacking insects. In addition, as long as the engineered bacteria were able to persist in the soil, they would continue to synthesize the insecticidal toxin, thus obviating the need for repeated sprayings of either biological or chemical insecticides. This approach has been tested on a small scale. The gene for the *B. thuringiensis* subsp. *kurstaki* insecticidal toxin was integrated into the chromosomal DNA of a strain of *Pseudomonas fluorescens* that colonizes corn roots. The integration of the toxin gene was achieved as follows (Fig. 15.6).

1. A transposon Tn5 element that had been cloned into a plasmid was genetically modified by altering portions of its left and right borders and deleting its transposase. Such an altered Tn5 element cannot be excised from the plasmid, even by exogenous transposase.
2. An isolated *B. thuringiensis* subsp. *kurstaki* insecticidal toxin gene was spliced into the middle of the altered Tn5 element on the plasmid and placed under the control of a constitutive promoter.
3. A wild-type Tn5 element was transposed into the chromosome of the root-colonizing strain of *P. fluorescens*.
4. The plasmid carrying the altered Tn5 element with the inserted toxin gene was introduced into the host bacterium that had the integrated wild-type Tn5 element.
5. Homologous recombination by means of a double crossover between the nontransposable Tn5 element on the plasmid that carried the toxin gene and the chromosomally integrated wild-type Tn5 led to the integration of the altered Tn5 with the toxin gene into the chromosomal DNA, with the concomitant loss of the wild-type Tn5 element.

In this form, the toxin gene is unlikely to be lost either during large-scale laboratory growth or after release of the engineered microorganism to the environment. Also, the probability of transfer of the toxin gene to other microorganisms in the environment is very low. Laboratory trials showed that the engineered *P. fluorescens* was toxic to tobacco hornworm larvae. On the basis of these initial results, investigators are ready to test this geneti-

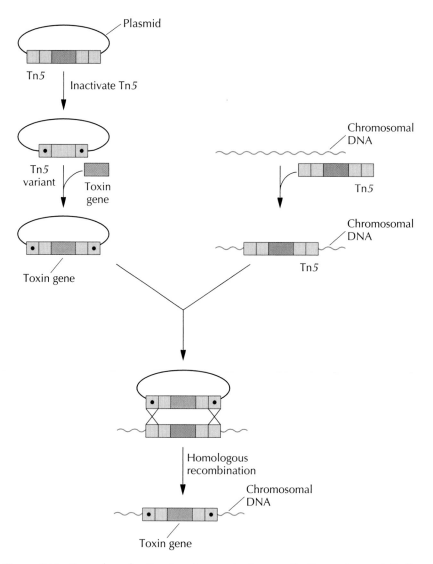

Figure 15.6 Procedure for the development of a genetically engineered *P. fluorescens* that carries a copy of the *B. thuringiensis* insecticidal toxin gene integrated into its chromosomal DNA. The *B. thuringiensis* insecticidal toxin gene is cloned into an excision-defective variant of Tn*5* on a plasmid. This construct is introduced into a *P. fluorescens* strain containing a wild-type Tn*5* sequence that has been integrated into its chromosomal DNA. By homologous recombination, the excision-defective Tn*5* element carrying the *B. thuringiensis* insecticidal toxin gene becomes integrated into the *P. fluorescens* chromosome.

cally manipulated microorganism for its ability to minimize root damage from insect predation in the greenhouse and in open-field trials.

In other laboratories, various *B. thuringiensis* insecticidal toxin genes have been introduced into the chromosomal DNA of a number of different microorganisms. For example, the *cryIA(c)* genes were introduced into a strain of *P. fluorescens* that protects sugarcane plants against the sugarcane borer, *Eldana saccharina*. As well, when this gene was used to transform *Clavibacter xyli* subsp. *cynodontis*, a bacterium that normally inhabits the xylem of Bermuda grass, the bacterium protected corn plants from damage caused by the European corn borer, *Ostrinia nubilalis*.

Baculoviruses as Biocontrol Agents

Mode of Action

Baculoviruses are rod-shaped, double-stranded DNA viruses that can infect and kill a large number of different invertebrate organisms. Subgroups of this viral family are pathogenic to several orders of insects, including the Lepidoptera, Hymenoptera, Diptera, Neuroptera, Trichoptera, Coleoptera, and Homoptera. In nature, some of these baculoviruses are important for the control of certain pest insects, and several have been registered for use as biological insecticides. Baculoviruses were used in North America against forest pests, including the spruce sawfly (*Neodiprion sertifer*), starting in the 1930s and ending with the advent of chemical pesticides in the 1960s. Baculoviruses continue to be used on a limited basis, mostly by the forestry industry in an effort to control the gypsy moth, *Lymantria dispar* (Table 15.3).

A baculovirus particle consists of a cylindrical nucleocapsid that surrounds the viral DNA. Often, in the nucleus of an infected cell, baculovirus particles are embedded in a crystalline protein matrix called an occlusion body. The occlusion body, which is called a polyhedron, is largely composed of the protein polyhedrin. When an infected insect dies, millions of polyhedra are released. Upon ingestion by an insect, the polyhedra move to the midgut, where the alkaline environment facilitates the dissolution of the polyhedrin protein coat, releasing infectious nucleocapsids. The nucleocapsids are taken up by the insect midgut cells and then migrate through the cytoplasm to the nucleus, where the nucleocapsid is removed. After viral replication, which takes place in the nucleus, and nucleocapsid assembly, some nucleocapsids are released by budding through the plasma membrane of infected cells into the circulatory system of the insect. Consequently, the infection spreads to other cells throughout the insect. It usually takes about 10 rounds of viral replication or about 5 to 6 days for the insect to die. At that stage, about 25% of the dry weight of the insect consists of polyhedra.

One positive feature of using baculoviruses as insect biocontrol agents is that, for the most part, they have limited host ranges and do not affect nontarget organisms. This characteristic means, however, that any particular baculovirus can be used to control only a limited number of insect pests. Also, since baculoviruses have coevolved with their insect hosts over many thousands of years, they are well adapted to avoid the insect's defense mechanisms. Therefore, resistance to these viruses develops only rarely, and much less frequently than resistance to *B. thuringiensis* develops. More-

Table 15.3 Some insect pests that are currently controlled with baculoviruses

Pest	Common name	Crop
Neodiprion sertifer	European pine sawfly	Pine
Lymantria dispar	Gypsy moth	Broad-leaved trees
Heliothis sp.	Cotton bollworm	Cotton, sorghum
Orgyia pseudotsugata	Tussock moth	Douglas fir
Cydia pomonella	Codling moth	Walnuts, apples
Trichoplusia ni	Cabbage looper	Brassicas
Oryctes rhinoceros	Rhinoceros beetle	Coconuts

over, baculovirus-resistant insects rapidly revert to susceptibility when they are no longer exposed to the virus.

Farmers and growers prefer to use a single insecticidal agent that can control many different insect pests rather than a number of different insecticides. In practice, this means that if baculoviruses are to be used more extensively, their limited host range needs to be expanded.

It has been known for some time that when insect cells are infected with two different strains of baculovirus at the same time, new viruses with slightly different specificities can form after the two starting viruses have replicated. These new viruses are the product of homologous recombination between the two starting viruses. Detailed analysis of this phenomenon revealed that a region of DNA that is only 79 nucleotides long and located within the p143 helicase gene was sufficient to permit homologous recombination between different baculoviruses. It is believed that this 79-nucleotide DNA segment is responsible for the host range of different baculoviruses. Therefore, alteration of some of the nucleotides within this 79-nucleotide DNA segment may eventually allow researchers to generate baculoviruses with modified (expanded) insect specificities.

Genetic Engineering for Improved Biocontrol

Baculoviruses are relatively slow in killing target insects. Depending on conditions, it can take from a few days to several weeks before the viral infection leads to the host's death. To remedy this ineffectiveness, several attempts have been made to enhance the virulence of baculoviruses by introducing foreign genes that will either severely impair or kill the targeted insect species (Table 15.4). One approach has been to use a gene that will disrupt the normal life cycle of the insect when it is expressed within the host insect cells.

During insect development, a reduction in the level of juvenile hormone in larvae initiates metamorphosis into pupae and leads to a cessation of larval feeding. The reduction in the juvenile hormone level is due to an increase in the amount of juvenile hormone esterase, an enzyme that converts the biologically active methyl ester form of juvenile hormone to an inactive acid form. Inhibition of juvenile hormone esterase activity leads to an in vivo accumulation of active juvenile hormone, so the larvae remain in the feeding stage longer and eventually become giant larvae. Therefore, researchers reasoned that an experimentally induced increase in the supply of juvenile hormone esterase should lower the endogenous level of active juvenile hormone and cause a premature cessation of feeding. Basically, their premise was that shortening the duration of larval feeding would curtail the extent of crop damage.

Table 15.4 Some genes that have been expressed in baculoviruses to increase insecticidal activity

Gene	Effect on host insect of introduced gene
Diuretic hormone	Reduced hemolymph volume
Juvenile hormone esterase	Feeding cessation
B. thuringiensis toxin	Feeding cessation
Scorpion toxin	Paralysis
Mite toxin	Paralysis
Wasp toxin	Premature melanization, low weight gain

Adapted from Maeda, *Curr. Opin. Biotechnol.* **6:**313–319, 1995.

To test this idea, the investigators first had to clone and express the gene for juvenile hormone esterase. This task was achieved by purifying the enzyme from the insect *Heliothis virescens* (the tobacco budworm), determining its amino acid sequence, synthesizing a DNA oligomer that corresponds to a portion of the esterase amino acid sequence, and then using this oligomer as a hybridization probe. The coding sequence for juvenile hormone esterase was isolated from an *H. virescens* complementary DNA (cDNA) library and inserted into the genome of a baculovirus under the control of baculovirus transcription signals. When *Trichoplusia ni* (the cabbage looper) at the first larval instar stage was treated with this genetically modified baculovirus, the amount of juvenile hormone in the insect was reduced by the cloned juvenile hormone esterase, and larval feeding and growth were dramatically curtailed relative to feeding and growth by the control larvae that were treated with native baculovirus.

This approach may have limited usefulness for enhancing baculoviruses as general biocontrol agents, however, because the reduction in larval feeding that is attributable to the effect of juvenile hormone esterase is confined to the first larval instar. Other stages of development are much less sensitive to this treatment. In other words, a baculovirus engineered to express juvenile hormone esterase would have to be applied exactly at the time when the majority of the target insect population was in its first larval instar stage, which, under natural conditions, is almost impossible to achieve.

Another approach for enhancing the effectiveness of baculoviruses as a pesticide is to incorporate into the viral genome an insect-specific toxin gene that, when expressed during the viral infection cycle, will yield a potent insect toxin. The gene that encodes the insect-specific neurotoxin produced by the North African scorpion, *Androctonus australis* Hector, was cloned into a baculovirus strain, and the genetically engineered virus was tested as a biological insecticide. This neurotoxin, which does not have any effect on mice, disrupts the flow of sodium ions in neurons of targeted insects and eventually leads to paralysis and death. Laboratory-raised insects that were infected with a baculovirus carrying the scorpion neurotoxin gene caused 50% less damage to the leaves of test plants than did insects that had been treated with wild-type baculovirus.

When the cDNA for the toxin from the Israeli yellow scorpion, *Leiurus quinquestriatus hebraeus*, was cloned and expressed in the baculovirus *Autographa californica* nuclear polyhedrosis virus, the time that it took to kill 50% of the insect larvae that were tested was reduced from 120 to 78 hours. Moreover, 120 hours after infection, the insect larvae treated with recombinant virus gained only one-third as much weight as larvae treated with wild-type virus. Thus, this engineered baculovirus not only hastened the demise of the infected insect larvae, it also significantly decreased the ability of the insects to damage plants.

Recently, a genetically engineered baculovirus, *A. californica* nuclear polyhedrosis virus, which expresses the insect-specific neurotoxin from *Androctonus australis*, was tested under field conditions. Interestingly, while laboratory studies had previously demonstrated a 25 to 50% reduction in the time that it took to kill the insect pest *T. ni* in controlled field tests, the modified baculovirus was even more effective (Fig. 15.7). In the field, the genetically engineered baculovirus killed the insect pests faster, decreased the damage to cabbage plants, and reduced the secondary cycle of infection compared to the wild-type virus.

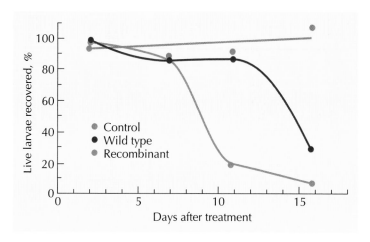

Figure 15.7 Survival of *T. ni* larvae after treatment of cabbage leaves with either wild-type or recombinant baculovirus expressing a scorpion neurotoxin gene. The control cabbage plants were treated only with insect larvae. The lower the percentage of live larvae recovered, the greater the killing of the larvae and the more effective the treatment. Plants were treated with baculovirus only once at the start of the experiment.

No matter how effective a particular genetically engineered baculovirus may be in small-scale experiments, a major hurdle to more widespread use is the difficulty and cost of propagating such viruses. Baculoviruses are obligate parasites; therefore, they must be grown either in living whole organisms or in insect cell culture. Consequently, the cost of baculovirus preparations, whether or not the virus has been genetically engineered, is not currently competitive with that of chemical insecticides. However, biological insecticides may become more appealing when the adverse environmental impact of chemical insecticides is factored into the cost-benefit analysis.

SUMMARY

Microbial insecticides are currently being developed as environmentally friendly biological substitutes for chemical pesticides. A number of subspecies of the bacterium *Bacillus thuringiensis* produce a protoxin as a part of a parasporal crystal, which, after ingestion, kills specific insects. The transition from insecticidal protoxin to toxin occurs in the gut of the target insect and is mediated by the pH and digestive proteases in the gut. The death of the insect is the consequence of the formation of membrane ion channels in the gut cells, which allow ATP to escape and in turn lead to decreased cellular metabolism, cessation of feeding, dehydration, and eventually death. The *B. thuringiensis* toxins are highly specific for a limited number of insect species, nontoxic to nontarget species, and biodegradable, and therefore they are unlikely to cause significant biological selection for resistant forms. These attributes make these biological insecticides good candidates for controlling insect damage to certain crops and preventing the proliferation of insects that act as vectors of human diseases.

The genes (*cry*) for various *B. thuringiensis* toxins have been cloned and characterized. By expressing a *B. thuringiensis* gene in a nonsporulating *Bacillus* strain, production during vegetative growth was achieved, bypassing the need for parasporal crystal formation.

To expand the specificity of a *B. thuringiensis* toxin to other pest insects, toxin genes from different subspecies were cloned into plasmids and introduced into a host strain, either on a broad-host-range plasmid or by integration into the chromosomal DNA of the host cell. In addition to dual toxicity, the bacteria with two different toxin genes sometimes showed an effect against a nontarget insect pest. In one instance, a fusion protein consisting of two toxin domains from different *B. thuringiensis* toxin genes was constructed by genetic manipulation, and the fusion protein retained both toxic activities. In another study, the receptor binding domain of one insecticidal toxin was combined with the toxin domain of another. It is thought that insect resistance is less likely to develop when such hybrid toxins are used.

To ensure that *B. thuringiensis* spraying for the control of mosquitoes is effective, the *B. thuringiensis* toxin genes have been cloned into various microorganisms that live near the surface of ponds and are eaten by mosquito larvae. These strategies appear to be effective means of delivering the *B. thuringiensis* toxin to the targeted insect. Also, rhizosphere bacteria that have been engineered with *B. thuringiensis* toxin genes lessen the damage caused by insects that attack the roots of plants.

Baculoviruses are pathogenic to many different species of insects. Each strain of baculovirus is specific to a small number of different insect species. Although baculoviruses kill their host organisms, the process is usually considered to be too slow to be effective for controlling insects that attack crop plants. However, when certain genes are cloned into different strains of baculovirus, the virus can act as a delivery system for a gene that produces an insecticidal protein during the viral life cycle. Several preliminary tests of this strategy has been successful in laboratory trials. In addition, in one case a gene for a neurotoxin that kills insects was cloned into a baculovirus, and this construct was shown to be effective in the field.

References

Ananda Kumar, P., R. P. Sharma, and V. S. Malik. 1996. The insecticidal proteins of *Bacillus thuringiensis. Adv. Appl. Microbiol.* **42**:1–43.

Baum, J. A., and T. Malvar. 1995. Regulation of insecticidal crystal protein production in *Bacillus thuringiensis. Mol. Microbiol.* **18**:1–12.

Bonning, B. C., and B. D. Hammock. 1992. Development and potential of genetically engineered viral insecticides. *Biotechnol. Genet. Eng. Rev.* **10**:455–489.

Bosch, D., B. Schipper, H. van der Kleij, R. A. de Maagd, and W. J. Stiekema. 1994. Recombinant *Bacillus thuringiensis* crystal proteins with new properties: possibilities for resistance management. *Bio/Technology* **12**:915–918.

Calogero, S., A. M. Albertini, C. Fogher, R. Marzari, and A. Galizzi. 1989. Expression of a cloned *Bacillus thuringiensis* delta-endotoxin gene in *Bacillus subtilis. Appl. Environ. Microbiol.* **55**:446–453.

Caramori, T., A. M. Albertini, and A. Galizzi. 1991. *In vivo* generation of hybrids between two *Bacillus thuringiensis* insect-toxin-encoding genes. *Gene* **98**:37–44.

Chejanovsky, N., N. Zilberberg, H. Rivkin, E. Zlotkin, and M. Guervitz. 1995. Functional expression of an alpha anti-insect scorpion neurotoxin in insect cells and lepidopterous larvae. *FEBS Lett.* **376**:181–184.

Chungjatupornchai, W. 1990. Expression of the mosquitocidal-protein genes of *Bacillus thuringiensis* subsp. *israelensis* and the herbicide-resistance gene *bar* in *Synechocystis* PCC6803. *Curr. Microbiol.* **21**:283–288.

Cory, J. S., M. L. Hirst, T. Williams, R. S. Hails, D. Goulson, B. M. Green, T. M. Carty, R. D. Possee, P. J. Cayley, and D. H. L. Bishop. 1994. Field trial of a genetically improved baculovirus insecticide. *Nature* **370**:138–140.

Crickmore, N., C. Nicholls, D. J. Earp, T. C. Hodgman, and D. J. Ellar. 1990. The construction of *Bacillus thuringiensis* strains expressing novel entomocidal δ-endotoxin combinations. *Biochem. J.* **270:**133–136.

Croizier, G., L. Croizier, O. Argaud, and D. Poudevigne. 1994. Extension of *Autographa californica* nuclear polyhedrosis virus host range by interspecific replacement of a short DNA sequence in the p143 helicase gene. *Proc. Natl. Acad. Sci. USA* **91:**48–52.

Ferré, J., M. D. Real, J. Van Rie, S. Jansens, and M. Peferoen. 1991. Resistance to the *Bacillus thuringiensis* bioinsecticide in a field population of *Plutella xylostella* is due to a change in a midgut membrane receptor. *Proc. Natl. Acad. Sci. USA* **88:**5119–5123.

Fuxa, J. R. 1991. Insect control with baculoviruses. *Biotechnol. Adv.* **9:**425–442.

Ge, A. Z., N. I. Shivarova, and D. H. Dean. 1989. Location of the *Bombyx mori* specificity domain on a *Bacillus thuringiensis* δ-endotoxin protein. *Proc. Natl. Acad. Sci. USA* **86:**4037–4041.

Gelernter, W., and G. E. Schwab. 1993. Transgenic bacteria, viruses, algae and other microorganisms as *Bacillus thuringiensis* toxin delivery systems, p. 89–104. *In* P. F. Entwistle, J. S. Cory, M. J. Bailey, and S. Higgs (ed.), Bacillus thuringiensis, *an Environmental Biopesticide: Theory and Practice.* John Wiley & Sons, Chichester, United Kingdom.

Hammock, B. D., B. C. Bonning, R. D. Possee, T. N. Hanzlik, and S. Maeda. 1990. Expression and effects of the juvenile hormone esterase in a baculovirus vector. *Nature* **344:**458–461.

Held, G. A., L. A. Bulla, E. Ferrari, J. Hoch, A. I. Aronson, and S. A. Minnich. 1982. Cloning and localization of the lepidopteran protoxin gene of *Bacillus thuringiensis* subsp. *kurstaki. Proc. Natl. Acad. Sci. USA* **79:**6065–6069.

Herrera, G., S. J. Snyman, and J. A. Thomson. 1994. Construction of a bioinsecticidal strain of *Pseudomonas fluorescens* active against the sugarcane borer, *Eldana saccharina. Appl. Environ. Microbiol.* **60:**682–690.

Kalman, S., K. L. Kiehne, N. Cooper, M. S. Reynoso, and T. Yamamoto. 1995. Enhanced production of insecticidal proteins in *Bacillus thuringiensis* strains carrying an additional crystal protein gene in their chromosomes. *Appl. Environ. Microbiol.* **61:**3063–3068.

Lampel, J. S., G. L. Canter, M. B. Dimock, J. L. Kelly, J. J. Anderson, B. B. Uratani, J. S. Foulke, Jr., and J. T. Turner. 1994. Integrative cloning, expression, and stability of the *cryIA(c)* gene from *Bacillus thuringiensis* subsp. *kurstaki* in a recombinant strain of *Clavibacter xyli* subsp. *cynodontis. Appl. Environ. Microbiol.* **60:**501–508.

Lereclus, D., H. Agaisse, M. Gominet, and J. Chaufaux. 1995. Overproduction of encapsulated insecticidal crystal proteins in a *Bacillus thuringiensis* spoOA mutant. *Bio/Technology* **13:**67–71.

Lereclus, D., A. Delécluse, and M. M. Lecadet. 1993. Diversity of *Bacillus thuringiensis* toxins and genes, p. 37–69. *In* P. F. Entwistle, J. S. Cory, M. J. Bailey, and S. Higgs (ed.), Bacillus thuringiensis, *an Environmental Biopesticide: Theory and Practice.* John Wiley & Sons, Chichester, United Kingdom.

Liu, J. T., M. J. Sui, D. D. Ji, I. H. Wu, C. C. Chou, and C. C. Chen. 1993. Protection from ultraviolet radiation by melanin of mosquitocidal activity of *Bacillus thuringiensis* var. *israelensis. J. Invertebr. Pathol.* **62:**131–136.

Liu, J. W., W. H. Yap, T. Thanabalu, and A. G. Porter. 1996. Efficient synthesis of mosquitocidal toxins in *Asticcacaulis excentricus* demonstrates potential of gramnegative bacteria in mosquito control. *Nat. Biotechnol.* **14:**343–347.

Maeda, S. 1995. Further development of recombinant baculovirus insecticides. *Curr. Opin. Biotechnol.* **6:**313–319.

McCutchen, B. F., P. V. Choudary, R. Crenshaw, D. Maddox, S. G. Kamita, N. Palekar, S. Volrath, E. Fowler, B. D. Hammock, and S. Maeda. 1991. Development of a recombinant baculovirus expressing an insect-selective neurotoxin: potential for pest control. *Bio/Technology* **9:**848–852.

Mettus, A. M., and A. Macaluso. 1990. Expression of *Bacillus thuringiensis* δ-endotoxin genes during vegetative growth. *Appl. Environ. Microbiol.* **56:**1128–1134.

Murphy, R. C., and E. S. Stevens. 1991. Cloning and expression of the *cryIVD* gene of *Bacillus thuringiensis* subsp. *israelensis* in the cyanobacterium *Agmenellum quadruplicatum* PR-6 and its resulting larvicidal activity. *Appl. Environ. Microbiol.* **58:**1650–1655.

Obukowicz, M. G., F. J. Perlak, K. Kusano-Kretzmer, E. J. Mayer, S. L. Bolten, and L. S. Watrud. 1986. Tn*5*-mediated integration of the delta-endotoxin gene from *Bacillus thuringiensis* into the chromosome of root-colonizing pseudomonads. *J. Bacteriol.* **168:**982–989.

Obukowicz, M. G., F. J. Perlak, K. Kusano-Kretzmer, E. J. Mayer, and L. S. Watrud. 1986. Integration of the delta-endotoxin gene of *Bacillus thuringiensis* into the chromosome of root-colonizing strains of pseudomonads using Tn*5*. *Gene* **45:**327–331.

Priest, F. G. 1992. Biological control of mosquitoes and other biting flies by *Bacillus sphaericus* and *Bacillus thuringiensis*. *J. Appl. Bacteriol.* **72:**357–369.

Schnepf, H. E., and H. R. Whiteley. 1981. Cloning and expression of the *Bacillus thuringiensis* crystal protein gene in *Escherichia coli*. *Proc. Natl. Acad. Sci. USA* **78:**2893–2897.

Stewart, L. M. D., M. Hirst, M. L. Ferber, A. T. Merryweather, P. J. Cayley, and R. D. Possee. 1991. Construction of an improved baculovirus insecticide containing an insect-specific toxin gene. *Nature* **352:**85–88.

Thanabalu, T., J. Hindley, S. Brenner, C. Oei, and C. Berry. 1992. Expression of the mosquitocidal toxins of *Bacillus sphaericus* and *Bacillus thuringiensis* subsp. *israelensis* by recombinant *Caulobacter crescentus*, a vehicle for biological control of aquatic insect larvae. *Appl. Environ. Microbiol.* **58:**905–910.

Thiery, I., L. Nicholas, R. Rippka, and N. Tandeau de Marsac. 1991. Selection of cyanobacteria isolated from mosquito breeding sites as a potential food source for mosquito larvae. *Appl. Environ. Microbiol.* **57:**1354–1359.

Tomalski, M. D., and L. K. Miller. 1991. Insect paralysis by baculovirus-mediated expression of a mite neurotoxin gene. *Nature* **352:**82–84.

Van Rie, J., W. H. McGaughey, D. E. Johnson, B. D. Barnett, and H. Van Mellaert. 1990. Mechanism of insect resistance to the microbial insecticide *Bacillus thuringiensis*. *Science* **247:**72–74.

Wood, H. A., and R. R. Granados. 1991. Genetically engineered baculoviruses as agents for pest control. *Annu. Rev. Microbiol.* **45:**69–87.

Yap, W. H., T. Thanabalu, and A. G. Porter. 1994. Expression of mosquitocidal toxin genes in a gas-vacuolated strain of *Ancylobacter aquaticus*. *Appl. Environ. Microbiol.* **60:**4199–4202.

REVIEW QUESTIONS

1. What are the advantages of biological insecticides over chemical insecticides?

2. Why is the *Bacillus thuringiensis* toxin not toxic to humans?

3. Outline a strategy that you would use to isolate an insecticidal protoxin gene from *B. thuringiensis* subsp. *israelensis*. How would you use this gene in a practical way?

4. How would you determine whether a particular insecticidal protoxin gene is present on a plasmid or part of the chromosome of a *B. thuringiensis* strain?

5. How would you use genetic engineering to improve the usefulness of a particular *B. thuringiensis* protoxin?

6. How would you genetically engineer baculoviruses so that they are more effective insecticides?

7. How can the species range of an insecticidal toxin be extended?

8. What information is provided when you know the class of a Cry protein?

9. How would you engineer a Cry protein to lessen or avoid the development of insect resistance to this toxin?

10. Why is the bacterium *Asticcacaulis excentricus* an attractive host organism for the expression of *B. thuringiensis* insecticidal toxin genes?

11. How might it be possible to expand the range of insects that are infected by a particular baculovirus?

16

CHAPTER

Large-Scale Production of Proteins from Recombinant Microorganisms

The production of commercial products that are synthesized by genetically engineered microorganisms requires the partnership of two kinds of experts. Molecular biologists are responsible for isolating, characterizing, modifying, and creating effectively expressed genes in microorganisms that can be used for industrial production, and biochemical engineers ensure that the genetically engineered form of a microorganism can be grown in large quantities under conditions that give optimal product yields. In the early days of molecular biotechnology, biologists naively thought that scale-up was simply a matter of multiplication; i.e., they believed that whatever conditions were found to be effective on a small scale would be equally effective on a large scale and that to achieve this it was merely necessary to use a larger reaction vessel with a correspondingly larger volume of medium.

This simplistic view is far from the reality of the situation. For example, good growth of aerobic microorganisms can usually be achieved in a standard 200-mL laboratory flask that is aerated with a mixer driven by a 300-watt motor. If the system were directly scaled up, a single 10,000-liter container would require a mixer with a 15-MW motor. Such a motor would be as large as a house, and the heat generated during stirring would boil the microorganisms. Although biochemical engineers may quibble about some aspects of this specific example, they all know that the industrial production of microorganisms is not merely a multiplication of bench-scale conditions. For a start, increasing the size of the reaction vessel (bioreactor; fermenter) is required for the large-scale growth of microorganisms, because it would be impractical to set up 50,000 individual 200-mL culture flasks to obtain 10,000 liters of cell suspension.

A number of parameters must be precisely regulated to obtain maximum yields from either small (1 to 10 liters) or large (>1,000 liters) bioreactors. These parameters include temperature, pH, rate and nature of mixing of the growing cells, and, with aerobic organisms, oxygen demand. Moreover, the optimal conditions generally change with each 10-fold increase in the volume of a bioreactor.

There are also other technical considerations. The design of the bioreactor is important. It should ensure adequate sterility and provide for appropriate levels of containment of the genetically engineered microorganism. The reactor should also include probes that permit the accurate and continuous on-line monitoring of as many critical reaction parameters as possible so that adjustments can be made rapidly and easily throughout the course of the fermentation reaction. In addition, because sterilization may alter the composition of the medium (e.g., by destroying vitamins), it is important to ascertain whether the medium composition is still optimal for maximal microbial growth following sterilization.

Generally, large-scale fermentation and product purification are stepwise processes (Fig. 16.1). A typical procedure begins with formulation and sterilization of the growth medium and sterilization of the fermentation equipment. The cells are grown first as a stock culture (5 to 10 mL), then in a shake flask (200 to 1,000 mL), and then in a seed fermenter (10 to 100

Figure 16.1 Generalized scheme for a large-scale fermentation process. The commercial product is usually in either the cell or cell-free fraction but not in both; consequently, one or the other of these fractions will be processed further (+) or discarded (−).

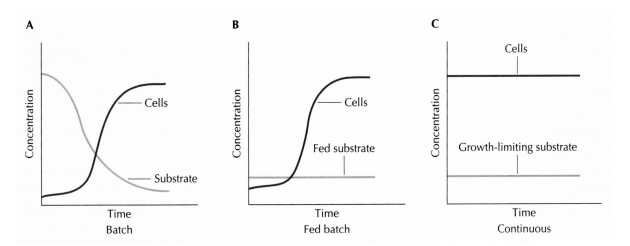

Figure 16.2 Schematic representation of the time course (progress curves) of cell concentration (mass) and substrate concentration in batch **(A)**, fed-batch **(B)**, and continuous **(C)** fermentations.

liters). Finally, the production fermenter (1,000 to 100,000 liters) is inoculated. After the fermentation step is completed, the cells are separated from the culture fluid by either centrifugation or filtration. If the product is intracellular, the cells are disrupted, the cell debris is removed, and the product is recovered from the debris-free fluid. If the product is extracellular, it is purified from the cell-free culture medium.

Principles of Microbial Growth

Microorganisms can be grown in batch, fed-batch, or continuous culture (Fig. 16.2). In batch fermentation, the sterile growth medium is inoculated with the appropriate microorganisms and the fermentation proceeds without the addition of fresh growth medium. In fed-batch fermentation, nutrients are added incrementally at various times during the fermentation reaction; no growth medium is removed until the end of the process. In the continuous fermentation process, fresh growth medium is added continuously during fermentation but there is also concomitant removal of an equal volume of spent medium containing suspended microorganisms. For each type of fermentation, oxygen (which is usually provided in the form of sterile air), an antifoaming agent, and, if required, acid or base are injected into the bioreactor as needed.

Batch Fermentation

During a batch fermentation, the composition of the culture medium, the concentration of microorganisms (biomass concentration), the internal chemical composition of the microorganisms, and the amount of either target protein or metabolite all change as a consequence of the state of cell growth, cellular metabolism, and availability of nutrients. Under these conditions, six typical phases of growth are usually observed: lag phase, acceleration phase, logarithmic (log) or exponential phase, deceleration phase, stationary phase, and death phase (Fig. 16.3).

Typically, there is no immediate increase in the numbers of cells after

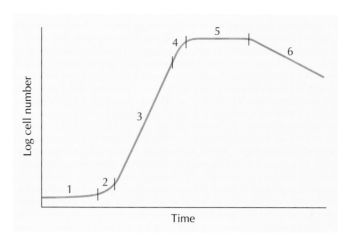

Figure 16.3 Pattern of microbial cell growth in a batch fermenter. The six phases of the growth cycle are (1) lag, (2) acceleration, (3) log (exponential), (4) deceleration, (5) stationary, and (6) death.

the inoculation into sterilized growth medium. This initial period is called the lag phase. During the lag period, the microbial cells adapt to the new environmental conditions. The cells may have to adjust to a different pH or to a new level of available nutrients. As part of an adaptive response, previously unexpressed metabolic pathways may be induced. A lag phase generally occurs whenever the cells of the inoculum are derived from a culture that has stopped growing (i.e., has entered stationary phase) because of substrate limitation or product inhibition. These cells need time to reset their metabolic systems to adjust to the new medium. The length of the lag phase corresponds to how long the inoculated cells were in stationary phase and the extent to which the previous growth medium of the starting cells differs from the new, fresh culture medium. Conversely, when the inoculum is a cell culture from a growing cell population in log phase, a discernible lag phase may not occur and growth may begin immediately. Following the lag phase, the brief period when the rate of cell growth increases until log phase growth is attained is called the acceleration phase.

During the log phase of growth, the cell mass undergoes several cell doublings and the specific growth rate of the culture remains constant. With excess substrate (nutrient supply) and no inhibition of growth by a compound that is present in the growth medium, the specific growth rate is independent of the substrate concentration. These changes and other related steps can be represented in mathematical form, making it possible for biochemical engineers to precisely model and then more easily scale up microbial cell growth. In this case, the rate of increase of the cell biomass with time, dX/dt, is the product of the specific growth rate, μ, and the biomass concentration, X:

$$dX/dt = \mu X$$

Similarly, the rate of increase of the cell number, dN/dt, is the product of the specific growth rate, μ, and the cell number, N:

$$dN/dt = \mu N$$

The specific growth rate, μ, is a function of the concentration of the limiting substrate (i.e., the carbon or nitrogen source), S, the maximum specific

growth rate, μ_{max}, and a substrate-specific constant, K_s. Both S and K_s are expressed in terms of concentration, e.g., in grams or moles per liter:

$$\mu = \mu_{max}S/(K_s + S)$$

Sometimes, scientists refer to the doubling or generation time, t, of a culture rather than to its specific growth rate μ, where $t = \ln 2/\mu$. The generation time of a culture is the length of time that it takes, under defined conditions, for the number of cells or the cell biomass to double. For the microorganisms that are commonly grown in culture, the value of μ_{max} varies from about 2.1 to 0.086 h-1 (reciprocal hours), which corresponds to doubling times of approximately 20 minutes to 8 hours.

When there is an excess of substrate (i.e., when $S \gg K_s$), then $\mu = \mu_{max}$ and the maximal rate of log-phase growth of the culture occurs. In practice, the value of K_s is often so low that substrate levels equivalent to K_s are rarely encountered during log-phase growth. For example, for *Escherichia coli*, while the K_s for glucose is approximately 1 mg/liter, the initial glucose level is usually around 10,000 mg/liter. However, as the culture nears the end of the log phase, the concentration of the remaining substrate, S, is depleted and may even fall below the value of K_s. Under conditions where $S < K_s$, the microorganisms rapidly enter the deceleration phase. However, because of the large cell population at the end of the log phase, the substrate may be so rapidly assimilated that the deceleration phase is short-lived and not observable.

Following either the depletion of a critical growth substance, such as the carbon source, from the medium or the accumulation of metabolic end products that inhibit growth, the increase in cell mass eventually ceases and the cells enter the stationary phase. During this phase, although the amount of biomass remains constant, cellular metabolism often changes dramatically; in some instances, compounds (secondary metabolites) that are of considerable commercial interest are synthesized. For example, antibiotics are usually produced during the stationary phase of the microbial growth cycle. The duration of the stationary phase depends on the particular organism and the conditions of growth.

In the death phase, the energy reserves of the cell are virtually exhausted and metabolic activity ceases. For most commercial processes, the fermentation reaction is halted and the cells are harvested before the death phase begins.

Fed-Batch Fermentation

In fed-batch fermentations, substrate is added in increments at various times throughout the course of the reaction. These additions prolong both the log and stationary phases, thereby increasing the biomass and the amount of synthesis of stationary-phase metabolites such as antibiotics. However, microorganisms in stationary phase often produce proteolytic enzymes or proteases, and these enzymes will attack any protein product synthesized by a genetically engineered microorganism. Therefore, when producing proteins from a recombinant microorganism, it is important to prevent the fermentation reaction from reaching this part of the growth cycle. Because it is often difficult to measure the substrate concentration directly during the fermentation reaction, other indicators that are correlated with the consumption of substrate such as the production of organic acids, changes in the pH, or the production of CO_2 can be used to estimate when additional substrate is needed. Generally, fed-batch fermentations require

more monitoring and greater control than batch fermentations and are therefore used to a lesser extent; however, since they may be advantageous in the development of systems for the production of proteins from recombinant microorganisms, they are becoming increasingly popular.

The periodic addition of substrate to a growing microbial culture prolongs the log phase of growth and delays the onset of the stationary phase, which initiates cellular stress responses, the production of proteases, and other metabolic changes that affect the yield of a recombinant protein. Nevertheless, with continued cell growth, an increasing amount of the incoming substrate is needed for maintenance of the host cell metabolism. This means that fewer cellular resources are used for the synthesis of the recombinant protein or commercial metabolite(s). To ensure that the synthesis and stability of a recombinant protein are not impaired, increasing amounts of nutrients must be added to the growing culture. This may be done by carefully monitoring the fermentation reaction and adding substrates (carbon and nitrogen sources together with trace elements) in increasing amounts as they are needed. Depending upon the particular microorganism, its genetic background, and the nature of the recombinant protein, a fed-batch fermentation strategy can increase the yield from 25% to more than 1,000% compared to batch fermentation.

Fed-batch processes are not limited to microbial cells but can also be used with mammalian and insect cells in culture. This is important because (1) these cell culture systems are increasingly being used for the production of human therapeutic proteins and (2) in the absence of fed-batch strategies, animal cells in batch culture are not very efficient in producing foreign proteins.

Continuous Fermentation

In a continuous fermentation, a steady-state condition, where $dX/dt = 0$, is attained when the total number of cells and the total volume in the bioreactor remain constant. In other words, under these conditions, the loss of cells due to outflow (product removal) is exactly balanced by the gain in new cells by growth (division). In more formal terms, for a continuous steady-state fermentation process, the dilution rate, D, is defined as the volumetric flow rate, F, divided by the constant liquid volume, V, in the bioreactor:

$$D = F/V$$

where D is equal to the specific growth rate, μ:

$$D = (dX/dt)(1/X) = \mu$$

To obtain hydrodynamically stable continuous cultures, the specific growth rate, μ, of the culture must be lower than the maximum attainable specific growth rate, μ_{max}. In practice, this condition is achieved by adjusting a pump that controls the volumetric flow rate, F, while keeping the volume of the culture within the bioreactor, V, constant.

The fundamental objective of industrial fermentations is to minimize costs and maximize yields. This goal can be achieved by attempting to develop the most efficient mode of fermentation for each particular process. Although continuous fermentation processes are not commonly used in industrial applications, primarily because of the greater experience that scientists have with growing cells in batch mode, the cost of producing a specific amount of cell biomass by continuous culture is potentially much

lower than that of producing the same amount by batch fermentation. The following factors account for the savings.

- Continuous fermentations use smaller bioreactors than do batch fermentations to produce the same amount of product.

- After a large-scale batch fermentation is completed, large-scale equipment is needed for cell harvesting, cell breakage, and subsequent downstream processing (purification) of the protein or metabolite product that is produced by the microorganism. Continuously grown cells, however, are produced "a little bit at a time," so that the equipment required for cell harvesting, cell breakage, and downstream processing can be much smaller.

- Continuous fermentation, by definition, avoids the "down time" between batch runs, during which the bioreactor is prepared for reuse. A common hindrance to efficient industrial fermentation is the loss of productivity due to the down time of the bioreactor for repairs, cleaning, or sterilization. Continuous fermentations have less down time, because a single reaction can be maintained for a much longer period.

- The physiological state of the cells during continuous fermentation is more uniform, so that yields of product tend to be more consistent. In batch fermentations, small differences in the timing of cell harvest, which coincides with the mid- to late-log phase of growth, can lead to significant physiological differences.

Continuous fermentation has been used for the commercial production of single-cell protein, antibiotics, and organic solvents.

Despite its merits, continuous fermentation has potential drawbacks that must be addressed before its use becomes more widespread.

- The duration of a continuous fermentation can be 500 to 1,000 hours, and therefore some cells might lose recombinant plasmid constructs. Cells that lack plasmids usually have a smaller energy burden and divide faster than do plasmid-containing microorganisms, so that yields may decline with time because fewer cells are synthesizing the product protein. Integration of the cloned gene into the genome of the host organism might avoid this problem.

- Maintenance of sterile conditions on an industrial scale for long periods is difficult. Furthermore, continuous processes need sterile backup equipment, a requirement that can greatly increase capital costs.

- The composition of culture medium that is used for industrial fermentations is not subject to the same level of quality assurance as that accorded laboratory medium components and therefore may vary from batch to batch. This variation can disrupt the physiology of the cells and decrease productivity.

Because batch fermentation has a proven history of reliability, there is a reluctance to switch to another type of fermentation system, even though a continuous mode of operation is generally regarded as the most efficient fermentation strategy. Nevertheless, a number of researchers have recently developed, on a laboratory (up to 10 liters) or pilot plant (up to 1,000 liters) scale, both continuous and fed-batch processes for the production of proteins from recombinant microorganisms. Therefore, it is probably only a

matter of time before the use of continuous and fed-batch fermentations becomes more widespread in industry.

Maximizing the Efficiency of the Fermentation Process

Regardless of the type of fermentation process that is used to grow cells, it is necessary to monitor and control culture parameters such as dissolved oxygen concentration, pH, temperature, and degree of mixing. Changes in any one of these parameters can have a dramatic effect on the yield of cells and the stability of the protein product.

Optimal growth of *E. coli* cells and many other microorganisms that are used as hosts for recombinant genes usually requires large amounts of dissolved oxygen. The maximal oxygen demand in a fermentation, Q_{max}, is dependent on the cell mass, X, the maximal specific growth rate, μ_{max}, and the growth yield based on oxygen consumed, Y_{O_2}, where

$$Q_{max} = X \cdot \mu_{max}/Y_{O_2}$$

Because oxygen is only sparingly soluble in water (0.0084 gram/liter at 25°C), it must be supplied continuously—generally in the form of sterilized air—to a growing culture. However, the introduction of air into a bioreactor produces bubbles, and if the bubbles are too large, the rate of transfer of oxygen to the cells is insufficient to support optimal growth. Thus, fermenter design should include provision for monitoring the dissolved oxygen level of the culture, providing oxygen to the culture, and adequately mixing the culture to efficiently disperse the bubbles.

Most microorganisms grow optimally between pH 5.5 and 8.5. However, during growth in a bioreactor, cellular metabolites are released into the growth medium, a process that can change the pH of the medium.

MILESTONE

Expression of Intracellular Hemoglobin Improves Protein Synthesis in Oxygen-Limited *Escherichia coli*
C. KHOSLA, J. E. CURTIS, J. DEMODENA, U. RINAS, and J. E. BAILEY
Bio/Technology **8**:849–853, 1990

Because of the low solubility of oxygen in water, the growth of aerobic bacteria often becomes limited by the amount of dissolved oxygen in the fermentation broth. This problem is especially acute at high cell densities or when fermentations are performed on a large scale. To address this problem, chemical engineers have attempted to increase the rate of transfer of introduced oxygen to the liquid of the growth medium. Their approaches have included (1) sparging the growth medium with pure oxygen rather than with air; (2) introducing

the air (or oxygen) under pressure; (3) adding chemicals such as perfluorocarbons to the fermentation broth to increase the solubility of oxygen; and (4) modifying the configuration of the fermentation vessel to optimize the aeration or agitation of the fermentation broth. While all of these solutions to the "oxygen problem" are somewhat effective, they are subject to a threshold effect beyond which it is impossible to introduce a sufficient amount of oxygen that would improve the final yield of the fermentation.

As an alternative to these "hardware" solutions, Bailey and coworkers designed a biological system in which the host organism was modified so that it would be more efficient at using the low levels of oxygen that are normally present in the growth medium. They cloned a gene encoding a hemoglobin-like molecule from the gram-negative bacterium *Vitreoscilla* sp. and introduced it into several different recombinant bacteria. The introduced bacterial hemoglobin bound oxygen from the environment and created a higher level of available oxygen within the cells, which resulted in an increase in growth and foreign gene expression. This approach provided a clever biological solution to what at first glance seemed to be an almost insurmountable engineering problem.

Therefore, the pH of the medium must be monitored and either acid or base must be added as needed to maintain a constant pH. Of course, the added acid or base must be well mixed into the fermentation broth so that the pH of the growth medium is the same throughout the entire reaction vessel.

Temperature is one of the parameters that is essential for the success of a fermentation reaction. Microorganisms grown at a temperature below the optimum grow slowly and have a reduced rate of cellular production (productivity). On the other hand, if the growth temperature is too high—but not lethal—there may be premature induction of the expression of the target protein, if it is under the control of a temperature-sensitive repressor, or induction of a heat shock (stress) response, which will produce cellular proteases that lower the yield of the protein product.

Adequate mixing of a microbial culture is essential for many aspects of a fermentation, including the assurance of an adequate supply of nutrients to the cells and the prevention of the accumulation of any toxic metabolic by-products in local, poorly mixed regions of the bioreactor. Effective mixing is relatively easy with small-scale cultures, but one of the major problems in increasing the scale of fermentation equipment is the difficulty in maintaining a level of mixing that ensures a homogeneous growth environment.

Agitation of the fermentation broth also affects other factors: the rate of transfer of oxygen from the gas bubbles to the liquid medium and then from the medium to the cells; efficient heat transfer; accurate measurement of specific metabolites in the culture fluid; and efficient dispersion of added solutions such as acids, bases, nutrients, or antifoaming agents. On these grounds, it might be concluded that the more mixing there is, the better the growth. However, excessive agitation of a fermentation broth can cause hydromechanical stress (shear), which may damage larger microbial or mammalian cells, and a temperature increase, which may also decrease cell viability. Thus, a balance must be struck between the need to provide thorough mixing and the need to avoid damage to the cells.

There is an additional consideration for scaled-up fermentations that has nothing to do with the technical aspects of the process but depends instead on whether a genetically engineered microorganism is being used. In most countries, specific rules and regulations must be followed when genetically engineered microorganisms are grown on a large scale. Although most recombinant microorganisms are not hazardous, it is nevertheless important to ensure that they are not inadvertently released into the environment. Therefore, fail-safe systems are used to prevent accidental spills of live recombinant organisms and to contain them if there is a spill. Furthermore, all recombinant microorganisms must be treated by a verified procedure to render them nonviable before they are discharged from the production facility. The spent culture medium must also be treated to ensure that it does not contain viable organisms and that its disposal does not create an environmental problem.

High-Density Cell Cultures

Generally, when foreign proteins are produced by recombinant *E. coli*, the greater the final cell density the greater the amount of product that is formed. In practice, cell concentrations of more than 50 (and in a few cases more than 100) grams of dry cell weight per liter of culture have been obtained with fed-batch cultures of recombinant *E. coli*. The dry weight of *E. coli* cells is approximately 20 to 25% of the wet weight.

One way to achieve a high density of *E. coli* cells is to optimize the growth medium. Some nutrients, including carbon and nitrogen sources, can inhibit cell growth if they are present at too high a concentration. Glucose is inhibitory above 50 grams per liter, ammonia is inhibitory above 3 grams per liter, iron is inhibitory above 1.15 grams per liter, magnesium is inhibitory above 8.7 grams per liter, phosphorus is inhibitory above 10 grams per liter, and zinc is inhibitory above 0.038 gram per liter. Therefore, merely increasing the amount of nutrients in the growth medium in batch culture does not generally yield a high cell density. In addition, since the nutrients in complex media, such as peptone or yeast extract, can vary from one batch of medium to another, fermentations that use complex media are not always reproducible.

Acetate, which can be inhibitory to cell growth, is produced by *E. coli* both when the cells are grown under oxygen-limiting conditions and in the presence of excess glucose. The acetate problem can be minimized by using glycerol instead of glucose as a carbon source, lowering the culture temperature, or using an *E. coli* strain that has been genetically engineered to shunt acetate into less toxic compounds (see Chapter 6).

Oxygen may also become limited in high-density cell cultures. To overcome this problem, the rate of introduction of air (sparging), the agitation rate, or both can be increased. Also, pure oxygen rather than air, which is only approximately 20% oxygen, can be introduced into growing cell cultures, or cells can be grown under pressure to increase the solubility of oxygen, which increases the rate of transfer of oxygen to the cells in the aqueous growth medium. Alternatively, expression in *E. coli* host cells of a *Vitreoscilla* hemoglobin gene significantly increases the uptake of oxygen to growing cells.

High-density cell cultures are most readily attained in fed-batch cultures. The addition of nutrients following the depletion of some of the original nutrients may be constant, stepwise, or exponential. With constant-rate feeding, nutrients are added at the same rate throughout the fermentation. However, under these conditions, the specific growth rate continually declines. With stepwise feeding, an increasing amount of nutrients is added at higher cell concentrations. In this case, the specific growth rate decline is largely compensated. With exponential feeding, nutrients are added as required to maintain constant cell growth. It is possible to automate the fed-batch addition of nutrients based on measuring the concentration of the growth-limiting substrate, such as glucose, in the culture medium during the fermentation process.

Bioreactors

A cursory examination of the biochemical engineering literature may give the impression that there is a limitless number of bioreactor designs. However, closer inspection reveals that virtually all of these designs fall into three fundamental classes:

- stirred-tank reactors (STRs), which have internal mechanical agitation (Fig. 16.4A);
- bubble columns, which rely on the introduction of air or another gas (sparging) for agitation (Fig. 16.4B); and
- airlift reactors, which have either an internal (Fig. 16.4C) or an external (Fig. 16.4D) loop; the mixing and circulation of the culture fluid

Figure 16.4 Simplified examples of bioreactor configurations. **A.** Stirred-tank reactor (STR). **B.** Bubble column. **C.** Internal-loop airlift reactor with a central draft tube. **D.** External-loop airlift reactor. The arrows within the bioreactors indicate the direction of the flow of the culture medium.

in these reactors are the result of the motion of an introduced gas (usually air), which causes density differences within the different parts of the bioreactor.

The traditional and by far the most commonly used bioreactor is the STR. This type of bioreactor has several advantages over the other types:

- It has highly flexible operating conditions.
- It is readily available commercially.
- It provides efficient gas transfer to the growing microbial cells. Or, in the words of fermentation engineers, the volumetric mass transfer coefficient, k_La, of STRs is high.
- It has been used extensively by fermentation engineers and microbiologists for growing a variety of microorganisms.

In an STR, gas (usually air) is added to the culture medium under pressure through a device called a sparger, which can be in the form of either a

ring with many small holes or a tube with a single orifice. Although sparging rings generate smaller bubbles and consequently create better initial gas distribution, sparging tubes are often preferred in many small-scale applications because they are less likely to become plugged. Thorough dispersion of the gas within the bioreactor is achieved by one or several impellers (agitators) and not by the sparger alone. Agitation of the culture medium by the impellers causes larger bubbles to break down into smaller ones, disperses the bubbles throughout the medium, and enhances the residence time of the bubbles in the bioreactor. At high levels of agitation, the mean size of the bubbles in large bioreactors is essentially independent of the size of the holes in the sparger. The type of impeller, its rotational speed, and the physicochemical properties of the liquid phase are important factors that give rise to efficient gas dispersion. In large bioreactors, however, if the initial gas distribution from the sparger is not uniform across the tank, even vigorous agitation may not create a homogeneous gas environment.

Because of the corrosive or abrasive nature of many culture media and sterilization procedures, STRs are usually constructed from stainless steel or glass. The glass units are usually limited to laboratory-scale bioreactors that have a capacity of less than 50 liters.

One limitation on bioreactor size is the ability to efficiently remove heat that is generated as a consequence of the metabolism of the growing cells and of the energy input by agitation. Too much heat raises the temperature and alters the physiological state of the cells and decreases the product yield. Heat can be removed by using a cooling jacket around the reaction vessel or by internal coils. Although internal cooling coils are more effective than jackets in keeping the fermentation reaction close to the desired temperature, they can become fouled (coated) with microorganisms, which prevents cooling, and they sometimes interfere with the proper agitation of the fermentation broth.

Contamination of a fermentation with fungi or bacteria is usually disastrous. Therefore, bioreactors are designed so that they can be sterilized, usually with pressurized steam. There should be no internal dead spaces or surfaces that escape contact with the steam during sterilization. All seals, probes, and valves also must be readily steam sterilizable. When a bioreactor is designed, a trade-off is often made between a full set of ports for probes—to monitor all of the possible fermentation parameters—and fewer ports. The latter design makes the maintenance of sterility easier.

The high level of agitation of the culture medium during a fermentation reaction often causes considerable foaming. Excessive foam can wet the filter at the port through which the introduced air exits the bioreactor, thereby decreasing the air flow or providing a pathway for the entry of contaminating cells. Either chemical antifoaming agents or mechanical foam breakers can be used to control foaming. However, the chemical agents can diminish the extent of microbial growth by preventing oxygen transfer or, in some cases, by inhibiting cellular enzymes. Furthermore, if these antifoaming compounds are not removed before downstream processing, they can contaminate the final product. Foaming can also be controlled by providing sufficient "head space" within the bioreactor, i.e., space above the liquid in which the bubbles can dissipate. In practice, then, the "working volume" or actual volume of the culture in an STR is typically only about 75% of the total volume of the bioreactor.

Many of the considerations that apply to STRs also apply to pneumatic reactors such as bubble columns and airlift bioreactors. Thus, for example,

the maintenance of sterility, constant pH, and constant temperature are key components of any fermentation, regardless of the precise configuration of the bioreactor.

The configurations of bubble columns and airlift bioreactors give them some distinct advantages over STRs. These pneumatic reactors are more energy efficient than STRs because agitation is provided by the injection of a stream of air—or another gas if anaerobic microorganisms are being grown—rather than by a mechanical stirrer. Also, the elimination of the need for the mixer shaft in these units means one less potential site of entry for contaminating organisms.

Pneumatic reactors generate a lower shear environment than do STRs, and in airlift reactors, the shear stress is more evenly distributed throughout the vessel than in STRs. The reduction of shear forces is important for the following reasons.

- Genetically engineered microorganisms are often more susceptible to lysis when exposed to shear stress than are unengineered organisms because the extra metabolic burden of synthesizing a foreign protein often causes genetically engineered microorganisms to form weakened cell walls.
- A frequent cellular response to the shear forces is a decreased synthesis of all proteins, including the recombinant protein.
- Shear stress can alter the physical and chemical properties of the cells so that the downstream processing steps become more difficult to perform. For example, the fermentation conditions can increase the amount of surface polysaccharides that a microorganism produces and, as a consequence, can change the conditions for effective harvesting and lysis of these cells, making it more difficult to purify the recombinant protein.

In bubble columns, the air is introduced under high pressure near the bottom but the smaller bubbles coalesce into larger ones as they rise through the column, leading to uneven gas distribution. In addition, the use of high-pressure air tends to cause excessive foaming of the medium. These disadvantages restrict the flexibility or effective range of operating conditions, as well as the potential size of bubble columns.

Airlift bioreactors, however, can be readily adapted for either pilot plant or large-scale fermentation processes. In an airlift reactor, the gas is introduced into the bottom of a vertical channel (riser). Both the gas and liquid flow up the riser until they reach an open space at the top (gas-liquid separator), where the gas is at least partially disengaged from the liquid. The degassed liquid, which is denser than the gassed liquid, descends in a separate vertical channel (downcomer) and flows along the base of the reactor until it reaches the bottom of the riser. In this way, the culture fluid and cells are continuously being circulated around the bioreactor.

There are two main types of airlift bioreactors: those that have a single container with internal baffles that create interior liquid circulation channels (internal-loop reactors) (Fig. 16.4C) and those that have an external loop so that the culture liquid circulates through separate, independent channels (internal-loop reactors) (Fig. 16.4D). Internal-loop airlift reactors are simple in design, but once they are constructed, both the volume and the circulation rate are fixed for all fermentation processes. In contrast, the external loop of external-loop airlift reactors can be changed or modified to suit the requirements of different fermentations.

Airlift bioreactors are generally more efficient than bubble columns, especially for denser or more viscous suspensions of microorganisms; in airlift reactors, mixing is generally better and bubble coalescence is not as big a problem as it is in bubble columns. In extremely large airlift fermenters, such as the 1,500,000-liter fermenter built by ICI in England for the production of single-cell protein, it takes a considerable amount of time for a cell to complete a full cycle through the reactor. To prevent the cells from becoming substrate depleted while they are traversing the bioreactor, multiple injection points for the introduction of substrate were added along the length of the unit.

Typical Large-Scale Fermentation Systems

Recombinant microorganisms have been used to overproduce protein products, such as pharmaceuticals like insulin. In addition, they have been used as "factories" to synthesize useful metabolites such as antibiotics. Protein levels are maximal during the mid-log to late log phase of the growth cycle, whereas metabolites are commonly synthesized during the deceleration and stationary phases. Clearly, these differences must be considered when a large-scale fermentation process is being developed.

The optimization of product synthesis is a scientific challenge. For protein production, it is generally preferable to use cloned genes that are under the control of strong promoters that can be regulated. Initially, it was thought that constitutive expression of a cloned gene would be sufficient to obtain reasonable quantities of the product. However, experience has shown that continuous transcription and translation of a cloned gene drains energy from essential cell functions and slows cell growth. With an inducible system, the expression of a cloned gene can be confined to a specific period in the growth cycle of the microorganism. On these grounds, for optimal protein production, the process should consist of two separate stages. First, the cells are grown under optimum conditions to a relatively high cell density. Second, depending on the nature of the promoter that drives the cloned gene, transcription is induced either by shifting the temperature or by adding a chemical inducer such as isopropyl-β-D-thiogalactopyranoside (IPTG) to the medium.

A two-stage system is not easy to implement in a large bioreactor (>100 liters) because it is technically very difficult either to raise the temperature quickly, typically from 30 to 42°C, or to ensure that the chemical inducer is rapidly and evenly mixed in a large vessel. As indicated in the following example, this problem can be overcome by using two connected bioreactors (two-stage fermentation) so that the cells are grown in the first vessel and the induction is carried out in the second vessel. Under these conditions, growth and induction are separately optimized, thereby increasing the overall amount of product formed per unit of time (productivity) of the fermentation.

Two-Stage Fermentation in Tandem Airlift Reactors

E. coli NM989, which carries the gene encoding the enzyme T4 DNA ligase under the transcriptional control of the p^L promoter and a temperature-sensitive *c*I repressor, was grown and induced in a two-stage airlift bioreactor. In this bacterial strain, the T4 DNA ligase gene was integrated in the chromosomal DNA, a location obviating any potential problems of plasmid instability that might occur during extended fermentation. The growth stage

was carried out at 30°C in an external-loop airlift bioreactor that had a 10-liter working volume. The T4 DNA ligase gene was not expressed under these conditions. A second external-loop airlift bioreactor, with a working volume of approximately 5 liters at 42°C, was used for the induction stage (Fig. 16.5). The two bioreactors were linked by a transfer tube with a pump that controlled the continuous flow of cell suspension from the growth stage bioreactor (first) into the induction stage vessel (second). In addition, cells suspended in culture medium were removed from the induction stage bioreactor at a specific rate and prepared for downstream processing.

The maximal specific growth rate (μ_{max}) of the microbial culture was approximately 0.66 reciprocal hour in the first bioreactor and 0.54 reciprocal hour in the second. These values correspond to cell doubling times of 63 and 77 minutes, respectively. Fresh medium was continuously added at a

Figure 16.5 Two-stage airlift reactor used for the temperature-dependent induction of a protein product. Cells from the growth stage (left) fermenter, which is at 30°C, are pumped into the induction stage fermenter (right), which is at 42°C. Each bioreactor has a double external loop that is fitted with valves. By changing the valve settings, working volumes best suited for different fermentations can be created.

rate of 2 liters per hour to the growth stage fermenter, and cell suspension (effluent) was simultaneously removed from the induction stage fermenter at the same rate. As a consequence of the liquid volumes of the two bioreactors, a cell spent about 5 hours in the growth stage bioreactor and 2 hours in the induction stage bioreactor. The different residence times in the two phases of this fermentation process were necessary to optimize both the number of cells produced and the yield and stability of the T4 DNA ligase. Generally, residence times can be altered as required by adjusting the relative working volumes in the two fermenters of a two-stage system and by adjusting the volumetric rate of input of nutrients into the first bioreactor.

The double-external-loop design of the airlift fermenter (Fig. 16.5) used in this work facilitated the adjustment of the working volumes of the two fermenters relative to one another. It also added versatility to the system, so that it was possible to obtain a variety of different growth conditions for different populations of recombinant cells. For T4 DNA ligase production, the best results were obtained when approximately 33 mL of cell suspension was transferred every minute from the growth stage to the induction stage bioreactor. Because this amount of cell suspension was equivalent to only 0.67% of the volume of the induction stage bioreactor, the incoming cells underwent a virtually instantaneous temperature shift from 30 to 42°C. Nutrients in a concentrated form were added at a specific rate to the induction stage bioreactor throughout the fermentation to keep the cells in this bioreactor in log phase. This action prevented the T4 DNA ligase from being degraded by proteolytic enzymes that are normally synthesized during the deceleration and stationary phases.

With this continuous two-stage bioreactor, induced *E. coli* NM989 can be grown to a density of approximately 4 grams (dry weight) of cells per liter of culture. After induction, about 4% of the cell protein is T4 DNA ligase, an amount that corresponds to approximately 25,000 units of enzyme activity per gram (dry weight). This process can produce approximately 100,000 units of enzyme activity per liter of culture, or about 4,800,000 units per day. Assuming that it is possible to recover about 20% of the initial activity following purification of the enzyme and that the enzyme sells for about $0.25 per unit, then the final yield of purified enzyme in 1 day will be worth about $240,000. Although these calculations do not consider all of the costs that go into the production of protein from a genetically engineered microorganism, it is clear that for highly valued products, small to moderately sized continuous-fermentation systems can generate significant returns on the initial capital investment.

Two-Stage Fermentation in a Single Stirred-Tank Reactor

The tripartite fusion protein AGβgal, which is used for immunological assays, was produced on a large scale in a single STR. The gene encoding AGβgal was constructed by recombinant DNA techniques and encodes the five immunoglobulin G (IgG) binding regions of *Staphylococcus aureus* protein A, two IgG binding regions from *Streptococcus* strain G148 protein G, and β-galactosidase from *E. coli*. The synthetic AGβgal gene was placed under the control of the bacteriophage lambda p^R promoter, which is regulated in the same manner as the p^L promoter, cloned into a plasmid that carries the gene for ampicillin resistance, and introduced by transformation into *E. coli*. The strain with the AGβgal DNA plasmid carries a second plasmid that has the genes for a temperature-sensitive *c*I repressor protein and a kanamycin resistance gene.

A 5-liter volume of these cells was grown in an STR at 30°C in the presence of both ampicillin and kanamycin—to provide selective pressure for the retention of both plasmids—and then used to initiate growth without antibiotics at 30°C in a 45-liter (working-volume) STR. The cell suspension in the 45-liter fermenter in turn served as an inoculum for a 600-liter STR, where the cells were grown at 30°C without antibiotics (Fig. 16.6). In general, to keep the cost of the process to a minimum, antibiotics are not added to large-scale microbial cultures. When the cell density in the 600-liter (working-volume) bioreactor reached the equivalent of about 4 grams (dry weight) per liter of growth medium, the temperature was shifted from 30 to 40°C to induce the expression of the AGβgal protein. Under these conditions, it takes about 1 hour to reach 40°C. A temperature of 40°C rather than 42°C was chosen because the lower temperature was found to yield the same level of Agβgal protein while allowing the cells to grow for a longer period. In other words, the low-temperature (40°C) conditions yielded a larger amount of protein product.

The specific activity of the AGβgal protein increased for 2 hours after the initiation of the induction and then decreased. This activity falloff was probably due to the synthesis of proteases by cells that had entered the deceleration and stationary phases of the growth cycle. In addition, 50% of the cells had lost their plasmids after growth for 4 hours at 40°C. These problems notwithstanding, after 4 hours at 40°C, the AGβgal protein was approximately 20% of the dry weight of the total biomass. Considering the very high yield of the target protein that was produced by this strategy, it is probably not necessary to integrate the genes for the AGβgal protein and the *cI* repressor into the chromosomal DNA of the *E. coli* host cell in an effort to increase the final yield.

Batch versus Fed-Batch Fermentation

In some instances a simple fed-batch strategy can be used to produce both a high cell density and a high level of expression of a target protein. For example, a plasmid carrying a gene encoding a hybrid protein that includes the insulin B peptide under the control of the *E. coli trp* promoter was introduced into a *trp* mutant strain of *E. coli* and the transformant was grown in media containing various amounts of tryptophan. At high levels of tryptophan, the synthesis of the target protein was repressed; following consumption of the tryptophan in the medium by the growing cells, synthesis of the target protein was induced. With this system, the addition of tryptophan to the medium resulted, in batch cultures, in increases in the amounts of both biomass and target protein produced. Moreover, the use of constant-rate fed-batch cultures resulted in additional increases in the amounts of both biomass and target protein (Table 16.1).

Harvesting Microbial Cells

The first step in the process of purification of a product that is synthesized during microbial fermentation is the separation of the cells from the culture medium. Recombinant and native microbial cells can both be harvested with the same type of equipment. However, as a consequence of physiological changes, such as alterations in cell size and the production of extracellular polysaccharides, conditions that have been established for nontransformed cells may not be optimal for recombinant cells expressing a foreign protein.

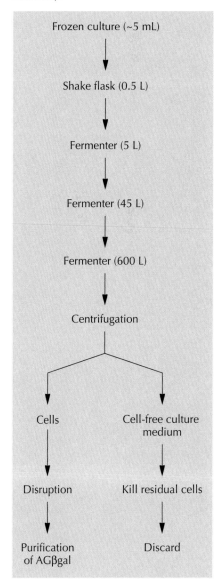

Figure 16.6 Scheme for the large-scale production of the protein AGβgal. The volumes in parentheses indicate how much culture medium was used at each step of the process. The culture medium occupied about 60 to 75% of the total volume of each of the bioreactors (fermenters).

Frozen culture (~5 mL)

↓

Shake flask (0.5 L)

↓

Fermenter (5 L)

↓

Fermenter (45 L)

↓

Fermenter (600 L)

↓

Centrifugation

Cells — Cell-free culture medium

Cells → Disruption → Purification of AGβgal

Cell-free culture medium → Kill residual cells → Discard

Table 16.1 Comparison of batch and fed-batch fermentation for the production of a fusion protein including the insulin B peptide

Product	Yield in fermentation system:		
	Batch	**Batch + Trp**	**Fed batch**
Biomass (g DW/L)	6.7	12	20
Fusion protein or total protein (%)	4.6	7.9	11
Total amount of fusion protein (g/L)	0.17	0.53	1.21
Plasmid-bearing cells (%)	86	62	90

Adapted from Gosset et al., *Appl. Microbiol. Biotechnol.* **39:**541–546, 1993.

Biomass is measured in grams (dry weight) per liter (g DW/L). In the "Batch + Trp" fermentation, 0.1 g of tryptophan was added. In the fed-batch fermentation, 0.1 g of tryptophan was added every 2 hours, for a total of five times during the course of the 10-hour fermentation. A higher level of tryptophan added to the batch fermentation did not increase the amount of either biomass or target protein produced.

For large volumes, either high-speed centrifugation, which is the current method of choice, or membrane microfiltration is used to separate cells from the growth medium. High-speed semicontinuous centrifuges have been specifically designed for the harvesting of microbial cells. The cell suspension is continuously fed into a running centrifuge, and the cells are concentrated within it while the clarified medium is collected in an external container. When the centrifuge chamber is full of packed cells, the run is stopped and the cells are removed. The need to stop and then restart the procedure periodically, especially when large volumes are being processed, can be a major inconvenience. Furthermore, the cost of both the equipment and the power to run it, the release of the microorganisms into the air as aerosols during the harvesting procedure, and the difficulty in removing all the microbial cells from the spent medium also limit the use of this separation procedure.

Membrane filtration is an alternative method of separating the cells from the culture medium. Unfortunately, with traditional (dead-end) filtration, the microbial cells accumulate on the surface of the polymeric membrane filter. Consequently, the flow rate of spent medium through the membrane decreases rapidly (Fig. 16.7A). Increasing the pressure on the membrane enhances the flow for a short time; however, the cells still accumulate on the surface of the membrane and may even form a more compact and less permeable layer as a result of the pressure.

An alternative filtration technique entails passing the cell suspension at a high speed across the surface of the membrane (cross-flow filtration) (Fig. 16.7B). Under these conditions, only a very small fraction of the circulating liquid actually goes through the membrane in any one pass. The remaining cell suspension acts to sweep the membrane clean of accumulated cells, so that the rate of liquid flow through the membrane does not decrease as rapidly as it does in dead-end filtration. In both dead-end and cross-flow filtration, the average pore size of the membranes is 0.2 to 0.45 μm. After many cycles in a cross-flow filtration system, almost all of the culture medium has passed through the membrane. The use of cross-flow filtration has so far been limited to laboratory-scale operations; industrial-scale operations still rely largely on centrifugation.

The next step in the purification process depends on both the nature and the location of the product. If it is a protein and is in the culture medium, the medium is concentrated and the target protein is purified by column chromatography or other standard procedures. If the product is a low-molecular-weight compound in the culture medium, it can be purified by the appropriate extraction procedures. Finally, if the product is in the cel-

A

Cell suspension

Cells

Membrane

Cell-free culture medium

B

Cells

Cell suspension

Cell concentrate

Membrane

Cell-free culture medium

Figure 16.7 Membrane filtration systems for concentrating microbial cells. **A.** Dead-end filtration. **B.** Cross-flow filtration. Arrows within each unit show the direction of the liquid flow.

lular fraction, the cells must be disrupted (lysed) before the steps leading to product purification are initiated.

Disrupting Microbial Cells

A large number of chemical, biological, and physical methods have been developed for disrupting microbial cells. All of these procedures represent a compromise, because they must be vigorous enough to break the micro-

bial cell walls yet gentle enough to ensure that the protein product is not denatured. There is no single set of conditions for cell wall lysis, because the cell walls of diverse microbial species are composed of different polymers.

- In gram-positive bacteria, the cell wall is external to the cytoplasmic membrane and consists of a thick peptidoglycan layer of *N*-acetyl-glucosamine and *N*-acetylmuraminic acid residues cross-linked by oligopeptides.
- The cell wall of gram-negative bacteria has an outer membrane, a thin peptidoglycan layer, and a cytoplasmic membrane.
- The yeast cell wall is composed of a thick layer of partially phosphorylated mannans and β-glucan.
- The lower fungi have multilayered cell walls, consisting of α- and β-glucan, glycoproteins, and chitin.

Cell wall composition and strength depend on culture conditions, cellular growth rate, the phase of the growth cycle when the cells are harvested, how the concentrated cells are stored, and whether the isolated microorganism was expressing a cloned gene. All these factors affect their susceptibility to disruption.

The chemical methods that disrupt microbial cell walls include treatment with alkali, organic solvents, or detergents. If the protein product is stable at pH values from about 10.5 to 12.5, bacterial cell lysis can easily be carried out on a large scale at low cost. For example, recombinant human growth hormone is efficiently released from *E. coli* by treatment with sodium hydroxide at pH 11. Few to no viable cells remain after alkali treatment, and so concerns about the inadvertent release of a genetically engineered microorganism from a production facility are considerably reduced. Treatment with organic solvents is a simple and inexpensive way to disrupt cells and has been used for the isolation of enzymes from yeasts. However, preliminary tests must be run to make sure that the proposed treatment does not denature the protein product. Detergents permeabilize bacterial cells by solubilizing cell membranes and membrane proteins. As a consequence of this activity, holes are formed and proteins and other molecules are released from the cells. Unfortunately, detergents are expensive, frequently denature the protein product, and are often retained as contaminants throughout the purification process.

The major biological method for disrupting microbial cells is enzymatic lysis. For example, the cell walls of gram-positive bacteria are readily hydrolyzed by the enzyme lysozyme, which is isolated from egg whites; the cell walls of gram-negative bacteria are hydrolyzed by lysozyme and ethylenediaminetetraacetic acid (EDTA); and those of yeasts are hydrolyzed by combinations of one or more of the enzymes β-1,3-glucanase, β-1,6-glucanase, mannanase, and chitinase. Enzymatic treatments are highly specific, and the conditions for lysis are mild. Currently, cost considerations limit the use of enzymes as cell lysis agents, but the use of genetically engineered microorganisms for large-scale production of enzymes that attack cell walls should make these enzymes less expensive.

Microbial cells can be physically disrupted either by nonmechanical methods that include osmotic shock and repeated cycles of freezing and thawing or by mechanical procedures such as sonication, wet milling, high-pressure homogenization, and impingement. Generally, after treatment by a nonmechanical method, many of the cells remain intact. In contrast, mechanical disruption is highly efficient, which makes it the preferred choice.

A sonicator that generates high-pressure sound waves that cause cell disruption by shear and cavitation (production of internal holes) is generally useful for small volumes.

Wet milling is quite commonly used for disrupting large quantities of cells. A concentrated cell suspension is pumped into the chamber of a high-speed agitator bead mill that is filled with an inert abrasive material such as small glass beads (<1 mm in diameter) and is fitted with a central shaft that has a number of attached blades. Most of the cell disruption occurs as a consequence of the shear forces generated by the high-speed motion of the glass beads. The conditions for optimizing cell disruption can be determined by experimenting with the number or configuration of the agitator disks, the agitator speed, the size of the glass beads, the number of glass beads, the cell concentration, the geometry of the grinding chamber, and the temperature. This type of apparatus has been successfully used to disrupt a wide range of different microbial cell types and can readily break recombinant as well as nonrecombinant cells.

In the high-pressure homogenization process, concentrated cells are pumped into a valve assembly under high pressure and the pressure is then rapidly decreased, causing the cells to lyse. This process can be customized for different microbial species and protein products by changing the operating pressure, the design of the valve, the temperature of the cell suspension, or the number of times the cell mass is treated.

Impingement is a cell disruption procedure in which a high-velocity stream of suspended cells under pressure hits either a stationary surface or a second fluid stream of suspended cells. The forces that are created at the point of contact disrupt the cells. With a device called a Microfluidizer, for example, two parallel streams of *E. coli* cells in suspension are impinged against one another; a high percentage of the cells are disrupted by a single passage through the unit. Additional passages may be required for other cell types. Unlike high-pressure homogenizers and high-speed agitator bead mills, which generally require highly concentrated cell suspensions, this device can be used to disrupt cells in either dilute or concentrated preparations. Preliminary reports indicate that the activities of cellular proteins are not significantly impaired as a consequence of cell breakage by this technique. When the cell suspension is pretreated with low levels of lysozyme and then disrupted in the Microfluidizer at much lower than normal pressure and fluid velocity, the activities of somewhat more labile proteins, which might be otherwise inactivated by the high pressure, are retained.

Downstream Processing

After cell disruption, cell debris is removed by either low-speed, high-capacity centrifugation or membrane microfiltration. The protein product is precipitated from either the crude lysate, the clarified lysate, or the cell-free culture medium with organic solvents (alcohol or acetone) or ammonium sulfate. Under these conditions, the target protein is usually enriched approximately two- to fivefold. Unfortunately, the cost of the precipitant can add significantly to the cost of the process. Alternatively, the crude protein mixtures are concentrated and fractionated by cross-flow ultrafiltration through membranes that have a smaller average pore size than those used for either cell concentration or debris removal (Fig. 16.7B). This approach, which is still in the developmental stage, can be used with vol-

umes ranging from 1 to several thousand liters, can be performed continuously (thereby limiting the size of the required apparatus), and can yield 10- to 100-fold enrichment, depending on the size and properties of the target protein.

The required degree of purity of the final protein product depends on its end use. In some cases, crude preparations are satisfactory, but, for other products, such as pharmaceutical proteins, additional purification procedures are required.

A number of proteins that are overproduced intracellularly are confined to insoluble particles (inclusion bodies) within the bacterium. After disruption, such inclusion bodies can be readily separated from the bulk of the remaining cell components. Initially, researchers found it difficult to solubilize inclusion bodies without irreversibly denaturing the protein, but protocols have now been devised by which the activity of the recombinant protein can be restored by a renaturation step. Of course, the additional steps involved in the isolation of active proteins from inclusion bodies increase the cost of the purification process.

Protein Solubilization

In some instances, overproduction of a recombinant protein can create both soluble and insoluble forms, which complicates the purification process. For example, when human insulin-like growth factor I (IGF-I), a 7.6-kilodalton (kDa) peptide, was expressed in *E. coli* cells, approximately 90% of the recombinant protein was localized in the *E. coli* periplasm and about 10% was found in the external medium. To recover both the soluble and insoluble forms of *IGF-I* from the periplasm, a high concentration of urea and dithiothreitol at alkaline pH was added to solubilize the insoluble forms of the peptide in situ. This treatment kills but does not lyse the cells. Consequently, the cytoplasmic proteins remain within the cells. This solubilization procedure produces a highly viscous solution, which precludes removing the cells and cell debris by centrifugation. Instead, an aqueous two-phase liquid extraction procedure which separates soluble and insoluble material was developed for this purpose. Both the in situ solubilization and aqueous two-phase liquid extraction procedure are highly efficient; 80 to 95% of the IGF-I is recovered from 10-, 100-, or 1,000-liter fermentations by these methods.

SUMMARY

The large-scale production of genetically engineered organisms in industrial-sized bioreactors (>1,000 liters) is not achieved merely by extrapolating directly from laboratory growth conditions (0.1 to 1.0 liter). The temperature, pH, rate and nature of mixing, oxygen demand for aerobic organisms, and nutrient levels must be taken into consideration when large bioreactors are being designed.

Microbial fermentations can be performed in several different ways. In batch fermentation, an inoculum of cells is added to fresh medium and the fermentation is allowed to proceed without supplementation until the maximum amount of the target product is synthesized. Under these conditions, the cell culture passes through six phases of growth: lag, acceleration, logarithmic (log), deceleration, stationary, and death. Protein production is optimal during the log phase, and peak production of many low-molecular-weight products occurs during the stationary phase. It is important to

monitor batch fermentations closely to ensure that the cells are harvested at the appropriate time. In fed-batch fermentation, growth medium is added at various intervals, usually to prolong the log phase of the fermentation process. Continuous fermentation entails adding fresh growth medium throughout the course of the fermentation and simultaneously removing cells and spent medium.

Each of these fermentation systems has disadvantages and advantages for large-scale production of recombinant products. Although continuous fermentation is still a relatively untried industrial process, this approach has some inherent benefits, suggesting that its use may become more widespread in the future.

One way to increase the amount of a recombinant protein is to grow transformants expressing that protein to as high a cell density as possible. The best way to obtain high-density cell cultures is to use a fed-batch fermentation strategy.

There are three basic bioreactor configurations: stirred-tank reactors (STRs); bubble columns, and airlift reactors. Currently, STRs are used most frequently in industry, but interest in airlift bioreactors is increasing. In STRs, mixing is achieved by mechanical agitation. In airlift reactors, both aeration and mixing are performed by a gas (usually air) introduced through a sparger at the bottom of the vessel, with either internal baffles or external loops causing the fluid to circulate within the vessel. Bubble columns are similar to airlift reactors but lack the construction features that cause the culture medium to circulate in the vessel. The problems of maintaining sterility, pH, temperature, and other fermentation parameters are overcome in different ways depending on the design of the bioreactor. Two-stage fermentation processes in either tandem airlift reactors or a single STR have been successfully used for the production of recombinant proteins.

When the product is present in the cells, the cells can be harvested by either centrifugation or cross-flow filtration and lysed chemically or mechanically. The product is then fractionated from the cell lysate. Various devices and processes have been created to ensure that these steps are conducted effectively. When the product is in the culture medium, the concentration and purification procedures are simpler.

References

Bailey, J. E., and D. F. Olis. 1977. *Biochemical Engineering Fundamentals.* McGraw-Hill, New York, N.Y.

Charles, M. 1985. Fermentation scale-up: problems and possibilities. *Trends Biotechnol.* **3:**134–139.

Datar, R. 1986. Economics of primary separation steps in relation to fermentation and genetic engineering. *Process Biochem.* **21:**19–26.

Engler, C. R. 1985. Disruption of microbial cells, p. 305–324. *In* C. L. Cooney, A. E. Humphrey, and M. Moo-Young (ed.), *Comprehensive Biotechnology,* vol. 2. Pergamon Press, Oxford, United Kingdom.

Giorgio, R. J., and J. J. Wu. 1986. Design of large scale containment facilities for recombinant DNA fermentations. *Trends Biotechnol.* **4:**60–65.

Gosset, G., R. de Anda, N. Cruz, A. Martínez, R. Quintero, and F. Bolivar. 1993. Recombinant protein production in cultures of an *Escherichia coli trp⁻* strain. *Appl. Microbiol. Biotechnol.* **39:**541–546.

Grund, G., C. W. Robinson, and B.R. Glick. 1991. Cross-flow ultrafiltration of proteins, p. 69–83. *In* M. D. White, S. Reuveny, and A. Shafferman (ed.), *Biologicals from*

Recombinant Microorganisms and Animal Cells: Production and Recovery. Verlag Chemie, Weinheim, Germany.

Hart, R. A., P. M. Lester, D. H. Reifsnyder, J. R. Ogez, and S.E. Builder. 1994. Large scale, *in situ* isolation of periplasmic IGF-I from *E. coli. Bio/Technology* **12**:113–117.

Kroner, K. H. 1986. Cross-flow filtration in the downstream processing of enzymes: current status. *Biotechnol. Forum* **3**:20–31.

Kroner, K. H., H. Nissinen, and H. Zeigler. 1987. Improved dynamic filtration of microbial suspensions. *Bio/Technology* **5**:921–926.

Lee, S. Y. 1996. High cell-density culture of *Escherichia coli. Trends Biotechnol.* **14**:98–105.

McKillip, E. R., A. S. Giles, M. H. Levner, P. P. Hung, and R. N. Hjorth. 1991. Bioreactors for large-scale t-PA production. *Bio/Technology* **9**:805–812.

Mendoza-Vega, O., C. Hebert, and S. W. Brown. 1994. Production of recombinant hirudin by high cell density fed-batch cultivations of a *Saccharomyces cerevisiae* strain: physiological considerations during the bioprocess design. *J. Biotechnol.* **32**:249–259.

Merchuk, J. C. 1990. Why use airlift bioreactors? *Trends Biotechnol.* **8**:66–71.

Park, T. H., J.-H. Seo, and H. C. Lim. 1991. Two-stage fermentation with bacteriophage λ as an expression vector in *Escherichia coli. Biotechnol. Bioeng.* **37**:297–302.

Paulson, D. J., R. L. Wilson, and D. D. Spatz. 1984. Cross-flow membrane technology and its applications. *Food Technol.* **Dec. 1984**:77–87.

Ramírez, D. M., and W. E. Bentley. 1995. Fed-batch feeding and induction policies that improve foreign protein synthesis and stability by avoiding stress response. *Biotechnol. Bioeng.* **47**:596–608.

Reuss, M. 1995. Stirred tank bioreactors, p. 207–255. *In* J. A. Asenjo and J. Merchuk (ed.), *Bioreactor System Design.* Marcel Dekker, Inc., New York, N.Y.

Robinson, D. K., C. P. Chan, C. Yu Ip, P. K. Tsai, J. Tung, T. C. Seamans, A. B. Lenny, D. K. Lee, J. Irwin, and M. Silberklang. 1994. Characterization of a recombinant antibody produced in the course of a high yield fed-batch process. *Biotechnol. Bioeng.* **44**:727–735.

Sauer, T., C. W. Robinson, and B. R. Glick. 1989. Disruption of native and recombinant *Escherichia coli* in a high-pressure homogenizer. *Biotechnol. Bioeng.* **33**:1330–1342.

Sayadi, S., M. Nasri, F. Berry, J. N. Barbotin, and D. Thomas. 1987. Effect of temperature on the stability of plasmid pTG201 and productivity of *xylE* gene product in recombinant *Escherichia coli*: development of a two-stage chemostat with free and immobilized cells. *J. Gen. Microbiol.* **133**:1901–1908.

Schügerl, K., and A. Lübbert. 1995. Pneumatically agitated bioreactors, p. 257–303. *In* J. A. Asenjo and J. Merchuk (ed.), *Bioreactor System Design.* Marcel Dekker, Inc., New York, N.Y.

Schütte, H., and M.-R. Kula. 1990. Pilot- and process-scale techniques for cell disruption. *Biotechnol. Appl. Biochem.* **12**:599–620.

Seigel, R., and D. D. Y. Ryu. 1985. Kinetic study of instability of recombinant plasmid pPLc23*trp*Al in *E. coli* using two-stage continuous culture system. *Biotechnol. Bioeng.* **27**:28–33.

Siegel, M. H., H. Hallaille, and J. C. Merchuk. 1988. Air-lift reactors: design, operation, and applications. *Adv. Biotechnol. Processes* **7**:79–124.

Strandberg, L., K. Köhler, and S.-O. Enfors. 1991. Large-scale fermentation and purification of a recombinant protein from *Escherichia coli. Process Biochem.* **26**:225–234.

Strandberg, L., L. Andersson, and S.-O. Enfors. 1994. The use of fed batch cultivation for achieving high cell densities in the production of a recombinant protein in *Escherichia coli. FEMS Microbiol. Rev.* **14**:53–56.

Strathman, H. 1985. Membranes and membrane processes in biotechnology. *Trends Biotechnol.* **3**:112–118.

Tanny, G. B., D. Mirelman, and T. Pistole. 1980. Improved filtration techniques for concentrating and harvesting bacteria. *Appl. Environ. Microbiol.* **40:**269–273.

Tutunjian, R. S. 1985. Scale-up considerations for membrane processes. *Bio/Technology* **3:**615–626.

Van Brunt, J. 1985. Scale-up: the next hurdle. *Bio/Technology* **3:**419–424.

Van Brunt, J. 1986. Fermentation economics. *Bio/Technology* **4:**395–401.

White, M. D., B. R. Glick, and C. W. Robinson. 1995. Bacterial, yeast and fungal cultures: the effect of microorganism type and culture characteristics on bioreactor design and operation, p. 47–87. *In* J. A. Asenjo and J. Merchuk (ed.), *Bioreactor System Design.* Marcel Dekker, Inc., New York, N.Y.

Whitney, G. D., B. R. Glick, and C. W. Robinson. 1989. Induction of T4 DNA ligase in a recombinant strain of *Escherichia coli. Biotechnol. Bioeng.* **33:**991–998.

Yamane, T. 1995. Bioreactor operation modes, p. 479–509. *In* J. A. Asenjo and J. Merchuk (ed.), *Bioreactor System Design.* Marcel Dekker, Inc., New York, N.Y.

REVIEW QUESTIONS

1. What are the differences between batch, fed-batch, and continuous fermentations?

2. What parameters need to be monitored and controlled in an optimized fermentation process?

3. How does the presence of a recombinant plasmid affect the growth of microbial cells?

4. How does the mixing of a growing microbial culture affect the transfer of oxygen from the growth medium to the cells?

5. What are the advantages of a high cell density during a large-scale fermentation? What conditions lead to a high cell density during a large-scale fermentation?

6. What are the relative advantages and disadvantages of using an STR or an airlift fermenter?

7. Compare the growth and induction of a recombinant microbial culture (a) using two reactors in tandem and (b) using a single reactor.

8. What is downstream processing?

9. What strategy would you use to purify a recombinant protein that is secreted into the growth medium?

10. What are the advantages and disadvantages of large-scale mechanical lysis of cells relative to chemical lysis?

11. How can microbial cells be concentrated after the fermentation stage of a biotechnological process? What are the advantages and disadvantages of these procedures?

12. How can the activity of a recombinant protein that is synthesized in an insoluble form be recovered?

Eukaryotic Systems

I N THE PAST, high-yielding strains of many different crop plants and farm animals have been successfully developed by selective cross-breeding. This time-consuming process has been, in part, superseded by the development of methods for the genetic engineering of higher organisms. Now, genes that determine specific traits can be introduced into plants and animals and then passed on from one generation to the next. How these transgenic plants and animals are formed is examined in Part III.

A number of transgenic plants have already been produced. Traits that have been engineered into plants include insecticidal activity, tolerance of viral infection, herbicide resistance, altered fruit ripening, altered flower pigmentation, improved nutritional quality of seed proteins, and self-incompatibility. Many of the transgenesis studies with animals are still in an early stage, so it is difficult to predict which genetic features will be introduced into the recipient species. However, transgenic mice have been developed as model systems for cancer induction, cystic fibrosis, Alzheimer disease, and other human diseases.

Recombinant DNA technology has also had a profound impact on the study of human genetic diseases and the development of therapies that are based on genetic engineering. It is now possible to find a chromosomal linkage between a human disease-causing gene, whose mode of action is totally unknown, and a marker site and then use the cloned sequence that identifies the nearby marker site as a start for a search for the defective gene. With this strategy, a number of genes for human genetic diseases have been identified and characterized. Subsequently, the mode of action of the normal gene and the malfunction of the defective version can be determined, thereby facilitating the development of effective therapies. Both human molecular genetics and human gene therapy are discussed in Part III.

Genetic Engineering of Plants: Methodology

Considerable effort has gone into developing varieties of plants that produce increased yields and have enhanced nutritional value. Although much of this endeavor has been directed toward the three major grains—corn, wheat, and rice—successful breeding programs for other food plants and horticultural species have also been established. Recombinant DNA technology, which has been used extensively with microbial systems, is also an important tool for the direct genetic manipulation of plants. There are a number of effective DNA-delivery systems and expression vectors that work with a range of plant cells. Furthermore, most plant cells are totipotent—meaning that an entire plant can be regenerated from a single plant cell—so fertile plants that carry an introduced gene(s) in all cells (i.e., transgenic plants) can often be produced from genetically engineered cells. If the transgenic plant flowers and produces viable seed, the desired trait is passed on to successive generations.

There are three major reasons for developing transgenic plants. First, the addition of a gene(s) often improves the agricultural, horticultural, or ornamental value of a crop plant. Second, transgenic plants can act as living bioreactors for the inexpensive production of economically important proteins or metabolites. Third, plant genetic transformation (transgenesis) provides a powerful means for studying the action of genes during development and other biological processes.

Some of the genetically determined traits that can be introduced into plants by a single gene or, possibly, a small cluster of genes include insecticidal activity, protection against viral infection, resistance to herbicides, delay of senescence, tolerance of environmental stresses, altered flower pigmentation, improved nutritional quality of seed proteins, and self-incompatibility. To date, numerous transgenic plants have been generated, including many crop and forest species. In the future, plant biotechnology will have an enormous impact on traditional plant breeding programs because it can significantly decrease the 10 to 15 years that it currently takes to develop a new variety using traditional plant breeding techniques; further, it will also be used to create plants with novel characteristics.

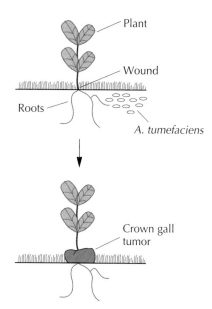

Figure 17.1 Infection of a plant with *A. tumefaciens* and formation of a crown gall.

Figure 17.2 Structures of the plant molecules acetosyringone and hydroxyacetosyringone. These compounds are released in response to wounding and can induce the *vir* genes of the Ti plasmid.

Acetosyringone

Hydroxyacetosyringone

Plant Transformation with the Ti Plasmid of *Agrobacterium tumefaciens*

The gram-negative soil bacterium *Agrobacterium tumefaciens* is a phytopathogen that, as a normal part of its life cycle, genetically transforms plant cells. This genetic transformation leads to the formation of crown gall tumors, which interfere with the normal growth of an infected plant (Fig. 17.1). This agronomically important disease affects only dicotyledonous plants (dicots), including grapes, stone-fruit trees, and roses.

Crown gall formation is the consequence of the transfer, integration, and expression of genes of a specific segment of bacterial plasmid DNA—called the T-DNA (transferred DNA)—into the plant cell genome. The T-DNA is actually part of the "tumor-inducing" (Ti) plasmid that is carried by most strains of *A. tumefaciens*. Depending on which bacterial strain the Ti plasmid comes from, the length of the T-DNA region can vary from approximately 12 to 24 kilobase pairs (kb). Strains of *A. tumefaciens* that do not possess a Ti plasmid cannot induce crown gall tumors.

The initial step in the infection process is the attachment of *A. tumefaciens* to a plant cell at the site of an open wound, often at the base of the stem (i.e., the crown) of the plant. Originally, it was thought that *A. tumefaciens* infected wounded plants because the physical barrier of the cell wall had been breached by injury, thereby facilitating entry of the bacterium. However, it is now believed that these bacteria respond to certain plant phenolic compounds such as acetosyringone and hydroxyacetosyringone (Fig. 17.2), which are excreted by susceptible wounded plants. These wound-response compounds resemble some of the products of phenylpropanoid metabolism, which is the major plant pathway for the synthesis of plant secondary metabolites such as lignins and flavonoids. These small molecules (i.e., acetosyringone, hydroxysyringone) act to induce the activity of the virulence (*vir*) genes that are encoded on the Ti plasmid (Fig. 17.3).

The *vir* genes are located on a 35-kb region of the Ti plasmid that lies outside of the T-DNA region. The products of the *vir* genes are essential for the transfer and integration of the T-DNA region (Fig. 17.3) into the genome of a plant cell. There are at least seven, and possibly eight, different *vir* genes.

After a Ti plasmid-carrying cell of *A. tumefaciens* attaches to a host plant cell and the *vir* genes are induced, the T-DNA is transferred by a process that is thought to be similar to plasmid transfer from donor to recipient cells during bacterial conjugation. In this model, the T-DNA is transferred as a linear, single-stranded molecule from the Ti plasmid, passes into the plant cell, and eventually becomes integrated in the plant chromosomal DNA. The 5' end of the single-stranded T-DNA carries the right border sequence, and the left border sequence is at the 3' end.

The formation of the single-stranded form of T-DNA is initiated by strand-specific cutting at both borders of the intact T-DNA region. The integration of the T-DNA into the plant genome is thought to depend on specific sequences that are located at the right border of the T-DNA. This border contains a repeating unit that consists of 25 base pairs (bp). Although the left border contains a similar 25-bp repeat, deletion studies have shown that this region is not involved in the integration process.

Most of the genes that are located within the T-DNA region are activated only after the T-DNA is inserted into the plant genome. The products

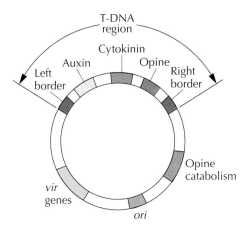

Figure 17.3 Schematic representation of a Ti plasmid. The T-DNA is defined by its left and right borders and includes genes for the biosynthesis of auxin, cytokinin, and an opine; these genes are only transcribed and translated in plant cells. Outside of the T-DNA region, there is a cluster of *vir* genes, a gene(s) that encodes an enzyme(s) for opine catabolism, and an origin of DNA replication (*ori*) which permits the plasmid to be stably maintained in *A. tumefaciens*. None of these features is drawn to scale.

of these genes are responsible for crown gall formation. The T-DNA region includes the genes *iaaM* and *iaaH*, which are also known as *tms1* and *tms2*, respectively. This pair of genes encodes enzymes that synthesize the plant hormone auxin (indoleacetic acid). Specifically, the gene *iaaM* codes for the enzyme tryptophan 2-monooxygenase, which converts tryptophan to indole 3-acetamide, and the gene *iaaH* contains the information for indole 3-acetamide hydrolase, which converts indole 3-acetamide to indoleacetic acid (Fig. 17.4A). In addition, the T-DNA region carries the *tmr* gene (also known as *ipt*), which encodes isopentenyl transferase. This enzyme adds 5'-AMP to an isoprenoid side chain to form the cytokinins isopentenyladenine and isopentenyladenosine (Fig. 17.4B). Hydroxylation of these two molecules by plant enzymes generates the cytokinin called transzeatin and transribosylzeatin, respectively. Both auxin and the cytokinins regulate plant cell growth and development. In excess, they can cause the plant to develop tumorous growths such as crown galls.

In addition to auxin and cytokinin biosynthesis genes, the T-DNA region from each specific Ti plasmid carries a gene for the synthesis of a molecule called an opine. Opines are unique and unusual condensation products of an amino acid and a keto acid, or an amino acid and a sugar. For example, the condensation product of arginine and pyruvic acid is called octopine; arginine with α-ketoglutaraldehyde is nopaline; and agropine is a bicyclic sugar derivative of glutamic acid (Fig. 17.5). The opines are synthesized within the crown gall and then secreted. They can be used as a carbon source (and sometimes also as a nitrogen source) by any *A. tumefaciens* that carries a Ti plasmid-borne gene for the catabolism of that particular opine (Fig. 17.3). The opine catabolism gene(s) is on the Ti plasmid but is not part of the T-DNA region. Most other soil microorganisms that have been tested are incapable of utilizing opines as a carbon source. Thus, a unique set of mechanisms has evolved whereby each strain of *A. tumefaciens* genetically engineers plant cells to be biological factories for the production of a carbon compound that it alone is able to use.

A

Tryptophan

O_2

CO_2 + H_2O

Indole-3-acetamide

NH_3

Indoleacetic acid

B

IPP

+

5'-AMP

PP_i

IPA

Figure 17.4 Biosynthesis of auxin and cytokinin by the enzymes encoded by the T-DNA genes of the Ti plasmid of *A. tumefaciens*. **A.** The auxin pathway involves the conversion of tryptophan to indole 3-acetamide by tryptophan monooxygenase and then indole 3-acetamide to indoleacetic acid by indole-3-acetamide hydrolase. **B.** The cytokinin synthesis reaction entails the attachment of an isopentenyl moiety from isopentenyl diphosphate (IPP) to 5'-AMP by the enzyme isopentenyltransferase to form isopentenyl adenosine monophosphate (IPA).

Figure 17.5 Chemical structures of three opines: octopine, nopaline, and agropine.

Ti Plasmid-Derived Vector Systems

The simplest way to exploit the ability of the Ti plasmid to genetically transform plants would be to insert a desired DNA sequence into the T-DNA region and then use the Ti plasmid and *A. tumefaciens* to deliver and insert this gene(s) into the genome of a susceptible plant cell. However, although the Ti plasmids are effective as natural vectors, they have several serious limitations as routine cloning vectors:

- The production of phytohormones by transformed cells growing in culture prevents them from being regenerated into mature plants. Therefore, the auxin and cytokinin genes must be removed from any Ti plasmid-derived cloning vector.
- A gene encoding opine synthesis is not useful to a transgenic plant and may lower the final plant yield by diverting plant resources into opine production. Therefore, the opine-synthesis gene should be removed.
- Ti plasmids are large (approximately 200 to 800 kb). For recombinant DNA experiments, however, a much smaller version is preferred, so large segments of DNA that are not essential for a cloning vector must be removed.
- Because the Ti plasmid does not replicate in *Escherichia coli*, the convenience of perpetuating and manipulating Ti plasmids carrying in-

serted DNA sequences in this bacterium does not exist. Therefore, in developing Ti plasmid-based vectors, an origin of replication that can be used in *E. coli* must be added.

Despite these constraints, a number of plant cell vectors have been constructed. All these Ti plasmid-based vectors are similarly organized and contain the following components:

- A selectable marker gene, such as neomycin phosphotransferase, that confers kanamycin resistance to transformed plant cells. Because the neomycin phosphotransferase gene (as well as many other marker genes used in plant transformation) is prokaryotic in origin, it is necessary to put it under the control of plant (eukaryotic) transcriptional regulation signals, including both a promoter and a termination-polyadenylation sequence, to ensure that it is efficiently expressed in transformed plant cells.
- An origin of DNA replication that allows the plasmid to replicate in *E. coli*. In some vectors, an origin of replication that functions in *A. tumefaciens* has also been added.
- The right border sequence of the T-DNA region. This region is absolutely required for T-DNA integration into plant cell DNA, although most cloning vectors include both a right and a left border sequence.
- A polylinker (multiple cloning site) to facilitate insertion of the cloned gene into the region between T-DNA border sequences.

Because these cloning vectors lack *vir* genes, they cannot by themselves effect the transfer and integration of the T-DNA region into recipient plant cells. Two different approaches have been used to meet these ends. In one approach, a binary vector system is used (Fig. 17.6A). The binary cloning vector contains both *E. coli* and *A. tumefaciens* origins of DNA replication but no *vir* genes; it is actually an *E. coli*–*A. tumefaciens* shuttle vector. All the cloning steps are carried out in *E. coli* before the vector is introduced into *A. tumefaciens*. The recipient *A. tumefaciens* strain carries a modified (defective, disarmed) Ti plasmid that contains a complete set of *vir* genes but lacks portions, or all, of the T-DNA region (so that this T-DNA cannot be transferred). In this system, the defective Ti plasmid synthesizes the *vir* gene products that mobilize the T-DNA region of the binary cloning vector plasmid. By providing the proteins encoded by the *vir* genes, the defective Ti plasmid is acting as a helper plasmid, enabling the T-DNA from the binary cloning vector to be inserted into the plant chromosomal DNA.

In the second approach, called the cointegrate vector system, the cloning vector recombines with a modified (disarmed) Ti plasmid which lacks the tumor-producing genes of the T-DNA within *A. tumefaciens*, and the entire cloning vector becomes integrated into the disarmed Ti plasmid (Fig. 17.6B). The cointegrate cloning vehicle and the disarmed helper Ti plasmid carry homologous DNA sequences that provide a shared site for in vivo homologous recombination; normally these sequences lie inside the T-DNA region. Following recombination, the cloning vector becomes part of the disarmed Ti plasmid, which provides the *vir* genes necessary for the transfer of the T-DNA to the host plant cells. The only way that this cloning vector can be maintained in *A. tumefaciens* is as part of a cointegrate structure. In this cointegrated configuration the genetically engineered T-DNA region can be transferred to plant cells.

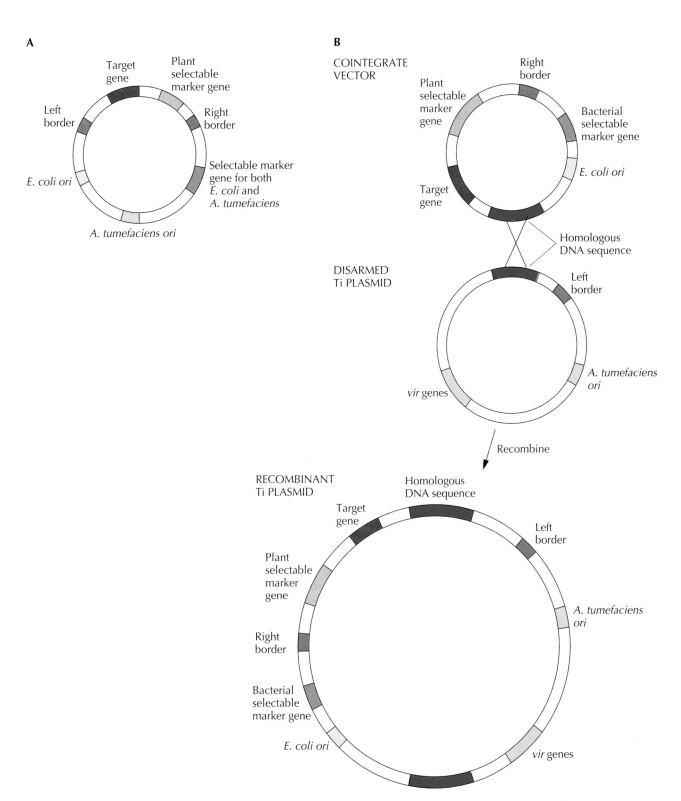

Figure 17.6 Two Ti plasmid-derived cloning vector systems. **A.** The binary cloning vector has origins of DNA replication (*ori*) for both *E. coli* and *A. tumefaciens* (or a broad-host-range origin), a selectable marker gene that can be used in either *E. coli* or *A. tumefaciens*, and both a target gene and a plant selectable marker gene inserted between the T-DNA left and right borders. **B.** The cointegrate cloning vector (top) carries only an *E. coli* origin of replication and cannot exist autonomously within *A. tumefaciens*. It also contains a selectable marker that can be used in either *E. coli* or *A. tumefaciens*, a T-DNA right border, a plant selectable marker (reporter) gene, a target gene, and a sequence of Ti plasmid DNA that is homologous to a segment on the disarmed Ti plasmid. The disarmed Ti plasmid (middle) contains the T-DNA left border, the *vir* gene cluster, and an *A. tumefaciens* origin of DNA replication (*ori*). Following recombination between the cointegrate cloning vector and the disarmed Ti plasmid, the final recombinant plasmid (bottom) has the T-DNA left and right borders bracketing the cloned and plant reporter genes.

Physical Methods of Transferring Genes to Plants

Although *A. tumefaciens*-mediated gene transfer systems are effective in several species, not all plants can be transformed in this way. In particular, monocotyledonous plants (monocots), including the world's major cereal crops such as rice, wheat, and corn, are not readily transformed by *A. tumefaciens*. However, by refining and carefully controlling conditions, protocols have been devised for the transformation of maize and rice, both monocots, by *A. tumefaciens* carrying Ti plasmid vectors. For example, immature maize embryos were immersed in an *A. tumefaciens* cell suspension for a few minutes and then incubated for several days at room temperature in the absence of selective pressure. The embryos were then transferred to a medium that contained a selective antibiotic that only allowed transformed plant cells to grow. These cells were maintained in the dark for a few weeks. Finally, the mass of transformed plant cells was transferred to a different growth medium and incubated in the light, which permitted regeneration of whole transgenic plants.

Table 17.1 lists some of the plant transformation methods that have been developed for monocots. A number of these methods require the removal of the plant cell wall to form protoplasts. Plant protoplasts can be maintained in culture as independently growing cells, or, with a specific culture medium, cell walls can be formed and whole plants can be regenerated from these cells. In addition, transformation methods have been developed that introduce cloned genes into a small number of cells of a plant tissue from which whole plants can be formed, bypassing the need for regeneration from a protoplast. At present, most researchers favor the use of either Ti plasmid-based vectors or microprojectile bombardment to deliver DNA into plant cells (Table 17.1). Using these various techniques, more than 50 different plants have been genetically transformed (Table 17.2).

Table 17.1 Plant cell DNA-delivery methods

Method	Comment
Ti plasmid-mediated gene transfer	An excellent and highly effective system that is limited to a few kinds of plants
Microprojectile bombardment	Used with a wide range of plants and tissues; easy and inexpensive
Viral vectors	Not an effective way to deliver DNA to plant cells
Direct gene transfer into plant protoplasts	Can be used only with plant cell protoplasts that can be regenerated into viable plants
Microinjection	Has limited usefulness because only one cell can be injected at a time; requires the services of a highly skilled individual
Electroporation	Generally limited to plant cell protoplasts that can be regenerated into viable plants
Liposome fusion	Can be used only with plant cell protoplasts that can be regenerated into viable plants

Table 17.2 Plants that have been genetically transformed

Alfalfa	Cucumber	Orchid	Rye
Apple	Eggplant	Papaya	Sorghum
Arabidopsis	Flax	Pea	Soybean
Asparagus	Grape	Peanut	Strawberry
Banana	Kiwi	Pearl millet	Sugar beet
Barley	Lettuce	Peony	Sugarcane
Bean	Licorice	Pear	Sunflower
Cabbage	Lily	Petunia	Sweet potato
Canola	Lotus	Plantain	Tall fescue
Carnation	Corn	Poplar	Tobacco
Carrot	Norway spruce	Potato	Tomato
Cotton	Oat	Red fescue	Wheat
Cranberry	Orchard grass	Rice	White spruce

Microprojectile Bombardment

Microprojectile bombardment, also called biolistics, is the most promising DNA delivery system for plants. Gold or tungsten spherical particles (approximately 0.4 to 1.2 μm in diameter) are coated with DNA that has been precipitated with $CaCl_2$, spermidine, or polyethylene glycol. The coated particles are accelerated to high speed (300 to 600 meters/second) with a special apparatus called a particle gun, which uses gunpowder, compressed air, or helium to provide the propelling force. At these speeds, the projectiles can penetrate plant cell walls and membranes. The particle density used does not significantly damage the cells.

Once inside a cell, the DNA, by some unknown process, integrates into the plant DNA. With the microprojectile bombardment system it is possible to transform a large number of different plant species, including monocots and conifers, plants that are not susceptible to *Agrobacterium*-mediated DNA transfer.

Microprojectile bombardment can be used to introduce foreign DNA into plant cell suspensions, callus cultures, meristematic tissues, immature embryos, protocorms, coleoptiles, and pollen in a wide range of different plants (Table 17.3). Furthermore, this method has also been used to deliver genes into chloroplasts and mitochondria, thereby opening up the possibility of introducing exogenous (foreign) genes into these organelles. Typically, plasmid DNA dissolved in buffer is precipitated onto the surface of the microprojectiles. Using this procedure it is possible to increase the transformation frequency by increasing the amount of plasmid DNA; however, too much plasmid DNA can be inhibitory.

With this technique, cells that appear to be transformed, based on the expression of a marker gene, often only transiently express the introduced DNA. Unless the DNA becomes incorporated into the genome of the plant, the foreign DNA will probably be lost when the transformed cells divide.

The configuration of the vector that is used to deliver foreign genes to the plant may have an influence on both the integration and expression in the plant of those genes. For example, transformation is more efficient when linear rather than circular DNA is used. Moreover, large plasmids (>10 kb) may become fragmented during microprojectile bombardment and there-

Table 17.3 Transgenic plants formed by microprojectile bombardment of various plant cells

Plant(s)	Cell source
Corn	Embryonic cell suspension, immature zygotic embryos
Rice	Immature zygotic embryos, embryogenic callus
Barley	Cell suspension, immature zygotic embryos
Wheat	Immature zygotic embryos
Turfgrass	Embryogenic callus
Rye	Meristems
Sorghum	Immature zygotic embryos
Pearl millet	Immature zygotic embryos
Orchid	Protocorms
Banana and plantain	Embryonic cell suspension
Poplar	Callus
Norway and white spruce	Somatic embryos
Pea	Zygotic embryos
Cucumber	Embryogenic callus
Sweet potato	Callus
Cranberry	In vitro-derived stem sections
Peony and lily	Pollen
Alfalfa	Embryogenic callus
Bean	Zygotic embryos
Cotton	Zygotic embryos
Grape	Embryonic cell suspension
Peanut	Embryogenic callus
Tobacco	Pollen

Adapted from Southgate et al., *Biotechnol. Adv.* **13**:631–651, 1995.

fore often result in lower levels of foreign gene expression than when smaller plasmids are used.

Use of Reporter Genes in Transformed Plant Cells

It is essential to be able to detect the foreign DNA that has been integrated into plant genomic DNA so that those cells that have been transformed can be identified. Furthermore, in studies of plant transcriptional regulatory signals and the functioning of these signals in specific plant tissues (such as in leaf, root, and flowers), it is often important to be able to quantify the level of expression of a gene with a readily identified product. Quantification and other applications require the use of reporter genes that either permit transformed cells to be selected or encode an activity that can be assayed. To these ends, a number of different genes have been tested as reporters for transformation, including both genes that can be used as dominant selectable markers and genes whose proteins produce a detectable response to a specific assay (Table 17.4). Many of these reporter genes are from bacteria and have been equipped with plant-specific regulatory sequences for expression in plant cells. Dominant marker selection provides a direct means of obtaining only transformed cells in culture. For example, in the presence of the antibiotic kanamycin, only plant cells with a functional neomycin phosphotransferase gene can grow.

Table 17.4 Plant cell reporter and selectable marker gene systems

Enzyme activity	Selectable marker	Reporter gene
Neomycin phosphotransferase	Yes	Yes
Hygromycin phosphotransferase	Yes	Yes
Dihydrofolate reductase	Yes	Yes
Chloramphenicol acetyltransferase	Yes	Yes
Gentamycin acetyltransferase	Yes	Yes
Nopaline synthase	No	Yes
Octopine synthase	No	Yes
β-Glucuronidase	No	Yes
Streptomycin phosphotransferase	Yes	Yes
Bleomycin resistance	Yes	No
Firefly luciferase	No	Yes
Bacterial luciferase	No	Yes
Threonine dehydratase	Yes	Yes
Metallothionein II	Yes	Yes
enol-Pyruvylshikimate-3-phosphate synthase	Yes	No
Phosphinothricin acetyltransferase	Yes	Yes
β-Galactosidase	No	Yes
Blasticidin S deaminase	Yes	Yes
Acetolactate synthase	Yes	No
Bromoxynil nitrilase	Yes	No

Adapted from Walden and Schell, *Eur. J. Biochem.* **192**:563–576, 1990, and Gruber and Crosby, p. 89–119, *in* B. R. Glick and J. E. Thompson (ed.), *Methods in Plant Molecular Biology and Biotechnology*, CRC Press, Boca Raton, Fla.

The needs of a particular experiment often dictate which reporter gene will be used. When the expression of a reporter gene interferes with normal plant functions, it cannot be used. In addition, in the perception of some observers of biotechnology, the presence of some reporter genes and their products may taint a commercial product. In this context, it is best not to introduce antibiotic resistance genes into any crop plants.

Some reporter gene products (e.g., β-D-glucuronidase and both firefly and bacterial luciferases) can be detected in intact plant tissues. The most popular of these systems is the *E. coli* β-D-glucuronidase (GUS) gene. It encodes a stable enzyme that is not normally present in plants and that catalyzes the cleavage of a range of β-D-glucuronides. The GUS activity in transformed plant tissues can be localized by observing the blue color that is formed after the hydrolysis of the uncolored substrate 5-bromo-4-chloro-3-indolyl β-D-glucuronic acid. Alternatively, GUS activity in plant extracts can be more quantitatively and sensitively assayed by a fluorometric analysis that involves the hydrolysis of the substrate 4-methylumbelliferyl β-D-glucuronide to form a fluorescent product.

Manipulation of Gene Expression in Plants

When genetic transformation of plants became routine, research efforts were directed towards introducing a wide range of plant and bacterial genes in plant cells. The transformed plants were assayed for the production of the foreign protein and studied physiologically to assess how the

MILESTONE

**Regeneration of Normal and Fertile Plants
that Express Octopine Synthase, from
Tobacco Crown Galls after Deletion of
Tumour-Controlling Functions**
H. De Greve, J. Leemans, J. P. Hernalsteens, L. Thia-Toong,
M. De Beuckeleer, L. Willmitzer, L. Otten,
M. Van Montagu, and J. Schell
Nature **300:**752–755, 1982

An efficient vector system is essential for the routine construction of transgenic plants. Most of the early efforts to develop such a vector focused on the Ti plasmid of the soil bacterium *A. tumefaciens* because, following infection of susceptible dicotyledonous plants, a portion of the Ti plasmid (the T-DNA) is inserted directly into the chromosomal DNA of the host plant cells. However, infection of plants with the Ti plasmid results in the transformed plant forming a crown gall tumor that interferes with the normal growth of the plant. Therefore, before the Ti plasmid could be used as a vector to transform plants, something had to be done to prevent crown gall tumor formation.

By studying the mRNAs that were transcribed from intact and modified T-DNAs, Schell and his colleagues determined that the tumor-inducing genes were a part of the T-DNA. Therefore, modified T-DNA in which the tumor-inducing genes were deleted could be constructed, and this modified T-DNA could then be introduced by homologous recombination into the Ti plasmid. Such a "disarmed" Ti plasmid could be introduced into plant cells and transfer its T-DNA into their chromosomal DNA in a stable manner without crown gall tumors being formed. The next logical step in the development of this system was the cloning of foreign marker and target genes into the T-DNA so that they could also be transferred to the chromosomal DNA of the host plant. This vector system utilizing the Ti plasmid has become the system of choice for creating transgenic plants and has been used successfully by thousands of laboratories around the world.

presence of an additional protein affected the whole plant. Many of these early experiments utilized promoters that were expressed constitutively in a range of plant cells. More recently, additional plant promoters have been isolated and characterized and used to express foreign proteins in specific cells at certain times during the growth and development of the plant. For example, instead of the strong constitutive 35S promoter from cauliflower mosaic virus, which is expressed in all plant tissues and throughout the life of the plant, researchers have used the promoter for the small subunit of the photosynthetic enzyme ribulosebisphosphate carboxylase, which is active only in photosynthetic tissues such as leaves. Similarly, plant promoters active only in specific tissues such as roots or flowers, or only during periods of environmental stress—e.g., the pathogenesis-related (PR) promoters—have been used to control the expression of some foreign genes.

While the vast majority of plant genes are found as part of the nuclear DNA, both the chloroplast and mitochondrion contain genes that encode a number of important and unique functions. However, not all of the proteins that are present in these organelles are encoded within the DNA. Some organellar proteins are encoded in the nuclear DNA, synthesized in the cell's cytoplasm, and then, by a special mechanism, imported into the appropriate organelle. Accordingly, there are two different ways in which a specific foreign protein can be introduced into the chloroplast or mitochondrion. On one hand, a fusion gene encoding the foreign protein and additional amino acids that direct the transport of the protein to the organelle can be inserted into the chromosomal DNA, and the recombinant protein can be transported into a particular organelle. On the other hand, the gene for the foreign protein can be inserted directly into either the chloroplast or mitochondrial DNA.

Isolation and Use of Different Promoters

Specialized vectors, called promoter tagging (labeling) vectors, have been used to isolate plant promoters from several plant species. This approach uses the *Agrobacterium*-mediated Ti plasmid transformation system. Briefly, a promoterless reporter gene is placed next to the right border of a Ti plasmid vector. After transfer of the T-DNA into a plant chromosome, the reporter gene from the vector is situated adjacent to the plant DNA. If the T-DNA is inserted at the promoter region of a functional gene, transcription of the reporter gene occurs. The neomycin phosphotransferase (*npt*) gene can be used as a reporter to identify plant promoters; expression of this gene is detected by selecting the kanamycin-resistant transformants. However, with this method it is difficult to identify (tag) a promoter that is active in only a certain developmental stage or is induced by a specific environmental factor. To ensure that transformed cells are selected, a hygromycin resistance gene, under the control of a constitutive promoter, is placed next to the promoterless reporter gene within the T-DNA. Hygromycin-resistant transformants are first selected, and then the transformants can be checked by an enzyme assay under different conditions for expression of the reporter gene. With this strategy, 5 to 30% of the transformed plant cells have the reporter gene under the control of an active promoter.

The cauliflower mosaic virus 35S promoter is frequently used as a strong promoter in plant systems, although the level of expression of a foreign protein under the control of this promoter is often lower than desired. To address this problem, it is necessary to test different promoter-gene constructs in plants to see if more effective promoters can be found. In addition to the promoter, several other elements may enhance foreign gene expression. These include enhancer sequences that are found from one to several hundred nucleotides upstream of the promoter sequence, introns that may stabilize mRNA, and transcription terminator sequences.

In one series of experiments, DNA constructs that contained all or some of the following elements were tested: the 35S promoter, the nopaline synthase gene transcription terminator, from one to seven tandemly repeated enhancer elements, and a DNA sequence from tobacco mosaic virus called Ω (omega) which is thought to increase gene expression at the translational level. The most active construct contained seven enhancer elements and directed a much higher level of foreign gene expression in both transgenic tobacco and rice plants than when the 35S promoter by itself was used (Table

Table 17.5 Testing promoter constructs in transgenic plants

Plant	Average gene expression		Maximum gene expression	
	35S promoter	Composite promoter	35S promoter	Composite promoter
Tobacco	1.0	2.8	2.8	18.3
Rice	1.0	14.4	7.2	47.1

Adapted from Mitsuhara et al., *Plant Cell Physiol.* **37**:49–59, 1996.

E. coli β-glucuronidase was the reporter gene. Enzyme activities were normalized to the average value per plant when the gene was under the control of the 35S promoter. The actual values in tobacco are approximately 30-fold higher than the values in rice. The composite promoter included the 35S promoter, the nopaline synthase gene transcription terminator, seven tandemly repeated enhancer elements, and the tobacco mosaic virus Ω sequence. Average gene expression is the mean value of a number of different transgenic plants. Maximum gene expression is the highest value observed in any transgenic plant with that promoter.

17.5). The various promoter constructs that were tested directed a wide range of foreign gene expression in transgenic plants. This variation is probably due to the site within the plant genome where the T-DNA is inserted. Nevertheless, this work shows that it is possible to engineer promoters that are much stronger than the naturally occurring 35S promoter. With this approach, it should be possible to engineer promoters that are tissue specific, developmentally regulated, and strong.

Targeting Foreign DNA to the Chloroplast Genome

Most higher plants have approximately 100 chloroplasts per leaf cell, and each chloroplast has about 100 copies of the chloroplast DNA genome. Stable genetic transformation of chloroplasts in order to modify chloroplast functioning requires insertion of the foreign DNA into the chloroplast genome rather than into the much larger chromosomal DNA—plant chromosomal DNA is generally around 10^4 to 10^5 times larger than chloroplast DNA. Moreover, the foreign DNA needs to be present in all of the approximately 10^4 chloroplast DNA genomes per leaf cell.

Initially, foreign DNA was introduced into the chloroplast genome on a plasmid vector with both the nonselectable foreign DNA and a selectable marker such as an antibiotic resistance gene, flanked by specific chloroplast DNA sequences (Fig. 17.7). While this strategy was effective, in many instances insertion of the selectable marker gene interfered with the expression of flanking chloroplast genes. To get around this problem, a protocol was devised in which the selectable marker and the foreign gene were not physically linked to one another. Tobacco plants were transformed with an equal mixture of two different plasmids, one with a selectable marker (a spectinomycin resistance gene) flanked by DNA from one region of the chloroplast, and the other with a foreign gene (a kanamycin resistance gene) flanked by DNA from a different region of the chloroplast (Fig. 17.8). Both genes had prokaryotic-like transcriptional signals enabling them to be transcribed in the chloroplast but not in the nucleus. The chloroplast DNA sequences on the plasmid were arranged so that recombination and insertion into the chloroplast genome would not disrupt any chloroplast genes. Following the introduction of these plasmids by microprojectile-mediated DNA transformation, transformed tobacco plants were selected on a medium that contained spectinomycin. Chloroplasts from the selected transformants were then assayed for the presence of the kanamycin resistance gene product (the unselected foreign gene). Surprisingly, approxi-

Figure 17.7 Plasmid vector used for integrating tandem genes into the chloroplast genome. Abbreviation: Spc^r, spectinomycin resistance.

Foreign gene Spc^r gene

Chloroplast DNA sequences

Figure 17.8 Plasmid vectors used to integrate two genes, one nonselectable and one selectable, into the chloroplast genome. Abbreviation: Spcr, spectinomycin resistance.

mately 30% of the spectinomycin-resistant transformants were also able to express the kanamycin resistance gene product. This experiment demonstrates the feasibility of the cotransformation protocol for introducing foreign genes into chloroplast DNA.

Production of Marker-Free Transgenic Plants

Usually, at the same time that a foreign gene is introduced into plants a selectable marker gene is also introduced. Although none of these genes has been shown to have an adverse effect on human, animal, or environmental safety, their inclusion in transgenic plants has raised concerns. For example, the products of some marker genes might be either toxic or allergenic; the antibiotic resistance genes that are used as selectable markers might be transferred to pathogenic soil microorganisms; and the presence of a selectable marker makes it technically more difficult to transform a transgenic plant with additional foreign genes since the same selectable marker cannot be used more than once. To allay these concerns, strategies for the production of transgenic plants without any marker genes have been developed.

One experimental approach to produce markerless transgenic plants includes cotransformation of plants with two separate DNAs, one carrying the marker gene and the other carrying the target foreign gene. In this case, approximately 30 to 80% of the plants contain both genes. However, the two genes are integrated at different sites on the chromosomal DNA. Following the selection of transformed plants, traditional breeding techniques can be used to rid the transgenic plant of the selectable marker. Thus, the two genes are separated by chromosome segregation during a few rounds of matings.

Alternatively, a selectable marker gene is cloned between plant transposable elements (Ds elements) and this combination is inserted into the T-DNA along with a transposase gene that excises the DNA between the Ds elements and integrates it at another chromosomal site (Fig. 17.9). Any sequence that is between two Ds elements can be mobilized to a new location in the genome—excising the gene from the original location—provided that

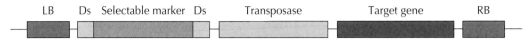

Figure 17.9 Schematic representation of the T-DNA of a vector. Following integration of the T-DNA into the plant chromosomal DNA, the transposase can excise the selectable marker gene and insert it into a different chromosomal location within the plant. Abbreviations: LB, the T-DNA left border; RB, the T-DNA right border; Ds, a plant transposable element. Note that the promoter and transcription termination sequences for the transposase gene, the target gene, and the selectable marker gene are not shown.

the appropriate transposase enzyme is available to catalyze the movement. During insertion of the T-DNA into the host plant DNA, about 90% of the time the selectable marker which is between the two Ds elements will be moved to another site on the chromosomal DNA. About half of the time the new location for the selectable marker will be far away from the original location. Thus, a selectable marker gene can be used to identify transformed plant cells, and subsequently it can be removed by breeding.

SUMMARY

Plants are genetically engineered by introducing a gene into plant cells that are growing in culture and then regenerating a fertile plant from selected transformed cells. Strains of the soil bacterium *Agrobacterium tumefaciens* can genetically engineer plants naturally. In this system, after responding to chemical signals from a surface wound, *A. tumefaciens* makes contact with an exposed plant cell membrane. A series of steps then occurs that results in the transfer of a segment (T-DNA) of a plasmid (Ti plasmid) from the bacterium into the nucleus of the plant cell. The T-DNA region becomes integrated into the plant genome, and subsequently the genes on the T-DNA region are expressed. The T-DNA region contains genes that encode enzymes for the production of phytohormones. These compounds cause the plant cells to enlarge and proliferate. Also, the plant cell becomes a factory for the production of an opine (encoded by a T-DNA gene), which can be catabolized only by *A. tumefaciens*. Thus, *A. tumefaciens* has evolved a mechanism that converts a plant cell into a production center for a carbon and nitrogen source (opine) that it can utilize exclusively.

The *A. tumefaciens*–Ti plasmid system has been modified for use as a delivery mechanism for cloned genes to some plant cells. In these vector systems, the phytohormone and opine metabolism genes have been removed from the T-DNA region, and this modified T-DNA sequence has been cloned into a plasmid that can exist stably in *E. coli*. A cloned gene that is inserted into this T-DNA region is part of the DNA that is transferred into the nucleus of a recipient plant cell. To achieve this transfer, *A. tumefaciens* is used as a delivery system. In one system, the shuttle vector with the T-DNA-cloned gene segment is introduced into an *A. tumefaciens* strain that carries a compatible plasmid with genes that are essential for transferring a T-DNA region into a plant cell (*vir* genes). In addition to this binary vector system, a cointegrate system has been designed so that, after the introduction of the shuttle vector into *A. tumefaciens*, it recombines with the *vir* gene-containing, disarmed Ti plasmid to give a single plasmid that has both *vir* gene functions and the T-DNA-cloned gene segment. The *A. tumefaciens* T-

DNA region has been used to produce a number of plants with introduced genes. Unfortunately, this system is not effective with all kinds of plants. However, microprojectile bombardment (biolistics) has been an effective procedure for delivering DNA to a wider range of plant cells.

Different plant promoters that are active only in specific plant tissues, or only at certain times during the life of the plant, are identified by the acquisition of expression of a promoterless reporter gene after it has integrated into plant chromosomal DNA. In addition, methods for the insertion of the foreign gene directly into either the chloroplast or mitochondrial DNA have been developed so that following expression the protein will be localized in the plastid. Finally, to allay a number of concerns, experimental protocols that remove marker genes from transgenic plants have been developed.

References

An, G., and Y. Kim. 1993. Techniques for isolating and characterizing plant transcription promoters, enhancers, and terminators, p. 155–166. *In* B. R. Glick and J. E. Thompson (ed.), *Methods in Plant Molecular Biology and Biotechnology.* CRC Press, Boca Raton, Fla.

Carrer, H., and P. Maliga. 1995. Targeted insertion of foreign genes into the tobacco plastid genome without physical linkage to the selectable marker gene. *Bio/Technology* **13:**791–794.

Christou, P. 1992. Genetic transformation of crop plants using microprojectile bombardment. *Plant J.* **2:**275–281.

Dale, E. C., and D. Ow. 1991. Gene transfer with subsequent removal of the selection gene from the host genome. *Proc. Natl. Acad. Sci. USA* **88:**10558–10562.

Goldsbrough, A. P., C. N. Lastrella, and J. I. Yoder. 1993. Transposition mediated re-positioning and subsequent elimination of marker genes from transgenic tomato. *Bio/Technology* **11:**1286–1292.

Gruber, M. Y., and W. L. Crosby. 1993. Vectors for plant transformation, p. 89–119. *In* B. R. Glick and J. E. Thompson (ed.), *Methods in Plant Molecular Biology and Biotechnology.* CRC Press, Boca Raton, Fla.

Halfter, U., P. C. Morris, and L. Willmitzer. 1992. Gene targeting in *Arabidopsis thaliana. Mol. Gen. Genet.* **231:**186–193.

Ishida, Y., H. Saito, S. Ohta, Y. Hiei, T. Komari, and T. Kumashiro. 1996. High efficiency transformation of maize (*Zea mays* L.) mediated by *Agrobacterium tumefaciens. Nat. Biotechnol.* **14:**745–750.

Jefferson, R. A., T. A. Kavanagh, and M. W. Bevan. 1987. GUS fusions: β-glucuronidase as a sensitive and versatile gene fusion marker in higher plants. *EMBO J.* **6:**3901–3907.

Klein, T. M., E. D. Wolf, R. Wu, and J. C. Sanford. 1987. High-velocity microprojectiles for delivering nucleic acids into living cells. *Nature* (London) **327:**70–73.

Krüger-Lebus, S., and I. Potrykus. 1987. A simple and efficient method for direct gene transfer to *Petunia hybridia* without electroporation. *Plant Mol. Biol. Rep.* **5:**289–294.

Miki, B. L., P. F. Fobert, P. J. Charest, and V. N. Iyer. 1993. Procedures for introducing foreign DNA into plants, p. 67–88. *In* B. R. Glick and J. E. Thompson (ed.), *Methods in Plant Molecular Biology and Biotechnology.* CRC Press, Boca Raton, Fla.

Mitsuhara, I., M. Ugaki, H. Hirochika, M. Ohshima, T. Murakami, Y. Gotoh, Y. Katayose, S. Nakamura, R. Honkura, S. Nishimiya, K. Ueno, A. Mochizuki, H. Tanimoto, H. Tsugawa, Y. Otsuki, and Y. Ohashi. 1996. Efficient promoter cassettes for enhanced expression of foreign genes in dicotyledonous and monocotyledonous plants. *Plant Cell Physiol.* **37:**49–59.

Ow, D. W., K. V. Wood, M. DeLuca, J. R. de Wet, D. R. Helinski, and S. H. How-

ell. 1986. Transient and stable expression of the firefly luciferase gene in plant cells and transgenic plants. *Science* **234**:856–859.

Paszkowski, J., M. Baur, A. Bogucki, and I. Potrykus. 1988. Gene targeting in plants. *EMBO J.* **7**:4021–4026.

Pauls, K. P. 1995. Plant biotechnology for crop improvement. *Biotechnol. Adv.* **13**:673–693.

Potrykus, I. 1990. Gene transfer to cereals: an assessment. *Bio/Technology* **8**:535–542.

Potrykus, I. 1991. Gene transfer to plants: assessment of published approaches and results. *Annu. Rev. Plant Physiol.* **42**:205–225.

Southgate, E. M., M. R. Davey, J. B. Power, and R. Marchant. 1995. Factors affecting the genetic engineering of plants by microprojectile bombardment. *Biotechnol. Adv.* **13**:631–651.

Vain, P., J. de Buyser, V. Bui Trang, R. Haicour, and Y. Henry. 1995. Foreign delivery into monocotyledonous species. *Biotechnol. Adv.* **13**:653–671.

Walden, R., and J. Schell. 1990. Techniques in plant molecular biology—progress and problems. *Eur. J. Biochem.* **192**:563–576.

Walden, R., and R. Wingender. 1995. Gene-transfer and plant-regeneration techniques. *Trends Biotechnol.* **13**:324–331.

Yoder, J. I., and A. P. Goldsbrough. 1994. Transformation systems for generating marker-free transgenic plants. *Bio/Technology* **12**:263–267.

Zambryski, P. 1988. Basic processes underlying *Agrobacterium*-mediated DNA transfer to plant cells. *Annu. Rev. Genet.* **22**:1–30.

Zambryski, P., J. Tempe, and J. Schell. 1989. Transfer and function of T-DNA genes from *Agrobacterium* Ti and Ri plasmids in plants. *Cell* **56**:193–201.

REVIEW QUESTIONS

1. Why is the Ti plasmid from *Agrobacterium tumefaciens* well suited for developing a vector to transfer foreign genes into plant chromosomal DNA?

2. How do (1) binary and (2) cointegrate Ti plasmid-based vector systems for plant transformation differ from one another?

3. What are reporter genes and how are they used when plant cells are transformed?

4. How can plants be transformed by microprojectile bombardment?

5. Describe in detail how you would isolate a root-specific plant promoter.

6. How is foreign DNA targeted for integration into chloroplast DNA?

7. How would you produce a transgenic plant that does not contain a marker gene?

8. How would you increase the activity of a plant promoter?

Genetic Engineering of Plants: Applications

The principal objective of plant biotechnology is to create new varieties of cultivated plants (cultivars). Many of the initial studies have focused on developing strains that give better yields without intentionally changing the food content of the plant. Genes that confer resistance to insects, viruses, herbicides, environmental stress, and senescence have been incorporated into various plants. Some of this work is discussed below. In addition, some work has gone into manipulating the biochemical pathway of flower pigmentation, modifying plant products, and using plants as bioreactors.

Development of Pathogen- and Herbicide-Resistant Plants

Insect-Resistant Plants

If crop plants could be genetically engineered to produce functional insecticides, then it might be possible to develop crops that would be intrinsically tolerant of insect predators and would not need to be sprayed (often six to eight times during a growing season) with costly and hazardous chemical pesticides. It has been estimated that in 1995 the amount spent on chemical insecticides worldwide was approximately $4 billion. The cost of maintaining such insect-resistant crops would be lower than that for nonresistant crops. Moreover, biological insecticides are usually highly specific for a limited number of insect species, and they are generally considered to be nonhazardous to humans and other higher animals.

Several different strategies have been used to confer resistance against insect predators. One approach involves a gene for an insecticidal protoxin produced by one of several subspecies of the bacterium *Bacillus thuringiensis* (see chapter 15). Another common strategy uses genes for plant proteins such as amylase inhibitors or protease inhibitors that have been shown to be effective against a wide variety of insects. After an insect ingests one of these inhibitors, it is not able to digest food (i.e., plants) because the inhibitor interferes with the hydrolysis of starch or plant proteins. Thus, the insect will feed less.

B. thuringiensis protoxin does not persist in the environment, nor is it

hazardous to mammals. Thus, it is a safe means of protecting plants. Unfortunately, a number of the economically significant pests of crop plants feed on internal plant tissues and are thus unlikely to be inhibited by *B. thuringiensis* preparations that has been sprayed onto the surface of the plant. To avoid this problem, genes for *B. thuringiensis* toxins can be expressed in plants. With these transgenic plants, there is no need to spray the insecticidal toxin. This limits the environmental distribution of the toxin and avoids problems associated with its limited environmental stability and the timing of the toxin application. The scientific challenge in utilizing the *B. thuringiensis* protoxin is to create a transgenic plant that expresses and synthesizes a functional form of this prokaryotic insecticide at sufficient levels to prevent damage by insect predation. The *B. thuringiensis* subsp. *kurstaki* insecticidal protein genes, *cryIA(a), cryIA(b)*, and *cryIA(c)*, are not particularly well expressed in plants (Table 18.1). This is problematic because high levels of expression of these insect control proteins are needed in order to produce commercially viable insect-resistant plants. In an effort to raise the level of the expressed protein, scientists truncated the gene so that only the N-terminal portion of the insecticidal toxin was produced and inserted a strong plant promoter to direct gene expression. Under these conditions, there was a significant increase in the level of insecticidal toxin produced, affording transgenic plants some protection against damage from insect predation.

To increase the level of expression, researchers sought the minimum sequence that encoded toxin activity. The amino acid sequences of protoxins from various strains of *B. thuringiensis* were compared to determine whether there is a common insecticidal domain. The analysis showed that the N-terminal portion of the protoxin molecule is highly conserved (~98%) among different strains of *B. thuringiensis* subsp. *kurstaki*, but the C-terminal region is more variable (~45% conserved). Further work showed that all of the insecticidal toxin activity resides within the first 646 amino acids from the N terminus of the 1,156-amino-acid protoxin. When the segment of the protoxin gene that encodes the highly conserved amino acid sequence was cloned and expressed in bacteria, the shortened protein was as active as the

Table 18.1 Expression of some *Bacillus thuringiensis* insecticidal toxin genes in transgenic plants

Plant(s)	Gene	% Expression	Insecticidal
Tobacco	*cryIA(b)*, full	0.0001–0.0005	No
Tobacco	*cryIA(b)*, truncated	0.003–0.012	Yes
Tobacco	*cryIA(a)*, full	Not detected	No
Tobacco	*cryIA(a)*, truncated	0.00125	Yes
Tobacco	*cryIA(c)*, truncated	<0.014	Yes
Tomato	*cryIA(b)*, truncated	0.0001	Yes
Cotton	*cryIA(b)*, truncated, WT	<0.002	No
Cotton	*cryIA(b)*, truncated, PM	0.05–0.1	Yes
Tomato, tobacco	*cryIA(b)*, truncated, WT	0.002	Yes
Tomato, tobacco	*cryIA(b)*, truncated, PM	0.002–0.2	Yes
Tomato, tobacco	*cryIA(b)*, truncated, FM	0.3	Yes

Adapted from Ely, p. 105–124, *in* Entwistle et al. (ed)., Bacillus thuringiensis, *an Environmental Biopesticide: Theory and Practice*, 1993.

Terms and abbreviations: full, the complete protoxin gene; truncated, a shortened version of the protoxin gene; WT, wild-type codons; PM, partially modified codons; FM, fully modified codons.

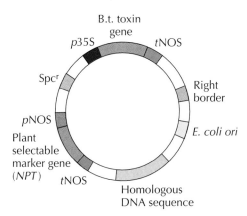

Figure 18.1 Cointegrate cloning vector carrying a *B. thuringiensis* (B.t.) insecticidal toxin gene. The toxin gene is under the control of the strong, constitutive 35S promoter (*p*35S) from cauliflower mosaic virus and the nopaline synthase transcription terminator–polyadenylation site (*t*NOS). The vector has an *E. coli* origin of DNA replication (*ori*); a spectinomycin resistance gene (Spcr), which allows the vector to be maintained and selected in *E. coli* cells; a T-DNA right border; a plant selectable marker gene; and a region of DNA that is homologous to DNA in the disarmed Ti plasmid, for integrating the two plasmids. The neomycin phosphotransferase gene (*NPT*), which acts as a plant reporter gene, is under the transcriptional control of nopaline synthase gene sequences (*p*NOS and *t*NOS) and is used to select for kanamycin-resistant transformed plant cells.

native form in protecting plants against lepidopteran insects in laboratory trials.

Transgenic tomato plants were produced to test whether the shortened protoxin gene would be able to protect the plants from damage by various insect pests. The shortened version of the protoxin gene was put under the transcriptional control of a strong constitutive promoter called the 35S promoter from cauliflower mosaic virus and the nopaline synthase transcription termination–polyadenylation site and was then cloned into the T-DNA region of a cointegrate-type Ti plasmid vector (Fig. 18.1). The vector contained a spectinomycin resistance (Spcr) gene that allowed it to be selected in either *Escherichia coli* or *Agrobacterium tumefaciens*, an *E. coli* origin of DNA replication, and a neomycin phosphotransferase gene that was under the control of the nopaline synthase promoter and transcription termination–polyadenylation sites and enabled selection of transformed plant cells in the presence of kanamycin. In addition, the cointegrate cloning vector had the right border of the T-DNA from a nopaline Ti plasmid and a segment of the octopine Ti plasmid that provides a region of homology for cointegrate formation by homologous recombination with a disarmed Ti plasmid. The plasmid was constructed and manipulated in *E. coli* before it was transferred by conjugation to a strain of *A. tumefaciens* that contained a disarmed Ti plasmid. After recombination in *A. tumefaciens*, the short form of the protoxin gene was transferred to the chromosomal DNA of tomato plants.

In both greenhouse and field trials, transgenic tomato plants that expressed the short form of the protoxin acquired some degree of protection against damage caused by tobacco hornworm (*Manduca sexta*), tomato fruitworm (*Heliothis zea*), and tomato pinworm (*Keiferia lycopersicella*) (Table 18.2). The extent of the protection is not the same for each of the insects, nor is it total. The transgenic plants were protected to some extent from damage

caused by tobacco hornworms and tomato fruitworms and to a lesser degree from damage by tomato pinworms. A combination of a low dose of chemical insecticide and production of the protoxin by the plants gave good protection in some cases. These levels of protection are encouraging, but more work is needed to determine how to lower even further the damage that is caused by these and other insects.

In an effort to dramatically increase the level of expression, two other approaches have been tried (Table 18.1). In the first attempt, an isolated insecticidal toxin gene was modified by site-directed mutagenesis to change any DNA sequences that might inhibit efficient transcription or translation in a plant host (in this case, both tobacco and tomato were used). This "partially" modified gene had a nucleotide sequence that was 96.5% unchanged from the wild-type gene and encoded the identical insecticidal toxin protein. Transgenic plants that expressed this partially modified sequence produced a 10-fold-higher level of insecticidal toxin protein than did plants that were transformed with the wild-type gene.

In the second attempt, a "fully" modified version of the insecticidal toxin gene was designed and chemically synthesized. This fully modified gene contained codons more commonly used by plants, as opposed to those favored by gram-positive bacteria. It was also modified to eliminate any potential messenger RNA (mRNA) secondary structure or chance plant polyadenylation sequences, which might decrease gene expression. It had a G+C content of 49% (the wild-type gene is 37% G+C) and a nucleotide sequence that was only 78.9% identical to the wild-type gene.

Transgenic plants that were transformed with this highly modified synthetic protoxin gene had an approximately 100-fold-higher level of toxin protein than did plants transformed with the wild-type gene. Moreover, this higher level of insecticidal toxin synthesis directly correlated with increased insecticidal activity. The success of this approach suggests that it might also be an effective way to increase the expression of a number of other foreign genes in plants.

In another approach to increasing the expression of the protoxin, one group of researchers expressed the fully modified protoxin gene under the control of the promoter for the gene that codes for the small subunit of the enzyme ribulose bisphosphate carboxylase and downstream from the chloroplast transit peptide sequence of this enzyme, so that the overproduced protoxin became localized within the chloroplast. This strategy led to a very high level of expression (nearly 1% of the total leaf protein) of the in-

Table 18.2 Susceptibility of wild-type and transgenic tomato plants to insect damage

Insect	% of plants or fruits damaged			
	Wild-type tomato plants		Transgenic tomato plants	
	No insecticide	Insecticide	No insecticide	Insecticide
Tobacco hornworm	47.5	3.75	1.25	0.00
Tomato fruitworm	20.1	ND	6.4	ND
Tomato pinworm	99.7	95.1	94.2	80.4

Adapted from Delannay et al., *Bio/Technology* 7:1265–1269, 1989.

The data are expressed as the percentage of tested plants (in the case of the tobacco hornworm) or fruits (for tomato fruitworm and tomato pinworm) that were damaged by insects when a chemical insecticide was or was not used. ND, not determined.

secticidal protoxin. Other researchers have introduced an insecticidal pro-
toxin gene directly into the chloroplast DNA of the host plant. This ap-
proach has a number of potential advantages. First, the protoxin gene does
not have to be modified, because the chloroplast transcriptional and trans-
lational apparatus are typically prokaryotic. Second, because there are
many chloroplasts per cell and many copies of chloroplast DNA per chloro-
plast, the protoxin gene will be present in multiple copies and therefore will
be more likely to be expressed at a high level. Third, chloroplasts are trans-
mitted only through the egg and not through pollen, which means that
plants receive all of their chloroplast DNA from their female parent. Con-
sequently, there is no risk of unwanted transfer of the protoxin gene to other
plants in the environment by pollen.

To date, some form of the gene for the protoxin has been introduced
and expressed in several plant species, including tomato, tobacco, potato,
rice, corn, apple, eggplant, canola, alfalfa, walnut, poplar, spruce, cran-
berry, and cotton; the prospects for the successful application of this ap-
proach to crop protection appear to be quite good. For example, transgenic
potato plants express a synthetic gene, designed with codons commonly
used by plants, based on an insecticidal toxin from *B. thuringiensis* subsp.
tenebrionis at high levels. These plants are highly resistant to the Colorado
potato beetle, a major insect pest of potatoes. Following several seasons of
successful field trials, these transgenic potatoes have recently been ap-
proved for commercial release in the United States. However, insect popu-
lations still have to be monitored to keep track of the frequency of resistant
organisms. Eventually, a more potent *B. thuringiensis* protoxin may be re-
quired or, more likely, other insecticidal genes that can act synergistically
with the *B. thuringiensis* protoxin will be identified and cloned into plants.

Experimental strategies are now being devised to prevent transgenic
plants that express the *B. thuringiensis* protoxin gene from acting as selec-
tion agents for resistant insects. In one approach, the expression of the in-
secticidal toxin in transgenic plants was temporally limited to a short pe-
riod. In that study, the gene for the *B. thuringiensis* protoxin was cloned
immediately downstream of the promoter of a gene from tobacco called the
pathogenesis-related protein 1a (PR-1a) gene. The expression of the PR-1a
gene is part of a natural defense mechanism that combats pathogens in to-
bacco. The PR-1a gene is normally induced by any one of a variety of
pathogenic organisms or by chemicals such as salicylic acid and polyacrylic
acid. Transgenic plants that carried the *B. thuringiensis* protoxin gene driven
by the PR-1a promoter were treated with a chemical inducer and were
found to synthesize detectable levels of insecticidal toxin within 1 day of ap-
plication, which subsequently protected the plants against insect attack.
Therefore, it is conceivable that the protoxin could be induced by the ad-
ministration of an inexpensive and safe chemical inducer only when it is re-
quired during the growing season. Such periodic production should lower
the selection pressure for resistant insects. This type of system could also be
useful in regulating the synthesis of a wide range of other foreign proteins
in transgenic plants.

No single *B. thuringiensis* protoxin is effective against a broad range of
insect species. This may limit the overall usefulness of these protoxins.
However, plants have evolved general insect defense mechanisms that are
sufficient for plant survival but not always effective enough to keep the
damage to a level that would be acceptable for crop plants. For example,
some plants produce protease inhibitors that, when ingested, prevent the

feeding insect from hydrolyzing plant proteins, thereby effectively starving the predator insect. Consequently, it seemed reasonable to isolate a plant gene for a protease inhibitor, add a strong promoter, and create transgenic crop plants that produce sufficiently high levels of the protease inhibitor to reduce damage from insect predation.

In one study, researchers isolated a clone that encodes cowpea trypsin inhibitor from a complementary DNA (cDNA) clone bank by using a chemically synthesized DNA probe based on the amino acid sequence of the cowpea trypsin inhibitor protein. The full-length cDNA was subcloned onto a Ti plasmid binary cloning vector (Fig. 18.2) and introduced into a strain of *A. tumefaciens* carrying a disarmed Ti plasmid that contained active *vir* genes. Following *A. tumefaciens* infection of tobacco leaf disks with this vector, cells that incorporated the cloned DNA were selected for growth on kanamycin, and transgenic plants were regenerated. The damage caused by *Heliothis virescens* (tobacco budworm) larvae to transgenic plants that expressed more than 2 μg of cowpea trypsin inhibitor per mg of protein was significantly less than the damage inflicted on nontransformed plants.

Cowpea seeds that contain approximately 2 μg of inhibitor per mg of plant protein are not toxic to either animals or humans. However, if the amount of protease inhibitor produced by a transgenic plant is determined to be a potential hazard, then it might be possible to limit the expression of the protease inhibitor to the plant tissues that the major insect pests prefer to attack but that are not used as food by humans or animals. In other words, a cloned protease inhibitor gene could be active in the leaves and roots of a plant but not in the commercially valuable fruit.

Introduction of the potato proteinase inhibitor II gene provides rice plants with protection against the pink stem borer (*Sesamia inferens*), a major insect pest of rice. Infestation of rice plants by pink stem borers causes severe damage to the plants, often resulting in a hollow stem and dead panicles with no seeds. A plasmid carrying the potato proteinase inhibitor II gene under the control of its own promoter and transcription termination region was constructed. The plasmid also contained the first intron from the

Figure 18.2 Binary cloning vector carrying a cowpea trypsin inhibitor gene. The vector contains a broad-host-range origin of DNA replication (*ori*) and a kanamycin resistance gene (Kanr), which function in both *E. coli* and *A. tumefaciens*. Between the T-DNA left and right borders, there are (1) a neomycin phosphotransferase gene (*NPT*) under the transcriptional control of nopaline synthase signals (*p*NOS and *t*NOS), which enables kanamycin-resistant transformed plant cells to be selected, and (2) the cowpea trypsin inhibitor gene, which is under the control of the 35S promoter (*p*35S) from cauliflower mosaic virus and the transcription terminator–polyadenylation region from the nopaline synthase gene (*t*NOS).

Figure 18.3 Plasmid vector carrying the potato proteinase inhibitor II gene. Designations: *Pin2* gene, the potato proteinase inhibitor II gene; 5′ end, the region of DNA preceding the gene; 3′ end, the region of DNA following the gene; *Act1* intron, the first intron from the rice actin 1 gene; *35S* 5′ end, the 35S promoter from cauliflower mosaic virus; *bar* gene, the bacterial phosphinothricin acetyltransferase gene; *nos* 3′ end, the region of DNA following the nopaline synthase gene. The *bar* gene serves as a selectable marker for transgenic plants, conferring resistance to the herbicide Basta (ammonium glufosinate).

rice actin gene inserted between the promoter and the potato proteinase inhibitor II coding region. This construct was introduced into rice suspension cells by microprojectile bombardment (Fig. 18.3), and transgenic plants were generated. When pink stem borer larvae were artificially applied to these transgenic rice plants, 70 to 100% of the wild-type plants were severely damaged by insect predation, while only 15 to 20% of the transgenic plants were damaged. Since plant proteinase inhibitors are common components of both human and animal food and are readily inactivated by cooking, their introduction into new crops can be regarded as safe.

Another strategy that is designed to increase the effectiveness of relatively low levels of *B. thuringiensis* insecticidal toxin activity entails combining the toxin with a serine protease inhibitor. In laboratory trials, investigators found that when the amount of purified *B. thuringiensis* insecticidal toxin that causes minimal insect mortality was mixed with a low concentration of protease inhibitor, the insecticidal activity of the mixture was 20-fold greater than that of the *B. thuringiensis* protoxin alone. To test whether this scheme would function in transgenic plants, a DNA fragment that encoded a fusion protein consisting of both a protease inhibitor and a truncated toxin was constructed. Transgenic tobacco plants that produced small amounts of this fusion protein were protected from insect attack.

Another way of imparting insect resistance to susceptible plants entails using a gene that encodes an α-amylase inhibitor. The cowpea weevil (*Callosobruchus maculatus*) and azuki bean weevil (*C. chinensis*) are seed-feeding beetles that both cause considerable economic loss of these crops, especially in developing countries. When larvae of these insects are fed a diet that includes the common bean (*Phaseolus vulgaris*), insect growth is inhibited. This growth inhibition is attributable to the presence of an α-amylase inhibitor in the seed proteins of the common bean. Accordingly, the gene for the α-amylase inhibitor from the common bean was isolated, placed under the transcriptional control of a strong seed-specific promoter for the bean phytohemagglutinin gene, and used to transform pea plants (*Pisum sativum*). Peas are usually quite susceptible to damage by both cowpea weevils and azuki bean weevils. However, transgenic pea plants that expressed the α-amylase inhibitor were resistant to both of these insects. The level of resistance to cowpea weevils was found to be proportional to the amount of α-amylase inhibitor that the transgenic plant produced (Fig. 18.4).

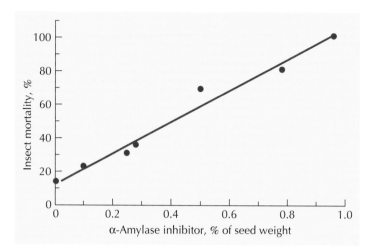

Figure 18.4 Mortality of cowpea weevil larvae reared on transgenic pea plants that produce different amounts of α-amylase inhibitor.

Another approach to developing insect-resistant transgenic plants makes use of a bacterial cholesterol oxidase gene. Cholesterol oxidase, which is present in a range of different bacterial genera, catalyzes the oxidation of 3-hydroxysteroids to ketosteroids and hydrogen peroxide. This enzyme is commonly used in assays to determine the levels of cholesterol in human serum. Low levels of this enzyme have a high level of insecticidal activity against larvae of the boll weevil (*Anthonomus grandis grandis*) (Fig. 18.5), a common and economically important insect (Coleoptera) pest of cotton and have lower levels of activity against some lepidopteran pests. Cholesterol oxidase probably acts by disrupting the insect's midgut epithelial membrane, thus killing the insect. A cholesterol oxidase gene encoding a protein with a molecular mass of approximately 55,000 daltons (and a length of 504 amino acids) plus a leader peptide of approximately 5,000 daltons (43 amino acids) was isolated from a strain of *Streptomyces* and cloned into a vector under the control of a plant virus (figwort mosaic virus) pro-

Figure 18.5 Effect of increasing amounts of cholesterol oxidase on the mortality of boll weevil larvae. ppm, parts per million. Adapted from Corbin et al., *Appl. Environ. Microbiol.* **60**:4239–4244, 1994.

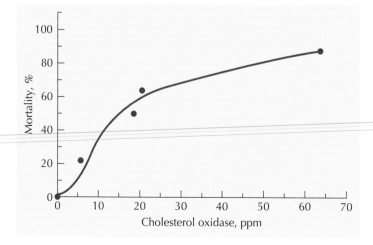

moter and a termination sequence from the 3' region of the *A. tumefaciens* nopaline synthase gene. When this construct was introduced into tobacco cell protoplasts, the transformed cells actively expressed the insecticidal enzyme. In the future, it is likely that this gene will be introduced into cotton plants. Then, either by itself or in combination with genes for other biological insecticides, it should be an effective means of protecting plants against damage from insect predation.

Virus-Resistant Plants

Plant viruses often cause considerable crop damage and significantly reduce yields. Therefore, in the absence of effective chemical treatments, plant breeders have attempted to transfer naturally occurring virus resistance genes from one plant strain (cultivar) to another. However, resistant cultivars often revert to virus sensitivity, and resistance to one virus does not necessarily confer resistance against other similar viruses. Natural virus resistance can be achieved in different ways: viral transmission can be blocked, establishment of the virus can be prevented, or viral symptoms can be bypassed or resisted. Recently, researchers have used the techniques of genetic engineering to develop nonconventional types of virus-resistant transgenic plants. These methods used "immunization" with viral coat protein genes, other viral genes, or viral gene antisense sequences to confer resistance.

When transgenic plants express the gene for the coat protein (which often is the most abundant protein of a virus particle) of a virus that normally infects those plants, the ability of the virus to subsequently infect the plants and spread systemically is often greatly diminished. Although the precise mechanism by which the presence of coat protein genes inhibits viral proliferation is not known, it is clear that the antiviral effect occurs early in the viral replication cycle and, as a result, prevents any significant amount of viral synthesis. This feature is an advantage because it decreases the probability of selecting for spontaneous viral mutants that can overcome this resistance and replicate in the presence of virus coat protein. The viral coat protein gene approach has been used to confer tolerance to a number of different plant viruses (Table 18.3). Using this approach, researchers have developed virus-resistant transgenic plants for a number of different crops. Although complete protection is not usually achieved, high levels of virus resistance have been reported. Moreover, a coat protein gene from one virus sometimes provides tolerance to a broad spectrum of unrelated viruses. The utility of this strategy is supported by the observation that transgenic plants do as well in field trials as in the laboratory studies.

In both eukaryotes and prokaryotes, an RNA molecule that is complementary to a normal gene transcript (mRNA) is called antisense RNA. The mRNA, being translatable, is considered to be a sense RNA. The presence of antisense RNA can decrease the synthesis of the gene product by forming a duplex molecule with the normal sense mRNA, thereby preventing it from being translated. The antisense RNA–mRNA duplex is also rapidly degraded, a response that diminishes the amount of that particular mRNA in the cell. Theoretically, it should be possible to prevent plant viruses from replicating and subsequently damaging plant tissues by creating transgenic plants that synthesize antisense RNA that is complementary to virus coat protein mRNA.

In one of many studies, the efficacies of the virus coat protein gene and antisense RNA approaches were compared by cloning the cDNA for the

Table 18.3 Some virus-resistant transgenic plants that contain cloned viral coat proteins

Plant(s)	Virus that provided the coat protein gene
Nicotiana benthamiana, N. clevelandii	Plum pox virus
N. benthamiana, squash	Watermelon mosaic virus 2
N. benthamiana, squash	Zucchini yellow mosaic virus
Papaya, tobacco	Papaya ringspot virus
Potato	Potato leafroll virus
Potato	Potato virus Y
Potato, *Nicotiana debneyii*	Potato virus S
Potato, tobacco	Potato virus X
Rice	Rice stripe virus
Tobacco	Arabis mosaic virus
Tobacco	Soybean mosaic virus
Tobacco	Tobacco etch virus
Tobacco	Tobacco streak virus
Tobacco	Tomato spotted wilt virus
Tobacco, alfalfa, tomato	Alfalfa mosaic virus
Tobacco, cucumber	Cucumber mosaic virus
Tobacco, *N. benthamiana*	Tobacco rattle virus
Tobacco, tomato	Tobacco mosaic virus
Tomato	Tomato mosaic virus

Adapted from Fitchen and Beachy, *Annu. Rev. Microbiol.* 47:739–763, 1993.

coat protein of cucumber mosaic virus (CuMV) into tobacco plants in two orientations (sense and antisense; one orientation per plant) and then testing transgenic plants for sensitivity to viral infection (Fig. 18.6). The genome of CuMV consists of three separate single-stranded pieces of RNA, each coding for a specific viral protein. In vivo, one of these pieces, RNA3, is processed to remove a portion of its sequence, thereby generating RNA4, which encodes the virus coat protein. To create transgenic plants that either produced normal mRNA and expressed the virus coat protein or produced its antisense RNA, the following steps were carried out:

1. Isolation of RNA4
2. In vitro enzymatic conversion of RNA4 into a double-stranded cDNA
3. Addition of linkers onto the cDNA
4. Insertion of the full-length cDNA sequences into cloning vectors in both orientations, with each oriented sequence under the control of the 35S promoter sequence from cauliflower mosaic virus and the termination-regulatory sequences from the plant gene for the small subunit of ribulose bisphosphate carboxylase
5. Formation of separate transgenic plants carrying the cDNA sequence in one of the two possible orientations

The Ti plasmid binary vector system was used to transfer both protein-producing sense and antisense RNA-producing cDNA sequences to separate tobacco cells, from which transgenic plants were regenerated (Fig. 18.7). The transgenic tobacco plants that expressed the CuMV coat protein were protected from viral particle accumulation and did not show symptoms of viral infection, regardless of whether the inoculum of the challenge virus

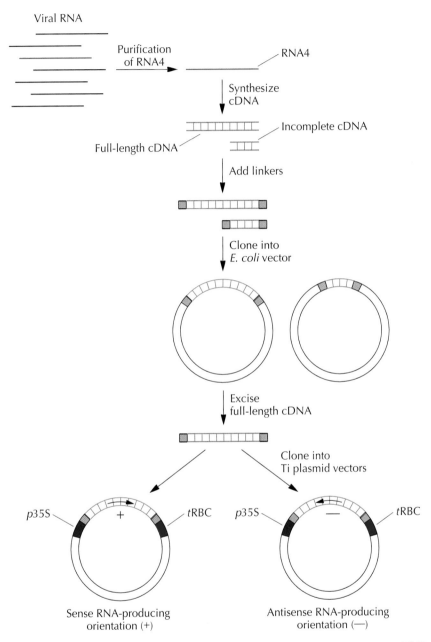

Figure 18.6 Procedure for introducing cucumber mosaic virus coat protein cDNA into plant cells. RNA4, which encodes the coat protein, is fractionated from a viral RNA preparation and used as the template for the synthesis of double-stranded cDNA. Linkers are added to the cDNA preparation, and the cDNAs are cloned into an *E. coli* plasmid vector. A full-length cDNA clone is identified, excised from the *E. coli* vector, and subcloned into a Ti plasmid cloning vector between the 35S promoter from cauliflower mosaic virus (*p*35S) and the transcription terminator from the gene for the small subunit of ribulose bisphosphate carboxylase (*t*RBC). This cloning step creates two orientations for the RNA4 cDNA. In one case, the RNA that is transcribed is translated into coat protein (sense RNA), and in the other case, the transcribed RNA is complementary to the mRNA for the coat protein (antisense RNA).

was high or low. However, the transgenic tobacco plants expressing the CuMV coat protein antisense RNA were protected only when the concentration of the challenge virus in the inoculum was low.

Several groups of scientists have constructed transgenic plants that synthesize antisense RNA copies of virus coat protein genes and tested whether these plants can withstand a viral challenge. In all instances, the plants were protected against the invading virus only when low concentrations of the virus were used. At high concentrations, plants were damaged by the virus. In addition, antisense RNA copies of virus coat protein genes generally afforded a much lower level of protection to transgenic plants than sense versions of the virus coat protein genes. Although it may be premature to give up on the antisense RNA approach as a means of creating virus-resistant plants, significant improvements are required before it will gain acceptance.

Often field crops are exposed to several different viruses, any one of which may damage the plant and lower the final yield. Ideally, transgenic plants should be resistant to more than one virus. With this in mind, Ti plasmid binary vectors expressing one or more coat protein genes for CuMV, zucchini yellow mosaic virus, and watermelon mosaic virus 2 were used to transform yellow crookneck squash (*Cucurbita pepo*) plants (Fig. 18.8). Transgenic plants that contained the coat protein genes from all three viruses were resistant to damage by all three viruses under laboratory conditions. Transgenic plants expressing coat protein genes for zucchini yellow mosaic virus and watermelon mosaic virus 2 were tested under field conditions by using aphids, which are small insects that naturally transmit these viruses to developing plants. The transgenic plants that expressed both coat protein genes were completely resistant to infection when these two viruses were applied at the same time (Fig. 18.9). On the other hand, while transgenic plants expressing only one of the two viral coat proteins were delayed

Figure 18.7 Ti plasmid binary cloning vectors containing either the protein-producing sense **(A)** or the antisense RNA-producing **(B)** orientation of the cucumber mosaic virus (CuMV) coat protein cDNA. Each cDNA sequence is under the control of the 35S promoter (*p*35S) from cauliflower mosaic virus and the transcription terminator–polyadenylation site (*t*RBC) from the gene for the small subunit of ribulose biphosphate carboxylase. The vector also contains a neomycin phosphotransferase (*NPT*) gene under the control of nopaline synthase transcription signals (*p*NOS and *t*NOS), a spectinomycin resistance gene (Spc^r), a T-DNA right border sequence, a T-DNA left border sequence, and a broad-host-range origin of DNA replication (*ori*). The protein-producing sense (+) orientation is shown by the A→Z arrow, and the antisense RNA-producing (−) orientation is shown by the Z→A arrow.

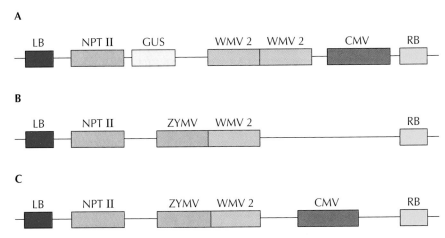

Figure 18.8 **A.** A T-DNA construct with a neomycin phosphotransferase gene (NPT II) as a selectable marker, a β-glucuronidase gene (GUS) as a reporter gene, two copies of the coat protein gene from watermelon mosaic virus 2 (WMV 2), and the coat protein gene from cucumber mosaic virus (CMV). The left and right borders of the T-DNA are indicated by LB and RB, respectively. **B.** Similar to **A** without CMV and NPT II, with one copy of WMV 2, and with the coat protein gene from zucchini yellow mosaic virus (ZYMV). **C.** Same as **B** with the addition of CMV. All of the genes in these constructs include both promoters and transcription terminator regions.

in succumbing to the viruses in comparison to the nontransformed control plants, all of these plants eventually developed severe symptoms of viral disease, making them unfit for sale to consumers. Clearly, using more than one viral coat protein gene is an effective strategy that should be useful in developing a range of transgenic plants that are resistant to all of the major viruses that normally inhibit their growth and development.

Some preliminary reports have suggested that transgenic plants that express viral genes other than the virus coat protein gene (e.g., virus satel-

Figure 18.9 Disease frequency in transgenic and nontransformed (wild-type) yellow crookneck squash in the field. Aphids were used to transmit a mixture of zucchini yellow mosaic virus (ZYMV) and watermelon mosaic virus 2 (WMV) to the squash plants. Adapted from Fuchs and Gonsalves, *Bio/Technology* **13:**1466–1473, 1995.

lite RNA or a virus replication gene) may also be protected against damage from infective viruses. However, at this time, it is not clear how effective and generally applicable these approaches will be.

In addition to "immunizing" plants against damage from viruses by expressing viral proteins in the plant cells, protection can be conferred by antiviral plant proteins. For example, pokeweed (*Phytolacca americana*) has three distinct antiviral proteins present in its cell wall: pokeweed antiviral protein (PAP), which is present in spring leaves; PAPII, which is found in summer leaves; and PAP-S, which appears in seeds. These three proteins can be readily purified from aqueous extracts of ground-up plant tissue. When small amounts of PAP were applied to the leaf surface of different plants, the plants were protected against infection by several viruses. Based on these observations, PAP is a good candidate for developing transgenic plants that are resistant to a broad spectrum of plant viruses.

After a cDNA encoding PAP was isolated, it was introduced into tobacco and potato plants by using binary Ti plasmid vectors. Transformants that expressed a high level of PAP (>10 ng/mg of protein) had a stunted and mottled appearance and were sterile. On the other hand, plants with a lower level of PAP (1 to 5 ng/mg of protein) were normal in appearance and fertile. These data suggest that above a certain level PAP interferes with normal cellular functioning. In transgenic plants expressing PAP, the major effect of the antiviral protein was to reduce the number of lesions; however, once a lesion formed, the plant became infected systematically, suggesting that PAP acts by inhibiting viral infection at an early stage. Nevertheless, when transgenic tobacco and potato plants expressing low levels of PAP were challenged with either potato virus X or Y, they developed significantly fewer lesions on their leaves than nontransformed control plants. Since PAP is effective at relatively low levels, it should be possible to develop transgenic plants that express a low level of PAP together with another means of generating virus resistance.

Herbicide-Resistant Plants

Approximately 10% of global crop production is lost through weed infestation every year, despite the expense of $10 billion spent on more than 100 different chemical herbicides. In addition, many herbicides do not discriminate weeds from crop plants, many must be applied early before the weeds take hold, and some persist in the environment. The creation of herbicide-resistant crop plants is one way to overcome some of these drawbacks.

A number of different biological manipulations that would cause a crop plant to be herbicide resistant can be envisioned:

- Inhibit the uptake of the herbicide.
- Overproduce the herbicide-sensitive target protein so that enough of it remains available for cellular functions despite the presence of the herbicide.
- Reduce the ability of a herbicide-sensitive target protein to bind to a herbicide.
- Endow plants with the capability to metabolically inactivate the herbicide.

Of these possibilities, the last three have been implemented to produce herbicide-resistant transgenic plants (Table 18.4).

Resistant plants have been developed for the herbicide glyphosate, which is "environmentally friendly" because it is readily degraded to nontoxic compounds in the soil. It acts as an inhibitor of 5-enolpyruvylshikimate-3-phosphate synthase (EPSPS), which is an enzyme in the shikimate

pathway. This enzyme plays an important role in the synthesis of aromatic amino acids in both bacteria and plants. An EPSPS-encoding gene from a glyphosate-resistant strain of *E. coli* was isolated, put under the control of plant promoter and transcription termination–polyadenylation sequences, and cloned into plant cells. Transgenic tobacco, petunia, tomato, potato, and cotton plants that produce an amount of the resistant *E. coli* EPSPS sufficient to replace the inhibited plant enzyme are resistant to the effects of glyphosate. Thus, in these cases, the crop plant would not be affected by glyphosate treatment, whereas the weeds would be.

Table 18.4 Some examples of gene-based herbicide resistance

Herbicide(s)	Mode of development of herbicide resistance
Triazines	Resistance is due to an alteration in the *psbA* gene, which codes for the target of this herbicide, chloroplast protein D-1.
Sulfonylureas	Genes encoding resistant versions of the enzyme acetolactate synthetase have been introduced into poplar, canola, flax, and rice.
Imidazolinones	Strains with resistant versions of the enzyme acetolactate synthetase have been selected in tissue culture.
Aryloxyphenoxypropionates, cyclohexanediones	These herbicides inhibit the enzyme acetyl coenzyme A carboxylase. Resistance, selected in tissue culture, is due either to an altered enzyme that is not herbicide sensitive or to the degradation of the herbicide.
Glyphosate	Resistance is from overproduction of EPSPS, the target of this herbicide. Resistance has been engineered by transforming soybean with the gene for a glyphosate-resistant EPSPS and tobacco with a glyphosate oxidoreductase gene, which encodes an enzyme that degrades glyphosate.
Bromoxynil	Resistance to this photosystem II inhibitor has been created by transforming tobacco and cotton plants with a bacterial nitrilase gene, which encodes an enzyme that degrades this herbicide.
Phenoxycarboxylic acids (e.g., 2,4-D and 2,4,5-T)	Resistant cotton and tobacco plants have been created by transformation with the *tfdA* gene from *Alcaligenes*, which encodes a dioxygenase that degrades this herbicide.
Glufosinate (phosphinothricin)	Over 20 different plants have been transformed with either the *bar* gene from *Streptomyces hygroscopicus* or the *pat* gene from *S. viridochromogenes*. The phosphinothricin acetyltransferase that these genes encode detoxifies this herbicide.
Cyanamide	Resistant tobacco plants were produced when a cyanamide hydratase gene from the fungus *Myrothecium verrucaria* was introduced. The enzyme encoded by this gene converts cyanamide to urea.
Dalapon	Tobacco plants transformed with a dehalogenase gene from *Pseudomonas putida* can detoxify this herbicide.

Bromoxynil 3,5-Dibromo-4-hydroxybenzoic acid

Figure 18.10 Detoxification of the herbicide bromoxynil by the enzyme nitrilase from *K. ozaenae*.

An example of resistance due to herbicide inactivation has been developed for bromoxynil (3,5-dibromo-4-hydroxybenzonitrile), a herbicide that acts by inhibiting photosynthesis. In this case, resistant plants were created by the introduction of a bacterial gene that encodes the enzyme nitrilase, which can inactivate bromoxynil before the herbicide can act (Fig. 18.10). The gene for nitrilase was isolated from the soil bacterium *Klebsiella ozaenae* and placed under the control of the light-regulated promoter from the small subunit of the enzyme ribulose bisphosphate carboxylase before it was transferred to tobacco plants. The transgenic plants expressed nitrilase activity and were resistant to the toxic effects of the herbicide.

Fungus- and Bacterium-Resistant Plants

Extensive damage and loss of crop productivity are caused by phytopathogenic fungi. It has been estimated that one fungal disease of one major crop (fungal rice blast, a disease that affects rice plants) costs farmers in Southeast Asia, Japan, and the Philippines about $5 billion per year. At present, the major way of controlling the damage and losses to crop plants that result from fungal infection is through the use of chemical agents that may persist and accumulate in the environment and subsequently be hazardous to animals and/or humans. It would therefore be beneficial if a simple, inexpensive, effective, and environmentally friendly nonchemical means of preventing fungal damage to crop plants could be found.

Plants often respond to pathogen invasion or other environmental stresses by synthesizing a group of proteins, commonly called pathogenesis-related (PR) proteins. The PR proteins include β-1,3-glucanases, chitinases, thaumatin-like proteins (thaumatin is a small, very sweet protein), and protease inhibitors. To develop plants resistant to fungal pathogens, researchers have engineered plants to constitutively express one or more PR proteins. For example, transgenic plants have often been designed to produce high levels of chitinase, which can hydrolyze the β-1,4 linkages of the N-acetyl-D-glucosamine polymer chitin, a major component of many fungal cell walls (Fig. 18.11).

Transgenic plants that have been engineered to constitutively express chitinase under the control of the cauliflower mosaic virus 35S promoter, include rice, tobacco, and canola. Moreover, a transgenic tobacco plant constitutively expressing both chitinase and β-glucanase was constructed. The doubly transgenic plant was created by genetically crossing one transgenic plant expressing chitinase with one expressing β-glucanase. Transgenic plants that expressed chitinase were more resistant to damage by fungal

Figure 18.11 A plasmid vector containing a rice chitinase gene cassette and a hygromycin resistance gene cassette (in both cases including transcriptional regulatory sequences) used to transform rice protoplasts. Rice cell protoplasts were transformed by polyethylene glycol treatment in the presence of this plasmid. Transformed cells were selected for their resistance to hygromycin. Later they were tested for the presence of chitinase genes by Southern hybridization and for chitinase by Western blot analysis; then they were used to regenerate transgenic plants.

pathogens than were control plants, even though the control plants could synthesize their own PR proteins in response to the fungal infection. In addition, while transgenic plants constitutively expressing chitinase were resistant to fungal pathogens, the beneficial fungus *Glomus mosseae* was not affected in its binding to the plant roots. This presumably reflects a difference in the cell wall compositions of different fungi. Importantly, a transgenic plant constitutively expressing chitinase has been found to be effective at resisting fungal damage under field conditions. In sum, the overexpression of PR proteins will probably be an effective strategy for protecting plants against damage from pathogenic fungi.

The annual worldwide losses to farmers from potato diseases caused by the pathogenic soil bacterium *Erwinia carotovora* are estimated to be approximately $100 million. Moreover, plant breeders have not identified any resistant traits that can be bred into commercial cultivars. To overcome this problem, one group of researchers developed transgenic potato plants that actively expressed bacteriophage T4 lysozyme. Furthermore, the lysozyme was targeted for secretion into the apoplast (the set of intercellular spaces which are inside the plant but outside the plant cells) in potato plants since this is the part of the plant where *E. carotovora* enters and spreads. More specifically, the T4 lysozyme gene was fused to the barley α-amylase signal peptide coding sequence and placed under the transcriptional control of the cauliflower mosaic virus 35S promoter, transcription terminator, and polyadenylation site. Although the T4 lysozyme gene was under the control of this strong promoter, only a very low level of lysozyme was synthesized. This result notwithstanding, under laboratory and greenhouse conditions transgenic plants with this construct were significantly protected from damage by high levels of *E. carotovora*. Since much lower levels of the pathogen than were used in these laboratory experiments are present in the field, this type of genetic construct should provide a high level of protection under natural conditions. Moreover, since lysozyme can lyse a wide range of both gram-positive and gram-negative bacteria, this approach should be useful for protecting many different plants from various bacterial pathogens.

Development of Stress- and Senescence-Tolerant Plants

Unlike many animals, plants cannot physically avoid adverse environmental conditions such as high levels of light, ultraviolet (UV) irradiation, heat, high salt concentrations, and drought, so they have evolved physiological strategies to cope with these stresses. At the molecular level, one of the un-

MILESTONE

Light-Inducible and Chloroplast-Associated Expression of a Chimaeric Gene Introduced into *Nicotiana tabacum* Using a Ti Plasmid Vector

L. Herrera-Estrella, G. Van den Broeck, R. Maenhaut,
M. Van Montagu, J. Schell, M. Timko, and A. Cashmore
Nature **310**:115–120, 1984

After researchers had established the Ti plasmid system as an effective means of transforming many different plants, their attention turned to the development of procedures for the expression of foreign genes in plants. Initially, most of the genes that were introduced into plant cells were under the transcriptional control of either the relatively strong constitutive 35S promoter from cauliflower mosaic virus or the nearly-as-strong constitutive promoter for the nopaline synthase gene that is encoded within some T-DNAs. However, the development of plants with new and modified useful traits often requires that a specific protein be expressed only in certain tissues, e.g., leaves or roots, or only at certain times in the life of the plant, e.g., during early seedling development, fruit formation, or high-temperature stress. As a first step toward the development of plants that express foreign genes in a tissue-specific or time-specific manner, Herrera-Estrella et al. constructed a chimeric gene that included the 5′-flanking region from the pea gene for the small subunit of ribulose bisphosphate carboxylase, the coding region of a bacterial chloramphenicol acetyl-transferase gene, and the 3′-flanking region from the nopaline synthase gene containing signals both for termination of transcription and for polyadenylation of the mRNA. Normally the gene for the small subunit of ribulose bisphosphate carboxylase is expressed only in green or photosynthetic tissue, and as expected, the chloramphenicol acetyltransferase gene was also expressed only in photosynthetic tissues. This work provided one of the first demonstrations that, despite their complexity, plant promoters could accurately and with tissue specificity direct the transcription of heterologous proteins. Since this study, researchers have used a wide range of different plant promoters to direct tissue- and development-specific heterologous gene expression in transgenic plants.

desirable consequences of physiological stress is the production of oxygen radicals. Thus, investigators reasoned that if they could create plants that were able to tolerate increased levels of oxygen radicals, these plants should also be able to withstand various forms of environmental stress.

Oxidative Stress

A common type of potentially damaging oxygen radical is the superoxide anion. Biologically, the enzyme superoxide dismutase detoxifies superoxide anion by converting it to hydrogen peroxide, which in turn is broken down to water by any of a number of cellular peroxidases or catalases (Fig. 18.12). In one experiment, tobacco plants that were transformed with a superoxide dismutase gene that was under the control of the 35S promoter from cauliflower mosaic virus expressed the enzyme superoxide dismutase and also acquired tolerance to oxygen radical damage.

Plants have several different isoforms of the enzyme superoxide dismutase. The Cu/Zn superoxide dismutases are found primarily in chloroplasts and to a lesser extent in the cytosol. The Mn superoxide dismutase is located in the mitochondria, and some plants also have an Fe form of superoxide dismutase. Transgenic tobacco plants that carried the cDNA for a chloroplast-localized Cu/Zn superoxide dismutase under the control of the 35S promoter from cauliflower mosaic virus were much more resistant to high-light damage than nontransformed plants. When they were tested, the transgenic plants retained 94% of their photosynthetic activity under conditions where nontransformed plants lost all of their activity. In another experiment, transgenic plants with cloned Mn superoxide dismutase targeted to their chloroplasts were three- to fourfold less sensitive to ozone damage than nontransformed plants.

An additional benefit of increasing the superoxide dismutase level in

Figure 18.12 Conversion of superoxide anion to hydrogen peroxide and then to water and oxygen.

higher plants is an increased resistance to the herbicide methyl viologen (paraquat) and a higher tolerance to light stress. A superoxide dismutase gene might also be used to maintain the quality of cut flowers during shipping. Worldwide, the cut flower industry is a multibillion-dollar-a-year business. Petal deterioration (wilting) is due in part to oxygen radicals produced during flower senescence. If the endogenous oxygen-scavenging capacity in cut flowers were increased by overexpression of a superoxide dismutase gene under the control of a flowering-specific promoter, then shelf life might increase as well.

Salt Stress

Many plants live in environments where growth can be severely impaired by either drought or high salinity. As the use of irrigation in agriculture has become more frequent, increased soil salinity has become a common problem. To proliferate under these conditions, many plants synthesize low-molecular-weight nontoxic compounds which are collectively called osmoprotectants. These compounds facilitate both water uptake and retention and also protect and stabilize cellular macromolecules from damage by high salt levels. Some well-known osmoprotectants include sugars, alcohols, proline, and quaternary ammonium compounds. In particular, the quaternary ammonium compound betaine is a highly effective osmolyte that accumulates in some plants during periods of water stress or high salinity.

Several important crops, including potatoes, rice, and tomatoes, do not accumulate betaine. However, the introduction of betaine-biosynthetic enzymes into these plants might enable them to withstand water stress and/or high salinity. Betaine is synthesized from choline in two steps in both plants and bacteria (Fig. 18.13). In plants such as spinach, choline is converted to betaine aldehyde by the enzyme choline monooxygenase and then to betaine by betaine aldehyde dehydrogenase. In bacteria such as *E. coli*, both steps of betaine biosynthesis are catalyzed by the same enzyme, choline dehydrogenase. Therefore, to create a more salt-tolerant tobacco, researchers used *A. tumefaciens* to transform plant cells with a Ti plasmid vector carrying the *E. coli betA* gene, which encodes choline dehydrogenase, under the control of the cauliflower mosaic virus 35S promoter. In laboratory tests, tobacco plants expressing the *E. coli betA* gene were up to 80% more tolerant of a high (i.e., 300 mM) salt concentration than were nontransformed tobacco plants. While it may be possible to improve the osmoprotection afforded by the *E. coli betA* gene by using a tissue-specific promoter to direct expression of this gene, this experiment is an important step in the development of plants that are more tolerant of high levels of salt.

Fruit Ripening

A major problem in fruit marketing is premature ripening and softening during transport. These changes are part of the natural aging (senescence)

Figure 18.13 Conversion of choline to glycine betaine.

process of the fruit. Some of the genes that are induced during ripening encode the enzymes cellulase and polygalacturonase. It has been postulated that by interfering with the expression of one or more of these genes, the ripening process might be delayed. This interference could be achieved by creating transgenic plants with antisense RNA-producing versions of these genes. In fact, when such an antisense RNA-producing gene for polygalacturonase was introduced into tomato plants, a $1.3 billion-a-year crop in the United States, both polygalacturonase mRNA and enzymatic activity were reduced by 90%. This genetically engineered tomato is known as the FLAVR SAVR (pronounced flavor saver) tomato. Thus, lowering polygalacturonase production inhibits fruit ripening in tomatoes, a useful idea that has recently become a practical reality, permitting tomatoes to ripen on the vine instead of being harvested while they are still green. On 18 May 1994, the U.S. Food and Drug Administration ruled that the FLAVR SAVR tomato is as safe as tomatoes that are bred by conventional means. And, because FLAVR SAVR tomatoes are essentially the same as other tomatoes, special labeling is not required.

The plant growth regulator ethylene induces the expression of a number of genes that are involved in fruit ripening and senescence. It is synthesized from *S*-adenosylmethionine by way of the intermediate compound 1-aminocyclopropane-1-carboxylic acid (ACC) (Fig. 18.14). Treatment of plants with chemical compounds that block ethylene production delays both fruit ripening and senescence. Thus, premature fruit ripening might be prevented by inhibiting the plant's ability to synthesize ethylene. This can be achieved in several different ways (Fig. 18.14). For example, transgenic plants that have been engineered to contain antisense RNA versions of ei-

Figure 18.14 Inhibition of ethylene biosynthesis by genetic manipulation. Normally, ACC is synthesized from *S*-adenosylmethionine by the enzyme ACC synthase, and then ACC oxidase converts ACC to ethylene. Ethylene synthesis may be inhibited in transgenic plants either by an antisense mRNA version of ACC synthase or ACC oxidase, which inhibits the synthesis of these enzymes, or by the enzyme ACC deaminase, which competes with ACC oxidase for the available ACC, producing ammonia and α-ketobutyrate rather than ethylene.

ther ACC synthase or ACC oxidase, enzymes whose activities are essential to the plant's synthesis of ethylene, have much lower than normal levels of ethylene, and therefore the fruit that is produced by those plants has an extended storage life.

Moreover, by screening, a large number of soil bacterial strains that degrade ACC were identified. From one of these strains, the gene for the enzyme ACC deaminase was isolated, fused to the 35S promoter from cauliflower mosaic virus, and expressed in tomato plants. The transgenic plants synthesized a lower level of ethylene than did normal plants, and again the fruit of the transgenic plants had a significantly longer shelf life. These bioengineered changes should result in fewer losses due to spoilage, presumably because of the much lower levels of ethylene. While most reports of transgenic plants in which ethylene levels have been lowered in an effort to delay fruit ripening have focused on tomato plants, one group of workers reported the construction of transgenic cantaloupe with lowered ethylene levels. This result indicates that the production of transgenic plants with lowered ethylene levels to delay fruit ripening should be an effective strategy with a range of different fruits.

Genetic Manipulation of Flower Pigmentation

The flower industry is continually attempting to improve flower appearance and postharvest lifetime. By traditional breeding techniques, over the years it has been possible to create thousands of new varieties that differ from one another in color, shape, and plant architecture. However, traditional plant breeding is a slow and painstaking procedure that is limited by the gene pool of a particular species; thus, for example, no one has been able to breed a blue rose. As an alternative to traditional breeding techniques, uniquely colored flowers can be developed by manipulating the genes for enzymes in the anthocyanin biosynthesis pathway. Anthocyanins, which are a class of flavonoids, are the most common type of flower pigment. They are synthesized from the amino acid phenylalanine by a series of enzyme-catalyzed reactions. The color of the flower is determined by the chemical side chain substitutions of different chemical structures, with the cyanidin derivatives producing more red and the delphinidin derivatives producing more blue (Fig. 18.15).

While the petunia enzyme dihydroflavonol 4-reductase can convert colorless dihydroquercetin to red cyanidin-3-glucoside and colorless dihydromyricetin to blue delphinidin-3-glucoside, it cannot use colorless dihydrokaempferol as a substrate (Fig. 18.15). However, when petunias were transformed with a dihydroflavonol 4-reductase gene from maize, the flowers of the transgenic plants were brick red-orange. This unique color, which had never been seen before in petunias, was due to the production of pelargonidin-3-glucoside by the transgenic plants.

Four plants (roses, carnations, tulips, and chrysanthemums) account for approximately 70% of the cut flower industry. Moreover, since genetic transformation protocols have been worked out for these four plants, they should become the focus of efforts to develop plants with unique colors. For example, transgenic chrysanthemums with both sense and antisense constructs of the chrysanthemum chalcone synthase cDNA have been produced. Chalcone synthase catalyzes the first step in anthocyanin biosynthesis (Fig. 18.15). It is expected that both the sense and the antisense cDNA will suppress chalcone synthase gene expression in transgenic plants.

Figure 18.15 Biosynthesis of anthocyanins. Abbreviations: CHS, chalcone syn-thase; CHI, chalcone isomerase; F3H, flavonone 3-hydroxylase; F3'H, flavonoid 3'-hydroxylase; F3'5'H, flavonoid 3',5'-hydroxylase; DFR, dihydroflavonol 4-reduc-tase; 3GT, UDP-glucose:flavonoid 3-*O*-glucosyltransferase. Petunia DFR can convert dihydroquercetin to cyanidin-3-glucoside and dihydromyricetin to blue delphinidin-3-glucoside. In addition to these conversions, maize DFR can convert dihydrokaempferol to pelargonidin-3-glucoside.

"Sense suppression," which is also called "cosuppression," occurs when an additional copy of an endogenous gene prevents the accumulation of the mRNA from the endogenous gene. At present, the molecular basis of this phenomenon is not understood. On the other hand, the antisense chalcone synthase RNA should block translation of endogenous chalcone synthase mRNA.

The sense and antisense constructs were placed under the control of the cauliflower mosaic virus 35S promoter on a binary Ti plasmid vector and then introduced into plant cells. Three of the 133 sense transformants and

three of the 83 antisense transformants produced white flowers, which indicated that endogenous chalcone synthase gene expression, and as a consequence anthocyanin synthesis, had been suppressed. The white-flowering plants were propagated vegetatively through cuttings, and approximately 90 to 98% of the plants continued to produce white, rather than pink, flowers when planted in the field. This work is an important step toward the development of commercial flower varieties with novel colors.

Modification of Plant Nutritional Content

Over the years, agronomists and plant breeders have been extremely successful in optimizing the useful properties (e.g., protein or oil content) and increasing the productivity (yield) of a large number of crop plants. However, traditional breeding approaches to crop improvement are both difficult and slow, and they are intrinsically limited by the existing genetic content of cross-breeding strains. Conversely, the use of genetic engineering techniques allows scientists both to dramatically speed up the process of developing plants with improved characteristics and to introduce traits that would otherwise be impossible to develop by traditional techniques. For example, on a laboratory scale, genetic engineering has been used to improve the nutritional quality of several different plants, including corn and pea, by modification of the amino acid content of some of their seed storage proteins; the fatty acid composition of both edible and nonedible oil-producing crops; and the taste of fruits and vegetables by the introduction of monellin, a sweet-tasting protein.

Amino Acids

Seed storage proteins, which are used as sources of both carbon and nitrogen during seed germination, contain a limited number of amino acids, which are organized into repeating units. Often, the nutritional value of these proteins is deficient because they lack one or more of the amino acids (generally lysine or methionine) that are essential for human health. The amino acid composition of the seed storage proteins can be altered to a limited extent by breeding programs, but recently genetic engineering strategies have also been used.

As a test case, the phaseolin gene (from broad bean), which encodes a seed storage protein composed of a broad spectrum of amino acids, was transferred into tobacco plants. The transgene was faithfully expressed, and in this experiment, the protein product was correctly compartmentalized. In addition, it has been possible to specifically alter in vitro the nucleotide sequence of seed storage protein genes to produce a protein with an improved amino acid composition. In transgenic plants, the altered proteins, unlike unaltered proteins, are not always properly compartmentalized into protein bodies in the seeds. However, recent studies have shown that if the desired amino acids are inserted near the hypervariable region of the C-terminal end of the seed storage protein molecule, the overall structure of the protein is not perturbed. Moreover, the product of these transgenes can be properly packaged during the development of the seed.

One novel way to increase the lysine content of seeds is to increase the production of lysine in transgenic plants by deregulating the lysine biosynthetic pathway. The amino acids lysine, threonine, methionine, and isoleucine are all derived from aspartic acid (Fig. 18.16). The first step in the conversion of aspartic acid to lysine is phosphorylation of aspartic acid by

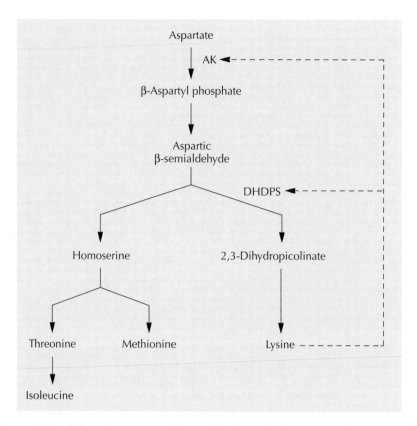

Figure 18.16 Schematic representation of the biosynthetic pathway for amino acids derived from aspartic acid. Not all of the steps and intermediates are shown. Feedback inhibition is shown by the dashed lines. Abbreviations: DHDPS, dihydrodipicolinic acid synthase; AK, aspartokinase.

aspartokinase (AK) to produce β-aspartyl phosphate. The condensation of aspartic β-semialdehyde with pyruvic acid to form 2, 3-dihydrodipicolinic acid, which is catalyzed by dihydrodipicolinic acid synthase (DHDPS), is the first reaction in the pathway that is committed to lysine biosynthesis. Both AK and DHDPS are feedback inhibited by lysine. Thus, to overproduce lysine, it is necessary to abolish the feedback inhibition of these two enzymes. This was accomplished by cloning lysine-feedback-insensitive genes for DHDPS and AK from *Corynebacterium* and *E. coli*, respectively; fusing each of these genes to a chloroplast transit peptide; placing each gene under the control of a seed-specific promoter; and then introducing the two genes on a Ti plasmid binary vector into canola and soybean plants (Fig. 18.17). Transgenic canola and soybean plants had more than a 100-fold increase in the free lysine in their seeds, with an overall doubling of the total seed lysine content in canola and a fivefold increase in the total lysine content in soybean.

Currently, when corn is used as an animal feed, it must be supplemented with soybean meal, purified lysine, or both. However, it may be possible to replace the use of expensive lysine with inexpensive transgenic soybean meal from soybean plants that overproduce lysine. Moreover, it may eventually be possible, by using the approach that has been successful with soybean, to engineer corn to overproduce lysine. High-lysine corn would be more nutritious for both animals and humans.

Figure 18.17 Schematic representation of a Ti plasmid vector used to transform soybean and canola with genes that modify their lysine content. Designations and abbreviations: Pv5', the promoter from the bean β-phaseolin gene; Pv3', the transcription termination region from the bean β-phaseolin gene; cts, the coding region for the chloroplast transit peptide from the small subunit of ribulosebisphosphate carboxylase; *dapA*, a gene from *Corynebacterium* which encodes a lysine-insensitive dihydrodipicolinic acid synthase; *lysC*M4, a mutant version of the *E. coli lysC* gene which encodes a lysine-insensitive aspartokinase. The left and right borders of the T-DNA are indicated by LB and RB, respectively.

Lipids

Annual global plant oil production was estimated to be worth about $45 billion in 1995 and is expected to grow to around $70 billion by 2010. More than 90% of this production is for human consumption in margarines, shortenings, salad oils, and frying oils. Together, soybean, palm, rapeseed (canola), and sunflower account for approximately 75% of worldwide plant oil production. For the most part, these oils consist of the following fatty acids: palmitic, stearic, oleic, linoleic, and linolenic acids (Table 18.5).

It is currently possible, by genetic engineering, to change the degree of unsaturation (i.e., the number of carbon–carbon double bonds) and modify the chain length of fatty acids in plants. A number of transgenic varieties of canola, each producing a different modified oil, have been created and are being field tested (Table 18.6). Each transgenic variety contains one additional gene. For example, the stearic acid variety contains an antisense copy of a *Brassica* stearate desaturase gene, which inhibits the expression of the normal canola gene and leads to the accumulation of stearic acid rather than

Table 18.5 Some important plant fatty acids

Common name	Abbreviation
Caprylic acid	$C_{8:0}$
Capric acid	$C_{10:0}$
Lauric acid	$C_{12:0}$
Myristic acid	$C_{14:0}$
Palmitic acid	$C_{16:0}$
Stearic acid	$C_{18:0}$
Petroselinic acid	$\varDelta 6C_{18:1}$
Oleic acid	$\varDelta 9C_{18:1}$
Linoleic acid	$\varDelta 9,12C_{18:2}$
Linolenic acid	$\varDelta 9,12,15C_{18:3}$
Ricinoleic acid	$12OH\varDelta 9C_{18:1}$
Erucic acid	$\varDelta 13C_{22:1}$

The first number after the C denotes the number of carbon atoms; the number after the colon is the degree of unsaturation, i.e., the number of C=C bonds; a $\varDelta$ followed by a number indicates the position of the first carbon atom that is involved in the C=C bond; a number followed by an OH indicates the position on the chain of a hydroxyl group. All C=C bonds are *cis*. The numbering system for fatty acids begins with the carboxyl group as 1.

Table 18.6 Transgenic canola varieties with modified seed lipid contents

Seed product	Commercial use(s)	First field trial
40% Stearic	Margarine, cocoa butter	1994
40% Lauric	Detergents	1994
60% Lauric	Detergents	1996
80% Oleic	Food, lubricants, inks	1995
Petroselinic	Polymers, detergents	Expected in 1998
"Jojoba" wax	Cosmetics, lubricants	1996
40% Myristate	Detergents, soaps, personal care items	1996
90% Erucic	Polymers, cosmetics, inks, pharmaceuticals	Expected in 1998
Ricinoleic	Lubricants, plasticizers, cosmetics, pharmaceuticals	1997

Adapted from Murphy, *Trends Biotechnol.* **14:**206–213, 1996.

the desaturation of stearic acid to oleic acid. Progress on the production of transgenic canola varieties with modified seed oil properties has been both rapid and impressive. Many of these varieties are likely to be used commercially within the next few years.

Modification of Food Plant Taste and Appearance

Preventing Discoloration

The postharvest discoloration of fruits and vegetables is a considerable problem for the food industry. A lack of acceptance of discolored foods by consumers has been dealt with by the food industry through the use of food additives in a wide range of foods. However, recently, the safety of some of these food additives (sulfites in particular) has been questioned.

The enzymes that are thought to be responsible for the initial step in the discoloration of fruits and vegetables, the oxidation of monophenols and *o*-diphenols to *o*-quinones, are polyphenol oxidases. These nucleus-encoded enzymes with a molecular weight of around 59,000 are localized in chloroplast and mitochondrial membranes.

The contention that inhibition of the enzyme polyphenol oxidase would decrease the extent of discoloration has been tested with transgenic potatoes carrying a number of different polyphenol oxidase cDNA constructs. Vectors were constructed with either the full-length or partial potato polyphenol oxidase cDNA in either the sense or antisense orientation under the control of the cauliflower mosaic virus 35S promoter, the granule-bound starch synthase promoter, or the patatin type I promoter (Fig. 18.18). The latter two promoters are specific for the potato tuber. The two commercial varieties of potato that were transformed with these constructs are considered to have a good level of resistance to black spot (enzymatic discoloration), so an increase in black spot resistance by genetic manipulation would be greater than what could be attained by traditional breeding techniques. Transgenic plants with the polyphenol oxidase cDNA constructs were deliberately bruised, and then the extent of black spot damage was assessed. Most of the transgenic potato plants with an antisense version of the polyphenol oxidase gene under the control of either the cauliflower mosaic virus 35S promoter or the granule-bound starch synthase promoter were significantly more resistant to black spot than the nontransformed potatoes. The patatin promoter, which may not be fully active in potato tubers, did

Figure 18.18 Sense and antisense polyphenol oxidase gene constructs. Transcription of either the sense or antisense gene is separately under the control of the cauliflower mosaic virus 35S promoter (*p*35S), the granule-bound starch synthase promoter (*p*GBSS), or the patatin type I promoter (*p*PATATIN) and the nopaline synthase transcription terminator region (*t*NOS). The abbreviations RB and LB represent the right and left borders of the T-DNA, respectively. Adapted from Bachem et al., *Bio/Technology* **12:**1101–1105, 1994.

not prevent polyphenol oxidase accumulation. The sense constructs all synthesized increased amounts of polyphenol oxidase and showed larger amounts of black spot than the control plants. Although these results are preliminary, this approach may be applicable to the reduction of enzymatic discoloration in a range of commercially important plants.

Sweetness

Even though a fruit or vegetable may have high nutritional value, if it is not tasty, humans usually will not eat it. Although palatability of food can be achieved by adding salt, sugar, flavors, or other ingredients during preparation, it would be advantageous to the food industry if certain foods could be made intrinsically more appetizing.

Monellin, a protein that is found in the fruit of an African plant with the unlikely name of serendipity (*Dioscorephyllum cumminsii* Diels), is approximately 100,000 times sweeter than sucrose on a molar basis. This feature makes monellin a candidate as a sugar substitute, with the added bonus that, because it is a protein, it would not have the same metabolic impact as sugar.

Monellin is a dimer with an A chain of 45 amino acid residues and a B chain of 50 residues; the chains are held together by weak noncovalent bonds. Unfortunately, the fact that monellin is composed of two separate polypeptide chains limits its usefulness as a sweetener because it is readily dissociated and consequently loses its sweetness when it is either heated during cooking or exposed to acid (e.g., lemon juice or vinegar). Also, the need to clone and express in a coordinated manner two separate genes complicates efforts to produce the protein in either transgenic plants or microorganisms. To circumvent this problem, a monellin gene that encodes both the A and B chains as a single peptide was chemically synthesized. The fusion protein was produced in transgenic tomato and lettuce plants. Two different promoters were used to express the monellin fusion protein gene. In the experiment with tomatoes, expression was directed by the tomato

fruit-specific promoter E8, which is activated at the onset of fruit ripening. The construct for the lettuce experiment was under the control of the 35S promoter from cauliflower mosaic virus. Each construct used the transcription termination–polyadenylation site from a Ti plasmid nopaline synthase gene. In each case, the synthetic monellin gene was introduced into plant cells by *A. tumefaciens* infection, using the Ti plasmid cointegrate vector system. Monellin was detected in ripe and partially ripe tomatoes and in lettuce leaves but not in green tomatoes. The monellin level in tomatoes could also be elevated by a burst of the plant hormone ethylene. Comprehensive taste tests have not yet been reported for these genetically sweetened foods. However, if these tests are positive, then this strategy for sweetening plants without sugar or chemical additives would be applicable to a wide range of fruits and vegetables.

Plants as Bioreactors

Plants are easy to grow and can generate considerable biomass. With these features in mind, research has been carried out to determine whether transgenic plants can be used for the production of commercial proteins and chemicals. Unlike recombinant bacteria, which are grown in large bioreactors (a process that requires highly trained personnel and expensive equipment), crops can be produced relatively inexpensively by less-skilled workers. Ultimately, the biggest hurdles to overcome in an endeavor of this sort will be the purification of the product of a transgene from the mass of plant tissue and the overall cost of producing a target protein in a transgenic plant compared with the cost of production in a microorganism. In small-scale trials, plants have been used to produce monoclonal antibodies, functional antibody fragments, and the polymer polyhydroxybutyrate, which can be used to make a biodegradable plastic-like material.

Antibodies

The production of antibodies and antibody fragments in transgenic plants has several potential advantages over their synthesis in recombinant microbial cells. For example, transformation of plants generally results in the stable integration of the foreign DNA into the plant genome, while most microorganisms are transformed with plasmids that can be lost during a prolonged or large-scale fermentation. In addition, the processing and assembly of foreign proteins in plants are similar to those in animal cells, whereas bacteria do not readily process, assemble, or posttranslationally modify eukaryotic proteins. Moreover, plants are inexpensive to grow on a large scale, and their production is not limited by fermentation capability. Finally, since most harvested plant tissues cannot usually be stored for long periods, foreign proteins might be produced in seeds where they will be stable for long periods under ambient conditions.

Polymers

It is costly to produce the polymer poly(3-hydroxybutyric acid), which is used in the synthesis of biodegradable plastics, by bacterial fermentation. Consequently, research has been conducted to determine if this polymer could be produced at less cost in plants. In bacteria such as *Alcaligenes eutrophus*, poly(3-hydroxybutyric acid) is synthesized from acetyl coenzyme A in three steps catalyzed by three enzymes (see Fig. 12.22) whose genes are organized on a single operon. Since plants are unable to process a transcript

of an operon with more than one gene, each of the three genes was cloned separately and targeted to the chloroplast of the plant *Arabidopsis thaliana*. The chloroplast was chosen as the site of synthesis since previous experiments had demonstrated that cytoplasmic synthesis of this polymer resulted in only low levels of polymer formation and produced highly stunted plants. Moreover, chloroplasts can accumulate high levels of starch, another biological polymer.

Each of the three poly(3-hydroxybutyric acid) biosynthesis genes was fused to a DNA fragment that encodes the chloroplast transit peptide of the small subunit of pea ribulose bisphosphate carboxylase and was placed under the transcriptional control of the cauliflower mosaic virus 35S promoter. Each of the biosynthetic genes was introduced into *A. thaliana* plants with Ti plasmid binary vectors. Two transgenic plants, each with a different foreign gene, were crossed to form a transgenic plant with two chloroplast-encoded foreign genes. Then the double-gene transgenic plant was crossed with a transgenic plant that carried the third foreign gene, and a transgenic plant carrying all three of the bacterial poly(3-hydroxybutyric acid) biosynthesis genes was selected. Mature leaves of some of the transgenic plants expressing all three of the bacterial poly(3-hydroxybutyric acid) biosynthesis genes produced more than 1 mg of poly(3-hydroxybutyric acid) per gram (fresh weight) of leaf. This work is an important first step in the development of crop plants that produce large amounts of poly(3-hydroxybutyric acid).

Foreign Proteins in Seeds

Oleosins, or oil body proteins, are found in the seeds of a wide range of plants. These proteins are quite hydrophobic and are mostly embedded within plant oil bodies, thus stabilizing the oil bodies as discrete organelles. However, the N- and C-terminal regions of oleosins are more hydrophilic than the rest of the protein and are exposed to the aqueous environment. It is therefore possible to engineer at the DNA level fusions between oleosins and water-soluble proteins (Fig. 18.19), with the expectation that the fusion protein will be targeted to plant oil bodies, making it relatively easy to purify the fusion protein. In this case, the water soluble target protein will not be embedded in the oil body but rather will be exposed to the aqueous environment. Depending upon the final use of the fusion protein, it is also possible to cleave the target protein from the fusion protein if necessary. Since it is relatively easy to purify oil bodies from plant seeds, the purification of the recombinant protein is simplified. Then the recombinant protein can be recovered by cleavage of the fusion protein. This system has the potential to significantly lower the costs of purifying target proteins produced in plants.

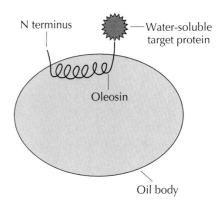

Figure 18.19 Schematic representation of a fusion protein including oleosin and a water-soluble target protein embedded in a plant seed oil body. The N- and C-terminal ends of the oleosin and the target protein are hydrophilic and are therefore found in the aqueous environment.

SUMMARY

By using a variety of techniques (see Chapter 17), it has become relatively straightforward to transform plants with foreign genes. Expression of foreign genes in plants makes it possible to produce a very wide range of new plant varieties. Plants have been engineered to be resistant to a range of environmental stresses, including insects, viruses, herbicides, pathogens, and oxidative and salt stress; to have flowers with modified colors; to have a modified nutritional content, including modifications in amino acids, lipids, discoloration, and sweetness; and to act as factories for the large-

scale production of important foreign proteins such as antibodies and therapeutics. Numerous transgenic plants with altered properties and contents have been successfully produced and tested in the laboratory and in some cases in the field. Although only a small number of genetically engineered plants have entered the marketplace thus far, it is likely that transgenic plants will become an important part of agricultural and horticultural practice.

References

Ananda Kumar, P., R. P. Sharma, and V. S. Malik. 1996. The insecticidal proteins of *Bacillus thuringiensis. Adv. Appl. Microbiol.* **42:**1–43.

Anderson, E. J., D. M. Stark, R. S. Nelson, N. E. Tumer, and R. N. Beachy. 1989. Transgenic plants that express the coat protein gene of TMV or A1MV interfere with disease development of non-related viruses. *Phytopathology* **12:**1284–1290.

Ayub, R., M. Guis, M. Ben Amor, L. Gillot, J.-P. Roustan, A. Latché, M. Bouzayen, and J.-C. Pech. 1996. Expression of ACC oxidase antisense gene inhibits ripening of cantaloupe melon fruits. *Nat. Biotechnol.* **14:**862–866.

Bachem, C. W. B., G.-J. Speckmann, P. C. G. van der Linde, F. T. M. Verheggen, M. D. Hunt, J. C. Steffens, and M. Zabeau. 1994. Antisense expression of polyphenol oxidase genes inhibits enzymatic browning in potato tubers. *Bio/Technology* **12:**1101–1105.

Beachy, R. N., S. Loesch-Fries, and N. E. Tumer. 1990. Coat protein-mediated resistance against virus infection. *Annu. Rev. Phytopathol.* **28:**451–474.

Bird, C. R., D. Grierson, J. A. Ray, and W. W. Schuch. October 1993. Tomato plants and cells containing pTOM36 antisense constructs. U.S. patent 5,254,800.

Boulter, D., A. M. R. Gatehouse, and V. Hilder. 1989. Use of cowpea trypsin inhibitor (CpTI) to protect plants against insect predation. *Biotechnol. Adv.* **7:**489–498.

Bowler, C., L. Slooten, S. Vandenbranden, R. De Rycke, J. Botterman, C. Sybesma, M. Van Montagu, and D. Inzé. 1991. Manganese superoxide dismutase can reduce cellular damage mediated by oxygen radicals in transgenic plants. *EMBO J.* **10:**1723–1732.

Brunke, K. J., and R. L. Meeusen. 1991. Insect control with genetically engineered crops. *Trends Biotechnol.* **9:**197–200.

Cheng, J., M. G. Bolyard, R. C. Saxena, and M. B. Sticklen. 1992. Production of insect resistant potato by genetic transformation with a δ-endotoxin from *Bacillus thuringiensis* var. *kurstaki. Plant Sci.* **81:**83–91.

Comai, L., D. Facciotti, W. R. Hiatt, G. Thompson, R. E. Rose, and D. M. Stalker. 1985. Expression in plants of a mutant *aroA* gene from *Salmonella typhimurium* confers tolerance to glyphosate. *Nature* **317:**741–744.

Corbin, D. R., J. T. Greenplate, E. Y. Wong, and J. Purcell. 1994. Cloning of an insecticidal cholesterol oxidase gene and its expression in bacteria and in plant protoplasts. *Appl. Environ. Microbiol.* **60:**4239–4244.

Courtney-Gutterson, N., C. Napoli, C. Lemieux, A. Morgan, E. Firoozabady, and K. E. P. Robinson. 1994. Modification of flower color in florist's chrysanthemum: production of a white-flowering variety through molecular genetics. *Bio/Technology* **12:**268–271.

Cuozzo, M., K. M. O'Connell, W. Kaniewski, R.-X. Fang, N.-H. Chua, and N. E. Tumer. 1988. Viral protection in transgenic tobacco plants expressing the cucumber mosaic virus coat protein or its antisense RNA. *Bio/Technology* **6:**549–557.

Day, A. G., E. R. Bejarano, K. W. Buck, M. Burrell, and C. P. Lichtenstein. 1991. Expression of an antisense viral gene in transgenic tobacco confers resistance to the DNA virus tomato golden mosaic virus. *Proc. Natl. Acad. Sci. USA* **88:**6721–6725.

Dekker, J., and S. O. Duke. 1995. Herbicide-resistant field crops. *Adv. Agron.* **54:**69–116.

Delannay, X., B. J. LaVallee, R. K. Proksch, R. L. Fuchs, S. R. Sims, J. T. Greenplate, P. G. Marrone, R. B. Dodson, J. J. Augustine, J. G. Layton, and D. A. Fischhoff. 1989. Field performance of transgenic tomato plants expressing the *Bacillus thuringiensis* var. *kurstaki* insect control protein. *Bio/Technology* **7**:1265–1269.

Duan, X., X. Li, Q. Xue, M. Abo-El-Saad, D. Xu, and R. Wu. 1996. Transgenic rice plants harboring an introduced potato proteinase inhibitor II gene are insect resistant. *Nat. Biotechnol.* **14**:494–498.

Düring, K., P. Porsch, M. Fladung, and H. Lörz. 1993. Transgenic potato plants resistant to the phytopathogenic bacterium *Erwinia carotovora*. *Plant J.* **3**:587–598.

Ely, S. 1993. The engineering of plants to express *Bacillus thuringiensis* δ-endotoxins, p. 105–124. *In* P. F. Entwistle, J. S. Cory, M. J. Bailey, and S. Higgs (ed.), Bacillus thuringiensis, *an Environmental Biopesticide: Theory and Practice.* John Wiley & Sons, Chichester, United Kingdom.

Falco, S. C., T. Guida, M. Locke, J. Mauvais, C. Sanders, R. T. Ward, and P. Webber. 1995. Transgenic canola and soybean seeds with increased lysine. *Bio/Technology* **13**:577–582.

Fiedler, U., and U. Conrad. 1995. High-level production and long-term storage of engineered antibodies in transgenic tobacco seeds. *Bio/Technology* **13**:1090–1093.

Fischhoff, D. A., K. S. Bowdish, F. J. Perlak, P. G. Marrone, S. M. McCormick, J. G. Niedermeyer, D. A. Dean, K. Kunsano-Kretzner, E. J. Mayer, D. E. Rochester, S. G. Rogers, and R. T. Fraley. 1987. Insect tolerant tomato plants. *Bio/Technology* **5**:807–813.

Fitchen, J. H., and R. N. Beachy. 1993. Genetically engineering protection against viruses in transgenic plants. *Annu. Rev. Microbiol.* **47**:739–763.

Fuchs, M., and D. Gonsalves. 1995. Resistance of transgenic hybrid squash ZW-20 expressing the coat protein genes of zucchini yellow mosaic virus and watermelon mosaic virus 2 to mixed infections by both potyviruses. *Bio/Technology* **13**:1466–1473.

Gray, J. E., S. Picton, J. J. Giovannoni, and D. Grierson. 1994. The use of transgenic and naturally occurring mutants to understand and manipulate tomato fruit ripening. *Plant Cell Environ.* **17**:557–571.

Grison, R., B. Grezes-Besset, M. Schneider, N. Lucante, L. Olsen, J.-J. Leguay, and A. Toppan. 1996. Field tolerance to fungal pathogens of *Brassica napus* constitutively expressing a chimeric chitinase gene. *Nat. Biotechnol.* **14**:643–646.

Hilder, V. A., A. M. R. Gatehouse, S. E. Sheerman, R. F. Barker, and D. Boulter. 1987. A novel mechanism of insect resistance engineered into tobacco. *Nature* **330**:160–163.

Hill, K. K., N. Jarvis-Eagan, E. L. Halk, K. J. Krahn, L. W. Liao, R. S. Mathewson, D. J. Merlo, S. E. Nelson, K. E. Rashka, and L. S. Loesch-Fries. 1991. The development of virus-resistant alfalfa, *Medicago sativa* L. *Bio/Technology* **9**:373–377.

Holton, T. A., F. Brugliera, D. R. Lester, Y. Tanaka, C. D. Hyland, J. G. T. Menting, C.-Y. Liu, E. Farcy, T. W. Stevenson, and E. C. Cornish. 1993. Cloning and expression of cytochrome P450 genes controlling flower color. *Nature* **366**:276–279.

Johnson, R., J. Narvaez, G. An, and C. Ryan. 1989. Expression of protease inhibitors I and II in transgenic tobacco plants: effects on natural defense against *Manduca sexta* larvae. *Proc. Natl. Acad. Sci. USA* **86**:9871–9875.

Jongedijk, E., A. A. J. M. de Schutter, T. Stolte, P. J. M. van den Elzen, and B. J. C. Cornelissen. 1992. Increased resistance to potato virus X and preservation of cultivar properties in transgenic potato under field conditions. *Bio/Technology* **10**:422–429.

Klee, H. J., M. B. Hayford, K. A. Kretzmer, G. F. Barry, and G. M. Kishore. 1991. Control of ethylene synthesis by expression of a bacterial enzyme in transgenic tomato plants. *Plant Cell* **3**:1187–1193.

Knutzon, D. S., G. A. Thompson, S. E. Radke, W. B. Johnson, V. C. Knauf, and J. C. Kridl. 1992. Modification of *Brassica* seed oil by antisense expression of a stearoyl-acyl carrier protein desaturase gene. *Proc. Natl. Acad. Sci. USA* **89**:2624–2628.

Lee, W. S., J. T. C. Tzen, J. C. Kridl, S. E. Radke, and A. H. C. Huang. 1991. Maize oleosin is correctly targeted to seed oil bodies in *Brassica napus* transformed with the maize oleosin gene. *Proc. Natl. Acad. Sci. USA* **88**:6181–6185.

Lilius, G., N. Holmberg, and L. Bülow. 1996. Enhanced NaCl stress tolerance in transgenic tobacco expressing bacterial choline dehydrogenase. *Bio/Technology* **14**:177–180.

Lin, W., C. S. Anuratha, K. Datta, I. Potrykus, S. Muthukrishnan, and S. K. Datta. 1995. Genetic engineering of rice for resistance to sheath blight. *Bio/Technology* **13**:686–691.

Ling, K., S. Namba, C. Gonsalves, J. L. Slightom, and D. Gonsalves. 1991. Protection against detrimental effects of potyvirus infection in transgenic tobacco plants expressing the papaya ringspot virus coat protein gene. *Bio/Technology* **9**:752–758.

Lodge, J. K., W. K. Kaniewski, and N. E. Tumer. 1993. Broad-spectrum virus resistance in transgenic plants expressing pokeweed antiviral protein. *Proc. Natl. Acad. Sci. USA* **90**:7089–7093.

Ma, J. K.-C., and M. B. Hein. 1995. Immunotherapeutic potential of antibodies produced in plants. *Trends Biotechnol.* **13**:522–527.

MacIntosh, S. C., G. M. Kishore, F. J. Perlak, P. G. Marrone, T. B. Stone, S. R. Sims, and R. L. Fuchs. 1990. Potentiation of *Bacillus thuringiensis* insecticidal activity by serine protease inhibitors. *J. Agric. Food Chem.* **38**:1145–1152.

Meyer, P., I. Heidmann, G. Forkmann, and H. Saedler. 1987. A new petunia flower colour generated by transformation of a mutant with a maize gene. *Nature* **330**:677–678.

Mol, J. N. M., T. A. Holton, and R. E. Koes. 1995. Floriculture: genetic engineering of commercial traits. *Trends Biotechnol.* **13**:350–355.

Mol, J. N. M., A. R. van der Krol, A. J. van Tunen, R. van Blokland, P. de Lange, and A. R. Stuitje. 1990. Regulation of plant gene expression by antisense RNA. *FEBS Lett.* **268**:427–430.

Murphy, D. J. 1996. Engineering oil production in rapeseed and other oil crops. *Trends Biotechnol.* **14**:206–213.

Nawrath, C., Y. Poirier, and C. Somerville. 1994. Targeting of polyhydroxybutyrate biosynthetic pathway to the plastids of *Arabidopsis thaliana* results in high levels of polymer accumulation. *Proc. Natl. Acad. Sci. USA* **91**:12760–12764.

Peñarrubia, L., R. Kim, J. Giovannoni, S.-H. Kim, and R. L. Fischer. 1992. Production of the sweet protein monellin in transgenic plants. *Bio/Technology* **10**:561–564.

Perlak, F. J., R. W. Deaton, T. A. Armstrong, R. L. Fuchs, S. R. Sims, J. T. Greenplate, and D. A. Fischhoff. 1990. Insect resistant cotton plants. *Bio/Technology* **8**:939–943.

Perlak, F. J., R. L. Fuchs, D. A. Dean, S. L. McPherson, and D. A. Fischhoff. 1991. Modification of the coding sequence enhances plant expression of insect control protein genes. *Proc. Natl. Acad. Sci. USA* **88**:3324–3328.

Powell, P. A., D. M. Stark, P. R. Sanders, and R. N. Beachy. 1989. Protection against tobacco mosaic virus in transgenic plants that express tobacco mosaic virus antisense RNA. *Proc. Natl. Acad. Sci. USA* **86**:6949–6952.

Quinn, J. P. 1990. Evolving strategies for the genetic engineering of herbicide resistance in plants. *Biotechnol. Adv.* **8**:321–333.

Rooijen, G. J. H., and M. M. Moloney. 1995. Plant seed oil-bodies as carriers for foreign proteins. *Bio/Technology* **13**:72–77.

Ryan, C. A. 1990. Protease inhibitors in plants: genes for improving defenses against insect and pathogens. *Annu. Rev. Phytopathol.* **28**:425–449.

Sen Gupta, A., R. P. Webb, A. S. Holaday, and R. D. Allen. 1993. Overproduction of superoxide dismutase protects plants from oxidative stress. *Plant Physiol.* **103**:1067–1073.

**Shade, R. E., H. E. Schroeder, J. J. Pueyo, L. M. Tabe, L. L. Murdock, T. J. V. Hig-

gins, and M. J. Chrispeels. 1994. Transgenic pea seeds expressing the α-amylase inhibitor of the common bean are resistant to bruchid beetles. *Bio/Technology* **12**:793–796.

Shah, D. M., C. M. T. Rommens, and R. N. Beachy. 1995. Resistance to diseases and insects in transgenic plants: progress and applications to agriculture. *Trends Biotechnol.* **13**:362–368.

Töpfer, R., N. Martini, and J. Schell. 1995. Modification of plant lipid synthesis. *Science* **268**:681–686.

Tricoli, D. M., K. J. Carney, P. F. Russell, J. R. McMaster, D. W. Groff, K. C. Hadden, P. T. Himmel, J. P. Hubbard, M. L. Boeshore, and H. D. Quemada. 1995. Field evaluation of transgenic squash containing single or multiple virus coat protein gene constructs for resistance to cucumber mosaic virus, watermelon mosaic virus 2, and zucchini yellow mosaic virus. *Bio/Technology* **13**:1458–1465.

Vaeck, M., A. Reynaerts, H. Höfte, S. Jansens, M. de Beuckeleer, C. Dean, M. Zabeau, M. Van Montagu, and J. Leemans. 1987. Transgenic plants protected from insect attack. *Nature* **328**:33–37.

Van Camp, W., H. Willekens, C. Bowler, M. Van Montagu, D. Inzé, P. Reupold-Popp, H. Sandermann, Jr., and C. Langebartels. 1994. Elevated levels of superoxide dismutase protect transgenic plants against ozone damage. *Bio/Technology* **12**:165–168.

Van Rie, J. 1991. Insect control with transgenic plants: resistance proof? *Trends Biotechnol.* **9**:177–179.

Vierheilig, H., M. Alt, J.-M. Neuhaus, T. Boller, and A. Wiemken. 1993. Colonization of transgenic *Nicotiana sylvestris* plants, expressing different forms of *Nicotiana tabacum* chitinase, by the root pathogen *Rhizoctonia solani* and by the mycorrhizal symbiont *Glomus mosseae*. *Mol. Plant-Microbe Interact.* **6**:261–264.

Williams, S., L. Friedrich, S. Dincher, N. Carozzi, H. Kessmann, E. Ward, and J. Ryals. 1992. Chemical regulation of *Bacillus thuringiensis* δ-endotoxin expression in transgenic plants. *Bio/Technology* **10**:540–543.

Zhu, Q., E. A. Maher, S. Masoud, R. A. Dixon, and C. J. Lamb. 1994. Enhanced protection against fungal attack by constitutive co-expression of chitinase and glucanase genes in transgenic tobacco. *Bio/Technology* **12**:807–812.

REVIEW QUESTIONS

1. A local crop is being ravaged by a nonenveloped virus with a single-stranded RNA genome (8,000 nucleotides long). The virus and its RNA can be readily isolated. In addition, you have antibodies against all four of the proteins encoded within the viral genome. Describe a strategy that you could use to protect the crop against viral infection and subsequent damage.

2. Suggest several different strategies for developing insect-resistant plants.

3. How is antisense RNA used to provide plants with resistance to specific viruses?

4. How can protease inhibitors, amylase inhibitor, and cholesterol oxidase each protect a plant against damage from insect predation?

5. Suggest a strategy for protecting a plant against damage from several different viruses.

6. What general strategies can be employed in genetically engineering plants to be resistant to herbicides?

7. How can plants be engineered to resist damage from pathogenic soil fungi?

8. How can plants be genetically engineered to be resistant to pathogenic bacteria?

9. Suggest a strategy that you could use to genetically engineer plants to be tolerant of high levels of salt.

10. You have been asked by an avocado grower to find a way to genetically engineer his crop to prevent it from ripening during shipping. What experimental approaches would you consider?

11. How can plants be genetically manipulated to produce flowers with unusual colors?

12. How can soybeans be genetically manipulated to increase their lysine content?

13. What is cosuppression? What is antisense suppression? Compare cosuppression and antisense suppression.

14. Suggest a strategy which would facilitate the large-scale purification of soluble proteins such as antibody fragments produced in plants.

CHAPTER 19

Transgenic Animals

Many generations of selective matings are required to improve live-stock and other domesticated animals genetically for traits such as milk yield, wool characteristics, rate of weight gain, and egg-laying frequency. At each successive generation, animals with superior performance characteristics are used as breeding stock. Eventually, high-production animals are developed as more or less pure breeding lines. This combination of mating and selection, although time-consuming and costly, has been exceptionally successful. Today, almost all aspects of the biological basis of livestock production can be attributed to this process. However, once an effective genetic line has been established, it becomes difficult to introduce new genetic traits by selective breeding methods. For example, a strain with a newly discovered, valuable gene may also carry deleterious genes that, after crossing, would diminish the existing genetically determined production levels. Thus, a completely new program of multigenerational crosses with rigorous selection procedures has to be initiated to ensure that a new breeding line retains both its original attributes and the new trait.

Until recently, selective breeding was the only way to enhance the genetic features of domesticated animals. However the combination of the successful transfer of genes into mammalian cells and the possibility of creating genetically identical animals by transplanting nuclei from embryonic tissue into enucleated eggs (nuclear transfer, nuclear cloning) led researchers to consider putting single, functional genes or gene clusters into the chromosomal DNA of higher organisms. Conceptually, the strategy used to achieve this end is simple.

- A cloned gene is injected into the nucleus of a fertilized egg.
- The inoculated fertilized eggs are implanted into a receptive female (because successful completion of mammalian embryonic development is not possible outside of a female).
- Some of the offspring derived from the implanted eggs carry the cloned gene in all of their cells.
- Animals with the cloned gene integrated in their germ line cells are bred to establish new genetic lines.

Table 19.1 Protein composition (grams/liter) of milk from cattle and sheep

Proteins	Cattle	Sheep
Casein		
α_{s1}-Casein	10.0	12.0
α_{s2}-Casein	3.4	3.8
κ-Casein	3.9	4.6
β-Casein	10.0	16.0
Major whey proteins		
α-Lactalbumin	1.0	0.8
β-Lactalbumin	3.0	2.8
Other proteins		
Serum albumin	0.4	Unknown
Lysozyme	Trace	Unknown
Lactoferrin	0.1	Unknown
Immunoglobulins	0.7	Unknown

This technique has many practical applications. If, for example, the product of the injected gene stimulates growth, animals that acquire this gene should grow faster and require less feed. An enhancement of feed efficiency by a few percent would have a profound impact on lowering the cost of production of either beef or pork.

During the 1980s—with considerable effort—the idea of genetically manipulating animals by introducing genes into fertilized eggs was converted into a reality. As with many new scientific enterprises, a set of terms was created to make communication easier. For example, an animal whose genetic composition has been altered by the addition of foreign (exogenous) DNA is said to be transgenic. The DNA that is introduced is called a transgene, and the overall process is called transgenic technology or transgenesis.

The genetic improvement of multicellular organisms by the introduction of relevant transgenes is only slowly being realized. In the meantime, however, transgenesis has become a powerful technique for studying fundamental problems of mammalian gene expression and development, for establishing animal model systems for human diseases, and for using the mammary gland to produce pharmaceutically important proteins in milk. With this last application in mind, the term "pharming" was coined to convey the idea that milk from transgenic farm ("pharm") animals can be a source of authentic human protein drugs or pharmaceuticals. There are a number of reasons why the mammary gland should be used in this way. Milk is a renewable, secreted body fluid that is produced in substantial quantities and can be collected frequently without harm to the animal. A novel drug protein that is confined to the mammary gland and secreted into milk should have no side effects on the normal physiological processes of the transgenic animal and should undergo posttranslational modifications that at least match closely those in humans. Finally, purification of a protein from milk, which contains only a small number of different proteins (Table 19.1), should be relatively straightforward.

Transgenic Mice: Methodology

Transgenic technology has been developed and perfected in the laboratory mouse. Since the early 1980s, hundreds of different genes have been introduced into various mouse strains. These studies have contributed to an un-

derstanding of gene regulation, tumor development, immunological specificity, molecular genetics of development, and other biological processes of fundamental interest. Transgenic mice have also played a role in examining the feasibility of the industrial production of human therapeutic drugs by domesticated animals and in the creation of transgenic strains that act as biomedical models for various human genetic diseases. For transgenesis, DNA can be introduced into mice by (1) retroviral vectors that infect the cells of an early stage embryo prior to implantation into a receptive female, (2) microinjection into the enlarged sperm nucleus (the male pronucleus) of a fertilized egg, or (3) introduction of genetically engineered embryonic stem cells into an early-stage developing embryo before implantation into a receptive female.

Retroviral Vector Method

Of the various gene transfer methods, the use of retroviral vectors (Fig. 19.1) has the advantage of being an effective means of integrating the transgene into the genome of a recipient cell. However, these vectors can transfer only small pieces (~8 kilobases [kb]) of DNA, which, because of the size constraint, may lack essential adjacent sequences for regulating the expression of the transgene.

There is a further major drawback to the use of retroviral vectors. Although these vectors are designed to be replication defective, the genome of the retroviral strain (helper virus) that is needed to create large quantities of the vector DNA can be integrated into the same nucleus as the transgene. Despite special precautions, helper strain retroviruses could be produced by the transgenic organism. Consequently, for applications in which either a commercial product is to be synthesized by the transgenic organism or the transgenic organism is to be used as food, it is absolutely necessary that there be no retroviral contamination. Therefore, because alternative methods are available, retroviral vectors are rarely used for creating transgenic animals that have a commercial end use.

DNA Microinjection Method

Because of the disadvantages of the retroviral vector method, microinjection of DNA is currently the preferred method for producing transgenic mice. This procedure is performed via the following steps (Fig. 19.2).

1. The number of available fertilized eggs that are to be inoculated by microinjection is increased by stimulating donor females to superovulate. They are given an initial injection of pregnant mare's serum and another injection, about 48 hours later, of human chorionic gonadotropin. A superovulated mouse produces about 35 eggs instead of the normal 5 to 10.
2. The superovulated females are mated and then sacrificed. The fertilized eggs are flushed from their oviducts.
3. Microinjection of the fertilized eggs usually occurs immediately after their collection. The microinjected transgene construct is often in a linear form and free of prokaryotic vector DNA sequences.

In mammals, after entry of the sperm into the egg, both the sperm nucleus (male pronucleus) and female nucleus exist separately. After the female nucleus completes its meiotic division to become a female pronucleus, nuclear fusion (karyogamy) can occur. The male pronucleus, which tends to be larger than the female pronucleus, can be located by using a dissecting

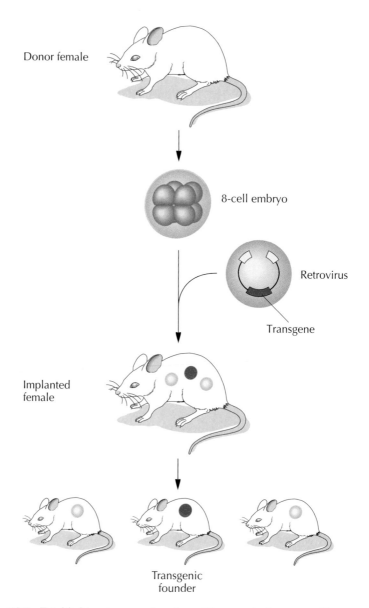

Donor female

8-cell embryo

Retrovirus

Transgene

Implanted
female

Transgenic
founder

Figure 19.1 Establishing transgenic mice with retroviral vectors. Cleavage stage embryos, usually the eight-cell stage, are infected with a defective retrovirus carrying a transgene. Implanted females (foster mothers) give birth to transgenic pups. Matings are carried out to determine which pups have the transgene in their germ line cells. Transgenic lines can be established from these founder transgenic animals.

microscope. The egg can then be maneuvered, oriented, and held in place while the DNA is microinjected. On a good day, several hundred male pronuclei can be inoculated.

After inoculation, 25 to 40 eggs are implanted microsurgically into a foster mother, which has been made pseudopregnant by being mated to a vasectomized male. In mice, copulation is the only known way to prepare the uterus for implantation. In this case, because the male mate lacks sperm, none of the eggs of the foster mother are fertilized. The foster mother will deliver pups from the inoculated eggs about 3 weeks after implantation.

To identify transgenic animals, DNA from a small piece of the tail can be assayed by either Southern blot hybridization or the polymerase chain

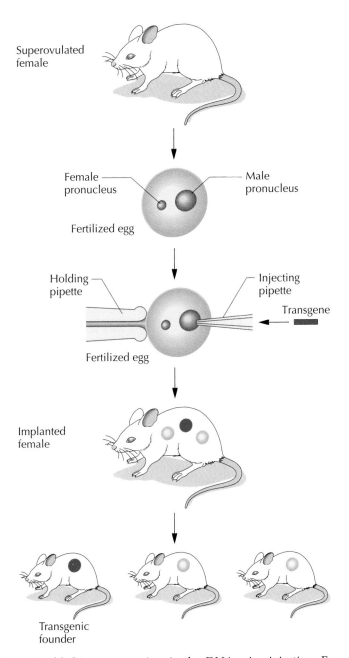

Superovulated
female

Female
pronucleus

Male
pronucleus

Fertilized egg

Holding
pipette

Injecting
pipette

Transgene

Fertilized egg

Implanted
female

Transgenic
founder

Figure 19.2 Establishing transgenic mice by DNA microinjection. Eggs are obtained from donor females that have been induced to superovulate and then mated with males. Purified samples of the transgene construct are microinjected into the male pronucleus of a fertilized egg. Implanted females (foster mothers) give birth to transgenic pups from which transgenic lines can be established.

reaction (PCR) for the presence of the transgene. A transgenic mouse can be mated to another mouse to determine if the transgene is in the germ line of the founder animal. Subsequently, progeny can be bred to generate pure (homozygous) transgenic lines.

The procedure, although apparently simple, requires the coordination of a number of experimental steps. Even a highly trained practitioner can expect, at best, only 5% of the inoculated eggs to develop into live trans-

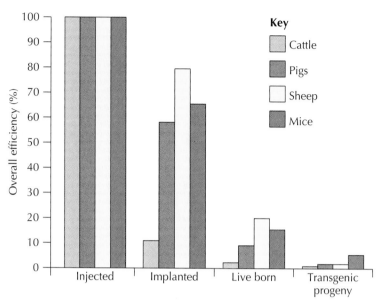

Figure 19.3 Overall efficiency of the transgenesis process after DNA microinjection. All the fertilized eggs (100%) of cattle, pigs, sheep, and mice are inoculated with a transgene, but the success of implantation and giving birth to offspring is much lower, and only 5% or fewer of the treated eggs become transgenic progeny.

genic animals (Fig. 19.3). None of the steps in the procedure is 100% efficient; consequently, large numbers of microinjected fertilized eggs must be used. With the mouse system, for example, about 66% of the fertilized eggs survive the injection procedure; about 25% of the implanted eggs develop into pups; and about 25% of the pups are transgenic. Thus, from 1,000 inoculated fertilized eggs, 30 to 50 transgenic pups are produced. Furthermore, with this method, the injected DNA integrates at random sites within the genome, and often multiple copies of the injected DNA are incorporated at one site. Not all of the transgenic pups will have the appropriate characteristic. In some individuals, the transgene may not be expressed because of the site of integration, and in others, the copy number may be excessive and may lead to overexpression, which disrupts the normal physiology of the animal. Notwithstanding the overall inefficiency, it has become routine to use DNA microinjection to generate lines of mice carrying functional transgenes.

Engineered Embryonic Stem Cell Method

Cells from the blastocyst stage of a developing mouse embryo can proliferate in cell culture and still retain the capability of differentiating into all other cell types—including germ line cells—after they are reintroduced into another blastocyst embryo. Such cells are called pluripotent embryonic stem (ES) cells. When in culture, ES cells can be readily engineered genetically without altering their pluripotency. With this system, for example, a functional transgene can be integrated at a specific site within a dispensable gene in the genome of ES cells. Then the genetically engineered cells can be selected, grown, and used to generate transgenic animals (Fig. 19.4). In this way, the randomness of integration that is inherent with the DNA microinjection and retroviral vector systems is avoided.

After transfection of ES cells in culture with a DNA vector that is designed to integrate within a specific chromosomal location, some cells will

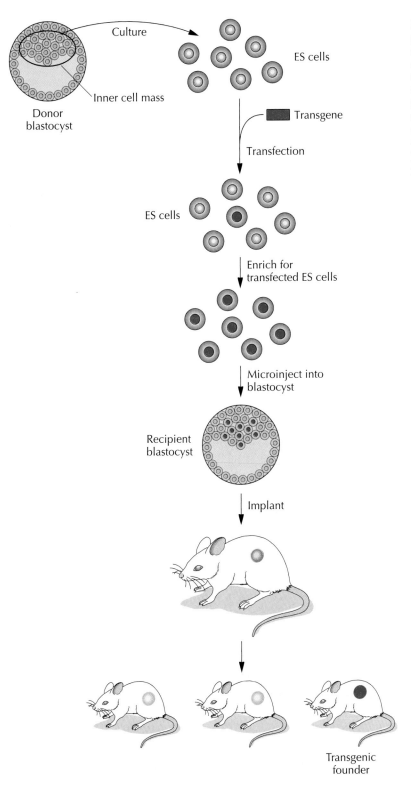

Figure 19.4 Establishing transgenic mice with genetically engineered embryonic stem (ES) cells. An ES cell culture is initiated from the inner cell mass of a mouse blastocyst. The ES cells are transfected with a transgene. After growth, the transfected cells are identified by either the positive-negative selection procedure or PCR analysis. Populations of transfected cells can be cultured and inserted into blastocysts, which are then implanted into foster mothers. Transgenic lines can be established by crosses from founder mice that carry the transgene in their germ lines.

Culture

ES cells

Inner cell mass

Donor blastocyst

Transgene

Transfection

ES cells

Enrich for transfected ES cells

Microinject into blastocyst

Recipient blastocyst

Implant

Transgenic founder

have DNA integrated at nontarget (spurious) sites, whereas in other cells, integration will occur at the target (correct) site. Moreover, in most of the ES cells, the input DNA will not be integrated at all. To enrich for the cells with DNA integrated at the target site, a procedure called positive-negative selection is implemented. This strategy uses positive selection for cells that have vector DNA integrated anywhere in their genomes and negative selection against the vector DNA sequence that is integrated at spurious sites.

The target site should be located in a section of genomic DNA that encodes no essential products, so that after integration of the input DNA, there is no interference with any developmental or cellular functions. Moreover, it is essential that the transgene be integrated into a part of the genome that does not prevent it from being transcribed. Researchers are continually looking for these kinds of sites.

A targeting DNA vector for the positive-negative selection procedure usually contains (1) two blocks of DNA sequences (HB1 and HB2) that are homologous to separate regions of the target site; (2) the transgene (TG), which will confer a new function on the recipient; (3) a DNA sequence that codes for resistance to the compound G-418 (Neo^r); and (4) two different genes for thymidine kinase (*tk1* and *tk2*) from herpes simplex virus types 1 and 2 (HSV-*tk1* and HSV-*tk2*) (Fig. 19.5A). The arrangement of these sequences is key to the positive-negative selection procedure. Between the two blocks of DNA that are homologous to the target site are the genes for the transgene and G-418 resistance (Neo^r gene). Outside of each of the homologous blocks are the genes for HSV-*tk1* and HSV-*tk2*. If integration occurs at a spurious site (that is, not at HB1 and HB2), either one or both of the HSV-*tk* genes has a high probability of being integrated along with the other sequences (Fig. 19.5A). Alternatively, if the integration event is due to homologous recombination by a double crossover at the target site, the HSV-*tk* genes will be excluded and only the transgene and the Neo^r gene will be incorporated into the genome (Fig. 19.5B). When transfected cells are grown in the presence of G-418, all the cells that lack the Neo^r gene will be killed. Therefore, only cells with integrated DNA will survive; i.e., these cells are positively selected. If the compound ganciclovir is added at the same time as G-418, the cells that express thymidine kinase will be killed because thymidine kinase converts ganciclovir to toxic compounds that kill cells; i.e., these cells are negatively selected. The cells most likely to survive this dual-selection scheme will be those that have DNA integrated at the target site. Although not foolproof, the positive-negative selection method enriches an ES cell population for cells that carry a transgene at a specific chromosomal location.

A more direct way to detect ES cells that carry a transgene at a targeted chromosomal site is to use PCR. In this case, the targeting DNA vector contains two blocks of DNA that are homologous to the target site, with one on either side of both the transgene and a cloned bacterial or synthetic (unique) DNA sequence that is not present in the mouse genome (Fig. 19.6). After transfection of ES cells, cells are pooled and samples are screened by PCR. One of the primers (P1) for PCR is complementary to a sequence within the cloned bacterial or synthetic (unique) DNA sequence of the integrating vector. The other (P2) is complementary to a DNA sequence that is part of the chromosome adjacent to the region of one of the homologous blocks of DNA. If integration is at a random site, the predicted amplified DNA product will not be synthesized (Fig. 19.6A). However, if site-specific integration occurs, the PCR amplifies a DNA fragment of known size (Fig. 19.6B). In

A

B

Figure 19.5 Positive-negative selection. **A.** Result of nonspecific integration. Both genes for thymidine kinase (*tk1* and *tk2*), the two DNA sequences that are homologous to a specific chromosomal region in the recipient cells (HB1 and HB2), a gene (Neor) that confers resistance to the cytotoxic compound G-418, and the transgene (TG) are incorporated into the chromosome. After transfection, cells are selected for resistance to both G-418 and the compound ganciclovir, which becomes cytotoxic to cells that synthesize thymidine kinase. Other nonhomologous integrations may occur and produce inserts with one or the other of the thymidine kinase genes. After treatment with G-418 and ganciclovir, all the cells with nonspecific integration of the input DNA that includes at least one of the thymidine kinase genes are killed. **B.** Result of homologous recombination. The product of a double crossover between homologous blocks (HB1 and HB2) of DNA on the vector DNA and on chromosomal DNA does not contain either of the two thymidine kinase genes (*tk1* and *tk2*). After treatment with G-418 and ganciclovir, only cells that have undergone homologous recombination survive.

this way, pools of cells with ES cells containing the desired gene at the targeted site can be identified. By subculturing from these pools, cell lines carrying the site-specific integration can be established.

ES cells carrying an integrated transgene can be cultured and inserted into blastocyst stage embryos, and these embryos can then be implanted into pseudopregnant foster mothers. From progeny in which the genetically engineered ES cells have contributed to the formation of germ line cells, transgenic lines can be established by first mating founder transgenic mice to animals from the same strain and then crossing trangenic littermates to create transgenic mice that are homozygous for the transgene.

Not only can a transgene be inserted into a specific chromosome site by homologous recombination in ES cells to provide a new function but also a specific mouse gene can be targeted for disruption by the incorporation of a DNA sequence, usually a selectable marker gene, into its coding region

Figure 19.6 Testing for nonspecific integration **(A)** and homologous recombination in transfected cells by PCR **(B)**. **A.** After nonspecific integration of the vector DNA, one of the primers (P2) is not able to hybridize to a chromosomal site that is a predetermined distance from the site of hybridization of P1, so that a DNA fragment with a specific size is not amplified. P1 hybridizes to a unique segment (US) of the input DNA that does not occur in the chromosomal DNA of the recipient cells. **B.** Homologous recombination between DNA sequences (HB1 and HB2) of the input DNA that are complementary to chromosomal sites (CS1 and CS2) creates hybridization regions for both P1 and P2 that are a predetermined distance apart. Amplification by PCR generates a DNA fragment of a specific size, which can be visualized by gel electrophoresis. In this case, the transgene (TG), which lies between the homologous blocks (HB1 and HB2), is integrated at a specific chromosomal location.

(Fig. 19.7). One of the aims of targeted gene disruption (gene knockout) is to determine the developmental and physiological consequences of inactivating a particular gene. Another is the hope that a transgenic line with a specific disabled gene can be used as a model system to study the molecular pathology of a human disease.

For example, inactivation of the mouse rhodopsin gene by targeted gene disruption leads to deterioration of the rod cells of the retina in transgenic mice that closely resembles the human disease retinitis pigmentosa. Thus, the progress of retinal degeneration and the effects of possible therapeutic agents that might either delay or block the genetically induced retinopathy can be studied by using the rhodopsin "knockout" mouse. To date, over 250 different types of knockout mice have been created as animal models for the study of various human abnormalities.

In principle, the procedures for establishing "gain-of-function" and "loss-of-function" transgenic animals are quite similar. Unfortunately, pluripotent ES cells comparable to those of the mouse have not yet been discovered for cattle, sheep, pigs, or chickens, but the search is under way.

Cloning by Nuclear Transfer

Pluripotency can be tested by tranferring a nucleus from a cultured cell to an enucleated egg and determining if the reconstituted ovum supports de-

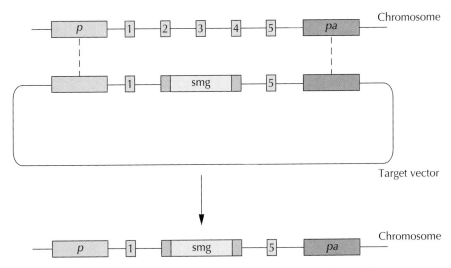

Figure 19.7 Gene disruption by targeted homologous recombinant. The target vector carries a selectable marker gene (smg) with flanking DNA sequences that are homologous to regions of the targeted gene. In this example, the targeted gene has five exons (1 to 5). Homologous recombination (dashed lines) disrupts (i.e., knocks out) the targeted gene.

velopment and the formation of a viable offspring. A number of laboratories have tested embryonic, fetal, and adult cell lines for pluripotency with varying success. Generally, although the efficiency is low, nuclei from embryonic cells support development to term. For example, viable cattle have been produced by nuclear transfer from short-term-cultured embryonic cells. In a highly publicized case, a sheep named Dolly was cloned by nuclear transfer of a mammary (udder) cell from an adult organism (Fig. 19.8). This was the first demonstration of pluripotency of a nucleus of a differentiated adult cell. There is the possibility, however, that Dolly's donor nucleus was from an undifferentiated cell that was present in the mammary epithelium of the source organism.

The development of Dolly from a differentiated cell nucleus and three other sheep from fetal cell nuclei is attributed to the transfer of nuclei from cells that have been induced to enter the quiescent stage (G_0) of the cell division cycle and, perhaps, to features that are specific to sheep embryogenesis. During the first three cell divisions of the sheep ovum, which takes several days, the DNA is replicated and no genes are expressed. It is thought that during these cell divisions the transferred DNA is stripped of cell-specific regulatory proteins and the appropriate embryonic development genes are bound by embryo-initiating protein factors from the cytoplasm of the egg.

If nuclear transfer is to be used for generating transgenic livestock, the key problem that must be resolved is whether pluripotency can be maintained during continuous cell culture. If it can be, it is a straightforward matter of genetically engineering these cells and creating transgenic organisms. However, because of species differences in the timing of early embryonic cell divisions and the onset of transcription during embryogenesis, it is not clear at this time whether matching the state of the donor nucleus with that of the ovum will lead to successful nuclear transfers with livestock other than sheep.

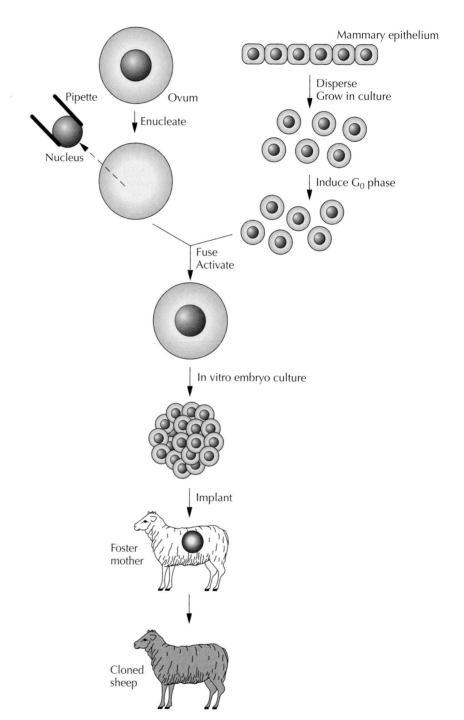

Figure 19.8 Cloning sheep by nuclear transfer. The nucleus of an ovum is removed (dashed arrow) with a pipette. Cells from the mammary epithelium of an adult are grown in culture and the G_0 state is induced by inhibiting cell growth. A G_0 cell and an enucleated ovum are fused, and the renucleated ovum is grown in culture or ligated oviducts until an early embryonic stage before it is implanted into a foster mother, where development proceeds to term. In the experiment described by Wilmut et al. (1997), 277 enucleated ova were fused with G_0 mammary cells and 1 of 29 transferred early-stage embryos produced a live lamb.

Yeast Artificial Chromosome Transgenesis

Generally, transgenes are cDNAs, small genes (≤20 kb), or parts of genes. Often, cDNAs are poorly expressed in mammalian cells. As well, when genomic DNA is used for transgenesis, important gene-specific regulatory sequences that lie either upstream or downstream of the gene are rarely retained as part of the insert. Moreover, complete genes and multigene complexes (≥100 kb) are too large for conventional vectors. For these reasons, yeast artificial chromosomes (YACs) that carry genomic DNA ranging in size from 100 to >1,000 kb have been used for transgenesis.

Transgenic mice have been produced by microinjection of the pronucleus of the fertilized egg or transfection of ES cells with YACs that carry either a number of related genes or a single large gene. For example, transgenic mice that carry the human β-globin gene cluster, which has five different functional globin genes and covers about 250 kb, express the genes of the β-globin ensemble in a tissue-specific and time-dependent manner that corresponds to the site-specificity and temporal pattern in humans. This concurrence is attributed to the flanking DNA that contains the promoter region and other important regulatory elements.

The production of mice that synthesize only human antibodies is a noteworthy example of YAC transgenesis. As noted in Chapter 10, monoclonal antibodies have the potential to be used as human therapeutic agents. However, it is impossible to generate human monoclonal antibod-

MILESTONE

Genetic Transformation of Mouse Embryos by Microinjection of Purified DNA
J. W. Gordon, G. A. Scangos, D. J. Plotkin, J. A. Barbosa, and F. H. Ruddle
Proc. Natl. Acad. Sci. USA **77:**7380–7384, 1980

Somatic Expression of Herpes Thymidine Kinase in Mice following Injection of a Fusion Gene into Eggs
R. L. Brinster, H. Y. Chen, M. Trumbauer, A. W. Senear, R. Warren, and R. D. Palmiter
Cell **27:**223–231, 1981

Gordon et al. were the first to show the feasibility of DNA transfer by microinjection into the pronucleus of mouse eggs. In their study, the procedure was tested by microinjecting several hundred eggs with a vector–gene construct that consisted of pBR322 carrying both the herpes simplex virus (HSV) thymidine kinase gene and a piece of the simian virus 40 genome. Of the 78 offspring that were obtained from the surrogate mothers, 2 retained some plasmid DNA. The authors concluded, "These data demonstrate that it is possible to use a recombinant plasmid as a vector to transfer foreign genes directly into mouse embryos, and that these embryos can maintain the foreign genes throughout development." Unfortunately, the plasmid DNA was not intact and the HSV thymidine kinase sequence did not become a transgene.

On the other hand, Brinster et al., who microinjected the pronuclei of a number of mouse eggs with a plasmid carrying the HSV thymidine kinase gene under the control of the promoter of the metallothionein-I gene, found that one of their transgenic mice expressed HSV thymidine kinase at a high level in its liver and kidney in comparison to three other transgenic mice, which produced low levels of this enzyme. As well, eight other transgenic animals carried the HSV thymidine kinase sequence but did not produce any active HSV thymidine kinase. Southern blot analysis showed that all of the transgenic mice contained multiple copies of the microinjected DNA.

These two studies laid the foundation for transgenesis of mice. Despite the technical complexity and relative inefficiency of the microinjection strategy, it has been exceptionally successful. Currently, scores of strains of mice with either foreign genes (transgenic mice) or endogenous genes that have been disrupted by the insertion of foreign DNA (knockout mice) are being used for studying gene regulation, mammalian development, viral pathogenesis, cancer, toxicology, and the mutagenicity of various agents, among other things. In addition, transgenic and knockout mice are of considerable biomedical importance as model systems for human diseases.

ies routinely. Also, unfortunately, rodent monoclonal antibodies are immunogenic to humans. Elaborate recombinant DNA strategies have been devised to "humanize" existing rodent monoclonal antibodies. These labor-intensive procedures produce Fv and Fab antibodies, which sometimes have a low affinity for a particular antigen. A more accessible method for generating complete human antibodies by using hybridomas would be a welcome technological advance.

The formation of natural antibodies is a biological marvel. An antibody is a tetrameric protein with two pairs of dissimilar chains. One of the chains is called the heavy (H) chain, and the other is a light (λ or κ) chain. The terms "heavy" and "light" refer to the difference of the molecular masses of the antibody subunits. The genetic information for a specific heavy chain is created by somatic DNA rearrangements of variable (V_H), diversity (D_H), joining (J_H), and constant (C_H) DNA segments (domains) in a B cell. There are two different types of antibody light chains (λ and κ) that are encoded after DNA rearrangements of their own variable ($V\lambda$, $V\kappa$); joining ($J\lambda$, $J\kappa$), and constant ($C\lambda$, $C\kappa$) domains. A single B cell synthesizes one kind of antibody molecule that has a unique set of combined segments that encodes an H chain and either a rearranged λ or κ chain.

The genetic repertoire for the formation of the vast numbers of different human H antibody chains consists of about 95 V_H segments, 30 D_H segments, 6 J_H segments, and 5 main constant ($C\alpha$, $C\gamma$, $C\delta$, $C\epsilon$, $C\mu$) domains. The κ gene locus has about 76 $V\kappa$ segments, 5 $J\kappa$ segments, and 1 constant ($C\kappa$) sequence (Fig. 19.9). The H and κ gene loci are each about 1 to 1.5 Mb in length. To create a transgenic mouse that is capable of synthesizing an extensive array of different human antibodies, the endogenous mouse H and light chain genes must be inactivated and YACs with a large number of H and light chain DNA elements from each human immunoglobulin gene must be inserted into the chromosome DNA of this mouse.

Figure 19.9 Schematic representation of the κ and H human immunoglobulin genes. **A.** Germ line version of the κ chain immunoglobulin gene. The dashed line represents a contiguous set of domains that are not shown. A functional κ chain gene such as Vκ8-Jκ4-Cκ is formed in a B cell after a series of DNA rearrangements that combine these immunoglobulin DNA domains. This final rearranged gene represents 1 of about 500 possibilities. **B.** Germ line version of the H chain immunoglobulin gene. The dashed lines represent contiguous sets of domains that are not shown. A functional H chain gene such as V_{H33}-D_{H26}-J_{H4}-Cα is formed in a B cell after a series of DNA rearrangements that combine these immunoglobulin DNA domains. This final rearranged gene represents 1 of about 140,000 possibilities. Although only one $C\gamma$ domain is shown here, there are, in reality, four types of $C\gamma$ domains ($C\gamma$1, $C\gamma$2a, $C\gamma$2b, and $C\gamma$3).

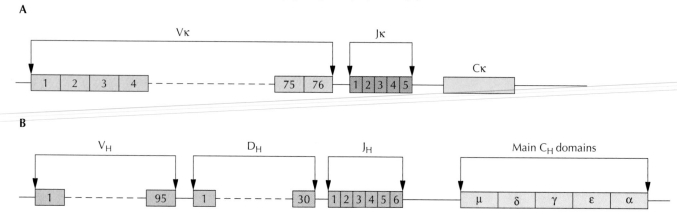

Initially, to meet this end, the mouse H and κ chain genes were replaced (knocked out) by a small portion of the human H chain gene cluster that included 4 V_H segments, 16 D_H segments, 6 J_H segments, $C\gamma$, and $C\mu$ and a human κ gene cluster that had 4 $V\kappa$ segments, 5 $J\kappa$ segments and $C\kappa$, respectively. Transgenic mice with this set of human antibody genes produced human antibody against some antigens, and hybridomas that secreted human monoclonal antibodies were generated. However, the variety of human antibodies produced by these transgenic mice was meager due to the limited set of H and κ chain variable segments. This drawback was overcome by constructing YACs with more human H and κ variable segments.

A 1,000-kb YAC that carries 66 V_H segments, about 30 D_H segments, 6 J_H segments, $C\mu$, $C\delta$, and $C\gamma$ was formed by combining segments from four other human H chain YACs. Similarly, an 800-kb YAC with 32 $V\kappa$ segments, 5 $J\kappa$ segments, and $C\kappa$ was created from three YACs that carried different $V\kappa$ segments. The YACs with H and κ chain segments were transfected separately into ES cells by cell fusion, cells with integrated YAC DNA were selected with a selectable marker, and the integrity of each insert was established by PCR assays. Blastocysts were injected separately with cells carrying either the H or κ chain inserts, and founder organisms were identified by PCR assays. Transgenic mice with H and κ chain inserts were bred separately to mice with inactivated mouse H and κ chain loci. Then the offspring were mated to each other to form mice without functional mouse H chain and κ genes that carried both of the human H and κ chain inserts.

The transgenic mice with increased numbers of human V_H and $V\kappa$ segments produce human antibodies. As a test, these transgenic mice were immunized with three different antigens, and in each case, hybridomas secreted a complete human monoclonal antibody that had a strong affinity for the immunizing antigen. It is very likely that this transgenic system will be used to generate human therapeutic monoclonal antibodies.

Transgenic Mice: Applications

Transgenic mice can be used both as model systems for human diseases and as test cases to determine if the production of a potential therapeutic agent is feasible. Whole-animal models simulate both the onset and progress of a human disease. However, a mouse is not a human, even though it is a mammal, and so the information gathered from transgenic models is not always medically relevant. In other instances, however, critical insights about the etiology of a complex disease can be discovered. With this in mind, mouse models for human genetic diseases such as Alzheimer disease, arthritis, muscular dystrophy, tumorigenesis, hypertension, neurodegenerative disorders, endocrinological dysfunction, coronary disease, and many others have been developed.

Alzheimer disease is a degenerative brain disorder that is characterized by the progressive loss of both abstract thinking and memory and is accompanied by personality change, language disturbances, and a slowing of physical capabilities. Clinical diagnosis of Alzheimer disease is poor, although 1% of the population between 60 and 65 years old and 30% of the population over 80 years old may develop it. In the neocortex and hippocampus of the brain of patients with Alzheimer disease, neurofibrillary tangles accumulate within the cell body of the neurons, dense extracellular aggregates called senile plaques develop at the ends of neurites, and brain

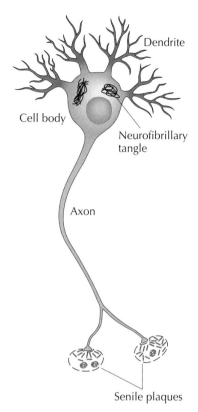

Dendrite

Cell body

Neurofibrillary
tangle

Axon

Senile plaques

Figure 19.10 Schematic representation of a neuron of the human cerebral cortex, showing some of the histopathological features of Alzheimer disease. Senile plaques containing amyloid deposits and apparent cellular debris accumulate at the synapses of the neuron. Within the cell body of the neuron, neurofibrillary tangles contain aggregated cytoskeletal and other proteins. Other changes that occur in the neurons are not depicted here.

cells (neurons) are lost (Fig. 19.10). In addition, aggregates called amyloid bodies are found in the blood vessels of the brain.

The principal component of senile plaques and amyloid bodies is a 4-kilodalton (kDa) protein called Aβ (amyloid β, β-protein, β-amyloid protein, β/A4). There are Aβ proteins with different numbers of amino acid residues, such as Aβ40 and Aβ42. All Aβ proteins are products of an internal proteolytic cleavage of the β-amyloid precursor protein (APP). It is not known what causes the accumulation of Aβ. Some families that have a high incidence of Alzheimer disease have mutations in the APP gene, a finding that implicates this gene in Alzheimer disease. Unfortunately, it is for the most part impossible to study the onset and pathogenesis of Alzheimer disease in human subjects. Accordingly, an animal model that mimics Alzheimer disease would be an invaluable research tool.

A number of transgenic mice have been generated with full or partial versions of the normal APP gene driven by neuron-specific promoters. In most cases, amyloid plaques, neurofibrillary tangles, neuronal cell death, or disturbed behavior was not observed. However, neurodegeneration that is similar to Alzheimer disease does occur with a transgene that is made up of the coding region for the terminal 100 amino acids of APP, which contains the Aβ protein sequence.

More convincing animal models for Alzheimer disease were created with transgenes that contain mutations of the APP gene that occur in some families with a high incidence of early onset ($\leq$50 years of age) of Alzheimer disease. In one set of these families, site 717 of APP (APP-717) contains phenylalanine instead of valine. In another group of families with

Alzheimer disease, sites 670 and 671 of APP (APP-670/671) contain asparagine and leucine instead of lysine and methionine.

A transgene with the APP-717 mutation was constructed from an APP cDNA. Modified introns were added between exons 6 and 7, 7 and 8, and 8 and 9 of the APP cDNA. The introns were introduced into the APP cDNA because experiments have shown that transgenes with introns have an increased rate of transcription in comparison to constructs without introns. The "APP cDNA–introns" construct is controlled by the platelet-derived growth factor β promoter, which is expressed in brain tissue (Fig. 19.11). The complete construct is called the PDAPP minigene. Aging transgenic mice (>6 months old) with about 40 copies of the PDAPP minigene display amyloid plaques, neuronal cell death, and memory defects. An APP-670/671 gene construct which is driven by a brain-specific promoter also produces in transgenic mice Alzheimer disease-like features including an excess of Aβ42. Interestingly, neither aging PDAPP minigene mice nor APP-670/671 transgenic mice have neurofibrillary tangles. Possibly, these structures are a secondary response to the overproduction of Aβ42 in humans.

Three other genes—*ApoE4*, presenilin 1 (*PS1*), and presenilin 2 (*PS2*)—have been implicated in Alzheimer disease in humans. The presence of the *ApoE4* allele of the *ApoE* locus, which plays a role in lipid transport, increases the likelihood that individuals who are at least 60 years old will develop Alzheimer disease. Mutations of the presenilin genes are found in families that show very early onset of Alzheimer disease. The actual role in the development of Alzheimer disease for each of these genes is problematic. Various studies indicate that the mutated presenilin genes increase the accumulation of Aβ42. For example, doubly transgenic mice that were generated by mating transgenic mice that carry a full-length human APP gene and those with a presenilin 1 mutation overproduce Aβ42. While there is much to learn about Alzheimer disease, the availability of an animal model will help answer some critical questions about the molecular basis of a disorder that affects about 4 million people in the United States at an annual cost of about $100 billion.

Transgenic mice have also been used as models for expression systems that are designed for secretion of the product of a transgene into milk. For example, large quantities of authentic cystic fibrosis transmembrane regulator (CFTR) protein are needed to study its function and to formulate potential therapies for treating cystic fibrosis (CF). This prevalent genetic disease affects about 1 in 2,500 live-born persons of European origin. The primary effect of a faulty CF gene is an alteration in the function of CFTR, which normally acts as a chloride channel. As a consequence of the disruption of the proper flow of chloride ions into and out of cells, mucus accu-

Figure 19.11 DNA construct for modeling Alzheimer disease in transgenic mice. The exons of the amyloid precursor protein cDNA are marked by numbers (1 to 18), and the introduced introns are indicated by letters (A to C). The regulatory sequences are the platelet-derived growth factor β promoter and simian virus 40 poly(A) sequence. The construct is called the PDAPP minigene.

Figure 19.12 Goat β-casein gene–CFTR cDNA expression construct. The full-length cDNA for CFTR was cloned between exon 2 (EX2) and exon 7 (EX7) of the goat β-casein gene. The promoter and transcription termination sequences and exons 1, 8, and 9 (EX1, EX8, and EX9) of the casein gene are retained.

mulates in the ducts of several organs, especially the lungs and pancreas. This mucus becomes the site of a bacterial infection that is difficult to control with antibiotics; as the bacteria die, their released DNA makes the mucus very thick. This thickened mucus prevents normal organ function by blocking the ducts and thereby exacerbates the effects of CF. Currently, the life expectancy for persons with CF is about 25 to 30 years.

A reliable supply of CFTR is required to learn more about its mode of action. With various in vitro cell expression systems, the yields of CFTR have been low, possibly as a result of the biological consequences of the accumulation in the cell membrane of transfected cells of CFTR, which is a transmembrane protein. This limitation would be remedied if plasma membranes were frequently shed from the host cell. With such a system, not only would a heterologous transmembrane protein be associated with the released fragments of plasma membrane but also concentrating and purifying the protein would be relatively easy. In fact, during lactation, the mammary gland cells use this procedure for the production of fat globules. In this instance, fat from within the mammary gland cell is encapsulated by plasma membrane and together they are secreted into milk as a globule.

To test the feasibility of this system, a full-length CFTR cDNA sequence was cloned into the middle of a defective goat β-casein gene from which the end of exon 2 to the beginning of exon 7 had been deleted (Fig. 19.12). The construct kept the promoter and termination sequences of the goat β-casein gene. The CFTR cDNA was cloned into a structural gene to provide introns for enhancing transcription of the transgene. The β-casein gene is actively expressed in mammary glands during lactation, and β-casein is a major milk protein.

Transgenic mouse lines carrying the CFTR sequence under the control of the β-casein gene regulatory sequences were established. As predicted, the milk of transgenic females contained the CFTR protein bound to the membrane of fat globules. There were no negative effects to either CFTR-transgenic lactating mothers or pups that were fed milk that contained CFTR. The CFTR protein was glycosylated and readily extracted from the fat-rich fraction of the milk. It remains to be determined whether the CFTR is authentic. However, the possibility of producing other membrane-bound proteins in milk has been established. Many other proteins that are potentially therapeutic for humans have also been synthesized by the mammary gland cells of lactating transgenic mice. Eventually, to obtain large quantities of CFTR, additional transmembrane proteins, and various human therapeutic proteins, the transgenic constructs will have to be incorporated into the genome of a much larger mammal such as a cow, sheep, or goat.

Transgenic Cattle

If the mammary gland is to be used as a bioreactor, dairy cattle, which each annually produce approximately 10,000 liters of milk, containing about 35 grams of protein per liter, are likely candidates for transgenesis. More specifically, if a recombinant protein was present at 1 gram per liter of milk and it could be purified with 50% efficiency, the yield from 20 transgenic cows would be about 100 kg per annum. Coincidentally, the annual global requirement for protein C, which is used for the prevention of blood clots, is about 100 kg. On the other hand, one transgenic cow would be more than sufficient for the production of the annual world supply of factor IX (plasma thromboplastin component), which is used by hemophiliacs to facilitate blood clotting.

A modified mouse transgenesis DNA microinjection protocol has been developed for cattle (Fig. 19.13). In brief, the essential steps entail

1. collecting oocytes from slaughterhouse-killed cows
2. in vitro maturation of oocytes
3. in vitro fertilization with bull semen
4. centrifugation of the fertilized eggs to concentrate the yolk, which in normal eggs prevents the male pronuclei from being readily seen under a dissecting microscope
5. microinjection of input DNA into male pronuclei
6. in vitro development of embryos
7. nonsurgical implantation of one embryo into one recipient foster mother in natural estrus
8. DNA screening of the offspring for the presence of the transgene

When this nonsurgical procedure was put to the test, two transgenic calves were obtained from a starting pool of 2,470 oocytes, a result indicating that the methodology is feasible although inefficient. As research in this area continues, the transgenesis procedure will be refined. For example, it should soon be possible to take a small number of cells from a developing embryo in vitro and assay for the transgene, because this loss of cells does not interfere with normal development. This test will ensure that only embryos carrying the transgene are implanted.

One of the goals of transgenesis of dairy cattle is to change the constituents of milk. The amount of cheese produced from milk is directly proportional to the κ-casein content. Increasing κ-casein production with an overexpressed κ-casein transgene is a reasonable likelihood. For a different end use, expression of a lactase transgene in the mammary gland could result in milk that is free of lactose. Such a milk would be welcomed by the many people who are lactose intolerant and experience severe indigestion after the consumption of milk or milk-containing foods. Although transgenesis of cattle holds promise, progress toward producing large numbers of genetically engineered animals will be slow because it takes about 2 years to produce a calf from a fertilized egg.

For livestock, in general, attempts will be made to create animals with inherited resistance to bacterial, viral, and parasitic diseases. For example, genetic resistance to bacterial diseases—such as mastitis (mammary gland abscesses) in dairy cattle, neonatal scours (dysentery) in swine, and fowl cholera—occurs in some breeds. If the basis of each of these resistances is a single gene, it may be possible to create transgenic animals that are resistant to these bacterial infections. Currently, infectious diseases of domestic ani-

Figure 19.13 Steps in the development of transgenic cattle.

mals are controlled by vaccination, drugs, physical isolation, and careful monitoring. The cost of disease prevention can be as much as 20% of the total production value.

Another approach that may be used to develop lines of animals that are resistant to infectious agents entails creating inherited immunological protection by transgenesis. A number of candidate genes that contribute to the immune system (e.g., major histocompatibility genes, T-cell receptor genes, and lymphokine genes) are being studied from this perspective. However, the most favorable preliminary results to date have come from research in which the genes encoding the H and light (L) chains of a monoclonal antibody have been transferred to mice, rabbits, and pigs. The rationale behind this strategy is to provide a built-in, inherited biological protection mechanism for the transgenic animal that gets rid of the need for immunization by vaccination.

The concept of introducing into a recipient organism the transgenes for an antibody that binds to a specific antigen has been called in vivo immunization. To test this idea, the genes for the immunoglobulin chains of a mouse monoclonal antibody against an antibody that binds to 4-hydroxy-3-nitrophenylacetate were cloned in tandem and microinjected into fertilized eggs of mice, rabbits, and pigs. In each case, monoclonal antibody activity was found in the serum of the transgenic animals. However, the level of monoclonal antibody containing both of the cloned H and L chains was low. Different transgene constructs will have to be examined to determine whether this problem can be resolved.

Transgenic Sheep, Goats, and Pigs

Transgenesis research with sheep and goats has, for the most part, been devoted to developing the mammary glands of these organisms as bioreactors for the production of pharmaceutical proteins. Although the quantity of milk produced by either a sheep or a goat is smaller than that produced by a cow, lactation in sheep and goats yields hundreds of liters of milk per year. With a method very similar to the one used for producing transgenic mice and with transgene constructs that have mammary gland-specific promoters driving human gene sequences (Table 19.2), investigators have created both transgenic sheep and goats that secrete human proteins into milk. The transgene-derived proteins were glycosylated and had biological activities comparable to those extracted from human sources. However, more detailed studies must be carried out to determine whether the proteins produced in animal mammary glands are equivalent to the native forms.

Expression of transgenes in the mammary glands of sheep or goats had no ill effects on either lactating females or nursing progeny. In contrast, when the transgene for bovine growth hormone—under the control of the metallothionein promoter—was introduced into pigs, adverse results were observed. Although the levels of growth hormone varied among these transgenic pigs, the group as a whole showed an increased ability to convert feed into body weight. However, this positive response was offset by pathological disturbances that included gastric ulceration, kidney dysfunction, lameness, inflammation of the lining of the heart, immobility of the joints, and susceptibility to pneumonia. The reasons for these symptoms are not known. Possibly, they were caused by long-term exposure to excess growth hormone. In these experiments, the transgene was produced more or less continuously. These studies show that no matter how appealing a concept

Table 19.2 Mammary gland transgenes, promoter sequences, and recipient organisms

Transgene	Promoter	Transgenic species
Longer-acting tissue plasminogen activator	Whey acidic protein	Goat
α_1-Antitrypsin	β-Lactoglobulin	Sheep
Clotting factor IX	β-Lactoglobulin	Sheep
Soluble CD4 protein	Whey acidic protein	Mouse
Lactoferrin	α_{s1}-Casein	Cattle
Urokinase	α_{s1}-Casein	Mouse
CFTR	β-Casein	Mouse
Interleukin-2	β-Casein	Rabbit

may be, there is no substitute for experimentation. In contrast to the experience with the growth hormone transgene in pigs, improved wool production due to enhanced growth of fleece in transgenic sheep was achieved by the overexpression of a construct that consisted of a mouse ultra-high sulfur keratin promoter driving a sheep insulin-like growth factor 1 cDNA without any apparent negative effects on the health of the animals.

As well, there have been successful experimental results with transgenic pigs. For example, with a construct that consisted of the regulatory region from the human β-globin gene joined to two human α_1-globin genes and one human β^A-globin gene, healthy transgenic pigs that expressed human hemoglobin in their blood cells were produced. After the human β-globin promoter was replaced with the pig β-globin promoter, higher levels of human hemoglobin were produced. By various chemical criteria, the human hemoglobin from the transgenic pigs had the same properties as natural human hemoglobin. As well, the transgenic human hemoglobin was separated from the pig hemoglobin by a simple chromatographic procedure.

Although preliminary, these results point to the possibility of replacing whole blood that is used for transfusions with human hemoglobin produced by transgenesis. However, free hemoglobin may not be the answer, because it does not conduct oxygen exchange as effectively as hemoglobin within erythrocytes in whole blood. Free hemoglobin also decomposes after administration to animals, and its breakdown products cause kidney damage. Thus, the formulation of a human blood substitute by transgenesis may be a long way off.

Animals are a potential source of organs for transplantation into humans, especially since human organs are generally in short supply. The significant impediment to organ transplantation between species is hyperacute rejection of the animal organ. Hyperacute rejection entails the binding of preexisting antibodies of the host organism to a carbohydrate epitope on the surface of the cells of the grafted organ. The bound antibodies elicit an inflammatory response (complement cascade) that destroys the antibody-coated cells and leads to the rapid loss of the transplanted organ.

Under natural conditions, proteins on the surface of the cells lining the blood vessels protect the cells from the inflammatory response. These complement-inhibiting proteins are species specific. Therefore, it was reasoned that if the donor animal carried one or more of the genes for a human complement-inhibiting protein, a transplanted organ would be protected from the initial inflammatory response. To this end, transgenic pigs with differ-

ent human complement inhibitor genes have been produced. Cells from one of these transgenic pigs were not affected to any great extent by components of the complement cascade system. In preliminary experiments, organs that were transplanted from transgenic pigs into primates showed reduced tissue damage, and organ rejection was delayed slightly. In the future, transgenic pigs with a human complement inhibitor gene(s) and possibly without the major pig cell surface protein that is responsible for hyperacute rejection may be a source of organs for transplantation into humans.

Transgenic Birds

Microinjection of DNA into fertilized eggs of birds to produce transgenic strains is not a simple procedure because of several features that are unique to avian reproduction and development. For example, during fertilization in birds, several sperm can penetrate the ovum instead of one, as usually occurs in mammals. Consequently, it is impossible to identify the male pronucleus that will fuse with the female pronucleus. Also, microinjection of DNA into the cytoplasm will not suffice, because this DNA does not become integrated into the genome of the fertilized egg. Finally, even if nuclear DNA microinjection were practicable, the technique would be difficult to implement because the avian ovum after fertilization becomes, in rapid succession, enveloped in a tough membrane, surrounded by large quantities of albumin, and enclosed in inner and outer shell membranes.

Notwithstanding these disadvantages, it is possible to inject a transgene into the region on the yolk (germinal disc) that contains the female and male pronuclei. The germinal disc is present before the eggshell is formed. After the administration of DNA to a germinal disc, each egg is cultured in vitro, and when an embryo forms, it is placed in a surrogate egg to produce a hatchling. One transgenic line of chickens has been established by this strategy. However, at present this method is inefficient and technically difficult to carry out on a routine basis.

By the time an avian egg outer shell membrane has hardened, the developing embryo (blastoderm stage) comprises two layers of 40,000 to 80,000 cells. In trial experiments, inoculation of the blastoderm stage with replication-defective retrovirus vectors carrying bacterial marker genes resulted in the production of germ line transgenic chickens and quail. Although in these cases some of the transgenic organisms did not produce virus, the use of retrovirus vectors to deliver genes for a product that is to be used as food would ultimately raise questions about its safety, whether real or imagined. Moreover, the size of the transgene that can be introduced into the recipient organism by retrovirus vectors is limited to ~8 kb, and, occasionally, integration at the initial site is not permanent. Consequently, alternative methods of transgenesis have been examined.

At present, no one has identified avian-specific ES cells; therefore, this approach cannot be used in birds. However, a procedure involving engineered cells from embryos offers possibilities. Blastoderm cells can be removed from a donor chicken, transfected with cationic lipid (liposome) transgene DNA complexes (lipofection), and reintroduced into the subgerminal space of embryos of freshly laid eggs (Fig. 19.14). Some of the progeny will consist of a mixture of cells, some of which are from the donor but most of which are from the recipient. Such a genetic mixture is called a chimera. In some of the chimeras, cells that were descended from trans-

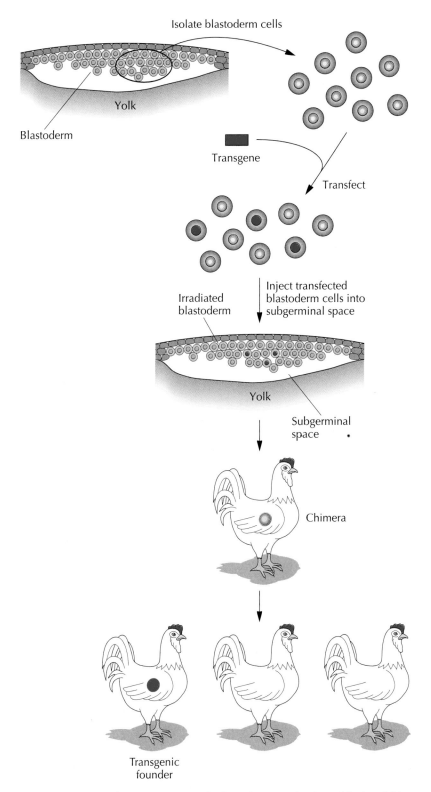

Figure 19.14 Establishing transgenic chickens by transfection of isolated blasto-derm cells. Cells from blastoderm donors are removed, transfected by lipofection with a transgene, and inserted into the subgerminal space of an irradiated recipient blastoderm. Some of the resulting chickens may be chimeric, and the chimeras that have the transgene in germ line cells can be bred to establish transgenic lines.

fected cells may become part of the germ line tissue and form germ cells. Transgenic lines can then be established from these chimeras by rounds of matings. The proportion of donor cells in chimeras can be increased to enhance the probability of obtaining germ line chimeras if the recipient embryos are irradiated with a dose of 540 to 660 rads for 1 hour before the introduction of the transfected cells. The radiation treatment destroys some but not all the blastoderm cells, thereby increasing the final ratio of transfected cells to recipient cells. It should be possible to produce transgenic chickens in this way, despite the inefficiency of the procedure.

Transgenic chickens could be used to improve the genetic makeup of existing strains with respect to built-in (in vivo) resistance to viral and coccidial diseases, better feed efficiency, lower fat and cholesterol levels in eggs, and better meat quality. Avian researchers have also suggested that the egg, with its high protein content, could be used as a source for pharmaceutical proteins. By analogy to the mammary gland of sheep, goats, and cows, the expression of a transgene in the cells of the reproductive tract of a hen that normally secretes large amounts of ovalbumin could lead to the accumulation of a transgene-derived protein that becomes encased in the eggshell. The pharmaceutical protein could then be fractionated from these natural storage cases.

Transgenic Fish

As natural fisheries become exhausted, production of this worldwide food resource will depend more heavily on aquaculture. Altering fish genetically by transgenesis is therefore a primary research objective. To date, transgenes have been introduced by microinjection or electroporation of DNA into the fertilized eggs of a number of fish species including carp, catfish, trout, salmon, and tilapia. In fish, the pronuclei are not readily seen under a microscope after fertilization; therefore, linearized transgene DNA is microinjected into the cytoplasm of either fertilized eggs or embryos that have reached the four-cell stage of development. Unlike mammalian embryogenesis, fish egg development is external; hence, there is no need for an implantation procedure. Development can occur in temperature-regulated holding tanks. The survival of fish embryos after DNA microinjection is high (35 to 80%), and the production of transgenic fish ranges from 10 to 70%. The presence of a transgene can be scored by PCR analysis of either nucleated erythrocyte samples or scale DNA. Founder fish can be mated and transgenic lines can be established.

Many of the initial studies with transgenic fish have focused on examining the effect of a growth hormone transgene on growth rate. In one study, a transgene consisting of the promoter region from the antifreeze protein gene of the fish called ocean pout, the growth hormone cDNA from salmon, and the termination–polyadenylation signals from the 3′ end of the antifreeze protein gene from the ocean pout was injected into eggs of Atlantic salmon. In general, the transgenic salmon were larger and grew faster than the nontransgenic controls. The expression system in this case was chosen to enhance the transcription of the growth hormone in cold waters and to be an "all-fish" construct to avoid possible biological incompatibilities that might arise from using a growth hormone gene from nonfish sources. In this context, sockeye salmon eggs were microinjected with an "all-salmon" growth hormone construct, and after about 1 year the transgenic salmon were about 11 times heavier than the nontransgenic salmon.

The physiological performance of these transgenic salmon lines under natural conditions will be of considerable interest. It is contemplated that genes for disease resistance, tolerance to environmental stress, and other biological features will eventually be introduced into both temperate and tropical fish.

SUMMARY

Genetic modification of animals by recombinant DNA technology (transgenesis) entails the introduction of a cloned gene(s) into the genome of a cell that might, after proliferation, be present in the germ lines of developed organisms, so that subsequently it may be possible to establish true-breeding lineages. Much of the research in this area has centered on developing transgenic mice. The commonly used protocol includes introducing a cloned gene into a fertilized egg by microinjection, implanting the treated fertilized egg into a receptive female, and then testing the offspring to determine whether any of them have the input gene in their cells. A cloned gene also can be delivered into a fertilized mouse egg by using retroviral vectors. Alternatively, mouse embryonic stem cells can be isolated and transfected with the cloned gene. In this method, the input DNA construct is formulated to integrate into the genome of embryonic stem cells. Subsequently, cells with the cloned gene at a specific chromosomal site are cultured, selected, and introduced into early mouse embryos. Mouse embryonic stem cells are pluripotent (totipotent) and consequently can develop into any type of cell including germ line tissue. Yeast artificial chromosomes (YACs) that carry large gene arrays have been used for transgenesis. A mouse that produces only human antibodies has been created by YAC transgenesis. Transgenic mice have been used to develop model systems for human genetic diseases such as Alzheimer disease.

By using similar experimental strategies, transgenic versions of cattle, sheep, goats, pigs, birds, and fish have been generated. It is hoped that transgenesis can be used to enhance existing genetic properties of livestock and provide the genetic basis for novel features. In addition, it is anticipated that the mammary glands—especially those of cows, sheep, and goats—will be used as a biological factory for cloned gene products that can be readily purified in large quantities from milk.

References

Brinster, R. L., J. M. Allen, R. R. Behringer, R. E. Gelinas, and R. D. Palmiter. 1988. Introns increase transcriptional efficiency in transgenic mice. *Proc. Natl. Acad. Sci. USA* **85:**836–840.

Citron, M., D. Westaway, W. Xia, G. Carlson, T. Diehl, G. Levesque, K. Johnson-Wood, M. Lee, P. Seubert, A. Davis, D. Kholodenko, R. Motter, R. Sherrington, B. Perry, H. Yao, R. Strome, I. Lieberburg, J. Rommens, S. Kim, D. Schenk, P. Fraser, P. St. George Hyslop, and D. J. Selkoe. 1997. Mutant presenilins of Alzheimer's disease increase production of 42-residue amyloid β-protein in both transfected cells and transgenic mice. *Nat. Med.* **3:**67–72.

Clark, A. J. 1996. Genetic modification of milk proteins. *Am. J. Clin. Nutr.* **63:** 633S–638S.

Damak, S., H. Su, N. P. Jay, and D. W. Bullock. 1996. Improved wool production in transgenic sheep expressing insulin-like growth factor 1. *Bio/Technology* **14:**185–188.

Devlin, R. H., T. Y. Yesaki, C. A. Blagl, E. M. Donaldson, P. Swanson, and W.-K. Chan. 1994. Extraordinary salmon growth. *Nature* **371:**209–210.

Diamond, L. E., K. R. McCurry, M. J. Martin, S. B. McClellan, E. R. Oldham, J. L. Platt, and J. S. Logan. 1996. Characterization of transgenic pigs expressing functionally active human CD59 on cardiac endothelium. *Transplantation* **61:**1241–1249.

DiTullio, P., S. H. Cheng, J. Marshall, R. J. Gregory, K. Ebert, H. M. Meade, and A. E. Smith. 1992. Production of cystic fibrosis transmembrane conductance regulator in the milk of transgenic mice. *Bio/Technology* **10:**74–77.

Fishwild, D. M., S. L. O'Donnell, T. Bengoechea, D. V. Hudson, F. Harding, S. L. Bernhard, D. Jones, R. M. Kay, K. M. Higgins, S. R. Schramm, and N. Lonberg. 1996. High-avidity human IgGκ monoclonal antibodies from a novel strain of minilocus transgenic mice. *Nat. Biotechnol.* **14:**845–851.

Fodor, W. L., B. L. Williams, L. A. Matis, J. A. Madri, S. A. Rollins, J. W. Knight, W. Velander, and S. P. Squinto. 1994. Expression of a functional human complement inhibitor in a transgenic pig as a model for the prevention of xenogeneic hyperacute organ rejection. *Proc. Natl. Acad. Sci. USA* **91:**11153–11157.

Games, D., D. Adams, R. B. Alessandrini, R. Barbour, P. Berthelette, C. Blackwell, T. Carr, J. Clemens, T. Donaldson, F. Gillespie, T. Guido, S. Hagopian, K. Johnson-Wood, K. Khan, M. Lee, P. Leibowitz, I. Lieberburg, S. Little, E. Masliah, L. McConlogue, M. Montoya-Zavala, L. Mucke, L. Paganini, E. Penniman, M. Power, D. Schenk, P. Seubert, B. Snyder, F. Soriana, H. Tan, J. Vitale, S. Wadsworth, B. Wolozin, and J. Zhao. 1995. Alzheimer-type neuropathology in transgenic mice overexpressing V717F β-amyloid precursor protein. *Nature* **373:**523–527.

Gong, Z., and C. L. Hew. 1995. Transgenic fish in aquaculture and developmental biology. *Curr. Top. Dev. Biol.* **30:**177–214.

Hsiao, K., P. Chapman, S. Nilsen, C. Eckman, Y. Harigaya, S. Younkin, F. Yang, and G. Cole. 1996. Correlative memory deficits, Aβ elevation, and amyloid plaques in transgenic mice. *Science* **274:**99–102.

Humphries, M. M., D. Rancourt, G. J. Farrar, P. Kenna, M. Hazel, R. A. Bush, P. A. Sieving, D. M. Sheils, N. McNally, P. Creighton, A. Erven, A. Boros, K. Gulya, M. R. Capecchi, and P. Humphries. 1997. Retinopathy induced in mice by targeted disruption of the rhodopsin gene. *Nat. Genet.* **15:**216–219.

Jänne, J., J.-H. Hyttinen, T. Peura, M. Tolvanen, L. Alhonen, R. Sinervirta, and M. Halmekytö. 1994. Transgenic bioreactors. *Int. J. Biochem.* **26:**859–870.

Johnson-Wood, K., M. Lee, R. Motter, K. Hu, G. Gordon, R. Barbour, K. Khan, M. Gordon, H. Tan, D. Games, I. Lieberburg, D. Schenk, P. Seubert, and L. McConlogue. 1997. Amyloid precursor protein processing and Aβ42 deposition in a transgenic mouse model of Alzheimer disease. *Proc. Natl. Acad. Sci. USA* **94:**1550–1555.

Krimpenfort, P., A. Rademakers, W. Eyestone, A. van der Schans, S. van den Broek, P. Kooiman, E. Kootwijk, G. Platenburg, F. Pieper, R. Strijker, and H. de Boer. 1991. Generation of transgenic dairy cattle using "in vitro" embryo production. *Bio/Technology* **9:**844–847.

Masliah, E., A. Sisk, M. Mallory, L. Mucke, D. Schenk, and D. Games. 1996. Comparison of neurodegenerative pathology in transgenic mice overexpressing V717F β-amyloid precursor protein and Alzheimer's disease. *J. Neurosci.* **16:**5795–5811.

McCurry, K. R., D. L. Kooyman, C. G. Alvarado, A. H. Cotterell, M. J. Martin, J. S. Logan, and J. L. Platt. 1995. Human complement regulatory proteins protect swine-to-primate cardiac xenografts from humoral injury. *Nat. Med.* **1:**423–427.

Mendez, M. J., L. L. Green, J. R. F. Corvalan, X.-C. Jia, C. E. Maynard-Currie, X. Yang, M. L. Gallo, D. M. Louie, D. V. Lee, K. L. Erickson, J. Luna, C. M.-N. Roy, H. Abderrahim, F. Kirschenbaum, M. Noguchi, D. H. Smith, A. Fukushima, J. F. Hales, M. H. Finer, C. G. Davis, K. M. Zsebo, and A. Jakobovits. 1997. Functional transplant of megabase human immunoglobulin loci recapitulates human antibody response in mice. *Nat. Genet.* **15:**146–156.

Oster-Granite, M. L., D. L. McPhie, J. Greenan, and R. L. Neve. 1996. Age-dependent neuronal and synaptic degeneration of mice transgenic for the C terminus of the amyloid precursor protein. *J. Neurosci.* **16:**6732–6741.

Petitte, J. N., M. E. Clark, G. Liu, A. M. Verrinder Gibbins, and R. J. Etches. 1990. Production of somatic and germline chimeras in the chicken by transfer of early blastodermal cells. *Development* **108**:185–189.

Pursel, V. G., C. A. Pinkert, K. F. Miller, D. J. Bolt, R. G. Campbell, R. D. Palmiter, R. L. Brinster, and R. E. Hammer. 1989. Genetic engineering of livestock. *Science* **244**:1281–1288.

Sang, H. 1994. Transgenic chickens—methods and potential applications. *Trends Biotechnol.* **12**:415–420.

Sharma, A., M. J. Martin, J. F. Okabe, R. A. Truglio, N. K. Dhanjal, J. S. Logan, and R. Kumar. 1994. An isologous porcine promoter permits high level expression of human hemoglobin in transgenic swine. *Bio/Technology* **12**:55–59.

Sims, M., and N. L. First. 1993. Production of calves by transfer of nuclei from cultured inner cell mass cells. *Proc. Natl. Acad. Sci. USA* **90**:6143–6147.

Swanson, M. E., M. J. Martin, K. O'Donn, K. Hoover, W. Lago, V. Huntress, C. T. Parsons, C. A. Pinkert, S. Pilder, and J. S. Logan. 1992. Production of functional human hemoglobin in transgenic swine. *Bio/Technology* **10**:557–559.

Weidle, U. H., H. Lenz, and G. Brem. 1991. Genes encoding a mouse monoclonal antibody are expressed in transgenic mice, rabbits, and pigs. *Gene* **98**:185–191.

Wilmut, I., A. E. Schnieke, J. McWhir, A. J. Kind, and K. H. S. Campbell. 1997. Viable offspring derived from fetal and adult mammalian cells. *Nature* **385**:810–813.

REVIEW QUESTIONS

1. How are transgenic mice created?

2. What is positive-negative selection, and how does it work?

3. What are knockout mice? How and why are they established?

4. What are the advantages and disadvantages of using transgenic mice as model systems for human diseases?

5. What is nuclear cloning?

6. Discuss how transgenesis was used to produce human monoclonal antibodies.

7. Discuss how the mammary gland could be used as a bioreactor for the production of commercial products.

8. Discuss how transgenesis could be used to improve organ transplantation.

9. What approaches have been developed to produce transgenic chickens?

10. Discuss how transgenesis might improve fish aquaculture.

Human Molecular Genetics

The overall quality of human health depends on a number of factors, including the environment, individual lifestyle, human biology, and access to a good health care system. As diagnosis, treatment, and prevention of infectious diseases have improved in the last 40 years, the impact of genetic disorders on human health has become more obvious, especially in developed countries. In Canada, for example, it is estimated that 5% of the population will exhibit a gene-related defect, disability, or handicap by the age of 25, and over a lifetime more than 50% of the population will have a disease that has, to some extent, a genetic basis. At present, over half of the admissions to children's hospitals are due to genetic diseases.

More than 1,000 human disorders are known to be inherited. Most of these are exceedingly rare (1 affected person in 100,000), whereas a few are relatively common (1 affected person in 10,000). Many human genetic diseases can be attributed to a nucleotide change (mutation) in a single gene, although a number of complex conditions, e.g., cancer, probably are the consequence of the presence of a number of different mutant genes. When researchers have a clear, accurate, and consistent definition of the symptoms (phenotype) of a disease and when the phenotype is determined by a single gene, the nature of the genetic basis of the disease can usually be ascertained from the pattern of inheritance within multigenerational families. There are four basic patterns of inheritance: autosomal dominant (Fig. 20.1), autosomal recessive (Fig. 20.2), X-linked dominant (Fig. 20.3), and X-linked recessive (Fig. 20.4). The term autosomal refers to the 22 pairs of non-sex-specific chromosomes (autosomes) in humans, and the term X-linked specifies a gene that is located on the X chromosome. A dominant condition is one in which the presence of a single mutant gene of the pair is sufficient to produce the disease phenotype, and a recessive disorder requires that both genes of the pair be defective. In males, most single X-linked genes, whether dominant or recessive, produce a phenotype because there is only one copy of an X chromosome per nucleus.

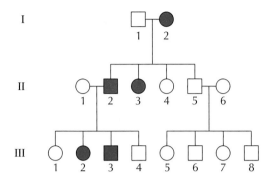

Figure 20.1 Pedigree showing autosomal dominant inheritance. Squares denote males, circles indicate females, solid symbols indicate affected individuals, and open symbols represent unaffected members of the family. A horizontal line joining a male and female symbol represents a marriage. A vertical line from a horizontal marriage line designates the offspring from a marriage. A number of vertical lines from a marriage line leading to symbols signifies a set of offspring (sibship). Within a sibship, the birth order is from left to right. The roman numerals (I, II, and III) indicate the generations, and the arabic numerals (1, 2, and 3) mark the members of each generation. A two-part code (e.g., II-3) is used to specify a particular individual within a family. Two key indicators for autosomal dominant inheritance are the following: (1) the trait occurs in successive generations if the penetrance is complete (i.e., every person that has the genotype shows the phenotype), and (2) males and females are equally affected.

Figure 20.2 Pedigree showing autosomal recessive inheritance. Key indicators are: (1) unaffected parents can have affected offspring; (2) males and females are equally affected; and (3) if both parents are affected, then all of their offspring will be affected (not shown).

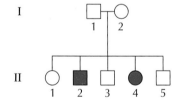

Pedigree studies are extremely helpful in determining the mode of inheritance of a specific condition, but they do not provide any information about the gene that determines the disorder, about the biological basis of the malfunction, or about the chromosome location if the condition is autosomal. Moreover, determination of whether a human disorder is genetically derived is not always straightforward. First, not all individuals who carry the defective gene show the symptoms of the disease (incomplete penetrance). Second, the symptoms (phenotype) can cover a range of defects from mild to severe (variable expressivity). Third, the same phenotype can occur due to defects in completely different genes (genetic heterogeneity, heterogeneity). Fourth, in some cases, alternative forms (alleles) of the same gene can produce different phenotypes. Fifth, small family size necessitates

Figure 20.3 Pedigree showing X-linked dominant inheritance. Key indicators are: (1) with complete penetrance, each generation shows an affected individual; (2) all daughters of an affected male are affected, and none of his sons is affected; (3) there is often a "male-to-female-to-male" pattern of inheritance in successive generations; and (4) there may be more affected females than males.

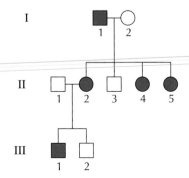

the collection of information from a large number of pedigrees to pin down the nature of a specific genetic disease.

The complete understanding of the relationship between a phenotype, whether it is normal or abnormal, and its corresponding genotype depends ultimately on the ability to isolate (clone) and characterize an individual gene. The DNA sequence of a gene can help explain the role that the gene product plays under normal conditions, how mutations disrupt its function, and the extent to which different mutations in different exons are responsible for the disease. With a cloned gene, experiments can be devised to test the actions and interactions of the gene product and how a mutant gene product undermines normal processes. As well, based on the DNA sequences of normal and mutated forms of a gene, diagnostic tests for specific gene mutations can be developed. As knowledge of the function of a disease gene accumulates, therapeutic treatments for the disease can be formulated.

The cloning of a human gene usually does not follow a precisely defined set of experimental procedures that are implemented in a stepwise manner. Rather, a repertoire of different tools and resources can be used as the circumstances dictate. For example, existing information on the gene product of a disease gene (disease-causing gene) determines how a gene search is initiated. Generally, both human linkage and physical maps are essential for identifying disease genes, and eventually they will facilitate the sequencing of the entire human genome. A genetic map (linkage map) displays the order of sites (loci) along a chromosome. The formation of comprehensive human linkage maps requires both frequently occurring alleles at many different loci on every chromosome and a ready way of determining the presence of a particular allele. On the other hand, a physical map is a set of ordered DNA clones that cover an entire chromosome or a chromosome region. In practice, these clones overlap each other to form a contiguous array which is called a contig. The breadth of coverage of a contig is expressed as base pairs of DNA. A physical map that is based on contigs provides an invaluable resource for generating the ultimate physical map, which is the complete DNA sequence of a chromosome.

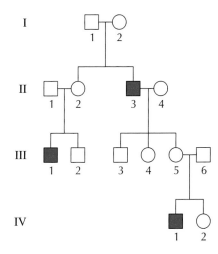

Figure 20.4 Pedigree showing X-linked recessive inheritance. Key indicators are: (1) unaffected parents can have affected offspring; (2) there is no direct transmission from father to son; and (3) there are more affected males than females.

Genetic Linkage and Gene Mapping

On the basis of his studies of garden peas in the 1850s, Gregor Mendel formulated some of the fundamental principles concerning the mechanisms of heredity. First, he deduced that the units of inheritance were discrete, occurred as pairs, and could exist in alternative forms. Later, by about 1905, the units of inheritance were called genes and each variant of a gene was called an allele. Second, Mendel found that only one member of each pair of genes is distributed to a sex cell (gamete). Third, he concluded that pairs of genes assort independently of one another, with the result that all possible genetic combinations will occur if enough offspring from a single genetically appropriate mating are sampled (Fig. 20.5). Unbeknownst to Mendel, this last conclusion depends on whether the pairs of genes are on separate chromosomes or, at least, extremely far apart on the same chromosome. In Mendel's experiments, none of the crosses (matings) involved pairs of genes that were close to one another on the same chromosome. If he had encountered such a genetic situation, he would have observed that these genes tended to be inherited as a linked set.

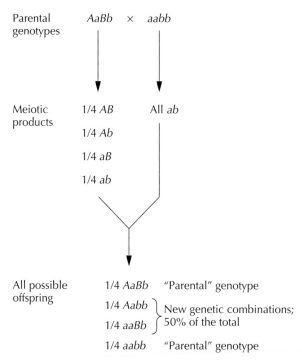

Figure 20.5 Independent assortment. When an organism with a doubly heterozygous genotype (*AaBb*) is crossed with an organism with a doubly homozygous recessive genotype (*aabb*)—and when the number of offspring is large enough to produce a reasonable sample size—then 50% of the offspring will have the parental genotypes (*AaBb, aabb*) and 50% will have new genotypic combinations (*Aabb, aaBb*). This result indicates that genes *A* and *B* have undergone independent assortment. With genetic crosses, the offspring class represents all the possible combinations of the meiotic products of each of the parents. It should be noted that the percentages of the parental and new combination genotypic classes depend on the genotypes of the parents; e.g., a cross with the genotypes *AABB* and *aabb* will give 100% of the offspring with a nonparental genotype (*AaBb*).

Theoretically, if genetic linkage were complete, then all genes along any chromosome would be passed on to the sex cells as an intact block, with no new genetic combinations on the chromosomes being formed during the meiotic process (Fig. 20.6). However, in most instances, linkage is not complete. Exchanges (recombination events, crossovers) occur between gene sites (loci) during meiosis, thereby creating new combinations of genes (Fig. 20.7). Because recombination tends to occur with increasing frequency as the distance between two specific gene loci increases, the percentage of recombination can be used to represent a measure of distance (map distance) between the two genes. Thus, by analyzing recombination frequencies among the progeny of parents that are heterozygous for a number of linked genes, a genetic map that places the genes in a linear array can be constructed. The distance between loci represents the frequency of occurrence of recombination and is not equivalent to a precise physical distance. However, approximations for the value of percentage recombination and number of base pairs of DNA have been deduced by correlating physical chromosome maps with genetic maps. In gene mapping terminology, 1 map unit is equal to 1 centimorgan (cM) and 1 cM is equal to 1% recombination, which, in humans, is roughly equivalent to 1×10^6 base pairs (bp) of DNA.

Some additional points about genetic linkage and gene mapping should be noted. First, to score the occurrence of new genetic combinations (recombinants), the genetic arrangement of one of the parents must be at least doubly heterozygous (*AB/ab* or *Ab/aB*). Second, doubly heterozygous genotypes can be in two configurations. When two linked genes on each chromosome are of the same type (i.e., both dominant, *AB*, or both recessive, *ab*), this arrangement is called the coupling (*cis*) phase. Alternatively, when two linked genes on each chromosome are of different types (i.e., one dominant and one recessive allele, *aB* or *Ab*), the arrangement is the repulsion (*trans*) phase. Third, recombination between two genes occurs regardless of the arrangement of the genes. From a genetic viewpoint, recombination between genes of a doubly homozygous individual (e.g., *Ab/Ab* or *AB/AB*) does not produce a new genetic combination, and as such, the recombination event cannot be scored even if it occurs. Fourth, 0% recombination denotes complete linkage, and a value of 50% recombination between two loci means that the genes are either on different chromosomes or so far apart on the same chromosome that they cannot be mapped. To resolve the problem of two genes being far apart but on the same chromosome, genes that lie between the original two genes must also be mapped to determine whether or not all of them make up a single linkage group.

To generate detailed genetic maps for eukaryotic organisms such as mouse, corn, fruit fly, nematode, and yeast, many genes, each with at least

Figure 20.6 Complete linkage. The genotype of the doubly heterozygous parent *AB//ab* is in the coupling (*cis*) phase and the other parent is doubly homozygous recessive, *ab//ab*. In the absence of any recombination events between the *A* and *B* loci, all the offspring will have parental genotypes, with half being *AB//ab* and the other half *ab//ab*. Nevertheless, complete linkage does not always prevent new genetic combinations from occurring; e.g., with the cross *Ab//aB* × *AB//ab*, all the offspring would have a new genetic makeup, namely, *Ab//AB*, *Ab//ab*, *aB//AB*, and *aB//ab*. However, in the absence of recombination, genes on the same chromosome will remain linked together forever. For convenience, genetic nomenclature uses a single horizontal line or a single slash instead of two lines or two slashes to show linkage relationships of members of the same pair of chromosomes (homologous chromosomes).

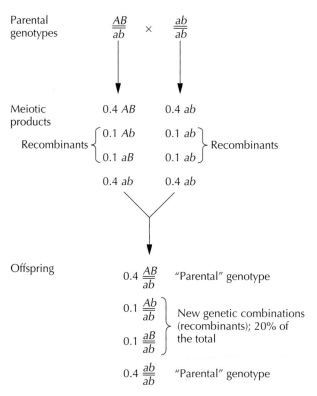

Figure 20.7 Incomplete linkage. During meiosis in this example, 20% (i.e., 0.1 + 0.1 = 0.2) of the products result from a recombination(s) event between the *A* and *B* loci. This recombination frequency occurs regardless of the genotypes of the parents. The doubly homozygous recessive parent produces only one type of gamete even though recombination occurs. In the test cross shown here, the recombinant meiotic products are readily scored phenotypically among the offspring.

two alleles, must be identified. Then, genetically defined crosses must be carried out and the frequency of recombinants among large numbers of offspring must be scored. The results indicate the extent of linkage between genes. Eventually, high-quality genetic maps can be produced by using multipoint (more than two linked gene pairs) crosses.

Detection and Estimation of Genetic Linkage in Humans

Prior to the advent of recombinant DNA technology in the early 1980s, the detection and estimation of genetic linkage in humans was difficult, arduous, and usually unsuccessful. In general, linkage studies in humans have inherent problems. The genetic phase of either parent is rarely known, which obscures the distinction between recombinant and nonrecombinant types among the progeny. Also, the statistical reliability of recombination fraction determinations is limited by the small size of most families.

Theoretically, the presence of a single X chromosome in human males should provide a good opportunity for estimating map distances between two or more gene loci. In this special circumstance, the alleles of loci on the X chromosome are registered directly as phenotypes in males. Mothers who are doubly heterozygous for X-linked loci will have sons who carry either recombinant or nonrecombinant X chromosomes. If the phase of the alleles of two gene loci in these mothers is known, then recombinant and nonre-

combinant types can be directly distinguished among the sons. The genotypes of the fathers, in these cases, are irrelevant because sons inherit X chromosomes only from their mothers. In some instances, the phase of a doubly heterozygous mother can be ascertained from the phenotype of her father. For example, if the mother's father (i.e., the grandfather) showed two X-linked recessive traits and she exhibited the dominant phenotype for both of these traits, then she would be a double heterozygote in *cis* phase, i.e., *AB/ab* (Fig. 20.8). The so-called "grandfather method" of linkage determination depends on scoring the phenotypes for two loci in sons of large numbers of different, phase-known doubly heterozygous mothers. In this case, the recombination fraction between two specific gene loci would be equivalent to the sum of the recombinant chromosomes (*R*) divided by the sum of the recombinant chromosomes plus the sum of the nonrecombinant chromosomes (*NR*), i.e.,

$$\frac{\Sigma(R)}{\Sigma(R) + \Sigma(NR)}$$

However, this strategy has drawbacks. Grandfathers are not always available for genetic typing. Consequently, the phase of a presumptive doubly heterozygous mother is unknown. Also, heterozygosity at the same two loci in a large number of different families is not a common occurrence. Despite much effort, prior to the 1980s it was not possible to assemble an ex-

Figure 20.8 Mapping the X chromosome with the "grandfather" method. With this method, the genetic phase of two or more X-linked loci in a daughter (Mother) is derived from knowledge of the X-linked alleles that were carried by her father (Grandfather). In turn, this information is used to determine which of the mother's sons (Sons) have received a nonrecombinant (NR) or recombinant (R) chromosome. In this example, the grandfather carries two recessive genes at loci *A* and *B* on his X chromosome and his daughter is a double heterozygote in the *cis* phase. The X chromosome is shown with alleles at the *A* and *B* loci, and the Y chromosome is the shorter solid line.

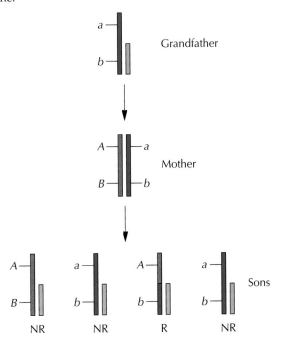

tensive, unequivocal linkage map of the human X chromosome by counting recombinant and nonrecombinant chromosomes. At that time, there were too few loci and not enough identifiable alleles.

The Logarithm of the Likelihood Ratio Method of Linkage Analysis: Lod Score

In addition to the direct method of counting large numbers of recombinants and nonrecombinants to determine the recombination fraction between two loci in humans, a broader-based, indirect method was needed that (1) could rigorously distinguish between independent assortment and linkage; (2) was not necessarily dependent on knowledge of the actual phase of the doubly heterozygous parent; (3) could combine the information from a number of different families; and (4) could give an estimate of the recombination fraction if linkage was evident. In 1955, N. E. Morton formulated such a method, which is now used extensively.

In human linkage studies, the recombination fraction is symbolized by the Greek letter theta (θ). Morton's method compares the likelihood [$L(\theta)$], for any recombination fraction (θ), that two loci within a group of siblings (sibship) are genetically linked (i.e., close to each other on the same chromosome) to the likelihood [$L(0.50)$] that the two loci are not linked (i.e., on separate chromosomes or very far apart from each other on the same chromosome). In the case of linkage, since the recombination fraction (θ) is unknown, it can have any value that is less than 0.50 and greater than or equal to zero ($0 \leq \theta < 0.50$). When two loci assort independently, the value of θ, by definition, is 0.50. In other words, when half of the gametes that are produced by a double heterozygote contain new genetic combinations, the two loci are either on nonhomologous chromosomes or so far apart on the same chromosome that they appear to be on different chromosomes. Consequently, if $L(\theta)$ equals $L(0.5)$, then the two loci are not linked. The common logarithm of the ratio of these two likelihoods, i.e., $\log[L(\theta)/L(0.50)]$, is the log-of-odds ratio which, in a contracted form, is called the lod (LOD) score. The letter Z is used to symbolize a lod score, and the term $Z(\theta)$ represents a lod score for a particular value of θ where $0 \leq \theta < 0.50$.

Part of the determination of $L(\theta)$ depends on the probability of occurrence of a particular combination of recombinant (R) and nonrecombinant (NR) chromosomes among the members of each sibship under study. The probability that an offspring will receive either nonrecombinant chromosome from a doubly heterozygous parent is $1/2 (1 - \theta) + 1/2 (1 - \theta)$, or $1 - \theta$. Similarly, the probability that an offspring will receive either recombinant chromosome from a doubly heterozygous parent is $1/2 \theta + 1/2 \theta$, or θ. Accordingly, the probability of each individual of a five-member sibship having a nonrecombined chromosome that originated from a doubly heterozygous parent is $K(1 - \theta)^5$, where $(1 - \theta)$ represents the frequency of receiving a nonrecombined chromosome; the exponential term (5) indicates the number of sibs with a nonrecombined chromosome, or, more precisely, the number of NR chromosomes in the sibship, and K is the coefficient for this example. When all the components of a term are the same, i.e., either all nonrecombinants or all recombinants, the coefficient equals 1 (i.e., 5!/5!0!, or $n!/n!0!$, where n is the number of sibs of the sibship). If all the individuals of a four-member sibship received a recombined chromosome from a doubly heterozygous parent, then the probability of this occurring would be $(\theta)^4$. Further, if a nine-member sibship had five NR types and four R types, then the overall probability of this combination is $K(1 - \theta)^5(\theta)^4$,

where K is 126; i.e., 9!/5!4!. The calculation of a lod score is based on a ratio where the coefficients of the terms of the numerator and denominator are the same and therefore factor out. Consequently, the coefficients are commonly disregarded in linkage analyses.

A single sibship will be used to illustrate how lod scores are calculated (Fig. 20.9). The notations B and O in Fig. 20.9 represent the alleles *ABO*B* and *ABO*O* of the ABO blood group, and the solid symbols denote the presence of a completely penetrant, autosomal dominant trait called nail-patella syndrome (NPS). The principal features of NPS are abnormal growth of the finger and toe nails and a reduced or absent knee cap (patella). The gene symbol for NPS is *NPS1*, and the normal and dominant alleles are *NPS1*N* and *NPS1*D*, respectively. NPS is a good trait for linkage studies because it is easily diagnosed, does not affect viability or fertility, and is present at birth.

Individual I-2 in Fig. 20.9 is heterozygous for NPS because among his offspring there are both affected and unaffected individuals. As well, I-2 is heterozygous at the ABO locus (*ABO*B/ABO*O*) because among his offspring there are both type B and type O phenotypes, which is consistent with his spouse (I-1) being *ABO*O/ABO*O*. Consequently, I-2 is a double heterozygote (*NPS1*D/NPS1*N; ABO*B/ABO*O*) for these two autosomal loci. If the *ABO* and *NPS1* loci were linked, then the phase of I-2 is unknown (phase-unknown). It could be either *ABO*B NPS1*D/ABO*O NPS*N* (phase 1) or *ABO*B NPS1*N/ABO*O NPS1*D* (phase 2). Or, with an abbreviated notation, the two possible phases would be *B D/O N* (phase 1) or *B N/O D* (phase 2).

If it is assumed that the *ABO* and *NPS* loci are linked and the genotype of I-2 is in phase 1 (*ABO*B NPS1*D/ABO*D NPS1*N*), then, among his progeny, individuals II-1, II-2, II-4, II-6, II-7, II-8, II-9, and II-10 would be nonrecombinant types because they received either the *ABO*B NPS1*D* or *ABO*O NPS1*N* chromosome from him (Fig. 20.10). All the offspring receive an *ABO*O NPS1*N* chromosome from I-1 because she is doubly homozygous (*ABO*O NPS1*N/ABO*O NPS1*N*). In this context, the genetic contribution of I-1 to her offspring is known and irrelevant for the analysis of linkage. Assuming a phase 1 genetic organization for individual I-2, individuals II-3, II-5, and II-11 have each received a recombinant chromosome. Hence, for this sibship, the probability of the combination of nonrecombinant and recombinant chromosomes is $(1 - \theta)^8(\theta)^3$.

The genotype of I-2 could equally as well be in phase 2, i.e., *ABO*B NPS1*N/ABO*O NPS1*D*. Accordingly, II-3, II-5, and II-11 would have received nonrecombinant chromosomes, and each of the remaining members

Figure 20.9 Nail-patella syndrome and ABO blood groups in a sample pedigree. Solid symbols represent individuals with nail-patella syndrome. Open symbols denote individuals who do not have nail-patella syndrome. The letters below each symbol designate the *ABO* blood group genotype. The *ABO* blood group alleles are abbreviated, with *O* representing *ABO*O* and *B* representing *ABO*B*.

Figure 20.10 The genetic organization of nail-patella and *ABO* blood group alleles of each member of the pedigree shown in Fig. 20.9 under the assumption that these two loci are linked. The *ABO* blood group allele designations are abbreviated, with *O* representing *ABO*O* and *B* representing *ABO*B*. The normal and dominant alleles for the nail-patella syndrome locus are presented as *N* and *D*, respectively. Individual I-2 could be in either of two genetic phases (Phase 1, Phase 2). The solid vertical bars represent the chromosomes of individual I-2 and the chromosomes inherited by the offspring from I-2. The stippled vertical bars represent the chromosomes of individual I-1 and the chromosomes inherited by the offspring from I-1. For phase 1 and phase 2, the chromosome that is inherited from individual I-2 by each of the offspring is identified as nonrecombinant (NR) or recombinant (R).

of the sibship would have inherited a recombined chromosome from I-2 (Fig. 20.10). The probability of this combination for the sibship is $(1 - \theta)^3(\theta)^8$.

Since the two possible genetic phases for I-2 are equally likely, then the overall likelihood [$L(\theta)$] for the sibship is $1/2(1 - \theta)^8(\theta)^3 + 1/2(1 - \theta)^3(\theta)^8$. The value for this expression can be computed for different values of θ. A commonly used set of θ values is 0, 0.001, 0.05, 0.10, 0.2, 0.3, 0.4, and 0.50. If time is not an issue, a full range of values for θ between 0 and 0.50 can be used. Next, the logarithm of the ratio of the likelihood of the sibship for each particular value of θ other than 0.50 to the likelihood for θ set to 0.50 is calculated. For example, with $\theta = 0.10$, the ratio $L(0.10)/L(0.50)$ is

$$\frac{\frac{1}{2}(1 - 0.10)^8(0.10)^3 + \frac{1}{2}(1 - 0.10)^3(0.10)^8}{\frac{1}{2}(1 - 0.50)^8(0.50)^3 + \frac{1}{2}(1 - 0.50)^3(0.50)^8} \text{ or } \frac{2.152 \times 10^{-4}}{4.883 \times 10^{-4}} \text{ which equals } 0.441.$$

The common logarithm of 0.441 is -0.356, which represents the lod score for this ratio. In other words, $Z(0.10)$ equals -0.356.

If the phase of I-2 were known (phase-known), then the likelihood for the sibship would also be known. For example, if I-2 had the phase 1 genotype (*ABO*B NPS1*D/ABO*O NPS1*N*), then, as noted above, the likelihood for the sibship would be $(1 - \theta)^8(\theta)^3$. The *Z* score with θ set to 0.10 would be

$$\log \frac{(1 - 0.10)^8(0.10)^3}{(1 - 0.50)^8(0.50)^3} \text{ or } \log \frac{4.305 \times 10^{-4}}{4.883 \times 10^{-4}}$$

which is equal to -0.055. Or, if I-2 had the phase 2 genotype (ABO^*B $NPS1^*N/ABO^*O\ NPS1^*D$), then the value for $Z(0.10)$ would be

$$\log \frac{(1-0.10)^3(0.10)^8}{(1-0.50)^3(0.50)^8} \text{ or } -4.826.$$

For the phase-unknown condition, the Z scores for the sibship shown in Fig. 20.9 range from -5.993 for a θ of 0.001 to $+0.029$ for a θ of 0.45 (Table 20.1). When at least one recombinant type is present in a sibship and θ is set to 0, mathematically the Z score is minus infinity ($-\infty$). The results in Table 20.1 indicate that the maximum lod score (Z_{max}) falls near a θ of 0.30. Additional calculations with a full range of θ values from 0.20 to 0.40 established that the Z_{max} is $+0.214$ at a θ of 0.276.

A Z_{max} of $+0.214$ is not sufficient to support the likelihood that the ABO and $NPS1$ loci are linked. By convention, a maximum lod score of $+3.000$ or greater is used as the criterion that two autosomal loci are linked because the odds in favor of linkage with this lod score are 1,000 to 1 or greater. For X-linked genes, because both loci are known to be on the same chromosome, the standard for linkage is set to a maximum Z score of $+2.000$ or greater, which represents odds of 100 to 1 or greater in support of linkage. A lod score of -2.000 is considered to be sufficient to exclude linkage between two loci because there is only 1 chance in 100 that the two loci are linked.

To detect and estimate linkage, the Z scores for the range of values for θ from a number of different sibships must be determined before the maximum Z score for a particular value of θ is obtained. The conversion of the ratio of the likelihoods for each sibship to the common logarithm allows each $Z(\theta)$ value to be summed. In a thorough study of the linkage between the ABO and $NPS1$ loci, 25 pedigrees, including several sibships, gave a $Z(0.10)$ of $+31.235$ (Table 20.2). Clearly, these data show that the two loci are linked. If they had not been linked, then all values of $Z(\theta)$ would have been less than $+3.000$.

The value of θ at Z_{max} gives a rough estimate of the recombination fraction between two linked loci. Although in the original study of the ABO and $NPS1$ loci a precise Z_{max} was not cited, the Z score with θ set to 0.10 was the highest for the range of recombination fractions that were calculated, which suggests that these two loci are approximately 10 cM apart. It should be stressed that with the various NPS pedigrees that were studied, different ABO alleles were linked to the $NPS1$ locus. In other words, there is no specific association between one particular ABO allele and the $NPS1$ locus. Unless proven otherwise, genetic linkage occurs between loci and not certain alleles. And, it should also be noted that the lod-score method does not reveal the autosomal location of two linked loci. Additional studies, described later in this chapter, were required to show that the ABO and $NPS1$ loci are located on the long arm (q) of chromosome 9 between regions 34 and 34.2, i.e., at 9q34–9q34.2.

Table 20.1 Z scores, based on the phase-unknown condition, for various values of θ for the sibship shown in Fig. 20.9

θ	0	0.001	0.01	0.05	0.10	0.20	0.30	0.40	0.45
Z score	$-\infty$	-5.993	-3.025	-1.071	-0.356	0.138	0.209	0.095	0.029

Table 20.2 Cumulative Z scores for various values of θ for the *ABO* and *NPS1* loci

θ	0.05	0.10	0.15	0.20	0.25	0.30	0.40
Z score	28.159	31.235	30.405	27.756	23.983	19.434	9.048

Adapted from Renwick and Schulze, *Ann. Hum. Genet.* **28**:379–392, 1965.

Genetic Mapping of Human Chromosomes

Genetic Polymorphism

The discovery of linkage between the ABO blood group and nail-patella syndrome was possible for two basic reasons. First, the presence of each major ABO allele (I^A, I^B, I^O) can be determined accurately with a simple laboratory test. As a result, the genotypes of all relevant parents and offspring are known. Second, each ABO allele occurs with a high frequency in the population at large, which increases the likelihood that a parent will be heterozygous. In Great Britain, where the ABO-NPS linkage study was initiated, the frequencies of I^O, I^A, and I^B alleles are about 0.66, 0.28, and 0.06, respectively.

The term "frequency of an allele" denotes the proportion of a particular allele of the total number of alleles at a locus in a population. For example, for a locus with two alleles (*A1*, *A2*) in a population of 13,000 people with 3,800 individuals that are *A1A1*, 6,400 that are *A1A2*, and 2,800 that are *A2A2*, the frequency of allele *A1* is

$$0.54, \text{i.e.,} \ \frac{(2 \times 3{,}800) + 6{,}400}{2 \times 13{,}000}$$

and the frequency of allele A2 is

$$0.46, \text{i.e.,} \ \frac{(2 \times 2{,}800) + 6{,}400}{2 \times 13{,}000}$$

For most loci, one allele is considerably more frequent (0.999 or more) than the other allele(s) (0.001 or less). In a large population, if the frequency of one allele at a locus is 0.999 and the other is 0.001, then the vast majority (99.8%) of the individuals would be homozygous for the more frequent allele, about 0.198% would be heterozygous, and 0.001% would be homozygous for the less frequent allele. Under these circumstances, it is virtually impossible to observe either segregation of the alleles of this locus or linkage with another locus, because most matings would be between homozygotes for the more frequent allele. If, on the other hand, the frequencies of two alleles at a locus were 0.99 and 0.01, then about 2% of the population would be heterozygous. In contrast to the previous example, the chance of observing segregation or linkage is increased because there are more heterozygotes at this locus in the population (Table 20.3). In short, human linkage studies depend on loci with frequently occurring alleles. When two or more alleles of a locus occur in a population with a frequency of 0.01 or greater, a genetic polymorphism exists and the locus is said to be polymorphic. Since genetic polymorphisms such as the alleles of the ABO locus are rare, extensive chromosome mapping projects could not be conducted until procedures were devised that readily detect large numbers of polymorphic sites.

Restriction Fragment Length Polymorphism

At the DNA level, a single base difference between two homologous genes is sufficient to create alleles. In many instances, a single base-pair change can cause a gene product to differ drastically from the normal product. There are also a large number of single-base-pair differences that have no effect on gene products, and others that occur without any biological consequences in segments of the DNA that do not code for proteins. These innocuous base-pair differences are dispersed throughout every chromosome and represent a reservoir of potential polymorphic sites (marker loci; genetic markers) that could be used for genetic mapping. The problem, then, is how to detect these polymorphic sites.

In 1980, D. Botstein, R. L. White, M. H. Skolnick, and R. W. Davis developed the theoretical framework for identifying polymorphic single-base-pair sites and using them as marker loci to map human chromosomes. Briefly, the rationale for this methodology is as follows. Restriction endonucleases cleave DNA at specific recognition sites. When a single base-pair change occurs within a recognition site, the sequence is no longer cut by the restriction endonuclease, whereas an intact site on another chromosome can be cleaved (Fig. 20.11A). In other words, one allelic form has a functional recognition site for a particular restriction endonuclease and the other does not. As a consequence of this allelic difference, DNA fragments of different sizes are produced when the DNA is digested with a particular restriction endonuclease. Thus, DNA hybridization with a probe that binds to DNA that only occurs at one chromosome location (single copy DNA) can be used to reveal the presence or absence of the polymorphic restriction endonuclease site.

For example, assume that a portion of a chromosome has three restriction endonuclease sites that are recognized by the restriction endonuclease *Hin*dIII (Fig. 20.11B). Further assume that in all individuals this chromosome has a functional *Hin*dIII recognition site at both sites A and B; i.e., within the population there are no other alleles for these sites. In other words, these sites are not polymorphic. However, for site 1, there is a high frequency of a single-base-pair change that makes this site resistant to cleav-

Table 20.3 Frequencies of alleles and genotypes in a large, random mating population

Allelic frequency		Genotypic frequency		
A1	*A2*	*A1A1*	*A1A2*	*A2A2*
1.0	0	1.0	0	0
0.999	0.001	0.998001	0.001998	0.000001
0.99	0.01	0.9801	0.0198	0.0001
0.90	0.10	0.81	0.18	0.01
0.75	0.25	0.5625	0.3750	0.0625
0.50	0.50	0.25	0.50	0.25
0.25	0.75	0.0625	0.3750	0.5625
0.10	0.90	0.01	0.18	0.81
0.01	0.99	0.0001	0.0198	0.9801
0.001	0.999	0.000001	0.001998	0.998001
0	1.0	0	0	1.0

The shaded area includes the allelic frequencies that produce genetic polymorphisms.

A

```
        +                        −
     AATGCTT                  ACTGCTT
     TTACGAA                  TGACGAA
        ↑                        ↑
        └────────────────────────┘
           Single base pair change
```

B

Site 1	Fragments produced	Fragments detected
+	X, Y	X
−	(X + Y)	(X + Y)

C

Genotypes	Fragments produced	Fragments detected
+ / +	X, Y	X
+ / −	X, (X + Y)	X, (X + Y)
− / −	(X + Y)	(X + Y)

D

Probe β

Site 1	Fragments produced	Fragments detected
+	X, Y	Y
−	(X + Y)	(X + Y)

Probe γ

Site 1	Fragments produced	Fragments detected
+	X, Y	X, Y
−	(X + Y)	(X + Y)

Figure 20.11 Use of restriction endonuclease sites as genetic markers. **A.** A single-base-pair change in a restriction endonuclease recognition site prevents cleavage. Functional and nonfunctional restriction endonuclease sites are shown by plus (+) and minus (−) signs, respectively. **B.** A segment of a single chromosome is represented with three restriction endonuclease sites (A, 1, and B) that are recognized by the same restriction endonuclease. The distance between sites A and 1 is X, that between sites 1 and B is Y, and that between sites A and B is (X+Y). In all DNA samples, sites A and B are functional (both +), whereas site 1 can be either a functional (+) or nonfunctional (−) cleavage site. With a functional site 1 (+), DNA fragments X and Y are produced after treating the DNA with the restriction endonuclease. With a nonfunctional site 1 (−), a single DNA fragment [(X+Y)] is produced after the DNA is treated with the restriction endonuclease. Probe α, after Southern blot

age by *Hind*III. Thus, in a population of individuals, two possibilities exist for a single chromosome with respect to this site. It can either be cleaved (+) or not cut (−).

If the distance from site A to site 1 is not the same as from site 1 to site B and each length is less than about 20 kb, and if a single copy DNA probe hybridizes to the DNA between site A and site 1 (Fig. 20.11B), then two possibilities can be identified by Southern DNA blotting after the *Hind*III-treated DNA is separated by agarose gel electrophoresis. First, if site 1 is cut, then two fragments are produced and the probe hybridizes to the DNA fragment from site A to site 1. Second, if site 1 is not cut, then the probe hybridizes to the DNA fragment from site A to site B (Fig. 20.11B).

The actual analysis of DNA samples from a group of individuals is slightly more complicated because chromosomes occur as pairs (Fig. 20.11C). However, each genotype (+/+, +/−, and −/−) produces a distinctive pattern of fragments after hybridization with the probe. As well, probes that hybridize to different regions between sites A and B are effective in distinguishing the presence or absence of the recognition site at site 1 (Fig. 20.11D). A mutated restriction endonuclease site that occurs frequently in a population and produces a distinctive pattern of DNA fragments is called a restriction fragment length polymorphism (RFLP). Polymorphic restriction endonuclease sites provide marker loci for each chromosome.

The genetic status of each RFLP locus on a single chromosome is called the haplotype. When a single site is examined there are two possible haplotypes (+ or −), with two different sites on a chromosome there are four haplotypes (++, + −, − +, and − −), and with n chromosome loci there are 2^n haplotypes. The determination of which alleles of RFLP loci (or any other polymorphic loci) are present on the chromosomes of an individual is called haplotyping (genotyping, DNA typing). RFLP loci are inherited in a strictly Mendelian manner, and the inheritance of an RFLP locus can be traced within a pedigree. When two or more RFLP loci are followed in a pedigree, it is possible to determine the occurrence of a recombination event. As shown in Fig. 20.12, the father (I-1) is heterozygous for three different RFLP loci on the same chromosome, and the mother (I-2) is triply homozygous and lacks the restriction endonuclease site of each locus on both chromosomes. The genetic status of the chromosome inherited from the father can be evaluated in each offspring by genotyping. In this example, offspring II-2 received a chromosome from his father that had undergone a crossover event; the other offspring inherited noncrossover chromosomes from their father.

In practice, DNA from a number of individuals is treated separately with different restriction endonucleases and then tested with cloned single copy DNA segments as probes to detect RFLPs. Hybridization of a DNA probe with a preparation of human metaphase chromosomes that are

analysis, will detect fragment X when site 1 is functional (+) and fragment (X+Y) when it is nonfunctional (−). **C.** The fragments that are produced after treating with the restriction endonuclease and those that are detected by probe α after Southern blot analysis are presented for each possible genotype at site 1 (+/+, +/−, and −/−). **D.** The fragments that are produced from a single chromosome after treating with the restriction endonuclease and those that are detected by DNA hybridization with either probe β or probe γ.

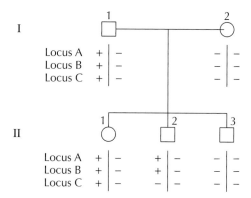

Figure 20.12 Detection of segregation and recombination of RFLP loci in a pedigree. The plus (+) and minus (−) signs represent functional and nonfunctional alleles at three RFLP loci (A, B, C) that are on the same chromosome. The genetic constitution of I-1 is based on his parents' genotypes, which were +++/+++ and −−−/−−− (not shown). Haplotyping of the offspring indicates that II-2 received a recombinant chromosome from I-1. Offspring II-1 and II-3 have noncrossover chromosomes from I-1. The vertical line separating each set of RFLP alleles under the symbol for each member of the pedigree depicts homologous chromosomes.

spread on a microscope slide (in situ hybridization) indicates whether the probe represents DNA from a single chromosome region (single copy DNA). A standardized nomenclature was devised for the thousands of RFLP loci. The notation D21S18 represents a locus that is identified by a DNA probe (D) that hybridizes to chromosome 21 (21) that is a single (S) copy sequence and was registered as isolate number 18 (18) by the DNA Committee of the International System for Human Linkage Maps (ISLM). A polymorphic marker site that falls within a known gene is designated by the name of the gene. For example, ADH represents a polymorphic locus within the alcohol dehydrogenase (*ADH1*) gene. There is no standardized designation for RFLP alleles. Some laboratories number the alleles in succession (D1S34*1, D1S34*2, etc.), and others use the length (kilobase pairs [kb]) of the fragment that is created by the presence or absence of a site(s) (D4S56*8, D4S56*12, D4S56*4, etc.).

The identification of thousands of RFLP loci expanded dramatically the number of alleles that could be used for genetic studies. Linkage between an RFLP locus (loci) and a disease gene can be established by a pairwise lod score analysis of haplotyped sibships within pedigrees that have the genetic disease under study. Similarly, linkage between RFLP loci can be determined using multigeneration pedigrees. The use of RFLP loci for mapping studies has some shortcomings. These loci are not uniformly distributed throughout every chromosome, the maintenance of the probes as clones is not convenient, and the requirement of haplotyping many individuals of a number of pedigrees by Southern DNA blotting is cumbersome. Fortunately, other polymorphic loci that consist of simple repeating units of two, three, or four nucleotide pairs (short tandem repeats [STRs]) are present in large numbers (>100,000) throughout the human genome and can be readily scored using the polymerase chain reaction (PCR).

Short Tandem Repeat Polymorphism

There are about 100,000 blocks of the dinucleotide repeat CA/GT [(CA)·(GT) (Fig. 20.13)] scattered throughout the human genome. These

CA CAC A
GT GTG T

Figure 20.13 A dinucleotide short tandem repeat DNA sequence. This (CA)·(GT) sequence has 24 repeating dinucleotides and is designated (CA)$_{24}$.

blocks contain from 1 to 40 repeating CA/GT units. Any one block at a particular chromosome location retains, more or less, the same number of repeating units through successive generations. A notation for describing a CA/GT repeat is (CA)$_n$, where n represents the number of CA repeats. In addition to CA dinucleotide repeats, other dinucleotide repeats [e.g., (AT)$_n$, etc.], as well as trinucleotide [e.g., (ATG)$_n$, etc.] and tetranucleotide repeats [e.g., (ATCG)$_n$, etc.], are also found dispersed among all human chromosomes. Collectively, these simple multicopy repetitive sequences are called short tandem repeats (STRs, or microsatellites).

The first step leading to the identification of polymorphic STR loci entails screening a small-insert (~1,000 bp) human genomic library with the appropriate hybridization oligonucleotide probe. Often, a 15-CA-unit oligonucleotide is used for isolating cloned (CA)n repeats. The insert of each positive clone is sequenced to determine both the length of the CA repeat and the sequences of the regions that flank the tandem repeat. In situ hybridization with probes that are based on sequences on either side of the repeated (CA)·(GT) block is used to determine if these flanking sequences represent single copy DNA. If the flanking sequences occur only once in the genome, then a pair of primers are synthesized that correspond to the flanking sequences and direct the amplification of the CA repeat. Next, PCR tests are run with DNA samples from a large number of individuals with this pair of primers. The PCR products, which are designed to be about 200 bp long, are analyzed with polyacrylamide gel electrophoresis. If the amplified DNA for a specific pair of primers has the same length in all the DNA samples, then that repeat is not polymorphic (Fig. 20.14A). Alternatively, if there are different PCR products, then an STR polymorphism (STRP) is indicated (Fig. 20.14B). Differences in the length of the CA repeats at the locus are responsible for the different PCR products. In other words, the different CA repeats represent alleles (Fig. 20.15). The CA repeat-alleles of a locus often have frequencies that are 0.20 or greater. The number of CA repeats of an allele is a relatively stable characteristic.

The naming convention for STRP loci is the same as for RFLP loci. To date, thousands of STRP loci have been registered. On the other hand, the names of STRP primers are idiosyncratic. Many STRP loci have been identified by French researchers with the financial support of the Muscular Dystrophy Association of France (Association Française contre les Myopathies). In recognition of this assistance, primer pairs for many STRPs have been given the prefix AFM followed by an identifying number (AFM349xc5). The primer pair name is often accompanied by the locus designation [AFM349xc5 (D3S2017)].

STRP loci have for the most part superseded RFLP loci as the choice for human gene mapping studies. In contrast to an RFLP probe, which must be maintained in a vector, purified, and labeled for genotyping, only the sequence information for a pair of STRP primers, which can be stored in a computer database, is required. Also, STRP loci are uniformly distributed in the human genome; STR allelic frequencies are very high, ensuring a high incidence of heterozygosity, and allelic differences can be clearly and easily identified after PCR amplification.

A

STR primer pair X

B

STR primer pair Y

Figure 20.14 DNA typing of STR loci. **A.** The DNA from different individuals (*n* = 7) is amplified by PCR with a pair of primers (X) that flank a (CA)·(GT) repeat. In all cases (lanes 1 to 7), the PCR product (solid rectangles) has the same size after poly-acrylamide gel electrophoresis, which signifies no differences in the length of the STR. Based on this sample, there is only one STR allele at this locus in this sample. **B**. Same as in **A**, except a different primer pair (Y) for another STR locus was tested. The occurrence of two different PCR products (solid rectangles) indicates that there are two STR alleles at this locus. Lanes 1 to 3 contain amplified DNA from individuals who are homozygous for one STR allele, lanes 4 and **5** contain amplified DNA from heterozygotes who carry two different STR alleles, and lanes 6 and 7 contain amplified DNA from individuals who are homozygous for the other STR allele. The lengths of the STR repeats are determined by DNA sequencing.

Figure 20.15 An example of two STR alleles. One allele (allele 1) has a (CA)$_{15}$ repeat and the other (allele 2) has a (CA)$_{10}$ repeat. The same single-copy (unique) sequences flank the repeated dinucleotide sequences of both alleles.

Allele 1

(CA)$_{15}$

CA CACACACACACACACACACACACACACA
GT GTG TG TG TG TG TG TG TG TG TG TG TG T

Unique
DNA sequence

Unique
DNA sequence

Allele 2

(CA)$_{10}$

CA CACACACACACACACACA
GT GTG TG TG TG TG TG TG T

Unique
DNA sequence

Unique
DNA sequence

Mapping of a Genetic Disease Locus to a Chromosome Location

Pedigree analysis, except for X-linked inheritance, does not reveal the chromosome location of the locus of a disease gene. However, the availability of RFLP and STRP probes (markers) that identify polymorphic loci throughout the genome makes it possible to test for linkage between a polymorphic and a disease locus using the lod score method. This approach works best when a genetic disease is clinically unambiguous, the mode of inheritance is unequivocal, and there is a reasonable estimate of the extent of penetrance.

A linkage study requires the collection of blood cells, with informed consent, from members of several two- and three-generation families or from individuals of a large multigeneration family with a particular genetic disorder. The blood cells are cultured and become a source of DNA for future procedures, which eliminates the need to draw blood repeatedly from the same subjects. The DNA from each individual is genotyped for a number of different polymorphic markers. In some studies, more than 250 polymorphic markers representing sites from all parts of all autosomes are used. A two-point (two-locus) lod score is calculated for each polymorphic locus and the genetic disease from all informative parent-offspring combinations. A Z value of +3.00 or more supports linkage, and one of −2.00 or less excludes it.

A large multigeneration pedigree with individuals who had benign familial neonatal convulsions (BFNC) was studied to determine the chromosome location of the disease (Fig. 20.16). This condition features violent episodes of uncontrolled jerking of the face, body, arms, and legs within the first six months of life. The symptoms stop in about 90% of the cases by the first birthday. Apparently, the seizures have no effect on subsequent neurological and intellectual functioning. BFNC is a rare disorder that is well defined clinically and is inherited as an autosomal dominant trait with high penetrance.

Of all the polymorphic markers that were tested, two probes (D20S19, D20S20) showed linkage with the *BFNC* locus within the large multigeneration pedigree under study (Table 20.4; Fig. 20.16). The alleles for the *D20S19* and *D20S20* loci of each genotyped member of the pedigree in Fig. 20.16 are shown as stacked numbers, and the vertical line separates the allelic constitution of each chromosome. The upper numbers are the alleles for the *D20S19* locus and the lower ones stand for the *D20S20* locus.

In this kindred, the D20S19 probe distinguishes among 10 different alleles of an RFLP locus, and the D20S20 probe identifies two alleles of its locus. For some RFLP loci, a number of restriction endonuclease recognition sites may be clustered in a small segment of DNA (∼20 kb) which collectively are considered as part of the same locus. Four closely spaced restriction endonuclease sites could give rise to 20 different fragment lengths that are identified by a single probe. Similarly, an STRP locus can have more than two alleles. In these cases, the lengths of the repeated nucleotide units differ from each other.

The (8,2) haplotype in the BFNC kindred cosegregated with the disease in many instances, which suggests that in this family the *BFNC* locus is on this particular chromosome. It was expected that individuals III-18, IV-4, and IV-14, who received an (8,2) chromosome from an affected parent, would also have been affected. These exceptions probably represent in-

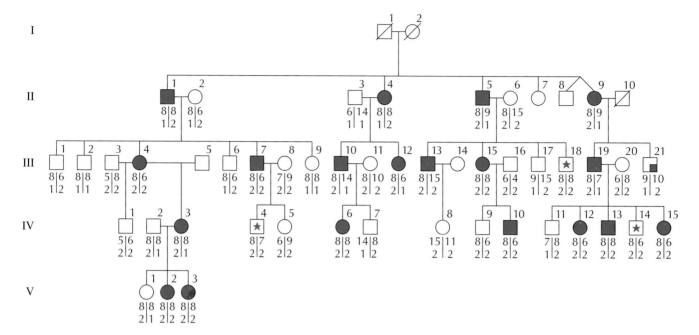

Figure 20.16 Haplotype analysis of two chromosome 20 polymorphic loci in a large kindred with BFNC. The haplotypes of the two polymorphic sites are represented by a number in which the allele of one locus is written above the allele of the second locus. The haplotypes of the two chromosomes of an individual are separated by a vertical line. The pedigree symbols with a diagonal slash denote deceased individuals. Solid symbols mark individuals affected by the disease. The symbol with a filled-in quadrant represents a case where the individual has a phenotype that is related but differs clinically from other affected members of the family. Individuals II-8 and II-9 are nonidentical (fraternal) twins. The stars mark cases of possible incomplete penetrance (III-18, IV-4, IV-14). The (8,2) allelic constitution in the affected members of the family cosegregates with the disease, and these individuals carry a chromosome that is a copy of one from a common ancestor. (Pedigree is adapted from Leppert et al., *Nature* [London] **337**:647–648, 1989.)

complete penetrance in this pedigree. In other words, these individuals have the *BFNC* disease gene but it was not expressed. In linkage studies with other traits, this kind of anomaly could be due to misdiagnosis, or the onset of the disease has not yet occurred in some of the individuals who carry the disease gene.

There are some instances in the BFNC family when an offspring of an affected parent has an (8,2) chromosome without showing the disease (e.g., IV-7 and IV-11). In both of these cases, it is possible to determine the origin of the (8,2) chromosome. For example, individual IV-7 inherited a (14,1)

Table 20.4 Pairwise lod scores between a BFNC locus and two chromosome 20 polymorphic loci

Locus	Recombination fraction (θ)					
	0.00	**0.05**	**0.10**	**0.20**	**0.30**	**0.40**
D20S20	3.12	3.05	2.88	2.32	1.57	0.69
D20S19	2.87	2.92	2.83	2.36	1.63	0.76

Adapted from Table 1 in Leppert et al., *Nature* (London) **337**:647–648, 1989.

MILESTONE

Construction of a Genetic Linkage Map in Man Using Restriction Fragment Length Polymorphism

D. Botstein, R. L. White, M. Skolnick, and R. W. Davis

Am. J. Hum. Genet. **32**:314–331, 1980

Abundant Class of Human Polymorphisms Which Can Be Typed Using the Polymerase Chain Reaction

J. L. Weber and P. E. May

Am. J. Hum. Genet. **44**:388–396, 1989

The incredible upsurge of human molecular genetics since the 1980s is due to the innovative insight of Botstein et al. They appreciated that human restriction fragment length polymorphisms (RFLPs) represented a potential source of polymorphic alleles (marker loci) that could be genetically mapped to human chromosomes. As they noted: "We describe a new basis for the construction of a genetic linkage map of the human genome. The basic principle of the mapping scheme is to develop, by recombinant DNA techniques, random single-copy DNA probes capable of detecting DNA sequence polymorphisms, when hybridized to restriction digests of an individual's DNA." Moreover, they realized that the chromosome location of a gene that is responsible for human disease could be identified by linkage with a DNA marker locus. Although not explicitly stated in their paper, the idea of using linkage between a DNA marker and a human disease gene as a means to isolate the disease gene flows directly from their concept. Botstein et al. were absolutely correct when they concluded that "The application of a set of probes for DNA polymorphisms to DNA available to us from large pedigrees should provide a new horizon in human genetics."

By 1992, hundreds of RFLP marker loci had been identified and mapped to various human chromosomes. A number of these DNA probes were instrumental in the isolation of genes for inherited diseases such as Duchenne muscular dystrophy and cystic fibrosis. However, despite their usefulness, highly polymorphic RFLP loci were not evenly spaced at close distances to one another along each human chromosome. As well, the RFLP assay procedure, which entails DNA hybridization with restriction endonuclease-treated DNA, was tedious and prone to misinterpretation. These limitations were overcome when Weber and May and others showed that large numbers of highly polymorphic di-, tri-, and tetranucleotide repeats (short tandem repeats, or STRs) are ubiquitously dispersed throughout the human genome and that alleles could be easily distinguished with PCR. As Weber and May noted, ". . . this class of polymorphisms is likely to find application in the study of many genetic-disease genes and should permit substantial improvement in the resolution of the human genetic map." STRs, especially dinucleotide tandem repeats, have superseded RFLPs as effective marker loci and are currently being used to construct high-density genetic maps of each human chromosome.

chromosome from the affected parent and the (8,2) chromosome with a normal *BFNC* gene from his unaffected mother. By contrast, the genotypes of individuals IV-9 and V-1 are not as easily explained. On the one hand, they may have inherited an (8,2) chromosome that came from outside the family that has the normal *BFNC* gene. For example, the (8,2) chromosome of IV-9 may have been derived through his mother (III-15) from his grandmother (II-6). On the other hand, the absence of the disease in IV-9 and V-1 could be due to incomplete penetrance if they each inherited an (8,2) chromosome with *BFNC*D* from an affected parent. It should be stressed that in other families with BFNC the *D20S19*8* and *D20S20*2* alleles are not necessarily linked to the BFNC disease allele. There is no biological significance between the *D20S19*8* and *D20S20*2* alleles and the dominant form of the *BFNC* gene. The polymorphic alleles merely happen to be on a chromosome that also carries the *BFNC*D* allele that was inherited from a common ancestor in this particular pedigree. In principle, loci, not alleles, are linked.

On the basis of the two-point lod scores (Table 20.4), the *D20S19* and *D20S20* loci are probably less than 5 cM ($<5 \times 10^6$ bp) from the *BFNC* locus. In general, linkage mapping is not able to resolve loci that are less than 1 to 2 cM apart. Since the *D20S19* and *D20S20* loci are located within region

13.2 to 13.3 of the long arm (q) of chromosome 20 (20q13.2–13.3), then the *BFNC* locus must be near or within this chromosome region. In all, the lod score approach with polymorphic markers has been used to map more than a hundred disease genes to specific chromosome sites.

Multilocus Mapping of Human Chromosomes

The availability of thousands of widely dispersed polymorphic markers has made it possible to determine both the order of loci and the map distances between these sites on each chromosome. A linkage map of polymorphic sites is an important resource for positioning a disease gene. The probes for the loci that closely flank the disease gene can then be used to isolate the disease gene.

Three-generation families with two sets of living grandparents and parents who have had a large number (>8) of children are ideal for mapping polymorphic loci. The genotypes of the grandparents delineate the genetic phase of each parent. Large sibships also increase the likelihood that recombination events will be observed. The Centre d'Etude du Polymorphisme Humain (CEPH) in Paris, which is now the Fondation Jean Dausset-CEPH, assembled a group of 65 families consisting, in most instances, of three generations with an average sibship size of 8.5 (e.g., Fig. 20.17). This reference panel of families (CEPH family panel) is a source of genotyped information for mapping projects in laboratories around the world. The panel actually consists of culture lymphoblastoid cell lines from most of the CEPH family members, which provide a ready source of DNA that can be used for mapping new polymorphic loci as they become available.

The construction of a multilocus human chromosome genetic (linkage) map is a difficult task that requires a specialized computer program to find the order of loci that is best supported by the recombination data. The ordering problem is compounded as the number of loci that are to be mapped increases. For N loci, there are $N!/2$ possible orders. With 10 loci, for example, there are 1,814,400 possible orders. Although some orders are clearly invalid, based on a visual inspection of the data, there are still a large number of likely orders. A commonly used strategy is to deduce the most likely order for a few linked loci at a time and then combine these "best" orders to generate a statistically supported linkage map with all of the loci that are being considered. The criterion for acceptance that one locus is next to another is a log-likelihood (lod, LOD) score of +3.00 or greater.

In general, map construction is a stepwise process. First, several polymorphic markers for a single chromosome are assembled. Next, DNA sam-

Figure 20.17 CEPH family K1331.

ples from a group of CEPH families are genotyped for each polymorphic marker. Because of the structure of the CEPH families, genotyping of the entire panel is not necessary. The quality of a map is not enhanced enough to justify the extra amount of work that is required to analyze additional families. Often 15 and occasionally 40 families are used. About 10,000 assays are required to genotype 40 CEPH families with 20 polymorphic markers. The genotype for each locus for each individual is entered into a database. At this stage, the database is checked for errors. Computer programs look for cases of impossible parent-offspring genotypes which would represent mistakes during data entry or, possibly, genotyping errors. In some instances, retyping is carried out to verify the original genotyping. Errors lead to incorrect locus ordering and inflated map distances. If possible, erroneous data are filtered out of the analysis. All pairwise lod scores and recombination fractions (θ) are determined from the genotyped CEPH families. Then, with either the lod score values or the recombination fractions, a particular computer program constructs the genetic (linkage) map.

Human chromosome linkage maps are continually being updated as additional polymorphic loci are identified. With more sites, a map becomes denser and the distance between loci decreases. By 1994, a set of CEPH families were genotyped for about 6,000 polymorphic markers, and a multilocus mapping analysis placed about 1,000 loci throughout the entire genome with an average distance of about 4 cM between loci. With additional polymorphic markers, the objective of large-scale mapping projects is to construct a map of every chromosome with loci that are 1 to 2 cM apart.

Inserting a Disease Gene into a Linkage Map

The members of pedigrees with a particular genetic disease are genotyped with polymorphic markers whose sites have been mapped and are on the same chromosome arm as the disease gene. The location of the disease gene is determined using the same principle as the two-point lod score for linkage analysis. In this case, the disease gene locus is arbitrarily positioned at sites among four ordered loci, and a lod score is determined for each position. For multipoint mapping, a lod score is the logarithm of the ratio of (1) the likelihood of four ordered polymorphic loci with the disease gene at a specific position on this four-point map to (2) the likelihood that the disease gene is not linked to any of the polymorphic loci. Four polymorphic loci are used because the computational load becomes excessive when additional loci are tested together. The disease gene can be placed preceding the first locus, at various intervals between the loci, or after the last locus. After lod scores for each of the various positions of the disease gene among the different sets of four loci are calculated, the highest lod score that is greater than +3.00 marks the most likely location of the disease gene.

Radiation Hybrid Mapping

Radiation hybrid (RH) mapping does not depend on assembling pedigrees and genotyping members of the CEPH family panel. Rather, it relies on somatic cell techniques and on screening (using PCR-based probes) a number of cell lines that maintain pieces (fragments) of a human chromosome. To date, RH mapping of either an entire chromosome or a chromosome region starts with a hybrid cell line that carries a single human chromosome. The cells of such a monochromosomal hybrid cell line are treated with a lethal dose of ionizing radiation from either a source of X or gamma (γ) rays. Both X and γ rays bombard living matter with particles that induce the formation

of ions which disrupt membranes, inactivate enzymes, and fragment chromosomes. The standard unit of biologically absorbed energy from ionizing radiation is called a rad (*radiation absorbed dose*). One rad is equivalent to 0.01 joule of energy per kilogram of tissue or 100 ergs per gram of tissue. Generally, cultured cells are killed by a dose of 3,000 rads. The larger the dose, the greater the cellular damage and the smaller the DNA fragments that are produced. A dose of 10,000 rads generates DNA fragments that are too small for RH mapping.

The crucial aspects of the RH mapping protocol are the rescue and perpetuation of the human DNA fragments that were produced by the radiation treatment. Both rescue and perpetuation are accomplished by fusing irradiated cells (donor cells) with nonirradiated rodent cells (recipient cells). Irradiated cells that fuse to each other, and those that do not fuse, will not be able to grow when cultured because of the radiation damage. Recipient cells that fuse to each other or remain unfused will not have a selectable marker which is present in the donor cells that would enable them to grow in the fusion culture medium. Consequently, only the donor-recipient cell fusions that carry the selectable marker will proliferate in the fusion culture medium. It is common for DNA fragments from the irradiated cells to be inserted into or translocated onto functional chromosomes of the recipient cells. The surviving fused cells are cultured en masse before single cell lines are established. These single cell lines are called radiation hybrids (RHs), and a group of radiation hybrids from one experiment is a radiation hybrid (RH) panel. Most of the human chromosome DNA from a monochromosomal cell hybrid line is retained as fragments in different members of an RH panel.

The DNA from a sample of each RH of the initial panel is screened with a number of chromosome-specific PCR-based probes. Many of these probes recognize polymorphic sites. However, polymorphism per se is not required for RH mapping. RH mapping relies on whether a chromosome site is either present or absent in the members of an RH panel. Consequently, PCR primers that amplify a unique (single copy) DNA sequence that is not polymorphic (i.e., is monomorphic) can also be used to screen an RH panel. Monomorphic PCR-identifiable, chromosome-specific sites are called sequence tagged sites (STSs). The presence (+) or absence (−) of a PCR-identified site is recorded for each probe with each member of an RH panel (Table 20.5). Any RH that is not amplified by any of the probes is removed from the panel. A panel of about 100 RHs from a monochromosomal hybrid cell line is required for effective RH mapping.

The theoretical basis of RH mapping closely resembles the rationale for meiotic mapping. The closer two sites are on a chromosome, the more likely they will be retained jointly on a radiation-induced DNA fragment. With meiotic mapping, the closer two sites are on a chromosome, the more likely they will not be separated by recombination. The basic assumptions of RH mapping are that (1) radiation-induced breakage between two sites is independent of marker retention and (2) the retention of a fragment with one marker is independent of the retention of any other fragment in the same cell.

The pattern (retention pattern, signature) of the presence (+) or absence (−) of each marker for each member of the RH panel is used to generate an RH map. A lod score is calculated from the logarithm of the ratio of the likelihood for a particular retention pattern for two sites to the likelihood that is expected if the two sites are always separated by radiation-induced

Table 20.5 Retention data for RH mapping

RH panel	Presence or absence of marker						
	A	B	C	D	E	F	G
1	+	−	−	+	−	−	+
2	+	+	−	−	+	+	−
3	−	−	+	+	−	+	−
4	+	−	+	+	+	−	−
5	−	−	+	−	−	−	−
6	+	+	−	−	−	+	−
7	−	−	+	+	−	−	+
8	−	+	−	−	−	−	−
9	+	+	−	+	+	−	+
10	−	−	−	−	−	+	−

The numbers and letters represent radiation hybrids and PCR-based markers, respectively. The plus and minus signs designate the presence or absence of a marker site in a radiation hybrid.

breakage. Unlike meiotic recombination, the range of θ values for radiation breakage frequency extends from 0 to 1. A value of zero signifies that two marker sites are never separated by a particular dose of radiation, and therefore are extremely closely linked. When $\theta = 1$, the markers are always separated by a particular dose of radiation; that is, they are not linked. A lod score of +3.00 or greater is support that two markers are linked. Computational methods have been devised for multisite ordering and determining the distance between sites of an RH map.

The units of distance between two sites on an RH map are called centiRays (cR), and 100 cR equals 1 Ray. The dose of radiation that was used to create an RH panel for an RH map must be cited because the size of radiation-induced fragments, more or less, varies inversely to the amount of irradiation. For example, a distance of 1 $cR_{8,000}$ represents a 1% occurrence of radiation-induced breakage between two markers with a dose of 8,000 rads.

There is no simple relationship between a centiRay and the number of base pairs it represents. Generally, the higher the dose in rads, the smaller the physical distance for a particular centiRay. For example, 1 $cR_{9,000}$, 1 $cR_{8,000}$, 1 $cR_{6,000}$, 1 $cR_{5,000}$, and 1 $cR_{3,000}$ are equivalent to approximately 50, 53, 26, 90 and 100 kb, respectively. By contrast, meiotic (genetic) mapping can, at best, resolve sites that are about 1 cM, or 1,000 kb, apart. In addition to higher resolution, RH maps tend to be more complete than genetic maps. Moreover, RH mapping is technically simpler than meiotic mapping, and new sites can be quickly added to an existing RH map. One limitation of RH mapping is that it cannot localize a disease gene to a specific chromosome region. Nevertheless, for the construction of multilocus human chromosome maps, RH mapping will probably supplant linkage mapping that relies on genotyping members of the CEPH families.

Physical Mapping of the Human Genome

Genetic and RH maps are linear representations of marker sites. The distances between sites on these maps are relative units that are based on either the frequency of recombination (cM) or the probability of coretention of two sites by an RH (cR). These units can be converted into approximate physical distances in base pairs. By contrast, a physical map records the ac-

tual distance (base pairs) between sites. In addition, a physical map of a whole chromosome or a chromosome region provides direct access to the DNA, which in turn facilitates the isolation and characterization of genes and the systematic sequencing of chromosome DNA.

The construction of a physical map depends on retrieving clones that have overlapping segments from a genomic DNA library. On the basis of the overlaps and other positional information, a continuum of clones can be established for a chromosome region, a whole chromosome, or the entire genome. Sets of contiguously ordered clones (contigs) have been generated from human YAC (yeast artificial chromosome), BAC (bacterial artificial chromosome), PAC (bacteriophage P1 artificial chromosome), bacteriophage P1 (P1), and cosmid libraries by using various procedures. In practice, the strategies for assembling large-insert contigs from a YAC, BAC, or PAC library differ slightly from those for forming P1 or cosmid contigs.

Assembling Contigs from YAC, BAC, and PAC Libraries

Sequence tagged site (STS) content mapping has become the preferred method for constructing physical maps of chromosome regions and whole chromosomes from large-insert (YAC, BAC, or PAC) genomic libraries. An STS is a short single copy segment of DNA (~100 to 300 bp) that can be specifically detected by PCR amplification with a unique set of primers. Large numbers of STSs with an STS every 50 to 100 kb are required to produce a continuous contig over a long span of a chromosome. By way of illustration, from 1,500 to 3,000 STSs are required for the physical mapping of a chromosome that consists of about 200 megabase pairs (Mb) of DNA. At least 30,000 STSs are needed for a high-resolution physical map of the complete human genome.

Various strategies have been devised to generate STSs. For example, DNA from a purified sample of a single human chromosome that was isolated by flow-sorting cytophotometry is treated with a restriction endonuclease and cloned into a small-insert (<1,000-bp) vector. Clones are selected at random, and the inserts are sequenced. Clones with inserts that are shorter than 100 bp and those that contain sequences from human repetitive DNA elements are discarded. The presence of repetitive DNA is determined by computer-based nucleotide matching of the insert sequence with all known human repetitive DNA sequences. Next, the primer sequences are determined for each appropriate clone. Each STS is tested to ensure that it amplifies a unique segment of chromosome DNA.

The STS content of individual clones of a large-insert library is ascertained by PCR screening. Then, on the basis of STSs that are shared between clones, a computer program is used to deduce a likely set of overlapping clones and, to a limited extent, the relative positions of some of the STSs (Fig. 20.18). With the development of RH mapping, it is possible, without excessive labor, to order sets of STSs, which makes it easier to deduce the members of a contig (Fig. 20.19). After the overlapping clones of a contig have been established, the extent of the overlaps, the continuity of the contig, and the total amount of DNA that it spans are determined by restriction endonuclease mapping with an electrophoresis system that separates DNA fragments larger than 10^5 bp (e.g., pulsed-field gel electrophoresis [PFGE]). Contigs for chromosome regions that range from 1 to over 20 Mb and, in a few instances, for an entire chromosome have been assembled. Eventually, large-insert coverage of the entire genome is anticipated.

A

Clone	STS														
	1	2	3	4	5	6	7	8	9	10	11	12	13	14	15
a		+	+	+		+				+	+L		+R	+	+
b	+	+L			+R					+			+	+	+
c			+	+R		+					+	+L			
d	+L				+		+	+R							
e							+L	+	+R						

B

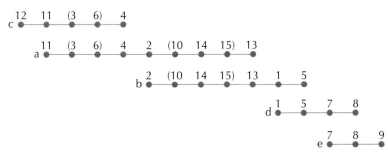

Figure 20.18 STS content mapping. **A.** STSs (1 through 15) that generate an ampli-fication product with a set of clones (a through e) are designated with a plus (+) sign. The letters L and R denote STSs at the 5′ and 3′ ends of the insert of a clone which represent the left and right ends of the insert, respectively. The STSs are un-ordered. **B.** Based on the overlaps from the results in part A, a contig consisting of five clones (a through e) can be arranged and an STS content map can be formulated. A dot with a number represents a specific STS. The data do not resolve the order of some adjacent STSs (bracketed numbers). The intervals between STSs are the same because actual distances are not known.

Figure 20.19 STS content mapping with ordered STSs. The numbers on the upper-most horizontal line are mapped STSs (20 through 31). The dots represent the STS content of each clone (f through j). Overlapping clones can be readily determined be-cause the STSs are ordered. The vertical dashed lines provide a visual guide joining each mapped STS with the corresponding STS in one or more of the clones of the contig.

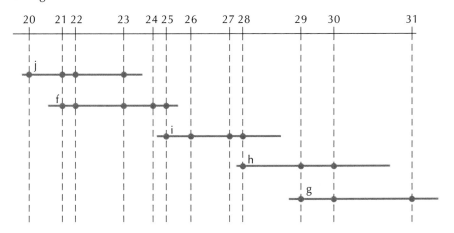

Assembling Contigs from Cosmid, P1, and λ Libraries

Small-insert contigs are often more amenable for the study of genes and large-scale DNA sequencing projects than are large-insert contigs. Cosmid libraries are used frequently as a source for assembling region-specific or whole-chromosome contigs. In most instances, overlapping cosmid clones are initially identified by DNA fingerprinting. Briefly, DNA is extracted from each clone and treated with a restriction endonuclease. The restriction endonuclease fragments are labeled, separated by electrophoresis, and visualized by autoradiography. Each clone produces a distinctive pattern of fragments which represents its unique DNA fingerprint; overlapping clones will share one or a few fragments.

The T4 DNA polymerase replacement reaction can be used for end-labeling DNA fragments regardless of whether the restriction endonuclease produces 5′ or 3′ extensions or blunt ends. In this case, T4 DNA polymerase and one labeled deoxyribonucleotide are added to a restriction endonuclease-treated cosmid DNA sample (Fig. 20.20). The 3′ exonuclease activity of the T4 DNA polymerase removes nucleotides from the 3′ ends of the fragments until a nucleotide that is complementary to the single nucleotide in the reaction mixture is reached. The free labeled nucleotide is linked to the 3′ end because the polymerase activity of T4 DNA polymerase supersedes the 3′ exonuclease activity. No additional incorporation of nucleotides occurs because no other nucleotides are present in the reaction mixture.

There are other strategies for labeling restriction endonuclease fragments. The recessed 3′ end of the restriction endonuclease fragments can be extended (filled in) by Klenow fragment, with the 5′ overhang as the tem-

Figure 20.20 End-labeling of double-stranded DNA molecules with T4 DNA polymerase. The 3′ exonuclease activity of T4 DNA polymerase removes nucleotides from the 3′ ends of DNA molecules that have blunt ends (**A**), 3′ extensions (**B**), or recessed 3′ ends (**C**). The 3′ exonuclease activity of T4 DNA polymerase continues until a base in the opposite strand is reached that is complementary to the deoxyribonucleotide in the reaction mixture; then the polymerase activity of T4 DNA polymerase links the deoxyribonucleotide from the reaction mixture to the available 3′ end. No further in vitro incorporation occurs because there are no other deoxyribonucleotides in the reaction mixture. Although not shown, both ends of a restriction endonuclease-digested DNA molecule are labeled. In this example, the incorporation of labeled dGTP (dGTP*) is depicted. Any deoxyribonucleotide can be used with the T4 DNA polymerase end-labeling procedure.

plate and one of the incorporated deoxyribonucleotides being labeled. Alternatively, labeled linkers can be added to the extensions of the restriction endonuclease fragments.

Because of the large numbers of cosmid clones that have to be analyzed for overlaps, the search for shared labeled restriction endonuclease fragments between pairs of clones has been computerized. The DNA fingerprint for each clone is captured by electronic scanning and digitized, and pairwise comparisons are conducted. Matches between bands are used to organize the clones into contigs. Overlaps of clones of a contig are confirmed by detailed restriction endonuclease mapping of the inserts. Gaps between contigs can be bridged by generating probes from the closest ends of adjacent contigs and screening a large-insert library for the missing section of DNA.

Transcriptional Mapping

The clones of a cDNA library are DNA copies of expressed gene transcripts that were present at the time the mRNA was extracted from a particular tissue. Assigning individual cDNA clones to chromosome regions provides researchers with possible candidates for disease genes. If a genetic linkage study maps a disease gene to the same chromosome region as a cDNA sequence(s), then the cDNA clone(s) can be used to determine if the clone(s) is derived from the gene that is responsible for the genetic disease.

The assignment of cDNA clones and other forms of expressed sequences to specific chromosome regions is called transcriptional (gene) mapping. Various methods have been devised for constructing transcriptional maps. With one approach, individual cDNA clones are partially sequenced and an STS is designed from the translated portion of each cDNA insert. This particular type of STS is called an expressed STS (eSTS; gene-based STS). Monochromosomal somatic cell hybrids and somatic cell hybrid cell lines that maintain a portion of a particular human chromosome (deletion panel) are used to determine the chromosome location of an eSTS. Briefly, PCR amplification with an eSTS of DNA from monochromosomal cell hybrids for each chromosome establishes the chromosome that carries the eSTS sequence. PCR amplification with DNA from members of a deletion panel then identifies the chromosome region that contains the eSTS (Fig. 20.21). Also, gene-based STSs can be added to existing RH maps. By 1996, about 20,000 human gene-based STSs were mapped to all autosomes and the X chromosome.

In addition to the mapping of cDNA clones to chromosome regions, a massive project of partial sequencing of the clones of cDNA libraries that are derived from virtually all human organs and tissues has been undertaken by The Institute for Genome Research (TIGR) of Rockville, Md., and other laboratories. One of the missions of this program is to create a catalog of short sequences (~150–300 nucleotides) for every expressed human gene. These short coding sequences are called expressed sequence tags (ESTs). The extent, diversity, and transcriptional activity of expressed human genes can be studied using ESTs. Furthermore, STSs can be created from ESTs for mapping studies and selecting genomic clones that contain the gene.

The partial sequencing of cDNA clones and the processing of the sequence data are, more or less, fully automated. Each new EST is analyzed to determine if it has been sequenced previously. If an EST is not redundant, it is entered into the EST database (dbEST). Further comparisons are con-

A

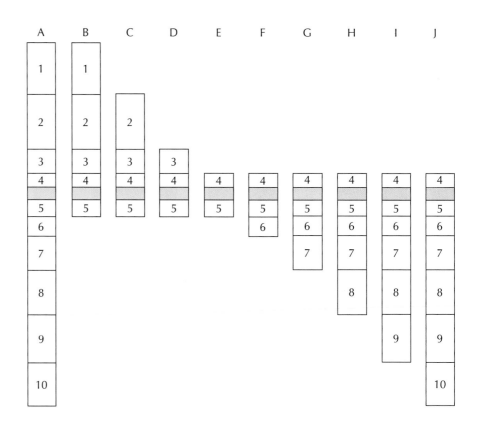

B

| Marker | Deletion panel | | | | | | | | | | Interval |
	A	B	C	D	E	F	G	H	I	J	
STS-a	+	+	+	+	−	−	−	−	−	−	3
STS-b	+	−	−	−	−	−	−	+	+	+	8
STS-b	+	+	+	−	−	−	−	−	−	−	2
STS-d	+	−	−	−	−	−	+	+	+	+	7
STS-e	+	−	−	−	−	−	−	−	−	+	10

Figure 20.21 Assigning a marker to a chromosome interval using a hybrid cell deletion panel. **A.** A schematic representation of the human chromosome content of a monochromosomal cell hybrid (A) and members (B through J) of a cell hybrid deletion panel. Intervals (1 through 10) are defined by the limits of the deletions of the chromosomes of the deletion panel. A solid rectangle marks the centromere of each chromosome. **B.** The results of PCR amplification with STS-based markers (STS-a through STS-e) with the cell hybrid lines (A through J) shown in panel **A**. The presence or absence of a PCR product is denoted by a plus or minus sign, respectively. The presence or absence of a PCR product from an STS marker with the members of the deletion panel is used to determine the chromosome interval that carries the STS. For example, STS-d is confined to interval 7 because a PCR product is always present when this region is carried by a member of the deletion panel.

ducted to ascertain if an EST resembles a known gene, gene family, or category of gene function or if it is unknown. By 1995, about 300,000 ESTs had been determined from 300 cDNA libraries derived from 37 organs and tissues. Of these ESTs, about 90,000 represent distinct human expressed sequences, with about 10,000 of them corresponding to genes with known cellular roles and the remaining 80,000 representing undiscovered genes.

Cloning Human Disease Genes

The cloning of a human disease gene is usually not accomplished by implementing a prescribed set of experimental protocols. Rather, methods and resources are used as the circumstances dictate. The initiation of a search for a disease gene depends on the existing information about its gene product. In some instances, the gene product is well known. In other cases, it is a matter of guesswork as to what the gene product might be. Finally, for many inherited diseases, the nature of a gene product is completely unknown. Strategies have been devised to deal with each of these conditions. In general, four approaches are used to hunt down a disease gene: functional gene cloning; candidate gene cloning; positional gene cloning (positional cloning); and positional-candidate gene cloning. Regardless of the approach, authentication of a disease gene is based on detecting a nucleotide change in the gene of affected individuals that does not occur in the same gene of unaffected individuals.

Detection of Mutations in Human Genes

A number of simple and inexpensive assays such as single-strand conformational (conformation) analysis (SSCA, also called single-strand conformational polymorphism [SSCP]), denaturing gradient gel electrophoresis (DGGE), heteroduplex analysis (HA), chemical mismatch cleavage (CMC), and the protein truncation test (PTT) have been devised to detect mutations.

Of these assays, SSCA is widely used. Briefly, some—or, if feasible, all—of the exons of a cloned gene are amplified individually by PCR of DNA from affected and unaffected individuals. Each pair of primers is determined from sequences that flank each exon or from the terminal ends of each exon. In addition, DNA sequence data can be used to develop PCR primers for the 5′ upstream region adjacent to the first exon, for the 3′ downstream region adjacent to the last exon of the gene, and across the splice junctions of exons and introns.

After separate PCR amplifications, the products of each reaction are denatured, rapidly cooled, and electrophoresed. Each denatured single-stranded DNA molecule assumes a three-dimensional conformation that is dependent on its primary nucleotide sequence. The conformation is the consequence of intrastrand base pairing and the formation of other bonds. Because of base complementarity, the two single strands of a double-stranded DNA molecule have different nucleotide sequences, and each has a different three-dimensional conformation. The different conformations have different rates of migration in a gel during electrophoresis. Hence, two bands can be visualized by either DNA-specific staining or autoradiography if the strands were labeled during PCR amplification. If two DNA molecules from different sources that represent the same segment of a gene differ by a single nucleotide pair, then the conformations of the two single DNA strands from one of the sources are very likely to be different from

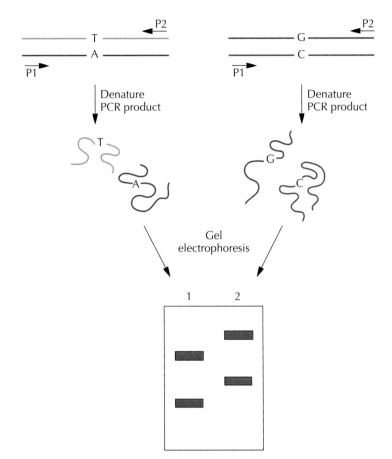

Figure 20.23 Functional gene cloning. The flow diagram highlights some of the steps that lead to the isolation and identification of a gene when information about the sequence of the gene product is known.

Figure 20.22 Single-strand conformation analysis (SSCA). Samples of DNA from two different sources that differ by a single nucleotide pair (A:T versus G:C) are amplified by PCR with the same primers (P1, P2). The PCR products are denatured and run in individual lanes (1, 2) of a nondenaturing gel. The distance that a single-stranded DNA molecule migrates into the gel depends on its conformation. A single nucleotide pair difference between DNA strands is often sufficient to produce different single-strand conformations. SSCA of amplified DNA from a heterozygote with a single nucleotide pair mutation will produce four bands (not shown).

those of the two strands from the other source. In other words, each of the four strands will migrate during electrophoresis at its own distinctive rate. With side-by-side comparisons between DNA from affected and unaffected individuals, the differences in the migration of single-stranded DNA molecules are easily detected (Fig. 20.22). The SSCA localizes a nucleotide alteration to a specific exon or region of a gene. The nature of the mutational difference is not revealed by the SSCA, but this information can be obtained by DNA sequencing. The SSCA is not foolproof. It detects about 90% of the single-base-pair alterations in PCR products that are 200 bp or less.

Functional Gene Cloning

A functional gene cloning project starts with a protein whose function is known. The amino acid sequence of the protein is used to deduce the possible DNA sequence of the coding region of the corresponding gene (target gene). A cDNA library that is derived from the tissue where the protein is

prevalent is screened with an oligonucleotide probe based on the coding sequence of the target gene. Alternatively, if the mRNA for the protein can be purified, then a full-length cDNA is synthesized and cloned. DNA sequencing confirms that the correct cDNA clone has been selected or synthesized.

The chromosome location of the target gene is determined by in situ hybridization with either the cDNA clone or a genomic clone that was selected by the cDNA clone. Alternatively, a monochromosome cell hybrid panel and then the appropriate deletion panel can be screened with either the cDNA or a selected genomic clone to pinpoint the chromosome site of the target gene.

A cosmid contig that covers the chromosome region where the target gene is located is screened for clones that hybridize to the cDNA clone. The selected genomic clones are sequenced and, with the cDNA sequence as a guide, the exons, introns, and 5′ and 3′ flanking regions of the gene can be delineated. If there is no cosmid contig that covers the chromosome location that contains the target gene, then a large-insert clone that spans the chromosome region is isolated with the cDNA or genomic probe. Small-insert subclones are established from the large-insert clone and are screened with the cDNA probe. These positive clones are sequenced to characterize the target gene (Fig. 20.23).

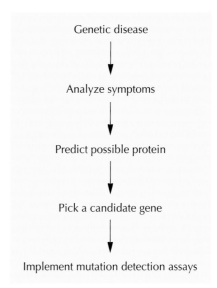

Figure 20.24 Candidate gene cloning. The flow diagram highlights some of the steps that lead to the isolation and identification of a disease gene by predicting, on the basis of the symptoms of the disease, which known gene is most likely to be the disease gene.

Candidate Gene Cloning

Although the candidate gene cloning approach is not a very efficient way to clone human genes, it can be useful in some instances. Briefly, after an analysis of the symptoms of a genetic disease, and in the absence of a specific protein, an informed guess is made about the kind of protein that might be responsible for the inherited condition. Then, the current repertoire of cloned genes is examined and a candidate(s) is chosen. A mutation detection assay is devised, based on the nucleotide sequence of the candidate gene, and is used to determine whether it is the actual gene (Fig. 20.24). Considering the large numbers of genes in the human genome, and how few have been characterized, it is not surprising that successful matches are infrequent. However, a negative outcome is helpful because the gene can be excluded as a major contributor to a particular genetic disorder.

Positional Gene Cloning

The positional gene cloning approach is mandated when nothing is known about the gene product that is responsible for a genetic disease, and there are no likely candidate genes (Fig. 20.25). In these instances, the disease gene is mapped to a chromosome location (position) and the actual gene is discovered by using various tools and resources ("gene hunting" technologies). A helpful starting point for a positional gene cloning project is if a chromosome rearrangement such as a translocation or a large deletion (>10 kb) is associated with some affected individuals. It is assumed that the gene that is responsible for an abnormal phenotype is damaged by the chromosome abnormality. With this clue, linkage studies can be focused on a specific chromosome region. Otherwise, with a large number of polymorphic markers, a whole genomic scan is carried out to find the position of the disease gene. After the linkage study localizes a gene to a chromosome region,

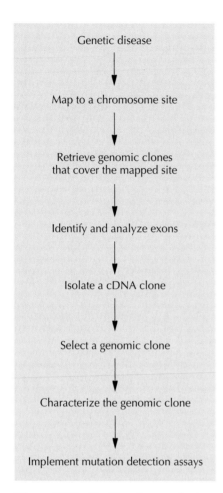

Genetic disease

↓

Map to a chromosome site

↓

Retrieve genomic clones
that cover the mapped site

↓

Identify and analyze exons

↓

Isolate a cDNA clone

↓

Select a genomic clone

↓

Characterize the genomic clone

↓

Implement mutation detection assays

Figure 20.25 Positional gene cloning. The flow diagram highlights some of the steps that lead to the isolation and identification of a disease gene when its gene product is unknown. Chromosome mapping and closely linked probes are used to track down the disease gene.

multipoint mapping with additional polymorphic probes is used to pinpoint the site of the disease gene and identify the closest flanking marker sites. At best, the level of resolution between mapped marker sites is 1 cM, which spans about 1×10^6 bp. On average, this amount of human DNA contains 20 and possibly 50 different genes. The objective of the positional gene cloning approach is to discover which of the possible genes between two marker sites is actually the disease gene.

Genomic clones that include the flanking marker sites and encompass the DNA between these sites are retrieved from a contig that covers the chromosome region that contains the disease gene. If a contig is not available, then probes for the closely linked marker sites are used to screen genomic DNA libraries for clones that are derived from the region with the disease gene. Between 1986 and 1990, when human gene hunting technologies and resources were rudimentary, chromosome jumping (Fig. 20.26) and walking (Fig. 20.27) were used to identify genomic clones that were close to or contained the disease gene. With the development of large-insert genomic libraries and the assembly of contigs, these strategies are no longer necessary.

Regardless of how the genomic clones from the chromosome region with the target gene are obtained, it is important to know which of these clones or subclones have exons. There are various indirect and direct methods, such as identification of CpG islands, interspecies DNA Southern blotting, hybrid selection, exon trapping, DNA sequencing, and computational gene identification, that can be used to determine whether a cloned insert has an exon(s).

The transcribed regions of vertebrate genomes are often preceded by clusters of nucleotides that are rich in cytosine and guanine residues. These clusters are called CpG islands. A set of CpG islands can be identified by the grouping on a restriction endonuclease map of *Eag*I, *Bss*II, and *Sac*II restriction endonuclease sites. These particular restriction endonuclease recognition sequences are GC-rich. If two or more of these sites fall within 5 to 10 kb of each other, it is a good indication of a CpG island. The occurrence of a CpG island is not a strict test for the presence of an exon, but it signifies that a transcribed gene is nearby.

Genomic clones or subclones can be hybridized by Southern DNA blotting to restriction-digested genomic DNA from different vertebrates such as mouse, rat, rabbit, monkey, cow, chicken, and fish (zoo blot, multispecies blot, Noah's ark blot). A positive cross-hybridization signifies, in all likelihood, that the clone has coding sequences, because many exons are evolutionarily conserved whereas repetitive DNA and noncoding nucleotide sequences, including introns, are not. A positive result with a zoo blot indicates that a clone contains an exon(s), but it does not reveal whether the clone represents the disease gene.

Hybrid selection is a rapid and effective procedure that identifies a ge-

Figure 20.26 Formation of a chromosome jumping library. Genomic DNA is partially digested with a rare-cutting restriction endonuclease, and fragments that are about 200 kb long are collected, ligated with the *supF* gene (~7 kb), and circularized. The pairs of letters—A and B, X and Y, and S and T—represent sequences that were originally 200 kb apart but, after circularization, are separated by 0.7 kb. The 200-kb circles contain many *Eco*RI sites. After digestion with this restriction endonuclease, some of the fragments that are produced contain supF and various flanking se-

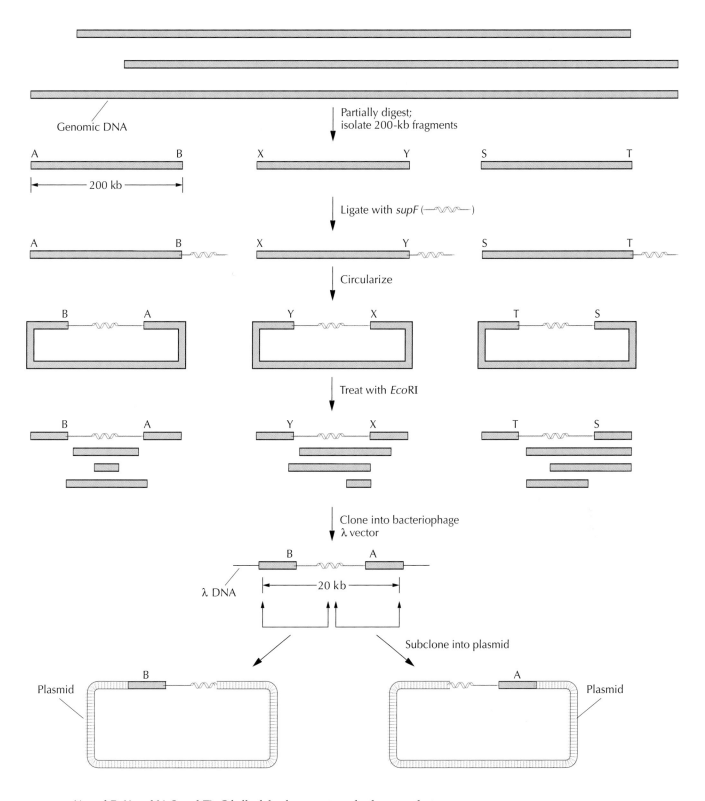

quences (A and B; X and Y; S and T). Of all of the fragments, only the ones that are about 20 kb long are cloned and packaged into a bacteriophage λ vector. Those vectors that carry the *supF* gene will grow on SupF⁻ host cells. The clone that hybridizes to a probe is identified. The subcloned half of this clone that does not hybridize to the probe contains DNA that was originally 200 kb away from the sequence that hybridizes to the probe. The different DNA features are not drawn to scale.

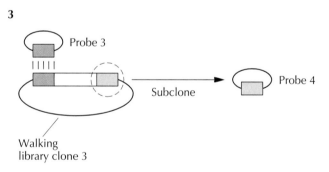

4 Extent of walk (from sequence 1 to sequence 4)

Figure 20.27 Chromosome walking. **(1)** A probe (probe 1) hybridizes to a cloned 40-kb piece of DNA. After subcloning and restriction endonuclease mapping, the sequence that is farthest away from the original hybridizing sequence is used to form a second probe (probe 2). **(2)** With probe 2, another clone from the library (other than library clone 1) is selected, and the sequence that is farthest away from it is used to form a third probe (probe 3). Walking library clones 1 and 2 together contain DNA that spans ~80 kb (minus the overlap between the two clones). **(3)** Same as steps 1 and 2, but using probe 3. With a third walking library clone, the walk has advanced another 40 kb. **(4)** Three overlapping DNA fragments that encompass ~120 kb of contiguous chromosomal DNA. Chromosome walking can be bidirectional, and restriction endonuclease mapping helps to determine the direction of the successive walks.

nomic clone with an exon(s) and concomitantly isolates the corresponding cDNA molecule. Hybrid selection can be conducted in a number of different ways. Basically, DNA of a genomic clone from the chromosome region that contains the disease gene is bound to a matrix, prehybridized with repetitive DNA, and then hybridized with linear vector-insert DNA molecules of a cDNA library constructed from the tissue that is most likely to express the target gene. The nonhybridized vector-cDNA molecules are washed off the filter, and the hybridized vector-cDNA molecules are eluted and amplified by PCR. The primers for the PCR are directed to vector sequences that flank the cDNA insert (Fig. 20.28). If the primary tissue of expression of the target gene is in doubt, then cDNA libraries from a number of tissues can be pooled and hybridized to individual genomic clones. The PCR product can be cloned and used to test whether it represents the coding region of the disease gene. For example, the cDNA can be sequenced and compared by computer analysis with the sequences of known genes. A significant sequence similarity provides information about the class of proteins represented by the cDNA sequence. If the cDNA sequence encodes a type of protein that is likely to be the gene product of the target gene, then the genomic clone(s) are sequenced and the exons, introns, and 5' and 3' flanking regions are identified. Mutation detection assays can be developed to ascertain if there is a nucleotide difference between affected and unaffected individuals. If the nucleotide sequence matching analysis fails to identify a likely candidate, then other genes in the region are sequenced and analyzed. In practice, to spare time and effort, a search for a target gene often entails partial characterization of a number of genes until a highly likely candidate is found that can be examined in detail, including the development of a mutation detection assay.

Exon trapping (exon entrapment, exon amplification) is a method that enables exons to be identified and cloned from subclones of genomic clones (Fig. 20.29). Briefly, DNA from a genomic clone is digested to produce pieces from 1 to 6 kb in length which are cloned into a specially designed vector. The multicloning site of this vector lies within an intron that is flanked by two exons (exon 1 and exon 2). This artificial gene (exon 1–intron–exon 2) is under the control of a strong eukaryotic promoter. The vector can be maintained in either *Escherichia coli* or cultured mammalian cells. After the introduction (transfection) of vector DNA without an insert into a mammalian cell, the exon 1–intron–exon 2 unit is transcribed, and then the intron is removed from the primary transcript. The presence of this processed RNA (exon 1–exon 2) can be assayed by reverse transcriptase PCR (RT-PCR). With RT-PCR, DNA is synthesized from an mRNA preparation with reverse transcriptase (first-strand synthesis). A second DNA strand is synthesized with a primer which, in this example, is complementary to a sequence from exon 1 of the first DNA strand. A second primer that is complementary to a sequence of the second DNA strand from exon 2 is added to the reaction mixture, and the DNA is amplified by PCR. The length of the PCR product is readily determined by gel electrophoresis.

If a restriction endonuclease fragment with an exon (exon A) flanked by intron sequences is cloned into an exon-trapping vector, then after transfection the processed transcript will have three exons (exon 1–exon A–exon 2). The size of the PCR amplification product will be greater than when there is no insert, when an insert does not have an exon with functional splice sites (splice donor and splice acceptor), or when an insert does not have an exon. If an insert has more than one exon, each with functional splice sites, then the processed transcript will contain all of the exons.

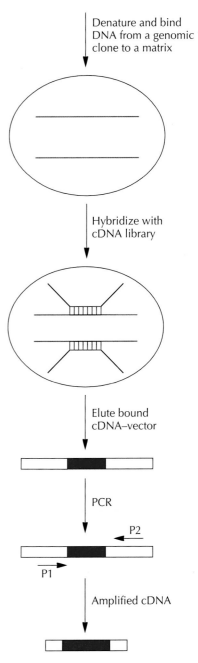

Figure 20.28 Hybrid selection. The DNA of a genomic clone is used to capture by hybridization a cDNA clone that subsequently is amplified, cloned, and analyzed.

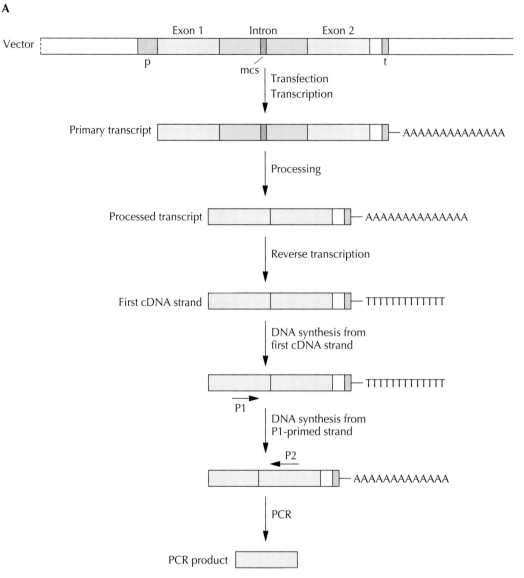

Figure 20.29 Exon trapping. **A.** An exon trapping vector consists of an artificial gene containing a promoter (p), two exons (exon 1 and exon 2) that are separated by an intron (intron) which carries a multiple cloning site (mcs), and the transcription termination sequence (t). The vector is transfected into eukaryotic cells. During the processing of the primary transcript of the artificial gene, the intron is removed. RT-PCR is used to produce a DNA molecule (PCR product) of a particular length that contains part of the coding region of both exons.

The PCR product carrying a trapped exon can be cloned if a restriction endonuclease recognition site is part of each primer sequence. A trapped exon can then be used as a probe to screen a cDNA library. With the nucleotide sequence of a trapped exon, a gene database can be searched for a similar sequence. When a trapped exon is identified as a likely part of the disease gene, genomic clones that cover the site of the disease gene are characterized and sequenced. Mutation detection assays are developed, and the DNA from affected and unaffected individuals is examined for mutations. Since the mutations that lead to an abnormal phenotype are not always uni-

Figure 20.29 (Continued) **B.** An insert is cloned into the multiple cloning site of an exon trapping vector and the vector-insert combination is transfected into eukaryotic cells. In this example, exon A, which is part of the insert, has functional acceptor and donor splice sites. During processing of the primary transcript, the intron sequences flanking exon A are removed and exon A is spliced between exons 1 and 2 in the processed transcript. The length of the RT-PCR product reveals whether an exon has been trapped between exons 1 and 2 and, accordingly, whether the insert contains an exon.

formly distributed among all exons, the larger the coding region of a putative gene that is scanned, the greater the likelihood that a mutation will be discovered.

Various computer programs (e.g., Gene Recognition and Analysis Internet Link [GRAIL]) can be used to identify exons within a DNA sequence. This form of detection depends on a number of features. One important indicator is based on the probabilities of nucleotide regularities that are expected for a coding region. If a laboratory is equipped for large-scale

DNA sequencing, then genomic clones covering the site of the disease gene can be sequenced and computationally analyzed for exons. The sequence of a predicted exon can be used to search a gene database for similar sequences, or an oligonucleotide probe can be synthesized for screening a cDNA library. Eventually, as is the case with all methods that identify exons in genomic clones, it is necessary to establish that a putative exon is part of the target gene.

Every positional gene cloning project is labor-intensive and time-consuming. It was an impressive accomplishment that between 1986 and 1995, over 50 human disease genes were discovered with this approach. In some instances, the search required only 1 to 2 years, whereas it took a consortium of research laboratories 10 years to find the gene for Huntington disease. However, as more gene cloning continues and high-density transcript maps for each chromosome are assembled, positional gene cloning will be superseded by positional-candidate gene cloning.

Positional-Candidate Gene Cloning

Positional-candidate gene cloning entails mapping a disease gene of unknown function to a chromosome location, and then examining up-to-date genetic and transcript maps to determine which coding regions (genes, gene-based ESTs) have been mapped to the same interval (bin) as the disease gene (Fig. 20.30). It is highly likely that one of these genes or sequences represents the disease gene. Mutation detection assays can be devised if any of these candidate genes have been characterized. Or, a likely EST(s) can be used as a probe to select a genomic clone that can be sequenced to provide the information for designing mutation detection assays. As physical and transcript maps become more complete, positional-candidate gene

Figure 20.30 Positional-candidate gene cloning. The flow diagram highlights some of the steps that lead to the isolation and identification of a disease gene when its gene product is unknown but it maps to the same chromosome region as some functional genes and ESTs. Of these genes and ESTs, those that are possible candidates for the disease gene are selected and used to determine which coding region corresponds to the disease gene.

cloning will become the predominant method for discovering human disease genes.

The Human Genome Project

On October 1, 1990, the Human Genome Project (HGP) was officially launched in the United States under the joint control of the Department of Energy and the National Institutes of Health. The project has as its ultimate goal the determination of the DNA sequence of the complete human genome. Overall, this vast array of genetic information will provide platforms for launching more detailed research programs into every simple genetic disorder, and act as a springboard for studying complex inherited human traits. It was estimated in 1990 that the HGP would take 15 years and cost $3 billion. In a short time, this effort became international in scope, with government-sponsored projects in the United Kingdom, France, Canada, Germany, and Japan. Now, there is significant cooperation and coordination among many national and international agencies, private companies, and not-for-profit research institutions. The HGP is an immense commitment that encompasses many different research ventures.

The U.S.-based component of the HGP is subdivided into discrete areas of study that include developing high-resolution genetic and physical maps; enhancing the cost-effectiveness and efficiency of massive-scale DNA sequencing; exploring new technologies for mapping genes and sequencing DNA; creating the computer capabilities for managing large amounts of data; and examining the ethical, legal, and social implications (ELSI) of the knowledge that would be gained from the HGP. The aim of the ELSI component is to establish guidelines for researchers and clinicians, as well as laying a foundation for government policies concerning the use of genetic information. As part of the "new technologies" component, the genome of the yeast *Saccharomyces cerevisiae* (150 Mb of DNA) has been completely sequenced by a consortium of research groups. The complete DNA sequences of other model organisms including a nematode (*Caenorhabditis elegans*, 100 Mb of DNA), a fruit fly (*Drosophila melanogaster*, 120 Mb of DNA), a bacterium (*E. coli*, 4.2 Mb of DNA), a plant (*Arabidopsis thaliana*, 100 Mb of DNA), and a mouse (*Mus musculus*, 3,000 Mb of DNA) are also being determined.

Some of the original goals of the 1990–1995 phase of the U.S. portion of the HGP were revised in 1993 because of the rapid progress in genetic mapping due to the introduction of microsatellite polymorphic markers and the development of fairly complete physical maps. By 1996, a number of the redefined targets were met. For example, in 1994 a human linkage (genetic) map that contained 5,826 loci covering 4,000 cM was published. Although only 908 of the loci were linked with odds of more than 1,000:1, the overall marker density was 0.7 cM (4,000 cM/5,826 loci), which is greater than the 2- to 5-cM map that was expected by 1995. An STS-based physical map with a resolution of 300 kb, that is, one STS site for every 300 kb of human DNA, was anticipated by 1996. However, in 1995, a complete physical map with 200 kb of resolution was assembled.

Headway in optimizing DNA sequencing technology has been steady, but not spectacular. The cost of sequencing decreased from about $5.00 (U.S.) per base pair in 1990 to about $0.30 (U.S.) in 1996. The rate of sequencing increased from about 10,000 bases per day in 1990 to 50,000 in 1996. It is estimated that 80 Mb, or 2.5% of the human genome, will be se-

quenced by 1998. Without further significant enhancements, a factory-scale effort of 30 automated DNA sequencing machines working 24 hours a day without any downtime, with a complete set of physical maps of cosmid clones, could generate about 3,000 Mb of DNA sequence data in about 6 years at a cost of about $900 million (U.S.). Additional time and cost would be required to check for errors and assemble the finished sequence. Before underwriting such a massive endeavor, research is being conducted to improve sequencing rates significantly with automated fluorescent-dye DNA sequencers using the Sanger method. Rapid "non-Sanger" DNA sequencing technologies are also being sought.

The HGP has allocated funds to a number of different research groups to handle various aspects of the overall project. In most instances, the responsibility of creating the genetic and physical maps of a particular chromosome has been given to large centers and small laboratories that cooperate with each other. Some very large research institutions have undertaken the prodigious task of assembling genetic and physical maps of the whole genome. Collectively, the human molecular genetics research projects have generated a massive amount of data about polymorphic probes; STSs; clones; the contents of genetic, physical, and integrated maps; restriction fragments; DNA fingerprints; and DNA sequences. These data must be amassed, organized, stored, interpreted, compared, combined, and made accessible in both raw and finished forms to other researchers. The comprehensive management of this information would not be possible without computer resources that include databases, database management systems, problem-solving algorithms, and programs that facilitate automation. Informatics is the name of the discipline that develops computer and statistical procedures for handling scientific information. Bioinformatics deals with the computer analysis and data management of biological information.

The bioinformatics component of the HGP has been extremely successful in creating the computer tools to cope with all aspects of genome data. Considerable information about the contents of the various chromosome maps including complete graphical representations, genome research resources, and software is available to both experts and the public at a variety of electronic sites. For example, the World Wide Web site of the U.S. National Center for Human Genome Research (http://www.nhgri.nih.gov/index.html) contains information about the HGP and many links to other genome centers.

Since its inception, the HGP has had a mandate to anticipate the ethical, legal, and social issues that might arise from the mapping and sequencing of the human genome and to develop guidelines, policies, and legislation that would ensure that human genetic knowledge would be used responsibly. Actually, the HGP does not raise any fundamentally new ethical, legal, or social issues that do not originate from medical genetics research in general. However, the HGP will lead to the identification of a large number of disease genes and to the determination of the sequences of many of these genes that can be used to develop DNA-based diagnostic tests.

This kind of knowledge generates many potential concerns. Will genetic information be used to discriminate against an individual who is applying for health insurance, employment, or immigration status? Will genetic information cause a person to be socially stigmatized? Are there sufficient safeguards to keep personal genetic information private? How should personal and societal needs be balanced? Do medical practitioners and clin-

icians have sufficient understanding of human genetics to explain the consequences of a particular genetic test to patients? Will counseling alleviate anxiety? Will genetic information disrupt family relationships? How is it possible to obtain truly informed consent from a patient? Should a DNA diagnostic test be offered if there is no cure for the condition? How can the public be educated to understand the implications of genetic information? There are no simple answers to these questions and the many others that emanate from human genetics research. In the United States, the ELSI program of the HGP has commissioned a large number of studies to examine these topics; established education programs, workshops, and exhibitions for students, teachers, clinicians, medical practitioners, the public, lawyers, and judges; investigated the feasibility of genetic testing for cystic fibrosis and inherited forms of breast, ovarian, and colon cancer; created two task forces (Genetic Information and Insurance; Genetic Testing) to study these particular issues in depth; and helped develop a proposal for U.S. federal legislation that safeguards the genetic information of identifiable individuals.

On the basis of the HGP-ELSI program and many other studies, five fundamental principles concerning the use of genetic information and the delivery of genetic services have been repeatedly emphasized: autonomy, privacy, justice, equity, and quality. In this context, the concept of autonomy stresses the requirement for personal freedom when an individual accesses genetic services. For example, genetic testing should be voluntary and carried out with informed consent; only high-risk individuals should be tested, and participants should be able to choose either to know or not to know the results of a genetic test. Knowledgeable counseling should be available so that all features of a procedure including its predictive capabilities, its medical consequences, and the nature of the therapeutic treatment, if it exists, are fully discussed and comprehended.

It is generally thought that genetic information differs from other kinds of personal information; and consequently special safeguards must be legislated to ensure privacy and confidentiality. Justice and equity are overlapping concerns. The consensus is that all those who need a genetic service should have access to it. As with other social and medical programs, adults with diminished mental capacity and children must be fully protected. In terms of quality, it is essential that genetic testing facilities be accredited and monitored to guarantee that the tests are conducted properly by professionally trained personnel and that the confidentiality of each participant is fully protected.

SUMMARY

The mode of inheritance of many human genetic diseases can be revealed by studying the pedigrees of multigeneration families with a clearly defined abnormality. On the basis of the pattern of inheritance within families, it can be determined whether an observed genetic condition is autosomal dominant, autosomal recessive, X-linked dominant, or X-linked recessive. With X-linked conditions, the gene causing the illness is, of course, on the X chromosome; for autosomal diseases, the specific chromosome location is not known. RFLP and STRP mapping can be used to identify marker sites that are linked to a disease gene, thereby localizing it to a specific chromosomal region. The likelihood of the odds (lod) method is used to establish

linkage between a marker site and a disease gene. The RFLP and STRP sites of a chromosome can be ordered by haplotype analysis of the inheritance patterns using a group of three-generation families with large sibships (CEPH families). In addition, the order of unique sites of a chromosome that are identified by PCR (STSs) can be ascertained by radiation hybrid mapping. Physical chromosome maps (contigs) are assembled from large-insert (YAC, BAC, and PAC) and small-insert (cosmid, P1, and λ) genomic libraries using STS-content mapping or other strategies including DNA fingerprinting. Transcript maps consist of cDNA sites and expressed sequence sites (ESTs) along a chromosome. The development of linkage, physical, and transcript maps facilitates the isolation and characterization of disease genes.

The discovery of a human gene is authenticated by establishing that individuals with a genetic defect have a change in a gene which is not found in unaffected individuals. Single-strand conformation analysis (SSCA) is commonly used to detect gene mutations.

In general, there are four approaches for isolating a human gene. First, functional gene cloning relies on information about the gene product which is used to create probes that, in turn, are used to screen a cDNA library. A positive cDNA clone carries the coding region of the target gene and can be used to select genomic clones for characterizing the complete gene. Second, candidate gene cloning is based on an educated guess of what gene might be responsible for a genetic disease. In this case, likely candidate genes from affected and unaffected individuals are examined with a mutation detection assay to determine which one, if any, is the disease gene. Third, positional gene cloning is used when nothing is known about the possible disease gene or its product. This approach is labor intensive and calls upon a repertoire of strategies and protocols. The site of the disease gene is mapped to a chromosome region with RFLP or STRP probes using pedigrees with the genetic disorder. Probes from sites that are closely linked to the disease gene are used to retrieve genomic clones that cover the mapped site of the disease gene. These genomic clones, or subclones derived from them, are tested for the presence of exons by zoo blotting, hybrid selection, or exon trapping. Mutation detection assays are devised from the DNA sequence information of various exons that are likely to correspond to the disease gene. Fourth, the positional-candidate gene cloning approach entails mapping a disease gene to a chromosome site, reviewing the functional genes and expressed sequence tags that have been mapped to the same chromosome interval, and deducing which of these genes is likely to be the disease gene. A mutation detection assay determines which of the candidate genes is the disease gene.

The Human Genome Project (HGP) is a massive, multidimensional research undertaking that has as its ultimate goal the complete sequencing of the human genome. The various facets of the HGP include developing high-resolution genetic and physical maps of all human chromosomes; sequencing the genomes of various model organisms such as *E. coli, C. elegans, S. cerevisiae, M. musculus,* and *A. thaliana*; optimizing the efficiency and rate of DNA sequencing; creating the computer capabilities for handling and analyzing the genetic, physical, and DNA sequence data generated by the HGP; informing the public about the consequences of acquiring human genetic information; and exploring the ethical, legal, and social implications that arise from sequencing the human genome. The HGP has made impressive

advances, and it is anticipated that the complete sequence of the human genome will be completed by the year 2005 or thereabouts.

References

Bellane-Chantelot, C., B. Lacroix, P. Ougen, A. Billault, S. Beaufils, S. Bertrand, I. Georges, F. Glibert, I. Gros, G. Lucotte, L. Susini, J. Codani, P. Gesnouin, S. Pook, G. Vaysseix, J. Lu-Kuo, T. Ried, D. Ward, I. Chumakov, D. Le Paslier, E. Barillot, and D. Cohen. 1992. Mapping the whole human genome by fingerprinting yeast artificial chromosomes. *Cell* **70**:1059–1068.

Botstein, D., R. L. White, M. Skolnick, and R. W. Davis. 1980. Construction of a genetic map in man using restriction fragment length polymorphisms. *Am. J. Hum. Genet.* **32**:314–331.

Chumakov, I. M., P. Rigault, I. Le Gall, C. Bellane-Chantelot, A. Billault, S. Guillou, P. Soularue, G. Guasconi, E. Poullier, I. Gros, M. Belova, J. Sambucy, P. Gervy, F. Glibert, S. Beaufils, H. Bui, C. Massart, M. De Tand, F. Dukasz, S. Lecoulant, P. Ougen, V. Perrot, M. Saumier, C. Soravito, R. Behouayila, A. Cohen-Akenine, E. Barillot, S. Bertrand, J. Codani, D. Caterina, I. Georges, B. Lacroix, G. Lucotte, M. Sahbatou, C. Schmit, M. Sangouard, E. Tubacher, C. Dib, S. Faure, C. Fizames, G. Gyapay, P. Millasseau, S. Nguyen, D. Muselet, A. Vignal, J. Morissette, J. Menninger, J. Lieman, T. Desay, A. Banks, P. Bray-Ward, D. Ward, T. Hudson, S. Gerety, S. Foote, L. Stein, D. C. Page, E. S. Lander, J. Weissenbach, D. Le Paslier, and D. Cohen. 1995. A YAC contig map of the human genome. *Nature* (London) **377**(Suppl.):175–297.

Collins, F. S. 1995. Positional cloning moves from perditional to traditional. *Nat. Genet.* **9**:347–350.

Collins, F., and D. Galas. 1993. A new five-year plan for the U.S. Human Genome Project. *Science* **262**:43–46.

Cooperative Human Linkage Center (CHLC): J. C. Murray, K. H. Buetow, J. L. Weber, S. Ludwigsen, T. Scherpbier-Heddema, F. Manion, J. Quillen, V. C. Sheffield, S. Suden, G. M. Duyk; Genethon: J. Weissenbach, G. Gyapay, C. Dib, J. Morissette, G. M. Lathrop, A. Vignal; University of Utah: R. White, N. Matsunami, S. Gerken, R. Melis, H. Albertsen, R. Plaetke, S. Odelberg; Yale University: D. Ward; Centre d'Etude du Polymorphisme Humain (CEPH): J. Dausset, D. Cohen, and H. Cann. 1994. A comprehensive human linkage map with centimorgan density. *Science* **265**:2049–2054.

Cox, D. R., M. Burmeister, E. R. Price, S. Kim, and R. M. Myers. 1990. Radiation hybrid mapping: a somatic cell genetic method for constructing high-resolution maps of mammalian chromosomes. *Science* **250**:245–250.

Fickett, J. W. 1996. Finding genes by computer: the state of the art. *Trends Genet.* **12**:316–320.

Green, E. D., and P. Green. 1991. Sequence-tagged site (STS) content mapping of human chromosomes: theoretical considerations and early experiences. *PCR Methods Appl.* **1**:77–90.

Grompe, M. 1993. The rapid detection of unknown mutations in nucleic acids. *Nat. Genet.* **5**:111–120.

Guyer, M. S., and F. S. Collins. 1995. How is the Human Genome Project doing, and what have we learned so far? *Proc. Natl. Acad. Sci. USA* **92**:10841–10848.

Gyapay, G., J. Morissette, A. Vignal, C. Dib, C. Fizames, P. Millasseau, S. Marc, G. Bernardi, M. Lathrop, and J. Weissenbach. 1994. The 1993–1994 Genethon human genetic linkage map. *Nat. Genet.* **7**:246–287.

Hudson, T. J., L. D. Stein, S. S. Gerety, J. Ma, A. B. Castle, J. Silva, D. K. Slonim, R. Baptista, L. Kruglyak, S. Xu, X. Hu, A. M. E. Colbert, C. Rosenberg, M. P. Peeve-Daly, S. Rozen, L. Hui, X. Wu, C. Vestergaard, K. M. Wilson, J. S. Bae, S. Maitra, S. Ganiatsas, C. A. Evans, M. M. DeAngelis, K. A. Ingalls, R. W. Nahf, L. T. Horton, M. O. Anderson, A. J. Collymore, W. Ye, V. Kouyoumjian, I. S. Zemsteva, J.

Tam, R. Devine, D. F. Courtney, M. T. Renaud, H. Nguyen, T. O'Connor, C. Fizames, S. Faure, G. Gyapay, C. Dib, J. Morissette, J. B. Orlin, B. W. Birren, N. Goodman, J. Weissenbach, T. L. Hawkins, S. Foote, D. C. Page, and E. S. Lander. 1995. An STS-based map of the human genome. *Science* **270**:1945–1954.

Knoppers, B. M., and R. Chadwick. 1994. The Human Genome Project: under an international ethical microscope. *Science* **265**:2035–2036.

Leppert, M., V. E. Anderson, T. Quattlebaum, D. Stauffer, P. O'Connell, Y. Nakamura, J.-M. Lalouel, and R. White. 1989. Benign familial neonatal convulsions linked to genetic markers on chromosome 20. *Nature* (London) **337**:647–648.

Ott, J. 1992. *Analysis of Human Genetic Linkage.* Johns Hopkins University Press, Baltimore, Md.

Poustka, A., T. M. Pohl, D. P. Barlow, A.-M. Frischauf, and H. Lehrach. 1987. Construction and use of human chromosome jumping libraries from NotI-digested DNA. *Nature* (London) **325**:353–355.

Schuler, G. D., M. S. Boguski, E. A. Stewart, L. D. Stein, G. Gyapay, K. Rice, R. E. White, P. Rodriguez-Tomé, A. Aggarwal, E. Bajorek, S. Bentolila, B. B. Birren, A. Butler, A. B. Castle, N. Chiannilkulchai, A. Chu, C. Clee, S. Cowles, P. J. R. Day, T. Dibling, N. Drouot, I. Dunham, S. Duprat, C. East, C. Edwards, J.-B. Fan, N. Fang, C. Fizames, C. Garrett, L. Green, D. Hadley, M. Harris, P. Harrison, S. Brady, A. Hicks, E. Holloway, L. Hui, S. Hussain, C. Louis-Dit-Sully, J. Ma, A. MacGilvery, C. Mader, A. Maratukulam, T. C. Matise, K. B. McKusick, J. Morissette, A. Mungall, D. Muselet, H. C. Nusbaum, D. C. Page, A. Peck, S. Perkins, M. Piercy, F. Qin, J. Quackenbush, S. Ranby, T. Reif, S. Rozen, C. Sanders, X. She, J. Silva, D. K. Slonim, C. Soderlund, W.-L. Sun, P. Tabar, T. Thangarajah, N. Vega-Czarny, D. Vollrath, S. Voyticky, T. Wilmer, X. Wu, M. D. Adams, C. Auffray, N. A. R. Walter, R. Brandon, A. Dehajia, P. N. Goodfellow, R. Houlgatte, J. R. Hudson, Jr., S. E. Ide, K. R. Iorio, W. Y. Lee, N. Seki, T. Nagase, K. Ishikawa, N. Nomura, C. Phillips, M. H. Polymeropoulos, M. Sandusky, K. Schmitt, R. Berry, K. Swanson, R. Torres, J. C. Venter, J. M. Sikela, J. S. Beckmann, J. Weissenbach, R. M. Myers, D. R. Cox, M. R. James, D. Bentley, P. Deloukas, E. S. Lander, and T. J. Hudson. 1996. A gene map of the human genome. *Science* **274**:540–546.

Taylor, K., N. Hornigold, D. Conway, D. Williams, Z. Ulinowski, M. Agochiya, P. Fattorini, P. De Jong, P. F. R. Little, and J. Wolfe. 1996. Mapping the human Y chromosome by fingerprinting cosmid clones. *Genome Res.* **6**:235–248.

Terwilliger, J. D., and J. Ott. 1994. *Handbook of Human Genetic Linkage.* Johns Hopkins University Press, Baltimore, Md.

REVIEW QUESTIONS

1. What is independent assortment?

2. Draw a four-generation pedigree with a disorder that is inherited as an autosomal dominant. The penetrance for the disorder is 70%. What is penetrance? How does incomplete penetrance affect the study of human genetic diseases?

3. What is a lod score? How is a lod score determined?

4. What is genetic polymorphism? Why are polymorphic loci essential for mapping human disease genes?

5. How are STRP probes used to locate a human disease gene?

6. What is the CEPH family panel? How is it used?

7. What is an STS? How is the chromosome location of an STS determined?

8. What is a contig? Describe in detail how a chromosome contig is assembled.

9. How would you establish that an insert of a genomic clone carries a coding region?

10. What is the rationale underlying the positional-candidate gene cloning approach?

Human Gene Therapy

For many years, the study of medical genetics was confined to determining whether a particular human disease had a genetic basis. Once the genetic nature of a disease was established, it was sometimes possible to devise a diagnostic test to identify the disease in newborns or prenatally. As a result of this type of biochemical analysis, genetic counseling was possible for families that had been shown to be at risk, so that they could at least be prepared for the occurrence of the disease among their future offspring. In addition, some form of therapy, such as medication, diet, or blood transfusions, could be sought to alleviate the impact of the gene defect.

Because genes have many different roles to play in the proper functioning of an organism, there is no single biological consequence of a mutated gene. A mutation that alters the activity of an enzyme may cause either an accumulation of a substrate that is toxic or a deficiency of a compound that is vital for normal cellular operations. If a mutated gene codes for a structural protein, then a cell, tissue, or organ may show severe architectural abnormalities. Moreover, a defect in a gene that is expressed in one tissue may have considerable impacts on other tissues and as a result produce a constellation of clinical symptoms. For example, a mutation in the gene for the liver enzyme phenylalanine dehydroxylase, a defect that prevents the conversion of the amino acid phenylalanine to tyrosine, has a profound effect on nervous tissue. When an individual is homozygous for the defective gene, either no phenylalanine dehydroxylase or an insignificant amount of it is produced, and excessive concentrations of endogenous phenylalanine accumulate in the blood. The high levels of phenylalanine cause the myelin sheath around neuronal axons of the central nervous system to be abnormally formed, a condition that, in turn, leads to severe mental retardation. This particular inborn error of amino acid metabolism occurs with a frequency of about 1 in 10,000 Caucasian newborns and is called phenylketonuria. Because each tissue has its own set of expressed genes, there are mutations that can lead to disorders in every part of the human body, including the muscles, eyes, liver, bones, kidneys, heart, nerves, skin, brain, stomach, intestines, and blood system.

Table 21.1 Current therapies for 10 single-gene human disorders

Gene product	Disease and symptoms	Frequency	Current therapy	Prognosis
Adenosine deaminase	Severe combined immunodeficiency Loss of T and B cells	1:1,000,000	Bone marrow transplant; adenosine deaminase replacement	Without therapy: fatal by 2 years of age With therapy: clinical improvement
LDL receptor	Familial hypercholesterolemia Elevated blood serum cholesterol level, coronary artery disease	1:500 (heterozygotes)	Diet, drugs, liver transplant	Some clinical improvement
Glucocerebrosidase	Gaucher disease Accumulation of glucocerebroside in macrophages causing liver, spleen, and bone damage	1:2,500 (Jewish populations); rare in non-Jewish populations	Symptomatic treatment: removal of spleen, antibiotics, repairing bone damage, bone marrow transplant, enzyme replacement	Some clinical improvement
Blood clotting factor VIII	Hemophilia A Altered plasma protein that causes defective blood clotting, chronic internal bleeding into joints, excessive bleeding after wounding	1:10,000 (males)	Concentrates of factor VIII by transfusion	Extended life expectancy, requires continual treatment, risk of viral infection from transfusions
Phenylalanine hydroxylase	Phenylketonuria Excess phenylalanine in the bloodstream of newborns causes mental retardation	1:10,000	Restriction of dietary phenylalanine	Good in many cases, if treatment is started early and maintained
α_1-Antitrypsin	Emphysema Deficiency of serum protein protease inhibitor, damage to the lungs, cirrhosis of the liver	1:3,500	Replacement therapy, lowering environmental risks	Progression of the disease is slowed but not stopped

The primary goal of human medical genetic research is to develop treatments for each of the many different genetic diseases. A few gene products, the characteristics of the diseases caused by mutations in the genes that encode these products, and the current treatments for these diseases are listed in Table 21.1. Because of the physiological complexity of these genetic diseases, curative measures have consisted mainly of treating the symptoms by administering drugs or conducting surgery. For some of the genetic diseases that cause errors of metabolism, restrictive diets have been devised to control the high endogenous levels of the toxicant that are a consequence of the gene mutation. For example, phenylketonuria, which can be identified at birth with a biochemical test, can be controlled by a phenylalanine-free diet. For other genetic diseases, a functional form of the defective protein that does not cause an adverse immune response is administered intravenously to alleviate the full impact of the disease. This "replacement therapy" is used for hemophilia, severe combined immunodeficiency (SCID), and Gaucher disease. Bone marrow and organ transplants are often carried out to provide blood cells or whole organs that will perform missing functions.

Table 21.1 (*continued*)

Gene product	Disease and symptoms	Frequency	Current therapy	Prognosis
Cystic fibrosis transmembrane regulator	Cystic fibrosis Multisystem disease, pancreatic insufficiency in some cases, intestinal blockage, blocked airways of the lungs	1:2,500 (Caucasians)	Antibiotics, physical clearing of the lungs, upgrading diet	Fatal by late 20s
Ornithine transcarbamylase	Hyperammonemia Urea cycle defect, ammonia accumulation arginine deficiency Early form: within 72 h of birth, lethargy, vomiting, coma, death, and, if survival, irreversible brain damage Late form: vomiting, lethargy, seizures	1:40,000	Restricted protein diet, arginine-supplemented diet, drugs, liver transplant	Late form: good Early (severe) form: symptoms diminished
Dystrophin	Duchenne muscular dystrophy Progressive muscle wasting	1:7,500 (males)	Only supportive treatments: good nutrition, aid in respiratory function, confinement to a wheelchair	Fatal by early 20s
β-Globin	Sickle-cell disease Chronic anemia, multisystem disease, damage to spleen, heart, kidneys, liver, and brain; heterozygotes have a mild form of the disease	1:500 (heterozygotes in populations of black African origin; less frequent in other populations)	Blood transfusions, drugs, analgesics, bone marrow transplant	Symptoms diminished; treatments suboptimal

Unfortunately, the most effective treatment for many genetic diseases can be administered only when the patient enters a crisis phase. Although these interventions are lifesaving, they are only temporary. Because genetic diseases often affect multiple systems and tend to be progressively debilitating, it has not been an easy task to develop effective therapies. Moreover, on the whole, the existing treatments tend to be inadequate, with very few of the patients having normal life spans or offspring. Although patients benefit from treatment, it is in most cases suboptimal, recurring, costly, and time-consuming. Consequently, new forms of therapies are continually being sought.

With the discovery of the molecular basis of DNA transformation in bacteria, in which a gene from one strain can be transferred from one strain to another in another strain, researchers have speculated that human genetic diseases might be cured in an analogous way by introduction of the normal gene into the appropriate somatic cell type. During the 1980s, the prospect of human somatic cell gene therapy became more likely and less theoretical as gene isolation techniques were developed, eukaryotic expression vectors were created, and transgenesis experiments with mice became

more routine. By 1990, after exhaustive reviews by various regulatory panels, the first human gene therapy trial was initiated with two young girls with SCID, which is the consequence of the absence of the enzyme adenosine deaminase (Table 21.1).

In this protocol, a cloned complementary DNA (cDNA) for adenosine deaminase (*ADA*) was introduced into lymphocytes that had been removed from each patient. After culturing, each subject received transfusions with her own genetically engineered lymphocytes, which synthesized adenosine deaminase, at regular intervals for 2 years. Four years after the onset of the trial, integrated vector and *ADA* gene expression was evident in both patients and the SCID symptoms were alleviated. However, it is not completely clear whether replacement therapy with a protected form of the enzyme adenosine deaminase (polyethylene glycol–adenosine deaminase) or gene therapy per se was responsible for this improvement. Regardless, the trial showed that this form of gene therapy was safe and helpful. Similar results were also observed with other SCID patients who received both types of therapies simultaneously. Consequently, these initial trials have been expanded to include more patients.

In the United States, the development of a therapeutic agent for humans usually passes through four clearly defined levels of study and testing before it can be sanctioned for use:

1. Preclinical trials that include extensive in vitro experiments and research on laboratory animals
2. Phase I trials with a small number (6 to 10) of human subjects who often are used to test the safety of the product
3. Phase II trials with an expanded number of human subjects to examine whether the product is helpful
4. Phase III trials that include a large human sample and fully comprehensive analysis of the safety and efficacy of the product, implementing the information gathered from the previous trials

Permission must be sought and granted by various regulatory agencies before preclinical trials and any of the other phases of study are initiated. Between 1990 and 1992, about a dozen phase I human gene therapy trials were approved, and by 1997 over 200 trials were under way to examine treatments for various cancers, hemophilia, AIDS, cystic fibrosis, hypercholesterolemia, amyotrophic lateral sclerosis, and other disorders (Table 21.2). A number of key considerations must be addressed before phase I trials can begin. For example, a proposed study must be directed toward developing a treatment for a recognized disease, the risk to patients should be minimal, and all existing rules and regulations for conducting biomedical experimentation must be observed.

Because gene therapy is so new and the number of diseases that could be treated this way are so diverse, a large number of different strategies are being examined. At present, all research on human gene therapy is directed toward correcting genetic defects of somatic cells, i.e., cells that do not contribute to the next generation (nonsex cells). For ethical, safety, and technical reasons, human germ line gene therapy, which entails introducing DNA into cells that can form part of a human germ line and be passed onto successive generations, is not being examined experimentally at this time.

In its broadest sense, human somatic cell gene therapy can be described as the introduction of a fully functional and expressible gene into a target

Table 21.2 Some human gene therapy trials that have been conducted since 1990

Gene therapy	Condition(s)	Target cells
Adenosine deaminase	Adenosine deaminase deficiency	Lymphocytes, bone marrow cells
Tumor necrosis factor	Melanoma	Tumor-infiltrating lymphocytes, autologous tumor cells
Interleukin-2	Melanoma, glioblastoma, renal cell cancer	Autologous tumor cells, tumor cells
Factor IX	Hemophilia B	Autologous skin fibroblasts
LDL receptor	Hypercholesterolemia	Autologous liver cells
Histocompatibility locus antigen class I-B7 plus β_2-microglobulin	Melanoma, colorectal cancer, renal cell cancer	Tumor cells
Herpes simplex virus thymidine kinase	Glioblastoma, AIDS, ovarian cancer	Tumor cells, T cells
Cystic fibrosis transmembrane conductance regulator	Cystic fibrosis	Nasal and airway epithelium
Multidrug resistance 1	Breast cancer	Blood CD34$^+$ cells
Granulocyte-macrophage colony-stimulating factor	Melanoma	Tumor cells
Interleukin-1 receptor antagonist	Arthritis	Autologous fibroblasts
Human CNTF	Amyotrophic lateral sclerosis	Encapsulated transduced xenogeneic cells
p53	Head and neck squamous carcinoma	Tumor cells
Fanconi anemia C	Fanconi anemia	Bone marrow cells

cell, with the result that a specific genetic disease will be corrected permanently. However, this simple characterization masks some critical biological considerations. For example, how will the cells that are to be targeted for correction be accessed? How will the remedial gene be delivered? What proportion of the target cells must acquire the input gene to counteract the disease? Does transcription of the input gene need to be precisely regulated to be effective? Will overexpression of the input gene cause alternative physiological problems? Will the cells with the input gene be maintained indefinitely, or will repeated treatments be required?

Although somatic cell gene therapy is at an early stage of development, some of these questions are beginning to be answered for certain genetic diseases. In addition, various strategies for implementing somatic cell gene therapy are emerging and can be grouped under two broad categories: ex vivo and in vivo gene therapy. In addition to gene therapy, other forms of nucleic acid therapeutic agents such as antisense oligonucleotides, genetically engineered RNA enzymes, and oligonucleotides that correct gene mutations in vivo are being developed.

Ex Vivo Gene Therapy

Usually, ex vivo gene therapy involves the following procedure (Fig. 21.1).

1. Collect cells from an affected individual.
2. Correct the genetic defect by gene transfer into the isolated cells.
3. Select and grow the genetically corrected cells.
4. Either infuse or transplant them back into the patient.

The use of a patient's own cells (autologous cells) ensures that no adverse immunological response occurs after infusion or transplantation.

It is important that the gene transfer protocol for ex vivo gene therapy be efficient and that the remedial (therapeutic) gene be both stably maintained and persistently expressed. Currently, with human subjects, vectors derived from mouse retroviruses are often used to meet these ends. However, because retroviruses can convert normal cells into cancerous ones, it is essential that this possibility be at least diminished and preferably completely abolished.

In general, a wild-type retrovirus particle carries two identical single-stranded RNA genomes that are each organized into six regions. From the 5' end, these regions consist of a long terminal repeat (5'-LTR) sequence; a noncoding sequence that is required for packaging the RNA strand inside a virus particle (encapsidation) and is designated psi$^+$ (ψ^+, packaging signal); three genes that code for an internal capsid structural protein (*gag*), reverse transcriptase and integrase functions (*pol*), and an envelope protein (*env*); and a 3'-LTR sequence (Fig. 21.2). Briefly, the life cycle of a retrovirus includes the following stages:

Figure 21.1 Schematic representation of ex vivo gene therapy. The steps in the procedure include (1) isolating cells with the gene defect from a patient; (2) growing the isolated cells in culture; (3) transfecting the isolated cells with a remedial gene construct; (4) selecting, growing, and testing the transfected cells; and (5) either transplanting or transfusing the transfected cells back into the patient.

Figure 21.2 Genetic map of a typical retrovirus. Designations: ψ⁺, encapsidation (packaging) signal sequence; *gag*, region encoding viral capsid protein; *pol*, region encoding enzymatic activities including reverse transcriptase and integrase; env, region encoding viral envelope protein. The 5′-LTR contains transcription initiation signals, and the entire genome is transcribed as a single RNA molecule. The 3′-LTR contains a polyadenylation signal sequence.

1. Infection of a target cell
2. Production of a DNA copy of the RNA genome by using the reverse transcriptase that is carried within the virion
3. Transport of viral DNA to the nucleus
4. Integration of the viral DNA into the host genome at one of a number of different chromosomal sites
5. Transcription of mRNA from the viral DNA that is driven by a strong promoter within the 5′-LTR sequence
6. Translation in the cytoplasm of the Gag, Pol, and Env proteins
7. Formation and packaging of the viral capsid with two RNA strands and reverse transcriptase molecules
8. Shedding of single-membrane-enveloped virions to the exterior of the cell

After an entire sequence of retroviral DNA is cloned into a plasmid, it can be constructed into a vector. Initially, with the use of restriction endonuclease sites and (if required) limited exonuclease digestion, most of the 3′ end of the *gag* gene and all of the *pol* and *env* genes are deleted, while the regions that contain part of the 5′ end of the *gag* gene and the 5′- and 3′-LTRs are retained. Then a remedial human gene whose transcription will be directed by the 5′-LTR promoter can be cloned adjacent to the ψ⁺region; a selectable marker gene with its own promoter can also be included (Fig. 21.3). This vector design allows both of the cloned genes to be synthesized. A variety of forms of retroviral vectors have been devised by using this basic scheme. Regardless of the format, the upper limit of the amount of cloned DNA that can be carried by a retroviral vector is ~8 kilobase pairs (kb).

Although retroviral vector DNA can be used by itself to transform cells, the efficiency of both delivery and integration is very low. Consequently, a methodology was devised to package retroviral vector RNA into virus particles. The intact particles deliver the complete vector RNA to a host cell at

Figure 21.3 Genetic map of a two-gene retroviral vector. Transcription of the remedial gene (Gene X) is driven by sequences within the 5′-LTR; an internal promoter *(p)* drives the transcription of a selectable marker gene (Neoʳ). The 3′-LTR sequence contains a polyadenylation signal sequence. The encapsidation region (ψ⁺II) is retained.

a high frequency, thereby virtually ensuring that its DNA equivalent will be integrated into the genome of a host cell. To this end, a cell line in culture was genetically engineered to carry at one chromosomal site within its genome a 5'-LTR–*gag*–3'-LTR fragment that lacks a functional ψ^+ sequence (designated $\Delta\psi^+$) and at another chromosomal site a 5'-LTR–*pol*–*env*–3'-LTR fragment which is also $\Delta\psi^+$. This is called a packaging cell line. Both of these segments are transcribed constitutively; because the ψ^+ region is absent and less-than-full-size viral RNA molecules are produced, empty virus particles are formed. However, when retroviral vector DNA is used to transfect these cells, it becomes integrated into the chromosomal DNA and is transcribed, thereby producing full-length retroviral vector RNA that has a ψ^+ sequence. Under these conditions, only the vector RNA is packaged into the virus capsids. These intact virus particles can be used to deliver the retroviral vector to the target cells at high efficiency (Fig. 21.4).

It is essential that no wild-type (replication-competent) retroviruses be produced by a packaging cell line when the retroviral vector particles are to be used for human somatic cell gene therapy, because such wild-type retroviruses might insert into genes and cause some cells to proliferate in an uncontrolled manner, i.e., become cancerous. Therefore, as a precaution, samples of packaged retroviral vectors are assayed routinely for the presence of wild-type retroviruses. Furthermore, because of the three separate chromosomal locations of the retroviral and vector sequences in a packaging cell line, the possibility of successive recombination events occurring to create a replication-competent retrovirus is extremely unlikely.

Retroviruses readily infect replicating cells, so actively growing target cells are either treated with purified packaged retroviral vector particles directly or cocultivated with the packaging cell line that is producing the packaged retroviral vector particles and then differentially selected to separate them from the packaging cells. In either case, the transduced target cells (those into which nonviral DNA was transferred by a virus) are tested to ensure that (1) the remedial gene product is synthesized, (2) replication-competent retroviruses are not being produced, and (3) the retroviral vector DNA has not been inserted into a site that either alters the growth properties of the cells or interferes with normal cellular functions. After these characterizations, the transduced cells are grown in culture, collected in large amounts, and then introduced into the patient at various intervals, with the hope that the cells will be maintained and the disease that is being treated will be corrected.

Individuals with genetic diseases that respond to bone marrow transplantation are likely candidates for ex vivo gene therapy (Fig. 21.5). The reason that bone marrow transplantation works as a therapeutic procedure for a diverse set of diseases is that bone marrow contains totipotent embryonic stem cells at a frequency of 10^{-4} to 10^{-5}. These cells can divide and differentiate into various important cell types, including B and T lymphocytes (B cells and T cells), macrophages, red blood cells, platelets, and bone cells (osteoclasts). Thus, for example, when a gene mutation interferes with the proper functioning of macrophages, bone marrow transplantation might provide the recipient with a steady supply of competent macrophages from its population of totipotent stem cells.

One of the preferred ways to carry out ex vivo gene therapy would be to engineer genetically totipotent stem cells and either infuse or transplant them into a patient, where they would continuously provide a missing cell type or gene product. For example, in humans, adenosine deaminase defi-

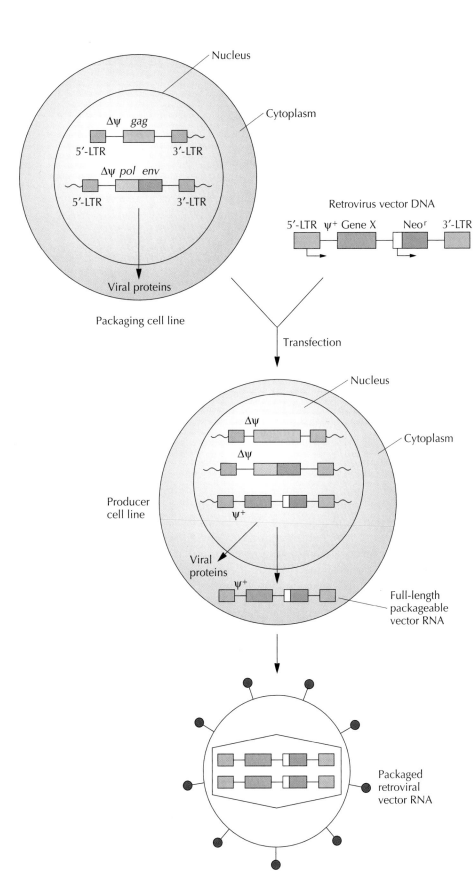

Figure 21.4 Production of packaged retrovirus vector RNA. The packaging cell line has two separate retroviral gene regions on its chromosomes; one contains the *gag* gene, and the other contains the *pol* and *env* genes. In these two regions, transcription is driven by sequences within the 5'-LTR region; both regions lack the encapsidation sequence ($\Delta\psi^+$). The packaging cell line synthesizes viral proteins, but because of the absence of the encapsidation sequence within either of the retroviral mRNAs, empty viral capsids are produced. After transfection of a packaging cell line with a retroviral vector, the viral proteins continue to be synthesized, while full-length RNAs from the retrovirus vector sequence that carries the encapsidation region (ψ^+) are transcribed and packaged into viral capsids. The released viral particles are replication defective and carry, in this example, a remedial gene (Gene X) and a selectable marker gene (Neor).

ADA-deficient severe combined immunodeficiency	Metachromatic leukodystrophy
Adrenoleukodystrophy	Non-ADA-deficient severe combined immunodeficiency
Chédiak-Higashi syndrome	Osteoporosis
Chronic granulomatous disease	Purine nucleotide phosphorylase deficiency
Fanconi anemia	
Gaucher disease	Reticular dysgenesis
gpL-115 deficiency	Sanfilippo syndrome
Granulocyte actin deficiency	Sickle-cell disease
Hunter disease	Thalassemia
Hurler syndrome	Wiskott-Aldrich syndrome
Infantile agranulocytosis	X-linked agammaglobulinemia
Maroteaux-Lamy syndrome	

Figure 21.5 Genetic diseases that can be treated by bone marrow transplantation.

ciency causes a buildup of adenosine and deoxyadenosine levels in the circulatory system. These compounds are toxic and lethal to both B and T lymphocytes and, as a consequence, cause severe immunodeficiency. Because B and T lymphocytes are both derived from totipotent stem cells, cloning a functional adenosine deaminase gene into stem cells, followed by reintroduction into a patient, should decrease the circulating levels of adenosine and deoxyadenosine and prevent B and T lymphocytes from being destroyed. The best chance of success for this procedure exists when totipotent stem cells originate from the patient. Alternatively, compatible donors, who are usually difficult to find, must be used as a source of stem cells.

Unfortunately, it is very difficult to isolate and culture totipotent stem cells from human bone marrow. In the meantime, trials have been carried out on retrovirus vector-engineered autologous T cells to determine whether ex vivo gene therapy might be useful for alleviating the symptoms of SCID due to adenosine deaminase deficiency. Because T cells have a limited lifetime, repeated infusions are required. In the initial trial, two girls received infusions of their own genetically corrected T cells over several months. Both girls showed clinical improvement, presumably because the adenosine deaminase-producing T cells lowered the toxic levels of adenosine and deoxyadenosine in their blood systems, thereby enabling B and T cells to persist.

In general, T cells are not the preferred delivery system for curing disease of bone marrow-derived (hematopoietic) cells. Alternatively, although not always available, umbilical cord blood, which contains stem cells, is a good source for ex vivo gene therapy of blood cell diseases. For example, newborn infants with adenosine deaminase deficiency were transfused once with CD34$^+$ cells from their own umbilical cord blood cells that had been transduced with human *ADA* cDNA. The *ADA* cDNA was maintained and expressed in white cells for at least 18 months.

One example of an unambiguously successful result with ex vivo gene therapy with recombinant autologous cells was achieved with a patient who is homozygous recessive for familial hypercholesterolemia. This individual has no low-density lipoprotein (LDL) receptors on her liver cells. Consequently, fat bodies containing LDL cholesterol are not processed by the liver and continually circulate in the bloodstream, leading to arterial blockage and severe heart disease. Such patients do not respond to drug

treatments, and the benefits of coronary artery bypass surgery are short-lived.

In this case, about 15% of the patient's liver was removed. Hepatocytes were released into culture medium and transduced with recombinant retrovirus carrying the cDNA for the LDL receptor. The transduced hepatocytes were infused into the patient's liver, where they became established and for at least 18 months expressed the LDL receptor cDNA and produced functional LDL receptors. The patient's condition, based on lipid profiles, showed significant improvement. In addition, she did not produce antibodies against the LDL receptor molecules.

Ex vivo gene therapy depends on genetically modified transplanted cells for the production of a therapeutic protein. Immunological rejection of these cells is curtailed by using the patient's own cells. However, dependence on autologous cells limits the range of possible applications for ex vivo gene therapy. In some instances, a patient may not have enough cells of the target tissue that can be retrieved and cultured in vitro. Also, the uptake of DNA by some somatic cells is inefficient and expression of the cloned gene is transient. For these reasons, systems that protect nonautologous (xenogenic, allogeneic) cells from immunological attack and allow the release of a therapeutic protein are being developed.

In brief, nonautologous ex vivo gene therapy entails isolating tissue-specific cells that grow well in culture, such as fibroblasts or keratinocytes from the skin, astrocytes from the brain, hepatocytes from the liver, or myo-

MILESTONE

Construction of a Retrovirus Packaging Mutant and Its Use To Produce Helper-Free Defective Retrovirus

R. MANN, R. C. MULLIGAN, and D. BALTIMORE
Cell **33:**153–159, 1983

Human gene therapy has always been a controversial subject. However, most scientists agreed with T. Friedmann and R. Roblin, who in 1972 (*Science* **175:**949–955) stated, "In our view, gene therapy may ameliorate some human genetic diseases in the future. For this reason, we believe that research directed at the development of techniques for gene therapy should continue." The essential features of human gene therapy are the delivery of a remedial gene and its expression in a cell type or tissue that cures a disease without risk to either those administering the therapeutic agent or those receiving it. Initially, vectors derived from human viruses seemed likely to be the principal mode for delivering remedial genes because they have specialized mechanisms for entering specific cells. In particular, of the various potential

viral vectors, those based on retroviruses were considered the most promising. However, a native retrovirus is an infectious agent that can cause cell damage and, in some instances, induce cancer. The most significant advance that made human gene therapy possible was the development of a system for packaging a remedial gene into a noninfectious virus particle that retains the capability of attaching to its host cell.

Mann et al. constructed the first retroviral packaging cell line. In essence, they integrated a viral genome from which they had removed a DNA segment that contained the packaging signal into a chromosome of a cell line. Under these conditions, the cell line produces noninfectious virus particles. However, after transfection of these cells with a DNA construct that has a packaging signal and

a remedial gene but no retroviral genes, the construct is packaged into virus particles that, in turn, can be used to deliver a remedial gene to a particular cell type. This clever strategy was immediately adopted by many other researchers who were working on human gene therapy. Over the years, the original retrovirus cell packaging line has been enhanced and the concept has been successfully applied to other viral vectors.

In 1985, as a result of the work of Mann et al. and others, W. F. Anderson, who has been a persistent advocate for human gene therapy, noted, "It now appears that effective delivery-expression systems are becoming available that will allow reasonable attempts at human gene therapy." On 22 May 1989, Anderson and his colleagues initiated the first clinical trial using a gene therapy strategy. To date, gene therapy in general has not been particularly successful; however, as more information accumulates from ongoing studies, it is inevitable that it will become the standard mode of treatment for many diseases.

blasts from muscle, and genetically modifying the cells with a therapeutic gene. The recombinant cells are encapsulated in an artificial, semipermeable, polymeric membrane that allows the recombinant protein to pass out and nutrients to enter. The nonimmunogenic membrane prevents the immune system of the patient from sensing and responding to the implanted cells. Various polymers, including alginate–poly-L-lysine–alginate, polyether sulfone, and poly(acrylonitrile-co-vinyl-chloride), have been used as the encapsulating material.

Preclinical trials with in vitro and animal systems have shown that encapsulated recombinant cells proliferate and express a high level of recombinant protein for long periods. A phase I trial with encapsulated cells that expressed human ciliary neurotrophic factor (CNTF) was undertaken to prevent the loss of motor neurons that is responsible for the neurodegenerative disorder amyotrophic lateral sclerosis. Unfortunately, the amyotrophic lateral sclerosis patients did not benefit from this treatment, although the procedure was found to be safe. However, when encapsulated CNTF-producing cells were implanted in the brains of animals with chemically induced damage that simulates Huntington disease, another fatal neurodegenerative disorder, brain cells were protected from destruction. Although this approach is at an early stage of development, the encapsulation of genetically engineered cells may provide an effective system for delivering remedial gene products for a variety of diseases.

In Vivo Gene Therapy

In vivo gene therapy entails the direct delivery of a remedial gene into the cells of a particular tissue of a prospective patient (Fig. 21.6). Retroviral vectors require that the target cells must be dividing in order to be infected. However, in many of the tissues that might be treated using gene therapy, the majority of the cells are quiescent. Consequently, a number of strategies, which include using viral and nonviral vector systems to deliver a therapeutic gene to cells of the target tissue, have been devised. The variety of gene delivery systems for in vivo therapy reflects both the diversity of potential target tissues such as skin, muscle, lung, brain, colon, spleen, liver, and blood cells and their location within the human body. The attributes of an ideal in vivo gene delivery system are high efficiency of uptake of the therapeutic gene by the target cells, transportation of the therapeutic gene to the nucleus of the target cell with a minimum of intracellular degradation, and sustained expression of the therapeutic gene at a level that alleviates the condition.

Viral Gene Delivery Systems

Retrovirus Vector System

The experience from a number of clinical trials with more than 200 patients has established that replication-defective retrovirus vectors are not responsible for any apparent adverse effects. Moreover, the safety features of retroviral vectors are continually being upgraded. For example, a plasmid that carries the retroviral *gag* and *pol* genes under the control of the 5'-LTR promoter, a therapeutic gene, and the *env* gene driven by the cytomegalovirus promoter has been constructed. After transfection, this plasmid directs the production of replication-defective virus particles. It is very unlikely that this construction, which is called a plasmovirus, will recom-

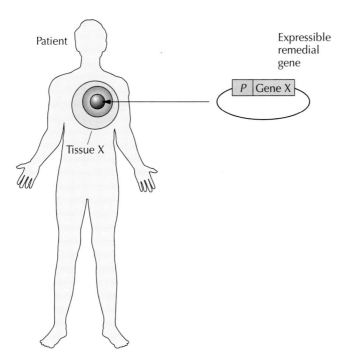

Figure 21.6 Schematic representation of in vivo gene therapy. A cloned, express-ible remedial gene is delivered to cells of a tissue of a patient with a genetic disease. The promoter (*p*) is tissue specific, and the remedial gene (Gene X) encodes a pro-tein that corrects the genetic defect of the patient.

bine to form replication-competent retroviruses. Although the DNA carry-ing capacity of a plasmovirus is only 3.5 kb, most potential therapeutic cDNAs and anticancer genes range from 0.5 to 2 kb in length.

Additional improvements of retroviral vectors that have been carried out include increasing the yield of virus particles, enhancing the efficiency of transduction, engineering them to infect nondividing cells, and specify-ing the cell type(s) that can be infected. In the last instance, a recombinant retrovirus vector genome that is packaged in the envelope protein of an-other virus will have the binding specificity and infection spectrum that are determined by the envelope protein. This phenomenon is called pseudo-type formation. A pseudotyped virus can be produced by cotransfecting a cell line that produces the *gag* and *pol* gene products with a recombinant retroviral vector and a second vector that expresses the envelope (*env*) gene of another virus. In this way, the infection spectrum can be limited to a spe-cific cell type. Alternatively, if necessary, the range of potential host cells can be extended, depending on the *env* gene that is used. Moreover, the *env* gene can be engineered to include a DNA element that encodes an amino acid sequence that binds to a cell receptor to create a recombinant retrovirus that infects a specific cell type. Finally, the specificity of expression of the therapeutic gene can be refined by putting it under the control of a cell-spe-cific promoter.

Adenovirus Vector System

Adenoviruses infect a wide range of nondividing human cells and have been used extensively as live vaccines against respiratory infections and gastroenteritis without side effects. These features make adenovirus a likely prospect for delivering genes to target cells.

For use as a vector for gene therapy, a cell line that produces the adenoviral E1 gene products, which are essential for adenoviral replication, is cotransfected with two portions of the adenovirus genome (Fig. 21.7). One of these segments, which is maintained in *Escherichia coli* as a plasmid, carries a therapeutic gene in place of the E1 region and is flanked by adenovirus DNA sequences. The E1 region is located near the 5′ end of the adenovirus genome. The second transfected DNA segment is an adenovirus DNA molecule that lacks part of its 5′ end, including the E1 region, but shares a region of adenoviral DNA with the plasmid that carries the therapeutic gene. A recombination event in the region of overlap between the two transfected pieces of DNA reconstitutes a full-length adenovirus genome that contains the therapeutic gene and lacks the E1 region. The E1 gene products that are supplied by the host cell initiate virus production; virus particles are assembled and then released after cell lysis. The DNA cloning capacity of this adenovirus vector system is about 7.5 kb. If recombination does not occur, both transfected DNA molecules are too small to be packaged properly. Moreover, recombination between the E1 DNA region in the genome of the host cell and the recombinant adenovirus DNA to form a replication-competent virus is extremely unlikely.

After infection of a target cell with a recombinant adenovirus, the DNA is passed into the cell nucleus, where the therapeutic gene is expressed. The recombinant DNA construct does not integrate into a chromosome, and consequently, it does not persist for long periods. Therefore, adenovirus-based gene therapy requires periodic administration with additional recombinant viruses.

Adenovirus gene delivery systems have been used in gene therapy trials for treating cystic fibrosis. The initial results were not encouraging. Very few of the patients' cells were transduced with the cystic fibrosis transmembrane regulating gene (*CFTR*), and multiple treatments with recombinant adenovirus triggered severe immunological responses in the patients. A low level of expression of some of the adenoviral genes after infection is responsible for the immunological response that destroys the transduced cells.

Various attempts have been made to overcome this problem. For example, a packaging cell line that carries the E1 region and some of the adenovirus genes that are expressed after transduction has been developed. Consequently, none of these adenoviral genes is present in transduced target cells. This approach has been extended so that no adenovirus genes form part of the transducing DNA. Briefly, an *E. coli* plasmid (28 kb) carrying one or more therapeutic gene expression constructs and no adenovirus genes is linearized, and a piece of DNA (4 kb) that contains the adenoviral origin of replication, packaging signal, and terminal sequence is ligated to each end. The length (36 kb) of this ligation product (ligation construct) is equivalent to a full-sized, packageable adenovirus genome. Next, the ligation construct and an adenovirus genome that lacks both the E1 region and a functional packaging signal are cotransfected into a host cell that expresses the genes of the E1 region. The DNA molecule that is both replication defective and packaging deficient supplies the genes for the production of the viral components. The ligation construct is replicated and packaged into virus particles. Over 99% of the released viral particles contain the DNA molecule with the therapeutic gene(s). The small number of viruses with replication-defective adenovirus genomes that are formed despite having a defective packaging signal can be separated by centrifugation

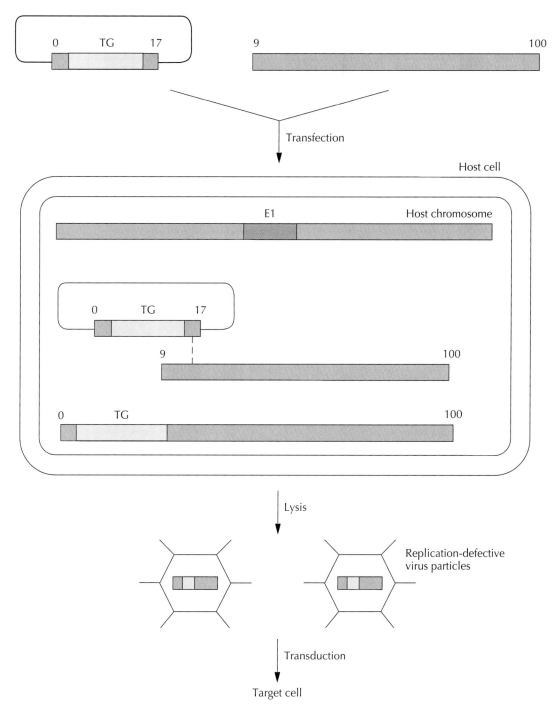

Figure 21.7 Adenovirus vector system. A plasmid with a therapeutic gene (TG) that is inserted into a segment of the adenovirus genome extending from map unit 0 to 17 and a linear adenovirus DNA molecule that consists of map units 9 to 100 are both introduced into a host cell that carries an integrated and functional adenovirus E1 gene (E1). The entire adenovirus genome is 100 map units. A recombination event (dotted line) between the shared DNA regions of the plasmid and linear adenovirus DNA molecule creates a DNA molecule that is equivalent to a full-length adenovirus genome. The recombinant DNA molecule with the therapeutic gene is packaged and released after cell lysis. The recombinant virus particles are replication defective. Plasmid DNA which is part of the final construct does not impact on the packaging of the recombinant DNA molecule (not shown).

from those with the therapeutic gene construct(s). This system has a DNA cloning capacity of 28 kb.

The efficiency of adenovirus-mediated gene delivery can be enhanced by developing a virus that preferentially infects a particular target cell. With this in mind, the gene for the attachment fibers of adenovirus can be engineered to include a DNA element that encodes a protein domain that binds to a cell-specific receptor.

Adeno-Associated Virus Vector System

Adeno-associated virus is a small, nonpathogenic, single-stranded human DNA virus (4.7 kb) that can integrate into a specific site on chromosome 19. Productive infection by adeno-associated virus depends on proteins from another virus (helper virus) such as adenovirus, hence the name adeno-associated virus. After the adeno-associated virus enters the nucleus, the polymerases of a host cell convert the adeno-associated virus genome to a double-stranded DNA that is then transcribed.

Its absence of pathogenicity makes adeno-associated virus a good candidate as a vector for the delivery of therapeutic genes. Recombinant adeno-associated virus is generated by cotransfection of two plasmids into a host cell that has been infected with adenovirus (helper virus) (Fig. 21.8). One of the plasmids carries a therapeutic gene flanked by the inverse terminal repeat sequences (~125 bp each) from the adeno-associated virus. The second plasmid contains the two genes (*rep* and *cap*) of the adeno-associated virus genome that are responsible for replication of the adeno-associated virus genome and the production of the adeno-associated virus capsid, respectively. After lysis of infected host cells, recombinant adeno-associated virus particles are purified from adenovirus by centrifugation and dialysis. Any residual adenovirus (helper virus) in the sample is killed by heat treatment. Recombinant adeno-associated virus can carry a cloned insert of 4.5 kb. In the absence of the *rep* gene, the recombinant adeno-associated virus DNA is unlikely to integrate into chromosome 19. On the other hand, without any adeno-associated virus genes, the recombinant vector will not evoke an immune response.

In one preclinical trial, in vivo transduction of mouse liver cells by intravenous injection with recombinant adeno-associated virus was enhanced 900-fold by prior nontoxic irradiation of the liver and the presence of nonrecombinant (wild-type) adeno-associated virus. The therapeutic gene that was tested in this case was the cDNA for human blood clotting factor IX, which was expressed for at least 5 months at a level that would be sufficient for a liver-based treatment of hemophilia in humans. Moreover, this amount of expression with other therapeutic genes would be effective for treating other disorders. In phase I clinical trials with cystic fibrosis patients, no inflammatory response was observed after administration of a *CFTR*-adeno-associated virus vector. The vector persisted for about 70 days. Further clinical trials will determine if therapeutic levels of the *CFTR* gene product are produced.

Herpes Simplex Virus Vector System

Although it is possible to engineer retrovirus and adenovirus vectors to infect specific cell types, some viruses naturally are directed to a particular type of cell. Of the latter category, herpes simplex virus type 1 infects and persists within nondividing neuronal cells. Herpes simplex virus type 1 is a common human pathogen that causes periodic cold sores and, rarely, fatal

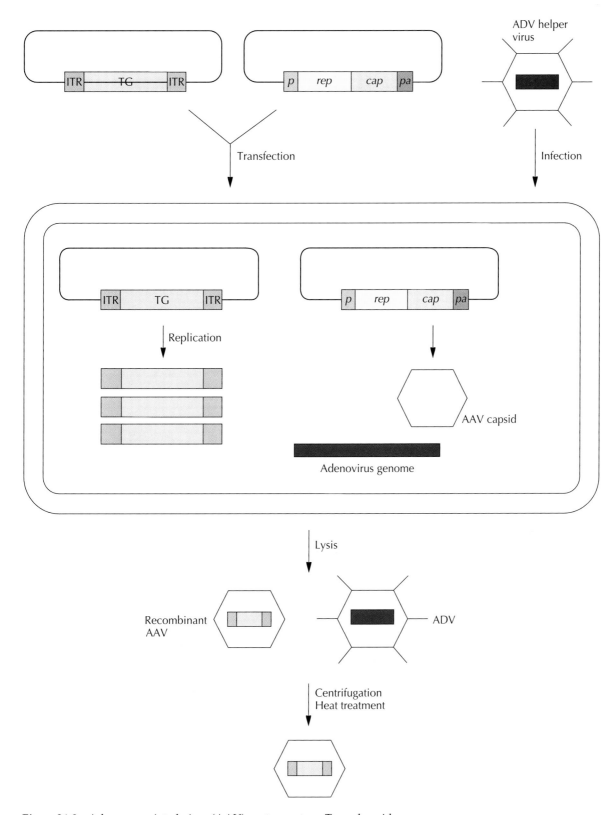

Figure 21.8 Adeno-associated virus (AAV) vector system. Two plasmids, one containing a therapeutic gene (TG) that is flanked by the inverted terminal repeats (ITR) of AAV and the other with the AAV genes for replication (*rep*) and capsid formation (*cap*) under the control of a promoter (*p*) and polyadenylation sequence (*pa*), are cotransfected into a host cell that is infected with a helper adenovirus (ADV). Both recombinant AAV and ADV particles are released after cell lysis. The recombinant AAV can be separated from ADV by centrifugation, and any remaining ADV is inactivated by heat treatment.

encephalitis. The virus usually remains latent in neurons. Reactivation of the lytic cycle is initiated by stress and hormonal changes.

There are many different disorders that affect the central and peripheral nervous systems, including tumors, metabolic and immunological defects, and neurodegenerative syndromes such as Alzheimer and Parkinson diseases. Neurological disorders are responsible for more hospitalizations and chronic care than almost all other diseases combined. Because of its preference for nerve cells, herpes simplex virus is a suitable candidate as a gene therapy vector for treating these kinds of diseases.

The genome of herpes simplex virus is a 152-kb-long, double-stranded DNA molecule. The virus fuses to the membrane of a neuron and is transported to the nucleus. The reproductive cycle consists of a lytic phase and a latent phase. During the former phase, the virus replicates and virus particles are produced. During the latter phase, the viral genome condenses and at least two viral promoters called latency-associated promoters are active.

About 30 kb of the herpes simplex virus genome can be replaced by cloned DNA without significant effects on replication, packaging, or infectivity. However, the large size of the herpes simplex virus makes genetic manipulations difficult. To overcome this problem, a scaled-down version of the herpes simplex virus genome consisting of the herpes simplex virus origin of replication and packaging signal has been incorporated into an *E. coli* plasmid that can, in addition, carry up to 8 kb of cloned DNA. These herpes simplex virus-based constructs are called amplicons (amplicon plasmids).

Most herpes simplex virus gene delivery systems rely on a helper herpes simplex virus to supply the viral proteins for replication and assembly of the viral components. The genome of the helper virus is modified so that it cannot be packaged and, therefore, infective virus particles cannot form. A herpes simplex virus amplicon plasmid is transfected into a host cell that has been infected with the helper virus. During DNA replication, which is similar to the mode of replication of bacteriophage λ, the inner circular strand of the amplicon acts as the template and DNA synthesis proceeds from the 3'-OH group that is made available by a nick in the outer strand. The growing strand continues around the intact circular template a number of times to produce a continuous, newly synthesized strand that is covalently linked to the original outer strand and comprises a linear array of multiple complementary copies of the inner strand. As the newly synthesized strand is displaced from the template, it acts as a template for the synthesis of a complementary strand. This type of replication is called rolling-circle DNA replication, and in this case, it generates a linear double-stranded molecule that has multiple copies of the amplicon. Since each amplicon is about 15 kb long, a set of 10 copies in tandem is equivalent to a full-size herpes simplex virus genome and is packaged into a herpes simplex virus capsid (Fig. 21.9)

An alternative procedure for forming recombinant herpes simplex viruses entails cotransfecting host cells that support the growth of herpes simplex virus with wild-type herpes simplex virus DNA and a plasmid carrying a therapeutic gene that is flanked by herpes simplex virus sequences that are derived from the nonessential (accessory) region of the herpes simplex virus genome. After replication of the wild-type herpes simplex virus DNA in the nucleus of the host cell, a recombination event between the herpes simplex virus genome and the herpes simplex virus sequences in the plasmid incorporates the therapeutic gene into a full-length herpes simplex virus genome. After packaging, both recombinant and wild-type virus par-

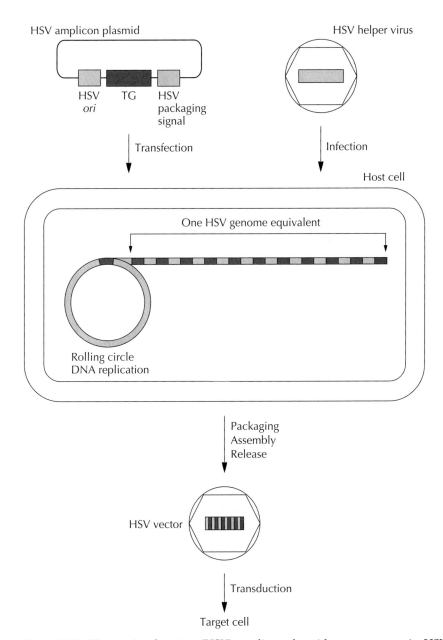

Figure 21.9 Herpes simplex virus (HSV) amplicon plasmid vector system. An HSV origin of replication (HSV *ori*), an HSV packaging signal, and a therapeutic gene (TG) are inserted into an *Escherichia coli* plasmid (HSV amplicon plasmid). The HSV amplicon plasmid is transfected into a host cell that is infected with helper HSV. The HSV amplicon plasmid DNA undergoes rolling circle DNA replication. One HSV genome equivalent of about 10 amplicons is packaged as a unit into an HSV capsid that is produced by the HSV helper virus. The HSV helper genome is not packaged. HSV particles carrying multiple copies of the therapeutic gene are released after cell lysis and used to transduce neurons.

ticles are released. Because there are so few recombinant viruses initially, the plaques are picked, the viruses are propagated, and subsequent plaques are assayed by the polymerase chain reaction or DNA hybridization for the presence of the therapeutic gene. A recombinant virus stock is identified and maintained free of wild-type herpes simplex virus (Fig. 21.10).

Figure 21.10 Formation of a recombinant herpes simplex virus (HSV) vector by recombination. A plasmid with a therapeutic gene (TG) that is flanked by HSV DNA sequences from a nonessential (accessory) region of the HSV genome is cotransfected with wild-type HSV genome DNA into a host cell. After rolling circle replication of the HSV genome in the nucleus of the host cell, recombination (dotted lines) can occur between the HSV sequences of the plasmid and the HSV DNA. Both full-length wild-type and recombinant HSV DNA molecules are packaged into virus particles and released. The viruses are propagated, and then plaques are screened to identify recombinant HSV vectors. Stocks of recombinant HSV vectors are maintained free of wild-type HSV.

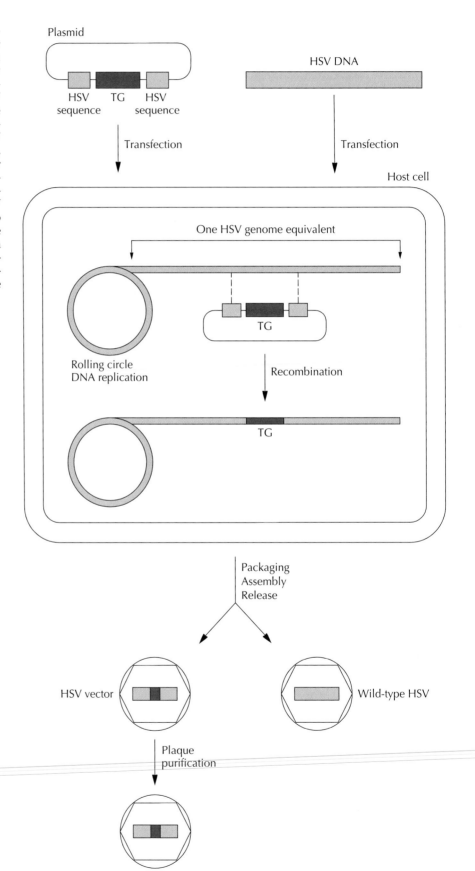

Preclinical trials with experimental animals have indicated that herpes simplex virus vectors can deliver genes to cells of both the brain and the peripheral nervous system that are expressed and maintained for long periods. However, more research is needed before recombinant herpes simplex virus can be tested in phase I trials with human subjects.

Nonviral Gene Delivery Systems

A virus-mediated gene delivery system utilizes cell receptor sites to enter a target cell, which avoids lysosomal degradation of the vector and leads to the deposition of the vector DNA into the nucleus of the cell. However, notwithstanding these important advantages, viral vectors are costly to maintain and in many instances have a limited cloning capacity, which precludes the use of tissue-specific regulatory sequences to drive the therapeutic gene. In addition, viral proteins may induce an inflammatory response that nullifies repeated administration of the vector. As an alternative to viral vectors for gene therapy, a number of nonviral gene delivery systems have been devised.

The least complicated nonviral gene delivery system is the introduction of pure DNA constructs directly into the cells of a target tissue. When plasmid DNA was injected into mouse skeletal muscle, some of the cells took up the DNA and, in this case, the reporter gene was expressed for more than 50 days. Although appealing in its simplicity, this approach is limited to accessible tissues and requires large amounts of DNA. Pure DNA constructs that cover the surface of gold particles (1 to 3 μm in diameter) have been propelled into skin cells and, after incision, into subcutaneous tumor cells with a gene gun. Therapeutic genes delivered in this way are expressed in the targeted tissues, and gene products are released into the circulatory system. In theory, secretion of a therapeutic protein into the circulatory system should facilitate its conveyance to a target tissue that is difficult to access directly. However, most proteins that are not normally part of the circulatory system are inactivated or degraded by cells or proteins in the blood. Special strategies need to be developed to overcome this problem.

Surrounding a DNA construct with artificial lipid layers to form a lipid sphere (liposome) with an aqueous core facilitates the passage of the DNA through the cell membrane. Various liposome formulations have been devised. For example, cationic liposomes have a positive charge on the outside and bind to negatively charged DNA molecules to form a lipid–DNA complex (lipoplex). Lipoplexes are simple to form, relatively nontoxic, and nonimmunogenic. Unfortunately, the efficiency of gene transfer by a lipoplex is not high, because after cellular internalization a large amount of the DNA is taken up by lysosomes and degraded. One clinical trial with cystic fibrosis patients who were treated with a liposome–*CFTR* gene complex showed no cytotoxic effects, although the frequency of transfection of nasal epithelial cells was low and the duration of expression was short.

DNA–molecular conjugates have been developed to deliver large DNA (>10-kb) constructs by the endosome cellular transport pathway, thereby avoiding lysosomal degradation of the DNA. With this approach, poly-L-lysine is first chemically linked (conjugated) to a molecule that binds to a specific cell receptor. Next, DNA is added and combines with the poly-L-lysine to form a tightly compacted, twisted, solid ring (toroid). The cell receptor binding sites are arrayed on the outside of the DNA–molecular conjugate (Fig. 21.11). Notwithstanding the specificity of a DNA–molecular

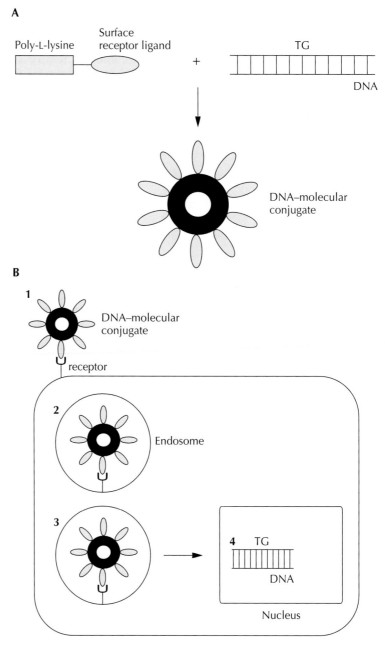

Figure 21.11 DNA–molecular conjugate delivery system. **A.** Poly-L-lysine that is conjugated to a cell surface receptor ligand combines with DNA containing a therapeutic gene to form a condensed DNA–polylysine ring with the cell surface receptor ligands facing outwards. **B.** A DNA–molecular conjugate binds to a specific cell receptor (1) and is engulfed by the cell membrane (2) to form an endosome (3) that protects the DNA–molecular conjugate from degradation by lysosomes. Some of the DNA within an endosome is released and passed into the cell's nucleus (4), where the therapeutic gene (TG) is expressed.

conjugate, in practice, the efficiency of transfection is low. In sum, the current generation of nonviral gene delivery systems has two major limitations. First, the frequency of transfection is often too low to create a therapeutic effect. Second, the duration of therapeutic gene expression is too brief to provide an effective treatment.

Theoretically, a human artificial chromosome would be an exciting therapeutic vector. It would have a very large DNA carrying capacity which would allow the inclusion of a complete set of regulatory elements for one or more therapeutic genes. Also, the genomic version of the therapeutic gene could be used; this might increase the efficiency of gene expression. Finally, both long-term stability and sustained expression of a therapeutic gene within either a proliferating or quiescent target cell would be ensured.

The three essential DNA elements of an artificial chromosome are the chromosomal ends (telomeres), a centromere, and origins of replication. The molecular features of the telomeres of human chromosomes are well known, whereas less information about centromeres and origins of replication is available. It was thought that without first-hand knowledge of each of these functional chromosome elements, it would be very difficult to construct a human artificial chromosome. Remarkably, stable linear human artificial chromosomes (microchromosomes) were formed and maintained after transfecting cells in culture with a long synthetic array (~1 megabase pair [Mb]) of a repetitive DNA element that occurs at the centromere region of the human Y chromosome, high-molecular-weight fragments of human genomic DNA, and human telomere DNA sequences. A neomycin resistance gene was ligated to the centromeric array before transfection so that the compound G418 could be used as a selection agent. Microchromosomes that ranged from about 6 to 10 Mb in length were found in some of the G418-resistant cells.

Two of the three kinds of microchromosomes that were observed arose by shortening (truncation) of an existing chromosome that either retained its original centromere or had its centromere replaced by the transfected centromeric array. The third type of microchromosome, which represents an entire human artificial chromosome, was formed by in vivo ligation of the three transfected DNA elements. Clearly, the prospect of creating a human artificial chromosome with a therapeutic gene(s) is feasible. However, the delivery of a DNA molecule as large as a human artificial chromosome to the nucleus of a target cell will be a major problem. Furthermore, the expression of genes that are retained within the blocks of DNA that make up the arms of a human artificial chromosome may have deleterious effects on target cells. Initially, as a therapeutic agent, cells with a genetically engineered human artificial chromosome might be encapsulated and implanted into tissues of patients.

Prodrug Activation Therapy

Despite intensive treatment with surgery, radiation therapy, and chemotherapy, cancer of all types is still a leading cause of death of humans. The combination of the herpes simplex virus thymidine kinase gene (HSV*tk*) and ganciclovir [GCV; 9-(1,3-dihydroxy-2-propoxymethyl)guanine] has been used to eradicate proliferating tumor cells. Briefly, the rationale for this form of therapy for proliferating cancer cells entails either transducing or transfecting tumor cells in vivo with the HSV*tk* gene under the control of an active promoter. Then, a few days later, ganciclovir is administered. The herpes simplex virus thymidine kinase phosphorylates ganciclovir to form monophosphate–GCV. Although host cell kinases do not efficiently phosphorylate GCV, they readily add phosphate groups to monophosphate–GCV to form triphosphate–GCV. Triphosphate–GCV inhibits DNA polymerase and terminates DNA synthesis, causing the death

of a proliferating cell. Moreover, triphosphate–GCV can pass into unmodified cells by cell-to-cell contact and kill these cells as well. Generally, one HSV*tk*-expressing tumor cell will destroy up to 10 unmodified cells. This phenomenon is called the "bystander effect."

When a gene under certain conditions causes the death of its own cell, it is called a "suicide" gene. The term "prodrug" refers to the inactive form of a therapeutic agent that is activated biologically after it is administered as part of a treatment. Other prodrug–activation gene combinations exist, but the GCV–HSV*tk* system is the one that is used most often.

In a number of preclinical trials, GCV–HSV*tk* suicide therapy has effectively reduced proliferating tumor cells. However, in the first phase I trial

Figure 21.12 Combining prodrug activation therapy and gene-based immunotherapy. Tumor cells are transduced in vivo with a herpes simplex virus thymidine kinase gene (HSV*tk*) and a cytokine gene. The transduced tumor cells that produce the cytokine (CYT) attract immune response cells, and those that produce thymidine kinase (TK) phosphorylate ganciclovir. As depicted by the solid bars, phosphorylated ganciclovir passes through contact sites between cells and kills a cell (X) after it is incorporated into chromosomal DNA.

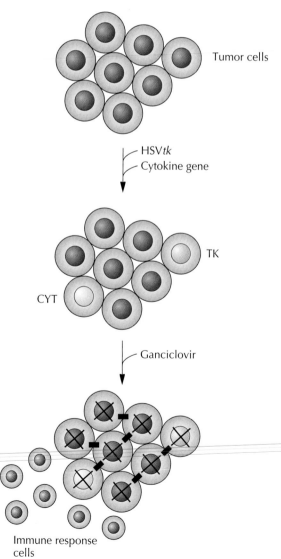

of the GCV–HSV*tk* system with terminally ill cancer patients, no tumor regression was observed. It is possible that in this case, notwithstanding the bystander effect, too few tumor cells were transduced with HSV*tk* to impede the growth of advanced tumors. Consequently, strategies are being devised both to increase the transduction frequency and to deliver the HSV*tk* gene to cells throughout a tumor.

A cancer therapy strategy using two different gene systems has been investigated. For example, GCV–HSV*tk* suicide gene therapy and gene-based immunotherapy have been combined to treat some cancers (Fig. 21.12). In this case, tumor cells are transduced separately with the HSV*tk* gene and a cloned cytokine cDNA or gene. Cytokines, such as interleukin-2, interleukin-12, and others, act as signals that mobilize immune cells and stimulate an immune response. It was reasoned that tumor cell peptides that are released after cell death caused by suicide gene therapy would interact with the immune cells that were attracted to the site of the tumor by the cytokine and subsequently initiate an immunological reaction directed against the tumor cells. As an added benefit, circulating tumor cell antibodies should prevent the initiation of cancer at other sites in the body due to the invasion of malignant cells from the primary tumor.

The idea of a two-gene cancer treatment was tested experimentally by implanting colon cancer cells that had been transduced separately with the HSV*tk* gene and a cytokine gene into the livers of animals. Treatment with GCV blocked the growth of the tumor in the liver. Furthermore, tumors did not form when other tissue locations of the treated animals were inoculated with nontransduced colon cancer cells. By contrast, under these conditions, control animals developed tumors at all the sites that were inoculated with nontransduced colon cancer cells. Notwithstanding these promising results, issues such as which tumors are responsive to this type of treatment and whether this treatment causes adverse side effects need to be resolved before either suicide gene therapy alone or various combination gene-based therapies can be tested in humans.

Nucleic Acid Therapeutic Agents

For the most part, protocols for ex vivo and in vivo gene therapy use a cloned DNA construct to supply a functional form of a protein which is either lacking or defective. However, a number of human disorders such as cancer, inflammatory conditions, and both viral and parasitic infections result from the overproduction of a normal protein. Therapeutic systems using nucleotide sequences are being devised to treat these types of conditions. Theoretically, a small single-strand nucleotide sequence (oligonucleotide) could hybridize to a specific gene or messenger RNA (mRNA) and diminish transcription or translation, respectively, thereby lowering the amount of protein that is synthesized. An oligonucleotide that is designed to bind to a gene and block transcription is called an antigene oligonucleotide, and one that base pairs with a specific mRNA is an antisense oligonucleotide. In addition, the binding of an oligonucleotide to a transcription factor that is responsible for the expression of a specific gene could lower both transcription and translation of the target gene. Moreover, double-stranded oligonucleotides that attach to DNA binding proteins could prevent the activation of transcription of specific genes. It is also conceivable that synthetic DNA molecules that bind to proteins that naturally are not DNA binding proteins and prevent them from functioning can be cre-

ated. Finally, ribozymes, which are natural RNA sequences that bind and cleave specific RNA molecules, could be engineered to target an mRNA and subsequently decrease the amount of a particular protein that is synthesized. The potential for nucleic acid therapeutic agents is considerable, with most of the current research efforts and clinical trials being focused on antisense sequences, and, especially, antisense oligonucleotides.

In Vivo Antisense RNA Production

To be an effective therapeutic agent, an antisense oligonucleotide must bind to a specified mRNA and prevent translation of the protein (Fig. 21.13). The possibility of using an expression vector to produce an antisense RNA that suppresses a pathogenic condition has been examined. For example, episomally based expression vectors that carry the cDNA sequence for either insulin-like growth factor 1 (IGF-1) or insulin-like growth factor 1 receptor (IGF-1R) were constructed with the cloned sequences oriented so the transcripts are antisense rather than mRNA (sense) sequences. IGF-1 is prevalent in malignant glioma, which is the most common form of human brain tumor. Excess production of IGF-1R occurs in prostate carcinoma, which is a significant type of cancer in males. In both vectors, the reverse-oriented cDNAs are under the control of the metallothionein promoter, which is induced by low levels of $ZnSO_4$.

Cultured glioma cells were transfected with the vector that produces the antisense version of the IGF-1 mRNA. In the absence of $ZnSO_4$, the tumorous properties were retained; in contrast, when $ZnSO_4$ was added to the culture medium, these distinctive features were lost. In another experiment, nontransfected glioma cells caused tumors after they were injected into rats, whereas glioma cells that had been transfected with antisense IGF-1 cDNA did not develop tumors.

When mice were injected with rat prostate carcinoma cells that were transfected with the IGF-1R cDNA in the antisense orientation, they developed either small or no tumors, whereas large tumors were formed when mice were treated with either nontransfected or control-transfected rat prostate carcinoma cells. It is assumed that in both cases the antisense RNA hybridizes with its complementary mRNA sequence and hinders translation of IGF-1 and IGF-1R, thus preventing the proliferation of the cancer cells.

Antisense Oligonucleotides as Therapeutic Agents

Sequence-specific effectiveness of chemically synthesized antisense oligodeoxynucleotides relies on hybridization to an accessible site on the target mRNA, resistance to degradation by cellular nucleases, and ready delivery into cells. Oligonucleotides with about 15 to 20 nucleotides have sufficient specificity to hybridize to a unique mRNA. Potential mRNA target sites are determined by testing a set of antisense oligonucleotides with cells in culture that produce the target mRNA. Those antisense oligonucleotides that diminish the translation of the specified protein are selected. Autoradiography after protein gel electrophoresis of cellular proteins that are radiolabeled during translation can be used to determine if the production of a particular protein is reduced in the presence of an antisense oligonucleotide. There are no general rules for predicting the best target sites in various RNA transcripts. Antisense oligonucleotides that are directed to the 5' and 3' ends of mRNAs, intron-exon boundaries, and regions that are naturally double stranded have been effective. Since oligodeoxynu-

Figure 21.13 Inhibition of translation of specific mRNAs by antisense (AS) nucleic acid molecules. The promoter and polyadenylation regions are marked by *p* and *pa*, respectively; the intron is indicated by the letter A; the exons are indicated by numbers (1 and 2). **A.** A cDNA (AS gene) is cloned into an expression vector in reverse orientation, and the construct is transfected into a cell, where the AS RNA is synthesized. The AS RNA hybridizes to the target mRNA, and translation is blocked. **B.** An AS oligonucleotide is introduced into a cell, and after it hybridizes with the target mRNA, translation is blocked.

cleotides are susceptible to degradation by intracellular nucleases, it was important to find ways to make these molecules resistant to attack by nucleases without affecting their ability to hybridize to a target sequence. To this end, the backbone, pyrimidines, and sugar moiety have been modified (Fig. 21.14). Currently, the most extensively used antisense oligonucleotide has a sulfur group in place of the free oxygen of the phosphodiester bond (Fig. 21.14B). This modification is called a phosphorothioate linkage. Phosphorothioate antisense oligonucleotides are water soluble, polyanionic, and resistant to endogenous nucleases. In addition, when a phosphorothioate antisense oligonucleotide hybridizes to its target site, the RNA–DNA duplex activates the endogenous enzyme ribonuclease (RNase) H, which cleaves the mRNA component of the hybrid molecule. Clinical trials with phosphorothioate antisense oligonucleotides, which are "first-generation" therapeutic agents, have been initiated. Some of the targets include RNAs of cytomegalovirus and human immunodeficiency virus as well as mRNAs for genes that are responsible for cancer, bowel disease, and other disorders.

Antisense oligonucleotides with phosphoramidite and polyamide (peptide) linkages have been synthesized (Fig. 21.14C and D). These molecules should be very resistant to nuclease degradation. Furthermore, chemical groups have been added to the 2′ carbon of the sugar moiety and the 5 carbon (C-5) of pyrimidines to both enhance stability and facilitate the binding of the antisense oligonucleotide to its target site (Fig. 21.14E and F). The potencies of these and other modifications are under study.

Free antisense oligonucleotides are not readily internalized by cells. However, cellular uptake occurs when an antisense oligonucleotide is encapsulated within a liposome. With an effective delivery system, low, nontoxic concentrations of the antisense molecule can be used to inhibit translation of the target mRNA. Moreover, liposomes that are conjugated with cell-specific binding sites direct an antisense oligonucleotide to a particular tissue.

Several preclinical trials have shown the usefulness of antisense oligonucleotides as therapeutic agents. For example, the narrowing (stenosis) of coronary and carotid arteries that leads to heart attacks and strokes, respectively, is often alleviated by angioplasty, which is a procedure that widens arteries by the insertion of an inflated balloon. However, arterial blockage recurs (restenosis) in about 40% of the patients within 6 months because angioplasty induces a healing reaction which stimulates the proliferation of smooth muscle cells and the secretion of an extracellular matrix in the inner layer of the artery at the site of the treatment. When phosphorothioate antisense oligonucleotides that targeted mRNAs for proteins that are essential for the mammalian cell cycle were applied to rat carotid arteries after angioplasty, restenosis was reduced by about 90%. In addition to postangioplasty restenosis, smooth muscle cell proliferation is implicated in atherosclerosis, hypertension, diabetes mellitus, and the failure of coronary bypass grafts. Presumably, these conditions might be controlled by similar antisense therapeutics.

In other studies, antisense oligonucleotides have curtailed viral infections and malaria. And, in a phase I clinical trial, an orally administered antisense oligonucleotide successfully reduced severe inflammatory enteritis (Crohn disease) without significant side effects. In this case, the target mRNA encodes intercellular adhesion molecule 1, which is produced in excess in patients with Crohn disease. The next round of trials will examine whether this antisense oligonucleotide is effective against other inflamma-

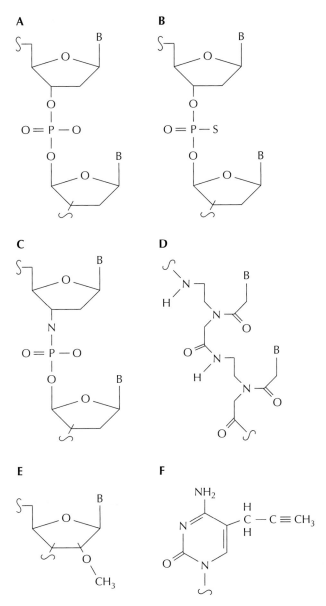

Figure 21.14 Modifications to antisense oligonucleotides. **A.** Phosphodiester linkage. **B.** Phosphorothioate linkage. **C.** Phosphoramidite linkage. **D.** Polyamide linkage (peptide nucleic acid). **E.** 2′-O-methyl ribose. **F.** C-5 propynylcytosine.

tory diseases, including rheumatoid arthritis, psoriasis, and ulcerative colitis.

In theory, an antisense oligonucleotide could form a triple helix with a chromosomal DNA target and block transcription. However, to date, the specificity of such an antigene oligonucleotide does not meet the standards that are expected for a therapeutic agent.

Nucleic Acid Pharmaceuticals

In addition to antisense therapy, the administration of an oligonucleotide that attaches to either a transcription or translation factor could prevent the expression of a target gene. However, the feasibility of this strategy has not

been rigorously examined to date. On the other hand, because nucleic acid molecules are capable of binding to proteins, it has been reasoned that one could devise an oligonucleotide that binds to a protein that does not normally bind a nucleic acid molecule and, as a consequence of this attachment, inhibit the functioning of the protein. This type of oligonucleotide is called an aptamer. For example, an aptamer that binds to thrombin might be an inexpensive way of inhibiting blood clotting after various surgical procedures.

With this use in mind, a potential antithrombin aptamer was isolated in the following way. A pool of oligonucleotides that each have two flanking regions of 18 bases (primer sequences) with a central section of 60 nucleotides with any of the four nucleotides at each of these 60 sites was chemically synthesized. Theoretically, this sample of oligonucleotides has about 1.3×10^{36} (4^{60}) different central sequences. Next, the pool of oligonucleotides is passed through a thrombin-bound column matrix. The oligonucleotides that bind to the thrombin are eluted. The aptamers in the initial sample are reselected three or more times by passage through fresh thrombin–matrix columns. The final set of thrombin aptamers is amplified by polymerase chain reaction (PCR) and cloned. The physical and biological properties of each aptamer are determined. Those aptamers which have high affinity, high specificity, and high inhibitory activity are retained for more detailed analysis.

An effective antithrombin aptamer was isolated in this way. However, this oligonucleotide is short-lived in vivo. Thus, it can only be used for procedures that require a transient inhibition of thrombin, such as cardiopulmonary bypass operations, in contrast to angioplasty, which requires a longer treatment with anticlotting agents. Clearly, the in vitro selection procedure for identifying thrombin aptamers can be used for any number of different target proteins.

Ribozymes as Therapeutic Agents

Ribozymes are naturally occurring catalytic RNA molecules (RNA enzymes) that have separate catalytic and substrate binding domains. The substrate binding sequence combines by nucleotide complementarity and, possibly, nonhydrogen bond interactions with its target sequence. The catalytic portion cleaves the target RNA at a specific site. The substrate domain of a ribozyme can be engineered to direct it to a specified mRNA sequence (Fig. 21.15).

In practice, an indirect strategy is necessary for creating a therapeutic ribozyme, since the large-scale production of synthetic RNA molecules is difficult and RNA molecules are susceptible to degradation after delivery to a target cell. One approach to overcome these drawbacks entails chemically synthesizing an oligodeoxynucleotide with a ribozyme catalytic domain (~20 nucleotides) flanked by sequences that hybridize to the target mRNA after transcription. The oligodeoxynucleotide is amplified by using the substrate binding sequences as primers. The amplification product is cloned into a eukaryotic expression vector, cells are transfected with the construct, and the transcribed ribozyme cleaves the target mRNA, thereby suppressing the translation of the protein that is responsible for the disorder. Various cancers and viral diseases could be treated with genetically engineered ribozymes.

To date, no naturally occurring DNA enzymes (deoxyribozymes) have been discovered. However, researchers have created oligodeoxynu-

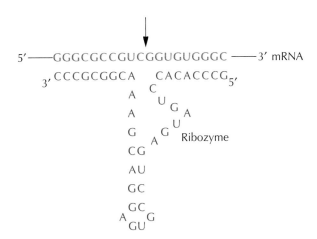

Figure 21.15 Ribozyme cleavage of an mRNA. A ribozyme with a genetically engineered substrate domain hybridizes to a target mRNA and cleaves it at a specific site (arrow). (Adapted from Tone et al., *In Vivo* **7**:471–476, 1993.)

cleotides with catalytic activity. As a therapeutic agent, a catalytic oligodeoxynucleotide has advantages over a ribozyme. For example, instead of using an expression vector to produce a ribozyme, a DNA enzyme could be chemically synthesized, packaged as a liposome, and delivered to the target cell. Although promising, the ability to generate effective DNA enzymes is at present rudimentary.

Oligonucleotide Correction of Genetic Conditions

The ability to convert a mutant base pair of a gene to the wild-type (normal, correct) version would reverse the consequences of many different genetic conditions. A strategy using a modified RNA–DNA oligonucleotide with 68 nucleotides (chimeric oligonucleotide) has been devised for this purpose. The composition of the chimeric oligonucleotide includes a single mixed oligonucleotide with ribonucleotides and deoxyribonucleotides in a duplex conformation with hairpin caps at the ends of the complementary segments and methylation of the oxygen of the 2′ carbon of the ribose sugars (Fig. 21.16). The rationale for this particular arrangement is based on various experimental observations. First, combined RNA–DNA molecules participate more readily than duplex DNA in homologous nucleic acid pairing reactions. Second, hairpin caps, which do not interfere with the pairing of homologous nucleic acid molecules, protect the molecule from exonucleases. Third, the 2′-O methylation of the ribose units shields the molecule from degradation by RNase H. In addition, the organization of the nucleotides of a chimeric oligonucleotide is important. Ten ribonucleotides flank a central core of five deoxyribonucleotides, and except for the correct base pair, this segment of the chimeric oligonucleotide has the same sequence as the target sequence.

The feasibility of base pair correction with a chimeric oligonucleotide was examined with both a mutated cDNA sequence carried by a plasmid and a mutant site within a chromosomal sequence. In both instances, with high frequencies, the mutant sites were replaced by the correct base pair. However, more studies are required before chimeric oligonucleotides will become effective therapeutic agents.

Target sequence

Figure 21.16 Correction of a single-base-pair mutation by a chimeric oligonu-cleotide. The double arrow points to the mutant site in the target sequence and the correct base pair in the chimeric oligonucleotide. The mutant and correct nu-cleotides are underlined. The uppercase letters represent deoxyribonucleotides, and the lowercase letters represent ribonucleotides. The nucleotides of the hairpin caps are shown in boldface. The vertical line and the 3′ and 5′ designations mark the 3′ and 5′ ends of the chimeric oligonucleotide. (Adapted from Yoon et al., *Proc. Natl. Acad. Sci. USA* **93**:2071–2076, 1996.)

Figure 21.17 Correction of a mutant splice site with an antisense oligonucleotide. **A.** The outcome of a mutation in a splice site is depicted. The numbers denote exons. The first intron is marked with the letter A, and the second intron contains a splice mutation (solid dot) which divides the intron into two parts (B1 and B2). The dotted lines span the RNA segments that are removed during RNA processing. There are two possible splicing events: pathway a leads to a functional mRNA, and pathway b leads to an RNA that includes part of the second intron (B2). **B.** An antisense oligonucleotide (AS) that binds to the mutant splice site prevents RNA processing at this site, and consequently only functional mRNA is produced.

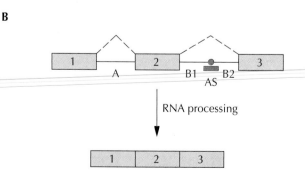

Aberrant splicing of an mRNA occurs when a mutation in an intron is recognized by the RNA processing system as an authentic splice site and, consequently, a portion of the intron is included as part of the processed mRNA (Fig. 21.17A). The presence of part of an intron disrupts the reading frame, and a truncated protein is produced. A disease condition may result from a diminished level of normal protein.

It was reasoned that an antisense oligonucleotide that targeted an aberrant splice site would prevent splicing at that site and increase the number of joining events between the correct intron–exon splice sequences. This notion was tested with a splice mutation in the second intron of the β-globin gene (Fig. 21.17B). This mutation is responsible for one form of β-thalassemia, which is an inherited blood disorder that leads to loss of red blood cells (anemia). After cells that are homozygous for the intron 2 splice site mutation (IVS2-654) were transfected with a 2'-O-methyl phosphorothioate antisense oligonucleotide that targeted the mutant splice site, the amount of normal β-globin chains was increased by about 50%, which theoretically would be beneficial to patients with this genetic defect. Further studies are required to determine if antisense rectification of splice site mutations is an effective therapeutic strategy for thalassemia and other conditions due to similar mutations.

SUMMARY

The development of effective treatments for genetic disease has been elusive because, in many instances, the appropriate gene product cannot be provided to a patient. However, when a normal version of a gene, usually in the form of a cDNA molecule, has been identified and cloned, it is possible that it can be used to correct the defect in affected individuals. Two main approaches for gene-based therapeutic treatment of various diseases have been developed: ex vivo gene therapy and in vivo gene therapy.

Ex vivo gene therapy entails adding a remedial gene to isolated cells of the affected individual by retroviral cloning vectors or other means. The transduced cells are cultured and infused back into the patient. Although this strategy avoids adverse immunological responses in the recipient, it is costly and difficult to implement. As an alternative, genetically engineered nonautologous cells that are encapsulated within a membrane that protects the cells from immunological attack and allows the release of the remedial gene product would enable ex vivo gene therapy to be used for a large number of patients with many different diseases.

For in vivo gene therapy, the remedial gene is introduced into the cells of the target tissue of an affected individual. Various strategies have been devised for delivering a gene to a target tissue. Viral vector systems based on retroviruses, adenovirus, adeno-associated virus, and herpes simplex virus have been developed. Nonviral gene delivery systems include injection of pure DNA, bombardment of a target tissue with DNA-coated particles, and cellular uptake of DNA that is enclosed within a lipid envelope (liposome). In addition, both viral and nonviral delivery systems that are cell specific are being devised.

Gene-based prodrug activation has considerable potential for treating proliferating cancers. In the most widely used system, the herpes simplex virus thymidine kinase gene (HSV*tk*) is introduced into tumor cells, and subsequent administration of ganciclovir (GCV) leads to formation of triphosphate–GCV, which is toxic to rapidly dividing cells. Moreover, un-

modified cells that are in contact with HSV*tk*-modified cells are also killed. The latter phenomenon is called the bystander effect.

In addition to gene-based therapies, oligonucleotides are being examined as therapeutic agents. For example, antisense therapy is designed to prevent or at least lower the expression of a gene that causes a disease condition. In one form of antisense therapy, a cloned gene is introduced into an expression vector in an orientation that allows transcription of the complementary DNA strand that is not normally transcribed. The antisense RNA transcript is complementary to the normal mRNA, can base pair with it, and, as a result, interferes with its translation. In the more prevalent strategy for antisense therapy, an oligonucleotide that is introduced into a target cell hybridizes to a specific mRNA and blocks translation. Modifications to the phosphodiester linkage, pyrimidines, and sugar moieties increase the effectiveness and longevity of antisense oligonucleotides. Researchers are also examining the therapeutic usefulness of oligonucleotides that cleave specific mRNAs (genetically engineered ribozymes), bind to specific proteins and block their function (aptamers), and correct single base pair and splice site mutations.

References

Agrawal, S. (ed.). 1996. *Antisense Therapeutics.* Humana Press Inc., Totowa, N.J.

Altman, S. 1993. RNA enzyme-directed gene therapy. *Proc. Natl. Acad. Sci. USA* **90:**10898–10900.

Anderson, W. F. 1992. Human gene therapy. *Science* **256:**808–813.

Blaese, R. M., K. W. Culver, A. D. Miller, C. S. Carter, T. Fleisher, M. Clerici, G. Shearer, L. Chang, Y. Chiang, P. Tolstoshev, J. J. Greenblatt, S. A. Rosenberg, H. Klein, M. Berger, C. A. Mullen, W. J. Ramsey, L. Muul, R. A. Morgan, and W. F. Anderson. 1995. T lymphocyte-directed gene therapy for ADA⁻ SCID: initial trial results after 4 years. *Science* **270:**475–480.

Bordignon, C., L. D. Notarangelo, N. Nobili, G. Ferrari, G. Casorati, P. Panina, E. Mazzolari, D. Maggioni, C. Rossi, P. Servida, A. G. Ugazio, and F. Mavilio. 1995. Gene therapy in peripheral blood lymphocytes and bone marrow for ADA⁻ immunodeficient patients. *Science* **270:**470–475.

Breaker, R. R. 1997. DNA enzymes. *Nat. Biotechnol.* **15:**427–431.

Burfeind, P., C. L. Chernicky, F. Rininsland, J. Ilan, and J. Ilan. 1996. Antisense RNA to the type I insulin-like growth factor receptor suppresses tumor growth and prevents invasion by rat prostate cancer cells *in vivo*. *Proc. Natl. Acad. Sci. USA* **93:**7263–7268.

Caplen, N. J., E. W. F. W. Alton, P. G. Middleton, J. R. Dorin, B. J. Stevenson, X. Gao, S. R. Durham, K. Jeffery, M. F. Hodson, C. Coutelle, L. Huang, D. J. Porteus, R. Williamson, and D. M. Geddes. 1995. Liposome-mediated *CFTR* gene transfer to the nasal epithelium of patients with cystic fibrosis. *Nat. Med.* **1:**39–46.

Cole-Strauss, A., K. Yoon, Y. Xiang, B. C. Bryne, M. C. Rice, J. Gryn, W. K. Holloman, and E. B. Kmiec. 1996. Correction of the mutation responsible for sickle cell anemia by an RNA-DNA oligonucleotide. *Science* **273:**1386–1389.

Cornetta, K., R. A. Morgan, and W. F. Anderson. 1991. Safety issues related to retroviral-mediated gene transfer in humans. *Hum. Gene Ther.* **2:**5–14.

Crystal, R. G. 1995. Transfer of genes to humans: early lessons and obstacles to success. *Science* **270:**404–410.

Culver, K. W., F. W. Anderson, and R. M. Blaese. 1991. Lymphocyte gene therapy. *Hum. Gene Ther.* **2:**107–109.

Culver, K. W., Z. Ram, S. Wallbridge, H. Ishii, E. H. Oldfield, and R. M. Blaese. 1992. In vivo transfer with retroviral vector-producer cells for treatment of experimental brain tumors. *Science* **256:**1550–1552.

Ellington, A. D., and R. Conrad. 1995. Aptamers as potential nucleic acid pharmaceuticals. *Biotechnol. Annu. Rev.* **1:**185–214.

Emerich, D. F., S. R. Winn, P. M. Hantraye, M. Peschanski, E.-Y. Chen, Y. Chu, P. McDermott, E. E. Baetge, and J. H. Kordower. 1997. Protective effect of encapsulated cells producing neurotrophic factor CNTF in a monkey model of Huntington's disease. *Nature* **386:**395–399.

Fender, P., R. W. H. Ruigrok, E. Gout, S. Buffet, and J. Chroboczek. 1997. Adenovirus dodecahedron, a new vector for human gene transfer. *Nat. Biotechnol.* **15:**52–56.

Flotte, T. R., and B. J. Carter. 1995. Adeno-associated virus vectors for gene therapy. *Gene Ther.* **2:**357–362.

Grossman, M., S. E. Raper, K. Kozarsky, E. A. Stein, J. F. Engelhardt, D. Muller, P. J. Lupien, and J. M. Wilson. 1994. Successful *ex vivo* gene therapy directed to liver in a patient with familial hypercholesterolaemia. *Nat. Genet.* **6:**335–341.

Harrington, J. J., G. V. Bokkelen, R. W. Mays, K. Gustashaw, and H. F. Willard. 1997. Formation of *de novo* centromeres and construction of first-generation human artificial microchromosomes. *Nat. Genet.* **15:**345–355.

Ishii-Morita, H., R. Agbaria, C. A. Mullen, H. Hirano, D. A. Koeplin, Z. Ram, E. H. Oldfield, D. G. Johns, and R. M. Blaese. 1997. Mechanism of 'bystander effect' killing in the herpes simplex thymidine kinase gene therapy model of cancer treatment. *Gene Ther.* **4:**244–251.

Jiao, S., P. Williams, R. K. Berg, B. A. Hodgeman, L. Liu, G. Repetto, and J. A. Wolff. 1992. Direct gene transfer into nonhuman primate myofibers in vivo. *Hum. Gene Ther.* **3:**21–33.

Kochanek, S., P. R. Clemens, K. Mitani, H.-H. Chen, S. Chan, and C. T. Caskey. 1996. A new adenoviral vector: replacement of all viral coding sequences with 28 kb of DNA independently expressing both full-length dystrophin and β-galactosidase. *Proc. Natl. Acad. Sci. USA* **93:**5731–5736.

Koeberl, D. D., I. E. Alexander, C. L. Halbert, D. W. Russell, and A. D. Miller. 1997. Persistent expression of human clotting factor IX from mouse liver after intravenous injection of adeno-associated virus vectors. *Proc. Natl. Acad. Sci. USA* **94:**1426–1431.

Lanza, R. P., J. L. Hayes, and W. L. Chick. 1996. Encapsulated cell technology. *Nat. Biotechnol.* **14:**1107–1111.

Leib, D. A., and P. D. Olivo. 1993. Gene delivery to neurons: is herpes simplex virus the right tool for the job? *BioEssays* **15:**547–554.

Lewis, J. G., K.-Y. Lin, A. Kothavale, W. M. Flanagan, M. D. Matteucci, R. B. De Prince, R. A. Mook, Jr., R. W. Hendren, and R. W. Wagner. 1996. A serum-resistant cytofectin for cellular delivery of antisense oligodeoxynucleotides and plasmid DNA. *Proc. Natl. Acad. Sci. USA* **93:**3176–3181.

Lyon, J., and P. Gorner. 1995. *Altered Fates: Gene Therapy and the Retooling of Human Life.* W. W. Norton & Company, Inc., New York, N.Y.

Michael, S. I., and D. T. Curiel. 1994. Strategies to achieve targeted gene delivery via the receptor-mediated endocytosis pathway. *Gene Ther.* **1:**223–232.

Miller, A. D. 1992. Human gene therapy comes of age. *Nature* **357:**455–460.

Mulligan, R. C. 1993. The basic science of gene therapy. *Science* **260:**926–932.

Pochon, N. A.-M., B. Heyd, N. Déglon, J.-M. Joseph, A. D. Zurn, E. E. Baetge, J. P. Hamming, M. Goddard, M. Lysaght, F. Kaplan, A. C. Kato, M. Schluep, L. Hirt, F. Regli, F. Porchet, and N. De Tribolet. 1996. Gene therapy for amyotrophic lateral sclerosis (ALS) using a polymer encapsulated xenogenic cell line engineered to secrete hCNTF. *Hum. Gene Ther.* **7:**851–860.

Putnam, D. A. 1996. Antisense strategies and therapeutic applications. *Am. J. Health-Syst. Pharm.* **53:**151–160.

Santoro, S. W., and G. F. Joyce. 1997. A general purpose RNA-cleaving DNA enzyme. *Proc. Natl. Acad. Sci. USA* **94:**4262–4266.

Sierakowska, H., M. J. Sambade, S. Agrawal, and R. Kole. 1996. Repair of thalassemic human β-globin mRNA in mammalian cells by antisense oligonucleotides. *Proc. Natl. Acad. Sci. USA* **93**:12840–12844.

Sun, W. H., J. K. Burkholder, J. Sun, J. Culp, J. Turner, X. G. Lu, T. D. Pugh, W. B. Ershler, and N.-S. Yang. 1995. *In vivo* cytokine gene transfer by gene gun reduces tumor growth in mice. *Proc. Natl. Acad. Sci. USA* **92**:2889–2893.

Tone, T., M. Kashani-Sabet, T. Funato, T. Shitara, E. Yoshida, B. I. Kashfian, M. Horng, O. Fodstadt, and K. J. Scanlon. 1993. Suppression of EJ cells tumorigenicity. *In Vivo* **7**:471–476.

Trojan, J., B. K. Blossey, T. R. Johnson, S. D. Rudin, M. Tykocinski, J. Ilan, and J. Ilan. 1992. Loss of tumorigenicity of rat glioblastoma directed by episome-based antisense cDNA transcription of insulin-like growth factor I. *Proc. Natl. Acad. Sci. USA* **89**:4874–4878.

Wagner, R. W. 1994. Gene inhibition using antisense oligodeoxynucleotides. *Nature* **372**:333–335.

Yoon, K., A. Cole-Strauss, and E. B. Kmiec. 1996. Targeted gene correction of episomal DNA in mammalian cells mediated by a chimeric RNA·DNA oligonucleotide. *Proc. Natl. Acad. Sci. USA* **93**:2071–2076.

REVIEW QUESTIONS

1. What is ex vivo gene therapy?

2. Describe how genetically engineered nonautologous cells could be used for ex vivo gene therapy. Why would these kinds of cells be important as therapeutic agents?

3. What is in vivo gene therapy?

4. Describe in detail at least two viral vector systems.

5. Describe in detail at least three nonviral gene delivery systems.

6. What is antisense oligonucleotide therapy?

7. What modifications have been used to increase the longevity and effectiveness of therapeutic antisense oligonucleotides?

8. Describe how the HSV*tk* gene is used to treat various cancers. What is the bystander effect?

9. Describe the changes that would be necessary to make a ribozyme into a therapeutic agent.

10. Describe how a therapeutic aptamer is isolated. What advantages would aptamers have over other kinds of drugs?

IV

Regulating and Patenting Molecular Biotechnology

INEVITABLY, REVOLUTIONARY technologies impact on societies in diverse ways. There can be economic, social, and ethical consequences that result from both the implementation of a new way of doing something and the displacement of traditional processes. Molecular biotechnology is no exception to this phenomenon. Moreover, because molecular biotechnology deals with the engineering of life forms, it impinges on a number of socially sensitive issues. As a result, there have been many questions about its propriety and how it is to be conducted in a safe and reliable manner. In Part IV, several socially important aspects of molecular biotechnology are examined. These include the development of recombinant DNA research guidelines, the regulation of genetically engineered food products, the development of policy for the deliberate release of genetically engineered organisms into the environment, the criteria for sanctioning human gene therapy trials, and the patenting of biotechnology inventions.

CHAPTER 22

Regulating the Use of Biotechnology

Major technological advances such as molecular biotechnology are seldom implemented without controversy. For some people, a new technology is a blueprint for unmitigated disaster. They believe that traditional ways and values will be thoroughly compromised and eroded. These individuals often feel that new initiatives have inherent hazards that can be avoided only if the development of the technology is prevented. Others view a new technology as a bountiful wellspring of untold benefits for society and argue that any obstacles will forestall, probably irreparably, essential advances. Often, those with the latter attitude assume that a new technology is fragile and must be given special concessions to enable it to reach fruition. A third, "neutral" group captures the middle ground between these two extremes. This group believes that nothing can be considered fundamentally new, and therefore, a new technology forms part of a continuum of discoveries. To this third group, existing regulatory systems are considered sufficient for handling the potential risks, and the expected benefits will inevitably occur as the new technology matures.

Because molecular biotechnology can potentially affect many aspects of modern society, including agricultural production and medical treatment, there are significant ethical, legal, economic, and social issues that need to be considered. For example, since its inception in 1973, serious doubts have been voiced by some individuals about the safety of recombinant DNA technology. These concerns prompted scientists to declare a self-imposed moratorium on certain types of recombinant DNA experiments until the adoption of official regulatory guidelines designed to ensure that recombinant microorganisms were unable to proliferate outside the laboratory and that laboratory workers were protected from any potential hazard. The formulation of these regulations took place in 1974 and 1975 in open meetings under the scrutiny of the press. Thus, the public became aware of the possibilities, both negative and positive, of genetically manipulating organisms. Nevertheless, in the late 1970s, apprehension persisted about the safety of this new technology. In particular, there was concern about the release (either accidental or deliberate) of genetically modified organisms into the environment and fear that they would become uncontrolled biological marauders of vulnera-

ble ecosystems. Additional specific guidelines seemed to be necessary to ensure that such rare possibilities would be even less likely to occur.

Moreover, there has been much discussion about the ethics of genetic manipulation of humans. The objective of these discussions has been to distinguish between inappropriate and acceptable procedures. There are no easy answers to the ethical, legal, and social questions raised by the various applications of molecular biotechnology. However, because the stakes are so high, many of these issues have been examined extensively.

Regulating Recombinant DNA Technology

In 1974, when it was realized that recombinant DNA technology could be used to engineer organisms with novel genes, concerns about safety, ethics, and unforeseen consequences were raised by scientists, the public, and government officials. Phrases such as "playing God," "manipulation of life," "the most threatening scientific research ever undertaken," and "man-made evolution" were often used by the popular press when describing recombinant DNA technology. The major apprehension was that, either inadvertently or possibly deliberately for the purposes of warfare, unique microorganisms that had never previously existed would be developed and would cause epidemics or environmental catastrophes. In response to a certain amount of public anxiety about these "doomsday bugs," a group of prominent molecular biologists called for a moratorium on several kinds of recombinant DNA research, especially experiments that involved pathogenic microorganisms.

In 1976, the U.S. National Institutes of Health (NIH), the primary U.S. research grant agency in the medical and health sciences, issued a set of guidelines for the conduct of NIH-supported research using recombinant DNA technology. These rules and regulations rigorously defined physical (laboratory) containment levels for the conduct of recombinant DNA experiments. They also required that biological containment be a component of any recombinant DNA experiment; i.e., the preferred hosts for foreign DNA would be those microorganisms considered least likely to proliferate outside of the laboratory or to transfer their DNA to other microorganisms. For research with known pathogenic organisms, for example, elaborate, negative-pressure, controlled, self-contained rooms were recommended. Conversely, research that was perceived to be less dangerous could be conducted in enclosed contained units equipped with high-quality filter systems. Although the NIH guidelines did not have legal status, most companies that were starting recombinant DNA technology research programs voluntarily complied. In addition, other countries, using the NIH guidelines as a model, adopted their own sets of restrictions for the conduct of recombinant DNA research.

The initial NIH guidelines were very stringent, and many scientists thought that the regulations were excessive. For example, the costs of the containment facilities required by the guidelines effectively prevented smaller companies and researchers with modest grant support from initiating programs using recombinant DNA technology. In anticipation of the need to modify the original guidelines, the NIH Recombinant DNA Advisory Committee (NIH-RAC) was created. This committee was charged with overseeing the developments in recombinant DNA research and, if necessary, refining the regulations. The NIH-RAC had to hold open meetings to discuss its decisions, it had to publish and distribute the minutes of its meet-

ings, and it had to allow any nonmember the opportunity to address the committee on any issue pertaining to recombinant DNA research. The membership of the NIH-RAC, although broad (including both ethicists and members of the general public), consisted mostly of scientists.

As part of the original NIH guidelines, one specific class of experiments that was "not to be initiated at the present time" under any circumstances was the "deliberate release into the environment of any organism containing a recombinant DNA molecule." However, it was inevitable that genetically engineered organisms (GEOs) that could function in natural settings would be developed.

By 1980, the original NIH guidelines were relaxed considerably by the NIH-RAC as a result of experience and specific experimental data from studies that the committee and NIH had sponsored. For example, it was established that the host organism (*Escherichia coli* K-12) that was most commonly used in recombinant DNA experiments was unable to proliferate to any significant extent outside the laboratory. In addition, microbiologists convinced molecular biologists and others that existing safety procedures for work with pathogenic organisms were of a high standard and more stringent rules were not required. Finally, it was conceded that it was extremely unlikely that a pathogenic organism would be created if the cloned gene had nothing to do with pathogenesis in the original organism. Most observers were satisfied that the safety of laboratory workers was ensured if good laboratory techniques were used. However, specific regulations were added to the guidelines to safeguard against accidental spills during large-scale fermentations with GEOs.

As a result of the easing of the containment requirements for more routine experiments, the use of recombinant DNA technology became more prevalent and flourished. The NIH-RAC and the NIH guidelines had effectively quelled, to some extent, the original concern about the potential dangers of recombinant DNA research. However, two significant problems remained. First, what kind of regulation was needed for foods that contained either GEOs or products that were derived from GEOs? Second, how were deliberate releases of GEOs into the environment to be managed?

A third potential problem area was avoided when pharmaceuticals that were made by recombinant DNA technology were deemed by regulatory authorities to be similar to pharmaceuticals produced by traditional means. In most countries, there was a strong consensus that existing regulations for the approval of pharmaceuticals for commercial use were sufficient to ensure both worker safety and public safety and that the process (i.e., recombinant DNA technology) by which a product was made was irrelevant. The axiom that the product alone should be evaluated for its safety and efficacy led to the approval of human drugs, vaccines, diagnostic devices, and other recombinant DNA products.

Regulating Food and Food Ingredients

In the United States, the Food and Drug Administration (FDA) is responsible for regulating the introduction of foods, drugs, pharmaceuticals, and medical devices into the marketplace. The safety of both crop foods and food ingredients that include flavors and additives must be thoroughly assessed before they can be licensed for human consumption. The FDA has had a well-established, although not foolproof, system for the approval of new foods and food products for some time. Critics of the FDA, however,

have argued that it tends to favor the interests of industry and is too lenient in enforcing its own regulations. Both the FDA and the food industry, which is represented by the International Food Biotechnology Council, have argued (somewhat convincingly) that new regulations are not required for foods and foodstuffs that are developed by recombinant DNA technology because any unlicensed food or food ingredient, regardless of how it is produced, must be assessed for safety by toxicity, allergenicity, and impurity testing. If genetic modifications by either selective breeding or recombinant DNA methodologies change the composition of an accepted food or foodstuff, then the company, after the safety of the product is proved, must inform the consumer through labeling that the new product differs from the traditional one.

Chymosin

A new food product is usually subjected to a large battery of different tests. However, in order to streamline the process and lower the costs of developing a food product, the similarity of the new product to the one that it is designed to replace is taken into consideration. For example, the FDA approved the enzyme chymosin, an agent produced by recombinant DNA technology for use in cheese making, without demanding a full range of tests. Chymosin, one of the key components of rennet, is a milk-clotting proteolytic enzyme that hydrolyzes the κ-casein protein of milk. This enzymatic cleavage creates curds, which in turn are processed into cheese. Traditionally, the milk-clotting agent for cheese making is derived from the fourth stomach of calves and consists of a mixture of substances that collectively is called rennet.

To ensure a reliable, convenient, and possibly cheaper industrial supply of chymosin, one of the chymosin genes was cloned and expressed, and the product was harvested from *E. coli* K-12. When a petition requesting permission to use recombinant chymosin for the commercial production of cheese was presented to the FDA, it was necessary to decide what criteria should be required for the approval process. Because there has been a long history of using rennet containing chymosin in the cheese-making industry, the FDA reasoned that, if the recombinant chymosin was identical to the naturally occurring chymosin, then excessive testing was not necessary. In essence, the petitioner had to show that the recombinant chymosin was identical to the chymosin of rennet. To substantiate this, restriction mapping, DNA hybridization, and DNA sequencing were used to establish that the cloned and native DNA sequences of the chymosin gene were identical. Moreover, the recombinant chymosin had the same molecular weight as purified calf chymosin, and the biological activities of the two forms of the enzyme were the same.

Next, it was essential to establish that the recombinant chymosin preparation was safe. The company showed that, as part of its purification process, recombinant chymosin is extracted from inclusion bodies and that the final preparation is free of whole bacterial cells, significant cell debris, and other impurities, including nucleic acids. Although the presence of minute amounts of *E. coli* K-12 cells in the final preparation of chymosin is undesirable, numerous studies have established that this strain is nontoxigenic and nonpathogenic to humans. To ensure that the recombinant chymosin preparations did not contain an unexpected toxin, the results of animal testing showed no adverse effects. After compiling all the information, the FDA concluded that the recombinant chymosin could be licensed for commercial use.

Tryptophan

For the most part, agencies in various countries that are responsible for regulating food and foodstuffs derived from recombinant DNA technology have adopted a case-by-case approach. Each submission is considered separately, and, depending on the judgment of the regulatory body, a series of tests is specified to ensure that the product is safe. Although the industry prefers and urges that government agencies create a single set of standards for all products derived by genetic engineering, there is considerable reluctance to go in that direction. Currently, the introduction of recombinant foodstuffs for human consumption is being handled with a degree of caution, especially since a false assumption that seems logical initially can cause unexpected and perhaps tragic results.

During 1989 and 1990 in the United States, an unusually high number of cases of the disease eosinophilia-myalgia syndrome (EMS) were reported. This generally rare disease causes severe, debilitating muscle pain and can be fatal as a result of respiratory arrest. A consistent feature among the occurrences of EMS was that the patients had been consuming large doses of the amino acid tryptophan as a food supplement. In each case, the source of the tryptophan was traced back to a single chemical company. The possible correlation between tryptophan and EMS was puzzling, because there had been no history of significant negative effects when tryptophan had previously been used as a dietary agent. Further investigation revealed that all of the suspected batches of the "tainted" tryptophan had been produced by a genetically engineered bacterial strain that had been designed to overproduce tryptophan. The company had assumed that the enhanced strain was identical to the previous one; therefore, no additional product safety tests were thought to be necessary. At the same time, what was thought to be a minor step in the purification process was changed, while the previous quality control measures that had been used to assay the purity of the final preparations were retained.

Chemical analyses of the commercial preparations that had been produced by the genetically engineered strain revealed that they contained novel metabolic derivatives of tryptophan, including 1-1'-ethylene-bis[tryptophan] (EBT). Initially, the presence of EBT was considered to be due to some metabolic quirk in the new strain. While research focused on determining whether EBT caused EMS, other studies showed that EBT was produced by wild-type strains as well. Toxicity studies established that EBT induced pathological changes in rats that are similar to EMS and, surprisingly, that even tryptophan, to a lesser extent, produced some EMS symptoms. Consequently, L-tryptophan, even without impurities, was banned for human consumption in the United States. It is not clear what actually caused EBT to appear in a product that had previously been safe. Most observers believe that the change in the purification process allowed EBT to contaminate the tryptophan. Presumably, although the company was unaware of it, the old method effectively removed it.

One of the lessons of this episode, even though genetic engineering may not have been the problem, is that biological equivalence between a strain and its genetically altered counterpart should not be assumed. This is as true for a strain produced by traditional methods as it is for one that has been genetically engineered. Furthermore, manufacturers are now more aware that a minor technical change in the purification phase can alter the nature of a product. However, what they do with this knowledge may be problematic; many companies do not want to run a complete battery of tox-

icity tests for a product that they believe has already been thoroughly tested. But, notwithstanding economic costs, many manufacturers are opting for a "better safe than sorry" approach.

Bovine Somatotropin

In addition to concerns about safety, there are other ways that a novel technology can affect society. The development of recombinant bovine somatotropin (bST), which is also known as bovine growth hormone, provides an example of an effective and safe treatment that is not necessarily a welcome innovation.

In the 1930s, it was shown that injection of bST into a dairy cow will increase its milk yield significantly. Because natural bST is both difficult and costly to accumulate in large quantities, it has not been used routinely as an augmenting agent by the dairy industry. However, with recombinant DNA technology, the gene for bST was cloned into *E. coli* and expressed. Then the recombinant bST was harvested and purified. Under trial conditions, as expected, milk production in dairy cows was increased by 20 to 25% after injection of the recombinant form of bST.

The safety of bST in milk has been studied exhaustively. In treated cows, the levels of bST in milk were not higher than those in control cows. Moreover, bST is not active in humans, and all toxicity trials have shown that there are no adverse effects on test organisms. The FDA, using all the research results that it could assemble, concluded that both the meat and milk of bST-treated cows were safe for human consumption. This conclusion was supported by the U.S. Office of Technology Assessment after its independent analysis of many of the bST studies.

However, an effective and powerful lobby group has been put together to block the approval of recombinant bST by the FDA. The principal reason for this opposition is based on the presumed economic consequences that recombinant bST will have on the dairy industry. The concern is that many small dairy farms will become unprofitable because fewer cows will be required to maintain current levels of milk production. In addition to the attrition of small dairy farmers, opponents of bST fear that the industry will become dominated by large corporate interests at the expense of the independent producers. These economic arguments may be justified, and certainly any group is entitled to protect itself against what it perceives to be a threat to its livelihood. However, the main issue in the advertising campaign waged against recombinant bST is that "genetically engineered hormones" in "hormone-laced milk" may be harmful and cause cancer in humans. Using recombinant DNA technology as a bogeyman has probably made this campaign emotionally effective.

In addition to economic concerns, the opponents of bST contend that its use will increase the incidence of bacterial infection of milk glands (mastitis) in dairy cattle. Consequently, it is argued that larger-than-normal amounts of antibiotics will have to be used in the maintaining of these animals, thereby resulting in increased levels of antibiotics in the cows' milk that may trigger allergic responses in some consumers. Moreover, increased antibiotic use could heighten the selection pressure for drug-resistant pathogens. However, the Veterinary Medicine Advisory Committee of the FDA and others have studied this issue and concluded that the frequency of mastitis is no greater in bST-treated cows than in untreated ones.

Recombinant bST was licensed for use in dairying in the United States in 1994. However, there are temporary prohibitions to the sale of milk from

bST-treated cows in a number of other countries. In each of these instances, the interim moratorium appears to be due to the potential socioeconomic impact of increasing milk production with recombinant bST rather than any possible adverse health effects.

On the whole, many of the initial apprehensions about regulating food-stuffs that are developed with recombinant DNA technologies have evaporated as the FDA and similar agencies in other countries have implemented traditional risk assessment procedures. The food industry prefers a minimal set of regulations to speed up the transition from the developmental phase to the marketplace. Governmental agencies, as representatives of the public, have a dual responsibility. They are entrusted with protecting public health and ensuring that new developments are not needlessly curtailed. As experience accumulates, the current requirements may be relaxed and a more standardized procedure for approval may be established. Presumably, these changes will not be made for the sake of expediency at the expense of safety.

Deliberate Release of Genetically Engineered Organisms

Despite its initial prohibition, by 1982 it became clear that the NIH-RAC would have to cope with requests for open-field testing of genetically engineered organisms, i.e., for their deliberate release into the environment. Uncharacteristically, neither guidelines nor protocols that advised applicants what information should be included in their submissions had been prepared. This initial reluctance to establish definitive regulations was due to a widely held belief among many molecular biologists that GEOs were not significantly different from their nonengineered progenitors. And, if a difference was present, it was thought that it would be readily detected by conventional biological testing.

Three applications for field trials of GEOs were received by the NIH-RAC in 1982. Two dealt with genetically modified plants (corn and tobacco). The third proposal was concerned with testing a genetically modified strain of the microorganism *Pseudomonas syringae* to determine if it could limit the extent of frost damage to plants. This particular submission became part of the landmark case for the development of regulatory procedures for the release of GEOs into the environment.

Ice-Minus *Pseudomonas syringae*

The genetic engineering portion of the *P. syringae* proposal involved removing a gene that coded for an ice nucleation protein from the organism and then testing whether the modified "ice-minus" strain, when sprayed onto the leaves of plants, could prevent frost damage. Under natural conditions, wild-type "ice-plus" *P. syringae*, which is usually found on the surface of plant leaves, secretes a protein that at low temperatures causes the formation of ice crystals, which, in turn, causes frost damage to the plant. The rationale for the deletion of the gene encoding the ice nucleation protein was that if a strain that lacked this protein were sprayed onto leaves before they became colonized with the wild-type strain, it might lower the temperature at which ice formation would occur, thereby preventing the leaves from being damaged by bacterially induced ice crystals. There is a significant economic incentive for such a novel treatment, because in the United States crop losses due to frost damage exceed a billion dollars per year.

In response to each of the requests for field testing of a genetically engineered organism, the NIH-RAC followed (more or less) the procedures it

had established for handling the regulation of recombinant DNA experimentation in the laboratory:

1. The submissions were announced in the U.S. Federal Register.
2. Information was sent to 3,000 "interested" persons.
3. A panel of experts reviewed the proposals.
4. A public meeting was called for discussion of each proposal.
5. At the same time the NIH-RAC was reviewing the proposals, the U.S. Department of Agriculture (USDA) also reviewed them.

After careful consideration, both the USDA and the NIH-RAC approved the "ice-nucleation gene deletion" proposal. In 1983, the director of NIH gave final endorsement to the NIH-RAC decision. On the same day that permission was granted to proceed with the field trial, a lawsuit to block the test was filed by an organization called the Foundation on Economic Trends, which is headed by Jeremy Rifkin, who strongly opposes all forms of genetic engineering. The lawsuit was upheld, with the judge noting that the NIH-RAC had not carried out a proper hearing in accordance with U.S. statutes and, more important, that it had failed to request an environmental impact statement.

This legal decision dramatically demonstrated that, notwithstanding the scientific opinion of NIH-RAC and its experts, the existing regulatory system for GEO field testing was inadequate. A prevalent opinion outside the confines of the NIH-RAC was that the release of a genetically modified organism into the environment could have far-reaching effects, because living microorganisms proliferate, persist, disperse, and sometimes transfer their DNA to other microorganisms. Some critics of the release of GEOs to the environment believed that, after its introduction into the environment, an engineered organism could displace an existing, important species from its ecological niche and as a result cause severe environmental damage. In addition, some opponents of release believed that genes could be transferred from an introduced GEO to indigenous strains, thereby creating (albeit inadvertently) an ecologically dangerous organism. Although these points of view presented worst-case, adverse-effect scenarios that might be exceedingly unlikely, it was essential that the regulatory protocol for field testing include a thorough assessment of the potential risk that an introduced organism might have on the environment.

Currently, the responsibility for assessing submissions for the deliberate release of GEOs in the United States resides with the U.S. Environmental Protection Agency (EPA) and the USDA. The NIH drew up an initial set of criteria for field tests with GEOs, but it relinquished its authority in this area to these other agencies.

The EPA decided to use two applications, both dealing with ice nucleation-defective bacteria, as prototype cases for developing an assessment process for the field testing of GEOs. Each proposal went through a series of reviews, which included appraisals of the environmental fate, ecological effects, and human health consequences of the test as well as product analysis, by the following groups:

- The Office of Pesticide Program Review of the EPA
- The Toxic Substances, Research and Development Policy Planning, and Evaluation Committees of the EPA
- The General Counsel of the EPA
- The USDA, FDA, and NIH

- A Science Advisory Panel that consisted of a microbiologist, a plant pathologist, and a community ecologist
- Open public meetings
- The various state agencies, which in this instance included the California Department of Agriculture

It was not envisioned that this elaborate, time-consuming, and often redundant process would become the routine mechanism for approving field testing of GEOs; rather, it was assumed that, with experience, the system would be trimmed without loss of effective assessment of the potential hazards of each trial. After what was thought to be a very thorough set of analyses, permission was granted for both of the field trials with ice nucleation-negative bacteria. However, in both instances, although the circumstances were different, local residents who were worried about the release of a GEO in their neighborhoods obtained court orders that temporarily blocked each of the field trials. As a consequence of this delay, both the EPA and the USDA implemented better methodologies for determining the risks of introducing GEOs into the environment. In a short time, the staffs at these agencies became more proficient at handling and analyzing the data submitted by the applicants. The scientific community, including ecologists, helped the process by initiating research programs that were designed to examine the consequences of the release of organisms into model environments, and scientific organizations formulated frameworks for deciding whether a particular GEO would have an adverse effect on the environment.

Eventually, in 1987, the field trials with ice nucleation-negative bacteria were conducted at sites in California. The results indicated that these GEOs were not dispersed to off-site locations, nor did they persist at the site of application. At one site, the freezing temperature of the test plants was lowered by 1°C. Currently, genetically engineered ice-minus bacteria are not used to any significant extent to protect crop plants from frost damage.

Open-Field Tests of Other Genetically Engineered Organisms

Since the first ice-minus bacteria trials, a number of other open-field tests of GEOs have been conducted. Overall, these studies have found that introduced microorganisms tend to remain confined to the test area, do not persist, do not appear to transfer genes to indigenous microorganisms, and have similar basic biological functions in natural and laboratory settings. Generally, because there can be many different adverse-effect possibilities for each GEO, a case-by-case approach has been adopted for granting permission to conduct a field test. These kinds of tests have been carried out in the United States, the United Kingdom, Australia, and other countries. However, biotechnology companies have been reluctant to develop genetically engineered microorganisms that can be used in the environment because the cost of field testing is high and because of the possibility that final approval can be denied despite successful test results. Nevertheless, there is a growing consensus that the environmental release of genetically engineered microorganisms after the appropriate laboratory and field tests will not be ecologically deleterious.

Regulators have been wary of the field testing of genetically engineered animals. For example, to test the ability of certain transgenic fish to thrive under simulated natural conditions, an elaborate impoundment facility has been built in the United States to ensure that the fish cannot escape and poachers cannot get in. In contrast, transgenic plants that have been devel-

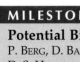

MILESTONE

Potential Biohazards of Recombinant DNA Molecules

P. BERG, D. BALTIMORE, H. W. BOYER, S. N. COHEN, R. W. DAVIS,
D. S. HOGNESS, R. ROBLIN, J. D. WATSON, S. WEISSMAN, and
N. D. ZINDER

Science **185**:303, 1974

After Cohen et al. (*Proc. Natl. Acad. Sci. USA* **70**:3240–3244, 1973) described the basic strategy for inserting a DNA molecule from any source into a plasmid, the safety and ethical implications of recombinant DNA technology were discussed at length in the scientific community. These informal discussions led to the formation of a temporary committee (Ad Hoc Committee on Recombinant Nucleic Acids) that was composed of leading molecular biologists who were charged with examining the concerns in more detail. In 1974, this group published a brief note, which was dubbed the "Berg letter," simultaneously in three major scientific journals: *Science*, *Nature*, and the *Proceedings of the National Academy of Sciences*. In this communication, Berg et al. recommended that researchers defer (1) creating microorganisms with novel genes for drug resistance and toxin production and (2) cloning cancer-causing genes from viruses into bacterial host cells. Moreover, they suggested that a government body, which would represent the interests of the public, should prepare guidelines and regulations for recombinant DNA technology and that an international meeting of scientists should be convened to "discuss appropriate ways to deal with the potential biohazards of recombinant DNA molecules." This letter had a significant impact. It led to a scientific meeting at the Asilomar Conference Center in California that attempted to define various recombinant DNA experiments in terms of their relative risk and suggested which laboratory safeguards should be implemented for experiments that were considered to have a minimal, low, moderate, or high risk. In turn, the Asilomar recommendations (Berg et al., *Science* **188**:991–994, 1975) contributed to the development of the 1976 NIH guidelines that established stringent regulations for all types of recombinant DNA experiments.

The Berg letter was responsible for the formulation of the regulatory framework for recombinant DNA technology. It focused public attention on the potential risks and benefits of recombinant DNA technology and ignited a heated public debate about recombinant DNA technology. Historically, it represents an interesting example of scientists trying to anticipate the possible risks of a new technology before there was any substantial evidence that there were any hazards.

oped for food or that have enhanced properties for conventional biological features are being handled more liberally. The prevalent, although not unanimous, view is that most genetically altered plants are no different from traditional plant strains (cultivars) that are derived from breeding experiments. In the United States, the same testing and licensing procedures are being applied to all plants that carry conventional genetic modifications, regardless of how these changes were introduced. On the other hand, field testing of transgenic plants that carry genes for insecticides, viral protection, or nonfood functions must satisfy some additional requirements.

In the United States, *Bacillus thuringiensis* (Bt)-transgenic plants, including corn, soybean, potato, and cotton, have been approved by the USDA, FDA, and EPA. By contrast, the release of Bt-rice into the Philippines was blocked by an international consortium of nongovernmental groups and the growing but not the consumption of Bt-corn has been banned in France. Many people, for various reasons including perceived risks, societal beliefs, and economic concerns, continue to be suspicious of GEOs.

The implementation of regulations for the deliberate release of GEOs has not been an easy task, and in practically all countries, it is still an ongoing process. The mandate of the regulatory agencies, as noted above, is to ensure the public safety and, specifically in these cases, to protect the environment from both short-term and long-term effects without constraining the development of worthwhile innovations. The Greek lawgiver Solon, who lived from about 638 to 559 B.C., once said, "Laws are like spiders' webs which, if anything small falls into them they ensnare it, but large

things break through and escape." In the context of the deliberate release of GEOs, no government wants a "large thing" to "break through and escape."

Human Gene Therapy

Human gene therapy, in a broad sense, is the addition of a functional gene(s) to a human cell with the aim of correcting a disorder that is due to a genetic defect. There are two possible kinds of treatments for inherited diseases. First, somatic (body) cells can be targeted for genetic transformation. With this strategy, "correction" of the genetic defect is confined to a specific organ or tissue. Second, germ line (sperm or egg) cells or a fertilized egg (zygote) could be genetically modified to create an individual who will carry the remedial gene(s) in all of his or her cells. As a consequence of germ line gene therapy, the genetic change is transmitted from one generation to the next.

Development of a Policy for Somatic Cell Gene Therapy

In 1980, representatives of the U.S. Catholic, Protestant, and Jewish communities wrote a joint letter to the President expressing deep concern about the possible ramifications of human genetic engineering. As a result, both a presidential commission and a congressional commission were established to examine the ethical and social implications of human gene therapy. These were important initiatives because, in the United States, significant public policy is often enacted on the basis of the recommendations of such commissions. In both final reports, a sharp line was drawn between somatic cell and germ line gene therapies. Somatic cell gene therapy was conceded to fall within the realm of standard medical interventions, not unlike organ transplantation. By contrast, germ line gene therapy was considered to be ethically problematic and too complicated technically to be of immediate concern. It was concluded that a regulatory framework for somatic cell gene therapy was required; however, development of guidelines for germ line gene therapy was premature. To avoid unauthorized trials, it was suggested that germ line gene therapy experiments be actively discouraged.

By 1985, NIH had prepared a document entitled "Points to Consider in the Design and Submission of Somatic-Cell Gene Therapy Protocols." This guideline tells an applicant what kinds of information must be included in a submission for permission to conduct a somatic cell human gene therapy trial on human subjects. The "Points to Consider" statement was, for the most part, crafted by combining the ideas that had been developed for regulating laboratory research using recombinant DNA technology with those for biomedical studies.

The biomedical guidelines had been updated in the 1970s in response to the revelation in 1972 about a 40-year human experiment that had been conducted by the U.S. Public Health Service in Alabama on about 400 rural, illiterate black men with syphilis. The study, which became known as the Tuskegee Study, was designed to monitor the natural development of this sexually transmitted disease, and no treatment was ever given or tested. Such a nontherapeutic experiment with uninformed subjects was appalling to many in the United States. Consequently, the U.S. Congress immediately ended the trial and initiated legislation to ensure that any similar projects would never be carried out again.

Some of the questions that have to be addressed in a somatic cell gene therapy application include the following:

- What is the disease that is to be treated?
- How serious is the disease?
- Do alternative, effective therapies exist?
- How dangerous is the proposed treatment for the patients?
- How effective is the treatment likely to be?
- How will the subjects be chosen for the trial?
- Is the choice of subjects fair and representative of the patient population?
- How will the patients be informed of the trial?
- What information will be conveyed to the patients?
- How will informed consent be obtained?
- How will confidentiality of both patients and medical information be maintained?

During the early stage of gene therapy experimentation, most of the applications were first reviewed by an ethics committee at the institution where the proposed experiment was to be conducted before being sent to the Human Gene Therapy Subcommittee of NIH-RAC. The latter group reviewed the proposal in terms of its basic science, medical content, adherence to the guidelines, coherence, and relevance. If unacceptable, the submission was returned with comments. The applicants could revise the proposal and resubmit it. If the proposal was approved by the subcommittee, then, using the same criteria, the NIH-RAC examined it in public meetings. After acceptance was granted at this level, the director of NIH considered the application and had to grant permission before the trial could proceed. In addition, because somatic cell gene therapy trials include the administration of novel DNA constructs, the FDA also examined and had to approve the application. The FDA is particularly interested in how the product was developed, what quality control measures are used to ensure the purity of the product, and what preclinical testing was conducted to ensure that the product is safe.

However, as the number of proposals grew and gene therapy became, as one commentator noted, "the hottest ticket in medicine," the original process was considered unnecessarily cumbersome and redundant. Consequently, after 1997, NIH will not be part of the human gene therapy regulatory system. In the future, if the NIH-RAC survives, it will be a forum for discussing the ethical issues pertaining to human gene therapy. This change removed the requirement that all gene therapy protocols must be examined publicly. The FDA, which is responsible for regulating biologic products, conducts approval reviews in private to secure the proprietary rights of the applicants. Human gene therapy is now considered to be a safe medical procedure, although not at present particularly efficacious. In this context, the initial apprehensions about human gene therapy have dissipated, and it has become a mainstream approach for treating human disease.

Most analysts believe that the approval process for somatic cell gene therapy trials in the United States is sufficient to ensure that patients are informed and fairly selected and that the treatments will be carried out, as well as can be expected for experimental procedures, without undue harm to either the patients or the population in general. Currently, other countries are developing their own guidelines for these kinds of trials. The implementation of the process in the United States has been the outcome of much deliberation. As one participant, Dr. Leroy Walters, the director of the

Center for Bioethics at Georgetown University in Washington, D.C., remarked during an NIH-RAC hearing in January 1989, "I don't know of any biomedical science or technique that has had the detailed scrutiny that human gene therapy has had."

Accumulation of Defective Genes in Future Generations

It is often contended that the treatment of genetic diseases by somatic cell gene therapy will eventually lead to the genetic deterioration of humans. The essence of this argument is that the frequency of a defective gene will increase in populations because a somatic cure will enable previously infertile individuals or those not expected to survive childhood to pass mutant genes on to the next generation. This concern about corrupting the human gene pool by an accumulation of defective genes is misplaced. Population geneticists have shown that a significant rise in the frequency of a rare deleterious or lethal gene due to effective treatment requires thousands of years. For example, for an uncommon genetic disorder that occurs once in 100,000 live births, it would take about 2,000 years after the initiation of a totally effective cure before the frequency of the disease doubles to 1 in 50,000.

Not only is the rate of increase of the frequency of a lethal gene extremely low, but as long as the treatment can be given to all those who require it, the genetic constitution of an individual is irrelevant. This point can be illustrated with an example from evolutionary history. Primates, including humans, are incapable of synthesizing vitamin C, which is essential for their survival. For these organisms, vitamin C must be supplied from external sources. Thus, all humans are genetically defective for the production of this vital molecule. By contrast, amphibians, reptiles, birds, and nonprimate mammals are capable of synthesizing their own vitamin C. Simply put, the vitamin C genetic deficiency did not interfere with the successful evolution of primates over millions of years. Similarly, human-initiated treatment of other human genetic defects will not create a significant burden for future generations.

Human Germ Line Gene Therapy

Although human germ line gene therapy is effectively forbidden at this time, it can be argued, somewhat convincingly, that certain genetic conditions may be treatable only in this way. Currently, the technical means to carry out human germ line gene therapy are lacking. However, with the development of genetic intervention in animals and the diagnostic testing of preimplantation embryos, knowledge that will eventually make human germ line gene therapy feasible will be gained. Moreover, as somatic cell gene therapy becomes a more routine form of medical treatment, public apprehension about human germ line gene therapy will lessen. Thus, at some time in the future, a credible request will be put forward for conducting a human germ line gene therapy trial. One can only hope that, by that time, all of the issues will have been fully discussed and the complete range of consequences, including social and biological issues, will have been addressed.

Currently, human gene therapy is being considered for the treatment of serious human diseases. Eventually, it might be capable of enhancements of physical or mental attributes, although it is questionable whether society will find this an acceptable use of this technology. In the meantime, like any new medical procedure, human somatic cell gene therapy raises a number of questions that need to be considered:

- What will be the financial costs of developing and delivering human somatic cell gene therapies?
- Should governments establish medical research priorities?
- Will the focus on human somatic cell gene therapy discourage the development of more conventional treatments?
- Will it be possible to provide a particular human somatic cell gene treatment to all of those who need it?
- Will any one individual or company own the exclusive rights to a certain genetic disease therapy?

Cloning Humans?

Public interest in the cloning of humans was stimulated in the 1960s by developmental biology research on frogs and toads. These studies established that a nucleus from an undifferentiated cell could replace the nucleus of the fertilized egg and support normal embryonic development. Experimentally, it was possible to inject undifferentiated nuclei from the same donor organism into a number of different fertilized eggs that had had their nuclei removed, thereby creating a set of offspring that all had the same genetic makeup. Such a group of organisms would be genetically identical. In other words, each of these organisms would be a genetic clone of the original donor. It was not too difficult, although far beyond existing technical capabilities at that time, to extrapolate from clones of toads to clones of humans. On this basis, science fiction writers and reporters created various scenarios. In one story, a novelist had her scientists use a cloning procedure to create multiple copies of the assassinated U.S. President John F. Kennedy. However, a more popular fictional theme was the cloning of evil men. These "cloning of humans" stories not only seemed far-fetched but also made the seriously fallacious assumption that the personality, character, and other attributes of an individual are due solely to his or her genetic makeup. In fact, all human beings are the product of both their genetic makeup and the environmental and cultural influences that they encounter during their development and maturation. For example, virulent racism, as exemplified by Hitler, is a learned behavior and not due to any gene or combination of genes. In a different environment with other cultural inputs, a clone of Hitler would not necessarily behave like the historical Hitler. Similarly, a clone of Mother Teresa would not necessarily devote her life to helping the poor and suffering of Calcutta.

Over the years, as research on both mammalian reproductive biology and transgenesis of mammalian species progressed, it appeared that the capability of cloning a human would be an eventuality sometime in the future. This perception became a reality in 1997, when a sheep called Dolly was cloned with a differentiated-cell nucleus from a donor ewe. The strategy that created Dolly could theoretically be used to form clones of any mammalian species, including humans. And, if the ovine protocol does not work for other species, it appears that not much experimentation will be required to devise the appropriate conditions for any particular mammalian species. As a result, notwithstanding the inefficiency of the ovine cloning procedure, the issue of human cloning instantly became a consequential part of any discussion of genetic and biomedical ethics.

Without question, human cloning is a highly charged and controversial topic. To some, the thought of creating a copy of an existing individual by experimental manipulation is odious. Others argue that a cloned person is

the same as an identical twin, despite the difference in age, and therefore, the process is not inherently evil, although it may be totally unnecessary. Others feel that human cloning may have therapeutic and social benefits that would allow it to be used under exceptional conditions. For example, in one scenario, it is asserted that parents should have the opportunity of creating a clone from a cell of a deceased child. For the most part, the legislative response to human cloning in many countries has been the passage of laws that ban all human cloning research. For the present time, these legislative restrictions should be sufficient to avert the cloning of humans. However, eventually, the question of whether human cloning is inevitable will have to be addressed.

SUMMARY

Significant technological advances, such as molecular biotechnology, are seldom implemented without controversy. The issues and concerns raised by the ability of scientists to genetically engineer organisms have had far-reaching implications and resulted in both the establishment of official guidelines for the conduct of research and the formulation of requirements for the introduction of recombinant DNA products into the marketplace. In this chapter, various aspects of the regulation of recombinant DNA technology, food products from genetically engineered organisms, the release of genetically engineered organisms into the environment, somatic and germ line cell gene therapies, and human cloning are discussed.

Guidelines for the proper conduct of recombinant DNA technology experiments were established by the U.S. National Institutes of Health in the late 1970s and were fine-tuned in the early 1980s. But two unresolved issues remained. First, how were genetically engineered products to be regulated? Second, how was the deliberate release of genetically engineered organisms into the environment to be managed? Industry, believing that no special regulations should be implemented for genetically engineered products, has taken the view that the nature of the product and its properties, not the process that was used to produce the product, are what matter. This view was adopted in the United States for pharmaceutical products. On the other hand, there has been more concern about genetically engineered food additives. Generally, the U.S. Food and Drug Administration, which is responsible for ensuring the safety of both pharmaceuticals and food additives, has taken a case-by-case approach to the problem of accepting genetically engineered food products as safe. Thus, depending on the product, a specific set of criteria must be met before it can be released for human consumption.

By contrast, a complete protocol had to be developed for those who wanted to release genetically engineered organisms into the environment. From the beginning of the public awareness of recombinant DNA technology, there have been concerns that scientists would produce, either by mistake or by design, an organism that would run rampant in the environment. As a consequence of public concerns, specific and rigorous regulations were formulated in the United States for the conduct of field trials with genetically engineered organisms.

The possibility of being able to genetically engineer humans has always been a cause for concern. Technically, human gene therapy has two aspects: somatic cell gene therapy and germ line gene therapy. Because germ line gene therapy has genetic consequences for future generations, it is currently not permitted. By contrast, somatic cell gene therapy is rapidly becoming a

potentially important mode of treatment for a number of diseases. The guidelines developed in the United States for the conduct of human somatic cell gene therapy combine the regulations for carrying out recombinant DNA technology research with the biomedical ethics requirements for human medical experiments.

With the nuclear cloning of the first mammal, a sheep, in 1997, human cloning has become a controversial topic of immediate concern. Currently, human cloning research is banned in a number of countries. However, notwithstanding this widespread prohibition, the question of whether a human will ever be cloned must be contemplated.

References

Anonymous. 1990. Biotechnologies and food: assuring the safety of foods produced by genetic modification. *Regul. Toxicol. Pharmacol.* **12:**S1–S196.

Berg, P., and M. Singer. 1995. The recombinant DNA controversy: twenty years later. *Bio/Technology* **13:**1132–1134.

Busch, L., W. B. Lacy, J. Burkhardt, and L. R. Lacy. 1991. *Plants, Power, and Profit: Social, Economic, and Ethical Consequences of the New Biotechnologies.* Blackwell Publishers, Cambridge, Mass.

Carmen, I. 1992. Debates, divisions, and decisions: recombinant DNA Advisory Committee (RAC) authorization of the first human gene transfer experiments. *Am. J. Hum. Genet.* **50:**245–260.

Cook-Deegan, R. M. 1994. *The Gene Wars: Science, Politics and the Human Genome.* W. W. Norton & Co. Inc., New York, N.Y.

Djerassi, C. 1993. Basic research: the gray zone. *Science* **261:**972–973.

Drahos, D. J. 1991. Field testing of genetically engineered microorganisms. *Biotechnol. Adv.* **9:**157–171.

Howard, T., and J. Rifkin. 1977. *Who Should Play God? The Artificial Creation of Life and What It Means to the Human Race.* Dell Publishing Co., New York, N.Y.

Juengst, E. T. 1990. The NIH "Points to Consider" and the limits of human gene therapy. *Hum. Gene Ther.* **1:**425–433.

Juskevich, J. C., and C. G. Guyer. 1990. Bovine growth hormone: human food safety evaluation. *Science* **249:**875–884.

Kamb, M. L., J. J. Murphy, J. L. Jones, J. C. Caston, K. Nederlof, L. F. Horney, L. A. Swygert, H. Falk, and E. M. Kilbourne. 1992. Eosinophilia-myalgia syndrome in L-tryptophan-exposed patients. *JAMA* **267:**77–82.

Kessler, D. A., M. R. Taylor, J. H. Maryanski, E. L. Flamm, and L. S. Kahl. 1992. The safety of foods developed by biotechnology. *Science* **256:**1747–1749.

Levin, M. A., and H. S. Strauss. 1991. *Risk Assessment in Genetic Engineering.* McGraw-Hill, Inc., New York, N.Y.

Lyon, J., and P. Gorner. 1995. *Altered Fates: Gene Therapy and the Retooling of Human Life.* W. W. Norton & Co. Inc., New York, N.Y.

Mayeno, A. N., and G. J. Gleich. 1994. Eosinophilia-myalgia syndrome and tryptophan production: a cautionary tale. *Trends Biotechnol.* **12:**346–352.

Miller, H. I. 1992. Putting the bST human-health controversy to rest. *Bio/Technology* **10:**147.

Tiedje, J. M., R. K. Colwell, Y. L. Grossman, R. E. Hodson, R. E. Lenski, R. N. Mack, and P. J. Regal. 1989. The planned introduction of genetically engineered organisms: ecological considerations and recommendations. *Ecology* **70:**298–315.

Walters, L. 1991. Human gene therapy: ethics and public policy. *Hum. Gene Ther.* **2:**115–122.

Wilson, M., and S. E. Lindow. 1993. Release of recombinant organisms. *Annu. Rev. Microbiol.* **47:**913–944.

Wivel, N., and L. Walters. 1993. Germ-line gene modification and disease prevention: some medical and ethical perspectives. *Science* **262:**533–540.

REVIEW QUESTIONS

1. What is NIH-RAC? What was its role in regulating recombinant DNA research?

2. What criteria are used by the FDA to determine if a recombinant protein is acceptable as a food or food additive?

3. Discuss the premise "Genetic engineering . . . is a radical new technology that violates the fundamental laws of nature."

4. Why and how are genetically engineered microorganisms that are designed to be released into the environment regulated?

5. Discuss the positive and negative aspects of licensing recombinant bovine somatotropin.

6. How is human somatic cell gene therapy regulated in the United States?

7. Discuss the premise "Somatic cell gene therapy will cause an increase in the frequency of defective genes in future generations."

8. Discuss why human germ line gene therapy research is prohibited.

9. Discuss the statement "Humans are what they are because of their genes."

10. Prepare arguments for both sides of a debate on the topic "Be it resolved that human cloning should be banned."

CHAPTER 23

Patenting Biotechnology Inventions

The principal objective of biotechnology is to produce commercial products for economic gain. However, no company will initiate high-risk, long-term projects without knowing that the results of its research efforts can be legally protected from competitors. At the same time, society at large has a stake in encouraging industrial innovation. A strategy that meets both of these objectives is for the government to grant inventors exclusive rights to the novel products or processes that they develop. Collectively, these sanctioned privileges are called intellectual property rights and include trade secrets, copyrights, trademarks, and patents. Trade secrets comprise private information about specific technical procedures and formulations that a company wishes to protect from others. Copyrights protect the authorship of published works from unauthorized use. Trademarks can be either words or symbols that identify a particular product or process of one company. For example, the name Gigapack has been legally designated to describe a bacteriophage λ in vitro packaging extract that is sold by the biotechnology company Stratagene. Other companies sell similar kits that have their own protected names.

For biotechnology, patents are the most important form of intellectual property. A patent is a legal document that gives the patent holder exclusive rights to implement the described invention commercially. Moreover, on the basis of the extent of the claims of the patent, the patent holder can develop other products that are directly derived from the original invention, while competitors would have to purchase the right to use the invention in order to develop a product based on it. On the other hand, a patent is a public document that must contain a detailed description of the invention, so it informs others about the nature and limits of the invention, allowing them to decide whether they should continue working in a particular direction or try to use the patented invention as a springboard to other possible innovations.

Patent decisions and laws vary from country to country, although there are ongoing attempts to develop international standards. For example, until recently, in the United States the duration of the exclusive rights of a

patent was 17 years after it was officially awarded, in Canada it was 20 years, and in many European countries it was 20 years from the date of the filing of the patent application. Now, under an international agreement, the life of a patent is limited to 20 years from the date of filing in most countries, including the United States. However, unlike many other countries, the U.S. government maintains that the date of the invention has precedence over the date of filing. Usually it takes 2 to 5 years following the filing of the initial patent application before a patent is granted. In every jurisdiction, the holding of a patent can be of considerable economic value, and it is not a trivial matter to receive one. For this reason, both the patent application and the invention must meet a very strict set of criteria.

Patenting

Generally, for either a product or a process to be patentable, it must satisfy four fundamental requirements:

1. The invention, after having been shown to work ("reduced to practice") must be "novel," meaning that the invention does not exist as another patent that is held by someone else in another country; is not an existing product or process; or, outside the United States, has not appeared in some published form prior to the submission of the patent application. In the United States, an inventor has 1 year following publication in which to apply for a patent.
2. A patent cannot be granted for something that was merely previously unknown; rather, the invention must contain, as judged by the patent office, an inventive step that was "not obvious" to other workers in the field.
3. The invention must be "useful" in some way, whether it is a process, an instrument, a compound, a microorganism, or a multicellular organism.
4. Every patent application must contain a description of the invention that is sufficiently thorough that a person knowledgeable in the same field can implement it.

Inventions and discoveries that are not patentable include scientific theories, mathematical methods, aesthetic creations, and therapeutic treatments of humans or animals. In addition, a consistent principle of patent law is that a patent cannot be granted for anything that is a "product of nature." The notion here is that it is not appropriate for society to give a monopoly to someone for something that occurs naturally and therefore belongs to the public. Often companies and individuals skirt around this constraint by applying for a patent that covers the process of purification of a product, thereby avoiding the direct question of ownership of either a natural substance or an organism that produces the product.

There is no simple, immediate system for the granting of a patent. The application must be prepared by an expert, normally a patent lawyer, and is organized in a defined pattern. In the United States, the application must have a title; an abstract describing succinctly the nature of the application; a section on the background of the invention that includes a full and open description of the current "state of the art" in the field of the invention; a comprehensive summary of the invention with, if considered helpful, figures and schematic representations; sections that explain the nature of the invention and describe how the invention works; and, finally, a list of

claims about the invention and how the invention may be used. The application is sent to the U.S. Patent and Trademark Office (PTO), where it is reviewed by an examiner for novelty, nonobviousness, utility, feasibility, and general acceptability as a patentable invention.

If an examiner agrees that the invention meets all the criteria for patentability, then a patent is awarded. However, the receipt of a patent is not a license to produce and sell the invention. All statutory regulations must be met before any product can be marketed. For example, if a patent is granted for a genetically engineered microorganism, the manufacturer must satisfy the recombinant DNA regulations for its production, distribution, and release. Protection of patent rights is the responsibility of the patent holder, and generally that means bringing a lawsuit(s) against those who are presumed to be infringing on the patent. These disputes are decided by the courts and not the patent office. Similarly, if a person or company feels that an awarded patent is inappropriate, the legitimacy of the patent can be challenged by a lawsuit.

If a patent application is rejected by an examiner, then the applicant can appeal the decision to a Patent Appeals Board. If this appeal is turned down, then the decision can be challenged legally. For some applicants, patenting can be a frustrating experience. There are a number of cases where the stakes were considered to be so high that costly court battles have gone on for years. Thomas Edison, who held more than a thousand patents in his lifetime, once described a patent as "an invitation to a lawsuit."

Product patents and process patents make up the two major categories of patents. Products include homogeneous substances, complex mixtures, and various devices; processes include preparative procedures, methodologies, or actual uses (Table 23.1) The patenting of biotechnology innovations has been based on the historical experience of patenting inventions by the agricultural, fermentation, pharmaceutical, and medical industries. For ex-

Table 23.1 Common types of patent categories, with examples from recombinant DNA technology

Categories	Examples
Product patents	
Substance	Cloned genes, recombinant proteins, monoclonal antibodies, plasmids, promoters, vectors, cDNA sequences, and monovalent vaccines
Compositions of matter	Multivalent vaccines, biofertilizers, bioinsecticides, pharmaceutical mixtures, microorganisms, and transgenic organisms
Devices	Pulsed-field gel electrophoresis apparatus, DNA sequencing apparatus, and microprojectile gene transfer machine
Process patents	
Process of preparation	DNA isolation, synthesizing double-stranded DNA, vector-insert construction, polymerase chain reaction (PCR) applications, and purification of recombinant protein
Method of working	Nucleic acid hybridization assays, diagnostic procedures, detection systems using PCR, and mutant assays
Use	Applying biofertilizers and bioinsecticides, fermentation of genetically modified organisms, and nontherapeutic animal treatment systems

ample, Louis Pasteur received a patent for a process for fermenting beer. Today, most but not all biotechnology patent applications are straightforward, and patents are granted without any significant problems. However, the first time a scientist attempted to patent a genetically modified microorganism that was engineered by the introduction of different plasmids, each of which carried the genes for a separate hydrocarbon degradative pathway, the case was highly controversial. This genetically modified bacterium, which was capable of breaking down many of the components of crude oil, was developed by A. Chakrabarty, who at the time worked for the General Electric Corporation. However, notwithstanding its potential usefulness in cleaning up oil spills, the patent application for this bacterium was rejected by the U.S. PTO on the grounds that microorganisms are products of nature and, as living things, are not patentable. In 1980, in a landmark decision, the U.S. Supreme Court decided that this organism was patentable according to the U.S. Patent Statute, arguing that "a live, human-made microorganism is patentable subject matter. . . as a manufacture or composition of matter."

The argument against patenting this genetically engineered microorganism tended to center on how the organism was developed. In the past, induced mutation followed by selection for novel properties was an acceptable way to create a patentable living organism. However, genetic engineering was considered by some to be "tampering with nature." Consequently, it was argued that no inventor should benefit from manipulating "products of nature." This position was not upheld. Thus, in the United States from 1980 onward and later in other countries, organisms, regardless of the means that were used to develop them, must be judged by the standard criteria of novelty, nonobviousness, and utility to determine if they are patentable.

Patenting in Different Countries

Patent offices in different countries often reach quite different conclusions about the same patent application. For example, in 1989 the biotechnology company Genentech applied for a patent in the United Kingdom (U.K.) for, among other things, the production of human tissue plasminogen activator (tPA) by recombinant DNA processes. This protein exists in small amounts in the human body and converts plasminogen to plasmin. Plasmin is an active enzyme that degrades the fibrin of a blood clot. Consequently, human tPA has been considered as a possible therapeutic agent for the prevention and treatment of coronary thrombosis. Genentech, after considerable effort, assembled a complete version of a human tPA complementary DNA (cDNA) and cloned this cDNA into *Escherichia coli* for the production of large amounts of pure tPA. In its patent application, Genentech claimed rights to human tPA as a product based on certain procedures of recombinant DNA technology that they developed, the cloning vector system, and the transformed microorganism. As a part of the "process" category, protection for the use of human tPA as a pharmaceutical agent was also sought by Genentech. A total of 20 claims were presented in the original patent application. Some of these were broad and others were narrow in scope. The patent was rejected by the U.K. patent office. Genentech then appealed to the U.K. Court of Appeals, which, after considerable deliberation, invalidated all of the claims for a variety of reasons. The judgment concluded that the patent was novel, but some of the judges argued that the submission was obvious; therefore, it could not be patented.

In contrast, Genentech was readily awarded a patent for human tPA in the United States. The U.S. patent not only protects the form of human tPA that was to be marketed by Genentech but also gives Genentech exclusive rights to all similar, but not identical, active forms of human tPA. Genentech won a lawsuit against two other biotechnology companies that were found to be infringing on its tPA patent, although they were selling nonidentical forms of tPA.

The Japanese version of Genentech's tPA patent is limited to the amino acid sequence of the human tPA that was cloned and patented by Genentech. In Japan, other companies can sell variant forms of human tPA. Thus, basically the same patent application was rejected, approved and given a broad interpretation, and approved and given a narrow interpretation by three different patent offices. For the present, at least, there are divergent views about what is or is not a patentable invention.

Patenting DNA Sequences

On the surface, any device, process, or compound that meets the criteria of novelty, inventiveness (nonobviousness), and utility should be patentable. Since 1980, over 5,000 patent applications for whole genes have been sub-

MILESTONE

Transgenic Non-Human Mammals
INVENTORS: P. LEDER and T. A. STEWART
Assignee: President and Fellows of Harvard College, Cambridge, Mass.
U.S. Patent 4,736,866
Date of Patent: 12 April 1988

In 1980, the U.S. Supreme Court defined a patentable invention as one that included "anything under the sun that is made by man." In 1988, a transgenic mouse was the first genetically engineered animal to be patented. In this case, the transgene consisted of a cancer-causing gene (oncogene) driven by a promoter in the long terminal repeat of the mouse mammary tumor virus (MMTV LTR). The oncogene was the *myc* gene from the chicken myelocytomatosis OK10 virus. The invention entailed cloning an MMTV LTR–*myc* fusion gene into a plasmid, injecting linearized plasmid DNA into the male pronuclei of fertilized one-celled mouse eggs, identifying offspring that expressed the *myc* gene, and establishing transgenic mouse lines. In some of these lines the *myc* gene was expressed in several different tissues, and in other lines it was limited to one or a few tissues. The in-

tegration of the MMTV LTR–*myc* gene construct, according to Leder and Stewart, "increases the probability of the development of neoplasms (particularly malignant tumors) in the animal." These transgenic organisms can be used to test whether a compound either causes or prevents cancer and as a source of cell lines from cells of various tissues such as the heart that are difficult to culture. Since 1989, Du Pont has been selling one of these lines of transgenic mice under the trade name OncoMice. More generically, others prefer to call this mouse line the "Harvard Oncomouse" or, for short, just "oncomouse."

The granting of U.S. patent 4,736,866 was contentious, with much of the concern directed at the ethical implications of such patents. Those who oppose the patenting of transgenic animals argue that this type of patent violates the sanctity of life,

threatens the integrity of species, and fosters inhumane treatment of animals. Notwithstanding these allegations, since 1988, a large number of patents have been granted in the United States for various transgenic organisms. For example, there are now, to name a few, patents for transgenic animals that act as models for benign prostatic disease, inflammatory disease, altered fat tissue metabolism, and thrombocytopenia. To date, neither the U.S. courts nor the U.S. government has suggested that, in principle, any of these patents is inappropriate. The patenting of transgenic organisms is no longer an issue in the United States. By contrast, in Europe and elsewhere, it remains a serious question that has not been completely resolved, although the Harvard Oncomouse has been patented by the European Patent Office. In their decision, the examiners concluded that the benefit to humankind of this transgenic system outweighed other factors that would have made it unacceptable for patenting. However, public interest groups and political parties are continuing to challenge this judgment.

mitted to patent offices throughout the world. Of these, about 1,500 have been approved. The most valuable human gene patent is for the production of recombinant erythropoietin, which had sales of more than $1 billion in 1996. Erythropoietin stimulates the formation of red blood cells and is used to prevent anemia in patients with kidney failure who require dialysis. Many of the other patented gene sequences are used as diagnostic probes (biomarkers).

With the advent of the Human Genome Project and, in particular, with the undertaking of the partial sequencing of thousands of human cDNA molecules from different tissues and organs, the patenting of partial DNA sequences became extremely contentious. In 1991, the issue of patenting gene fragments was broached when the U.S. National Institutes of Health (NIH) filed for the patent rights for 315 partially sequenced human cDNAs (expressed sequence tags [ESTs]). Two additional filings brought the total number of partial sequences to 6,869. In 1994, in a preliminary ruling, the U.S. PTO notified NIH that it would reject the patent application on the grounds that the functions of the sequences were not known. In other words, partial sequences by themselves did not fulfill the requirement of utility and were not patentable. The NIH opted not to appeal this ruling, and the question of patenting ESTs is still unresolved.

In the meantime, by 1997, over 350 patent applications for more than 500,000 partial DNA sequences had been filed, mostly by private companies. One of these applications seeks patent protection for about 18,500 ESTs. The U.S. PTO has not yet ruled on any of these applications. However, in the future, both to expedite the examination process and to oblige the applicants to be more selective, no more than 10 sequences can be included in a single application.

Individuals who oppose the patenting of DNA fragments of unknown function contend that, although eventually some of these sequences might be useful, it is premature and speculative to award patent protection now. In addition, it is argued that granting of such patents would not only give the patent holders too much control but would act as a constraint against the development of various diagnostic and therapeutic agents. In this context, the thousands of ESTs are considered to be "means to ends" and not the actual end points. On the other side, those who favor patenting ESTs maintain that these collections are novel because they define the normal messenger RNA (mRNA) complement of various tissues and organs and consequently have utility because each collection can be used as a diagnostic assay to determine the extent to which a disease alters the normal complement of mRNAs in various organs. In addition, these individuals assert that, historically, the existence of patents has not deterred the development of new products but, to the contrary, has stimulated the process.

The outcome of the EST-based patent applications is uncertain. It may take some time before the issue is completely resolved, especially if the U.S. PTO rejects one or more of these applications and the applicants appeal to the courts. Moreover, as research continues, the full-length versions of many of the partial cDNAs will be determined and potential applications will be deduced. Then the primary question will be whether a previously described partial sequence undermines the patentability of the complete cDNA sequence.

Until recently, in the United States, it was difficult to patent newly discovered genes even though the usefulness requirement was satisfied. In

these cases, the U.S. PTO had taken the position that the applications failed the test of nonobviousness. These inventions were considered obvious and therefore not patentable. Briefly, the U.S. PTO examiners asserted that the isolation of a gene or cDNA with synthetic probes based on a published partial or complete amino acid sequence to screen gene libraries was common practice and apparent to anyone working in the field. Therefore, obtaining a gene in this way was not an inventive process. In two cases, judges of the U.S. Court of Appeals for the Federal Circuit ruled that knowledge of an amino acid sequence did not reveal the actual cDNA or gene sequence due to the degeneracy of the genetic code. Consequently, when a researcher starts a gene search with an amino acid sequence, neither the cDNA nor the gene sequence is obvious. In other words, as one of the rulings noted, "What cannot be contemplated or conceived cannot be obvious." Although this issue has been resolved, the U.S. PTO maintains an exacting approach to patenting genes and has refused at least one recent application because the detailed methods that were used to isolate the gene sequence were considered routine and obvious. Patent attorneys point out that patentability does not depend on how an invention was made but whether the invention meets the requirements for patenting. The courts will be asked to consider whether the use of established gene isolation procedures nullifies nonobviousness. Since the U.S. PTO uses a case-by-case approach for biotechnology patent applications, it appears that there will be no absolute standard for defining nonobviousness and the courts will play an integral role in settling disputes concerning the patentability of genes.

Patenting Multicellular Organisms

The patenting of multicellular organisms continues to raise ethical and social concerns. However, there is nothing intrinsically new about the exclusive ownership of living material. In the past, microorganisms were routinely patented, and specific laws were promulgated to give plant breeders the right to own various plant varieties. To date, a transgenic mouse ("oncomouse") carrying an activatable gene that makes it susceptible to tumor formation has been patented in both the United States and Europe. In addition, genetically manipulated plants have also been accepted as patentable inventions.

It is likely that the most vigorous challenges to patenting transgenic animals will be put forward on moral grounds. In other words, the issue is whether society considers this form of patenting unacceptable. From a historical perspective, it is unlikely that a position based on ethical considerations will be completely successful in preventing the patenting of all transgenic animals. For example, if an invention purports to facilitate a new treatment for human disease, the currently prevalent view is that human rights and needs supersede those of animals. However, patenting is not an absolute right, and governments, by passing specific laws, can determine what can or cannot be patented. If an invention is considered by various special interest groups to have a potentially negative economic impact on an existing agricultural practice, then it is quite possible that a law preventing the implementation of the new technology could be passed. For example, the Danish parliament has passed a statute that exempts animals from becoming intellectual property. Therefore, they are not subject to patent protection.

Patenting and Fundamental Research

Not everyone believes that patenting is worthwhile. Some opponents argue that awarding a monopoly restricts competition, leads to higher prices, curtails new inventions, and favors large corporations at the expense of individual inventors or small companies. Notwithstanding these concerns, the patent system is well established and is here to stay. Moreover, patent ownership does not appear to prevent significant research and development by other researchers and companies. Indeed, it might be argued that if patents were serious impediments to innovation, then U.S. patent 4,237,224, which was granted to Stanley Cohen and Herbert Boyer in 1980 for recombinant DNA technology for both the use of viral and plasmid vectors and the cloning of foreign genes, should have seriously constrained the development of recombinant DNA technology (Fig. 23.1). Obviously, no such hindrance has occurred.

In the past, patenting and patent enforcement were rarely of interest to academic researchers working in the biological sciences. Now, however, there is a view within the academic scientific community that patents and the consequences of patenting may be detrimental to established scientific values. Traditionally, science, especially university-based research, has been an open system with a free exchange of ideas and materials through publications and personal communications. The ideas of others have been respected, and contribution to the technical development of an area of study has, in many instances, been a shared enterprise. However, more recently, some scientists have begun to feel that the integrity of traditional scientific inquiry has become secondary to self-interest, where public recognition and financial gain from innovations are the prime motivations for conducting scientific research. It is argued that research is often carried out secretly and has created elite, noncooperating research groups. In the past, there was a tendency to avoid secrecy in basic research. The belief was that scientific knowledge would grow if research results were published as articles in journals that could be read by anyone, thereby enabling researchers to direct their studies in appropriate directions and to benefit from the discoveries of others. With secrecy, time and effort may be wasted on repeating experiments that, unbeknownst to the researcher, have already been done. Now scientists are advised by patent lawyers to keep their work secret until a patent is filed. Consequently, the lure of patenting has made a large number of scientists reluctant to talk about their work, at least until after the patent application has been filed.

Furthermore, due to funding cutbacks by governments, nonprofit institutions and especially universities have sought other sources of revenue. Licensing fees and royalties from patents can be sources of new income. An exceptional example is the Cohen-Boyer patent for recombinant DNA, which, during its lifetime from 1980 to 1997, may have earned a total of about $45 million for Stanford University and the University of California. Also, the Massachusetts Institute of Technology files more than 100 patents annually in all research fields and generates about $5.5 million per year from licensing patent rights. Most universities have established patent policies and offices that facilitate both patenting and the transfer of technology, at a price, to industry. Faculty members usually receive a portion of the income from their inventions. Clearly, entrepreneurial activity is a fact of life at many U.S. and other universities. The challenge may be to prevent this legitimate function from dominating all aspects of academia.

A method for replicating a biologically functional DNA, which compromises: transforming under transforming conditions compatible unicellular organisms with biologically functional DNA to form transformants; said biologically functional DNA prepared in vitro by the method of: (a) cleaving a viral or circular plasmid DNA compatible with said unicellular organism to provide a first linear segment having an intact replicon and termini of a predetermined character; (b) combining said first linear segment with a second linear DNA segment, having at least one intact gene and foreign to said unicellular organism and having termini ligatable to said termini of said first linear segment, wherein at least one of said first and second linear DNA segments has a gene for a phenotypical trait, under joining conditions where the termini of said first and second segments join to provide a functional DNA capable of replication and transcription in said unicellular organism; growing said unicellular organisms under appropriate nutrient conditions; and isolating by means of said phenotypical trait imparted by said biologically functional DNA.

Figure 23.1 The first claim of U.S. patent 4,237,224, granted to S. Cohen and H. Boyer on 2 December 1980 and entitled "Process for producing biologically functional molecular chimeras."

In sum, the enthusiasm for patenting and patent protection has elicited the perception that traditional science may become hostage to patent holders and that research will become less fruitful. Others feel that the traditional way of doing science is an outmoded, inefficient, and indulgent exercise and that patent ownership and the drive for ownership will spur new discoveries. This controversy will not be readily resolved. It is clear that the emergence of molecular biotechnology has raised far-reaching considerations, even including how scientific inquiries ought to be conducted.

SUMMARY

Historically, there are a number of reasons for issuing patents. For example, the patent holder obtains exclusive rights for the invention, rights that are a reward for developing an innovative procedure, compound, or apparatus; the public learns about the details of an invention, knowledge that will both prevent loss of time and energy in the pursuit of something that has already been invented and stimulate further research; and the opportunity for financial gain drives companies and researchers to create and develop new ideas. Patents are intended to spur innovation and to recognize the rights of the innovators. For a patent to be granted, an invention must be novel, not obvious, and useful. In addition, the invention should not be a "product of nature."

The patenting of DNA molecules in the United States has been controversial. The U.S. Patent and Trademark Office (PTO) has refused to grant patents for collections of partial cDNA sequences (expressed sequence tags), because there is no concrete information concerning the usefulness of the sequences, or for genes that were isolated using published amino acid sequences to generate hybridization probes. In the latter cases, the courts have overruled the PTO and concluded that the degeneracy of the genetic code precludes knowing the nucleotide sequence of a cDNA based on an amino acid sequence; consequently, in the view of the courts, the nonobviousness requirement for patenting was not negated.

The key case that established that genetically engineered microorganisms were patentable was brought forward by A. Chakrabarty. In 1980, the

U.S. Supreme Court ruled that a bacterium that had been created by a form of genetic manipulation could be patented. More recently, U.S. patents have been granted for a cancer-prone transgenic mouse and some transgenic plants. However, in some other countries, there is a reluctance to patent genetically engineered animals.

With the development of molecular biotechnology, questions about whether private industries should be allowed to own or patent genetically engineered organisms have been raised. On one hand, it has been argued that without such proprietary rights, biotechnology companies would not have the incentive to develop and market novel products. On the other hand, some critics find this type of privilege to be morally unacceptable, and they contend that patents of this sort inhibit research and constrain innovation. And, for a variety of reasons, patenting has had an impact on how university-based biotechnology research is conducted.

References

Adler, R. 1984. Biotechnology as an intellectual property. *Science* **224**:357–363.

Belcher, M., and A. G. Sheard. 1993. Profiting from inventions in academia: American and British perspectives. *Ann. Clin. Biochem.* **30**:1–10.

Bizley, R. E. 1991. Patenting animals in Europe. *Bio/Technology* **9**:619–622.

Caskey, C. T. 1996. Gene patents—a time to balance access and incentives. *Trends Biotechnol.* **14**:298–302.

Chahine, K. G. 1997. Patenting DNA: just when you thought it was safe. *Nat. Biotechnol.* **15**:586–587.

Crespi, R. S. 1997. Biotechnology patents and morality. *Trends Biotechnol.* **15:** 123–129.

Eckenswiller, C., and J. Morrow. 1996. Why patent life forms? *Policy Options* **17**:11–15.

Johnson, E. 1996. A benchside guide to patents and patenting. *Nat. Biotechnol.* **14**:288–291.

Marshall, E. 1997. Companies rush to patent DNA. *Science* **275**:780–781.

Poste, G. 1995. The case for genomic patenting. *Nature* **378**:534–536.

Saliwanchik, R. 1986. Legal protection for biotechnology, p. 389–401. *In* A. L. Demain and N. A. Solomon (ed.), *Manual of Industrial Microbiology and Biotechnology.* American Society for Microbiology, Washington, D.C.

Yablonsky, M. D., and W. J. Hone. 1995. Patenting DNA sequences. *Bio/Technology* **13**:656–657.

REVIEW QUESTIONS

1. What are the essential requirements for patenting an invention?

2. What are process patents? Product patents? Give examples.

3. What information must be present in a patent application?

4. Why is a patent helpful to researchers who do not hold the patent?

5. What is the purpose of a patent?

6. Discuss the issues concerning to the patenting of ESTs.

7. Discuss the differences between the U.S. PTO and the Court of Appeals

for the Federal Circuit with regard to the patentability of genes that were isolated by using amino acid sequences to generate hybridization probes.

8. Prepare arguments for both sides of the debate "Be it resolved that patenting of genetically engineered multicellular organisms should be banned."

9. Discuss how patenting might impact research in nonprofit institutions.

10. Contact the U.S. PTO web site (http://patents.uspto.gov/), conduct a Boolean search for recombinant DNA patents (e.g., recombinant AND DNA), and summarize the inventions in the five most recent recombinant DNA patents.

Glossary

A Adenine residue in either DNA or RNA.

Ab *See* Antibody

Absolute linkage *See* Complete linkage

ACC In plants, 1-aminocyclopropane-1-carboxylate (ACC) is the immediate precursor of ethylene.

ACC deaminase A bacterial enzyme that can cleave ACC to ammonia and α-ketobutyrate.

ACC oxidase A plant enzyme that catalyzes the oxidation of ACC to ethylene. Formerly called ethylene-forming enzyme.

ACC synthase A plant enzyme that catalyzes the synthesis of ACC from *S*-adenosylmethionine. Its activity is stimulated by indoleacetic acid.

Actin A major protein component of skeletal muscle.

Activation Enhancing the rate of transcription.

Activator (1) A substance or physical agent that stimulates transcription of a specific gene or operon. (2) A protein that binds to an operator and enhances the rate of transcription. Also called activator protein.

Activator site A DNA sequence to which an activator protein binds. Also called activating site.

Acyl carrier protein A low-molecular-weight protein that forms part of a larger complex for either fatty acid or polyketide biosynthesis.

Adaptor (1) A synthetic double-stranded oligonucleotide that is blunt ended at one end and at the other has a nucleotide extension that can base pair with a cohesive end created by cleavage of a DNA molecule with a specific type II restriction endonuclease. After blunt-end ligation of the adaptor to the ends of a target DNA molecule, the construct can be cloned into a vector by using the cohesive ends of the adaptor. (2) A synthetic single-stranded oligonucleotide that, after self-hybridization, produces a molecule with cohesive ends and an internal restriction endonuclease site. When the adaptor is inserted into a cloning vector by means of the cohesive ends, the internal sequence provides a new restriction endonuclease site.

Adenine One of the organic bases found in either DNA or RNA.

Aerobe A microorganism that requires oxygen for growth.

Affinity tag A short sequence of amino acids that is engineered as part of a recombinant protein and binds to a specific element, compound, or macromolecule which facilitates the identification or purification of the recombinant protein. Also peptide tag, protein tag.

Ag *See* Antigen

Airlift fermenter A cylindrical fermentation vessel in which the cells are mixed by air that is introduced at the base of the vessel and rises through the column of culture medium. The cell suspension circulates around the column as a consequence of the gradient of air bubbles in different parts of the reactor.

Alginate A polysaccharide polymer produced by different seaweeds and bacteria that is comprised of β-D-mannuronate and α-L-guluronate.

Allele Alternative form of a gene. Alleles of a specific gene occupy the same location on homologous chromosomes.

Allelic frequency The ratio of the occurrence of one particular allele at a locus to the occurrence of all of the alleles of the locus in a large number of individuals of a population.

Allogeneic Having different antigens (an attribute of cell types). Also called allogenic.

Allosteric control *See* Allosteric regulation

Allosteric regulation A catalysis-regulating process in which the binding of a small effector molecule to one site on an enzyme affects the activity at another site of the enzyme.

Alternative splicing Cell-specific removal of an exon(s) during processing of a primary transcript that leads to a functional mRNA.

Amino acid A building block of a protein.

Aminoacyl site The portion of a ribosome where the anticodon-codon interaction takes place during translation. Also called an A site.

Aminoacyl-tRNA A charged tRNA; a tRNA with its specific amino acid attached to its 3' end.

Amplicon A herpes simplex virus type 1 plasmid vector. Also called amplicon plasmid.

Amylolytic Agents that are capable of breaking down starch into sugars.

Anaerobe A microorganism that grows in the absence of oxygen.

Annealing The process of heating (denaturing step) and slowly cooling (renaturing step) double-stranded DNA to allow the formation of hybrid DNA or DNA–RNA molecules.

Antibiosis The prevention of growth or development of an organism by a substance or another organism.

Antibiotic A biological substance that is produced by one organism and that can inhibit the growth of or kill another organism.

Antibody A protein (immunoglobulin) that is synthesized by a B lymphocyte and recognizes a specific site on an antigen. The basic immunoglobulin molecule consists of two identical heavy and two identical light chains.

Anticodon A set of three contiguous nucleotides in a tRNA molecule that are complementary to a set of three contiguous nucleotides (codon) in an mRNA.

Antifreeze protein A type of protein that binds to ice crystals and depresses the freezing temperature of the crystal below its melting temperature. Antifreeze proteins have been found in fish, insects, plants, fungi, and bacteria and protect cells from being damaged by ice crystals.

Antigen A compound that induces the production of antibodies.

Antigenic determinant *See* Epitope

Anti-idiotype antibody An antibody that has the properties of an antigen.

Antiparallel orientation The arrangement of the two strands of a duplex DNA molecule, which are oriented in opposite directions so that the 5'-phosphate end of one strand is aligned with the 3'-hydroxyl end of the complementary strand.

Antisense DNA (1) The sequence of chromosomal DNA that is transcribed. (2) A DNA sequence that is complementary to all or part of a functional RNA (mRNA).

Antisense RNA An RNA sequence that is complementary to all or part of a functional RNA.

Antisense therapy The in vivo treatment of a genetic disease by blocking translation of a protein with a DNA or an RNA sequence that is complementary to a specific mRNA.

Antiserum The fluid portion of the blood that contains the antibodies of an immunized organism.

Aphid A plant-sucking insect of the family Aphididae.

Aptamer A synthetic nucleic acid that binds to a protein that normally does not bind to a nucleic acid.

Arabidopsis thaliana A plant with a very small genome that is used as a model organism for the study of plant growth and development.

ARS Autonomous replicating sequence. Any cloned DNA sequence that initiates and supports extrachromosomal replication of a DNA molecule in a host cell; often used in yeast cells. Also called autonomous(ly) replicating segment, autonomously replicating sequence.

Attenuated vaccine A virulent organism that has been modified to produce a less virulent form but nevertheless retains the ability to elicit antibodies against the virulent form.

Authentic protein A recombinant protein that has all the properties, including any posttranslational modifications, of its naturally occurring counterpart.

Autologous cells Cells that are taken from an individual, cultured (or stored), and possibly genetically manipulated before being infused back into the original donor.

Autonomous(ly) replicating sequence *See* ARS

Autoradiography A technique that captures the image formed in a photographic emulsion as a result of the emission of either light or radioactivity from a labeled component that is placed next to unexposed film.

Autosomal dominant inheritance When a single allele of a gene pair on a chromosome other than the sex chromosomes is sufficient to produce a phenotypic effect.

Autosomal recessive inheritance When two alleles of a gene pair must be present on homologous chromosomes other than sex chromosomes to produce a phenotypic effect.

Autosomes All members of a chromosome set except the sex chromosomes. Human somatic cells have 22 pairs of autosomes and 1 pair of sex chromosomes.

Auxin *See* IAA

B cells Lymphocytes that produce antibodies and are derived from bone marrow cells.

BAC *See* Bacterial artificial chromosome

Bacmid A shuttle vector based on the AcMNPV genome that can be propagated in both *E. coli* and insect cells.

Bacterial artificial chromosome A vector system based on the *Escherichia coli* F factor plasmid that is used for cloning large (100- to 300-kb) DNA inserts.

Bacteriocin A compound produced by one bacterium that can kill cells of another bacterial species.

Bacteriophage A virus that infects bacteria. Also called phage.

Baculovirus A virus that infects insects.

Bank *See* Gene bank

Base pair The term *base pair* represents complementary nucleotides; in DNA adenine (A) is hydrogen bonded with the base thymine (T) and guanine (G) is hydrogen bonded with cytosine (C). A thousand base pairs is often called a kilobase (kb).

Base pair substitution Permanent replacement in chromosomal DNA of a nucleotide pair with another nucleotide pair.

Batch culture *See* Batch fermentation

Batch fermentation A process in which cells or microorganisms are grown for a limited time. At the beginning of the fermentation, an inoculum is introduced into fresh medium, and no addition or removal of medium is done for the duration of the process.

Bench-scale process A small-scale (or laboratory-scale) process; often refers to fermentation.

Betaine A low-molecular-weight compound that acts as a methyl group donor for methionine biosynthesis.

Bifunctional vector *See* Shuttle vector

Binary fission Asexual cell division that produces equal-sized daughter cells.

Binary vector system A two-plasmid system in *Agrobacterium* for transferring a T-DNA region that carries cloned genes into plant cells. The virulence genes are on one plasmid, and the engineered T-DNA region is on the other plasmid.

Bioaccumulation Concentration of a chemical agent (e.g., DDT) in increasing quantities in the organisms of a food chain.

Biocontrol Any process using living organisms to restrain the growth and development of pathogenic organisms.

Biodegradation The breakdown of a compound to its chemical constituents by living organisms.

Bioinformatics *See* Informatics

Biolistics Delivery of DNA to plant and animal cells and organelles by means of DNA-coated pellets that are fired under pressure at high speed. Also called microprojectile bombardment.

Biological aging *See* Senescence

Biomarker A biological feature that is used to measure either the progress of a disease or the effect of a treatment.

Biomass (1) The cell mass produced by a population of living organisms. (2) The organic mass that can be used either as a source of energy or for its chemical components.

Biomass concentration The amount of biological material in a specific volume.

Biopolymer Any large polymeric molecule (protein, nucleic acid, polysaccharide, lipid) produced by a living organism.

Bioreactor A vessel in which cells, cell extracts, or enzymes carry out a biological reaction. Often refers to a growth chamber (fermenter, fermentation vessel) for cells or microorganisms.

Bioremediation A process that uses living organisms to remove contaminants, pollutants, or unwanted substances from soil or water.

Biotin labeling (1) The attachment of biotin to another molecule. (2) The incorporation of a biotin-containing nucleotide into a DNA molecule.

Blot *See* Northern blotting; Southern blotting

Blotting Transfer of a macromolecule by capillary action from a gel to a membrane.

Blunt end The end of a DNA duplex molecule in which neither strand extends beyond the other. Also called flush end.

Blunt-end cut To cleave phosphodiester bonds in the backbone of duplex DNA between the corresponding nucleotide pairs on opposite strands. This cleavage process produces no nucleotide extensions on either strand. Also called flush-end cut.

Blunt-end ligation Joining (ligation) of the nucleotides that are at the ends of two DNA duplex molecules, neither of which has an extension.

Boll weevil An insect pest of cotton plants.

Box A short DNA sequence that plays a role in regulating, facilitating, enhancing, or silencing transcription.

bp *See* base pair

Brewer's yeast Strains of yeast, often *Saccharomyces cerevisiae,* that are used in the production of beer.

Broad-host-range plasmid A plasmid that can replicate in a number of different bacterial species.

Bubble column fermenter A fermentation vessel, or bioreactor, in which the cells or microorganisms are kept suspended in a tall cylinder by rising air that is introduced at the base of the column.

Bystander effect The death of an unmodified cell by a cytotoxic metabolite that is produced by a genetically modified cell and acquired by cell-to-cell contact.

C Cytosine residue in either DNA or RNA.

C terminus The last amino acid of a protein. Sometimes denotes the final amino acids of a protein. Also called carboxy(l) terminus, carboxy(l)-terminal end.

Calibrator lane The lane of a gel that contains electrophoretically separated size markers.

Callus A mass of undifferentiated plant tissue from tissue cuttings or from individual plant cells, grown in culture on defined medium.

cAMP *See* cyclic AMP

Cancer Uncontrolled growth of the cells of a tissue or an organ in a multicellular organism.

Candidate gene A coding sequence that has some characteristics, including chromosome location and function, that make it likely that it could be responsible for a specific genetic disorder.

Candidate gene cloning A strategy for isolating a disease gene that is based on an informed guess about the possible gene product.

Canola A plant whose seed is used to produce high-quality cooking oil. Formerly called rapeseed. Canola is Canada's most economically important crop.

Cap *See* G cap

Capsid A structure that is composed of the coat protein(s) of a virus and is external to the viral nucleic acids. The capsid often determines the shape of the virus.

Carrier In genetics, an individual that has one mutant allele and one normal allele of a gene and whose phenotype is normal.

Cassette A combination of DNA elements that performs a specific function and is maintained as a clonable unit.

CD molecules Designation for surface molecules on various cells of the immune system; e.g., CD4 is present on the surface of helper T cells.

cDNA Complementary DNA. A double-stranded DNA complement of an mRNA sequence; synthesized in vitro by reverse transcriptase and DNA polymerase.

cDNA clone A double-stranded DNA molecule that is carried in a vector and was synthesized in vitro from an mRNA sequence by using reverse transcriptase and DNA polymerase.

cDNA library A collection of cDNA clones that were generated in vitro from the mRNA sequences of a single tissue or cell population.

CDR Complementarity-determining region. A part of the variable (V) regions of light and heavy antibody chains that makes contact with the antigen. The amino acid sequences of CDRs are highly variable from one antibody of the same class to another.

Cell line A cell lineage that can be maintained in culture.

Cell-free protein synthesis *See* In vitro translation

Cell-mediated immune response The activation of T cells of the immune system in response to the presence of a foreign antigen.

Cellulose A high-molecular-weight polysaccharide of unbranched chains of (1,4)-linked β-D-glucose units that, as part of lignocellulose, contributes to the structural framework of plant cell walls.

Cellulosome A multiprotein aggregate that is present in some cellulolytic microorganisms and contains multiple copies of all the enzymes required to completely break down cellulose. This complex is often found on the outer surface of cellulolytic microorganisms.

Centimorgan A unit of genetic distance between two gene loci, abbreviated cM. Named after T. H. Morgan (1866–1945), who used the fruit fly *Drosophila melanogaster* to establish the principles of genetic linkage and gene mapping. In humans, 1 cM is approximately 10^6 base pairs of DNA.

CEPH family panel A set of about 65 two- or three-generation families with sibships of eight or more that are used for the genetic mapping of the human genome. The blood cells from members of these families that provide the source of DNA for genotyping are maintained by the Fondation Jean Dausset Centre d'Etude Polymorphisme Humain in Paris, France.

Cephem-type antibiotic An antibiotic that shares the basic chemical structure of cephalosporin.

Chagas' disease A parasitic disease caused by the protozoan *Trypanosoma cruzi*.

Charged tRNA A transfer RNA molecule that is coupled to its specific amino acid. Also aminoacylated tRNA, amino acyl tRNA.

Chemiluminescence The emission of light from a chemical reaction.

Chimera Usually, a plant or animal that has populations of cells with different genotypes. Sometimes it refers to a recombinant DNA molecule that contains sequences from different organisms.

Chimeric protein *See* Fusion protein

Chitinase An enzyme produced by plants in response to infection by fungal pathogens that hydrolyzes the chitin components in fungal cell walls. Some bacteria can produce chitinase.

Chromogenic substrate A compound or substance that contains a color-forming group.

Chromosomal integration site A chromosomal location where foreign DNA can be integrated, often without impairing any essential function in the host organism.

Chromosome A physically distinct unit of the genome.

Chromosome jumping A technique that allows two segments of duplex DNA that are separated by thousands of base pairs (about 200 kb) to be cloned together. After subcloning, each segment can be used as a probe to identify cloned DNA sequences that, at the chromosome level, are roughly 200 kb apart.

Chromosome walking A technique that identifies overlapping cloned sequences that are about 40 kb long that form one continuous segment of a chromosome.

Cistron A sequence of DNA that encodes a polypeptide chain.

Claims A section of a patent that states, in detail, the uses and possible applications of the invention described in the patent.

Cleave To break phosphodiester bonds of duplex DNA, usually with a type II restriction endonuclease. Also called cut, digest.

Clone (1) A population of cells or organisms that are genetically identical as a result of asexual reproduction, breeding of purebred (isogenic) organisms, or forming genetically identical organisms by nuclear transplantation. (2) A population of cells that all carry a cloning vehicle with the same insert DNA molecule. (3) To insert a DNA segment into a vector or host chromosome.

Clone bank *See* Gene bank

Cloning Incorporating a DNA molecule into a chromosomal site or a cloning vector.

Cloning site A location on a cloning vector into which DNA can be inserted.

Cloning vector A DNA molecule that can carry inserted DNA and be perpetuated in a host cell. Also called cloning vehicle, vector, or vehicle.

Cloning vehicle *See* Cloning vector

cM *See* Centimorgan; Map distance

Coding triplet A set of three contiguous nucleotides of the nontranscribed DNA strand of the coding region of a structural gene that is complementary to a transcribed triplet.

Codon A set of three nucleotides in mRNA that specifies a tRNA carrying a specific amino acid which is incorporated into a polypeptide chain during protein synthesis.

Codon optimization An experimental strategy in which codons within a cloned gene that are not the ones generally used by the host cell translation system are changed to the preferred codons without changing the amino acids of the synthesized protein.

Codon usage The mean frequency of occurrence of each codon determined from a large sample of structural genes of an organism.

Cofactor A low-molecular-weight compound that is a required component in an enzymatic reaction.

Cofermentation The simultaneous growth of two microorganisms in one bioreactor.

Cohesive ends Complementary single-strand extensions on the ends of duplex DNA molecules. Also called sticky ends. *See also* cos ends

Cointegrate vector system A two-plasmid system for transferring cloned genes to plant cells. The cloning vector has a T-DNA segment that contains cloned genes. After introduction into *Agrobacterium,* the cloning vector DNA undergoes homologous recombination with a resident disarmed Ti plasmid to form a single plasmid carrying the genetic information for transferring the genetically engineered T-DNA region to plant cells.

Coleoptile A protective organ that covers the youngest leaves of a plant.

Colony hybridization A technique that uses a nucleic acid probe to identify a bacterial colony with a vector carrying a specific cloned gene(s).

Combinatorial library During the ligation reaction with cDNAs of light and heavy antibody chains into a bacteriophage λ vector, many novel combinations consisting of one heavy and one light chain coding region are formed. The library comprises these combinations, each in a separate vector.

Competence The ability of bacterial cells to take up (usually plasmid) DNA molecules.

Complement A group of serum proteins that is activated by an antibody–antigen complex and has enzymatic and other biological activities such as degrading antibody–antigen complexes, lysing cells, modulating antibody production, stimulating immune cells to migrate to a site of complement activity, and inducing the release of histamine.

Complement cascade The series of sequential activations and enzymatic reactions by serum proteins that is launched in response to the formation of an antibody–antigen complex.

Complementarity (1) One of a pair of nucleotide bases that form hydrogen bonds with each other. Adenine (A) pairs with thymine (T) (or with uracil [U] in RNA), and guanine (G) pairs with cytosine (C). (2) One of a pair of segments or strands of nucleic acid that will hybridize (join by hydrogen bonding) with each other.

Complementarity-determining region *See* CDR

Complementary base pairs In double-stranded DNA, adenine forms hydrogen bonds with thymine and cytosine forms hydrogen bonds with guanine. In double-stranded regions of RNA molecules, and in both RNA–RNA and DNA–RNA strand interactions, adenine forms hydrogen bonds with uracil and cytosine forms hydrogen bonds with guanine.

Complementary DNA *See* cDNA

Complementary homopolymeric tailing The process of adding complementary nucleotide extensions to different DNA molecules—e.g., dG (deoxyguanosine) to the 3′-hydroxyl ends of one DNA molecule and dC (deoxycytidine) to the 3′-hydroxyl ends of another DNA molecule—to facilitate, after mixing, the joining of the two DNA molecules by base pairing between the complementary extensions. Also called dG·dC tailing, dA·dT tailing.

Complementation *See* Genetic complementation

Complete linkage Two or more adjacent gene loci on the same chromosome that are always inherited together. Also called absolute linkage.

Complete penetrance All the individuals with a mutant allele(s) at a gene locus show the same abnormal (mutant) phenotype.

Concatemer A tandem array of repeating unit-length DNA elements.

Conformation The shape of a molecule or any other object.

Conjugation The unidirectional transfer of DNA from one bacterium to another, involving cell-to-cell contact.

Conjugative functions Plasmid-based genes and their products that facilitate the transfer of a plasmid from one bacterium to another.

Constant domains Regions of antibody chains that have the same amino acid sequence in different members of a particular class of antibody molecules.

Constitutive synthesis Continual production of RNA or protein by an organism.

Contig A set of overlapping contiguous clones that cover a chromosome region or a whole chromosome.

Continuous fermentation A process in which cells or microorganisms are maintained in culture in the exponential growth phase by the continuous addition of a fresh medium that is exactly balanced by the removal of cell suspension from the bioreactor.

Copy number The average number of a specific type of plasmid molecules in a cell.

Corepressor A low-molecular-weight compound that combines with an inactive repressor protein to form a complex that binds to an operator region and prevents transcription.

Cosegregation When two genetic conditions appear to be inherited together.

cos ends The 12-base, single-strand, complementary extensions of bacteriophage λ DNA. Also called cos sites.

Cosmid A vector that uses the cos end sequences of bacteriophage λ and in vitro bacteriophage packaging to form, after injection of the vector into a host cell, a plasmid that can carry as much as 45 kb of insert DNA.

Cosuppression The transformation of a plant with a gene in the sense orientation that the plant already possesses. This results in the down regulation of both the endogenous and introduced genes. Also called sense suppression.

Cotransfection The introduction of two different DNA molecules into a eukaryotic cell. In baculovirus expression systems, the procedure by which the baculovirus and the transfer vector are simultaneously introduced into insect cells in culture.

Coupling The phase state in which either two dominant or two recessive versions of two different genes occur on the same chromosome. Also called *cis* configuration. *See also* Repulsion

CpG islands Clusters of GC-rich regions that precede many transcribed vertebrate genes.

Cross In genetic studies, the mating of two individuals. Also called mating.

Crossover (1) The site of recombination. A single crossover represents one reciprocal breakage-and-reunion event. A double crossover requires two simultaneous reciprocal breakage-and-reunion events. (2) The reciprocal exchange of DNA between two chromosomes or DNA molecules by a breakage-and-reunion process. Also called recombination, recombination event.

Crown gall A bulbous growth that occurs at the base of certain plants and is due to infection of the plant by a member of the bacterial genus *Agrobacterium*. Also called crown gall tumor.

Crown gall tumor *See* Crown gall

Crucifer A plant, such as cabbage, that is a member of the family Cruciferae.

Cryptic site A functional macromolecular sequence in an unlikely location. Also used, in some instances, to denote a macromolecular sequence whose function is unknown.

Cultivar A variety of plant that is (1) below the level of a subspecies taxonomically and (2) found only under cultivation.

Culture A population of cells or microorganisms that are grown under controlled conditions.

Culture medium A solid or liquid mixture that is used to grow microorganisms, organisms, or cells.

Cut *See* Cleave

Cyclic AMP Adenosine 3′,5′-cyclic phosphoric acid, an important regulatory molecule. Also called cAMP.

Cytokinin A plant hormone that stimulates cell division.

Cytosine One of the organic bases found in either DNA or RNA.

dA·dT tailing *See* Complementary homopolymeric tailing

Dehalogenation The removal of halogen atoms (chlorine, iodine, bromine, fluorine) from molecules, usually during biodegradation.

Deletion Loss of an internal portion of chromosomal DNA.

Denaturation (1) Separation of duplex nucleic acid molecules into single strands. (2) Disruption of the conformation of a macromolecule without breaking covalent bonds.

Denatured DNA Duplex DNA that has been converted to single strands by breaking the hydrogen bonds of complementary nucleotide pairs.

Deoxyribonuclease I *See* DNase I

Deoxyribonucleic acid *See* DNA

Deoxyribose The 5-carbon sugar component of DNA.

Deoxyribozyme A DNA molecule that has catalytic activity.

Derepression Displacement of a repressor protein from a promoter region of DNA; the "turning on" of a gene. When attached to the DNA, the repressor protein prevents RNA polymerase from initiating transcription.

dG·dC tailing *See* Complementary homopolymeric tailing

Diagnostic procedure A test or assay used to determine the presence of an organism, substance, or nucleic acid sequence alteration that represents a disease or pathogenic condition.

Diaminopimelic acid The immediate precursor to L-lysine in bacteria and plants, and a component of some bacterial cell walls.

Diazotroph An organism that can fix nitrogen.

Dicistronic vector A mammalian cloning vector that is specially designed to carry two functional genes.

Dicotyledon A class of plants that has two seed leaves. Also called dicot.

Dideoxynucleotide A nucleoside triphosphate that lacks hydroxyl groups on both the 2′ and 3′ carbons of the pentose sugar. Also referred to as ddNTP.

Digest *See* Cleave

Dihydrofolate reductase An enzyme that catalyzes the formation of tetrahydrofolic acid.

Diploid A cell or organism that has a set of all pairs of its chromosomes.

Directed mutagenesis The process of generating nucleotide changes in cloned genes by any one of several procedures, including site-specific and random mutagenesis. Also called in vitro mutagenesis.

Disarm Delete from a plasmid or virus those genes that are cytotoxic or induce tumor formation.

Dithiothreitol A low-molecular-weight thiol-containing reducing agent. It is added to buffers in low concentrations to prevent protein sulfhydryl groups from being oxidized. At higher concentrations, it is used to reduce disulfide linkages in proteins.

DNA Deoxyribonucleic acid; the genetic material of living things.

DNA codon A set of three contiguous deoxyribonucleotide pairs of the coding region of a structural gene where the bases of one strand are transcribed into a codon.

DNA construct A cloning vector with a DNA insert.

DNA delivery system Generic term for any procedure that facilitates the uptake of DNA by a recipient cell.

DNA fingerprint A set of DNA fragments that are characteristic for a particular source of DNA such as an insert of a clone. In some cases, restriction endonuclease DNA fragments are visualized by hybridization after gel electrophoresis. In other instances, the polymerase chain reaction (PCR) is used to generate a distinctive pattern of DNA bands that are evident after gel electrophoresis.

DNA fingerprinting A comparative diagnostic technique that characterizes the DNA of an organism or a sample.

DNA hybridization The pairing of two DNA molecules, often from different sources, by hydrogen bonding between complementary nucleotides. This technique is frequently used to detect the presence of a specific nucleotide sequence in a DNA sample.

DNA polymerase An enzyme that links an incoming deoxyribonucleotide, which is determined by complementarity to a deoxyribonucleotide in a template DNA strand, with a phosphodiester bond to the 3′-hydroxyl group of the last incorporated nucleotide of the growing strand during replication.

DNA probe A segment of DNA that is labeled (tagged) so that, after a DNA hybridization reaction, any base pairing between the probe and a complementary base sequence in a DNA sample can be detected.

DNA transformation *See* Transfection; Transformation

DNA typing *See* Genotyping

DNase I An enzyme that degrades DNA. It is used to remove DNA from RNA preparations and from cell free extracts. Also called deoxyribonuclease I.

Domain A segment of a protein that has a discrete function or conformation. At the protein level, a domain can be as small as a few amino acid residues and as large as half of the entire protein.

Dominant (1) When a heterozygous and homozygous genotype determines the same phenotype. (2) An allele that produces the same phenotype whether the genotype is heterozygous or homozygous. Also called dominance.

Dominant gene One of a pair of alleles that is sufficient to produce a phenotype in a heterozygote.

Dominant marker selection A gene encoding a product that enables only the cells that carry the gene to grow under certain conditions. For example, plant and animal cells that express the introduced Neor gene are resistant to the compound G418, and cells that do not carry the Neor gene are killed by G418. Also called positive selection, positive selectable marker.

Double crossover Two simultaneous reciprocal breakage-and-reunion events between two DNA molecules.

Double heterozygote Two different gene loci, each with two different alleles.

Doubling time *See* Generation time

Downstream (1) In molecular biology, the stretch of nucleotides of DNA that lie in the 3′ direction of the site of initiation of transcription, which is designated as +1. Downstream nucleotides are marked with plus signs, e.g., +2, +10. Also refers to the 3′ side of a particular gene or sequence of nucleotides. (2) In chemical engineering, those phases of a manufacturing process that follow the biotransformation stage. Refers to recovery and purification of the product of a fermentation process. Also called downstream processing.

Downstream processing *See* Downstream

Drug *See* Therapeutic agent

Dry weight The weight of a sample of biological material that has been dried in an oven to remove the water.

Ds element A plant transposable element.

Duplex DNA Double-stranded DNA.

Effector A low-molecular-weight compound that modifies the function of a regulatory protein.

Effector cells Cells of the immune system that degrade antigens.

Electrophoresis A technique that separates molecules (often DNA, RNA, or protein) on the basis of relative migration in a strong electric field.

Electroporation Electrical treatment of cells that induces transient pores, through which DNA is taken into the cell.

Electrotransfer Transfer of a macromolecule by an electric field from a gel to a membrane.

ELISA *See* Enzyme-linked immunosorbent assay

Elongation Sequential addition of one monomer at a time to a polymer.

ELSI The ethical, legal, and social implications (ELSI) that arise from genomic and other kinds of research.

Embryonic stem cells Cells of an early embryo that can give rise to all differentiated cells, including germ line cells.

Encode To specify, after decoding by transcription and translation, the sequence of amino acids in a protein.

Endoprotease An enzyme that cleaves the peptide bonds between amino acids within a protein. Cleavage is usually at one or more specific sites.

Endotoxin A component of the cell wall of gram-negative bacteria that elicits, in humans, an inflammatory response and fever.

End-product inhibition The inhibition of the activity of an enzyme by a metabolite. The enzyme is sometimes the first enzyme in a biosynthetic pathway, and the metabolite is generally the product of the last step in the pathway.

Enhancer A DNA sequence that increases the transcription of a eukaryotic gene when they are both on the same DNA molecule. Also called enhancer element, enhancer sequence.

Enolase An enzyme that catalyzes the conversion of 2-phosphoglycerate to phosphoenolpyruvate.

Enoylreductase An enzyme involved in the synthesis of polyketide antibiotics.

Enterotoxin A bacterial protein that, after its release into the intestine, causes cramps, diarrhea, and nausea.

Enzyme-linked immunosorbent assay A technique for detecting specific molecules in a mixed sample. An antibody (primary) is bound to the target molecule; another antibody (secondary), which binds to the primary antibody, is added later. The secondary antibody has attached to it an enzyme that can convert a colorless substrate into a colored product. If the target molecule is not present in the sample, washing steps will remove both antibodies and no colored product will be produced. Also called ELISA.

Epithelial cells Cells covering a surface or lining a cavity, e.g., gut epithelial cells.

Epitope A specific chemical domain on an antigen that is recognized by an antibody. Each epitope on a molecule such as a protein elicits the synthesis of a different antibody. Also called antigenic determinant.

Epitope tag An affinity tag that is recognized by an antibody.

ES cells *See* Embryonic stem cells

EST *See* Expressed sequence tag

Established cell line A population of cells that is grown in vitro and can be subcultured indefinitely.

Ethylene A gaseous compound that acts as a plant hormone. It is important in fruit ripening, flower senescence, seed germination, rooting of cuttings, root elongation, and the response of the plant to environmental stress.

Eukaryotes Organisms, including animals, plants, fungi, and some algae, that have (1) chromosomes enclosed within a membrane-bounded nucleus and (2) functional organelles, such as mitochondria and chloroplasts, in the cytoplasm of their cells.

Excision The natural or in vitro enzymatic release (removal) of a DNA segment from a chromosome or cloning vector.

Excrete *See* Export

ExoIII *See* Exonuclease III

Exogenous Derived externally, or foreign.

Exogenous DNA DNA that has been derived from a source organism and has been cloned into a vector and introduced into a host cell. Also referred to as foreign or heterologous DNA.

Exon A segment of a gene that is transcribed as part of the primary transcript and is retained, after processing, with other exons to form a functional mRNA molecule.

Exon trapping A protocol for the isolation of exons from genomic clones. Genomic DNA (1 to 6 kb) is cloned into an intron that is flanked by two exons (artificial gene) in a vector. When an exon is present in the cloned DNA, reverse transcriptase polymerase chain reaction (RT-PCR) shows that the mRNA from the artificial gene is longer than expected. The cloned DNA can be amplified, subcloned, and characterized. Also called exon entrapment, exon amplification.

Exonuclease III An *Escherichia coli* enzyme that removes nucleotides from the 3'-hydroxyl ends of double-stranded DNA. Also called ExoIII, exodeoxyribonuclease III.

Exopolysaccharide A high-molecular-weight polymer that is composed of sugar residues and is secreted by a microorganism into the surrounding environment.

Export To transport a protein out of a cell. Also to secrete, to excrete.

Expressed sequence tag A sequence tagged site derived from a cDNA clone. Also called EST, expressed sequence tagged site, eSTS.

Expression Transcription and translation of a gene.

Expression library A population of different DNA molecules cloned into one kind of expression vector.

Extension A single-stranded DNA region consisting of one or more nucleotides at the end of a strand of duplex DNA. Also called protruding end, sticky end, overhang, cohesive end.

3' extension A short single-stranded nucleotide sequence on the 3'-hydroxyl end of a double-stranded DNA molecule. Also called 3' protruding end, 3' sticky end, 3' overhang.

5' extension A short single-stranded nucleotide sequence on the 5'-phosphate end of a double-stranded DNA molecule. Also called 5' protruding end, 5' sticky end, 5' overhang.

Extracellular matrix The organized carbohydrate and protein structure that is secreted by animal cells when they are part of a tissue.

Extrachromosomal DNA A replicatable DNA element that is not part of a chromosome.

Ex vivo gene therapy The delivery of a gene(s) to the isolated cells of an individual. After culturing, the transformed cells are introduced back into the individual by transfusion, infusion, or injection, to alleviate a genetic disorder.

F_1 First generation of offspring from a cross. Also called first filial generation.

F_2 Second generation of offspring from a cross with at least one F_1 parent. Also called second filial generation.

False negative A negative assay result that should have been positive.

False positive A positive assay result that should have been negative.

Fed-batch fermentation Growth of cells or microorganisms during which nutrients are added periodically to the bioreactor.

Feedback inhibition *See* End-product inhibition

Fermentation (1) In chemical engineering, the growth of cells or microorganisms in specialized vessels (fermenters, bioreactors). (2) In biochemistry, the breakdown of carbon compounds by cells or organisms to ATP without using molecular oxygen.

Fermenter *See* Bioreactor

Ferredoxin An iron–sulfur protein that acts as an electron carrier.

Fertile Capable of breeding and reproduction.

Flavonoid A member of a class of plant phenolic compounds containing two aromatic rings connected by a 3-carbon bridge.

Fluorescein A fluorescent dye often used to label antibodies so that they may be visualized after they have reacted with antigens in cells.

Fluorography Detection of the emission of light on X-ray film from a labeled source.

Fluorophore That portion of a molecule that can fluoresce.

Flush end *See* Blunt end

Flush-end cut *See* Blunt-end cut

Foreign DNA A DNA molecule that is incorporated into either a cloning vector or a chromosomal site.

Fouling The coating or plugging by materials or microorganisms of equipment that prevents it from functioning properly.

Founder animal In transgenesis research, an organism that carries a transgene in its germ line and can be used in matings to establish a pure-breeding transgenic line, or one that acts as a breeding stock for transgenic animals.

Four-cutter A type II restriction endonuclease that binds and cleaves DNA at sites that contain four nucleotide pairs.

Frameshift mutation In chromosomal DNA, an insertion or deletion of base pairs that changes the reading frame of a gene.

Functional gene cloning A strategy for isolating a gene that depends on information about its gene product.

Fusion protein The product of two or more coding sequences from different genes that have been cloned together and that, after translation, form a single polypeptide sequence. Also called hybrid protein, chimeric protein.

G Guanine residue in either DNA or RNA.

G cap The 5′-terminal methylated guanine nucleotide that is present on many eukaryotic mRNAs; it is joined, after transcription, to the mRNA in a 5′-to-5′ linkage.

Gamete A cell with a haploid chromosome content. In animals, sperm or egg; in plants, pollen or ovum.

Gap A missing internal segment of one strand of duplex DNA

Gapped DNA A duplex DNA molecule with one or more internal single-stranded regions.

Gel matrix A semisolid macromolecular lattice that is used for the electrophoretic fractionation of macromolecules.

Gelatinization Steam cooking of milled grain, a process that increases the surface area of the starch and converts the original mash to a material with a gel-like consistency.

Gene A segment of nucleic acid that encodes a functional protein or RNA. The unit of inheritance.

Gene bank A population of organisms, each of which carries a DNA molecule that was inserted into a cloning vector. Ideally, all of the cloned DNA molecules represent the entire genome of another organism. Also called gene library, clone bank, bank, library. This term is sometimes also used to denote all of the vector molecules, each carrying a piece of the chromosomal DNA of an organism, prior to the insertion of these molecules into a population of host cells.

Gene cloning Insertion of a gene into a DNA vector (often a plasmid) to form a new DNA molecule that can be perpetuated in a host cell. Also called recombinant DNA technology, genetic engineering, gene splicing, gene transplantation, molecular cloning, cloning.

Gene library *See* Gene bank

Gene map The linear array of genes of a chromosome.

Gene therapy Use of a gene or cDNA to treat a disease.

Generation time The time that it takes for a population of single-celled organisms to double its cell number. Also called doubling time.

Genetic code The complete set of 64 codons that code for all 20 amino acids and 3 termination codons.

Genetic complementation When two genomes or DNA molecules that are in the same cell produce a function that neither genome or DNA molecule can supply on its own. Also called complementation.

Genetic engineering *See* Gene cloning

Genetic heterogeneity The condition or state of a population in which different mutant genes produce the same phenotype. Also called heterogeneity.

Genetic immunization Delivery of a cloned gene that encodes an antigen to a host organism. After the cloned gene is expressed, it elicits an antibody response that protects the organism from infection by a virus, bacterium, or other disease-causing organism.

Genetic linkage *See* Linkage

Genetic map The linear array of genes on a chromosome based on recombination frequencies. Also called linkage map.

Genetic mapping Determining the linear order of marker sites along a chromosome. Also called mapping.

Genetic polymorphism When two or more alleles of a locus in a population of individuals occur at a frequency of 1% or greater. Often, in the appropriate context, it is simply called polymorphism.

Genome (1) The entire complement of genetic material of an organism, virus, or organelle. (2) The haploid set of chromosomes (DNA) of a eukaryotic organism.

Genomics The study and development of genetic and physical maps, large-scale DNA sequencing, gene discovery, and computer-based systems for managing and analyzing genomic data.

Genotype (1) The genetic constitution of an organism. (2) The alleles at a genetic locus.

Genotyping The determination of the alleles of a chromosome of an individual. Also called DNA typing, haplotyping.

Germ line cells Cells that produce gametes.

Germ line gene therapy The delivery of a gene(s) to a fertilized egg or an early embryonic cell. The transferred gene(s) is present in all nuclei of the cells of the mature individual, including the reproductive cells, and alters the phenotype of the developed individual.

β-1,3-Glucanase A plant enzyme produced in response to infection by fungal pathogens that hydrolyzes some components in fungal cell walls. Some bacteria can produce β-1,3-glucanase.

Glycosylation The covalent addition of sugar or sugar-related molecules to proteins or polynucleotides.

Gram-negative organism Any prokaryotic organism that does not retain the first stain (crystal violet) used in the Gram technique. It does retain the second stain (safranin O) and therefore has a pink color when viewed under a light microscope. Retention of the stain is due to the structure of the cell wall.

Gram-positive organism Any prokaryotic organism that retains the first stain (crystal violet) used in the Gram technique, which gives a purple-black color when viewed under a light microscope. Retention of the stain is due to the structure of the cell wall.

Grandfather method Knowledge of the X-linked alleles of a father is used to determine the genetic phase of a daughter whose sons can be examined directly as recombinant or nonrecombinant types for the genetic mapping of X-linked gene loci.

GRAS An acronym for "generally regarded as safe." In the United States, this designation has been given to foods, drugs, and other materials that have been used for a considerable period of time and have a history of not causing illness to humans, even though extensive toxicity testing has not been conducted. More recently, certain host organisms for recombinant DNA experimentation have been given this status.

Gratuitous inducer A substance that can induce transcription of a gene(s) but is not a substrate for the induced enzyme(s).

Guanine One of the organic bases found in either DNA or RNA.

HAC *See* Human artificial chromosome

Haploid One copy of each autosome and one sex chromosome.

Haplotype The alleles of the loci of a chromosome. Derived by combining *haplo* from haploid and *type* from genotype.

Haplotyping *See* Genotyping

Heat shock proteins The proteins synthesized in the nearly universal response of organisms to environmental stress such as high temperature.

Helper cells T cells that assist other cell types to respond immunologically to the presence of an antigen. Also called helper T lymphocytes.

Helper plasmid A plasmid that provides a function(s) to another plasmid in the same cell. Some helper plasmids are used to mobilize nonconjugative plasmids from a donor cell into a recipient cell.

Helper virus A virus that provides a function(s) to another virus or viral genome in the same cell.

Heterogeneity *See* Genetic heterogeneity

Heterologous From a different source, as in heterologous DNA.

Heterologous probe A DNA probe that is derived from one organism and used to screen for a similar DNA sequence in a clone bank derived from another organism.

Heterologous protein *See* Recombinant protein

Heteromer A protein with two or more different protein chains. Also called heteromeric polypeptide, heteromeric protein.

Heterozygote An individual that has different alleles at the same locus in its two homologous chromosomes.

HGP *See* Human Genome Project

High-resolution map Closely spaced sites throughout a genetic or physical map.

Holoenzyme A catalytically active enzyme containing all of the necessary cofactors and subunits.

Homodimer A protein with two identical polypeptide chains.

Homologous From the same source, or having the same evolutionary function or structure.

Homologous chromosomes Chromosomes that normally synapse during zygotene.

Homomer A protein with two or more identical protein chains. Also called homomeric polypeptide, homomeric protein.

Homopolymer A nucleic acid strand that is composed of one kind of nucleotide.

Homopolymeric tailing *See* Tailing

Homozygote An individual that has identical alleles at the same locus in its two homologous chromosomes.

Homozygous dominant A genotype with two dominant alleles at a gene locus.

Homozygous recessive A genotype with two recessive alleles at a gene locus.

Host A microorganism, organism, or cell that maintains a cloning vector.

***Hpa*II tiny fragments** Small DNA fragments that are generated from a genomic clone after treatment with the restriction endonuclease *Hpa*II, which cleaves within the CpG islands. Also called *Hpa*II-induced tiny fragments, HTFs.

HTFs *See Hpa*II tiny fragments

Human artificial chromosome A chromosome that is assembled from telomere, centromere, and human genomic DNA sequences. Also called HAC, microchromosome.

Human Genome Project An international research effort dedicated to developing high-resolution human genetic and physical maps and, eventually, the complete genomic DNA sequences of humans and other organisms. Also called HGP.

Human minisatellite DNA Human DNA that is noncoding and generally G+C rich and contains tandem repeats of short (9- to 40-base-pair) stretches of DNA.

Humanized Having segments of a cloned (antibody) gene, usually from mice, replaced by comparable regions of human DNA. Such a recombinant protein is less likely to be recognized as a foreign protein when it is used as a human therapeutic agent.

Humoral immune response The production of antibody by B cells of the immune system in response to the presence of a foreign antigen.

Hybrid gene The combination of two genes or the parts of two genes in the correct reading frame that encodes a single protein that has amino acid sequences from both genes.

Hybrid selection A protocol for determining which genomic clones hybridize to a cDNA or mRNA molecule.

Hybridization The pairing of two polynucleotide strands, often from different sources, by hydrogen bonding between complementary nucleotides.

Hybridoma The product of the fusion of a myeloma cell with an antibody-producing lymphocyte. This cell combination (hybridoma) can continue to divide in cell culture and secrete a single type of antibody.

Hydrogen uptake positive A term describing a microorganism that is capable of assimilating (or taking up) hydrogen gas. Also called Hup$^+$.

Hydromechanical stress *See* Shear

3′-hydroxyl end The hydroxyl group that is attached to the 3′ carbon atom of the sugar (ribose or deoxyribose) of the terminal nucleotide of a nucleic acid molecule.

Hypervariable region The parts of both the heavy and light chains of an antibody molecule that enable it to bind to a specific site on an antigen.

Hypervariable segment A region of a protein that varies considerably between strains or individuals.

IAA The plant hormone indole-3-acetic acid.

Ice-minus bacteria Bacteria that do not synthesize ice nucleation proteins.

Ice nucleation protein A protein around which ice crystals form.

Immediate-early gene A viral gene that is expressed promptly after infection.

Immune response The processes, including the synthesis of antibodies, that are used by vertebrates to respond to the presence of a foreign antigen.

Immunoaffinity chromatography A purification technique in which an antibody is bound to a matrix and is subsequently used to bind a specific protein and separate it from a complex mixture.

Immunoassay A protocol that uses antibody specificity to detect the presence of a particular compound in a biological sample.

Immunogen A substance that induces an antibody response. Also called antigen.

Immunoglobulin *See* Antibody

Immunosuppression A substance, an agent, or a condition that prevents or diminishes the immune response.

Immunotherapeutic procedure The use of an antibody or a fusion protein containing the antigen binding site of an antibody to treat a disease and enhance the well-being of a patient.

Immunotoxin A fusion protein that has separate domains with antibody and toxin activity. The antibody portion of the molecule facilitates binding to a target molecule or cell, and the toxin inactivates the target molecule or kills the cell.

Impeller An agitator that is used for mixing the contents of a bioreactor.

Inactivated agent A virus, bacterium, or other organism that has been treated to prevent it from causing a disease.

Inclusion body A protein that is overproduced in a recombinant bacterium and forms a crystalline array inside the bacterial cell.

Incompatibility group A classification scheme indicating which plasmids can coexist within a single cell. Plasmids must belong to different incompatibility groups to coexist within the same cell. Plasmids that belong to the same incompatibility group are unstable when placed into the same cell. A plasmid cloning vector should always belong to an incompatibility group different from that of the host bacterium's endogenous plasmids.

Incomplete penetrance When some individuals in a population have a specific genotype that causes an abnormality but are not affected.

Independent assortment The formation of all possible gene combinations in gametes with genes on different chromosomes, followed by the random joining of male and female gametes. Also called Mendel's second law of inheritance.

Indole-3-acetic acid *See* IAA

Inducer A low-molecular-weight compound or a physical agent that interacts with or alters a repressor protein and prevents it from blocking transcription.

Induction Turning on transcription of a specific gene or operon. The consequences of the action of an inducer.

Infectious agent Often, a proliferating virus, bacterium, or parasite that causes a disease in plants or animals.

Informatics The study of the management of information using computer and statistical techniques. The term bioinformatics is used by some authors when the information is from a biological source(s).

Initiation The start of the biosynthesis of a polymeric macromolecule.

Initiation codon The codon AUG, which specifies the first amino acid (methionine [N-formylmethionine in prokaryotes]) of a protein. Also called initiator, translational start codon, translational initiation signal.

Initiation complex The fMet-tRNA$_F^{Met}$–mRNA–small ribosomal subunit–large ribosomal subunit combination in prokaryotes or the Met-tRNAMet–mRNA–small ribosomal subunit–large ribosomal subunit combination in eukaryotes that is ready for the elongation phase of translation.

Initiator *See* Initiation codon

Initiator element *See* Response element

Initiator tRNA The fMet-tRNA$_F^{Met}$ in prokaryotes and Met-tRNAMet in eukaryotes that starts translation.

Insecticide A substance or living organism that kills insects.

Insert A DNA molecule that is incorporated into a cloning vector.

Integrating vector A vector that is designed to integrate cloned DNA into the host cell chromosomal DNA.

Integration Insertion of a DNA molecule (usually by homologous recombination) into a chromosomal site.

Integration–excision (I/E) region The portion of bacteriophage λ DNA that enables bacteriophage λ DNA to be inserted into a specific site in the *Escherichia coli* chromosome and excised from this site.

Interleukin-2 A lymphokine secreted by certain T lymphocytes that stimulates T-cell proliferation.

Internal ribosomal entry site A nontranslated sequence following a coding region of a polycistronic RNA that binds to a small ribosomal subunit and forms an initiation-of-translation complex.

Intervening sequence *See* Intron

Intron A segment of a gene that is transcribed but is then excised from the primary transcript during processing to a functional RNA molecule. Also called intervening sequence.

In vitro mutagenesis *See* Directed mutagenesis

In vitro translation Protein synthesis that is directed by either purified DNA with bacterial extracts or mRNA with wheat germ or rabbit reticulocyte extracts that provide ribosomes, tRNAs, and protein synthesis factors. The reaction mixture is often supplemented with ATP, GTP, and amino acids.

In vivo gene therapy The delivery of a gene(s) to a tissue or organ of an individual to alleviate a genetic disorder.

Ion channel An integral protein, within a cell membrane, that facilitates selective ion transport.

IPTG Isopropyl-β-D-thiogalactopyranoside, an inducer of the *lac* (lactose) operon. In recombinant DNA technology, IPTG is often used to induce cloned genes that are under the control of the *lac* repressor–*lac* promoter system.

IRES *See* Internal ribosomal entry site

Jumping library *See* Chromosome jumping

kb *See* Kilobase pair

k_{cat} The catalytic rate constant that characterizes an enzyme-catalyzed reaction. The higher the k_{cat}, the faster the conversion of substrate into product.

k_{cat}/K_m The catalytic efficiency of an enzyme-catalyzed reaction. The greater the value of k_{cat}/K_m, the more rapidly and efficiently the substrate is converted into product.

β-Ketoreductase An enzyme involved in the synthesis of polyketide antibiotics.

Ketosynthase A low-molecular-weight enzyme involved as part of a larger complex in polyketide biosynthesis.

Kilobase pair 1,000 base pairs; a unit of length of DNA. Also called kb.

Kindred A group of individuals who are related to each other either genetically or by marriage. Also called kinship.

Klenow fragment A product of proteolytic digestion of the DNA polymerase I from *E. coli* that retains both polymerase and 3′-exonuclease activities but not 5′-exonuclease activity.

K_m A dissociation constant that characterizes the binding of an enzyme to a substrate. The lower the K_m, the tighter the binding of the enzyme to the substrate. Also called Michaelis constant.

Knockout The targeted disruption of a gene by homologous recombination. Also gene knockout.

Label A compound or atom that is either attached to or incorporated into a macromolecule and is used to detect the presence of a compound, substance, or macromolecule in a sample. Also called tag.

Large ribosomal subunit The larger component of a ribosome.

Latent agent Usually a virus that is present in a host organism but does not produce any symptoms.

Leader peptide *See* Signal sequence

Leader sequence A sequence of nucleotides at the 5′ end of an mRNA that is not translated into protein.

Library *See* Gene bank

Ligase chain reaction A technique for determining the presence or absence of a specific nucleotide pair within a target gene. Also called LCR.

Ligation Joining of two DNA molecules by the formation of phosphodiester bonds. In vitro, this reaction is usually catalyzed by the enzyme T4 DNA ligase.

Lignocellulose The combination of lignin, hemicellulose, and cellulose that forms the structural framework of plant cell walls.

Linkage The occurrence of two or more genes on the same chromosome.

Linkage map *See* Genetic map

Linker A synthetic double-stranded oligonucleotide that carries the sequence for one or more restriction endonuclease sites.

Lipase An enzyme that degrades lipids.

Lipofection Delivery into eukaryotic cells of DNA, RNA, or other compounds that have been encapsulated in an artificial phospholipid vesicle.

Lipopolysaccharide A compound containing lipid bound to a polysaccharide; often a component of microbial cell walls. Also called LPS.

Liposome A circular collection of lipid molecules in which the hydrophobic portions of the molecule are facing inward; a lipid vesicle with an aqueous interior that can carry nucleic acids, drugs, or other therapeutic agents.

Liquefaction Enzymatic digestion (often by α-amylase) of gelatinized starch to form lower-molecular-weight polysaccharides.

Live vaccine (1) A living, nonvirulent form of a microorganism or virus that is used to elicit an antibody response that protects the inoculated organism against infection by a virulent form of the microorganism or virus. (2) A living, nonvirulent microorganism or virus that expresses a foreign antigenic protein and is used to inoculate a human or animal. The latter type of organism is also called a live recombinant vaccine.

Locus The site on the chromosome where a specific gene is located.

Lod score The logarithm of the ratio of the likelihood of the odds that two loci are linked with a recombination fraction equal to or greater than 0 and less than 0.5, to the likelihood of the odds for independent assortment. Also called LOD score, Z, Z score.

Long template A DNA strand that is synthesized during the polymerase chain reaction, has a primer sequence at one end, and is extended beyond the site that is complementary to the second primer at the other end.

Long terminal repeats Similar blocks of genetic information that are found at the ends of the genomes of retroviruses. Also called LTRs.

LTR *See* Long terminal repeats

Lymphoma Cancer originating in the lymph nodes, spleen, and other lymphoreticular sites.

Lysis The destruction or breakage of cells by (1) viruses or (2) chemical or physical treatment.

Lysogeny A condition in which a bacteriophage genome (prophage) survives within a host bacterium either as part of the host chromosome or as part of an extrachromosomal element and does not initiate lytic functions.

Lytic cycle The process of viral production that usually leads to cell lysis.

M13 + strand The single-stranded DNA molecule that is present in infective bacteriophage M13.

MAb *See* Monoclonal antibody

Macromolecule A molecule with a high molecular mass such as a nucleic acid, protein, or polysaccharide.

Macrophage A large phagocytic white blood cell.

Maltose binding protein An abundant bacterial protein located within the periplasmic space and involved in the uptake of maltose. Fusion with other proteins often results in the fusion protein being secreted to the periplasm.

Map distance The number of map units or centimorgans between two gene loci.

Map unit A measure of genetic distance between two linked genetic loci; usually 1% recombination corresponds to one map unit.

Marker An identifiable DNA sequence on a chromosome. Also called marker site, marker locus, genetic marker.

Marker density The spacing of marker sites along a chromosome.

Marker peptide A portion of a fusion protein that facilitates its identification or purification.

Mb *See* Megabase pair

MCS *See* Polylinker

Megabase pair 1,000,000 base pairs; a unit of length of DNA. Also called Mb.

Meristematic tissue Plant tissue that is actively dividing. In young plants, it is usually found at the tips of the stem and the roots.

Mesophile A microorganism that is able to grow within the temperature range of 20 to 50°C; optimal growth often occurs at about 37°C.

Messenger RNA An RNA molecule carrying the information that, during translation, specifies the amino acid sequence of a protein molecule. Also called mRNA.

Metabolite (1) A low-molecular-weight biological compound that is usually synthesized by an enzyme. (2) A compound that is essential for a metabolic process.

Methylation The addition of a methyl group(s) to a macromolecule. For example, the addition of a methyl group to specific cytosine and, occasionally, adenine residues in DNA.

Michaelis constant *See* K_m

Microchromosome *See* Human artificial chromosome

Microinjection The introduction of DNA or other compounds into a single eukaryotic cell with a fine, microscopic needle.

Microprojectile bombardment *See* Biolistics

Mismatch The lack of base pairing between one or more nucleotides of two hybridized nucleic acid strands.

Missense mutation A genetic mutation that changes a codon for one amino acid into a codon specifying another amino acid.

Mobilizing functions The genes on a plasmid that facilitate the transfer of either a nonconjugative or conjugative plasmid from one bacterium to another.

Modification (1) Enzymatic methylation of a restriction enzyme DNA recognition site. (2) Specific nucleotide changes in DNA or RNA molecules.

Monochromosomal Refers to the presence of a single human chromosome in a somatic cell hybrid line.

Monoclonal antibody A single type of antibody that is directed against a specific epitope (antigenic determinant) and is produced by a hybridoma cell line, which is formed by the fusion of a lymphocyte with a myeloma cell. Some myeloma cells synthesize single antibodies naturally. Also called MAb.

Monocotyledon A class of plants that has one seed leaf. Also called monocot.

Monomer A unit of a polymer.

mRNA *See* Messenger RNA

Multigene RNA An RNA transcript of an operon.

Multiple cloning site *See* Polylinker

Multipoint linkage analysis The determination of the order and map distances of many loci on a chromosome at one time. Also called multilocus mapping, multipoint mapping, multilocus linkage analysis.

Multivalent vaccine A single vaccine that is designed to elicit an immune response either to more than one infectious agent or to several different epitopes of a molecule.

2μm plasmid A naturally occurring, double-stranded, circular DNA plasmid (6,318 bp) found in the nuclei of *Saccharomyces cerevisiae.* Many yeast plasmid vectors are derived from the 2μm plasmid. Also called 2μm circle, 2μ plasmid, 2-micron plasmid.

Must The juice from crushed grapes before fermentation; used in wine making.

Mutagenesis Chemical or physical treatment that changes the nucleotides of the DNA of an organism.

Mutant An organism that differs from the wild type because it carries one or more genetic changes in its DNA. Also called a variant.

Mutant complementation *See* Genetic complementation

Mutation A change of one or more nucleotide pairs of a DNA molecule.

Mutation detection assay A protocol that identifies the difference of one or a few nucleotide pairs between the same DNA molecules from different sources.

Mycelium A mass of interwoven thread-like filaments of a fungus or bacteria.

Myeloma A plasma cell cancer.

N terminus The first amino acid(s) of a protein. Also called amino terminus, amino-terminal end.

Narrow-host-range plasmid A plasmid that can replicate in one, or at most a few, different bacterial species.

Native protein The naturally occurring form of a protein.

Negative control A system of regulation of transcription that requires the removal of a repressor from an operator.

Neomycin phosphotransferase An enzyme that inactivates the antibiotics neomycin and kanamycin. This enzyme is often used as a selective marker for transgenic plants.

Nested primers Two or more sets of pairs of polymerase chain reaction primers that bind to neighboring DNA sequences.

Nick (1) To break a phosphodiester bond in the backbone of one of the strands of a duplex DNA molecule. (2) A break in the backbone of one of the strands of duplex DNA.

Nitrogen fixation The conversion of atmospheric nitrogen to ammonia. Biological nitrogen fixation is catalyzed by the enzyme nitrogenase, which is found only in prokaryotes.

Nod box A DNA sequence that controls the transcriptional regulation of *Rhizobium* nodulation genes.

Nodulation The formation of nodules by symbiotic bacteria on the roots of plants.

Nonautologous From a different species or individual.

Nonvirulent agent *See* Attenuated vaccine

Northern blotting Similar to Southern blotting, except that RNA that has been separated by gel electrophoresis is transferred from a gel onto a matrix such as a nitrocellulose or nylon membrane, and the presence of a specific RNA molecule is detected by DNA–RNA hybridization.

Nuclear cloning Production of an organism by placing a nucleus from a somatic cell into an enucleated fertilized egg. Also nuclear transfer.

Nucleoside A base (purine or pyrimidine) that is covalently linked to a five-carbon (pentose) sugar. When the sugar is ribose, the nucleoside is a ribonucleoside; when it is deoxyribose, the nucleoside is a deoxyribonucleoside.

Nucleotide A nucleoside with one or more phosphate groups linked to the 5′ carbon of the pentose sugar. Ribose-containing nucleosides are often called ribonucleoside monophosphate (NMP), ribonucleoside diphosphate (NDP), or ribonucleoside triphosphate (NTP). When the nucleoside contains the sugar deoxyribose, the nucleotides are called deoxyribonucleoside mono-, di-, or triphosphates (dNMP, dNDP, or dNTP).

Off-state When a gene is not being transcribed.

Oleosins Hydrophobic oil body proteins associated with plant seeds.

Oligonucleotide A short molecule (usually 6 to 100 nucleotides) of single-stranded DNA. Oligonucleotides are sometimes called oligodeoxyribonucleotides or oligomers and are usually synthesized chemically.

Oligonucleotide ligation assay A diagnostic technique for determining the presence or absence of a specific nucleotide pair within a target gene which indicates whether a gene is wild type (normal) or mutant (defective). Also called OLA.

Oligonucleotide-directed mutagenesis *See* Site-specific mutagenesis

On-state When a gene is being transcribed.

Oncogene A gene that plays a role in the cell division cycle. Often mutated forms of oncogenes cause a cell to grow in an uncontrolled manner.

Open reading frame A sequence of nucleotides in a DNA molecule that encodes a peptide or protein. This term is often used when, after the sequence of a DNA fragment has been determined, the function of the encoded protein is not known. Also called ORF.

Operator The region of DNA that is upstream from a prokaryotic gene(s) and to which a repressor or activator binds.

Operon A cluster of genes that are coordinately regulated.

Opine The condensation product of an amino acid with either a keto acid or a sugar.

ori *See* Origin

Origin The nucleotide sequence at which DNA synthesis is initiated. Also called origin of replication, *ori*.

Origin of replication *See* Origin

Osmolyte A compound that regulates the osmotic pressure within a cell.

Outflow The volume of growing cells that is removed from a bioreactor during a continuous fermentation process.

Overhang *See* Extension

P site *See* Peptidyl site

P1 artificial chromosome A plasmid vector system based on bacteriophage P1 that uses electroporation for introducing a vector with a large DNA insert (100 to 300 kb) into *Escherichia coli.* Also called PAC.

P1 cloning system A plasmid vector system based on bacteriophage P1 that uses in vitro bacteriophage P1 packaging for introducing a vector with a large DNA insert (80 to 100 kb) into *Escherichia coli.*

PAC *See* P1 artificial chromosome

Packaging cell line A cell line that is designed to produce viral particles that do not contain infective nucleic acid. This process has been described as "putting a sheep in wolf's clothing."

Palindromic sequences Complementary DNA sequences that are the same when each strand is read in the same direction (e.g., 5′ to 3′). These types of sequences serve as recognition sites for type II restriction endonucleases.

Panicle A pyridimal, loosely branched flower cluster.

Parasporal crystal Tightly packaged insect protoxin molecules that are produced by strains of *Bacillus thuringiensis* during the formation of resting spores.

Partial digest Treatment of a DNA sample with a type II restriction endonuclease under conditions that result in a limited number of cuts in each DNA molecule to yield many possible combinations of cleaved pieces in the final sample.

Passaging Subculturing cells that are growing in vitro.

Passive immunity (1) Natural acquisition of antibodies by the fetus or newborn from the mother. (2) The artificial introduction of specific antibodies by the injection of serum from an immune animal. In both cases, it confers temporary protection on the recipient.

Patatin A storage protein commonly found in potatoes.

Patent A government-issued document that allows the holder the exclusive right to manufacture, use, or sell an invention for a defined period, usually 20 years.

Pathogenesis-related promoter The promoter for a plant gene whose transcription is activated upon infection of the plant by pathogens.

PCR *See* Polymerase chain reaction

Pedigree A diagrammatic representation of the history of a trait in a multigeneration family.

Penetrance A certain fraction, usually expressed as a percentage, of individuals with a mutant gene(s) at a locus that show the expected abnormal (mutant) phenotype. Also called incomplete penetrance, nonpenetrance, reduced penetrance.

Peptide A short chain of amino acids that are linked with peptide bonds.

Peptide bond The covalent bond between the free carboxyl group of the α carbon of one amino acid and the free amino group of the α carbon of an adjacent amino acid in a peptide or protein.

Peptide vaccine A short chain of amino acids that can induce antibodies against a specific infectious agent.

Peptidyl site The portion of a ribosome where the tRNA with the peptide chain participates in peptide bond formation with the aminoacyl-tRNA during translation. Also called P site.

Peptidyl–tRNA The tRNA that has a growing peptide chain attached to it during translation.

Periplasm The space (periplasmic space) between the cell (cytoplasmic) membrane of a bacterium or fungus and the outer membrane or cell wall.

Phage *See* Bacteriophage

Pharmaceutical agent *See* Therapeutic agent

Phase state The coupling or repulsion of two linked genes.

Phase-known A double heterozygous genotype that is known to be in either a *cis* or *trans* configuration.

Phase-unknown The gene arrangement of a double heterozygous genotype that is not known.

Phenocopy A nongenetic condition that closely resembles a genetic disorder.

Phenotype An observable feature or set of traits that is determined by a gene or combination of genes of an organism.

Phenylketonuria An autosomal recessive disorder in humans that is due to the lack of the liver enzyme phenylalanine hydroxylase and that causes phenylalanine to accumulate. Also called PKU.

5′-phosphate end The phosphate group that is attached to the 5′ carbon atom of the sugar (ribose or deoxyribose) of the terminal nucleotide of a nucleic acid molecule.

Phosphodiester bond The linkage of a phosphate group to the 3′ carbon of one nucleotide and the 5′ carbon of another nucleotide; the linkage between nucleotides of the same nucleic acid strand.

Phosphorothioate linkage The linkage between nucleotides after a sulfur group replaces an available oxygen of a phosphodiester linkage.

Photosynthetic Able to convert atmospheric carbon dioxide into organic compounds. Nearly all plants, most algae, and some bacteria are photosynthetic.

Physical map A map of the positions of chromosome sites such as restriction endonuclease recognition or sequence tagged sites on a chromosome. The distance between sites is measured in base pairs.

Phytohormone A substance that stimulates growth or other processes in plants; a plant hormone, e.g., auxin, cytokinin, gibberellin, ethylene, or abscisic acid.

Phytopathogen An organism, such as a fungus, bacterium, or virus, that causes disease in plants.

PKU *See* Phenylketonuria

Plaque A clear area that is visible in a bacterial lawn on an agar plate and is due to lysis of the bacterial cells by bacteriophage.

Plasmid An autonomous, self-replicating extrachromosomal DNA molecule.

Plasmovirus A combined retrovirus packaging-and-vector system that is confined to a plasmid.

Plastid In plants, a double-membrane-bound organelle such as a chloroplast.

Pluripotent *See* Totipotent

Pneumatic reactor *See* Airlift fermenter

Pollen Microspores of plants that carry male gametes.

Poly(3-hydroxybutyric acid) The best studied and characterized of the polyhydroxyalkanoates.

Polyadenylation The addition of adenine residues to the 3′ end of eukaryotic mRNAs. Also called poly(A) tailing. The adenine-rich 3′-terminal segment is called a poly(A) tail.

Polyadenylation signal A sequence that terminates transcription and provides a recognition site at the end of an mRNA for the enzymatic addition of adenine residues.

Poly(A) tail *See* Polyadenylation

Polycistronic RNA An mRNA that encodes two or more proteins.

Polyclonal antibody A serum sample that contains antibodies that bind to different antigenic determinants of one antigen.

Polyhedron A structure in which baculovirus genomes are inside a cylindrical nucleocapsid, which is enclosed in an occlusion body composed of the protein polyhedrin.

Polyhydroxyalkanoate Biodegradable polymers produced by microorganisms as a carbon and energy storage material.

Polyketide A class of antibiotics

Polyketide synthase An enzyme involved in the biosynthesis of polyketide antibiotics

Polylinker A synthetic DNA sequence that contains a number of different restriction endonuclease sites. Also called multiple cloning site, MCS.

Polymer A macromolecule made up of a series of covalently linked monomers.

Polymerase chain reaction A technique for amplifying a specific segment of DNA by using a thermostable DNA polymerase, deoxyribonucleotides, and primer sequences in multiple cycles of denaturation, renaturation, and DNA synthesis. Also called PCR.

Polymorphic probe An assay system that identifies a chromosome site with two or more allelic DNA sequences that each occur with a frequency of 1% (0.01) or greater in a large population.

Polymorphic site A chromosome location that has two or more identifiable allelic DNA sequences that each occur with a frequency of 1% (0.01) or greater in a large population. Also called polymorphic locus.

Polynucleotide A linear series of 20 or more nucleotides linked by phosphodiester bonds.

Polypeptide A linear series of amino acids linked together with peptide bonds. Also called protein, protein chain.

Positional gene cloning A strategy for isolating an unknown disease gene. The disease gene is mapped to a chromosome site. A contig or genomic clone that covers the site of the disease gene is tested for exons. Examination of exons and mutation detection assays establish which gene is the disease gene.

Positional-candidate gene cloning A strategy for isolating an unknown disease gene. The disease gene is mapped to a chromosome site. The genes and expressed sequence tags that have been mapped to the same interval (bin) as the disease gene become candidates for the disease gene. Examination of exons and mutation detection assays establish which gene is the disease gene.

Positive control A system of regulation of transcription that requires the addition of a protein activator to an activator site on the DNA.

Positive selectable marker *See* Dominant marker selection

Positive selection *See* Dominant marker selection

Positive-negative selection A protocol that both selects for cells that carry a DNA insert integrated at a specific targeted chromosomal location (positive selection) and selects against cells that carry a DNA insert integrated at a nontargeted chromosomal site (negative selection).

Posttranslational modification The specific addition of phosphate groups, sugars (glycosylation), or other molecules to a protein after it has been synthesized.

Preventive immunization Injection of an antigen to elicit an antibody response that will protect the organism against future infections. Also called vaccination.

Primary antibody The antibody that binds to the target molecule in an ELISA or other immunological assay.

Primary cell culture A population of growing cells that is started directly from a tissue or cells of an organism.

Primary transcript Unprocessed RNA that is transcribed from a eukaryotic structural gene that has exons and introns.

Primer A short oligonucleotide that hybridizes with a template strand and provides a 3′-hydroxyl end for the initiation of nucleic acid synthesis.

Primer walking A method for sequencing long (>1-kb) cloned pieces of DNA. The initial sequencing reaction reveals the sequence of the first few hundred nucleotides of the cloned DNA. On the basis of these data, a primer that contains about 20 nucleotides and is complementary to a sequence near the end of sequenced DNA is synthesized and used for sequencing of the next few hundred nucleotides of the cloned DNA. This procedure is repeated until the complete nucleotide sequence of the cloned DNA is determined.

Probe (1) For diagnostic tests, the agent that is used to detect the presence of a molecule in a sample. (2) A DNA sequence that is used to detect the presence of a complementary sequence by hybridization with a nucleic acid sample.

Prodrug An inactive compound that is converted into a pharmacological agent by an in vivo metabolic process.

Productivity The amount of product that is produced in a bioreactor within a given period of time.

Progeny The offspring of a mating.

Prokaryotes Organisms, usually bacteria, that have neither a membrane-bound nucleus enclosing their chromosomes nor functional organelles such as mitochondria and chloroplasts.

Promoter A segment of DNA to which RNA polymerase attaches. It usually lies upstream of (5′ to) a gene. A promoter sequence aligns the RNA polymerase so that transcription will initiate at a specific nucleotide.

Prophage A repressed or inactive state of a bacteriophage genome that is maintained in a bacterial host cell as part of the chromosomal DNA.

Protease An enzyme that hydrolyzes peptide bond linkages and cleaves proteins into smaller peptides. Also called proteinase or proteolytic enzyme.

Protease inhibitor A protein that can form a tight complex with a protease and block its activity.

Protein *See* Polypeptide

Protein chain *See* Polypeptide

Protein drug *See* Therapeutic agent

Proteolysis Enzymatic degradation of a protein.

Proteolytic Hydrolytic breakage of a peptide bond of a protein.

Protocorm Precursor form of a corm, an underground plant propagule.

Protoplast A bacterial, yeast, or plant cell that has had its cell wall removed either chemically or enzymatically.

Protoxin A latent, nonactive precursor form of a toxin.

Pseudomonas A common gram-negative bacterial genus that is widely distributed. Many of the soil forms produce a pigment that fluoresces under ultraviolet light, hence the descriptive term fluorescent pseudomonads.

Pseudotype formation The packaging of the genome of one virus in the envelope or capsid protein of another virus. Also called phenotypic mixing.

Psychrophile A microorganism that can grow at temperatures as low as 0 to 5°C.

Purine Fusion of a pyrimidine and an imidazole ring, e.g., adenine or guanine.

Pyrimidine A heterocyclic ring, e.g., thymine, cytosine, or uracil.

Pyrogen A bacterial substance that causes fever in humans.

Quencher The portion of a molecule that can quench fluorescence.

RACE Rapid amplification of cDNA ends. Polymerase chain reaction methods are used to generate double-stranded DNA (cDNA) from the 3′ end (3′ RACE) or the 5′ end (5′ RACE) of a specific mRNA.

Random amplified polymorphic DNA A diagnostic procedure in which chromosomal DNA (usually from plants but sometimes from microorganisms or animals) is characterized by the DNA fragments that are synthesized when PCR is initiated after the addition of a single primer to the reaction mixture. Also called RAPD.

Random mutagenesis A nondirected change of a nucleotide pair(s) in a DNA molecule.

Random primer method A protocol for labeling DNA in vitro. A sample of random oligonucleotides containing all possible combinations of nucleotide sequences is hybridized to a DNA probe. Then, in the presence of a DNA polymerase and the four deoxyribonucleotides (one of which is labeled), the 3′-hydroxyl ends of the hybridized oligonucleotides provide initiation sites for DNA synthesis that uses the separated strands of the probe DNA as a template. This reaction produces labeled copies of portions of the probe DNA.

RAPD *See* Random amplified polymorphic DNA

Reading frame A series of codons that code for amino acids in a nucleotide sequence.

Read-through Transcription or translation that proceeds beyond the normal stopping point because the transcription or translation termination signal of a gene is absent or mutated.

RecA A protein, found in most bacteria, that is essential for DNA repair and DNA recombination.

Recessive gene An allele that does not demonstrably contribute to the phenotype in a heterozygote.

Recognition site *See* Restriction site

Recombinant An individual with two or more linked genes that are a consequence of one or more crossover events.

Recombinant DNA technology *See* Gene cloning

Recombinant protein A protein whose amino acid sequence is encoded by a cloned gene.

Recombinant toxin A single multifunctional toxic protein that has been created by combining the coding regions of various genes.

Recombinant vaccine *See* Subunit vaccine

Recombination *See* Crossover

Recombination frequency The number of recombinant individuals (or chromosomes) among a set of offspring divided by the total number of individuals (or chromosomes) that constitute the offspring population. In human genetics, it is represented by the Greek letter theta (θ).

Regulatory protein A protein that plays a role in either turning on or turning off transcription.

Relative molecular mass The sum of the atomic weights of all the atoms in a molecule; the mass of a molecule relative to the dalton. Molecular weight is a synonym for relative molecular mass.

Remedial *See* Therapeutic

Renaturation The reassociation of two nucleic acid strands after denaturation.

Replacement therapy The administration of metabolites, cofactors, or hormones that are deficient as the result of a genetic disease.

Replica plating The transfer of cells from bacterial colonies on one petri plate to another petri plate; the locations of the colonies that grow on the second plate correspond to those on the original (master) petri plate.

Replicatable biological unit Any biological entity that is capable of being reproduced.

Replication The process of DNA synthesis.

Replicative form The molecular configuration of viral nucleic acid that is the template for replication in a host cell. Also called RF.

Reporter gene A gene that encodes a product that can readily be assayed. For example, reporter genes are used to determine whether a particular DNA construct has been successfully introduced into a cell, organ, or tissue.

Repression Inhibition of transcription by preventing RNA polymerase from binding to the transcription initiation site; a repressed gene is "turned off."

Repressor A protein that binds to the operator or promoter region of a gene and prevents transcription by blocking the binding of RNA polymerase.

Repulsion The phase state in which a dominant version and a recessive version of two different genes occur on the same chromosome. Also called *trans* configuration. *See also* Coupling

Response element A sequence of deoxyribonucleotides of a gene that acts as a binding site for a protein (transcription factor) that regulates transcription. Also called initiator element, signal region.

Restenosis Recurring narrowing of a biological opening, tube, or canal.

Restriction endonuclease (type II) An enzyme that recognizes a specific duplex DNA sequence and cleaves phosphodiester bonds on both strands between definite nucleotides.

Restriction fragment length polymorphism The occurrence of variations in the lengths of certain DNA fragments that are produced after cleavage with a type II restriction endonuclease. The differences in DNA lengths are due to the presence or absence of a specific restriction endonuclease recognition site(s) and are detected by DNA hybridization with DNA probes after separation by gel electrophoresis. Also called RFLP.

Restriction map The linear array of restriction endonuclease sites on a DNA molecule.

Restriction site The sequence of nucleotide pairs in duplex DNA that is recognized by a type II restriction endonuclease. Sometimes called restriction enzyme site, restriction endonuclease site, or recognition site.

Retrovirus A class of eukaryotic RNA viruses that can form double-stranded DNA copies of their genomes; the double-stranded forms can integrate into chromosomal sites of an infected cell.

Reverse transcriptase An RNA-dependent DNA polymerase that uses an RNA molecule as a template for the synthesis of a complementary DNA strand.

Reverse transcription–polymerase chain reaction A two-step protocol for synthesizing cDNA molecules. First, cDNA strands are synthesized in vitro by reverse transcriptase with oligo(dT) as a primer and mRNA as the template. Second, a specific cDNA strand is amplified by the polymerase chain reaction (PCR), with one primer directed to a sequence of the first cDNA strand and the other to a sequence of the complementary cDNA strand (second strand) that is synthesized during the first PCR cycle. Also called RT–PCR.

RFLP *See* Restriction fragment length polymorphism

Rhizobacterium A microorganism whose natural habitat is near, on, or in plant roots.

Rhizosphere The zone in the immediate vicinity of growing plant roots.

Ribonucleic acid *See* RNA

Ribose The 5-carbon sugar component of RNA.

Ribosomal binding site A sequence of nucleotides near the 5′-phosphate end of a bacterial mRNA that facilitates the binding of the mRNA to the small ribosomal subunit. Also called Shine-Dalgarno sequence.

Ribosomal RNA The RNA molecules that form part of the large and small ribosomal subunits. Also called rRNA.

Ribosome The subcellular structure that contains both RNA and protein molecules and mediates the translation of mRNA into protein. Ribosomes contain both large and small subunits.

Ribozyme An RNA molecule that has catalytic activity.

Ribulose bisphosphate carboxylase The most abundant enzyme in the world, it is found in all green plants and is responsible for the fixation of carbon dioxide in photosynthesis. Sometimes called Rubisco.

RNA Ribonucleic acid; a polynucleotide that has ribose as its pentose sugar and uracil as one of its pyrimidines.

RNA polymerase An enzyme that links an incoming ribonucleotide, which is determined by complementarity to a base in a template DNA strand, with a phosphodiester bond to the 3′-hydroxyl group of the last incorporated ribonucleotide of the growing RNA strand during transcription.

Rolling circle A mode of DNA replication that produces concatemeric duplex DNA.

Root nodule A small round mass of cells that is located on the roots of plants and contains nitrogen-fixing bacteria.

rRNA *See* Ribosomal RNA

RT-PCR *See* Reverse transcription–polymerase chain reaction

Saccharifaction Hydrolysis of polysaccharides, after liquefaction, by glucoamylase to maltose and glucose.

Scale-up Conversion of a process such as fermentation of a microorganism from a small scale to a larger scale.

SCP *See* Single-cell protein

SDS *See* Sodium dodecyl sulfate

SDS-PAGE Sodium dodecyl sulfate-polyacrylamide gel electrophoresis.

Secondary antibody In an ELISA or other immunological assay system, the antibody that binds to the primary antibody. The secondary antibody is often conjugated with an enzyme such as alkaline phosphatase.

Secondary metabolite A compound that is not necessary for growth or maintenance of cellular functions but is synthesized, generally for the protection of a cell or microorganism, during the stationary phase of the growth cycle.

Secrete *See* Export

Secretion The passage of a molecule from the inside of a cell through a membrane into the periplasmic space or the extracellular medium.

Selectable Having a gene product that, when present, enables a researcher to identify and preferentially propagate a particular cell type.

Selection (1) A system for either isolating or identifying specific organisms in a mixed culture. (2) Survival of a more reproductively fit organism.

Self-incompatibility In plants, the inability of the pollen to fertilize ovules (female gametes) of the same plant. Also called self-sterility.

Self-replicating elements Extrachromosomal DNA elements that have origins of replication for the initiation of their own DNA synthesis.

Self-sterility *See* Self-incompatibility

Senescence The last stage in the postembryonic development of multicellular organisms, during which loss of functions and degradation of biological components occur. Also called biological aging.

Sensitivity For diagnostic tests, the ability of an assay to detect the smallest amount of the target molecule.

Sequence tagged site A short (200- to 500-bp) DNA sequence that occurs once in the genome and is identified by PCR amplification. Also called STS.

Sequence tagged site content mapping Markers that are based on unique polymerase chain reaction primers that are used to determine shared sites among clones of a library, which in turn facilitates the assembly of a contig.

Serotype Classification of an organism or protein on the basis of its interaction with antibodies.

Shear The sliding of one layer across another, with deformation and fracturing in the direction parallel to the movement. This term usually refers to the forces that cells are subjected to in a bioreactor or a mechanical device used for cell breakage.

Shine-Dalgarno sequence *See* Ribosomal binding site

Short tandem repeat A DNA sequence with a sequential repeating set of two (di-), three (tri-), or four (tetra-) nucleotide pairs. Also called STR.

Short tandem repeat polymorphism Two or more short tandem repeats that differ from each other by the number of repeated nucleotide units with each allelic DNA sequence occurring at a frequency of 1% or more in a large population. The difference in the lengths of allelic short tandem repeats is detected by gel electrophoresis following the polymerase chain reaction with primers that are complementary to unique DNA sequences that flank a short tandem repeat. Also called STRP.

Short template A DNA strand that is synthesized during the polymerase chain reaction and has a primer sequence at one end and a sequence complementary to the second primer at the other end.

Shuttle vector A plasmid cloning vehicle, usually a plasmid, that can replicate in two different organisms because it carries two different origins of replication. Also called bifunctional vector.

Sibs Brothers and/or sisters; full-sibs and half-sibs who have both biological parents or one biological parent in common, respectively. Also called siblings.

Sibship All the sibs who have at least one common biological parent.

Siderophore A low-molecular-weight substance that binds very tightly to iron. Siderophores are synthesized by a variety of soil microorganisms and plants to ensure that the organisms can obtain sufficient amounts of iron from the environment.

Sigma factor An accessory bacterial protein(s) that directs the binding of RNA polymerase to specific promoters.

Signal peptide *See* Signal sequence

Signal region *See* Response element

Signal sequence A segment of about 15 to 30 amino acids at the N terminus of a protein that enables the protein to be secreted (pass through a cell membrane). The signal sequence is removed as the protein is secreted. Also called signal peptide, leader peptide.

Signal-to-noise ratio Comparison of the extent of the response to an assay when the target entity is present (signal) in a sample to the extent when it is absent (noise) from the sample.

Silage Cattle feed that has been allowed to ferment.

Simplicity For diagnostic tests, the ease with which an assay can be implemented.

Single-cell protein Dried mass of a pure sample of a protein-rich microorganism, which may be used either as feed (for animals) or as a food (for humans). Also called SCP.

Single-site mutation A change in one base pair in DNA. Also point mutation.

Single-strand conformation analysis A mutation detection assay that is based on the conformation of single strands of DNA. If there is a nucleotide difference between the DNA molecules from two different sources, then following denaturation and gel electrophoresis the locations of the single strands in the two lanes of the gel will be different. Also called single-strand conformational analysis, single-strand conformation(al) polymorphism, SSCA, SSCP.

Site-specific mutagenesis A technique to change one or more specific nucleotides in a cloned gene in order to create an altered form of a protein with a specific amino acid change(s). Also called oligonucleotide-directed mutagenesis.

Six-cutter A type II restriction endonuclease that binds and cleaves DNA at sites that contain six nucleotide pairs.

Size markers A set of macromolecules of known relative molecular masses that are used to calculate the relative molecular masses of electrophoretically fractionated macromolecules.

Small ribosomal subunit The smaller component of a ribosome.

Sodium dodecyl sulfate An anionic detergent that denatures proteins.

Somatic cell Any cell of a multicellular organism that does not produce gametes.

Somatic cell gene therapy The delivery of a gene(s) to a tissue other than reproductive cells of an individual, with the aim of correcting a genetic defect.

Somatic cell hybrid panel A set of derived chromosome-specific hybrid cell lines that each carry a different portion of a particular chromosome. The members of such a panel have chromosomal deletions and, in some cases, carry translocated chromosomes that retain a segment of a particular chromosome. Ideally, the retained portions of the cell lines of a panel cover the entire chromosome. Also called somatic cell hybrid mapping panel, somatic cell deletion panel.

S1 nuclease An enzyme that specifically degrades single-stranded DNA.

Sonication Disruption of cells or DNA molecules by high-frequency sound waves. Also called ultrasonication.

Source DNA The DNA from an organism that contains a target gene; this DNA is used as starting material in a cloning experiment.

Source organism An organism (e.g., a bacterium, plant, or animal) from which DNA is purified and used in a cloning experiment.

Southern blotting A technique for transferring denatured DNA molecules that have been separated electrophoretically from a gel to a matrix (such as a nitrocellulose or nylon membrane) on which a hybridization assay can be performed.

Sparger A device that introduces air into a bioreactor in the form of separate, fine streams.

Specificity For diagnostic tests, the ability of a probe to react precisely with a specific target molecule.

Splice site The nucleotides at (1) the end of an exon and the beginning of an intron and (2) the end of an intron and the beginning of the next exon that are required for the joining of two exons and removal of an intron during the processing of a primary transcript to a functional mRNA.

Splice site mutation Loss or gain of a functional splice site that affects the proper removal of introns during mRNA processing.

Spore A small, protected reproductive form of a microorganism, often produced when nutrient levels are low.

Sporulation Formation of spores or resting structures, usually after the near depletion of nutrients from the growth medium, by some bacteria or fungi.

Staggered cuts Symmetrically cleaved phosphodiester bonds that lie on both strands of duplex DNA but are not opposite one another.

Start codon *See* Initiation codon

Steady state In a continuous fermentation process, the condition when the number of cells that are removed with the outflow is exactly balanced by the number of newly synthesized cells.

Stem cell A precursor cell that undergoes division and gives rise to lineages of differentiated cells.

Stenosis Narrowing of a biological opening, tube, or canal.

Sticky ends *See* Cohesive ends

Stirred tank fermenter A growth vessel in which cells or microorganisms are mixed by mechanically driven impellers.

STR *See* Short tandem repeat

Strain A microorganism or multicellular organism that is a genetic variant of a standard parental stock.

Strand A linear series of nucleotides that are linked to each other by phosphodiester bonds.

STRP *See* Short tandem repeat polymorphism

Structural gene A sequence of DNA that encodes a protein.

STS *See* Sequence tagged site

Subcloning Splicing part of a cloned DNA molecule into a different cloning vector.

Subspecies Population(s) of organisms sharing certain characteristics that are not present in other populations of the same species.

Substrate (1) A compound that is altered by an enzyme. (2) A food source for growing cells or microorganisms.

Subtilisin A proteolytic enzyme usually found in *Bacillus subtilis.*

Subunit vaccine An immunogenic protein(s) either purified from the disease-causing organism or produced from a cloned gene.

Sucrose density gradient centrifugation A procedure used to fractionate mRNAs or DNA fragments on the basis of size.

Suicide gene A plasmid-borne, inducible sequence that produces a protein that directly or indirectly kills the host cell.

Superbug Jargon for the bacterial strain of *Pseudomonas* developed by Dr. A. Chakrabarty, who combined hydrocarbon-degrading genes carried on different plasmids into one organism. Although this genetically engineered microorganism is neither "super" nor a "bug," it represents a landmark example because it showed how genetically modified microbial strains could be used in a novel way and because it was the basis for the precedent-setting legal decision that, in the United States, declared that genetically engineered organisms were patentable.

Symbiosis A close biological relationship between two organisms in which neither organism is extremely harmful to the other. In some cases, the relationship is mutually beneficial.

Syndrome A constellation of features that together make up the symptoms of a disorder or disease.

T Thymine residue in DNA.

T cells Lymphocytes that pass through the thymus gland during maturation. Different kinds of T cells play important roles in the immune response.

T-DNA The segment of a Ti plasmid that is transferred and integrated into chromosomal sites in the nuclei of plant cells.

T4 DNA ligase An enzyme from bacteriophage T4-infected cells that catalyzes the joining of duplex DNA molecules and repairs nicks in DNA molecules. The enzyme joins a 5'-phosphate group to a 3'-hydroxyl group.

T4 DNA polymerase end labeling DNA that has been cut with a restriction endonuclease(s) is mixed with T4 DNA polymerase and one labeled deoxyribonucleotide. The 3' exonuclease activity of the T4 DNA polymerase removes deoxyribonucleotides from the 3' ends of the DNA fragments. Immediately after a deoxyribonucleotide that is the same as the deoxyribonucleotide in the reaction mixture is cleaved off, the T4 DNA polymerase activity incorporates a labeled deoxyribonucleotide from the reaction mixture. No further incorporation of deoxyribonucleotides occurs because there is only one kind of deoxyribonucleotide in the reaction mixture.

Tag *See* Label

Tailing The in vitro addition of the same nucleotide by the enzyme terminal transferase to the 3'-hydroxyl ends of a duplex DNA molecule. Also called homopolymeric tailing.

Tandem array Usually a DNA molecule that contains two or more identical nucleotide sequences in series.

Target For diagnostic tests, the molecule or nucleic acid sequence that is being sought in a sample.

Target gene A descriptive term for a gene that is to be either cloned or specifically mutated.

Targeting vector A cloning vector carrying a DNA sequence capable of participating in a crossover event at a specified chromosomal location in the host cell.

TATA box The DNA sequence, to which RNA polymerase binds, that lies upstream from the site of initiation of transcription and ensures that transcription starts at a specified nucleotide. Also called a Pribnow box in prokaryotes and a Hogness box in eukaryotes after the researchers who discovered the function of the TATA box in prokaryotes and eukaryotes, respectively.

Temperature-sensitive protein A protein that is functional at one temperature but loses function at another (usually higher) temperature.

Template strand The polynucleotide strand that a polymerase uses for determining the sequence of nucleotides during the synthesis of a new nucleic acid strand.

Termination The cessation of the biosynthesis of a polymeric macromolecule.

Termination codon A naturally occurring codon that does not base pair with the anticodon of any tRNA. Generally, the three codons in this class (UAA, UAG, and UGA) are used to terminate translation, although in some rare instances one of these codons does code for an amino acid. Also called translational stop signal.

Terminator A sequence of DNA at the 3′ end of a gene that stops transcription. Also called transcription terminator.

Thaumatin A plant protein that has a sweet taste. It is also synthesized in some plants in response to infection by pathogens.

Therapeutic A treatment of a disease.

Therapeutic agent A compound that is used for the treatment of a disease and for improving the well-being of an organism. Also called pharmaceutical agent, drug, protein drug.

Thermophile A microorganism that grows optimally at high temperatures, usually above 50°C. Some thermophiles can grow at temperatures of 90 to 100°C.

Thermosensitivity Loss of activity of a protein at high temperature.

Thermostability Retention of activity at high temperature.

θ The Greek letter theta, used for the value of the recombination fraction with studies of human gene linkage.

Thioredoxin A small protein that acts as an electron carrier.

Thrombin A blood protein that plays a role in blood clotting.

Thrombus A blood clot.

Ti plasmid A large extrachromosomal element that is found in strains of *Agrobacterium* and is responsible for crown gall formation.

Thymidylate synthase An enzyme that catalyzes the methylation of the uracil moiety in dUMP to convert it to dTMP.

Thymine One of the organic bases found in DNA.

Tissue plasminogen activator A protein involved in dissolving blood clots. Also called tPA.

Totipotent Generally, when a cell is able to respond to any one of a number of different stimuli and, subsequently, to develop into any one of a number of differentiated cell types. Also called pluripotent.

tPA *See* Tissue plasminogen activator

Tracking dye A low-molecular-weight, visible colored compound that moves with the ion front during gel electrophoresis.

Transcribed triplet A set of three contiguous nucleotides of the transcribed DNA strand of the coding region of a structural gene that determines a codon in the mRNA.

Transcript An RNA molecule that has been synthesized from a specific DNA template.

Transcription The process of RNA synthesis that is catalyzed by RNA polymerase that uses a DNA strand as a template.

Transcription factor A protein that facilitates RNA synthesis by binding to a specific DNA sequence or another transcription factor that is bound to a specific DNA sequence.

Transcription mapping Assigning gene transcripts, in the form of cDNA clones or expressed sequence tags, to specific chromosome regions by fluorescence in situ hybridization, hybridization, polymerase chain reaction (PCR), analysis of somatic cell hybrid mapping panels, or other strategies. Also called transcript mapping, transcriptional mapping.

Transduction The transfer of nonviral DNA by a virus to a cell.

Transfection The transfer of DNA to a eukaryotic cell.

Transfer RNA The RNA molecules that decode the sequence information contained in an mRNA molecule during the translation process. Also called tRNA.

Transformation (1) The uptake and establishment of DNA in a bacterium or yeast cell, in which the introduced DNA often changes the phenotype of the recipient organism. (2) Conversion, by various means, of animal cells in tissue culture from controlled to uncontrolled cell growth.

Transformation efficiency The number of cells that take up foreign DNA as a function of the amount of added DNA; expressed as transformants per microgram of added DNA.

Transformation frequency The fraction of a cell population that takes up foreign DNA; expressed as the number of transformed cells divided by the total number of cells in a population.

Transgene A gene from one source that has been incorporated into the genome of another organism. Often refers to a gene that has been introduced into a multicellular organism.

Transgenesis The introduction of a gene(s) into animal or plant cells that leads to the transmission of the input gene (transgene) to successive generations.

Transgenic animal A fertile animal that carries an introduced gene(s) in its germ line.

Transgenic plant A fertile plant that carries an introduced gene(s) in its germ line.

Transient Of short duration.

Translation The process of protein (polypeptide) synthesis in which the amino acid sequence of a protein is determined by mRNA mediated by tRNA molecules, and carried out on ribosomes.

Translational initiation signal *See* Initiation codon

Translational start codon *See* Initiation codon

Translational stop signal *See* Termination codon

Translocation (1) The movement of peptidyl-tRNA and mRNA from the aminoacyl site to the peptidyl site on the ribosome during the elongation phase of translation; this movement opens the aminoacyl site for the next codon. (2) The transfer of chromosome material from one chromosome to another. (3) The movement of compounds through a plant.

Transposable element *See* Transposon

Transposase An enzyme that is encoded by a transposon gene and that facilitates the insertion of the transposon into a new chromosomal site and excision from a site.

Transposon A DNA sequence (mobile genetic element) that can insert randomly into a chromosome, exit the site, and relocate at another chromosomal site. For example, Tn5 is a bacterial transposon that carries the genes for resistance to the an-

tibiotics neomycin and kanamycin and the genetic information for its insertion and excision. Also called transposable element.

Tribrid protein A fusion protein that has three segments, each encoded by parts of different genes.

Trichloroethylene An organic compound, used as a solvent and degreasing agent, that often persists in the environment.

Tripartite mating A process in which conjugation is used to transfer a plasmid vector to a target cell when the plasmid vector is not self-mobilizable. When (1) cells that have a plasmid with conjugative and mobilizing functions are mixed with (2) cells that carry the plasmid vector and (3) target cells, mobilizing plasmids enter the cells with the plasmid vector and mobilize the plasmid vector to enter into the target cells. Following tripartite mating, the target cells with the plasmid vector are separated from the other cell types in the mixture by various selection procedures.

tRNA *See* Transfer RNA

U Uracil residue in RNA.

Universal donor cells Cells that, after introduction into a recipient, will not induce an immune response that would lead to their rejection.

Upstream (1) In molecular biology, the stretch of DNA base pairs that lie in the 5′ direction from the site of initiation of transcription. Usually, the first transcribed base is designated +1 and the upstream nucleotides are indicated with minus signs, e.g., −1, −10. Also, to the 5′ side of a particular gene or sequence of nucleotides. (2) In chemical engineering, those phases of a manufacturing process that precede the biotransformation step. Refers to the preparation of raw materials for a fermentation process. Also called upstream processing.

Upstream processing *See* Upstream

Uracil One of the organic bases found in RNA.

Vaccination *See* Preventive immunization

Variable domains Regions of antibody chains that have different amino acid sequences in different antibody molecules. These regions are responsible for the antigen binding specificity of the antibody molecule.

Variable expressivity Variation in the phenotype caused by different alleles of the same gene or by the presence of nonallelic genes. Also called expressivity.

Variant An organism that is genetically different from the wild-type organism. Also called mutant.

Vector *See* Cloning vector

Vegetative Referring to the normal growth cycle of a microorganism.

Vehicle *See* Cloning vector

***vir* genes** A set of genes on a Ti plasmid that prepare the T-DNA segment for transfer into a plant cell.

Virion An infectious virus particle.

Virulence The degree of pathogenicity of an organism.

V_{max} The maximal rate of an enzyme-catalyzed reaction. V_{max} is the product of E_0 (the total amount of enzyme) and the value of k_{cat} (the catalytic rate constant).

Washout The loss of the slower-growing microorganism when two organisms are being grown together.

Western blotting Transfer of protein from a gel to a membrane.

Wild type A genetic term that denotes the most commonly observed phenotype, or the normal state, in contrast to a mutant condition.

X linkage *See* X linked

X linked When a gene is present on the X chromosome.

Xenobiotic A chemical compound that is not produced by living organisms; a manufactured chemical compound.

Xenogeneic From a different species or individual (an attribute of cells or tissue). Also called xenogenic.

Xylose A 5-carbon sugar that is a major constituent of hemicellulose.

YAC *See* Yeast artificial chromosome

Yeast A single-celled, usually rounded fungus.

Yeast artificial chromosome A yeast-based vector system for cloning large (>100-kb) DNA inserts. Also called YAC.

Yeast episomal vector A cloning vector for the yeast *Saccharomyces cerevisiae* that uses the 2μm plasmid origin of replication and is maintained as an extrachromosomal nuclear DNA molecule.

Z *See* Lod score

Z$_{max}$ The maximum lod score for a range of recombination fractions from 0 to 0.5.

Zinc finger Sequence-specific DNA binding proteins that contain domains that bind Zn^{2+}.

Zoo blot Hybridization of a cloned human DNA sequence to DNA from various vertebrate organisms to determine whether the cloned DNA is evolutionarily conserved.

Index